Michael A. Johnson

Mohammad H. Moradi

PID Control

New Identification and Design Methods

Michael A. Johnson and Mohammad H. Moradi (Editors)

With

J. Crowe, K.K. Tan, T.H. Lee, R. Ferdous, M.R. Katebi, H.-P. Huang,
J.-C. Jeng, K.S. Tang, G.R. Chen, K.F. Man, S. Kwong, A. Sánchez,
Q.-G. Wang, Yong Zhang, Yu Zhang, P. Martin, M.J. Grimble and
D.R. Greenwood

PID Control

New Identification and Design Methods

With 285 Figures

Michael A. Johnson, PhD
Industrial Control Centre
University of Strathclyde
Graham Hills Building
50 George Street
Glasgow
G1 1QE
UK

Mohammad H. Moradi, PhD
Electrical Engineering Group
Faculty of Engineering
Bu-Ali Sina University
Hamadan
Iran

British Library Cataloguing in Publication Data
PID control : new identification and design methods
  1.PID controllers
  I.Johnson, Michael A., 1948- II.Moradi, Mohammad H.
  (Mohammad Hassan), 1967-
  629.8
ISBN 1852337028

Library of Congress Cataloging in Publication Data
PID control : new identification and design methods / Michael A. Johnson
  (editor), Mohammad H. Moradi (editor); with J. Crowe ... [et al.]
    p. cm.
  Includes bibliographical references and index.
  ISBN 1-85233-702-8
  1. PID control–Design and construction. I. Johnson, Michael A., 1948- II. Moradi,
Mohammad H. (Mohammad Hassan), 1967- III. Crowe, J.

TJ223.P55P53 2005
629.8--dc22
                                          2004057797

Apart from any fair dealing for the purposes of research or private study, or criticism or review, as permitted under the Copyright, Designs and Patents Act 1988, this publication may only be reproduced, stored or transmitted, in any form or by any means, with the prior permission in writing of the publishers, or in the case of reprographic reproduction in accordance with the terms of licences issued by the Copyright Licensing Agency. Enquiries concerning reproduction outside those terms should be sent to the publishers.

ISBN-10: 1-85233-702-8
ISBN-13: 978-1-85233-702-5
Springer Science+Business Media
springeronline.com

© Springer-Verlag London Limited 2005

MATLAB® and SIMULINK® are registered trademarks of The Mathworks Inc., 3, Apple Hill Drive, Natick, MA 01760-2098, USA.
  http://www.mathworks.com/
LabVIEW™ is a registered trademark of National Instruments Corporation, 11500, N. Mopac Expwy., Austin, TX 78759-3504, USA.
SIMATIC® is a registered trademark of Siemens AG, Germany.
DeltaV™ is a trademark of Emerson Process Management.
Plantweb® is a registered trademark of Emerson Process Management.

The use of registered names, trademarks, etc. in this publication does not imply, even in the absence of a specific statement, that such names are exempt from the relevant laws and regulations and therefore free for general use.

The publisher makes no representation, express or implied, with regard to the accuracy of the information contained in this book and cannot accept any legal responsibility or liability for any errors or omissions that may be made.

Typesetting: Ian Kingston Publishing Services, Nottingham, UK
Printed in the United States of America
69/3830-543210 Printed on acid-free paper SPIN 10894354

For the gift of loving parents and family and for my grandchildren, Ethan and Teigan
Michael A. Johnson

To my wife, Mehri, and my sons, Aref and Ali, for their understanding and consideration; To my parents for their love over many years and to my family for their support.
Mohammad H. Moradi

Preface

The industrial evidence is that for many control problems, particularly those of the process industries, the Proportional, Integral and Derivative (PID) controller is the main control tool being used. For these industrial problems, the PID control module is a building block which provides the regulation and disturbance rejection for single loop, cascade, multi-loop and multi-input multi-output control schemes. Over the decades, PID control technology has undergone many changes and today the controller may be a standard utility routine within the supervisory system software, a dedicated hardware process controller unit or an input–output module within a programmable electronic system which can be used for control system construction.

With such a well-developed industrial technology available it is not surprising that an academic colleague on learning that we planned a book on PID control exclaimed, "Surely not! Is there anything left to be said?". Of course, the short answer is that technology does not stand still: new solution capabilities are always emerging and PID control will evolve too. Indeed, the Ziegler–Nichols rules have been famous for over sixty years and the Åström and Hägglund relay experiment has been around for twenty years, so it would be disappointing if some new approaches to PID control had not emerged in the meantime. However, that is not to claim that all the methods discussed in this book will replace existing technologies; nor is this book a definitive survey of all that has taken place in the developments of PID control since, say, 1985. The book was originally conceived as a set of chapters about new ideas that are being investigated in PID control; it might be more accurately subtitled "*Some* new identification and design methods".

The first proposals for this book were constructed using a classification scheme based on the extent to which a method used a model, then what type of model and then whether the method used optimisation principles or not; a very academic approach. Such a scheme does work, but, as one reviewer remarked, it is perhaps unnecessarily rigid. However, another objective of the Editors was to incorporate into the text a set of contributions from international authors, and this is more difficult to achieve with a very strict classification framework. Consequently, the finished book has a more relaxed structure but retains an inherent methodological agenda.

The book opens with two basic chapters about PID controllers. Industrial technology is examined using discussions, examples and pictures in Chapter 1. Two interesting industrial product reviews significantly add to the value of this chapter. Chapter 2 is constructed around a set of useful concepts which say more about the PID notation and conventions than anything else. The material in these two opening chapters is descriptive and informative; some of it is theory, but it is selective. It is designed to be partly a repository of existing technology and expertise and partly an introduction to some of the terminology and concepts that will be used in subsequent chapters. The sections in these two chapters are written with some repetition of material to enable individual sections to be read in isolation when using the text in reference mode.

This is followed by 11 chapters that make different contributions to ideas for identification for PID control, and to the tuning of PID controllers. Almost all of the contributions arise from problems and research issues which have intrigued the various authors, and the chapters describe some answers to these problems. This is not just the leavening of a set of the various authors' published papers but a fully explained presentation of the investigative directions being followed by the contributors. The Editors hope that the reader will find the presentations quite readable and be able to follow up the research literature directly.

The underlying continuity in the book is that, as the chapters follow each other, the quality of model information used by the problem formulation and solution increases. This agenda starts at Chapter 3, where the methods assume no model information at all. The next group of chapters, numbers 4 to 7, use nonparametric models. Because the reaction curve method is historically associated with nonparametric methods, Chapter 8 on extensions to the reaction curve method is placed next. In the gap between nonparametric and parametric model-based methods, Chapters 9 and 10 report on the genetic algorithms and fuzzy model approach and on a so-called subspace identification method, respectively. Finally, methods based on parametric models take the stage in the final three chapters of the book. The last of these chapters looks at the idea of predictive PID control.

The emphasis within each chapter varies depending on what is important to the method being described. For example, a chapter might describe how to obtain the appropriate model information for a PID control design method, or how to use the appropriate model information in PID control design algorithm; sometimes both aspects of identification and design are treated. At no point can it be claimed that existing PID tuning methods are treated systematically; rather, the book has some chapters that explain some new ideas, whilst in other chapters existing techniques are given and then extended, reinterpreted and renovated. The book is most certainly not a cookbook for PID control tuning recipes, and to return to our colleague's surprised, "Is there anything left to be said?", the book now written shows clearly that the PID control still has many avenues to be explored.

This is also the place to give thanks to various people who have been so helpful in the compilation, construction and production of the book. All the contributors are very gratefully thanked for agreeing to participate and for their patience during the editorial period. It was an extraordinary pleasure to meet some of them at their home institutions in Singapore and Taipei, Taiwan, in 2000, and others at various recent international control conferences.

Professor M. J. Grimble is gratefully thanked for allowing Emeritus Professor M. A. Johnson the use of facilities at the Industrial Control Centre, University of Strathclyde in Glasgow, during the writing of this book.

Finally, the Editors would like to thank the publishing staff at Springer-Verlag London, Oliver Jackson and Anthony Doyle, and at Springer Verlag's New York offices, Jenny Wolkowicki, for their kind encouragement, and patience during the gestation period of the book. Also the copy editor and typesetter, Ian Kingston, is thanked for his thoroughness with the manuscript and for the excellent modern typographical interpretation of the text.

Michael A. Johnson and Mohammad H. Moradi
December 2004

How to Use This Book

In many cases the production of a contributed book leads to an opus which looks like set of collected papers from the authors. In the case of this book, care has been taken to have sufficient explanation introduced so that the book might also be used constructively. With this agenda, the typical structure for a chapter is:

- Learning objectives
- Introductory material
- Main algorithms described
- Worked examples and case studies
- Conclusions and discussion
- References and bibliography

Thus it is hoped that this book can be used for:

- Support material for possible advanced course study
- Self study by industrial and academic control engineers
- A source for future research ideas and projects
- A reference resource and a source of references
- Contacting researchers working on particular PID control topics

To assist the reader in navigating the various approaches and methods it is useful to have a map of the book. Firstly, the design approaches are given broad definitions, and a tree diagram of the book structure follows. The broad classification scheme that has been used to organise this book is based on the quality and type of the process model information used and then whether or not optimisation concepts have been used to generate the PID controller tunings. A brief description of the main categories follows next.

- *Model-free methods*: the method does *not* use the explicit identification of significant model points or a parametric model *per se*.

- *Nonparametric model methods*: the method uses the explicit identification of significant model points or a nonparametric model, but does not use a parametric model *per se*.

- *Data-intensive methods*: these methods are halfway between the nonparametric and parametric model-based methods. They are characterised by the use of process data, as in the subspace method, or grey knowledge, as in the case of fuzzy-logic methods.
- *Parametric model methods*: the method straightforwardly depends on the use of a parametric model; usually a transfer function model.

The second categorisation depends on whether the tuning method uses optimisation concepts. As an example, many optimisation-based methods use the appropriate classic linear quadratic cost function over a deterministic or stochastic problem formulation. Figure 1 shows the tree diagram map of the book chapters.

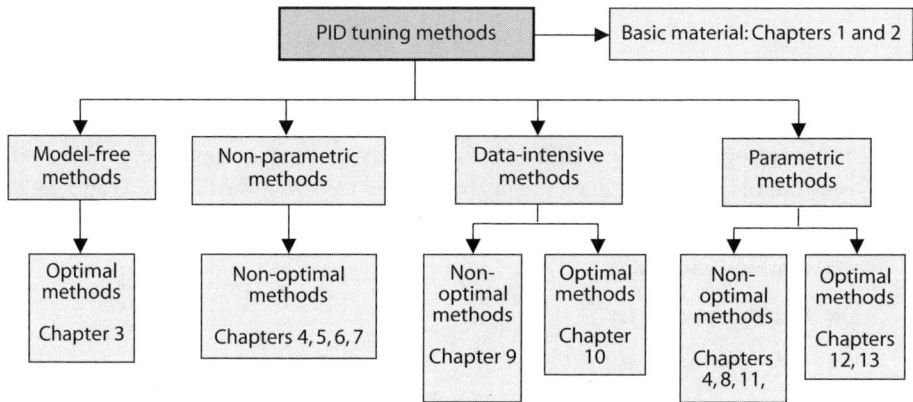

Figure 1 Book chapters: tree diagram map.

Contents

Editorial Responsibilities xix

Notation xxv

1   PID Control Technology 1
    Learning Objectives 1
    1.1   Basic Industrial Control 2
        1.1.1   Process Loop Issues – a Summary Checklist 6
    1.2   Three-Term Control 7
        1.2.1   Parallel PID Controllers 9
        1.2.2   Conversion to Time constant PID Forms 10
        1.2.3   Series PID Controllers 12
        1.2.4   Simple PID Tuning 14
    1.3   PID Controller Implementation Issues 17
        1.3.1   Bandwidth-Limited Derivative Control 18
        1.3.2   Proportional Kick 22
        1.3.3   Derivative Kick 24
        1.3.4   Integral Anti-Windup Circuits 26
        1.3.5   Reverse-Acting Controllers 29
    1.4   Industrial PID Control 29
        1.4.1   Traditional Industrial PID Terms 30
        1.4.2   Industrial PID Structures and Nomenclature 32
        1.4.3   The Process Controller Unit 33
        1.4.4   Supervisory Control and the SCADA PID Controller 35
    Acknowledgements 46
    References 46

2   Some PID Control Fundamentals 47
    Learning Objectives 47
    2.1   Process System Models 48
        2.1.1   State Space Models 49
        2.1.2   Convolution Integral Process Models 52
        2.1.3   Laplace Transfer Function Models 53

		2.1.4	Common Laplace Transform Process Models	55
	2.2	Controller Degrees of Freedom Structure		57
		2.2.1	One Degree of Freedom Control	57
		2.2.2	Two Degree of Freedom Control	57
		2.2.3	Three Degree of Freedom Structures	59
	2.3	PID Control Performance		60
		2.3.1	Controller Performance Assessment – General Considerations	60
		2.3.2	Controller Assessment – the Effectiveness of PID Control	66
		2.3.3	Classical Stability Robustness Measures	73
		2.3.4	Parametric Stability Margins for Simple Processes	79
	2.4	State Space Systems and PID Control		88
		2.4.1	Linear Reference Error Feedback Control	88
		2.4.2	Two Degree of Freedom Feedback Control System	90
		2.4.3	State Feedback With Integral Error Feedback Action	91
		2.4.4	State Space Analysis for Classical PI Control Structure	95
	2.5	Multivariable PID Control Systems		99
		2.5.1	Multivariable Control	100
		2.5.2	Cascade Control Systems	103
	Acknowledgements			106
	References			106
3	**On-line Model-Free Methods**			109
	Learning Objectives			109
	3.1	Introduction		110
		3.1.1	A Model-Free Control Design Paradigm	110
	3.2	Iterative Feedback Tuning		114
		3.2.1	Generating the Cost Function Gradient	114
		3.2.2	Case Study – a Wastewater Process Example	117
		3.2.3	Some Remarks on Iterative Feedback Tuning	122
	3.3	The Controller Parameter Cycling Tuning Method		124
		3.3.1	Generating the Gradient and Hessian – Some Theory	125
		3.3.2	Issues for a Controller Parameter Cycling Algorithm	131
		3.3.3	The Controller Parameter Cycling Algorithm	135
		3.3.4	Case Study – Multivariable Decentralised Control	136
	3.4	Summary and Future Directions		143
	Acknowledgements			144
	Appendix 3.A			144
	References			145
4	**Automatic PID Controller Tuning – the Nonparametric Approach**			147
	Learning Objectives			147
	4.1	Introduction		148
	4.2	Overview of Nonparametric Identification Methods		149
		4.2.1	Transient Response Methods	149
		4.2.2	Relay Feedback Methods	150
		4.2.3	Fourier Methods	150
		4.2.4	Phase-Locked Loop Methods	151

	4.3 Frequency Response Identification with Relay Feedback	152
	4.3.1 Basic Idea	153
	4.3.2 Improved Estimation Accuracy	155
	4.3.3 Estimation of a General Point	161
	4.3.4 Estimation of Multiple Points	164
	4.3.5 On-line relay tuning	164
4.4	Sensitivity Assessment Using Relay Feedback	166
	4.4.1 Control Robustness	166
	4.4.2 Maximum Sensitivity	167
	4.4.3 Construction of the $\lambda - \phi$ Chart	168
	4.4.4 Stability Margins Assessment	170
4.5	Conversion to Parametric Models	171
	4.5.1 Single and Multiple Lag Processes	172
	4.5.2 Second-Order Modelling	174
4.6	Case Studies	174
	Case Study 4.1: Improved Estimation Accuracy for the Relay Experiment	176
	Case Study 4.2: Estimation of a General Point	177
	Case Study 4.3: Estimation of Multiple Points	177
	Case Study 4.4: On-line Relay Tuning	179
	Case Study 4.5: Sensitivity Assessment	179
	References	180

5 Relay Experiments for Multivariable Systems — 183

Learning Objectives — 183

- 5.1 Introduction — 184
- 5.2 Critical Points of a System — 185
  - 5.2.1 Critical Points for Two-Input, Two-Output Systems — 185
  - 5.2.2 Critical Points for MIMO Systems — 186
- 5.3 Decentralised Relay Experiments for Multivariable Systems — 187
  - 5.3.1 Finding System Gains at Particular Frequencies — 188
  - 5.3.2 Decentralised Relay Control Systems – Some Theory — 190
  - 5.3.3 A Decentralised Two-Input, Two-Output PID Control System Relay-Based Procedure — 191
- 5.4 A Decentralised Multi-Input, Multi-Output PID Control System Relay-Based Procedure — 197
- 5.5 PID Control Design at Bandwidth Frequency — 202
- 5.6 Case Studies — 207
  - 5.6.1 Case Study 1: The Wood and Berry Process System Model — 207
  - 5.6.2 Case Study 2: A Three-Input, Three-Output Process System — 210
- 5.7 Summary — 210

References — 211

6 Phase-Locked Loop Methods — 213

Learning Objectives — 213

- 6.1 Introduction — 214
  - 6.1.1 The Relay Experiment — 215
  - 6.1.2 Implementation Issues for the Relay Experiment — 216
  - 6.1.3 Summary Conclusions on the Relay Experiment — 220
- 6.2 Some Constructive Numerical Solution Methods — 221

		6.2.1 Bisection Method	222
		6.2.2 Prediction Method	224
		6.2.3 Bisection and Prediction Method – a Comparison and Assessment	227
	6.3	Phase-Locked Loop Identifier Module – Basic Theory	229
		6.3.1 The Digital Identifier Structure	230
		6.3.2 Noise Management Techniques	242
		6.3.3 Disturbance Management Techniques	248
	6.4	Summary and Discussion	255
		References	256

7 Phase-Locked Loop Methods and PID Control — 259
Learning Objectives — 259
7.1 Introduction – Flexibility and Applications — 260
7.2 Estimation of the Phase Margin — 260
7.3 Estimation of the Parameters of a Second-Order Underdamped System — 261
7.4 Identification of Systems in Closed Loop — 265
    7.4.1 Identification of an Unknown System in Closed Loop with an Unknown Controller — 265
    7.4.2 Identification of an Unknown System in Closed Loop with a Known Controller — 268
7.5 Automated PI Control Design — 270
    7.5.1 Identification Aspects for Automated PID Control Design — 271
    7.5.2 PI Control with Automated Gain and Phase Margin Design — 275
    7.5.3 PI Control with Automated Maximum Sensitivity and Phase Margin Design — 286
7.6 Conclusions — 294
References — 295

8 Process Reaction Curve and Relay Methods Identification and PID Tuning — 297
Learning Objectives — 297
8.1 Introduction — 298
8.2 Developing Simple Models from the Process Reaction Curve — 302
    8.2.1 Identification Algorithm for Oscillatory Step Responses — 303
    8.2.2 Identification Algorithm for Non-Oscillatory Responses Without Overshoot — 305
8.3 Developing Simple Models from a Relay Feedback Experiment — 310
    8.3.1 On-line Identification of FOPDT Models — 312
    8.3.2 On-line Identification of SOPDT Models — 314
    8.3.3 Examples for the On-line Relay Feedback Procedure — 315
    8.3.4 Off-line Identification — 317
8.4 An Inverse Process Model-Based Design Procedure for PID Control — 320
    8.4.1 Inverse Process Model-Based Controller Principles — 320
    8.4.2 PI/PID Controller Synthesis — 323
    8.4.3 Autotuning of PID Controllers — 325
8.5 Assessment of PI/PID Control Performance — 329
    8.5.1 Achievable Minimal IAE Cost and Rise Time — 329
    8.5.2 Assessment of PI/PID Controllers — 332
References — 336

9 Fuzzy Logic and Genetic Algorithm Methods in PID Tuning — 339
Learning Objectives — 339

	9.1	Introduction	340
	9.2	Fuzzy PID Controller Design	340
		9.2.1 Fuzzy PI Controller Design	342
		9.2.2 Fuzzy D Controller Design	343
		9.2.3 Fuzzy PID Controller Design	344
		9.2.4 Fuzzification	345
		9.2.5 Fuzzy Control Rules	346
		9.2.6 Defuzzification	346
		9.2.7 A Control Example	349
	9.3	Multi-Objective Optimised Genetic Algorithm Fuzzy PID Control	350
		9.3.1 Genetic Algorithm Methods Explained	351
		9.3.2 Case study A: Multi-Objective Genetic Algorithm Fuzzy PID Control of a Nonlinear Plant	353
		9.3.3 Case study B: Control of Solar Plant	354
	9.4	Applications of Fuzzy PID Controllers to Robotics	355
	9.5	Conclusions and Discussion	357
		Acknowledgments	358
		References	358
10	**Tuning PID Controllers Using Subspace Identification Methods**		**361**
	Learning Objectives		361
	10.1	Introduction	362
	10.2	A Subspace Identification Framework for Process Models	363
		10.2.1 The Subspace Identification Framework	363
		10.2.2 Incremental Subspace Representations	366
	10.3	Restricted Structure Single-Input, Single-Output Controllers	368
		10.3.1 Controller Parameterisation	369
		10.3.2 Controller Structure and Computations	370
	10.4	Restricted-Structure Multivariable Controller Characterisation	371
		10.4.1 Controller Parameterisation	371
		10.4.2 Multivariable Controller Structure	372
	10.5	Restricted-Structure Controller Parameter Computation	372
		10.5.1 Cost Index	373
		10.5.2 Formulation as a Least-Squares Problem	373
		10.5.3 Computing the Closed-Loop System Condition	374
		10.5.4 Closed-Loop Stability Conditions	375
		10.5.5 The Controller Tuning Algorithm	375
	10.6	Simulation Case Studies	376
		10.6.1 Activated Sludge Wastewater Treatment Plant Layout	377
		10.6.2 Case study 1: Single-Input, Single-Output Control Structure	378
		10.6.3 Case Study 2: Control of Two Reactors with a Lower Triangular Controller Structure	379
		10.6.4 Case Study 3: Control of Three Reactors with a Diagonal Controller Structure	382
		10.6.5 Case Study 4: Control of Three Reactors with a Lower Triangular Controller Structure	385
		References	387
11	**Design of Multi-Loop and Multivariable PID Controllers**		**389**
	Learning Objectives		389

11.1 Introduction 390
    11.1.1 Multivariable Systems 390
    11.1.2 Multivariable Control 391
    11.1.3 Scope of the Chapter and Some Preliminary Concepts 392
11.2 Multi-Loop PID Control 394
    11.2.1 Biggest Log-Modulus Tuning Method 394
    11.2.2 Dominant Pole Placement Tuning Method 395
    11.2.3 Examples 404
11.3 Multivariable PID Control 408
    11.3.1 Decoupling Control and Design Overview 409
    11.3.2 Determination of the Objective Loop Performance 412
    11.3.3 Computation of PID Controller 421
    11.3.4 Examples 422
11.4 Conclusions 426
References 427

12 Restricted Structure Optimal Control 429
Learning Objectives 429
12.1 Introduction to Optimal LQG Control for Scalar Systems 430
    12.1.1 System Description 431
    12.1.2 Cost Function and Optimisation Problem 432
12.2 Numerical Algorithms for SISO System Restricted Structure Control 436
    12.2.1 Formulating a Restricted Structure Numerical Algorithm 436
    12.2.2 Iterative Solution for the SISO Restricted Structure LQG Controller . . 439
    12.2.3 Properties of the Restricted Structure LQG Controller 440
12.3 Design of PID Controllers Using the Restricted Structure Method 441
    12.3.1 General Principles for Optimal Restricted Controller Design 442
    12.3.2 Example of PID Control Design 443
12.4 Multivariable Optimal LQG Control: An Introduction 444
    12.4.1 Multivariable Optimal LQG Control and Cost Function Values 448
    12.4.2 Design Procedures for an Optimal LQG Controller 451
12.5 Multivariable Restricted Structure Controller Procedure 453
    12.5.1 Analysis for a Multivariable Restricted Structures Algorithm 454
    12.5.2 Multivariable Restricted Structure Algorithm and Nested Restricted
           Structure Controllers 458
12.6 An Application of Multivariable Restricted Structure Assessment – Control of the
Hotstrip Finishing Mill Looper System 461
    12.6.1 The Hotstrip Finishing Mill Looper System 461
    12.6.2 An Optimal Multivariable LQG Controller for the Looper System . . . 463
    12.6.3 A Controller Assessment Exercise for the Hotstrip Looper System . . . 465
12.7 Conclusions 470
Acknowledgements 471
References 472

13 Predictive PID Control 473
Learning Objectives 473
13.1 Introduction 474

	13.2	Classical Process Control Model Methods	475
		13.2.1 Smith Predictor Principle	475
		13.2.2 Predictive PI With a Simple Model	477
		13.2.3 Method Application and an Example	480
	13.3	Simple Process Models and GPC-Based Methods	485
		13.3.1 Motivation for the Process Model Restriction	485
		13.3.2 Analysis for a GPC PID Controller	486
		13.3.3 Predictive PID Control: Delay-Free System $h = 0$	490
		13.3.4 Predictive PID Control: Systems with Delay $h > 0$	491
		13.3.5 Predictive PID Control: An Illustrative Example	493
	13.4	Control Signal Matching and GPC Methods	500
		13.4.1 Design of SISO Predictive PID Controllers	500
		13.4.2 Optimal Values of Predictive PID Controller Gains	503
		13.4.3 Design of MIMO Predictive PID controllers	515
Acknowledgements			524
Appendix 13.A			524
		13.A.1 Proof of Lemma 13.1	524
		13.A.2 Proof of Theorem 13.1	526
		13.A.3 Proof of Lemma 13.2	527
References			529

About the Contributors ... 531

Index ... 539

Editorial Responsibilities

The concept of this book originated with Michael A. Johnson and Mohammad H. Moradi, who were also responsible for the overall editorial task.

Chapter Contributions

Chapter 1 PID Control Technology
M.A. Johnson
> *Industrial Control Centre, University of Strathclyde, Glasgow, G1 1QE, Scotland, UK*

This chapter is an introduction to the basic technology of PID control. It opens with a brief look at the way in which control engineers solve process control problems in different stages. These stages influence the different methods that the engineer uses. A presentation on the different forms of three-term controllers follows. This concentrates on notation, conversion between the types and the structural flexibility of the controllers. Simple implementation issues for the terms of the controller are given next. The chapter concludes with a section on industrial PID control including material on the process controller unit, supervisory control and the SCADA PID controller setup. Two industrial examples, kindly provided by Siemens AG and Emerson Process Management, are included in this section.

Chapter 2 Some PID Control Fundamentals
M.A. Johnson
> *Industrial Control Centre, University of Strathclyde, Glasgow, G1 1QE, Scotland, UK*

M.H. Moradi
> *Electrical Engineering Group, Faculty of Engineering, Bu-Ali Sina University, Hamadan, Iran*

This chapter archives some basic control and PID control properties and definitions used and referred to in later chapters. The fundamentals covered are process models, controller degrees of freedom structures, classical performance measures, a look at PID in state space formalism, and common structures for multivariable system PID-based controllers.

Chapter 3 On-line Model Free Methods
J. Crowe
> *British Energy Generation (UK) Ltd, 3 Redwood Crescent, Peel Park, East Kilbride, G74 5PR, Scotland, UK*

M.A. Johnson
Industrial Control Centre, University of Strathclyde, Glasgow, G1 1QE, Scotland, UK
The idea of model-free control design is slowly gaining credence. Papers quite often appear claiming a model-free approach. However, it is thought that the only real model-free method available is Iterative Feedback Tuning (IFT). This uses the process to generate optimisation information directly without the use of any model computations at all. In this chapter, IFT is given a simple tutorial presentation emphasising the main ideas of the method. A wastewater process example is then used to demonstrate the method. The Controller Parameter Cycling tuning method (due to J. Crowe) is a new addition to the model-free paradigm. This method is presented in the second half of the chapter.

Chapter 4 Automatic PID Controller Tuning – The Nonparametric Approach
K.K. Tan, T.H. Lee and R. Ferdous
Department of Electrical and Computer Engineering, National University of Singapore, 4 Engineering Drive 3, Singapore, 117576
This chapter first gives an overview of the nonparametric identification methods currently available and used for automatic tuning of PID controllers. Following the overview, the chapter focuses on the relay feedback method and its variants for the purpose of realising efficient autotuning solutions. These variants are mainly due to the joint work of the authors and their co-workers. Relay feedback is useful not just for control design, but also for process and control monitoring. The next part of the chapter reports on the use of the relay-based method for control robustness assessment and evaluation. It is shown how nonparametric results can be converted into parametric transfer function models. Finally, case studies on laboratory setups are presented to consolidate the materials presented giving an applications perspective.

Chapter 5 Relay Experiments for Multivariable Systems
M.H. Moradi
Electrical Engineering Group, Faculty of Engineering, Bu-Ali Sina University, Hamadan, Iran
M.R. Katebi
Industrial Control Centre, University of Strathclyde, Glasgow, G1 1QE, Scotland, UK
The relay experiment for multivariable processes has been given in several different forms. This chapter opens with a brief survey of these methods and considers the definition of critical points for a multivariable process. The decentralised relay experiments for multivariable systems are then investigated in detail and extended from the familiar two-input, two-output form to a multi-input multi-output relay procedure. The remainder of the chapter concentrates on a relay method based on PID control design at bandwidth frequency. Two case studies are given to demonstrate the method. The results are based on the research of M. H. Moradi and M. R. Katebi.

Chapter 6 Phase-Locked Loop Methods
J. Crowe
British Energy Generation (UK) Ltd, 3 Redwood Crescent, Peel Park, East Kilbride, G74 5PR, Scotland, UK
M.A. Johnson
Industrial Control Centre, University of Strathclyde, Glasgow, G1 1QE, Scotland, UK
The chapter opens with a look at the possible problems that can arise with the relay experiment. This leads naturally to posing the question: can the process identification data obtained from a relay experiment be produced without using relays? A first answer reported in this chapter used a constructive numerical route, but this was subsequently replaced by a new nonparametric identification method based on a phase-locked loop concept. The remainder of the chapter is devoted to presenting the basic

theory of the phase-locked loop identification procedure and demonstrating the potential of the method. The results are based on the research of J. Crowe and M.A. Johnson.

Chapter 7 PID Tuning Using a Phase-Locked Loop Identifier

J. Crowe
> *British Energy Generation (UK) Ltd, 3 Redwood Crescent, Peel Park, East Kilbride, G74 5PR, Scotland, UK*

M.A. Johnson
> *Industrial Control Centre, University of Strathclyde, Glasgow, G1 1QE, Scotland, UK*

The phase-locked loop identifier can be packaged as a self-contained module for use with various identification tasks. This chapter reports on the potential for identification flexibility. The first part demonstrates how the identifier can be used to estimate the phase margin, estimate the parameters of a second-order underdamped system, and identify a system in closed loop. A target use of the phase-locked loop identifier is to achieve automated on-line PI control design. Algorithms to achieve the on-line PI design for a gain and phase margin specification pair and for a maximum sensitivity and phase margin design specification are demonstrated in the remainder of the chapter. The results are based on the research of J. Crowe and M.A. Johnson.

Chapter 8 Process Reaction Curve and Relay Methods – Identification and PID Tuning

H.-P. Huang and J.-C. Jeng
> *Department of Chemical Engineering, National Taiwan University, Taipei, Taiwan 106*

This chapter was inspired by the classical work on process reaction curve methods for the tuning of three-term controllers. Author Hsiao-Ping Huang has spent many years developing extensions and new ways of approaching this field of engineering science. A special emphasis has been to develop techniques for use in the process and chemical industries. The chapter reports some of the author's more recent research on developing simple models from step response and relay feedback experiments and perform autotuning for PID controllers. As the emphasis is on the use of models to derive IMC-PID controllers or other related model-based controllers, alternative simple models other than the most popular FOPDT models may be required. This chapter provides systematic procedures to obtain these simple models for design purposes. An inverse model-based synthesis of PID controllers is also presented. It is especially simple and effective for application to FOPDT and overdamped SOPDT processes. Finally, in this chapter, a method to assess the performance status of a PID controller is given. This will be helpful to determine whether the PID controller is well tuned or not. The material in this chapter will help engineers to understand the methods of identification and autotuning from time-domain data without resorting to the traditional least square methods, and will enable them to understand and assess the performance that a PID controller can achieve.

Chapter 9 Fuzzy Logic and Genetic Algorithm Methods in PID Tuning

K.S. Tang, G.R. Chen, K.F. Man and S. Kwong
> *Department of Electrical Engineering, City University of Hong Kong, Tat Chee Avenue, Kowloon, Hong Kong*

The potential of fuzzy logic in solving controller tuning problems was realised in the mid-1960s. However, its application to tuning the classical PID controller emerged much later in the 1980s. Since that time, the authors of this chapter have spent some years with world-leading experts developing extensions and new ways of approaching this field of Fuzzy PID controller design. A special emphasis has been to develop techniques that use the power of genetic algorithms. This chapter demonstrates just how effective the genetic algorithm approach is and reports some of the authors' recent research.

Chapter 10 Tuning PID Controllers using Subspace Identification Methods

A. Sanchez

Departamento de Automatización y Control Industrial, Escuela Politécnica Nacional, P.O. Box 17-01-2759, Quito, Ecuador

M.R. Katebi and M.A. Johnson

Industrial Control Centre, University of Strathclyde, Glasgow, G1 1QE, Scotland, UK

This chapter presents a general framework for using subspace identification methods to tune PID controllers. The restricted controller concept is used to give the tuning procedure generality, but the specific application is for tuning PID controllers. The method does not use a parametric model directly, but the model is formulated using the subspace identification paradigm. Thus, data is organised to provide a data-based model for the iterative PID tuning algorithm. The applications of the PID tuning algorithm are from the wastewater industry and use simulations of a sequential wastewater treatment process. Various scalar and multivariable system examples are given. This method has an industrial origin, since Dr M.R. Katebi recently led the Strathclyde team in an EU-supported wastewater control system development project (the SMAC project) and A. Sanchez was Research Fellow on the project. The problems and solutions reported in the chapter were inspired by wastewater industry problems and are the research results of A. Sanchez, M.R. Katebi and M.A. Johnson.

Chapter 11 Design of Multi-Loop and Multivariable PID Controllers

Q.-G. Wang

Department of Electrical and Computer Engineering, The National University of Singapore, 4 Engineering Drive 3, Singapore, 117576

Yong Zhang and Yu Zhang

Global Research Shanghai, GE China Technology Center, 1800 Cai Lun Road, Zhang Jiang Hi-tech Park, Pudong New Area, Shanghai 201203, China

A major assumption for this chapter is that the designer has a multi-input, multi-output transfer function model for the process to be controlled. This will be used to show the difficulty in multivariable control arising from the interactions between the multiple process elements. For multi-loop PID control, the BLT tuning method will be outlined first. Improved settings may be obtained by assigning the dominant pole in suitable locations with the help of some computer graphs. For multivariable PID control, a decoupling design is adopted and the characteristics of decoupled systems are developed to reach achievable optimal target closed loop and produce the desired PID settings. These new methods are the work of Qing-Guo Wang in collaboration with his former PhD students, Yu Zhang and Yong Zhang. Professor Wang has spent many years developing extensions and new ways of designing PID controllers for multivariable processes. The authors would like to thank Professor K. J. Åström for his collaboration on the dominant pole placement method; we learnt much from him.

Chapter 12 Restricted Structure Optimal Control

P. Martin

BAE Systems Avionics, Crewe Toll, Ferry Road, Edinburgh, EH5 2XS, Scotland, UK

M.J. Grimble, D. Greenwood and M.A. Johnson

Industrial Control Centre, University of Strathclyde, Glasgow, G1 1QE, Scotland, UK

This chapter continues the theme of model-based design methods for PID controllers. An optimal restricted structure control concept is presented for both single-input and single-output systems and multivariable systems. M. J. Grimble and M.A. Johnson have worked with the Linear Quadratic Gaussian paradigm for over two decades and the work presented in this chapter is a response to industrial engineers who wish to use classical PID controllers and implement simple physically based multivariable controllers. The methods are demonstrated on applications from the marine and steel industries. P.

Martin and M. J. Grimble developed and wrote the first part of the chapter, whilst the second part of the chapter was devised and written by D. R. Greenwood and M.A. Johnson.

Chapter 13 Predictive PID Control

M.H. Moradi

Electrical Engineering Group, Faculty of Engineering, Bu-Ali Sina University, Hamadan, Iran

M.A. Johnson and M.R. Katebi

Industrial Control Centre, University of Strathclyde, Glasgow, G1 1QE, Scotland, UK

The last chapter of the book seeks to explore possible links between model-based PID control design and predictive control methods. The presentation is in three parts. Firstly, there is a section on the predictive PID control design method due to Tore Hägglund (Lund University, Sweden). This method uses the Smith predictor framework and was devised for systems with long dead times. Secondly, there is a look at the predictive PID control concept due to Kok Kiong Tan and his colleagues based on a Generalised Predictive Control formulation. In both cases, some simple simulation results are given. Using the published literature, M.A. Johnson wrote the presentations for these first two methods.

Finally, the control signal matching method for predictive PID control design is reported in the extended final section of the chapter. This method involves defining an M-step ahead PID controller and selecting the controller coefficients to match the control signal from a Generalised Predictive Control problem solution. The matching process leads to a formula for computing the PID gains. The other degree of freedom is the choice of forward prediction horizon M. This selection is used to ensure closed loop stability and good fidelity with the GPC control signal. The method has the potential to incorporate other features such as actuator constraints and integral anti-windup modifications. The results presented are the research results of M.H. Moradi and M.R. Katebi.

About the Contributors

Biographical sketches for all the contributors can be found in this section at the end of the book.

Notation

The area of PID control and the various related control techniques is still developing. Consequently, the engineering notation used by the many authors, research workers and control engineers working in this area is not always consistent. In fact, there is much creativity in the notation alone, because different authors wish to emphasise different aspects of the subject and as a result use some highly personalised notation. In this edited book, an attempt has been made to regularise some of the notation used without interfering with the notational colour introduced by some authors.

Complex Numbers

Cartesian representation: $s = \alpha + j\beta$, $\alpha, \beta \in \Re$
Polar representation: $s = r(\cos\phi + j\sin\phi)$
where $r = |s|$ and $\phi = \arg\{s\}$ with $0 \leq r$ and $-\pi < \phi \leq \pi$

Single-Input, Single-Output Systems

Process model: $G_p(s) \in \Re(s)$ or $W(s) \in \Re(s)$
Controller: $G_c(s) \in \Re(s)$ or $C(s) \in \Re(s)$
Forward path transfer: $G_{fp}(s) = G_p(s)G_c(s)$
For a single input, single output unity feedback system:

Closed loop transfer function $G_{cl}(s)$ and equivalently the complementary function $T(s)$

$$G_{cl}(s) = T(s) = \left[\frac{G_p(s)G_c(s)}{1 + G_p(s)G_c(s)}\right] = \left[\frac{G_{fp}(s)}{1 + G_{fp}(s)}\right]$$

Sensitivity function $S(s)$,

$$S(s) = \left[\frac{1}{1 + G_p(s)G_c(s)}\right] = \left[\frac{1}{1 + G_{fp}(s)}\right]$$

Nyquist Geometry

For SISO unity feedback systems, the loop transfer function is $G_L(s) = G_{fp}(s)$
Frequency variable ω where $0 \leq \omega < \infty$
Cartesian representation: $G_L(j\omega) = \alpha(\omega) + j\beta(\omega)$, where $\alpha: \Re^+ \to \Re$ and $\beta: \Re^+ \to \Re$
Polar representation: $G_L(j\omega) = r(\omega)(\cos\phi(\omega) + j\sin\phi(\omega))$, where the radial function is $r(\omega) = |G_L(j\omega)|$ and the angular function is $\phi(\omega) = \arg\{G_L(j\omega)\}$
In general for physical processes, the angular function $\phi(\omega)$ satisfies $-\infty < \phi(\omega) \leq 0$

Important Nyquist Geometry Points

Stability point: $s = -1 + j0$

Gain Margin Condition

Phase crossover frequency ω_{pco} is the frequency at which $\phi(\omega_{pco}) = \arg\{G_L(j\omega_{pco})\} = -\pi$. Alternative notation for this frequency is $\omega_{-\pi}$ and ω_π. Gain margin is defined as

$$GM = \frac{1}{|G_L(j\omega_{pco})|}$$

An alternative notation for gain margin is G_m.

In the context of the sustained oscillation phenomenon, these quantities are often referred to by the use of the term *ultimate* with related notation:

Ultimate frequency: $\omega_u = \omega_{pco}$

Ultimate gain: $k_u = \dfrac{1}{|G_L(j\omega_{pco})|} = \dfrac{1}{|G_L(j\omega_u)|}$

Ultimate period: $P_u = 2\pi/\omega_u$

Phase Margin Condition

The gain crossover frequency ω_{gco} is the frequency at which $r(\omega_{gco}) = |G_L(j\omega_{gco})| = 1$. An alternative notation for this frequency is ω_1.

The phase margin in degrees is given as

$$PM° = 180° + \arg\{G_L(j\omega_{gco})\}°$$

In radians, this is given by

$$\phi_{PM} = \pi + \arg\{G_L(j\omega_{gco})\}$$

Maximum sensitivity:

$$M_S = \max_{0 \leq \omega < \infty} |S(j\omega)|$$

Block Diagram Conventions

At block diagram summing points, the default for a signal entering the summing point is "+". If the signal is to be subtracted then a minus sign will appear along side the signal arrow. An illustration is given below. No "+" signs will appear in block diagram figures.

Across the summing point, $T = +\alpha + \beta - \chi$

Common Control Abbreviations

SISO	Single-input, single-output
TITO	Two-input, two-output
MIMO	Multi-input, multi-output
FOPDT	First-order plus dead time
SOPDT	Second-order plus dead time
IMC	Internal model control
MPC	Model predictive control
MBPC	Model-based predictive control
GPC	Generalised predictive control
FFT	Fast Fourier Transform
DFT	Discrete Fourier Transform
SNR	Signal-to-Noise Ratio
RGA	Relative Gain Array
IAE	Integral of Absolute Error Performance Criterion
ITAE	Integral of Time × Absolute Error Performance Criterion
LQG	Linear Quadratic Gaussian
GA	Genetic Algorithm

1 PID Control Technology

Learning Objectives

1.1 Basic Industrial Control

1.2 Three-Term Control

1.3 PID Controller Implementation Issues

1.4 Industrial PID Control

Acknowledgements

References

Learning Objectives

PID control is a name commonly given to three-term control. The mnemonic *PID* refers to the first letters of the names of the individual terms that make up the standard three-term controller. These are *P* for the proportional term, *I* for the integral term and *D* for the derivative term in the controller.

Three-term or PID controllers are probably the most widely used industrial controller. Even complex industrial control systems may comprise a control network whose main control building block is a PID control module. The three-term PID controller has had a long history of use and has survived the changes of technology from the analogue era into the digital computer control system age quite satisfactorily. It was the first (only) controller to be mass produced for the high-volume market that existed in the process industries.

The introduction of the Laplace transform to study the performance of feedback control systems supported its technological success in the engineering community. The theoretical basis for analysing the performance of PID control is considerably aided by the simple representation of an Integrator by the Laplace transform,

$$\left[\frac{1}{s}\right]$$

and a Differentiator using [s]. Conceptually, the PID controller is quite sophisticated and three different representations can be given. First, there is a symbolic representation (Figure 1.1(a)), where each of the three terms can be selected to achieve different control actions. Secondly, there is a time domain operator form (Figure 1.1(b)), and finally, there is a Laplace transform version of the PID controller (Figure 1.1(c)). This gives the controller an *s*-domain operator interpretation and allows the link between the time domain and the frequency domain to enter the discussion of PID controller performance.

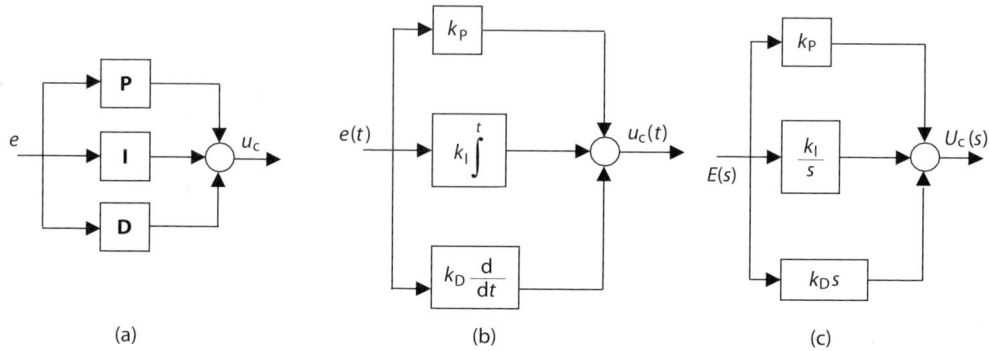

Key	Controller input e (system error)		PID control signal u_c	
Symbolic forms	e, u_c	Time domain forms	$e(t), u_c(t)$	Laplace domain forms	$E(s), U_c(s)$
Proportional gain	k_P	Integral gain	k_I	Derivative gain	k_D

Figure 1.1 PID controller representations.

This chapter concentrates on some basic structural features of the controller and reports on some industrial and implementation aspects of the controller.

The learning objectives for the chapter are to:

- Explain the process background for control loop components and signals
- Introduce the forms of the three terms in the PID controller
- Discuss the engineering implementation of the different PID terms
- Examine current industrial PID control technology

1.1 Basic Industrial Control

As can be seen from the typical industrial control loop structure given in Figure 1.2, even simple process loops comprise more than four engineering components.

The main components can be grouped according to the following loop operations:

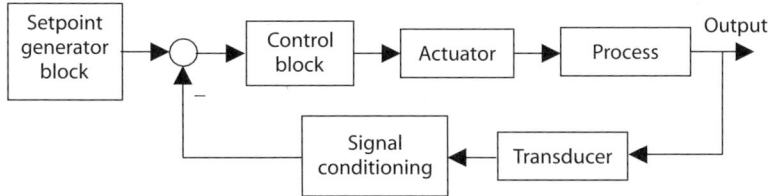

Figure 1.2 Components in a typical industrial control loop.

- **Process:** this is the actual system for which some specific physical variables are to be controlled or regulated. Typical process industry examples are boilers, kilns, furnaces and distillation towers.

- **Actuation:** the actuator is a process unit that supplies material or power input to the process. The actuator can be considered to act through amplification. For example, the control signal could be a small movement on a valve stem controlling a large flow of natural gas into a gas-fired industrial boiler.

- **Measurement:** the common adage is that *without measurement there will be no control*. Typically, the measurement process incorporates a transducer and associated signal processing components. The transducer will comprise a sensor to detect a specific physical property (such as temperature) and will output a representation of the property in a different physical form (such as voltage). It is quite possible that the measured output will be a noisy signal and that some of that noise will still manage to pass through the signal-conditioning component of the measurement device into the control loop.

- **Control:** the controller is the unit designed to create a stable closed-loop system and also achieve some pre-specified dynamic and static process performance requirements. The input to the controller unit is usually an error signal based on the difference between a desired setpoint or reference signal and the actual measured output.

- **Communications:** the above units and components in the control loop are all linked together. In small local loops, the control system is usually hardwired, but in spatially distributed processes with distant operational control rooms, computer communication components (networks, transmitters and receivers) will possibly be needed. This aspect of control engineering is not often discussed; however, the presence of communication delays in the loop may be an important obstacle to good control system performance.

To specify the controller, the performance objectives of the loop must be considered carefully, but in many situations it is after the loop has been commissioned and in use for a longer period of production activity that new and unforeseen process problems are identified. Consequently industrial control engineering often has two stages of activity: (i) control design and commissioning and (ii) post-commissioning control redesign.

Control Design and Commissioning

It is in constructing the first control design that explicit process modelling is likely to be used. Almost all physical processes are nonlinear in operation, but fortunately, many industrial processes run in conditions of steady operation to try to maximise throughput. This makes linearisation of the nonlinear process dynamics about steady state system operation a feasible analysis route for many processes. Subsequently, nonlinearity at different operating conditions can be overcome by gain scheduling or adaptation techniques. Although linearity is the route to simple model construction, even a straightforward loop can lead to a complicated model if all the components are considered, and the result could be a block diagram model as shown in Figure 1.3.

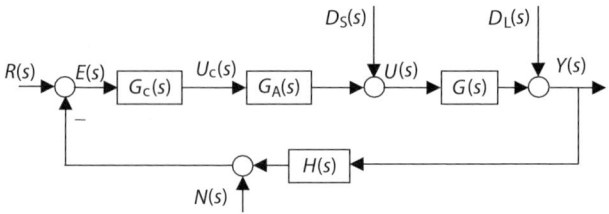

Key

Components		Loop signals	
$G(s)$	Process model	$Y(s)$	Process output
		$D_L(s)$	Load disturbance signal
$H(s)$	Measurement process model	$N(s)$	Measurement noise
$G_c(s)$	Controller unit	$R(s)$	Setpoint or reference signal
		$E(s)$	Process error input to controller
		$U_c(s)$	Controller output
$G_A(s)$	Actuator unit model	$U(s)$	Actuator ouput to process
		$D_S(s)$	Supply disturbance signal

Figure 1.3 Loop model diagram using Laplace transform terms.

In most cases, this finely detailed block diagram is usually only a step on the way to the standard control engineering textbook block diagram. It is usual to manipulate, refine, approximate and reduce the models to reach the standard control block diagram of Figure 1.4. At this point, many standard textbooks can be used to determine a control design (Doebelin, 1985; Seborg et al., 1989; Luyben, 1990; Åström and Hägglund, 1995; Dutton et al., 1997; Tan et al., 1999; Yu, 1999; Wilkie et al., 2002).

Using the standard block diagram, the performance objectives will include:

1. The selection of a controller to make the closed-loop system stable.

2. Achieving a reference-tracking objective and making the output follow the reference or setpoint signal.

3. If a process disturbance is present, the controller may have disturbance rejection objectives to attain.

4. Some noise filtering properties may be required in the controller to attenuate any measurement noise associated with the measurement process.

5. A degree of robustness in the controller design to model uncertainty may be required. In this case, the construction of a model and the subsequent model reduction may provide information about the probable level of uncertainty in the model. This information can be used when seeking control loop robust stability and robust control loop performance.

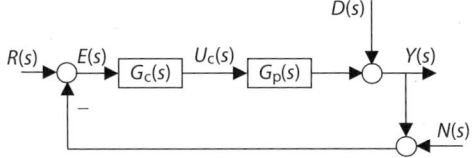

Key			
Composite process model	$G_p(s)$	Reference signal	$R(s)$
Controller unit	$G_c(s)$	Noise signal	$N(s)$
Disturbance signal	$D(s)$		

Figure 1.4 Standard control block diagram.

Post-Commissioning Control Redesign

Industrial processes always seem to have the ability to surprise the operators and control engineers by introducing a few physical effects that the process designer did not think of. Usually these take the form of either supply disturbances or load disturbances, or even both. These disturbances may be predictable or unpredictable; they may or may not be measurable. It is usually after these effects have manifested themselves in the process and upset process operation by degrading the product quality that the control engineer is asked to re-examine the controller settings. On the real plant, it is always essential to understand what is causing process disturbances to arise, since this may affect how to deal with them. For example, it could be that the control engineer has omitted an important physical process effect from the modelling step. Alternatively, it could be that a disturbance effect was erroneously removed in the modelling reduction stages. It may even mean that upstream processes need to be investigated and controlled better, or that a much better disturbance rejection control design is required. In the case of complex industrial processes, it may even be necessary to redesign the control system.

Some Post-Commissioning Control Issues

There are more opportunities for reconsidering the control design solution once the experience of running an industrial process is available. The control engineer may wish to re-examine the following issues:

1. *The controller design specification*
   Reference tracking performance is a common control textbook exercise, but in many industrial situations either the control design should only have used a disturbance rejection specification or the controller should have introduced a reasonable amount of disturbance rejection. It is also possible that insufficient measurement noise filtering is present in the loop. It is even possible that the structure of the control loop needs to be completely changed to eradicate poor control performance.

2. *Model-based techniques*
   With the availability of actual process data, the possibility of updating the modelling database arises. It is also possible that the modelling exercise did not fully consider the existence of potential process disturbances. In both cases, the structure and the data of the block diagrams of Figures 1.3 and 1.4 can be updated and controller redesign considered. The availability of verified models for the loop will enable model-based control design techniques to be used more accurately. The options for the control engineer include control redesign based on a more appropriate reference tracking – disturbance rejection loop specification, an assessment to ensure that PID control can meet the loop

performance specification, the use of a feedforward control design if the disturbances are measurable, or even the use of an intermediate measurement in a classical cascade control design.

3. *On-line methods*

   In most industrial plants there are hundreds of loops, and the plant technical staff may not have sufficient time or expertise to pursue a detailed control engineering exercise to optimise the performance of each loop. In these cases, some form of automated tuning procedure installed in the process control device or the SCADA system software may be the option to follow. These procedures tend to perform a process identification followed by a rule-based computation to determine the controller parameters. The rule used will achieve a specific control performance, so it is necessary to ensure that the computed controller matches the type of control needed by the actual process.

1.1.1 Process Loop Issues – a Summary Checklist

In control engineering it is very useful to consider the process operation carefully and to try to anticipate the likely behaviour of the process including all potential process disturbances. This knowledge can then be used in constructing a control design specification. A short checklist of key issues follows.

1. *System nonlinearity*

   The process relationship between input and output is often nonlinear for large input changes but locally linear for small input changes. This is a key assumption for the use of linear control around the base operating or linearising point, combined with gain scheduling and adaptation to accommodate the wider range of nonlinear dependencies. It is always useful to try to have a measure of the nonlinearity of the process.

2. *Input disturbances*

   Possible sources of disturbances to the material and energy flow into the process should be listed and incorporated in the model of the process. These are the process input disturbances.

3. *Output disturbances*

   Loading disturbances on the process often caused by downstream process demands should be investigated and modelled.

4. *Measurement noise disturbance*

   The measurement system should be examined to determine the levels of noise that can be expected on the measured variables in the control loop.

5. *Closed-loop requirements*

   Closed-loop requirements for the system should be enumerated. This has an influence on the type of control system to be installed and provides a basis for asking the question: *Can the desired performance be attained*? Closed-loop requirements are readily partitioned into:

   (a) Closed-loop stability – an essential closed-loop control system requirement

   (b) Closed-loop performance – comprising some, or all of:

   Reference tracking performance

   Supply disturbance rejection

   Load disturbance rejection

   Measurement noise rejection

   Robustness

   These different requirements use a mixture of time and frequency domain indices to delineate the specification. For example, reference tracking performance might use parameters like rise time t_r, Percentage overshoot $OS(\%)$, steady state error e_{ss} and settle time t_s, whilst measurement noise rejection might use a frequency domain performance specification for controller roll-off and robustness

might use the classical frequency domain specifications of gain margin *GM*, phase margin *PM* and maximum sensitivity M_S.

Why PID Control Is Important

PID control remains an important control tool for three reasons: past record of success, wide availability and simplicity in use. These reasons reinforce one another, thereby ensuring that the more general framework of digital control with higher order controllers has not really been able to displace PID control. It is really only when the process situation demands a more sophisticated controller or a more involved controller solution to control a complex process that the control engineer uses more advanced techniques. Even in the case where the complexity of the process demands a multi-loop or multivariable control solution, a network based on PID control building blocks is often used.

1.2 Three-Term Control

The signal framework of inputs and outputs for the three-term controller is shown in Figure 1.5 and is used to discuss the three terms of the PID controller.

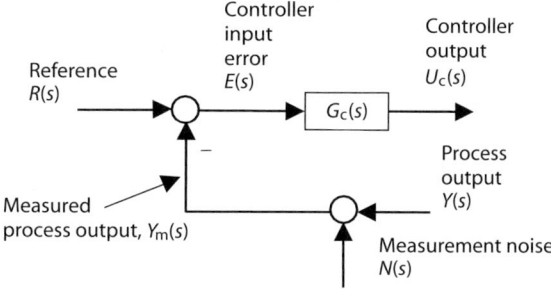

Figure 1.5 Controller inputs and outputs.

Proportional control

Proportional control is denoted by the P-term in the PID controller. It used when the controller action is to be proportional to the size of the process error signal $e(t) = r(t) - y_m(t)$. The time and Laplace domain representations for proportional control are given as:

Time domain $\qquad u_c(t) = k_P e(t)$

Laplace domain $\qquad U_c(s) = k_P E(s)$

where the proportional gain is denoted k_P. Figure 1.6 shows the block diagrams for proportional control.

Figure 1.6 Block diagrams: proportional control term.

Integral control

Integral control is denoted by the I-term in the PID controller and is used when it is required that the controller correct for any steady offset from a constant reference signal value. Integral control overcomes the shortcoming of proportional control by eliminating offset without the use of excessively large controller gain. The time and Laplace domain representations for integral control are given as:

Time domain $\quad u_c(t) = k_I \int^t e(\tau) d\tau$

Laplace Domain $\quad U_c(s) = \left[\dfrac{k_I}{s}\right] E(s)$

where the integral controller gain is denoted k_I. The time and Laplace block diagrams are shown in Figure 1.7.

Figure 1.7 Block diagrams: integral control term.

Derivative control

If a controller can use the rate of change of an error signal as an input, then this introduces an element of prediction into the control action. Derivative control uses the rate of change of an error signal and is the D-term in the PID controller. The time and Laplace domain representations for derivative control are given as:

Time domain $\quad u_c(t) = k_D \dfrac{de}{dt}$

Laplace domain $\quad U_c(s) = [k_D s] E(s)$

where the derivative control gain is denoted k_D. This particular form is termed *pure* derivative control, for which the block diagram representations are shown in Figure 1.8.

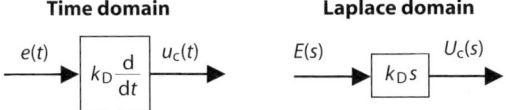

Figure 1.8 Block diagrams: derivative control term.

To use derivative control more care is needed than when using proportional or integral control. For example, in most real applications a pure derivative control term cannot be implemented due to possible measurement noise amplification and a modified term has to be used instead. However, derivative control has useful design features and is an essential element of some real-world control applications: for example, tachogenerator feedback in d.c. motor control is a form of derivative control.

Proportional and derivative control

A property of derivative control that should be noted arises when the controller input error signal becomes constant but not necessarily zero, as might occur in steady state process conditions. In these circumstances, the derivative of the constant error signal is zero and the derivative controller produces no control signal. Consequently, the controller is taking no action and is unable to correct for steady state offsets, for example. To avoid the controller settling into a somnambulant state, the derivative control term is always used in combination with a proportional term. This combination is called proportional and derivative, or PD, control. The formulae for simple PD controllers are given as:

Time domain $\quad u_c(t) = k_P e(t) + k_D \dfrac{de}{dt}$

Laplace domain $\quad U_c(s) = [k_P + k_D s] E(s)$

where the proportional gain is k_P and the derivative gain is k_D. The block diagrams for simple PD controllers are given in Figure 1.9.

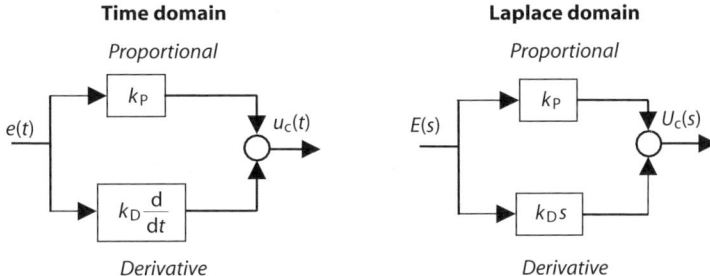

Figure 1.9 Block diagrams: proportional and derivative control.

1.2.1 Parallel PID Controllers

The family of PID controllers is constructed from various combinations of the proportional, integral and derivative terms as required to meet specific performance requirements. The formula for the basic parallel PID controller is

$$U_c(s) = \left[k_P + k_I \dfrac{1}{s} + k_D s \right] E(s)$$

This controller formula is often called the textbook PID controller form because it does not incorporate any of the modifications that are usually implemented to give a working PID controller. For example, the derivative term is not usually implemented in the pure form due to adverse noise amplification properties. Other modifications that are introduced into the textbook form of PID control include those used to deal with the so-called *kick* behaviour that arises because the textbook PID controller operates directly on the reference error signal. These modifications are discussed in more detail in Section 1.3.

This parallel or textbook formula is also known as a *decoupled* PID form. This is because the PID controller has three decoupled parallel paths, as shown in Figure 1.10. As can be seen from the figure, a numerical change in any *individual* coefficient, k_P, k_I or k_D, changes only the size of contribution in the path of the term. For example, if the value of k_D is changed, then only the size of the derivative action changes, and this change is *decoupled* and independent from the size of the proportional and integral terms. This decoupling of the three terms is a consequence of the *parallel* architecture of the PID controller.

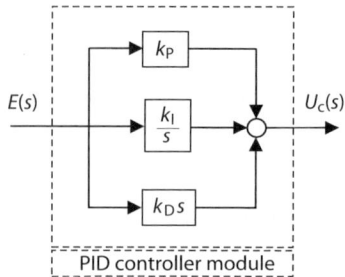

Figure 1.10 Parallel architecture for PID controller.

1.2.2 Conversion to Time constant PID Forms

In the parallel form of the PID controller, three simple gains k_P, k_I or k_D are used in the decoupled branches of the PID controller. The parallel PID control architecture can be given the following equivalent time-domain and Laplace s-domain mathematical representations:

Time-domain PID controller formula
$$u_c(t) = k_P e(t) + k_I \int^t e(\tau) d\tau + k_D \frac{de}{dt}$$

Transfer function PID controller formula
$$U_c(s) = \left[k_P + \frac{k_I}{s} + k_D s \right] E(s)$$

In the PID formulae, k_P is the proportional gain, k_I is the integral gain, k_D is the derivative gain, and the controller operates on the measured reference error time signal, $e(t) = r(t) - y_m(t)$, or equivalently, in the Laplace domain, $E(s) = R(s) - Y_m(s)$.

However, industrial representations of the PID controller often use a *time constant* form for the PID parameters instead of the decoupled form. This time constant form is easily derived from the parallel form and the analysis for the time domain formula is presented next. Begin from the parallel time domain form of the PID controller:

$$u_c(t) = k_P e(t) + k_I \int^t e(\tau) d\tau + k_D \frac{de}{dt}$$

The proportional gain k_P is first factored out to give

$$u_c(t) = k_P \left(e(t) + \frac{k_I}{k_P} \int^t e(\tau) d\tau + \frac{k_D}{k_P} \frac{de}{dt} \right)$$

Define two new time constants, viz

$$\tau_i = \frac{k_P}{k_I} \quad \text{and} \quad \tau_d = \frac{k_D}{k_P}$$

Then

$$u_c(t) = k_P \left(e(t) + \frac{1}{\tau_i} \int^t e(\tau) d\tau + \tau_d \frac{de}{dt} \right)$$

In this new time constant form, k_P is the proportional gain, τ_i is the integral time constant and τ_d is the derivative time constant.

A similar analysis follows for the transfer function expressions, viz

$$U_c(s) = \left[k_P + \frac{k_I}{s} + k_D s\right] E(s)$$

is rearranged to give

$$U_c(s) = k_P \left[1 + \frac{k_I}{k_P s} + \frac{k_D}{k_P} s\right] E(s)$$

Define

$$\tau_i = \frac{k_P}{k_I} \quad \text{and} \quad \tau_d = \frac{k_D}{k_P}$$

Then

$$U_c(s) = k_P \left[1 + \frac{1}{\tau_i s} + \tau_d s\right] E(s)$$

This analysis has the same definitions for k_P, τ_i and τ_d as the new time domain formula above. The parallel form and the industrial time constant forms are given in Table 1.1.

Table 1.1 Parallel and time constant forms for the PID controller.

	Time domain	Laplace s-domain
Parallel	$u_c(t) = k_P e(t) + k_I \int^t e(\tau)d\tau + k_D \frac{de}{dt}$	$U_c(s) = \left[k_P + \frac{k_I}{s} + k_D s\right] E(s)$
Time constant	$u_c(t) = k_P \left(e(t) + \frac{1}{\tau_i} \int^t e(\tau)d\tau + \tau_d \frac{de}{dt}\right)$	$U_c(s) = k_P \left[1 + \frac{1}{\tau_i s} + \tau_d s\right] E(s)$

The industrial time constant form of the PID controller has led to industry developing a specific PID terminology that involves terms like proportional band and reset action. It is also useful to note that the decoupled property of the parallel PID formula has been lost in the time constant PID formulae. When analysis proofs are being developed it is often more convenient to use the decoupled PID form. The following two simple examples show the effect of the changes.

Example 1.1

Consider the effect in the time constant PID formula of changing the value of k_P. Such a change directly affects the size of the contribution of *all* three terms in the controller; thus unlike the decoupled PID, the change in k_P now interacts with both the I-term and the D-term. It is important to note this effect when setting up a PID controller tuning procedure based on the time constant form.

Example 1.2

Consider how the I-term is *switched out* in the decoupled form. This is simply a matter of setting $k_I = 0$; as a limiting process this would be the operation $k_I \to 0$. However, in the time constant PID formula, switching out the I-term requires the integral time constant τ_i to be made extremely large, or the limiting operation $\tau_i \to \infty$ has to be considered; this might not be so useful in analytical problems or even in an on-line tuning procedure.

1.2.3 Series PID Controllers

Some early PID controllers used pneumatic hardware for which a series transfer function representation was an appropriate mathematical description. To maintain continuity in later analogue PID devices, some manufacturers retained this series structure. However, current modern PID controllers are likely to be digital and in parallel form, but these series PID formulae are still found in some industrial PID controller manuals. A block diagram of the series PID structure is found in Figure 1.11, and this can be used to determine the overall $G_{\text{series}}(s)$ transfer function for this type of PID structure.

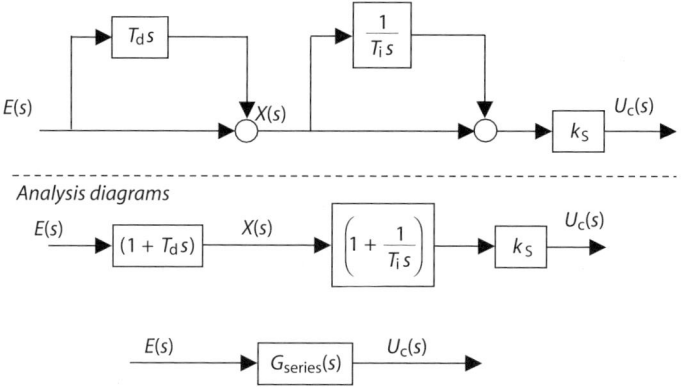

Figure 1.11 Series PID controller block diagram.

From Figure 1.11, the basic series PID control law is given in terms of a product of transfer functions as

$$U_c(s) = [G_{\text{series}}(s)]E(s) = \left[k_S\left(1 + \frac{1}{T_i s}\right)(1 + T_d s)\right]E(s)$$

The series PID controller can be converted to the parallel and then the time constant PID forms.

From the Series PID Controller to the Parallel PID Formula

First, the series transfer function form is multiplied out and then the terms of the parallel PID architecture identified as follows:

$$U_c(s) = \left[k_S\left(1 + \frac{1}{T_i s}\right)(1 + T_d s)\right]E(s)$$

becomes

$$U_c(s) = \left[k_S\left(\left(1 + \frac{T_d}{T_i}\right) + \frac{1}{T_i s} + T_d s\right)\right]E(s)$$

and hence

$$U_c(s) = \left[k_S\left(1 + \frac{T_d}{T_i}\right) + \frac{k_S}{T_i s} + k_S T_d s \right] E(s)$$

Comparison with the parallel form

$$U_c(s) = \left[k_P + \frac{k_I}{s} + k_D s \right] E(s)$$

yields

$$k_P = k_S\left(1 + \frac{T_d}{T_i}\right), \quad k_I = \frac{k_S}{T_i} \quad \text{and} \quad k_D = k_S T_d$$

The time domain form for the series form is easily identified from the above as

$$u_c(t) = k_S\left(1 + \frac{T_d}{T_i}\right) e(t) + \frac{k_S}{T_i} \int_0^t e(\tau) d\tau + k_S T_d \frac{de}{dt}$$

Thus, the series PID form is linked to the parallel PID architecture and has a time domain PID form. It is a short step from this to a connection with the time constant PID structure.

From the Series PID Controller to the Time constant PID Formula

Using the expanded form for the series PID controller, a rescaling is used to arrive at the usual time constant form. Begin from

$$U_c(s) = \left[k_S\left(\left(1 + \frac{T_d}{T_i}\right) + \frac{1}{T_i s} + T_d s \right) \right] E(s)$$

Then

$$U_c(s) = k_S\left(1 + \frac{T_d}{T_i}\right)\left[1 + \left(1 + \frac{T_d}{T_i}\right)^{-1} \frac{1}{T_i s} + \left(1 + \frac{T_d}{T_i}\right)^{-1} T_d s \right] E(s)$$

and

$$U_c(s) = k_S\left(\frac{T_i + T_d}{T_i}\right)\left[1 + \frac{1}{(T_i + T_d)s} + \left(\frac{T_i T_d}{T_i + T_d}\right) s \right] E(s)$$

Comparison with the usual time constant PID form yields

$$U_c(s) = \left[k_P\left(1 + \frac{1}{\tau_i s} + \tau_d s\right) \right] E(s)$$

to give

$$k_P = k_S\left(\frac{T_i + T_d}{T_i}\right), \quad \tau_i = T_i + T_d \quad \text{and} \quad \tau_d = \left(\frac{T_i T_d}{T_i + T_d}\right)$$

Even though there are links between the series and time constant PID formulae, the series form is still used and cited. Some claim that the series form is easier to tune manually, so it is included here for completeness.

1.2.4 Simple PID Tuning

The original technology for industrial PID controllers was analogue, and these controllers usually had a very simple interface for manually tuning the controller. Indeed, it was not unknown to have a process controller interface comprising three simple dials marked P, I and D respectively! This reflected the fact that for many of the simple processes found in the process industries a manual procedure was often quite adequate. One of the great strengths of PID is that for simple plant there are straightforward correlations between plant responses and the use and adjustment of the three terms in the controller. There are two parts to the tuning procedure: one is how to choose the structure of the PID controller (namely which terms should be used) and the second is how to choose numerical values for the PID coefficients or tune the controller.

Choosing the Structure of a PID Controller

For the structure of the controller, industrial personnel often relied on knowledge of the heuristic behaviour of PID terms as captured in Table 1.2.

Table 1.2 Tuning effects of PID controller terms.

	Reference tracking tuning — Step reference		Disturbance rejection tuning — Constant load disturbance	
	Transient	**Steady state**	**Transient**	**Steady state**
P	Increasing $k_P > 0$ speeds up the response	Increasing $k_P > 0$ reduces but does not eliminate steady state offset	Increasing $k_P > 0$ speeds up the response	Increasing $k_P > 0$ reduces but does not eliminate steady state offset
I	Introducing integral action $k_I > 0$ gives a wide range of response types	Introducing integral action $k_I > 0$ eliminates offset in the reference response	Introducing integral action $k_I > 0$ gives a wide range of response types	Introducing integral action $k_I > 0$ eliminates steady state offsets
D	Derivative action $k_D > 0$ gives a wide range of responses and can be used to tune response damping	Derivative action has no effect on steady state offset	Derivative action $k_D > 0$ gives a wide range of responses and can be used to tune response damping	Derivative action has no effect on steady state offset

Table 1.2 is very similar to a toolbox where each term of the controller can be selected to accomplish a particular closed-loop system effect or specification. Consider the following two examples.

Example 1.3

If it is necessary to remove steady state offsets from a closed-loop output response, the table indicates that a D-term will not do this and that a P-term will reduce the offset if the proportional gain k_P is increased, but that an I-term will eliminate the offset completely. Therefore an integral term would be chosen for inclusion in the controller structure.

Example 1.4

Suppose that the control engineer has elected to use integral action to eliminate constant steady state process disturbance offset errors but now wishes to speed up the closed-loop system response. The table shows that increasing the proportional gain k_P will have just this effect, so both proportional (P) action and integral (I) action would be selected, giving a PI controller solution.

The examples show how the PID controller structure is selected to match the desired closed-loop performance. This simple selection process can be followed as a flow diagram, as shown in Figure 1.12.

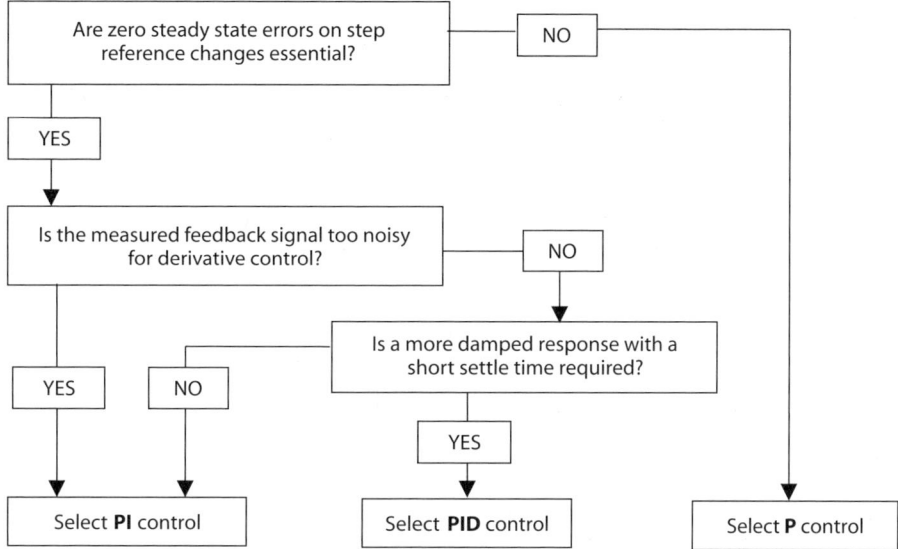

Figure 1.12 PID term selection flowchart.

The selection flowchart shows that PID control is actually a family of controllers, with the labels, P, PI, PD and PID indicating particular controller structures from the PID family. Table 1.3 shows all the members of the PID control family giving both the time domain and transfer function formulae.

Tuning the PID Controller
The second part of setting up a PID controller is to tune or choose numerical values for the PID coefficients. Many industrial process companies have in-house manuals that provide guidelines for the tuning of PID controllers for particular process plant units. Thus for simple processes it is often possible to provide rules and empirical formulae for the PID controller tuning procedure. Some of these manuals base their procedures on the *pro forma* routines of the famous Ziegler–Nichols methods and their numerous extensions of the associated rules (Ziegler and Nichols, 1942). The two Ziegler–Nichols methods use an on-line process experiment followed by the use of rules to calculate the numerical values of the PID coefficients. In the 1980s, when analogue control was being replaced by digital processing hardware, industrial control companies took the opportunity to develop new PID controller methods for use with the new ranges of controller technology appearing. Consequently, the Ziegler–Nichols methods became the focus of research and have since, become better understood. New versions of the Ziegler–Nichols procedures were introduced, notably the Åström and Hägglund relay experiment (Åström and Hägglund, 1985). In many applications, the implicit underdamped closed-loop

Table 1.3 Members of the PID control family.

	Time domain form	Laplace domain form
P	$u_C(t) = [k_P]e(t)$	$U_C(s) = [k_P]E(s)$
I	$u_C(t) = k_I \int^t e(\tau)d\tau$	$U_C(s) = \left[\dfrac{k_I}{s}\right]E(s)$
D	$u_C(t) = k_D \dfrac{de}{dt}$	$U_C(s) = [k_D s]E(s)$
PI	$u_C(t) = k_P e(t) + k_I \int^t e(\tau)d\tau$	$U_C(s) = \left[k_P + \dfrac{k_I}{s}\right]E(s)$
PD	$u_C(t) = k_P e(t) + k_D \dfrac{de}{dt}$	$U_C(s) = [k_P + k_D s]E(s)$
PID	$u_C(t) = k_P e(t) + k_I \int^t e(\tau)d\tau + k_D \dfrac{de}{dt}$	$U_C(s) = \left[k_P + \dfrac{k_I}{s} + k_D s\right]E(s)$

performance inherent in the original Ziegler–Nichols design rules was found to be unacceptable. The result was an extensive development of the rule-base for PID controller tuning. O'Dwyer (1998a,b) has published summaries of a large class of the available results.

Continuing competitive pressures in industry have led to a constant need for continual improvements in control loop performance. One result of these trends is that industry is much better at being able to specify the type of performance that a control system has to deliver. Table 1.4 shows some widely used control loop performance indices.

Table 1.4 Common indices for control loop performance specification.

Time domain performance indices	
Closed-loop time constant τ_{CL}	Closed-loop pole positions s_i, $i = 1,\ldots,n$
Rise time t_r (10%, 90%)	Settle time t_s (2%) or t_s (5%)
Percentage overshoot $OS(\%)$	Steady state offset e_{ss}, e_{Rss} or e_{Dss}
Damping ratio ζ	Disturbance peak value y_{Dpeak}
Natural frequency ω_n	Disturbance settle time D_{ts} (2%) or D_{ts} (5%)
Frequency domain indices	
Gain margin GM	Phase Margin PM
Maximum sensitivity S_{max}	Delay margin DM

Some of these performance indices are given typical values; for example, closed-loop system damping is often specified to lie in the range 0.6 to 0.75, whilst phase margin is often taken to lie in the range 45° to 60°. Other control performance specifications will depend on the particular type of system, and past engineering experience may be used to provide numerical values. In some cases, a specification may be especially critical and specific numerical values that have to be achieved will be given. Table 1.5 shows some typical examples based on past industrial experience (Wilkie *et al.*, 2002).

Table 1.5 Simple control system performance specifications.

Ship autopilot control	Specification	Design solution
No overshoot to a step demand in heading, but good speed of response	$OS(\%) = 0$ $e_{ss} = 0$	Design for critical damping, $\zeta = 1$. Use Proportional and Integral Control
Liquid tank level system	**Specification**	**Design solution**
Good speed of response, but steady state accuracy not required	$\tau_{CL} \ll \tau_{OL}$ e_{ss} small	Design for first-order response type. Use Proportional Control
Gas turbine temperature control	**Specification**	**Design solution**
Load changes frequent. Steady state accuracy essential. Fast disturbance rejection needed within five minutes	Disturbance rejection design $e_{Dss} = 0$ $D_{ts} < 5$ minutes	Underdamped response design for speed of response. Use Proportional and Integral Control

When the closed-loop specification involves the more precise system performance indices, the need for systematic, accurate and reliable procedures to select the PID controller coefficients emerges. Textbooks that give scientific PID coefficient calculation procedures include those due to Åström and Hägglund (1995), Tan *et al.* (1999) and Wilkie *et al.* (2002).

1.3 PID Controller Implementation Issues

Industrial PID control usually comes in a packaged form, and before attempting a tuning exercise, it is invaluable to understand *how* the PID controller has been implemented. This usually means a detailed examination of the manufacturer's User Manual, and possibly a meeting and discussion with the controller manufacturer's personnel. Even then, many of the manufacturer's innovations in PID control may remain commercially sensitive, since for a number of the problems arising in industrial PID control manufacturers have introduced customised features, and details of these may not be available to the user or installer. However, there are several common problems in the implementation of the terms of the PID controller and it is useful to examine general solutions and terminology even if specific industrial details are not available. Table 1.6 shows some common process control problems and the appropriate PID implementation solution.

To perform well with the industrial process problems of Table 1.6, the parallel PID controller requires modification. In this section, detailed consideration is given to the bandwidth-limited derivative term, proportional and derivative kick, anti-windup circuit design and reverse acting control.

Table 1.6 Process control problems and implementing the PID controller.

Process control problem	PID controller solution
Measurement noise	
■ Significant measurement noise on process variable in the feedback loop ■ Noise amplified by the pure derivative term ■ Noise signals look like high frequency signals	■ Replace the pure derivative term by a bandwidth limited derivative term ■ This prevents measurement noise amplification
Proportional and derivative kick	
■ P- and D-terms used in the forward path ■ Step references causing rapid changes and spikes in the control signal ■ Control signals are causing problems or outages with the actuator unit	■ Move the proportional and derivative terms into feedback path ■ This leads to the different forms of PID controllers which are found in industrial applications
Nonlinear effects in industrial processes	
■ Saturation characteristics present in actuators ■ Leads to integral windup and causes excessive overshoot ■ Excessive process overshoots lead to plant trips as process variables move out of range	■ Use anti-windup circuits in the integral term of the PID controller ■ These circuits are often present and used without the installer being aware of their use
Negative process gain	
■ A positive step change produces a wholly negative response ■ Negative feedback with such a process gives a closed-loop unstable process	■ Use the option of a reverse acting PID controller structure

1.3.1 Bandwidth-Limited Derivative Control

The Problem

The measured process variable of the feedback loop may contain excessive noise. Such noise is modelled as a high-frequency phenomenon and this will be amplified by the pure derivative term in the three-term controller. Figure 1.5 shows the situation of the controller in the loop, and from this figure an analysis of the noise filtering characteristics is derived. The controller error in the time domain is

$$e(t) = r(t) - y_m(t) = (r(t) - y(t)) - n(t)$$

and in the Laplace domain is

$$E(s) = (R(s) - Y(s)) - N(s)$$

Clearly, the operation of the controller on the error signal comprises a component due to a noise-free error and one due to the measurement noise, viz

$$\begin{aligned}U_c(s) &= [G_c(s)]E(s)\\ &= [G_c(s)]((R(s) - Y(s)) - N(s))\end{aligned}$$

and

$$U_c(s) = U_c^{nf}(s) - U_{noise}(s)$$

where $U_c^{nf}(s)$ is the noise-free control term and $U_{noise}(s) = [G_c(s)]N(s)$.

Consider now a decomposition of the noise component where the controller $G_c(s)$ takes the time constant PID form:

$$\begin{aligned} U_{noise}(s) &= [G_c(s)]N(s) \\ &= \left[k_P \left(1 + \frac{1}{\tau_i s} + \tau_d s \right) \right] N(s) \\ &= [G_P(s) + G_I(s) + G_D(s)]N(s) \end{aligned}$$

Hence

$$U_{noise}(s) = U_{noise}^P(s) + U_{noise}^I(s) + U_{noise}^D(s)$$

where

$$U_{noise}^P(s) = [G_P(s)]N(s), \; U_{noise}^I(s) = [G_I(s)]N(s), \; U_{noise}^D(s) = [G_D(s)]N(s)$$

and

$$G_P(s) = k_P, \; G_I(s) = \frac{k_P}{\tau_i s}, \; G_D(s) = k_P \tau_d s$$

Bode plots of the three components $G_P(s), G_I(s), G_D(s)$ of the PID controller will show how the P-, I- and D-terms respond to a noise signal input. This is shown in Figure 1.13, which has been constructed for $k_P = 5, \tau_i = 5$ and $\tau_d = 5$.

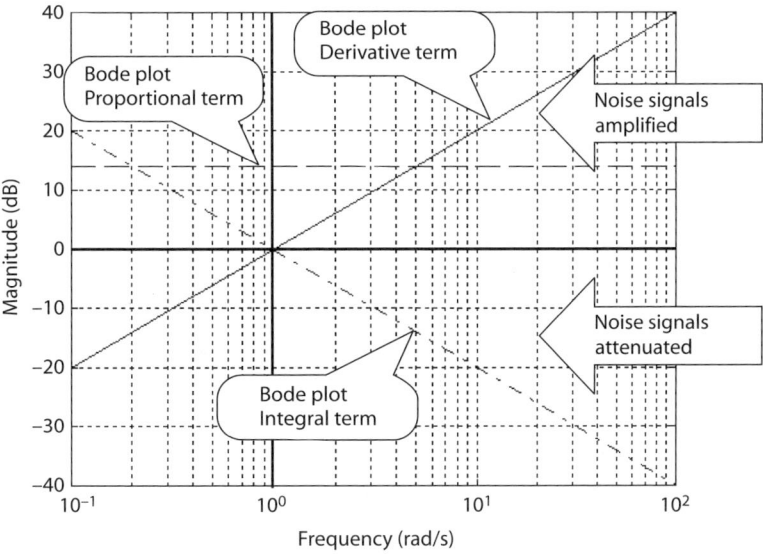

Figure 1.13 Bode plot of terms in PID controller.

This figure clearly shows that the effects of the PID terms on high-frequency measurement noise are as follows:

- **Proportional term:** the given proportional gain is $k_P = 5 = 13.98$ dB, and in this case noise amplification occurs but is fixed and constant across the frequency range.

- **Integral term:** the magnitude Bode plot for the integral term shows a roll-off of 20 dB per decade. Thus, high-frequency measurement noise is attenuated by the integral term.

- **Derivative term:** the magnitude Bode plot for the derivative term shows a *roll-on* of gain at +20 dB per decade. This produces increasing amplification as the noise frequency increases. It can also be seen from the Bode plot that a pure derivative term has no effect on steady (constant) signals, since in the low-frequency range the plot shows near zero gain and hence the attenuation of low-frequency signals occurs.

The Remedy

Measurement noise is quite likely to be present in practical applications and the undesirable noise amplification property of pure derivative action is prevented by using a low-pass filter in the D-term; the result is a bandwidth-limited derivative term.

A low-pass filter has a transfer function form

$$G_f(s) = \left[\frac{1}{\tau_f s + 1} \right]$$

This is included into the derivative term as follows:

$$U^D_{noise}(s) = [G_f(s)][G_D(s)]N(s)$$

$$= \left[\frac{1}{\tau_f s + 1} \right] [k_P \tau_d s] N(s)$$

$$= \left[k_P \left(\frac{\tau_d s}{\tau_f s + 1} \right) \right] N(s)$$

and

$$U^D_{noise}(s) = [G_{mD}(s)]N(s)$$

where the modified derivative term is defined by

$$G_{mD}(s) = \left[k_P \left(\frac{\tau_d s}{\tau_f s + 1} \right) \right]$$

The high-frequency gain of the modified derivative term is

$$k_D^\infty = \lim_{\omega \to \infty} |G_{mD}(j\omega)| = k_P \frac{\tau_d}{\tau_f}$$

In some commercial PID controllers, the following re-parameterisation of the filter time constant is often used. Set $\tau_f = \tau_d/n$ and the modified derivative term becomes

$$G_{mD}(s) = \left[k_P \left(\frac{\tau_d s}{(\tau_d/n)s + 1} \right) \right]$$

from which the high-frequency gain is given by

$$k_D^\infty = k_P \frac{\tau_d}{\tau_f} = k_P n$$

Typical commercial PID controller devices select n in the range $5 \leq n \leq 20$. Figure 1.14 shows the new noise filtering characteristics of the modified derivative form. Using the values $k_P = 5, \tau_d = 0.2$ and $n = 5$, the high-frequency signal amplification is now limited to $k_D^\infty = 25 = 27.95$ dB, as can be seen in the figure.

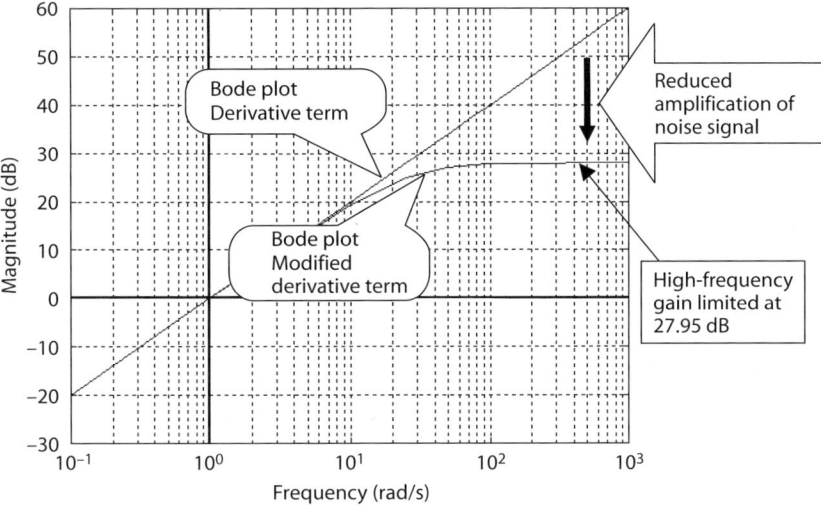

Figure 1.14 Magnitude Bode plot for the pure differential term and the modified derivative term for the parameter values $k_D^\infty = 25 = 27.95$ dB.

Inserting the new modified derivative term gives a new PID controller form:

$$U_c(s) = [G_P(s) + G_I(s) + G_{mD}(s)]E(s)$$
$$= \left[k_P + k_P \tau_i s + k_P \left(\frac{\tau_d s}{(\tau_d/n)s + 1}\right)\right] E(s)$$
$$= \left[k_P \left(1 + \frac{1}{\tau_i s} + \frac{\tau_d s}{(\tau_d/n)s + 1}\right)\right] E(s)$$
$$= G_c(s)E(s)$$

and the controller is given by

$$G_c(s) = k_P \left(1 + \frac{1}{\tau_i s} + \frac{\tau_d s}{(\tau_d/n)s + 1}\right)$$

Often different manufacturers use different notation for the PID controller involving different forms for the modified D-term and different values for n. For example, an industrial user manual might show the PID controller as

$$U_c(s) = [G_{pid}(s)]E(s) = G\left[1 + \frac{1}{TIs} + \frac{TDs}{TFs + 1}\right] E(s)$$

where G, TI, TD and TF are used to denote the terms $k_P, \tau_i, \tau_d, \tau_f$ respectively.

1.3.2 Proportional Kick

The Problem

Proportional kick is the term given to the observed effect of the proportional term in the usual parallel PID structure on rapid changes in the reference signal. Recall first the parallel PI controller structure as shown in Figure 1.15.

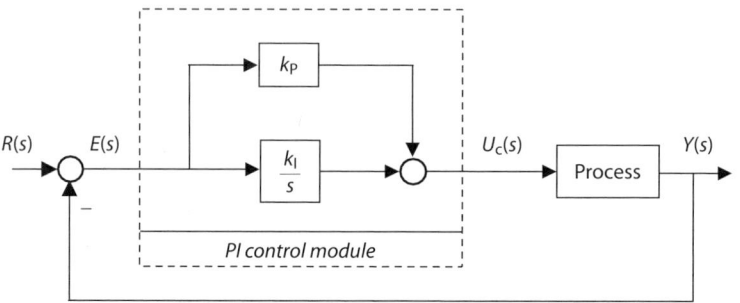

Figure 1.15 Parallel PI control structure.

Using Figure 1.15, if the process is under control and the outputs of the system are steady then the error signal $E(s) = R(s) - Y(s)$ will be close to zero. Consider now the effect of a step change in the reference input $R(s)$. This will cause an immediate step change in $E(s)$ and the controller will pass this step change directly into the controller output $U_c(s)$ via the proportional term $k_P E(s)$. In these circumstances, the actuator unit will experience a rapidly changing command signal that could be detrimental to the operation of the unit; the actuator will receive a proportional *kick*. A typical sharp spike-like change in the control signal is seen in Figure 1.16, which shows output and control signals for this proportional kick problem.

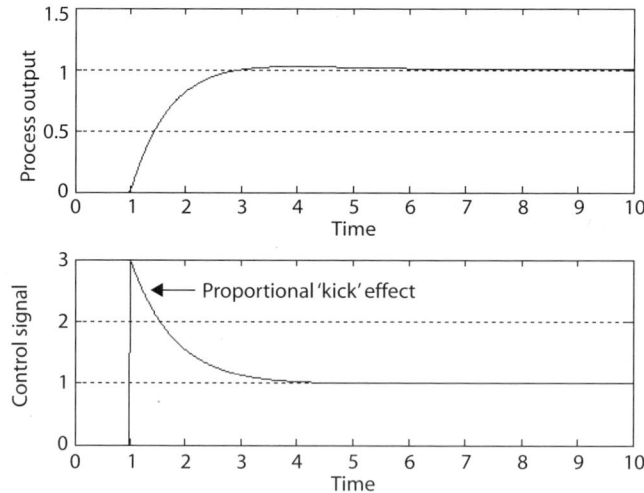

Figure 1.16 Process output and control signals showing proportional kick effects due to unit step change in reference signal at $t = 1$.

The Remedy

The remedy for proportional kick is simply to restructure the PI controller, moving the proportional term into the feedback path, as shown in Figure 1.17.

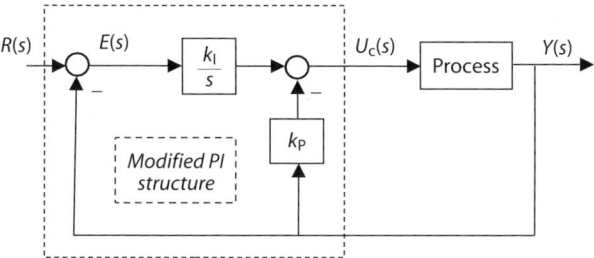

Figure 1.17 Restructured PI controller removing proportional kick effects.

The step response and control signal for this modified PI structure typically look like those of Figure 1.18. The spike on the control signal has been removed and the control signal is no longer an aggressive-looking signal. Meanwhile, the process output signal is now a little slower.

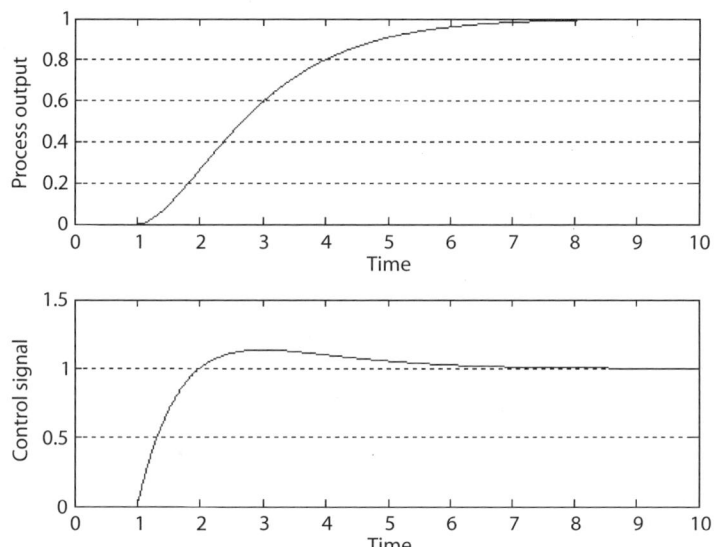

Figure 1.18 Typical output and control signals for the restructured PI controller (removing proportional kick).

The equation for the restructured form of the PI controller is

$$U_c(s) = \left[\frac{k_I}{s}\right] E(s) - [k_P] Y(s)$$

This structure shows the integral (I) term to be on the setpoint error signal and the proportional (P) term to be on the measured output or process variable signal. This has lead to the industrial terminology where this structure is called I–P, meaning I on error and P on process variable. Clearly, a new set of PID controllers is possible by restructuring the controller in this way.

1.3.3 Derivative Kick

The Problem

Derivative kick is very similar to proportional kick (Section 1.3.2). Figure 1.19 shows a parallel ID–P control system. This structure is read as "integral (I) and derivative (D) on error and proportional (P) on process variable". The derivative term is also the modified derivative term from Section 1.3.1. Thus with this particular form of three-term controller, the proportional (P) on process variable has eliminated proportional kick and the presence of the modified derivative term has reduced high-frequency noise amplification.

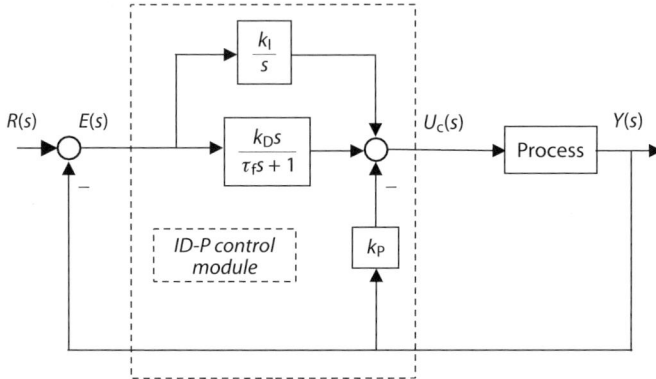

Figure 1.19 Three-term ID–P control system, with modified derivative term.

The equation for the ID-P control action is

$$U_c(s) = \left[\frac{k_I}{s} + \frac{k_D s}{(\tau_f s + 1)}\right] E(s) - [k_P] Y(s)$$

If the output of the process is under control and steady then the setpoint error signal $E(s) = R(s) - Y(s)$ will be close to zero. A subsequent step change in the reference signal $R(s)$ will cause an immediate step change in the error signal $E(s)$. Since the proportional term of the controller operates on the process output, proportional kick will not occur in the control signal; however, the output of the derivative term

$$\frac{k_D s}{(\tau_f s + 1)} E(s)$$

must be considered. Differentiating a step change will produce an impulse-like spike in the control signal and this is termed derivative kick. Figure 1.20 shows typical output and control signals for this problem. Note the very sharp spike-like change in the control signal. This control signal could be driving a motor or a valve actuator device, and the kick could create serious problems for any electronic circuitry used in the device.

The Remedy

If the derivative term is repositioned so that the reference signal is not differentiated, then derivative kick is prevented. The ID–P controller transfer function is

$$U_c(s) = \left[\frac{k_I}{s} + \frac{k_D s}{(\tau_f s + 1)}\right] E(s) - [k_P] Y(s)$$

1.3 PID Controller Implementation Issues

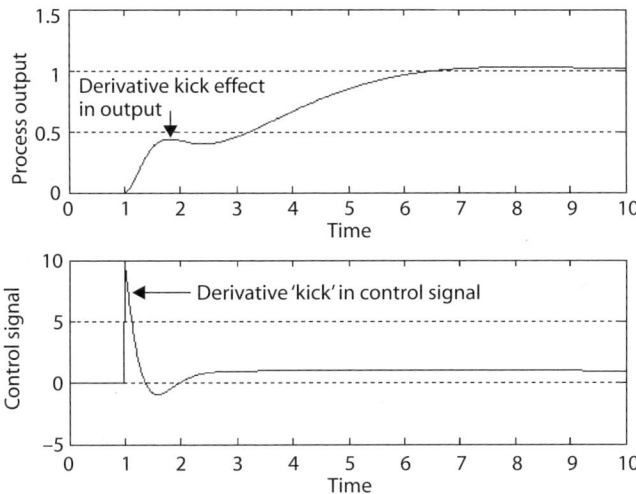

Figure 1.20 Output and control signals showing derivative kick in the control signal (unit step change in reference at $t = 1$).

and hence removing the operation of the derivative term on the reference gives

$$U_c(s) = \left[\frac{k_\mathrm{I}}{s}\right] E(s) - \left[k_\mathrm{P} + \frac{k_\mathrm{D} s}{(\tau_f s + 1)}\right] Y(s)$$

This new I-PD controller is shown in Figure 1.21. In this case, the I-PD terminology denotes Integral term on error and Proportional and Derivative terms on process variable or measured output.

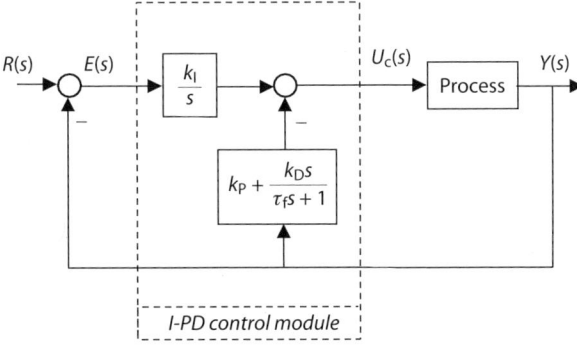

Figure 1.21 Three-term I-PD control for preventing derivative kick (and proportional kick).

Typical step response and control signals for the modified I-PD control structure are shown in Figure 1.22. In the figure, it can be seen that the spike on the control signal due to derivative kick has been removed and that no proportional kick is present either.

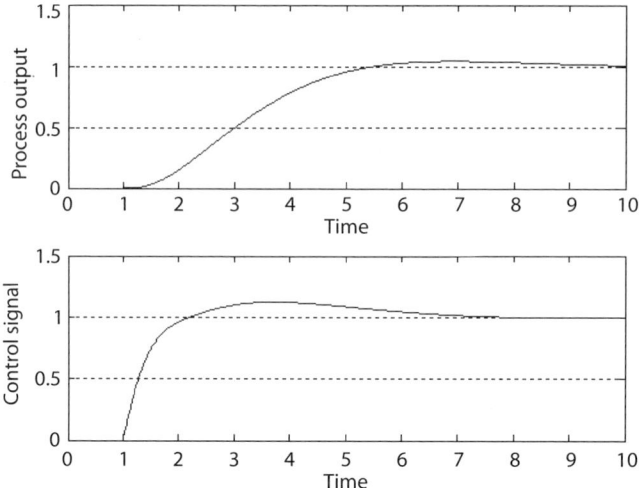

Figure 1.22 Typical output and control signals for the three-term I-PD controller showing that derivative kick has been removed.

1.3.4 Integral Anti-Windup Circuits

The Problem

Nonlinearity occurs in industrial plant in several ways. The process plant may be nonlinear, so that different operating conditions will have different process models and different dynamics. The most common solution to maintain good control performance over a range of nonlinear operating conditions is to schedule a set of PID controllers where each has been designed to achieve good performance for a specific operating point. Many PID controller units will offer gain or controller schedule facilities.

A very different problem for PID control arises from the nonlinear behaviour of process actuators. Many of these actuator devices have a limited range of input and output operation. For example, valves have a fully open position, a fully closed position and a flow characteristic in-between that could be linear or nonlinear. A typical characteristic is shown in Figure 1.23.

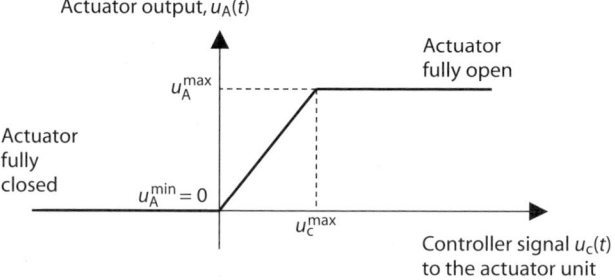

Figure 1.23 Typical actuator saturation characteristic.

The fully open position imposes a real limit on the effectiveness of any control action, because if the actuator output is in the saturation region u_A^{max} then the system is in open loop control because $u_C^{max} < u_C(t)$ and the control signal is having no effect on the actuator output. Actuator saturation is a common industrial nonlinearity that creates a windup effect.

To understand the integral wind up effect, a simple simulation of a switched step response of a pure I-controller configured with a saturating actuator can be used. This setup is shown in Figure 1.24.

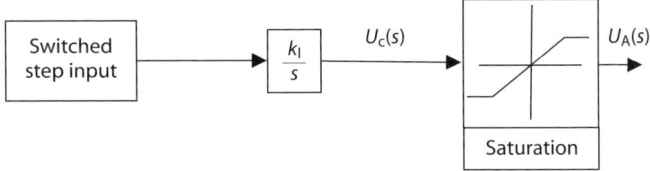

Figure 1.24 Pure I-control in cascade with actuator saturation characteristic.

Figure 1.25 comprises three plots showing the input switched step, I-control signal and actuator signal for control setup.

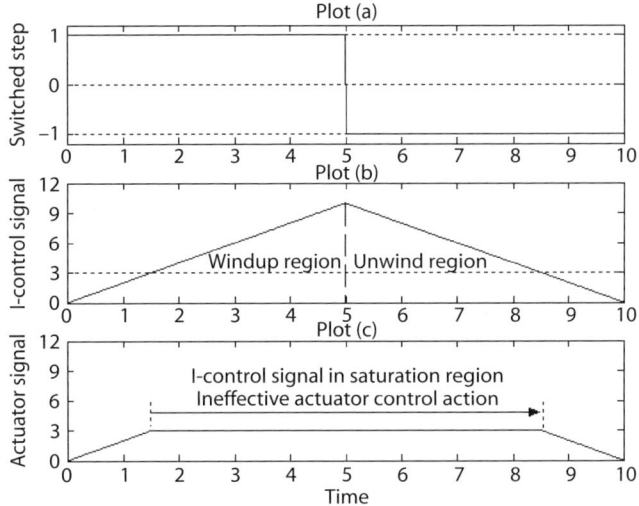

Figure 1.25 Input switched step, I-control and actuator signals.

From Figure 1.25, plot (a) shows the input switch step beginning at +1 from $t = 0$ and switching to −1 when $t = 5$. Plot (b) shows the I-control signal $u_c(t)$ and plot (c) shows the actual actuator output signal $u_A(t)$. As can be seen from plot (b), when the step input is at +1, the I-control signal $u_c(t)$ ramps up to a peak of 10 at $t = 5$. This is the windup period of the integral term. In the same time interval $0 \leq t \leq 5$, the actuator output $u_A(t)$ viewed in plot (c) has gone into the saturation region and a constant step signal is driving the process. When the step input switches to −1 at $t = 5$, the I-control signal $u_c(t)$ begins to decrease in value and the integral term unwinds, as shown in plot (b). Nevertheless, during most of this unwind period the actuator output $u_A(t)$ remains in the saturation region and the process continues to be driven by a constant step input signal (plot (c)). There are several practical consequences of this windup behaviour.

1. If the process is under closed-loop control and the control signal enters a saturation region, then the process reverts to *open* loop control and this is potentially dangerous if the system is open loop unstable.

2. In Figure 1.25, Plot (c) shows that the applied control signal at the actuator output, $u_A(t)$, spends significant periods as a constant signal. This may cause the process to exhibit excessive overshoot in the process output.

3. While the actuator signal $u_A(t)$ is in the constant signal period, the actual control signal $u_c(t)$ is requesting very different control action. This action does not begin to reach the process until the integral has first wound up and then unwound, when the controller becomes effective once more. Thus integrator windup delays the effective action of the control signal.

The Remedy

The remedy for integrator windup is to switch off the integral action as soon as the control signal enters the saturation region and switch the integral action back on as soon as the controller re-enters the linear region of control. This switching is implemented using an anti-windup circuit. In most commercial PID controllers an anti-windup circuit will be present, but the details of the circuit will not usually be available to the end-user. It is simply that many engineers have used their ingenuity on this problem and found different ways to design an anti-windup circuit. Some of these designs are commercially proprietary to particular controllers. For the user, it is usually sufficient to know that anti-windup protection is present. For the interested reader, a simple anti-windup circuit is given in Figure 1.26 (Wilkie et al., 2002).

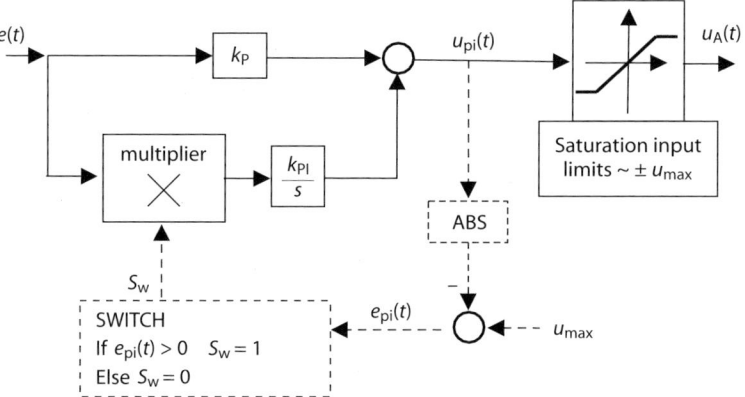

Figure 1.26 Simple anti-windup circuit for PI control.

The principle of the circuit is quite simple:

- The saturation input limits are denoted $\pm u_{max}$.
- Compute $e_{pi}(t) = u_{max} - |u_{pi}(t)|$.
- If $e_{pi}(t) > 0$ then $S_w = 1$ else $S_w = 0$.

The value of S_w is used to switch the integral term in the PI controller on and off. Simple simulation exercises show that this anti-windup circuit successfully reduces excessive overshoot caused by the control action remaining in the saturation region for prolonged periods (Wilkie et al., 2002).

1.3.5 Reverse-Acting Controllers

The Problem

Some processes have difficult dynamics and will produce a so-called *inverse response*. One example is the class of non-minimum phase systems. Figure 1.27 shows a typical system inverse response.

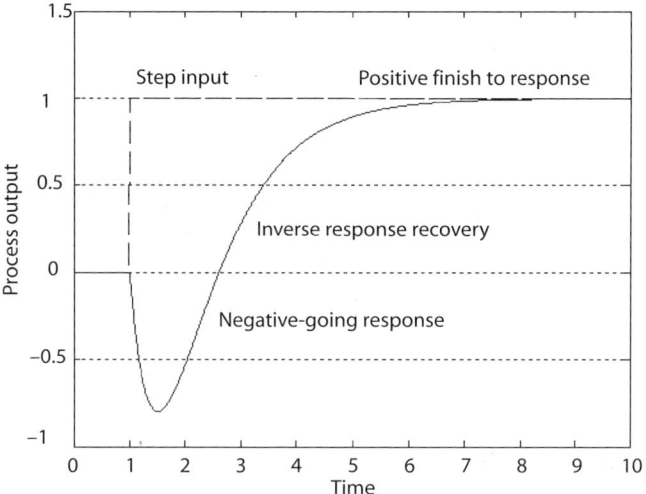

Figure 1.27 Typical inverse system response.

An inverse response occurs when a positive step change at the process input causes the output response first to go negative and then to recover to finish positive. This process behaviour usually has a physical origin where two competing effects, a fast dynamic effect and a slow dynamic effect, conspire to produce the negative start to the response before the step recovers to settle at a positive steady state value (Wilkie *et al.*, 2002).

If a process has a negative gain then a positive-going step change at the process input produces a negative going step response that remains negative. This is quite different from the system that produces an inverse response; it also leads to a requirement for a reverse acting controller. A typical step response for a system with negative gain is shown in Figure 1.28.

The control problem for a process with negative gain is that if negative feedback is used then the negative process gain generates positive feedback and closed-loop stability cannot often be found.

The Remedy

To remedy this problem, an additional gain of [−1] is placed at the output of the controller to maintain a negative feedback loop as shown in Figure 1.29. This combination of a controller and the [−] block is called a reverse-acting controller.

1.4 Industrial PID Control

The long history of PID control usage in industry has led to the development of an industrial PID control paradigm. This development has survived and adapted to the changes of controller technology from the analogue age into the digital computer control system era. It is therefore useful to know

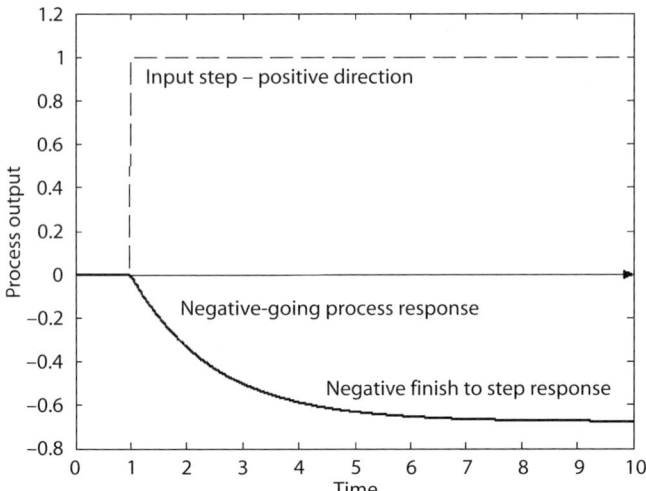

Figure 1.28 System with negative gain: the step response.

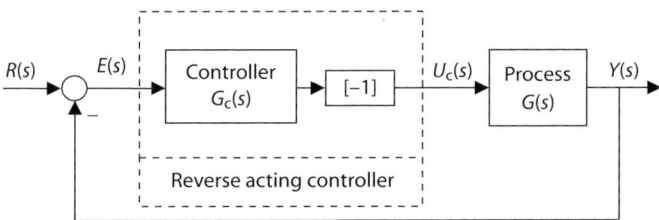

Figure 1.29 Reverse-acting controller for process with negative gain.

something of the industrial terminology and the look of current PID controller technology. This last section provides some insights to these industrial areas.

1.4.1 Traditional Industrial PID Terms

Proportional Band (PB)

The term *proportional band* is sometimes used in the process industries for the specification of the proportional action of the PID controller. In the PID controller, the proportional term can be written as the usual relation between signal values, namely

$$u_c(t) = k_P e(t)$$

Alternatively, this can be given in terms of a linear relation between signal changes for the error and control signals as

$$\Delta u_c = k_P \Delta e$$

It is usual that process hardware has limited signal ranges determined by the specific type of actuators and measurement devices used. Proportional band terminology is motivated by this physical setup of the PID controller and the linear change relation. Define the range e_R for the input error signal to the controller unit as

$$e_R = e_{max} - e_{min}$$

Similarly, introduce the range u_R for the output control signal as

$$u_R = u_{max} - u_{min}$$

Proportional band is denoted *PB* and is a percentage value. Proportional band is defined as the percentage of the error signal range e_R that gives rise to a 100% change in the control signal range, u_R. From the definition above, it is derived using the linear change relation as follows:

$$100\% u_R = k_P(PB\% \times e_R)$$

which rearranges to give

$$PB\% = \left(\frac{1}{k_P}\right)\left(\frac{u_R}{e_R}\right) \times 100\%$$

An alternative form which uses the size of signal changes follows by using the change relation to eliminate the proportional gain from the equation, namely

$$PB\% = \left(\frac{\Delta e}{\Delta u_c}\right)\left(\frac{u_R}{e_R}\right) \times 100\%$$

Sometimes a proportional band is referred to as *narrow* or *wide*. For example, a proportional band of just 20% would be termed *narrow* when compared with a proportional band of 80%, which would be termed *wide*.

When the ranges e_R, u_R are scaled to be 100% each, the proportional band formula reduces to

$$PB\% = \left(\frac{1}{k_P}\right) \times 100\%$$

This form is easily inverted to give a formula for the proportional gain as

$$k_P = \left(\frac{100}{PB}\right)$$

Some values for the inverse numerical correlation between proportional band and proportional gain are given Table 1.7, where it can be seen that a large proportional band corresponds to a small proportional gain value.

Table 1.7 Proportional band and corresponding proportional gain values.

Proportional band (PB%)	400	200	100	50	25	10	1
Proportional gain (k_P)	0.25	0.5	1	2	4	10	100

Reset Time, Reset Rate and Pre-Act Time

These terms use the industrial time constant form for the PID controller:

$$U_c(s) = \left[k_P\left(1 + \frac{1}{\tau_i s} + \frac{\tau_d s}{(\tau_d/n)s + 1}\right)\right]E(s)$$

Reset time is an alternative name for the integral time constant τ_i. Some manufacturers refer to the integral time constant in time per repeat, whilst others use the inverse of this and scale the integral term using *reset rate*. The units of reset rate are repeats per time. *Pre-act time* is an industrial name given to the derivative time constant τ_d.

These traditional terms have their origin in the pre-1980s analogue technology and an industrial need for descriptive labels for the effects of the different terms in PID controllers. It is thought that the use of traditional terms will diminish as the PID tuning process becomes more automated.

1.4.2 Industrial PID Structures and Nomenclature

In industrial implementations, PID control is usually available in an easy-to-use format. Thus PID controllers may be a process controller hardware unit or are viewed via a PID tuning interface window in an industrial SCADA or supervisory control system. Many control textbooks (this one for example) still follow the academic tradition of using symbols from the Greek alphabet in describing control and its supporting analysis. Such facilities are rarely available or even desirable in implemented industrial systems, where simplicity in representation and operation are overriding concerns. This has led to some representation differences that are described in this section.

Control engineers and process engineers tend to use slightly different terms for some system quantities, as shown in Table 1.8. These are minor differences in terminology, but it is useful to be aware of the possible different usage.

Table 1.8 General control engineering and process industry terms.

General control engineering		Process control industries	
System variables			
Term	**Notation**	**Term**	**Notation**
Output	$y(t)$, $Y(s)$	Process variable	PV, PV(s)
Reference	$r(t)$, $R(s)$	Setpoint	SP, SP(s)
Error	$e(t)$, $E(s)$	Setpoint error	E, E(s)
PID controller terms			
Term	**Notation**	**Term**	**Notation**
Proportional gain	k_P	Proportional gain	G
Integral time constant	τ_i	Reset time	TI
Derivative time constant	τ_d	Pre-act time	TD

In earlier sections, it was shown how the PID controller can be modified to overcome practical implementation problems like integral windup and measurement noise amplification. These changes do not affect the structure of the PID controller itself. It was to solve the problems associated with proportional and derivative kick that industrial engineers exploited the structural flexibility in the PID framework. For example, to avoid both proportional and derivative kick, a PID controller is restructured as "I on setpoint error with P and D on process variable". To introduce an apt and shorthand notation for this, the name for this PID controller is hyphenated and given as the I-PD controller. In this shorthand, the letters before the hyphen refer to terms acting on the setpoint error, and the letters after the hyphen refer to terms acting on the process variable. As a second example, recall that to avoid derivative kick alone, the PID controller is structured as "P and I on setpoint error and D on process variable". The shorthand name for this PID controller is the PI-D controller, where the PI before the hyphen indicates the PI terms act on the setpoint error, and the D after the hyphen shows that the D-term acts on the process variable. These structural variations of the PID controller can be assembled systematically to include the controllers P, I, PI, PD, PID, PI-D, I-P and I-PD. A full list, as might appear in an industrial control manual, is given in Table 1.9. Both the parallel (decoupled) control engineering forms and the industrial process control forms have been used in the table.

Table 1.9 Summary of transfer function formulae for the PID controller family.

Structure	General notation	Process control industries
P	$U_C(s) = [k_P]E(s)$	$U_C(s) = [G]E(s)$
I	$U_C(s) = \left[\dfrac{k_I}{s}\right]E(s)$	$U_C(s) = \left[G\left(\dfrac{1}{T_I s}\right)\right]E(s)$
PI	$U_C(s) = \left[k_P + \dfrac{k_I}{s}\right]E(s)$	$U_C(s) = \left[G\left(1 + \dfrac{1}{T_I s}\right)\right]E(s)$
PD	$U_C(s) = \left[k_P + \dfrac{k_D s}{(\tau_f s + 1)}\right]E(s)$	$U_C(s) = \left[G\left(1 + \dfrac{T_D s}{T_F s + 1}\right)\right]E(s)$
PID	$U_C(s) = \left[k_P + \dfrac{k_I}{s} + \dfrac{k_D s}{(\tau_f s + 1)}\right]E(s)$	$U_C(s) = \left[G\left(1 + \dfrac{1}{T_I s} + \dfrac{T_D s}{T_F s + 1}\right)\right]E(s)$
PI-D	$U_C(s) = \left[k_P + \dfrac{k_I}{s}\right]E(s) - \left[\dfrac{k_D s}{(\tau_f s + 1)}\right]Y(s)$	$U_C(s) = \left[G\left(1 + \dfrac{1}{T_I s}\right)\right]E(s) - \left[G\left(\dfrac{T_D s}{T_F s + 1}\right)\right]PV(s)$
I-P	$U_C(s) = \left[\dfrac{k_I}{s}\right]E(s) - [k_P]Y(s)$	$U_C(s) = \left[G\left(\dfrac{1}{T_I s}\right)\right]E(s) - [G]PV(s)$
I-PD	$U_C(s) = \left[\dfrac{k_I}{s}\right]E(s) - \left[k_P + \dfrac{k_D s}{(\tau_f s + 1)}\right]Y(s)$	$U_C(s) = \left[G\left(\dfrac{1}{T_I s}\right)\right]E(s) - \left[G\left(1 + \dfrac{T_D s}{T_F s + 1}\right)\right]PV(s)$

1.4.3 The Process Controller Unit

Many process control companies produce a range of hardware process controllers. Such units usually offer a limited range of PID controller capabilities. Anti-windup circuits and structures avoiding derivative kick are likely to be standard. Figure 1.30 shows a typical interface panel that might be found on the front of a modern process controller.

Standard features might include:

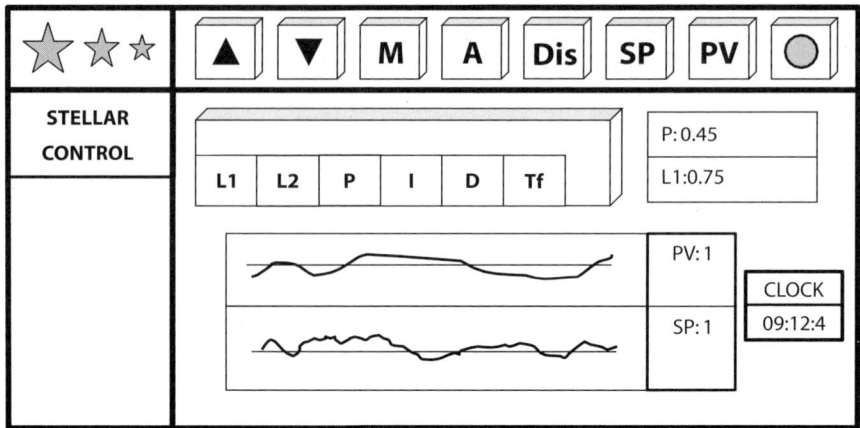

Figure 1.30 Typical process control unit front panel layout.

1. **Trend display facilities:** in the figure, the display shows the process variable PV 1 and the setpoint, SP 1. The button *Dis* in conjunction with the buttons *PV* and *SP* is used to display the trend for the process variables PV and setpoint SP respectively. The button marked *SP* can be used to set up the desired value of the setpoint level and the buttons marked "▲" and "▼" would be used to increase or decrease numerical inputs such as numerical values for the setpoint, or numerical coefficients for the PID control law used.

2. **Manual override facilities:** the operator (perhaps using the in-company tuning guidelines) would use these to allow tuning of the PID controller. The *M* button is used to activate manual tuning and the PID parameters can be set up via the *P*, *I*, and *D* buttons for control loops L1 and L2.

3. **Autotune facilities:** many process controller units have an autotune facility or an automatic tuning procedure. This would be initiated using the *A* button. Usually, the operator would have the ability to read and then accept or reject the PID parameters obtained.

4. **Control of several loops:** the process controller unit in Figure 1.30 can be used to control two loops: loop 1 and loop 2. The buttons *L1* and *L2* allow the operator to enter PID coefficients for these two loops respectively.

5. **Loop alarms:** some process controller units, but not the one shown in Figure 1.30, have facilities to input process alarm settings. These are made using specified *ALARM* entry ports.

The more advanced features that a process controller unit might have include:

1. **Gain schedule facilities:** to accommodate the requirement to control processes at different operating points with differing dynamics, facilities to store and switch in different sets of PID controller parameters may be available.

2. **Self-tuning facilities:** as an alternative to gain scheduling, the process controller unit may have an on-line self-tuning capability. This would be used to track changing system dynamics as the operating conditions change.

3. **Standard control loop structures:** common process control structures are the cascade control system, ratio control, feedforward control and even multiple nested loops. The unit may have facilities to set up and tune some of these control structures.

An important concluding observation is that the enthusiastic industrial control engineer will find the process control unit User's Handbook essential reading. If manual tuning is to be followed then it is important to fully understand which form of PID controller is used by the unit and the background to the different tuning procedures available.

1.4.4 Supervisory Control and the SCADA PID Controller

So many operational aspects of industry are now under computer control. The range of applications is very wide. In the primary industries, such as iron and steel production, electric power generation and distribution, in the utilities (water and gas) and in industries like paper and pulp production, the process line will invariably be under computer control. Computer control is also widespread in the secondary industries, such as manufacturing, food processing, pharmaceuticals, transport and healthcare. To the industrial control engineer one fascination of the process industries is that almost all of the operations are very complex and it is a considerable challenge to establish successful high-performance computer control over these operations. A complete industrial computer control system installation (hardware and software) is usually called a Distributed Computer System (DCS system), a Distributed Computer Control System (DCCS system) or a Supervisory Control and Data Acquisition System (SCADA system). The use of a particular term for a computer control system is often industry-dependent. Engineers in the heavy industries like the iron and steel industry or the electric power industry generally refer to the overall control system as a *DCS system*. In the oil production fields and in the petrochemical chemical industries virtually the same type of computer control system is more likely termed a *SCADA system*. In fields where monitoring and telemetry are important components of the computer control system then the overall system is also very likely to be called a *SCADA system*. Despite the fact that, on close examination, the technological differences between a so-called DCS system and a SCADA system might not be significant, different industries always seem to develop different terminology and jargon for the processes, operations and procedures particular to the industry.

The operator interfaces of these computer supervisory systems use the latest techniques of graphical representation and supporting computer hardware to ensure safe and efficient operation of the process systems. Other developments include software tools able to assist with the tuning of control loops and investigating the effectiveness of the installed control loop settings. Two examples from industry are used to illustrate typical facilities.

Example 1.5

Siemens AG (Germany) has kindly supplied the next few figures (1.31, 1.32 and 1.34) as examples of state of the art computer supervisory control system interfaces. DCS software usually contains sets of schematics where each tableau shows different categories of information about the control of a process. A simple sequence of layers relating to the process control is Process Plant Overview, Process Regulator Overview, and Regulator Status and Data View. The schematics supplied by Siemens AG follow this pattern.

Process Plant Overview

Figure 1.31 shows a typical modern process plant schematic for a drying process.

The top-level schematic shows the main components in a Drier (Trockner) process which should be read from left to right. On the left is a Burner (Brenner) unit supplying the drying processes. In the one separation stage, coarse material (Grobgut) is removed. Note the nice pictorial representation of waste transportation for the coarse material as a little wagon. The material enters a drying drum process. As will be seen in the next schematic, showing the regulator structure, a control objective is to maintain

36 PID Control Technology

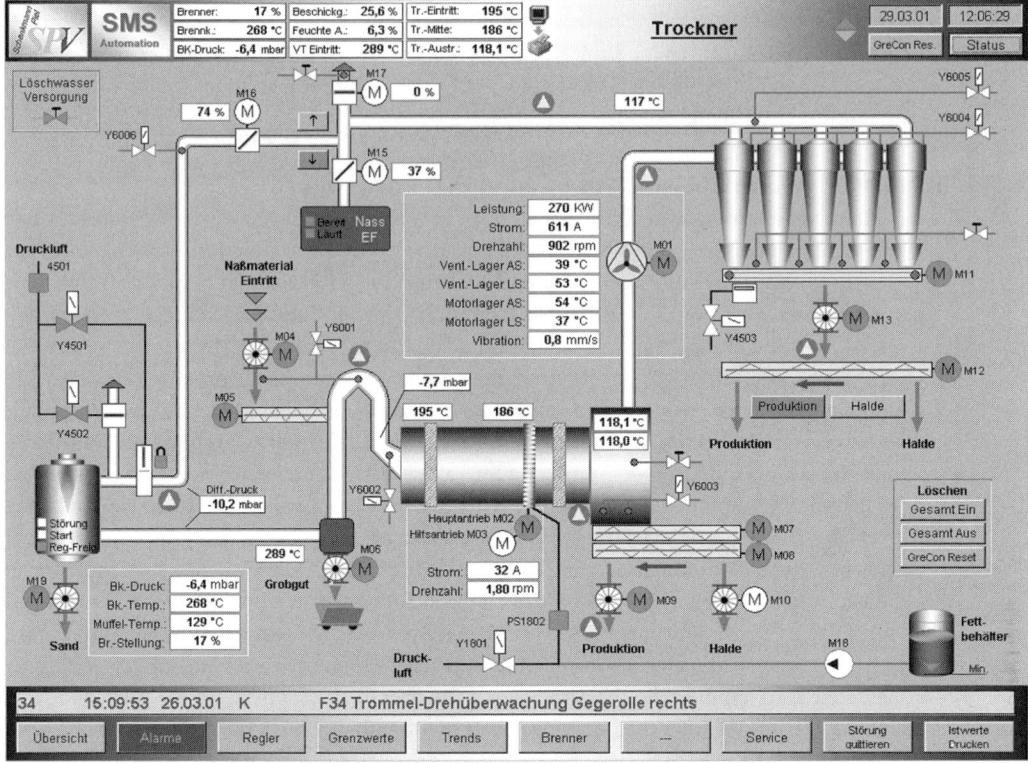

Figure 1.31 Modern process plant schematic for a drying process (courtesy of Siemens AG).

the exit temperature of the drying drum at a given setpoint. Some outfall from the drum drier is transported to a another drying stage. Thus the top-level schematic, the Process Plant Overview, provides an overall view of the status of the process operation. However, it is important to note that the Process Plant Overview does not detail either the control objectives or the control system to achieve those objectives. To find details of the control system it is necessary to look into a schematic from a lower level of the supervisory system. To do this the overview schematic has a task bar along the bottom where various options are available; Table 1.10 lists these options. The *Regler* button will lead to the regulator levels.

Process Regulator Overview
Figure 1.32 shows the schematic for the regulator structure (Regler Struktur) in the drier process control system.

This shows an example of a process control system that uses a network of PID controller modules to construct the control system. In fact, the scheme shown is a classical cascade control system. There are two loops in this scheme and the objective is to control the exit temperature of the drum drier (Temperatur Trommelaustritt Istwert). The outer loop, primary loop or master loop is given the name Führungsregler (variable command control) in this schematic. The inner loop, secondary loop or slave loop is given the name Folgeregler (follow-up control) in this schematic. The inner loop uses the temperature at the centre of the drum (Temperatur Trommelmitte Istwert) as the measured variable. The setpoint for the exit temperature of the drum drier (Temp. Trommelaustritt Sollwert) is in the top right-hand corner. The process sequence under control is the burner process, which is an actuator to

Table 1.10 Task bar labels for the Process Plant Overview schematic.

Task bar label (left to right)	Task available
Übersicht	Overview of process
Alarme	Process Alarms
Regler	Controller structure schematic
Grenzwerte	Variable boundary values
Trends	View variable trend data
Brenner	Burner data and details
Service	Service records
Störung quittieren	Confirm the presence of process disturbances
Istwerte Drucken	View instantaneous values of pressures

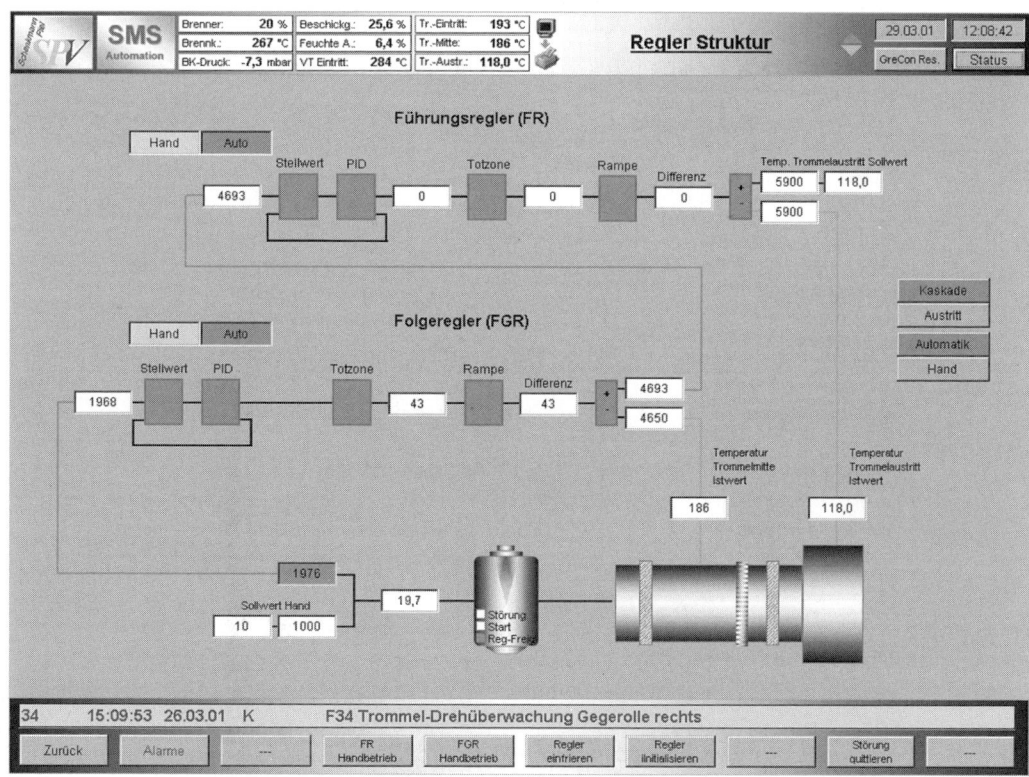

Figure 1.32 Process Regulator Overview schematic (courtesy of Siemens AG).

the rotating drum process. This is at the bottom right corner of the schematic. An industrial control engineer might possibly be far more familiar with the control scheme in the block diagram form of Figure 1.33. This clearly shows the two PID controllers in a cascade loop structure.

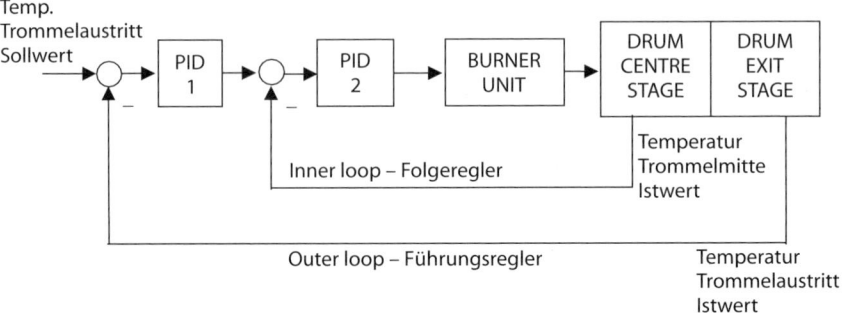

Figure 1.33 Block diagram for the Process Regulator Overview schematic.

For an operator, this Process Regulator Overview allows the structure of the control loops to be viewed along with the values of intermediate process variables. To find details of the controller parameters it is again necessary to view another schematic at this level of the supervisory system. However, to allow the control engineer to work with controller design options, the process regulator overview schematic also has a task bar along the bottom where various options are available; these are listed in the Table 1.11.

Table 1.11 Task bar labels for the Process Regulator Overview schematic.

Task bar label (left to right)	Task available
Züruck	Return
Alarme	Process Alarms
FR Handbetrieb	FR Controller in manual operation
FGR Handbetrieb	FGR Controller in manual operation
Regler einfrieren	Freezing or stopping the controller
Regler initialisieren	Initialising the controller parameters
Störung quittieren	Confirm the presence of process disturbances

Regulator Status and Data View

To find details of the controller parameters it is necessary to view another schematic, shown in Figure 1.34.

Figure 1.34 shows the data for four PID controllers. Reading from left to right, these are for the four process variables: Ventilatorleistung (Air intake unit power), Druck Brennkammer (Combustion chamber pressure), Temp. Trommelmitte (Drum centre temperature) and Temp. Trommelaustritt

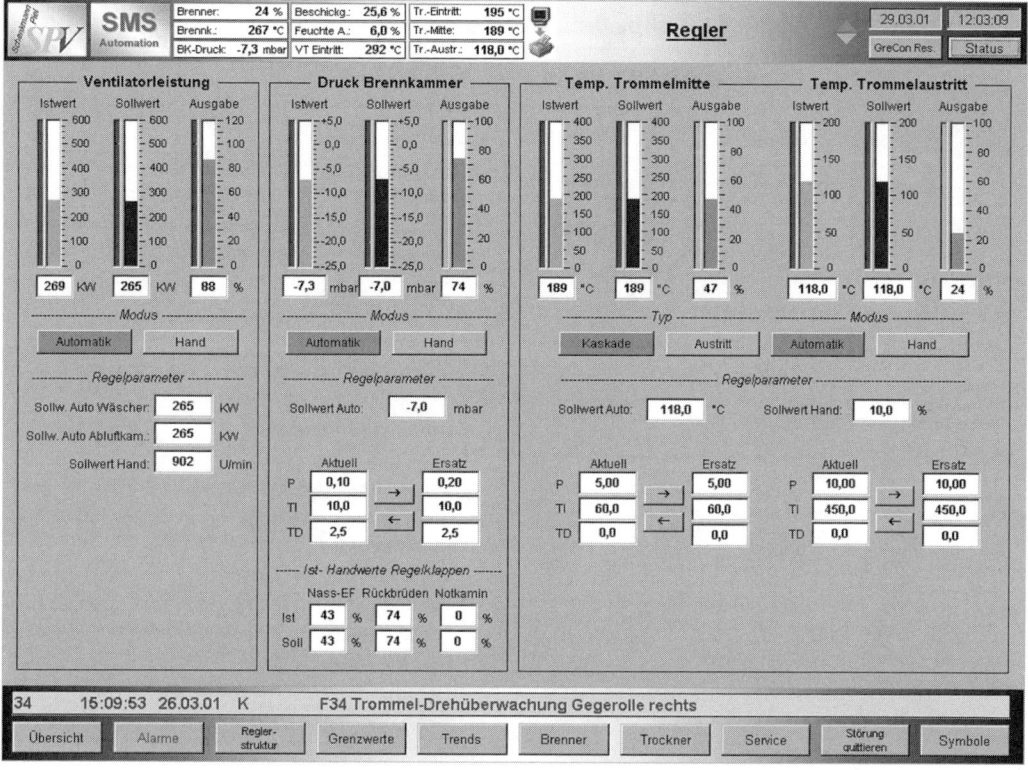

Figure 1.34 Regulator status and data view (courtesy of Siemens AG).

(Drum exit temperature). For each of the variables are three column icons giving readings for instantaneous value (Istwert), setpoint value (Sollwert) and output (Ausgabe). Beneath the data columns are the details for the control loops present. These are first given according to the classifiers: Typ (type of control loop), Modus (method or procedure used to obtain the controller parameters), for which there are the options, Kaskade (cascade), Austritt (exit), Automatik (automatic) and Hand (manual tuning). The actual PID controller parameters are denoted P, TI and TD, and with this notation, it seems very likely that the PID controller has the form

$$G_{\text{PID}}(s) = P\left(1 + \frac{1}{TIs} + TDs\right)$$

A look at the process control system handbook would be needed to confirm this point. Finally, the straightforward mechanism for replacing controller parameters can be seen as using the arrowed buttons lying between the parameter columns headed Aktuell (current value) and Ersatz (replacement value).

Finally, as with the other interface windows, transfer to other schematics is via the task bar at the bottom of the window. Table 1.12 lists the options available on this schematic.

The demand from industry for economic gains from the control of manufacturing and utility processes ensures that the development of supervisory systems for complex processes continues to progress. Recent ideas for computer–human interfaces in supervisory control have been published by Wittenberg (2004) among others.

Table 1.12 Task bar labels for the Regulator Status and Data View schematic.

Task bar label (left to right)	Task available
Übersicht	Overview of process
Alarme	Process Alarms
Regler struktur	Controller structure schematic
Grenzwerte	Variable boundary values
Trends	View variable trend data
Brenner	Burner data and details
Trockner	Drier process overview schematic
Service	Service records
Störung quittieren	Confirm the presence of process disturbances
Symbole	Key to the symbols

Example 1.6

The Emerson Process Management company has a coordinated system for control and supervision of production and manufacturing processes. The system covers all aspects of installation, commissioning, operation, monitoring and control.

The DeltaV™ System

The overall system comprises hardware and software aspects. The digital automation of processes is achieved using the Plantweb® architecture that allows the integration of process hardware, field devices and the necessary communications. Within this architecture, the DeltaV software provides the coordination, the control, the monitoring, the alarms and the data interrogative services. Strong links with Microsoft have resulted in a system with Windows functionality and this is evident in the system displays. The PID controller within this system is captured as a function block.

The PID Function Block

The provision of a PID function block necessarily involves taking decisions on which control facilities should be available to the users. These can be divided into two categories, the input–output signal processing and the PID control capabilities. As in this particular PID function block, the input–output signal processing typically includes analogue input channel processing, analogue output channel processing, signal scaling and limiting, override tracking, alarm limit detection, and signal status propagation.

The complexity of the PID control provision is sensibly guided by incorporating all the tools that an industrial process control engineer might wish to use. This PID function block offers two types of formula (termed standard and series) to perform proportional-integral-derivative (PID) control with the option for nonlinear control (including error-squared and notched gain), feedforward control, cascade control, reverse acting and direct acting control. Operation can be automatic or manual. Most interestingly for control educationalists, (1) the manual is written in terms of Laplace transfer function notation, but, as might be expected, the implementation uses discrete forms and (2) industrial terms like reset time (seconds), derivative time constant (rate in seconds), bumpless transfer, reset limiting, setpoint(SP), process variable (PV) and so on are frequently used.

The PID function block has some parameters that enable the PID equation to be configured for different structures. These parameters determine which of the three PID actions (Proportional, Integral and Derivative) are active and how the actions are applied. The structures available include:

- *PID action on Error* – proportional, integral and derivative action are applied to the error (where error = SP – PV).

- *PI action on Error, D action on PV* – proportional and integral action are applied to error; derivative action is applied to PV. A setpoint change will exhibit a proportional kick.

- *I action on Error, PD action on PV* – integral action is applied to error; proportional and derivative action are applied to PV.

- *PD action on Error* – proportional and derivative actions are applied to error; there is no integral action.

- *P action on Error, D action on PV* – proportional action is applied to error; derivative action is applied to PV; there is no integral action.

- *ID action on Error* – integral and derivative action are applied to error; there is no proportional action.

- *I action on Error, D action on PV* – integral action applied to error; derivative action applied to PV; there is no proportional action.

In each case, the manual indicates the outcomes of application of a particular structure, for example whether proportional kick, derivative kick or steady state offsets to constant reference changes will occur. This is more evidence that these types of control properties should be firmly in the basic control engineering course.

To conclude the discussion of this industrial PID function block it is interesting to note that there are two structure parameters available that allow a two degrees of freedom PID controller to be configured. This is particularly useful when tuning a control loop for disturbance rejection, since the setpoint response often exhibits considerable overshoot. This is particularly true when there is derivative action required and the derivative action is taken only on PV (to avoid large bumps in output as the result of modest setpoint changes). The two degrees of freedom structure provided by the DeltaV PID function block allows shaping of the setpoint response.

Tuning the PID Controller

An interesting feature of the DeltaV system is the approach adopted for using advanced process control. A determined effort has been made to develop a set of easy to use dedicated control modules. Each module is concerned with automating the application of a particular control methodology. These modules include DeltaV Tune, DeltaV Fuzzy, DeltaV Predict and DeltaV Simulate. The presence of a simulation capability is compatible with control engineering educational trends where simulation with tools such as MATLAB forms a strong part of the teaching methods used today (Wilkie et al., 2002). The DeltaV modular approach has been explored and described in detail by Blevins et al. (2003). A discussion of the DeltaV Tune module follows.

The DeltaV Tune Module

The idea behind the DeltaV Tune module is to provide a simulation capability and some analysis tools to allow loop performance to be predicted before new controller parameters are updated. The module is part of the DeltaV system and is initiated by a mouse-click to provide an on-demand control loop tuning facility. An automated process test procedure is available and good graphical interfaces aim to make the tuning of PID loops easier.

Technically, the DeltaV Tune uses the relay experiment followed by the application of tuning rules and computations. The Åström-Hägglund algorithm (1985) has been enhanced to define a first-order process model with dead time. This is a closed-loop test procedure so that the process remains secure during the loop tuning exercise. Tunings are then devised for user-specified performance requirements ranging from no overshoot to very aggressive (fast response) or tunings based on process type. The DeltaV Tune user interface can be viewed in Figure 1.35.

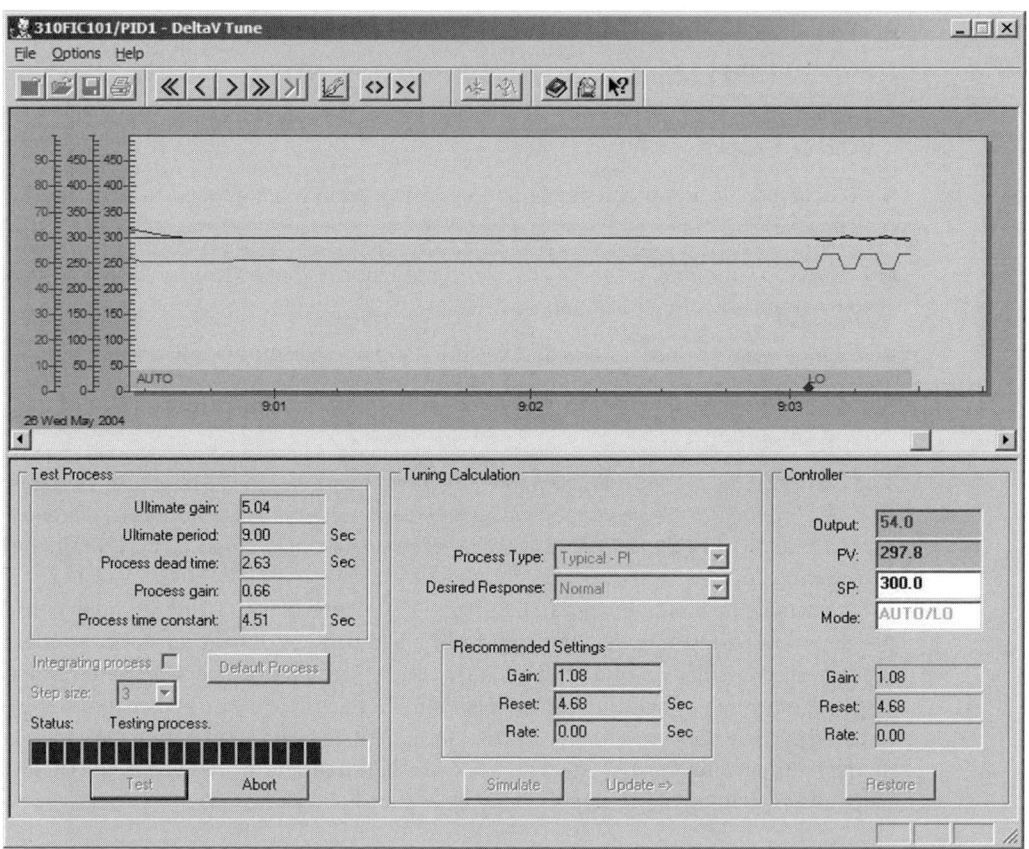

Figure 1.35 DeltaV Tune user interface screen (courtesy of Emerson Process Management).

DeltaV Tune is equipped with an Expert option that leads to controller tunings based on more advanced methods. The methods supported include:

- *Modified Ziegler Nichols rules for PI control* – this control design is based on the Ziegler–Nichols (1942) tuning rules but with modifications to minimise process overshoot.

- *Phase and gain margin rules for PID control* – this controller tuning has a default phase margin of 45°. In most cases, this phase margin will cause very little overshoot. A slower response with less overshoot for most processes can be achieved with a greater gain margin and a 60° phase margin.

- *Lambda-based tuning rules* – this method for PI control design allows the desired closed-loop response time constant-to-open-loop time constant ratio (defined as λ) to be specified through the lambda factor. For non-self-regulating level loops with PI control, the Lambda-Averaging Level design method is available. When the process dead time is greater than the process time constant, a Smith Predictor template is used in a Lambda-Smith Predictor design method.

- *Internal Model Control (IMC) Tuning* – this design method assumes a first-order process with a time delay and provides controller settings for PID control. The process model is identified during tuning test where the procedure for identifying the model is an Emerson Process Management patented technique. The IMC design is especially useful when a process has a delay longer than half the process time constant. The process time delay and the process time constant are shown in the Process Test Results panel of the Tuner user interface screen (see Figure 1.35).

Simulation is a powerful tool for understanding the present and possible future control-loop performance. Once recommended controller parameters have been found, the user can view simulated loop responses as shown in Figure 1.36. This module also allows an assessment of closed-loop stability using a robustness plot displaying process gain and phase margin. To compute new controller parameters for a different robustness specification, a mouse-click on the plot initiates this computation and produces the new simulated loop responses.

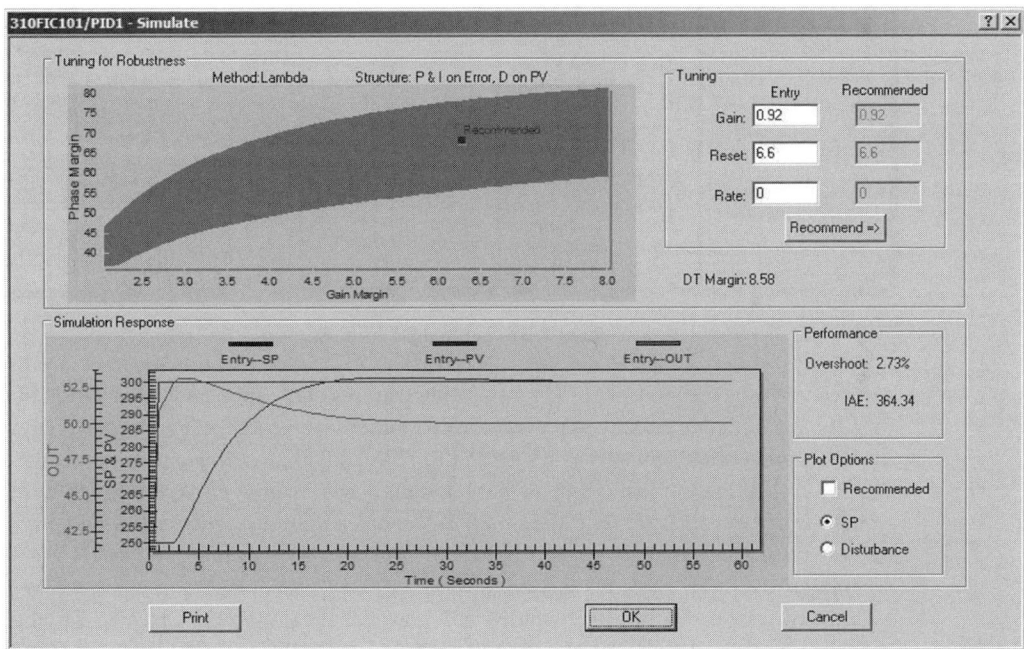

Figure 1.36 DeltaV user interface to view simulated closed-loop responses (courtesy of Emerson Process Management).

Some General PID Control Interface Features for SCADA Systems

The products reviewed enable some general features of PID control interfaces to be identified and enumerated. In the DCS or SCADA system, access to the setup and tuning of the PID control loops will usually be via a control engineer's interface with Windows-like functionality. Since the process industry involves some processes that are capable of causing catastrophic damage it will be important to guarantee staff and plant safety from the possible outcomes of simple operational errors or even vandalism or sabotage to the control loop setup. Consequently, access to the controller level of the supervisory control system is usually accompanied by password entry and security checks. This is to prevent unauthorised alterations to key process operational parameters. For further discussion, a typical (but hypothetical) example of a control engineer's interface for the setup and tuning of process control loops is shown in Figure 1.37.

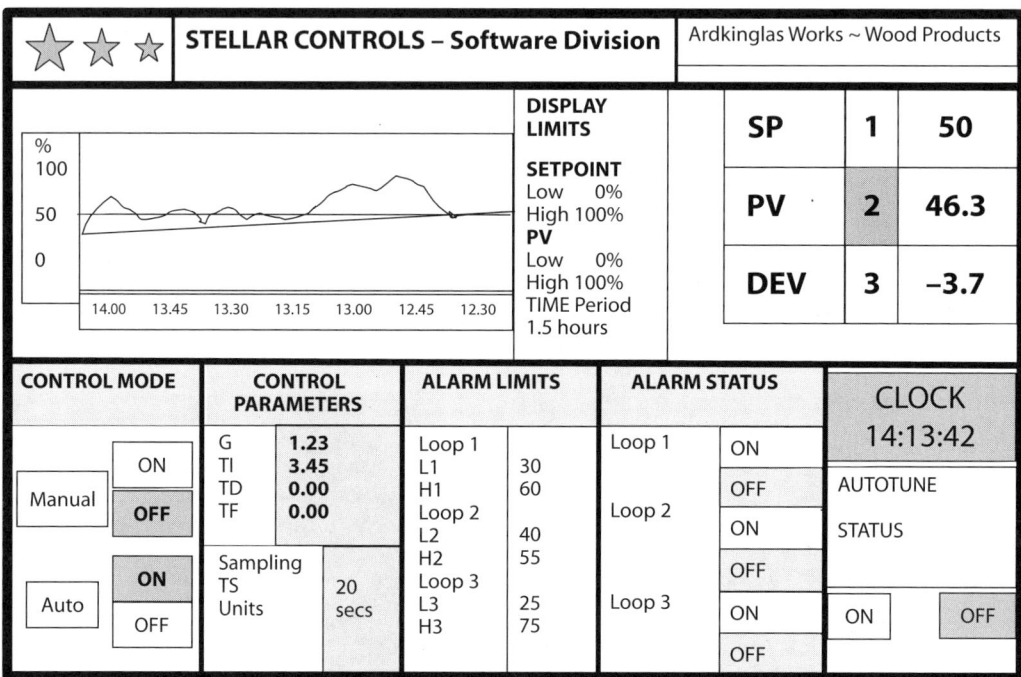

Figure 1.37 Typical DCS PID tuning schematic.

As can be seen, the interface of Figure 1.37 has features of notation and presentation in common with those on the front panel of the process controller unit. However, a SCADA system usually controls many loops in a process and will have much greater functionality than a simple process control unit. Since the SCADA system is a sophisticated software system it will be able to page through different windows which display information on different control loops, different process units and even the complete process plant. The different types of schematic possible were exemplified by the Siemens SIMATIC schematics given earlier in this section.

In the typical SCADA interface window of Figure 1.37 the setpoint and process variables are denoted as *SP* and *PV* respectively. The deviation of the process variable from the setpoint value is given as the additional variable *DEV*. To the left of the *SP*, *PV* and *DEV* display, a graphical presentation of the setpoint and process data is given. Just next to the trend display is a small table of graphical setup data. The *DEV* reading gives the operator a current value for the instantaneous error of the process variable from the setpoint. This has the advantage of immediacy over having the operator doing some mental arithmetic, and then possibly getting it wrong! The visual trend data is useful since this allows the operator to view the historic variability of the process operation for the previous 1.5 hours. Such a visual presentation of data is possibly more meaningful to an operator than formal mean and variance data. However, the formal mean and variance data for process variability might be of more interest to a control engineer looking at the long-term performance of a control loop.

Looking now along the bottom of the schematics, from left to right, the following items can be seen:

- *Control mode*

  The control mode options are manual or automatic operation. In manual operation, the control engineer is able to disable the PID controller and use a constant fixed control signal level so that the plant is in open loop operation. In automatic mode, the PID control action will be computed using the PID parameters supplied and the computed control signal automatically applied.

- *Control parameters*

  The PID parameters are stored and edited in this section of the interface. For this particular interface, cursor editing is used to change the controller parameters. The figure shows the PID parameters as $G = 1.23$; $TI = 3.45$; $TD = 0.00$; $TF = 0.00$. The sampling interval is given below the PID controller parameters as $TS = 20.0$ secs. The interpretation of these parameters is as follows.

Parameter symbol	PID controller term to which parameter is related
G	Proportional gain
TI	Integral term
TD	Derivative term
TF	Filter in the derivative term
TS	Specifies sample interval; Relates to all three terms of PID controller

  To find out the form of the PID controller it is necessary to refer to the User's Manual as provided by the DCS/SCADA system manufacturer. In this particular example, the presence of the parameter TF probably indicates that noise protection on the derivative term has been provided and the presence of TS specifying the sample interval probably means that the PID controller has been used in discrete or sampled-data form.

- *Alarm limits and alarm status*

  The three loops Loop 1, Loop 2 and Loop 3 have been provided with high (H1, H2, H2) and low (L1, L2, and L3) alarm limits respectively. Thus if the process variable exceeds a high limit or falls below a low limit then the Alarm status changes from *OFF* to *ON*. This change could be accompanied by flashing indicators, indicators changing colour and even possibly an audio alarm signal. The intimation is that the operator should be investigating the process situation and possibly taking some action. One local Scottish process plant refers such alarm activation to a real-time on-line expert system for further advice.

- *Clock*

  This is a simple real-time display of time using the 24-hour clock.

- *Autotune*

  The last facility is an autotune option. This is an automated PID controller tuning facility. The basic outline of the method used will probably be described in the manufacturer's User's Manual. Most of the autotune products used in the process industry are based on on-line identification followed by a rule based PID controller calculation. J. G. Ziegler and N. B. Nichols first devised the original concepts for this approach in the 1940s. Although a manufacturer might follow this well-established template there is bound to be manufacturer expertise engineered into the actual procedure, and these additional features will almost certainly be commercially sensitive. This is where product differentiation arises in the competitive field of PID control engineering! An important lesson for the industrial control engineer is to know as much as possible about the automated routine implemented since an inappropriate application of an autotune techniques can easily lead to poor loop control and excessive process variability.

Acknowledgements

The author, Michael Johnson, would like to acknowledge that the presentation of the material in this chapter has benefited enormously from many previous discussions with Dr Jacqueline Wilkie of the Industrial Control Centre, University of Strathclyde, Glasgow. The author would also like to thank Dr L. Giovanini (Industrial Control Centre, Glasgow) for his review of the chapter.

The author would like to acknowledge the permission of Siemens AG to publish the three SIMATIC® process control schematics used in this chapter. Permission was granted though the kind assistance of Manfred Baron, Promotion Manager SIMATIC HMI Software, Siemens AG, Nürnberg, Germany. The author would also like to thank his colleague Dr Marion Hersh (University of Glasgow) for her kind help with the German translations that were essential for gaining an understanding of the information content of the various sheets.

The author would like to acknowledge the kind permission of Emerson Process Management to publish the two DeltaV Tune display schematics used in this chapter. The kind assistance of Travis Hesketh, Emerson Process Management (UK) in reviewing the text describing the DeltaV system and the PID components is very gratefully acknowledged.

References

Åström, K.J. and Hägglund, T. (1985) *US Patent 4,549,123: Method and an apparatus in tuning a PID regulator.*
Åström, K.J. and Hägglund, T. (1995) *PID Controllers: Theory, Design and Tuning.* ISA Publishers, Research Triangle Park, NC.
Blevins, T.L., McMillan, G.K., Wojsznis, W.K. and Brown, M.W. (2003) *Advanced Control Unleashed: Plant Performance Management for Optimum Benefit.* ISA Publishers, Research Triangle Park, NC.
Doebelin, E.O. (1985) *Control System Principles and Design.* John Wiley & Sons Ltd, New York.
Dutton, K., Thompson, S. and Barraclough, R. (1997) *The Art of Control Engineering.* Addison Wesley Longman, Harlow.
Luyben, W.L. (1990) *Process Modelling, Simulation, and Control for Chemical Engineers.* McGraw-Hill, New York.
O'Dwyer, A. (1998a) PI and PID controller tuning rules for time delay processes: a summary. Part 1: PI controller tuning rules. *Preprints ACSP Symposium*, Glasgow, pp. 331–338.
O'Dwyer, A. (1998b) PI and PID controller tuning rules for time delay processes: a summary. Part 2: PID controller tuning rules. *Preprints ACSP Symposium*, Glasgow, pp. 339–346.
Seborg, D.E., Edgar, T.F. and Mellichamp, D.A. (1989) *Process Dynamics and Control.* John Wiley & Sons Ltd, New York.
Tan, K.K., Wang, Q.G. and Hang, C.C., with Hägglund, T. (1999) *Advances in PID Control.* Springer-Verlag, London.
Wilkie, J., Johnson, M.A. and Katebi, M.R. (2002) *Control Engineering: An Introductory Course.* Palgrave, Basingstoke.
Wittenberg, C. (2004) A pictorial human–computer interface concept for supervisory control. *Control Engineering Practice*, **12**(7), 865–878.
Yu, C.C. (1999) *Autotuning of PID Controllers.* Springer-Verlag, London.
Ziegler, J.G. and Nichols, N.B. (1942) Optimum settings for automatic controllers. *Trans. ASME*, **64**, 759–768.

2 Some PID Control Fundamentals

Learning Objectives

2.1 Process System Models

2.2 Controller Degrees of Freedom Structure

2.3 PID Control Performance

2.4 State Space Systems and PID Control

2.5 Multivariable PID Control Systems

Acknowledgements

References

Learning Objectives

Process models and control structures are essential to any discussion of PID control. Process models can use at least four different mathematical notations and that is for the continuous time domain alone. These process model types are given in this section including those of special interest for PID control studies, the low-order standard transfer function models. Perhaps less well known are the control degrees of freedom system structures. For example, the two degrees of freedom structure is often cited but it is not so usual to find a presentation of the full set of possible structures.

Creating a specification for the desired controller performance is an important control engineering task, and some aspects of this task are covered in this chapter. Later chapters will use the concepts of classical stability and performance robustness, and these concepts receive an early overview here. By way of contrast, there is little state space analysis used with PID control in this book and only a short section is given in this chapter.

The notation and structures for multivariable control are summarised in the concluding section of the chapter. This provides a useful opportunity to preview cascade, multi-loop, decentralised, multivariable control system layouts.

Thus the learning objectives of the chapter are to:

- Study process model types.
- Understand an overview of controller degrees of freedom structures.
- Learn about basic controller performance issues and specifications.
- Gain insight from a state space analysis of PID control.
- Review multivariable control system structures and notation.

2.1 Process System Models

Closed-loop control system design has several objectives: to achieve a stable closed-loop system, to ensure that the process output is close to a desired setpoint value, to reject output variable fluctuations due to process disturbance inputs, and to filter out residue process measurement noise from the control signal. The control design usually has to be completed with inaccurate or uncertain knowledge about the process to be controlled and the disturbances present in the system. There are two main categories of controller tuning procedures:

- Trial and error method
- Systematic tuning methods

Trial and Error Method
The trial and error tuning of PID controllers is said to be widespread in industrial practice. The method starts from some intelligently guessed PID controller parameters. Using these initial controller parameters, the output of the closed-loop system is observed and the controller parameters are modified until the desired output is obtained. Success is highly dependent on the tuning skill of the industrial control engineer and the knowledge available about the process to be controlled.

Systematic Tuning Methods
Systematic tuning methods introduce engineering science into the controller tuning procedure. This usually means using some form of model for the real process. The multitude of different control design methods available in control engineering textbooks and the literature arise from the different assumptions made for the type of process model and the actual information available for the system model. Once a model scheme has been determined, some consideration of what the controller has to achieve is captured as a control system specification. Then, using the specification and the process model, a procedure will be devised to calculate the controller parameters to attain the required controller specification.

In this book some quite different process model descriptions are used to generate tuning procedures for PID control systems. Indeed, that is one aim of the book: to explore some of these new tuning methods. However, it is useful have a statement of the conventional modelling framework and terminology from which to begin. Conventional systematic controller parameter tuning generally uses two different process models:

- *Steady state gain model*
  A steady state gain model is a description of the relationship between the process input and output to achieve desired steady state levels. This model is often obtained by using different open loop or closed-loop experiments to determine the range of the process inputs required to reach the desired output range. The model relationship between output and input in steady state may be represented by a function or may be described by means of a look-up table. In practice, control strategies based on

steady state gain models of the process do not usually give adequate shaping of the closed-loop process *dynamic* response, even if the desired steady state is obtained.

- *Dynamic model*
  Whereas a static model shows the relationship between the steady state input and output, a dynamic model describes the complete relationship between the input and output of a system. This includes the transient part of the process response as well as the steady state part of system behaviour. An important tool in the modelling of the complete dynamic behaviour of a linear system is the state space representation. Linear state space system models can be used as a framework for several other mathematical modelling descriptions, as shown in Figure 2.1. In fact, Figure 2.1 shows how the conventional classical process modelling mathematical framework arises. The modelling forms and terminology – linear state space, convolution integral, Laplace transform, discrete linear state space, discrete convolution form, z-transform – are fundamental in any modern control engineering discourse.

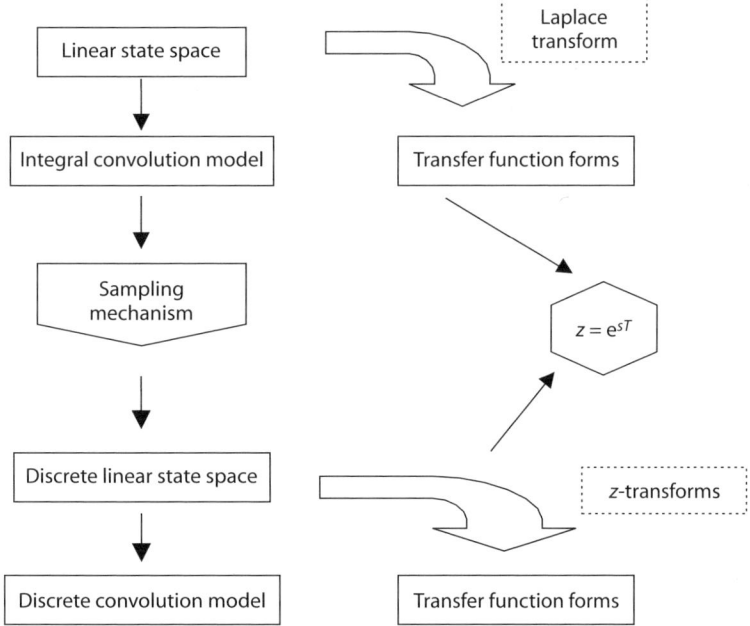

Figure 2.1 Modelling framework.

2.1.1 State Space Models

A state space model is constructed using a set of system variables which define the status of a process at any instant in time. In general, system behaviour changes with time, and the information about this evolution of system status usually resides in the rate-of-change variables within a system or in combinations of these variables and their derivatives. These status variables are known as the state variables of the system and the set of state variables which describe the behaviour of a system is termed the system state. The n state variables are collected together as a state vector and given the vector notation

$$x(t) = \begin{bmatrix} x_1(t) \\ \vdots \\ x_n(t) \end{bmatrix} \in \Re^n$$

The Process Input Variables

Whilst state variables define how the internal behaviour of the system evolves, a process will also have physical variables which can be described as inputs to the system. These external inputs are an additional mechanism by which the state variables are caused to evolve in value. The inputs to a process comprise two categories. Firstly there are those which can be manipulated and used to change the system state variables; these are termed controls. Secondly there are those inputs which cannot be manipulated or controlled; these are the disturbance inputs to the system. The m controls or system inputs which can be manipulated are usually given the notation $u_i(t)$, $i = 1,\ldots,m$. As with the state vector, the m control variables are given vector form as

$$u(t) = \begin{bmatrix} u_1(t) \\ \vdots \\ u_m(t) \end{bmatrix} \in \Re^m$$

The Process Output Variables

The variables that provide a view of the state variables are the output variables. These are denoted as $y_i(t), i = 1,\ldots,r$, where it is assumed that there are r output variables. As with the state vector and the control inputs, these output signals are given vector form as

$$y(t) = \begin{bmatrix} y_1(t) \\ \vdots \\ y_r(t) \end{bmatrix} \in \Re^r$$

Output variables can be classified according to four different types. One property of an output variable is whether or not it is measured. A second property of an output variable is whether or not it can be controlled or affected by the available control inputs. Thus there are four different types of output variable, as shown in Table 2.1.

Table 2.1 Classification of output variables.

Output property	Measured, m	Unmeasured, um
Controlled, c	$y_{c,m}$	$y_{c,um}$
Uncontrolled, uc	$y_{uc,m}$	$y_{uc,um}$

The real interpretative power of the state space representation comes from writing the relations for the equations between the input, the state variables and the outputs using vector–matrix notation, since the representation is generically multivariable in nature. Thus the state space description provides a general system modelling framework for the discussion of inputs, state variables and outputs, as depicted in Figure 2.2.

Linear State Space Equations

Typically physical, kinetic and chemical relationships between the various state variables use differential equation descriptions. Different types of differential equation leads to different types of state space

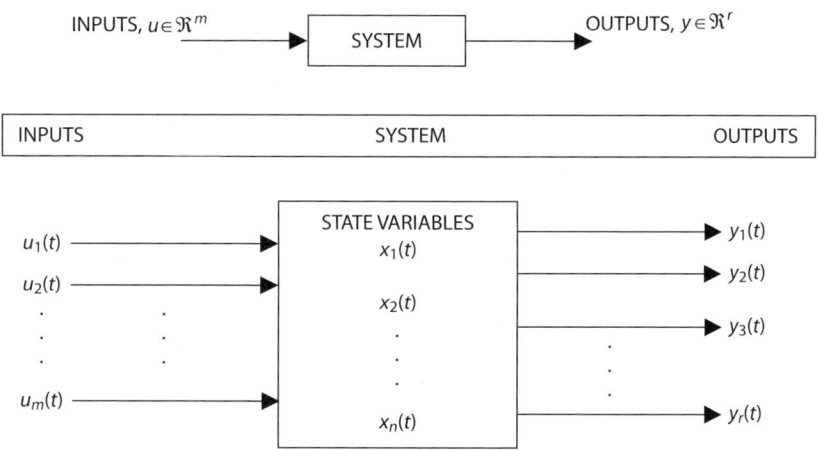

Figure 2.2 General system modelling framework.

system description for the system where the main distinction is between linear and nonlinear system descriptions. Commonly the physical differential equation system for a system is manipulated into a linear system of constant coefficient ordinary linear differential equations. A linear state variable description would then use a set of n first-order linear differential equations to generate a set of n state variables. In vector–matrix form a linear state space system description results. Consider a system with m inputs, n state variables and r outputs; then the linear state space system is given by the $ABCD$ model as

$$\dot{x}(t) = Ax(t) + Bu(t)$$
$$y(t) = Cx(t) + Du(t)$$

where $A \in \Re^{n \times n}$ is the system matrix, $B \in \Re^{n \times m}$ is the input matrix, $C \in \Re^{r \times n}$ is the output matrix, and $D \in \Re^{r \times m}$ is the direct-feed through matrix.

The matrix D in the output represents any direct connections between the input variables and the output variables. In practical systems such connections do not occur very often, so in many cases the D matrix is zero. However, there are theoretical situations where a non-zero D matrix occurs, and thus a four matrix $ABCD$ state space model provides complete generality.

The four matrix $ABCD$ state space model diagram is shown in Figure 2.3. To reflect the common situation where in many examples the D matrix will be zero, the lines connecting the D matrix are shown as dashed. In the cases where the D matrix is zero this connection will not be present in the diagram.

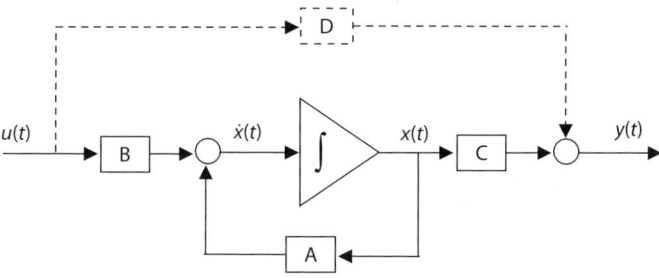

Figure 2.3 $ABCD$ state space model diagram.

The *ABCD* model represents a set of first-order differential equations that can be integrated to find the values of the system states. To be able to solve the differential equations a set of initial conditions, $x(0) \in \Re^n$, are required. In many cases these initial conditions are zero and they do not appear in the Figure 2.3. An example of the four matrix *ABCD* state space model diagram where the *D* matrix is set to zero, and showing the initial condition vector $x(0) \in \Re^n$ is given in Figure 2.4.

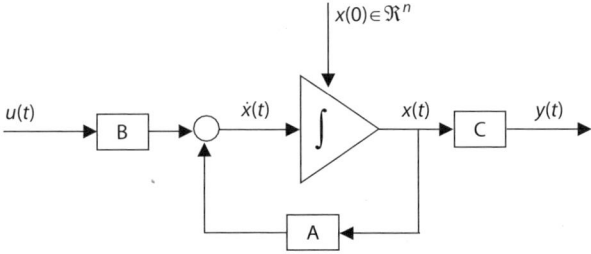

Figure 2.4 Example of the state space model diagram with $D = 0$ and initial condition vector $x(0) \in \Re^n$.

2.1.2 Convolution Integral Process Models

The linear state space equations can be integrated to give several convolution integral model forms. These convolution integral model forms essentially complete the framework of theoretical model structures that can be obtained as shown in Figure 2.1. They are of particular value when sampled data systems are being derived. The analysis for the convolution integral models begins from the state space equations:

$$\dot{x}(t) = Ax(t) + Bu(t)$$
$$y(t) = Cx(t) + Du(t)$$

State Integral Convolution Form

Firstly, evaluate the state equation at the time value, $t = \tau$, where $0 \leq \tau < t$ and the initial condition is given by $x(0) \in \Re^n$, to obtain

$$\dot{x}(\tau) = Ax(\tau) + Bu(\tau)$$

Pre-multiply both sides of the equation by $e^{-A\tau}$, and rearrange as

$$e^{-A\tau}\dot{x}(\tau) - e^{-A\tau}Ax(\tau) = e^{-A\tau}Bu(\tau)$$

This may be written as

$$d\{e^{-A\tau}x(\tau)\} = e^{-A\tau}Bu(\tau)d\tau$$

Integrating this expression over the interval $0 \leq \tau < t$ yields

$$[e^{-A\tau}x(\tau)]_0^t = \int_0^t e^{-A\tau}Bu(\tau)d\tau$$

and hence

$$e^{-At}x(t) - x(0) = \int_0^t e^{-A\tau}Bu(\tau)d\tau$$

The state integral convolution form then follows as

$$x(t) = e^{At}x(0) + \int_0^t e^{A(t-\tau)}Bu(\tau)d\tau$$

Common Terminology

The free-state response is $x_{\text{free}}(t) = e^{At}x(0)$

The forced state response is $x_{\text{forced}}(t) = \int_0^t e^{A(t-\tau)}Bu(\tau)d\tau$

The state response decomposes as $x(t) = x_{\text{free}}(t) + x_{\text{forced}}(t)$

Output Integral Convolution Form

The analysis uses the output equation

$$y(t) = Cx(t) + Du(t)$$

Assuming, as is commonly found in practical situations, $D = 0$, then

$$y(t) = Cx(t)$$

Substituting for the state integral convolution yields

$$y(t) = Cx(t) = C\left\{e^{At}x(0) + \int_0^t e^{A(t-\tau)}Bu(\tau)d\tau\right\}$$

and hence the output convolution form follows as

$$y(t) = Ce^{At}x(0) + \int_0^t Ce^{A(t-\tau)}Bu(\tau)d\tau$$

Common Terminology

The free output response is $y_{\text{free}}(t) = Ce^{At}x(0)$

The forced output response is $y_{\text{forced}}(t) = \int_0^t Ce^{A(t-\tau)}Bu(\tau)d\tau$

The output response decomposes as $y(t) = y_{\text{free}}(t) + y_{\text{forced}}(t)$

2.1.3 Laplace Transfer Function Models

The Laplace transform of a signal is given by

$$X(s) = \int_0^\infty x(t)e^{-st}dt$$

This transform relationship can be applied to the state space system description to give a link to the transfer function description of systems. For complete generality it is necessary to introduce vectors of transfer function representations of the control, state and output signals in an obvious way as

$$U(s) = \int_0^\infty u(t)e^{-st}dt, \text{ where } u(t) \in \Re^m \text{ and } U(s) \in \Re^m(s) \text{ with } U_i(s) = \int_0^\infty u_i(t)e^{-st}dt, \quad i=1,\ldots,m$$

$$X(s) = \int_0^\infty x(t)e^{-st}dt, \text{ where } x(t) \in \Re^n \text{ and } X(s) \in \Re^n(s) \text{ with } X_i(s) = \int_0^\infty x_i(t)e^{-st}dt, \quad i=1,\ldots,n$$

$$Y(s) = \int_0^\infty y(t)e^{-st}dt, \text{ where } y(t) \in \Re^r \text{ and } Y(s) \in \Re^r(s) \text{ with } Y_i(s) = \int_0^\infty y_i(t)e^{-st}dt, \quad i=1,\ldots,r$$

State Transfer Functions

The analysis begins from the linear state space description:

$$\dot{x}(t) = Ax(t) + Bu(t)$$

For the state transfer functions, multiply the state equation by e^{-st} and integrate out the time variable t over the infinite time interval:

$$\int_0^\infty \dot{x}(t)e^{-st}dt = A\int_0^\infty x(t)e^{-st}dt + B\int_0^\infty u(t)e^{-st}dt$$

Then, using the vector transfer function definitions gives

$$\int_0^\infty \dot{x}(t)e^{-st}dt = AX(s) + BU(s)$$

The vector version of the formula for the Laplace transform of the derivative of a time function is

$$\int_0^\infty \dot{x}(t)e^{-st}dt = sX(s) - x(0)$$

Using this relationship then yields

$$sX(s) - x(0) = AX(s) + BU(s)$$

Rearrangement of this matrix–vector equation gives the state transfer function relationship as

$$X(s) = (sI - A)^{-1}x(0) + (sI - A)^{-1}BU(s)$$

Common Terminology

The Laplace transfer function for free-state response is $X_{\text{free}}(s) = (sI - A)^{-1}x(0)$

The Laplace transfer function for forced state response is $X_{\text{forced}}(s) = (sI - A)^{-1}BU(s)$

The state response transfer function decomposes as $X(s) = X_{\text{free}}(s) + X_{\text{forced}}(s)$

The input-to-state transfer function matrix is given by $G(s) = (sI - A)^{-1}B \in \Re^{n \times m}(s)$

Output Transfer Functions

The analysis uses the output equation

$$y(t) = Cx(t) + Du(t)$$

Assuming, as is commonly found in practical situations, $D = 0$, then

$$y(t) = Cx(t)$$

For the output response transfer function, multiply the output equation by e^{-st} and integrate over the infinite time interval

$$\int_0^\infty y(t)e^{-st}dt = C \int_0^\infty x(t)e^{-st}dt$$

Introduce the vector of output signal transfer function representations to give

$$Y(s) = CX(s)$$

Substituting for the state transfer function relationship gives

$$Y(s) = CX(s) = C\{(sI - A)^{-1}x(0) + (sI - A)^{-1}BU(s)\}$$

and hence

$$Y(s) = C(sI - A)^{-1}x(0) + C(sI - A)^{-1}BU(s)$$

Common Terminology

The Laplace transfer function for the free output response is $Y_{\text{free}}(s) = C(sI - A)^{-1}x(0)$

The Laplace transfer function for the forced output response is $Y_{\text{forced}}(s) = C(sI - A)^{-1}BU(s)$

The output response transfer function decomposes as $Y(s) = Y_{\text{free}}(s) + Y_{\text{forced}}(s)$

The input–output transfer function matrix is denoted $G_p(s) = C(sI - A)^{-1}B \in \Re^{r \times m}(s)$

The above analysis shows the important theoretical links from time-domain state space system model descriptions through to matrix Laplace transform function models. The real power of these results lies in their interpretative generality.

2.1.4 Common Laplace Transform Process Models

The parameterised first-order differential equation that is commonly obtained in simple process modelling exercises is given by

$$\tau \dot{y}(t) + y(t) = Ku(t)$$

where K represents the steady state gain of the plant and τ represents the time constant of the plant. The step response of this system with zero initial conditions is the familiar exponential rise curve. In many modelling exercises, it is experimental exponential rise time response data which leads to the choice of an empirical model based on the first-order differential equation.

Whilst the first-order model has an exponential rise step response, to obtain an oscillatory characteristic in the step response requires a second-order process model. Thus the simple parameterised second-order differential equation which is widely studied for this is given by

$$\ddot{y}(t) + 2\zeta\omega_n\dot{y}(t) + \omega_n^2 y(t) = K\omega_n^2 u(t)$$

where K represents the steady state gain of the plant, ω_n represents the natural frequency of the process and ζ represents the process damping ratio. The introduction of the oscillatory characteristic is related to the value of the process damping ratio ζ as shown in Table 2.2.

The use of this simple modelling route based on experimental step response data leading to a model-fitting exercise with a first- or second-order system response is common in the process

Table 2.2 Oscillatory characteristic in second-order system responses.

Nature of oscillatory characteristic	Value of damping ratio ζ
Purely oscillatory	$\zeta = 0$
Underdamped	$0 < \zeta < 1$
Critically damped	$\zeta = 1$
Overdamped	$1 < \zeta$

industries. For this reason the properties of the first- and second-order models and their time-delayed forms have been thoroughly investigated and common control engineering phraseology is based on this study. It would be superfluous to repeat a development that has been treated in so many introductory control engineering textbooks (Wilkie *et al.*, 2002, for example). However, it is useful to list the main simple process transfer function models arising for the general system equation

$$Y(s) = [G_p(s)]U(s)$$

where the system output is $Y(s)$, the system control input is $U(s)$ and the process transfer function is $G_p(s)$.

First-Order (FO) Process Transfer Function

$$G_p(s) = \left[\frac{K}{\tau s + 1}\right]$$

where K represents the steady state gain of the plant and τ represents the time constant of the plant.

First-Order Plus Dead Time (FOPDT) Process Transfer Function

$$G_p(s) = \left[\frac{K e^{-s T_D}}{\tau s + 1}\right]$$

where K represents the steady state gain of the plant, τ represents the time constant of the plant and T_D represents the process time delay.

Second-Order (SO) Process Transfer Function

$$G_p(s) = \frac{K}{(s^2/\omega_n^2) + (2\zeta s/\omega_n) + 1}$$

where K represents the steady state gain of the plant, ω_n represents the natural frequency of the process and ζ represents the process damping ratio.

Second-Order Plus Dead Time (SOPDT) Process Transfer Function

$$G_p(s) = \frac{K e^{-s T_D}}{(s^2/\omega_n^2) + (2\zeta s/\omega_n) + 1}$$

where K represents the steady state gain of the plant, ω_n represents the natural frequency of the process, ζ represents the process damping ratio and T_D represents the process time delay.

2.2 Controller Degrees of Freedom Structure

There is a generic system of degrees of freedom in the structure of the control loop. The usual classical control loop found in most control textbooks uses the single degree of freedom structure. Not so well known is the two degrees of freedom structure, where the control has two components, one concerned with closed-loop stability and one which can be used to shape the closed-loop response. The idea of feedforward control is well known in the process control community; perhaps less well known is its location in the three degrees of freedom controller structure. The analysis for each of these controller structures is given below.

2.2.1 One Degree of Freedom Control

The single degree of freedom structure is shown in Figure 2.5.

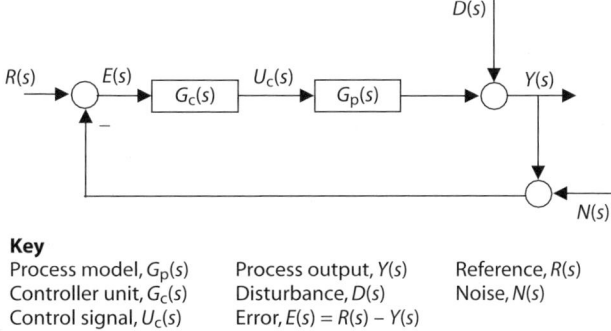

Key
Process model, $G_p(s)$ Process output, $Y(s)$ Reference, $R(s)$
Controller unit, $G_c(s)$ Disturbance, $D(s)$ Noise, $N(s)$
Control signal, $U_c(s)$ Error, $E(s) = R(s) - Y(s)$

Figure 2.5 Single degree of freedom controller block diagram.

The expression for the closed loop is given by

$$Y(s) = \left[\frac{G_p(s)G_c(s)}{1 + G_p(s)G_c(s)}\right](R(s) - N(s)) + \left[\frac{1}{1 + G_p(s)G_c(s)}\right]D(s)$$

The single controller, $G_c(s)$ in this structure, has to:

- Establish closed-loop stability

- Shape the dynamic and the static qualities of the output response $[Y/R]$ and the disturbance response $[Y/D]$

- Attenuate the effect of the measurement noise, $N(s)$

Generally, the outcome of a single degree of freedom design taxes the control engineer's ingenuity to achieve all these objectives.

2.2.2 Two Degree of Freedom Control

Although the advantages of the two degree of freedom controller are often exploited in academic papers and solutions it rarely seems to be mentioned as being used in industrial systems; perhaps it is an industrial trade secret when it is used. One documented industrial example was in the marvellously innovative Toshiba EC 300 autotune controller series which appeared in the 1980s and 90s. A two degree of freedom structure is shown in Figure 2.6.

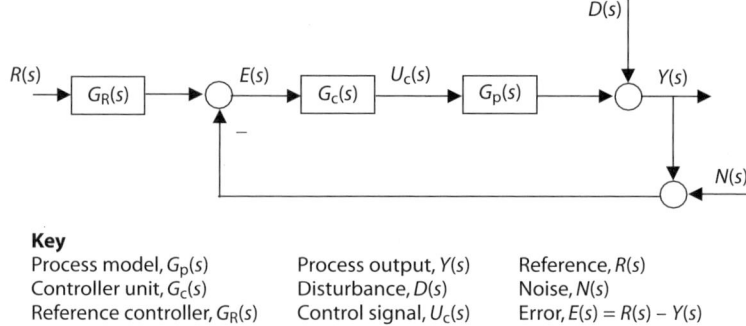

Key
Process model, $G_p(s)$ Process output, $Y(s)$ Reference, $R(s)$
Controller unit, $G_c(s)$ Disturbance, $D(s)$ Noise, $N(s)$
Reference controller, $G_R(s)$ Control signal, $U_c(s)$ Error, $E(s) = R(s) - Y(s)$

Figure 2.6 Two degree of freedom controller block diagram.

The expression for the closed loop is given by

$$Y(s) = \left[\frac{G_p(s)G_c(s)G_R(s)}{1+G_p(s)G_c(s)}\right]R(s) - \left[\frac{G_p(s)G_c(s)}{1+G_p(s)G_c(s)}\right]N(s) + \left[\frac{1}{1+G_p(s)G_c(s)}\right]D(s)$$

The controller $G_c(s)$ in this structure has to:

- Establish closed-loop stability
- Shape the dynamic and the static qualities of the disturbance response [Y/D]
- Attenuate the effect of the measurement noise $N(s)$

The two controllers, $G_c(s)$ and $G_R(s)$, in this structure are available to

- Shape the dynamic and the static qualities of the output response [Y/R]

As can be seen, whilst a single controller, $G_c(s)$, is used to shape the disturbance response [Y/D], an additional degree of freedom from the reference controller, $G_R(s)$, is introduced to shape [Y/R]; hence the designation *two* degree of freedom controller.

It might be thought that if the controller structure was split into two component controllers, $G_{c1}(s)$ in the forward path and a new controller $G_{c2}(s)$ in the feedback path, as shown in Figure 2.7, that an extra degree of freedom might accrue.

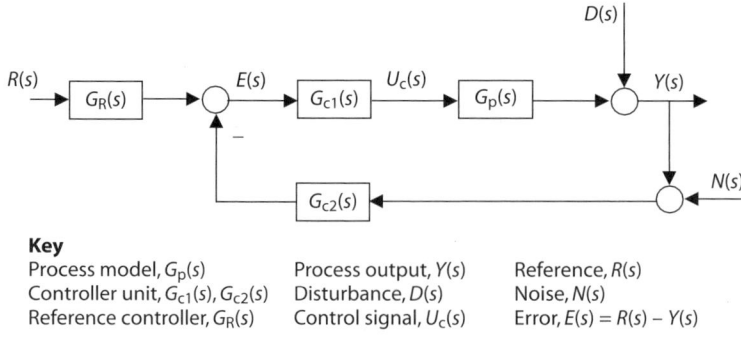

Key
Process model, $G_p(s)$ Process output, $Y(s)$ Reference, $R(s)$
Controller unit, $G_{c1}(s)$, $G_{c2}(s)$ Disturbance, $D(s)$ Noise, $N(s)$
Reference controller, $G_R(s)$ Control signal, $U_c(s)$ Error, $E(s) = R(s) - Y(s)$

Figure 2.7 Two degree of freedom controller block diagram.

The expression for the closed loop is given by

$$Y(s) = \left[\frac{G_p(s)G_{c1}(s)G_R(s)}{1+G_p(s)G_{c1}(s)G_{c2}(s)}\right]R(s) - \left[\frac{G_p(s)G_{c1}(s)G_{c2}(s)}{1+G_p(s)G_{c1}(s)G_{c2}(s)}\right]N(s) + \left[\frac{1}{1+G_p(s)G_{c1}(s)G_{c2}(s)}\right]D(s)$$

As can be seen from this analysis, the composite controller $G_{c1}(s)G_{c2}(s)$ simply replaces the role that the controller $G_c(s)$ took in Figure 2.6. Thus the composite controller $G_{c1}(s)G_{c2}(s)$ in this structure has to

- Establish closed-loop stability
- Shape the dynamic and the static qualities of the disturbance response, $[Y/D]$
- Attenuate the effect of the measurement noise, $N(s)$

Along with the controller $G_{c1}(s)G_{c2}(s)$, the reference controller $G_R(s)$ provides an additional degree of controller freedom to

- Shape the dynamic and the static qualities of the output response $[Y/R]$

Thus the structure of Figure 2.7 remains a two degree of freedom controller.

2.2.3 Three Degree of Freedom Structures

To obtain the three degree of freedom controller it is assumed that the disturbance signal $D(s)$ in the loop is measurable. A three degree of freedom structure is shown in Figure 2.8.

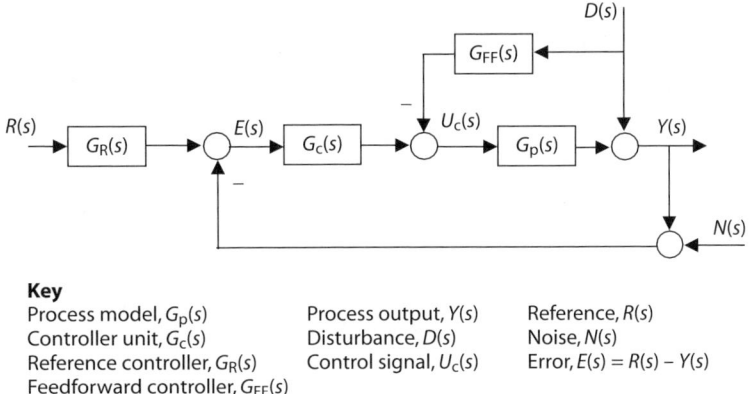

Key
Process model, $G_p(s)$
Controller unit, $G_c(s)$
Reference controller, $G_R(s)$
Feedforward controller, $G_{FF}(s)$
Process output, $Y(s)$
Disturbance, $D(s)$
Control signal, $U_c(s)$
Reference, $R(s)$
Noise, $N(s)$
Error, $E(s) = R(s) - Y(s)$

Figure 2.8 Three degree of freedom structure controller.

The expression for the closed loop is given by

$$Y(s) = \left[\frac{G_p(s)G_c(s)G_R(s)}{1+G_p(s)G_c(s)}\right]R(s) - \left[\frac{G_p(s)G_c(s)}{1+G_p(s)G_c(s)}\right]N(s) + \left[\frac{1-G_p(s)G_{FF}(s)}{1+G_p(s)G_c(s)}\right]D(s)$$

In this structure, the controller $G_c(s)$ has to

- Establish closed-loop stability
- Attenuate the effect of the measurement noise, $N(s)$

The two controllers, $G_c(s)$ and $G_R(s)$, in this structure are available to

- Shape the dynamic and the static qualities of the output responses [Y/R]

The controller $G_{FF}(s)$ in this structure is available to

- Eliminate the disturbance $D(s)$ if it can be chosen so that $1 - G_p(s)G_{FF}(s) = 0$.

This disturbance elimination requires $G_{FF}(s) = G_p(s)^{-1}$, which may not be feasible. In the case where the system is minimum phase, the introduction of a filter to ensure that $G_{FF}(s)$ is strictly proper is the usual modification made, and significant disturbance attenuation is still practical. Thus there are three degrees of freedom in the controller structure.

Sometimes an alternative structure is seen for the three degrees of freedom controller, as shown in Figure 2.9.

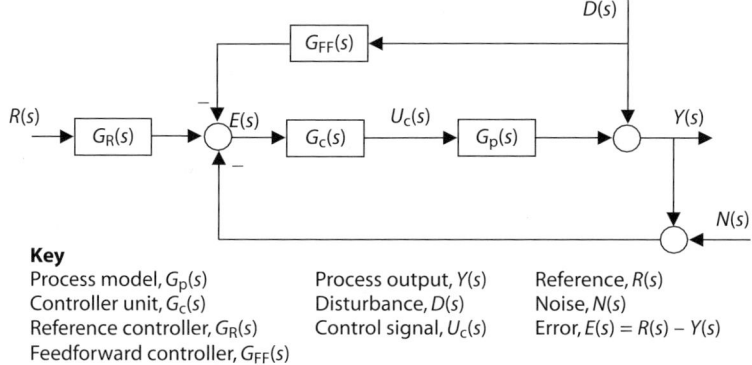

Key
Process model, $G_p(s)$
Controller unit, $G_c(s)$
Reference controller, $G_R(s)$
Feedforward controller, $G_{FF}(s)$

Process output, $Y(s)$
Disturbance, $D(s)$
Control signal, $U_c(s)$

Reference, $R(s)$
Noise, $N(s)$
Error, $E(s) = R(s) - Y(s)$

Figure 2.9 Alternative three degree of freedom controller block diagram.

The expression for the closed loop is given by

$$Y(s) = \left[\frac{G_p(s)G_c(s)G_R(s)}{1 + G_p(s)G_c(s)}\right]R(s) - \left[\frac{G_p(s)G_c(s)}{1 + G_p(s)G_c(s)}\right]N(s) + \left[\frac{1 - G_p(s)G_c(s)G_{FF}(s)}{1 + G_p(s)G_c(s)}\right]D(s)$$

In this structure, the feedforward controller $G_{FF}(s)$ can eliminate the disturbance, $D(s)$, if it can be chosen so that $1 - G_p(s)G_c(s)G_{FF}(s) = 0$. This requires $G_{FF}(s) = [G_p(s)G_c(s)]^{-1}$. Although this will achieve the objective of eliminating the disturbance, it has the obvious disadvantage that every time the feedback controller $G_c(s)$ is redesigned the feedforward controller also has to be updated.

2.3 PID Control Performance

Designing a control system is a circular process (see Figure 2.10) and at some point in the design cycle, the control engineer decides that the design is satisfactory and then leaves the cycle. In itself, the design cycle is also a learning process, so that each step of the cycle often yields insight into the behaviour of the process to be controlled and into the other steps of the design process.

Specifying and determining the appropriate control performance can be the most demanding part of the cycle.

2.3.1 Controller Performance Assessment – General Considerations

In an industrial control problem, understanding the system and its disturbances is an essential prerequisite for a satisfactory design exercise. This first step is assisted by a general system framework which

2.3 PID Control Performance

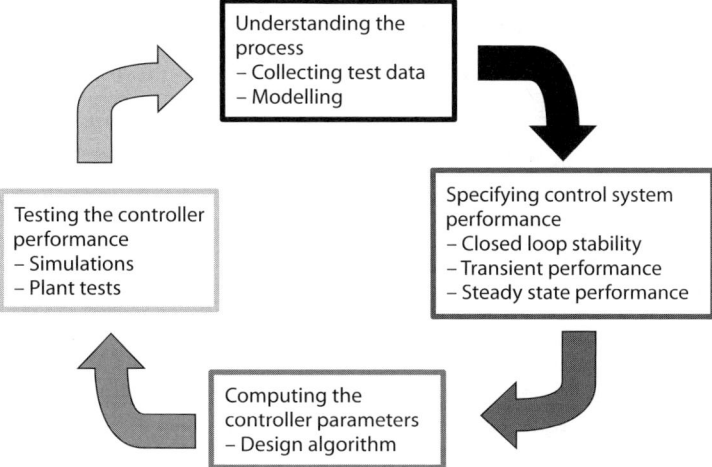

Figure 2.10 The control design cycle.

can be used almost as a checklist to identify the important features of the process. The framework is shown in Figure 2.11.

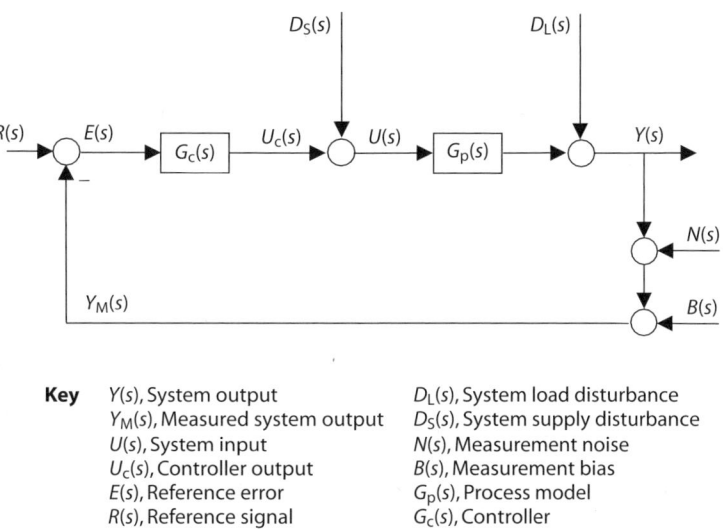

Key
- $Y(s)$, System output
- $Y_M(s)$, Measured system output
- $U(s)$, System input
- $U_c(s)$, Controller output
- $E(s)$, Reference error
- $R(s)$, Reference signal
- $D_L(s)$, System load disturbance
- $D_S(s)$, System supply disturbance
- $N(s)$, Measurement noise
- $B(s)$, Measurement bias
- $G_p(s)$, Process model
- $G_c(s)$, Controller

Figure 2.11 General system framework.

The basic process is represented by the model equation

$$Y(s) = [G_p(s)]U(s) + D_L(s)$$

where the system output is labelled $Y(s)$, the system input is $U(s)$, the system load disturbance is $D_L(s)$ and $G_p(s)$, represents the process model. The control system is designed so that output $Y(s)$ follows the reference signal $R(s)$. In the case that the reference signal is zero, the control design is that of a regulator design. Following the reference signal is termed reference tracking performance.

The process operation is upset by

(a) System load disturbance, denoted $D_L(s)$; this represents the effect of different process loadings on the system output.

(b) System supply disturbance, denoted $D_S(s)$; this represents the effect of variations in the quality of the energy or material supply at the input to the system.

The control design seeks to reject these disturbance inputs. Attenuating the effect of the disturbance inputs is termed disturbance rejection performance.

The controller equation is given by

$$U_c(s) = [G_c(s)]E(s)$$

where the controller transfer function is denoted $G_c(s)$, the controller input is the reference error signal, $E(s)$ and the controller output signal is labelled $U_c(s)$. The reference error $E(s)$ is formed as the difference between the reference signal $R(s)$ and the measured output signal, $Y_M(s)$. The measured signal comprises the sum of the process output signal, $Y(s)$, the process measurement bias signal, $B(s)$ and the process measurement noise signal, $N(s)$. The effect of the measurement bias on the control loop must be examined. Also note that measurement noise is one of the inputs to the controller, and one of the tasks of the controller design is to attenuate measurement noise in the control loop, otherwise the control signal itself contains these noise components. Attenuating the effect of the noise inputs to the controller is termed the noise rejection performance of the controller.

A transfer function analysis of the control loop is extremely useful in identifying clearly the performance objectives of the control problem. The analysis starts at the process output and goes round the loop to give

$$Y(s) = \left[\frac{G_p(s)G_c(s)}{1+G_p(s)G_c(s)}\right]R(s) - \left[\frac{G_p(s)G_c(s)}{1+G_p(s)G_c(s)}\right]B(s) - \left[\frac{G_p(s)G_c(s)}{1+G_p(s)G_c(s)}\right]N(s)$$
$$+ \left[\frac{1}{1+G_p(s)G_c(s)}\right]D_L(s) + \left[\frac{1}{1+G_p(s)G_c(s)}\right][G_p(s)]D_S(s)$$

This can be succinctly represented as a five-branch diagram showing the types of performance required for each branch (Figure 2.12).

Performance can now be considered in several categories.

Closed-Loop Stability

There are two basic expressions in the closed-loop transfer function equation for output $Y(s)$. One is the complementary sensitivity function or the closed-loop transfer function:

$$T(s) = G_{CL}(s) = \left[\frac{G_p(s)G_c(s)}{1+G_p(s)G_c(s)}\right]$$

and the other is the sensitivity function:

$$S(s) = \left[\frac{1}{1+G_p(s)G_c(s)}\right]$$

Closed-loop stability is determined from the rationalised forms of the two equations.

Introduce the transfer function form for the process:

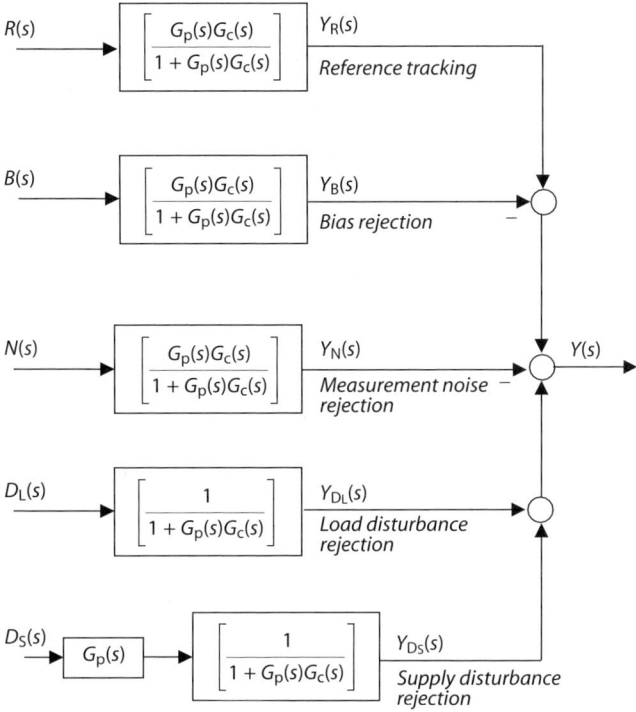

Figure 2.12 System performance framework.

$$G_p(s) = \left[K \frac{n(s)}{d(s)} \right]$$

where $n(0)=1$, $d(0)=1$, $\deg(n)=n_{num}$, $\deg(d)=n_{den}$, $n_{num} < n_{den}$ and $\mathit{diff} = n_{den} - n_{num} > 0$, and the controller

$$G_c(s) = \left[\frac{n_c(s)}{d_c(s)} \right]$$

Then the closed-loop transfer function or complementary function becomes

$$G_{CL}(s) = \left[\frac{Kn(s)n_c(s)}{d(s)d_c(s) + Kn(s)n_c(s)} \right] = \left[\frac{Kn(s)n_c(s)}{\rho_{CL}(s)} \right]$$

and the sensitivity is

$$S(s) = \left[\frac{d(s)d_c(s)}{d(s)d_c(s) + Kn(s)n_c(s)} \right] = \left[\frac{d(s)d_c(s)}{\rho_{CL}(s)} \right]$$

where the closed-loop characteristic expression is given by

$$\rho_{CL}(s) = d(s)d_c(s) + Kn(s)n_c(s)$$

Given that the process has no unstable hidden modes, closed-loop stability is obtained if the roots of the equation $\rho_{CL}(s) = d(s)d_c(s) + Kn(s)n_c(s) = 0$ all lie in the open left half of the complex plane. This is a condition assuming good model and physical process fidelity. The problem of robust stability begins from this point, for when the model representation is not accurate the designed controller still has to guarantee closed-loop stability. The subsequent development of robust stability then depends on how the control engineer is able to specify the model inaccuracy. In later sections the classical measures of robust stability, the gain and phase margin, are discussed along with some work on specifying parametric stability margins for control systems based on simple process models.

Reference Tracking Performance

The reference tracking performance of the system lies in the expression

$$Y_R(s) = \left[\frac{G_p(s)G_c(s)}{1 + G_p(s)G_c(s)}\right] R(s)$$

One part of the control design problem is to choose a controller to shape the transient part of the reference tracking response. The second and easier part is to control the steady state behaviour. For a PID control system, subject to step reference signals, the following analysis for the steady state behaviour is of interest. Assuming a PID controller can be selected to stabilise the closed loop, then define the process model as

$$G_p(s) = \left[K\frac{n(s)}{d(s)}\right]$$

where $n(0) = 1$, $d(0) = 1$, $\deg(n) = n_{num}$, $\deg(d) = n_{den}$, $n_{num} < n_{den}$ and $diff = n_{den} - n_{num} > 0$; the PID controller with $k_I > 0$ as

$$G_c(s) = \left[\frac{k_I + k_P s + k_D s^2}{s}\right]$$

and the step reference signal as

$$R(s) = \frac{r}{s}$$

Then use of the final value theorem gives

$$y_R(\infty) = \lim_{t \to \infty} y_R(t) = \lim_{s \to 0} s Y_R(s) = \lim_{s \to 0} s \times \left[\frac{G_p(s)G_c(s)}{1 + G_p(s)G_c(s)}\right] R(s)$$

Write

$$G_p(s)G_c(s) = \left[\frac{Kn(s)}{d(s)}\right]\left[\frac{k_I + k_P s + k_D s^2}{s}\right] = \left[\frac{Kn(s)(k_I + k_P s + k_D s^2)}{sd(s)}\right]$$

and

$$y_R(\infty) = \lim_{s \to 0} s \times \left[\frac{G_p(s)G_c(s)}{1 + G_p(s)G_c(s)}\right] R(s) = \lim_{s \to 0} s \times \left[\frac{Kn(s)(k_I + k_P s + k_D s^2)}{sd(s) + Kn(s)(k_I + k_P s + k_D s^2)}\right]\left(\frac{r}{s}\right) = \frac{Kk_I n(0)}{Kk_I n(0)} r = r$$

This analysis shows that the reference level is attained *provided* integral control is present. Note that the result does not depend on correct knowledge – or indeed any knowledge – of the process parameters. This is the important practical robust reference tracking performance result for the application of integral control. It is a guaranteed robust performance result, which is why integral control is such an important control technique in practical applications.

2.3 PID Control Performance

Measurement Bias Rejection

The measurement bias rejection property lies in the expression

$$Y_B(s) = \left[\frac{G_p(s)G_c(s)}{1+G_p(s)G_c(s)}\right]B(s)$$

If it is assumed that the bias is a constant step signal, written as $B(s)=b/s$, then an analysis identical to that for steady state reference tracking performance with the assumption of integral action in the controller yields

$$y_B(\infty) = \lim_{s\to 0} s \times \left[\frac{G_p(s)G_c(s)}{1+G_p(s)G_c(s)}\right]B(s) = \lim_{s\to 0} s \times \left[\frac{Kn(s)(k_I + k_P s + k_D s^2)}{sd(s)+Kn(s)(k_I + k_P s + k_D s^2)}\right]\left(\frac{b}{s}\right) = \frac{Kk_I n(0)}{Kk_I n(0)}b = b$$

This result has important consequences since it implies that in steady state the reference level, r, will be incorrect by the amount of the measurement bias, b. The inescapable conclusion is that it is imperative that any measuring device bias must be removed by manual means, since it will not be compensated for by controller design.

Measurement Noise Rejection

The measurement noise rejection performance lies in the expression

$$Y_N(s) = \left[\frac{G_p(s)G_c(s)}{1+G_p(s)G_c(s)}\right]N(s)$$

It is often the case that process noise is assumed to be represented by high-frequency signals and hence the measurement noise rejection performance resides in the high-frequency properties of the complementary sensitivity function. This is most commonly shaped in the high-frequency region by incorporating filters in the controller to increase the controller high-frequency roll-off rate. This will also have the effect of increasing the high-frequency roll-off rate of the complementary sensitivity function. For cases where PID control is used and there is a high measurement noise component in the measured output it is essential not to use the derivative control term or at worst to ensure that a filtered derivative term is used. This was discussed in Chapter 1.

Load Disturbance Rejection

The load disturbance rejection performance lies in the expression

$$Y_{D_L}(s) = \left[\frac{1}{1+G_p(s)G_c(s)}\right]D_L(s)$$

In the time domain, the particular controller selected will determine the nature of the transient response to a particular form of load disturbance signal $D_L(s)$. In many cases the load disturbance is a low-frequency phenomenon which can be modelled by the step signal model $D_L(s)=d_L/s$. Using the general system and PID controller descriptions from above, the final value theorem approach yields a steady state result

$$y_{D_L}(\infty) = \lim_{s\to 0} s \times \left[\frac{1}{1+G(s)G_c(s)}\right]D_L(s) = \lim_{s\to 0} s \times \frac{1}{1+[Kn(s)(k_I + k_P s + k_D s^2)/sd(s)]}\left(\frac{d_L}{s}\right)$$

$$= \lim_{s\to 0} s \times \left[\frac{sd(s)}{sd(s)+Kn(s)(k_I + k_P s + k_D s^2)}\right]\left(\frac{d_L}{s}\right) = 0$$

In this analysis, it is the integral term in the PID controller that ensures this load disturbance rejection result. The result is also a guaranteed property of integral control.

Supply Disturbance Rejection

The supply disturbance rejection performance lies in the expression

$$Y_{D_S}(s) = \left[\frac{1}{1 + G_p(s)G_c(s)}\right][G_p(s)]D_S(s)$$

If the supply disturbance phenomenon is fluctuating, then the nature of the controller selected will determine the shape of the transient response to a particular form of supply disturbance signal $D_S(s)$. In some cases the supply disturbance signal is a low-frequency disturbance which can be modelled by the step signal, $D_S(s) = d_S/s$. Using the general system and PID controller descriptions above, the final value theorem approach yields the following steady state result:

$$y_{D_S}(\infty) = \lim_{s \to 0} s \times \left[\frac{G_p(s)}{1 + G_p(s)G_c(s)}\right]D_S(s) = \lim_{s \to 0} s \times \left[\frac{(Kn(s)/d(s))}{1 + [Kn(s)(k_I + k_P s + k_D s^2)]/sd(s)}\right]\left(\frac{d_S}{s}\right) = 0$$

As with the load disturbance rejection result, it is again the presence of the integral term in the PID controller that secures this supply disturbance rejection result. The result is also a guaranteed property of integral control.

It should be noted that these results can be given a refined presentation for a more general system description where there are open loop poles $s^l = 0$, $0 \leq l$, namely,

$$G_p(s) = \left[\frac{Kn(s)}{s^l d(s)}\right]$$

where $n(0) = 1$, $d(0) = 1$, $\deg(n) = n_{num}$, $0 \leq l$, $\deg(d) = n_{den}$ and $n_{num} < n_{den} + l$.

2.3.2 Controller Assessment – the Effectiveness of PID Control

PID control is used in many industrial process systems because it is all that is needed to achieve effective control performance for a large number of mundane control problems. Furthermore, this success has been reinforced by its ease of application, its well-known properties and its wide availability. But the wide availability of PID control as a menu module in the controller toolbox can also lead to its inappropriate use. PID control is not the answer to every control problem, and it is important to know just when it is necessary to use a more advanced control strategy. If other controller modules were commonly available to the industrial control engineer as very simple plug-in controllers with well-defined properties, then it is very likely that these would also find common application. Indeed, there have been attempts by the vendors of control SCADA and DCS software to unleash the potential of advanced control in this way (Blevins et al., 2003).

The objective of this section is to initiate a diagnostic framework to determine just when PID control is appropriate. To achieve this, the performance of PID control is linked to various types of system properties and a set of facts covering the successes and limitations of PID control is created. The framework uses facts established using the general unity feedback control system as shown in Figure 2.13.

The output expression is given by

$$Y(s) = \left[\frac{G_p(s)G_c(s)}{1 + G_p(s)G_c(s)}\right](R(s) - N(s)) + \left[\frac{1}{1 + G_p(s)G_c(s)}\right]D_L(s)$$

2.3 PID Control Performance

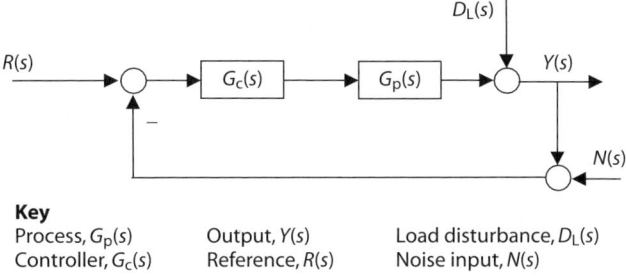

Figure 2.13 General unity feedback control system.

The definitions of the complementary sensitivity $T(s)$ and sensitivity $S(s)$ can be given and the output expression rewritten. The complementary sensitivity is

$$T(s) = \left[\frac{G_p(s)G_c(s)}{1 + G_p(s)G_c(s)} \right]$$

and the sensitivity is

$$S(s) = \left[\frac{1}{1 + G_p(s)G_c(s)} \right]$$

Then the output expression is

$$Y(s) = [T(s)](R(s) - N(s)) + [S(s)]D_L(s)$$

From this relationship the performance diagram of Figure 2.14 can be constructed.

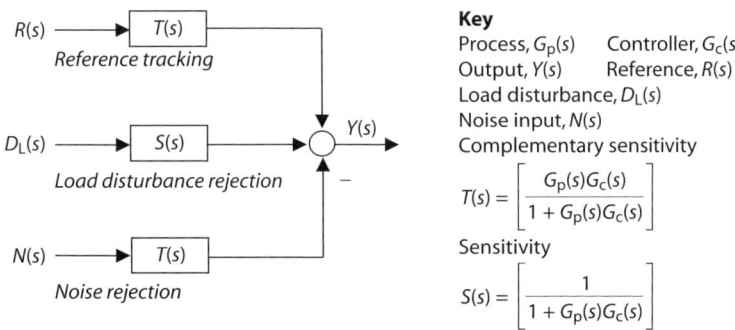

Figure 2.14 Performance branches for control system.

Using Figure 2.14, the application of PID control for different types of system characteristics are investigated for the categories closed-loop stability, reference tracking, load disturbance rejection and measurement noise attenuation. Define the numerator and denominator forms for the process and controller as follows:

Process model: $\quad G_p(s) = K \dfrac{n(s)}{d(s)}$

with $n(0) = 1, d(0) = 1, \deg(n) = n_{num}, \deg(d) = n_{den}, n_{num} < n_{den}$ and $0 < n_{den} - n_{num} = \mathit{diff}$.

Controller: $$G_c(s) = \frac{n_c(s)}{d_c(s)}$$

Stability: Potential Closed-Loop Pole Assignment
Using the process and controller definitions, the closed-loop system polynomial is

$$\rho_{CL}(s) = d(s)d_c(s) + Kn(s)n_c(s)$$

The potential for closed-loop pole location is determined by the flexibility available to compute acceptable three-term controller coefficients to achieve arbitrarily chosen stable closed-loop pole positions.

Tracking Reference Steps and Load Disturbance Rejection – Steady State Performance
Assuming that the closed loop is stable, the steady state performance can be determined as follows. The performance is examined for step reference changes $R(s) = r/s$ and for low-frequency load disturbances $D_L(s)$. For the latter a step signal model is used $D_L(s) = d_L/s$. Omitting the noise term from the output expression, the final value theorem then gives

$$y_{ss} = \lim_{s \to 0} sY(s) = \lim_{s \to 0} s[T(s)]R(s) + \lim_{s \to 0} s[S(s)]D_L(s)$$

Hence, assuming limits exist,

$$y_{ss} = [T(0)]r + [S(0)]d_L$$

Thus if $T(0) = 1$ perfect steady state tracking is achieved for step reference signals, and if $S(0) = 0$ then for low frequency load disturbances (modelled by a step signal) perfect steady state load disturbance rejection occurs.

Effect on Measurement Noise
To determine the effect on the measurement noise $N(s)$ the following component of the output signal is relevant:

$$Y_N(s) = \left[\frac{G_p(s)G_c(s)}{1 + G_p(s)G_c(s)}\right]N(s) = [T(s)]N(s)$$

Consequently, if noise is considered to be a high-frequency phenomenon, the noise attenuation of the proposed control system is assessed by examining the high-frequency behaviour of $T(s)$.

This analytical and diagnostic framework can be used with various process model types and controllers from the three-term control family. Some results are given in Table 2.3.

Table 2.3 shows clearly the immediate improvement attainable in reference tracking and disturbance rejection performance whenever integral action is included in the three-term controller. However, over P and PD control, PI and PID control increase the dynamic order of the closed-loop characteristic equation and may make it more difficult to stabilise the closed loop.

PID Control of Some Special System Features
There are some classes of system for which the control solution needs further consideration than just the blind application of an easy-to-use module like PID control. In most of these cases, the system feature is directly attributable to some particular physical aspect of the system.

Systems With Purely Oscillatory Modes
Purely oscillatory modes or extremely lightly damped modes are often found in processes which have a rotating physical element or repetitive system action. Examples include a steel strip rolling mill stand, where the slightly non-circular rolls produce an almost oscillatory output effect in strip gauge. In electrical power systems, the voltage control problem can have situations where the system becomes very lightly damped and exhibits an oscillating phenomenon. Ship models are often very lightly damped to

Table 2.3 General assessment of three-term controller performance for different system model types.

First-order process – P control

Process	Controller	Closed-loop stability
$G_p(s) = \left[\dfrac{K}{\tau s + 1}\right]$	$G_c(s) = k_P$	$\rho_{CL}(s) = \tau s + (1 + Kk_P)$

First-order dynamics, wide flexibility to assign single real closed-loop pole

Reference tracking	Load disturbance rejection	Noise rejection performance	Comments
$T(0) = Kk_P / (1 + Kk_P)$	$S(0) = 1/(1 + Kk_P)$	$T(s) \sim 1/s$	Wide range of simple processes approximately first-order
Gain-dependent steady state tracking performance	Gain dependent rejection	atten 20 dB/decade	

First-order process – PI control

Process	Controller	Closed-loop stability
$G_p(s) = \left[\dfrac{K}{\tau s + 1}\right]$	$G_c(s) = k_P + \dfrac{k_I}{s}$	$\rho_{CL}(s) = \tau s^2 + (1 + Kk_P)s + Kk_I$

Second-order dynamics, wide flexibility to assign closed-loop poles

Reference tracking	Load disturbance rejection	Noise rejection performance	Comments
$T(0) = 1$	$S(0) = 0$	$T(s) \sim 1/s$	Wide range of simple processes approximately first-order
Perfect steady state tracking	Perfect rejection	atten 20 dB/decade	

Standard second-order process – P control

Process	Controller	Closed-loop stability
$G_p(s) = \left[\dfrac{K\omega_n^2}{s^2 + 2\zeta\omega_n s + \omega_n^2}\right]$	$G_c(s) = k_P$	$\rho_{CL}(s) = s^2 + 2\zeta\omega_n s + \omega_n^2(1 + Kk_P)$

Second-order dynamics, limited flexibility to assign closed-loop poles

Reference tracking	Load disturbance rejection	Noise rejection performance	Comments
$T(0) = Kk_P / (1 + Kk_P)$	$S(0) = 1/(1 + Kk_P)$	$T(s) \sim 1/s^2$	Poor tracking and rejection performance
Gain-dependent steady state tracking performance	Gain-dependent steady state rejection performance	atten 40 dB/decade	Good measurement noise attenuation

Table 2.3 (continued)

Standard second-order process – PI control

Process

$$G(s) = \left[\frac{K\omega_n^2}{s^2 + 2\zeta\omega_n s + \omega_n^2}\right]$$

Controller

$$G_c(s) = k_P + \frac{k_I}{s}$$

Closed-loop stability

$$\rho_{CL}(s) = s^3 + 2\zeta\omega_n s^2 + \omega_n^2(1 + Kk_P)s + \omega_n^2 Kk_I$$

Third-order dynamics, limited flexibility to assign closed-loop poles

Reference tracking

$T(0) = 1$

Perfect steady state tracking

Load disturbance rejection

$S(0) = 0$

Perfect rejection

Noise rejection performance

$T(s) \sim 1/s^2$

atten 40 dB/decade

Comments

Good measurement noise attenuation

Standard second-order process – PD control

Process

$$G_p(s) = \left[\frac{K\omega_n^2}{s^2 + 2\zeta\omega_n s + \omega_n^2}\right]$$

Controller

$$G_c(s) = k_P + k_D s$$

Closed-loop stability

$$\rho_{CL}(s) = s^2 + (2\zeta\omega_n + \omega_n^2 Kk_D)s + \omega_n^2 Kk_P$$

Second-order dynamics, wider flexibility to assign closed-loop poles

Reference tracking

$T(0) = Kk_P/(1 + Kk_P)$

Offset in reference tracking

Load disturbance rejection

$S(0) = 1/(1 + Kk_P)$

Offset in disturbance rejection

Noise rejection performance

$T(s) \sim 1/s$

atten 20 dB/decade

Comments

Poor tracking and disturbance rejection performance

Standard second-order process – PID control

Process

$$G_p(s) = \left[\frac{K\omega_n^2}{s^2 + 2\zeta\omega_n s + \omega_n^2}\right]$$

Controller

$$G_c(s) = k_P + \frac{k_I}{s} + k_D s$$

Closed-loop stability

$$\rho_{CL}(s) = s^3 + (2\zeta\omega_n + \omega_n^2 Kk_D)s^2 + (1 + Kk_P)\omega_n^2 s + \omega_n^2 Kk_I$$

Third-order dynamics, wider flexibility to assign closed-loop poles

Reference tracking

$T(0) = 1$

Perfect steady state tracking

Load disturbance rejection

$S(0) = 0$

Perfect rejection

Noise rejection performance

$T(s) \sim 1/s$

atten 20 dB/decade

Comments

Integral action recovers good tracking and rejection properties

2.3 PID Control Performance

Table 2.3 (continued)

Special second-order process – Type 1 model – P control

Process	Controller	Closed-loop stability	
$G_p(s) = \left[\dfrac{K}{s(\tau s + 1)}\right]$	$G_c(s) = k_P$	$\rho_{CL}(s) = \tau s^2 + s + Kk_P$ Second-order dynamics, limited assignment of closed-loop poles available	

Reference tracking	Load disturbance rejection	Noise rejection performance	Comments
$T(0) = 1$ Perfect steady state tracking	$S(0) = 0$ Perfect rejection	$T(s) \sim 1/s^2$ atten 40 dB/decade	Integral action within the process providing good properties. Might be difficult to obtain good transient properties

Special second-order process – Type 1 model – PI control

Process	Controller	Closed-loop stability	
$G_p(s) = \left[\dfrac{K}{s(\tau s + 1)}\right]$	$G_c(s) = k_P + \dfrac{k_I}{s}$	$\rho_{CL}(s) = \tau s^3 + s^2 + Kk_P s + Kk_I$ Third-order dynamics, limited flexibility to assign closed-loop poles	

Reference tracking	Load disturbance rejection	Noise rejection performance	Comments
$T(0) = 1$ Perfect steady state tracking	$S(0) = 0$ Perfect rejection	$T(s) \sim 1/s^2$ atten 40 dB/decade	Might be difficult to stabilise process

Special second-order process – Type 1 model – PD control

Process	Controller	Closed-loop stability	
$G_p(s) = \left[\dfrac{K}{s(\tau s + 1)}\right]$	$G_c(s) = k_P + k_D s$	$\rho_{CL}(s) = \tau s^2 + (1 + Kk_D)s + Kk_P$ Second-order dynamics, wide flexibility to assign closed-loop poles, control over damping through k_D	

Reference tracking	Load disturbance rejection	Noise rejection performance	Comments
$T(0) = 1$ Perfect steady state tracking	$S(0) = 0$ Perfect rejection	$T(s) \sim 1/s$ atten 20 dB/decade	Motor control systems often use this model Good performance properties

Table 2.3 (continued)

High-order systems – non-dominant second-order process – P control		
Process	**Controller**	**Closed-loop stability**
$G_p(s) = \left[K \dfrac{n(s)}{d(s)} \right]$, $d(0) = 1 = n(0)$ $\deg(d) = n_{den}$, $\deg(n) = n_{num}$ $\text{diff} = (n_{den} - n_{num}) > 0$	$G_c(s) = k_P + \dfrac{k_I}{s}$	$\rho_{CL}(s) = sd(s) + Kn(s)(k_P s + k_I)$ (n_{den})th-order dynamics, limited flexibility to assign closed-loop poles
Reference tracking	**Load disturbance rejection**	**Noise rejection performance**	**Comments**
$T(0) = Kk_P/(1 + Kk_P)$ Gain-dependent steady state tracking performance	$S(0) = 1/(1 + Kk_P)$ Gain dependent steady state rejection performance	$T(s) \sim 1/s^{\text{diff}}$ attenuation is $20 \times \text{diff}$ dB/dec	May be difficult to stabilise process or obtain suitable transient response

High-order systems – non-dominant second-order process – PI control		
Process	**Controller**	**Closed-loop stability**
$G_p(s) = \left[K \dfrac{n(s)}{d(s)} \right]$, $d(0) = 1 = n(0)$ $\deg(d) = n_{den}$, $\deg(n) = n_{num}$ $\text{diff} = (n_{den} - n_{num}) > 0$	$G_c(s) = k_P + \dfrac{k_I}{s}$	$\rho_{CL}(s) = sd(s) + Kn(s)(k_P s + k_I)$ $(n_{den} + 1)$th-order dynamics, limited flexibility to assign closed-loop poles
Reference tracking	**Load disturbance rejection**	**Noise rejection performance**	**Comments**
$T(0) = 1$ Perfect steady state tracking	$S(0) = 0$ Perfect rejection	$T(s) \sim 1/s^{\text{diff}}$ attenuation is $20 \times \text{diff}$ dB/dec	May be difficult to stabilise process or obtain suitable transient response

replicate the oscillatory effect of wave motions on the vessel. Stationary marine vessels like drill-ships exhibit this type of oscillatory effect due to wave forces. In all of these examples, it is necessary to give very careful consideration to the type of control system to be employed. Industrial solutions may range from the use of PID control with some add-ons like a notch filter. Other solutions may involve the use of significant modelling and estimation input along with the use of advanced control methods.

Systems With Non-Minimum Phase Zeros

It is also interesting to look at the physical situation in a process if a system model shows the presence of non-minimum phase or right half-plane zeros. Firstly, such zeros often have a distinct physical origin (Wilkie et al., 2002). They may indicate some internal mechanism at work in the system or a deficiency in the actuator or measurements points in the system. Secondly, in the case of a single input, single output system they will present a limitation to performance achievable by any control system. An extended discussion of the implications of the presence of non-minimum phase zeros in a system model has been given by Skogestad and Postlethwaite (1996).

Systems With Time Delays

Two key models in process control are the simple system models which have time delays.

First-Order Plus Dead Time (FOPDT) Process Transfer Function

$$G_p(s) = \left[\frac{Ke^{-sT_D}}{\tau s + 1} \right]$$

Second-Order Plus Dead Time (SOPDT) Process Transfer Function

$$G_p(s) = \frac{Ke^{-sT_D}}{(s^2/\omega_n^2) + (2\zeta s/\omega_n) + 1}$$

The time delay, which is a non-minimum phase phenomenon, usually represents a transport delay in the physical system. Consequently, in the process control literature such delay models are quite common and there is a wealth of control material for which such a model is the first assumption. Another way in which this has affected the literature is that slow-reacting first-order types of response are often represented by an FOPDT model whether, or not the physical process actually has a time delay.

In general a time delay will tend to destabilise a closed-loop system and hence make it much more difficult to control. For this reason it is always useful to look again at the physical process to determine the origin of the time delay effect. Possibly it is due to measurements being made too far away from the variable or physical property to be controlled. The repositioning of a sensor could reduce or eliminate a long measurement delay, making the process much easier to control. Any process with a long time delay will be difficult to control, and advanced control methods like model-based predictive control may be required.

2.3.3 Classical Stability Robustness Measures

Closed-loop stability is usually expressed in terms of the concept of absolute stability. Namely, in a transfer function description of the closed-loop system where there are no unstable hidden poles, the closed loop is stable if all the closed-loop poles have negative real parts. To introduce a stability specification for performance, the concept of relative stability is used. Relative stability measures, specifies or determines just how stable the closed loop actually is. Thus measures of relative stability are used to specify or determine the distance of a closed-loop system from instability. A different interpretation of measures or margins of relative stability links them to the amount of model error or uncertainty permissible before the closed-loop system is unstable. In this interpretation of the available robustness within the closed-loop control design, the measures are termed stability robustness measures. Introductory control engineering textbooks almost always use the method of Routh–Hurwitz to investigate absolute stability, and in such textbooks, gain margin and phase margin in the frequency domain are given as the measures of relative stability or classical robustness.

The Nyquist Stability Criterion is an application of Cauchy's theorem to the problem of determining whether a closed-loop system is stable. The result depends on the number of encirclements made by an s-domain function, $F(s) = 1 + G_p(s)G_c(s)$, as the complex variable, s, traverses the right half-plane Nyquist D-contour. But for a significant class of practical systems the technicalities of the encirclements reduce to a simple situation, as shown in the Nyquist geometry of Figure 2.15.

In this simple situation encirclement of the stability point $(-1, 0)$ implies that the closed-loop system will be unstable. Furthermore, encirclement or non-encirclement is equivalent to finding the stability point on the right-hand side or left-hand side respectively of the frequency response plot of $G_p(j\omega)G_c(j\omega)$ as frequency ω travels from $\omega = 0$ to $\omega = \infty$. In Figure 2.15, plot A represents a system $G_p(s)G_A(s)$ whose closed loop is stable whilst plot B represents a system $G_p(s)G_B(s)$ whose closed loop is unstable. Examined closely, as ω travels from $\omega = 0$ to $\omega = \infty$ (shown by the direction of the arrows on the

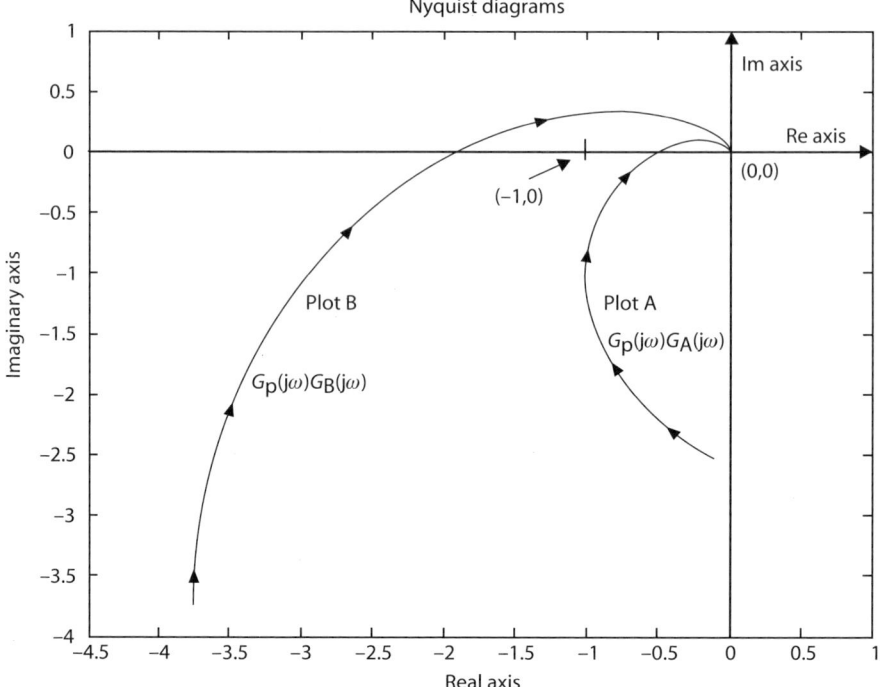

Figure 2.15 Simple Nyquist geometry.

figure), for plot A the stability point $(-1,0)$ is on the left-hand side, equivalent to no encirclement and no right half-plane poles in the closed-loop system, and for plot B the stability point $(-1,0)$ is on the right-hand side, equivalent to an encirclement and the presence of a right half-plane pole in the closed-loop system. This is sometimes called the *Left Hand Rule* test for relative stability. This geometric situation can be used to make a specification for a desired degree of relative stability. These two measures are the classical robustness measures of gain margin and phase margin.

Gain Margin

Gain margin, denoted *GM*, is simply a specification of where the frequency response should cross the negative real axis of the Nyquist response plane. It is a *gain margin* because it defines the gain multiplier that would cause the closed loop to become unstable.

Using Figure 2.16, the gain margin is defined through the condition

$$GM \times |G_p(j\omega_{-\pi})G_c(j\omega_{-\pi})| = 1$$

so that

$$GM = \frac{1}{|G_p(j\omega_{-\pi})G_c(j\omega_{-\pi})|}$$

The frequency $\omega_{-\pi}$ where the gain margin is calculated is known as the gain margin frequency or the phase crossover frequency. The term *phase crossover frequency* relates to the property that this is the frequency at which the system phase is $-180°$. The robustness interpretation is that if the process model is incorrect, then the steady state system gain can be multiplied by the gain margin before the closed loop becomes unstable. Thus the control design is robust to within the gain margin factor.

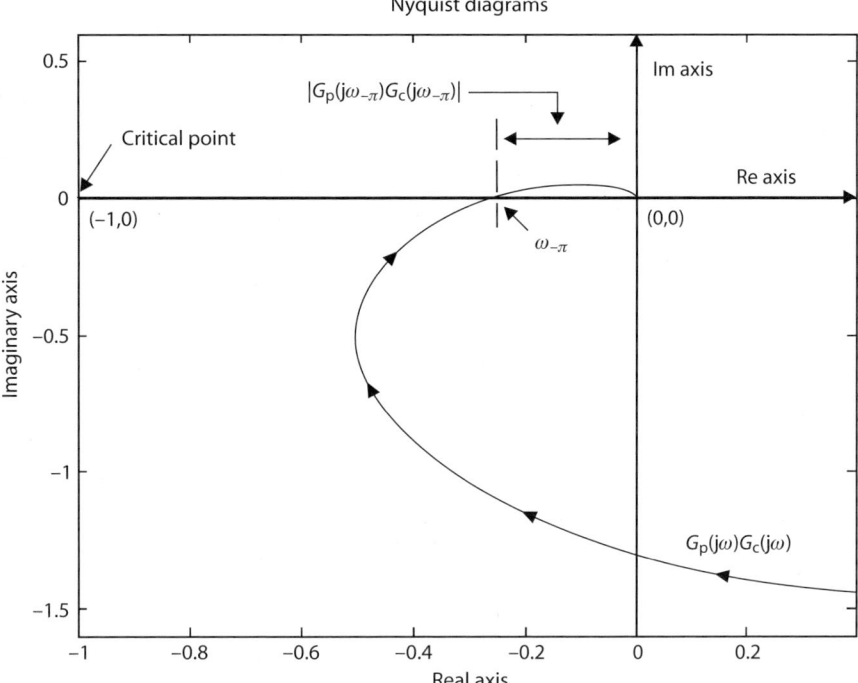

Figure 2.16 Gain margin geometry.

Design values for the gain margin can be given as gain values or in dB:

Absolute gain margin ranges: $2 < GM < 2.5$

dB gain margin ranges: $6\,\text{dB} < GM\,\text{dB} < 8\,\text{dB}$

Phase Margin

Phase margin, denoted *PM*, is simply a specification of how much phase lag can be inserted thereby rotating the frequency response so that it passes through the stability point. It is called a *phase margin* because it defines the amount of phase shift that can be tolerated before a closed-loop system becomes unstable.

Using the Figure 2.17, the phase margin is defined through the angle condition given in degrees as

$$-PM° + \arg\{G_p(j\omega_1)G_c(j\omega_1)\}° = -180°$$

so that

$$PM° = 180° + \arg\{G(j\omega_1)G_c(j\omega_1)\}°$$

The frequency ω_1 where the phase margin is calculated is known as the phase margin frequency or the gain crossover frequency. The term gain crossover frequency relates to the property that this is the frequency at which the system gain passes through an absolute value of unity. The robustness interpretation is that if the process model is incorrect by a lag unit, then the system can experience an additional phase lag of $PM°$ before the closed loop becomes unstable. Thus the control design is robust to within a phase shift of $PM°$ at frequency ω_1.

Design values for phase margin are usually given in degrees:

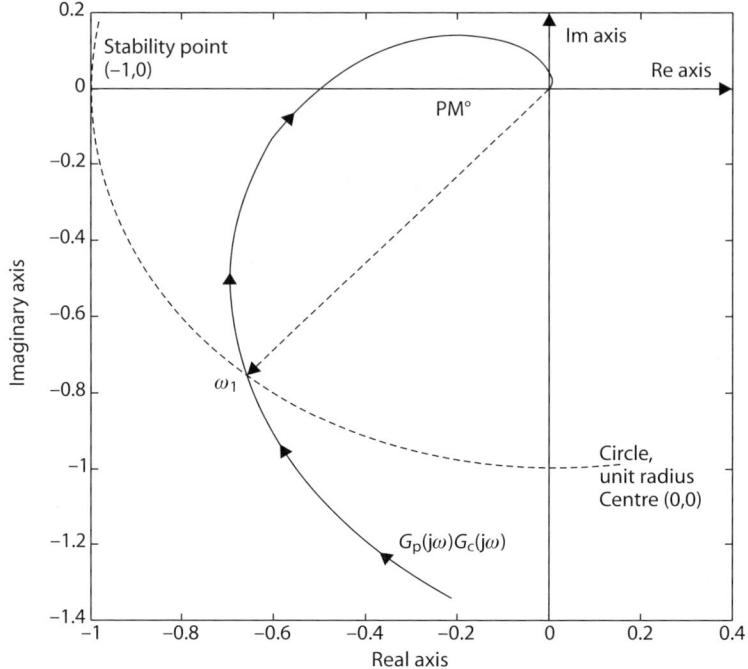

Figure 2.17 Phase margin geometry.

Phase margin ranges: $30° < PM < 60°$

Phase margin can also be usefully interpreted as a delay margin, denoted *DM*. The delay margin can be computed from the phase margin as

$$DM = \frac{PM°}{180°}\frac{\pi}{\omega_1}$$

The delay margin represents the delay that can occur in the forward path before the closed-loop system becomes unstable. For example, this might represent the amount of delay available to compute new control signal values before the closed-loop system becomes unstable. As such, it is quite useful to know the delay margin in closed-loop control studies.

Maximum Sensitivity Specification

The sensitivity function, $S(s)$, appeared in the load and supply disturbance terms in the generic analysis of control loop performance. The two expressions were:

Load disturbance term: $Y_{D_L}(s) = \left[\dfrac{1}{1 + G_p(s)G_c(s)}\right] D_L(s) = S(s) D_L(s)$

Supply disturbance term: $Y_{D_S}(s) = \left[\dfrac{1}{1 + G_p(s)G_c(s)}\right] \tilde{D}_S(s) = S(s) \tilde{D}_S(s)$

where $\tilde{D}_S(s) = [G_p(s)] D_S(s)$.

2.3 PID Control Performance

Disturbance rejection is a performance objective for control design. It is possible to give this disturbance rejection issue a frequency domain interpretation using these expressions. Set $s = j\omega$ and consider the magnitude content of output components, $Y_{D_L}(s)$ and $Y_{D_S}(s)$; hence

$$|Y_{D_L}(j\omega)| = |S(j\omega)| \times |D_L(j\omega)|$$

and

$$|Y_{D_S}(j\omega)| = |S(j\omega)| \times |\tilde{D}_S(j\omega)|$$

If it is assumed that the disturbance signals are low frequency in nature then in a frequency range of interest, $0 < \omega < \omega_D$, the transmission of the disturbance into the system output can be bounded by

$$|Y_{D_L}(j\omega)| = |S(j\omega)| \times |D_L(j\omega)| \leq \left\{ \max_{0 < \omega < \omega_D} |S(j\omega)| \right\} \times |D_L(j\omega)|$$

and

$$|Y_{D_S}(j\omega)| = |S(j\omega)| \times |\tilde{D}_S(j\omega)| \leq \left\{ \max_{0 < \omega < \omega_D} |S(j\omega)| \right\} \times |\tilde{D}_S(j\omega)|$$

Thus the disturbance rejection problem can be related to the maximum sensitivity M_S, computed as

$$M_S = \max_{0 < \omega < \omega_D} |S(j\omega)|$$

A further step can be taken since this bound can be minimised by choice of controller, $G_c(s)$ to give the minimum–maximum sensitivity value as

$$M_S^\circ = \min_{G_c \in G_{cstable}} \{ M_S(G_c) \} = \min_{G_c \in G_{cstable}} \{ \max_{0 < \omega < \omega_D} |S(j\omega)| \}$$

This is one way to use the maximum sensitivity expression. But a look at the Nyquist geometry for the maximum sensitivity leads to its separate use as a stability robustness measure.

Recall the definition of the maximum sensitivity as

$$M_S = \max_{0 < \omega < \omega_D} |S(j\omega)| = \max_{0 < \omega < \omega_D} \left| \frac{1}{1 + G_p(j\omega)G_c(j\omega)} \right|$$

Then consider the quantity

$$M_S(j\omega) = \frac{1}{1 + G_p(j\omega)G_c(j\omega)}$$

and hence

$$\frac{1}{M_S(j\omega)} = 1 + G_p(j\omega)G_c(j\omega)$$

Thus, as shown in Figure 2.18, the vector for $1/M_S(j\omega)$ is readily found using the vector summation in the expression $1 + G_p(j\omega)G_c(j\omega)$.

Examining the maximisation over the range of frequency ω; if

$$|M_S(j\omega_A)| = \left| \frac{1}{1 + G_p(j\omega_A)G_c(j\omega_A)} \right| < \left| \frac{1}{1 + G_p(j\omega_B)G_c(j\omega_B)} \right| = |M_S(j\omega_B)|$$

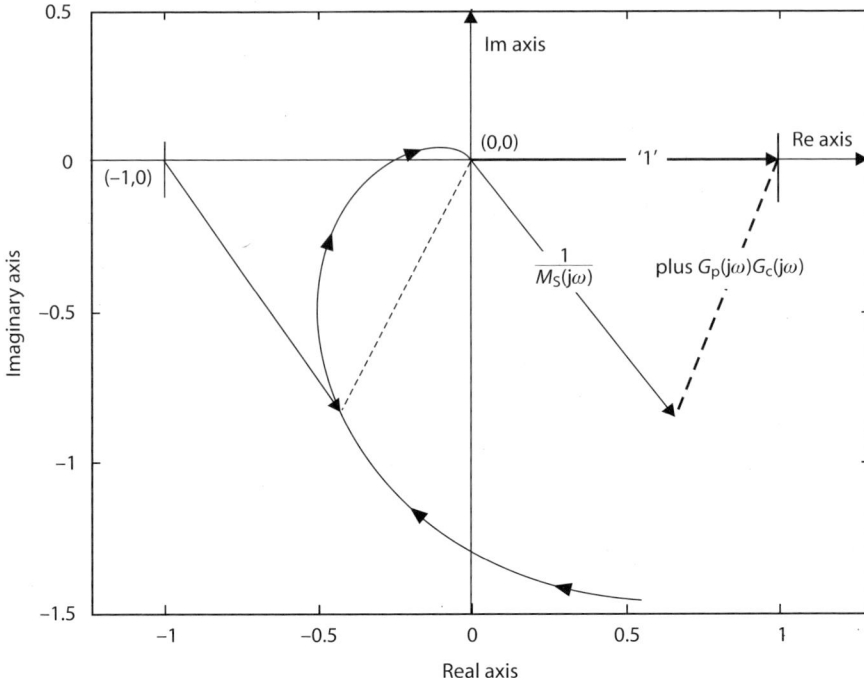

Figure 2.18 Nyquist geometry for sensitivity measures.

then

$$\frac{1}{|M_S(j\omega_A)|} > \frac{1}{|M_S(j\omega_B)|}$$

so that maximising the quantity

$$|M_S(j\omega)| = \frac{1}{|1+G_p(j\omega)G_c(j\omega)|}$$

is equivalent to minimising the quantity

$$\frac{1}{|M_S(j\omega)|} = |1+G(j\omega)G_c(j\omega)|$$

Consequently the maximum sensitivity, M_S, will occur with a circle of radius $1/M_S$, centred on the stability point $(-1, 0)$ which is tangential to the forward path frequency response $G_p(j\omega)G_c(j\omega)$. This is shown in Figure 2.19.

This geometrical condition can be reversed to set up a specification for controller design, since a desired design maximum sensitivity value can be selected and the circle of radius $1/M_S$, centred on the stability point $(-1, 0)$ which is tangential to the forward path frequency response $G_p(j\omega)G_c(j\omega)$ then becomes a control design specification to be met by the selection of an appropriate controller $G_c(s)$. It is not difficult to see from Figure 2.19 that the specification of a maximum sensitivity circle is closely related to stability robustness since it is another way of bounding the forward path transfer function away from the stability point $(-1, 0)$.

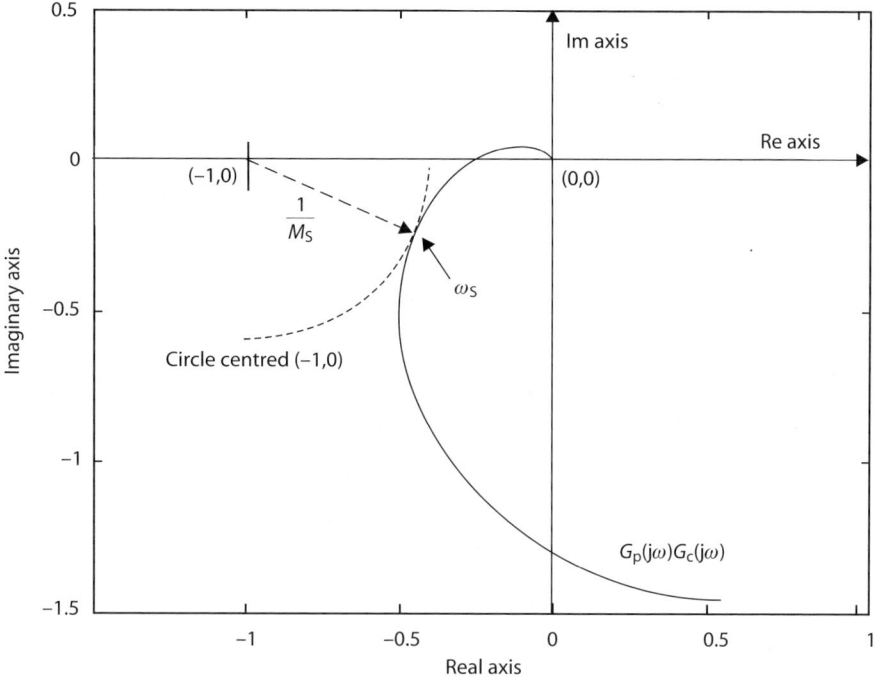

Figure 2.19 Nyquist geometry for maximum sensitivity.

Design values for maximum sensitivity are usually given in absolute terms or in dB:

Maximum sensitivity ranges: $M_S < 2$

Maximum sensitivity ranges: $M_S \, dB < 6 \, dB$

Skogestad and Postlethwaite (1996) give the following relationships between gain and phase margin and the value of maximum sensitivity:

$$GM \geq \frac{M_S}{M_S - 1} \quad \text{and} \quad PM \geq 2\arcsin\left(\frac{1}{2M_S}\right) > \frac{1}{M_S}$$

where PM has units of radians.

Thus $M_S = 2$ yields $GM \geq 2$ with $PM \geq 29°$ and hence the maximum sensitivity specification can be considered a composite gain and phase margin condition. However, gain and phase margins are useful specifications of stability robustness for single input, single output control loops, since they are simple to understand and interpret and are widely used. The main disadvantage is that the two measures are decoupled from one another and can be over conservative as design specifications. An alternative approach to stability robustness measures is reported next.

2.3.4 Parametric Stability Margins for Simple Processes

In industrial process control applications, three-term or PID control continues to dominate at the process control-loop level, whilst model-based predictive control is often found at supervisory level. Industrial process engineers also now better appreciate robustness as a concept and software tools incorporating some form of robustness tests are often an integrated part of commercial process operating packages (Blevins *et al.*, 2003). For example, the process control software might include a robustness map based on single loop gain and phase margins for tuning PID controllers (Gudaz and Zhang, 2001). The main drawback of these classical robustness measures is that they do not necessarily describe robust stability when acting together; they are essentially non-simultaneous measures. Gain margin describes how far the system is from instability based on a system gain change without a phase change and the phase margin is a stability margin measure based on a phase change without a gain change.

An alternative measure or margin for robust stability is given by the parametric stability margin (*psm*). This approach assumes a parametric based model for the system and the *psm* margin supplies an answer to the question of what is the largest *simultaneous* change that can occur to all the system parameters whilst the closed-loop system remains closed-loop stable. The parameters of the closed-loop system can be considered in several different ways; these can be the parameters of the open loop system model $G_p(s)$, the parameters of the controller $G_c(s)$ or the parameters of the composite closed-loop system model $G_{CL}(s)$. Thus parametric stability margin assessment can lead to four types of stability robustness problem:

1. Closed-loop stability robustness with only changes in the process parameters.

2. Closed-loop stability robustness with only changes in the controller parameters.

3. Closed-loop stability robustness with changes in both process and controller parameters.

4. Closed-loop stability robustness with changes in the composite parameters (those which subsume both process and controller parameters) of the closed-loop system transfer function.

In this section, the parametric stability margin is investigated for parametric changes for problems of type 1. However, the methods proposed could easily be extended to the other three classes of problem. Thus the case to be studied is that of the parametric stability margin for the closed-loop system *after* the controller has been designed for the nominal model and assuming that the controller remains immutable. This specifies the worst-case margin available in the accuracy of the process model parameters before closed-loop instability results.

Defining the Parametric Stability Margin
The parametric stability margin is to be defined for a system in the presence of parametric uncertainty around a nominal transfer function model. Define a *generic* system transfer function model which is dependent on a vector of n_p uncertain parameters q_i as

$$G_m(s,q) = \tilde{G}_m(s, q_1, \ldots, q_{n_p}), \text{ with } q \in \Re^{n_p}$$

and $0 < q_i^{min} \leq q_i \leq q_i^{max} < \infty$ for $i = 1, 2, \ldots, n_p$

It is important to note that the transfer function $G_m(s,q)$ is a *generic* system transfer function model which could, for example, be the closed-loop system transfer function. Define the nominal parameter vector as $q^0 \in \Re^{n_p}$, then the nominal system transfer function is given by $G_m(s,q^0) = \tilde{G}_m(s, q_1^0, \ldots, q_{n_p}^0)$. The uncertainty in each q_i parameter can be given the normalised description $\beta_i = (q_i - q_i^0)/q_i^0$, so that $q_i = q_i^0(1 + \beta_i)$ and nominal conditions occur at $\beta_i = 0$. The normalised parameter changes β_i, $i = 1, \ldots, n_p$ correspond to common engineering usage where the potential uncertainty in a process parameter is known in fractional terms, and the conversion to percentage form is $\beta_i^{per\ cent} = \pm \beta_i \times 100\%$. For example, a process parameter q_i might be known to the accuracy of $\beta_i = \pm 0.1$ or in percentage terms $\beta_i^{per\ cent} = \pm 10\%$. The range condition for parameters q_i readily yields the range relationship for β_i as $-1 < \beta_i^{min} \leq \beta_i \leq \beta_i^{max} < \infty$ for $i = 1, 2, \ldots, n_p$, where

2.3 PID Control Performance

$$\beta_i^{\min} = \frac{q_i^{\min} - q_i^0}{q_i^0} \quad \text{and} \quad \beta_i^{\max} = \frac{q_i^{\max} - q_i^0}{q_i^0}$$

Thus the transfer function models satisfy the relations

$$G_m(s,q) = G_{m\beta}(s,\beta) = \tilde{G}_{m\beta}(s,\beta_1,\ldots,\beta_{n_p})$$

with $\beta \in \Re^{n_p}$ and $-1 < \beta_i^{\min} \leq \beta_i \leq \beta_i^{\max} < \infty$ for $i = 1, 2, \ldots, n_p$, and nominal conditions are given by $G_m(s,q^0) = G_{m\beta}(s,0)$, $0 \in \Re^{n_p}$ so that the nominal model is positioned at the origin of the β-space.

Define the set of parameter changes $\beta_S \subset \Re^{n_p}$ such that if $\beta_{Si} \in \beta_S \subset \Re^{n_p}$ then $G_{m\beta}(s,\beta_{Si})$ lies on the stability–instability interface. Using the stability set β_S, the parametric stability margin can be defined as

$$psm = \min_{\beta \in \beta_S} \|\beta\|_p$$

where p characterises the β-norm.

If $p = 2$, then

$$\|\beta\|_2 = \sqrt{\sum_{i=1}^{n_p} \beta_i^2}$$

and the parametric stability margin is the largest radius β-hypersphere where simultaneous changes of *all* the parameters at a point on the surface of the hypersphere puts the system transfer function $G_{m\beta}(s,\beta)$ at the verge of instability. The centre of the hypersphere will be at the origin $\beta = 0 \in \Re^{n_p}$ of the β-space. Similarly, if $p = \infty$, then

$$\|\beta\|_\infty = \max_i |\beta_i|$$

and the parametric stability margin is measured in terms of the largest β-hypercube, rather than the largest β-hypersphere. It is this case, $p = \infty$, where the parametric stability margin corresponds to the standard engineering usage of a maximal fractional or percentage change in all the parameters whilst the system transfer function $G_{m\beta}(s,\beta)$ just remains stable. This situation is given a diagrammatic interpretation in Figure 2.20 for $p = \infty$ and $n_p = 2$.

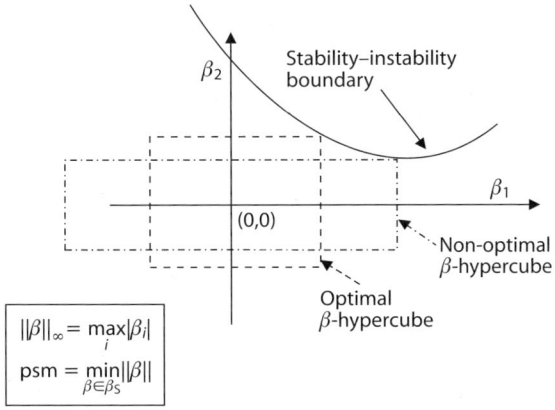

Figure 2.20 Parametric stability margin for $p = \infty$ and $n_p = 2$.

In the literature, the parametric stability margin is often computed using some very unintuitive polynomial algebraic methods and the presentation is usually for general system descriptions which do not easily relate to the simple process models commonly found in process control applications. Another problem in these polynomial approaches is the treatment of time delays, since this is often achieved by replacing the time delay with a Padé approximation to create polynomial expressions suitable for the polynomial analysis methods used to compute the parametric stability margin. By way of contrast, the method described here exploits the presence of a time delay term in the model and does not require a Padé approximation for the delay term.

Computing the *psm* for Simple Process Models

The results are given for the parametric stability margin of a process described by one of the two common process system models with the system controlled by a three-term controller. The process setup is shown in unity feedback structure of Figure 2.21, where the controller has the usual industrial time constant PID form:

$$G_c(s) = k_p \left(1 + \frac{1}{\tau_i s} + \tau_d s\right)$$

The process model is defined by one of the common process control transfer function models:

(a) First-Order Plus Dead Time (FOPDT) Model

$$G_p(s) = \frac{K e^{-T_D s}}{\tau s + 1}$$

where the static gain is K, the time constant is τ and the time delay is T_D.

(b) Second-Order Plus Dead Time (SOPDT) Model

$$G_p(s) = \frac{K e^{-T_D s}}{(s^2/\omega_n^2) + (2\zeta s/\omega_n) + 1}$$

where the static gain is K, the damping is ζ, the natural frequency is ω_n and the time delay is T_D.

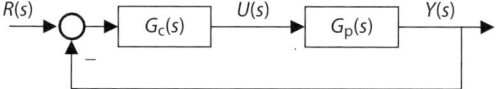

Figure 2.21 Process control loop.

The generic transfer function to be used in the case of the simple process models is the closed-loop transfer function:

$$G_m(s,q) = G_{CL}(s) = \left[\frac{G_p(s)G_c(s)}{1 + G_p(s)G_c(s)}\right]$$

In the robustness to be investigated the controller transfer function $G_c(s)$ is considered to be fixed at the design values obtained using the nominal process transfer function model $G_p(s)$. The parameter vector $q \in \Re^{n_p}$ takes two cases appropriate to the two models adopted for the process.

(a) First-Order Plus Dead Time (FOPDT) Model
The number of parameters is $n_p = 3$ and the q-vector is a 3-tuple, where $q_1 = K$ is the static gain, $q_2 = \tau$ is the time constant and $q_3 = T_D$ is the time delay.

(b) Second-Order Plus Dead Time (SOPDT) Model

The number of parameters is $n_p = 4$ and the q-vector is a 4-tuple, where $q_1 = K$ is the static gain, $q_2 = \zeta$ is the damping, $q_3 = \omega_n$ is the natural frequency and $q_4 = T_D$ is the time delay.

The creation of the β_i parameters follows from these definitions so that

$$\beta_i = \frac{q_i - q_i^0}{q_i^0}, \quad i = 1, 2, \ldots, n_p$$

and the closed-loop transfer function models satisfy

$$G_{CL}(s) = G_m(s, q) = G_{m\beta}(s, \beta)$$

with $\beta \in \Re^{n_p}$ and $-1 < \beta_i^{\min} \leq \beta_i \leq \beta_i^{\max} < \infty$ for $i = 1, 2, \ldots, n_p$, where

$$\beta_i^{\min} = \frac{q_i^{\min} - q_i^0}{q_i^0} \quad \text{and} \quad \beta_i^{\max} = \frac{q_i^{\max} - q_i^0}{q_i^0}$$

For either of the two model cases the brute force calculation of the parametric stability margin could be achieved by forming a grid in β-space, and testing whether each point $\beta \in \beta_{\text{grid}} \subset \Re^{n_p}$ is in the set β_S, and if so then computing $\|\beta\|_\infty$. The calculated norm value would only be stored if its value was smaller than any other previous computed norm value. The difficulty with this proposal is the step of testing for $\beta \in \beta_S \subset \Re^{n_p}$. To surmount this difficulty it should be noted that in the above two-model case, the parameter $q_{n_p} \in \Re^{n_p}$ is the time delay T_D of the process model. Thus, for given values of the parameters $q_1, \ldots, q_{n_p-1}$ and equivalently $\beta_1, \ldots, \beta_{n_p-1}$, the existing theory due to Walton and Marshall (1987) can be used to compute the size of the smallest time delay which destabilises the closed-loop system. From this it is a simple step to use the nominal model time delay to compute the value of the remaining element β_{n_p}. Clearly, the computed full β-vector is *constructed* to satisfy $\beta \in \beta_S \subset \Re^{n_p}$ and in this way it is possible to search through elements lying on the stability–instability interface; the computational algorithm is given below.

Algorithm 2.1: Calculation of *psm* for closed-loop systems with time delay processes

Step 1   Initialisation

Set up step sizes for a grid in $(n_p - 1)$ space for the parameter variations $\beta_1, \ldots, \beta_{n_p-1}$

Set *psm* = largevalue

Step 2   Loop step

Set value of $\beta_1, \ldots, \beta_{n_p-1}$

Find smallest time delay $q_{n_p}^{\min}$ which destabilises the closed-loop system using

$$G_{fp}(s) = \tilde{G}_{m\beta}(s, \beta_1, \ldots, \beta_{n_p-1}) e^{-q_{n_p} s} G_c(s)$$

Compute $\beta_{n_p} = \dfrac{q_{n_p}^{\min} - q_{n_p}^0}{q_{n_p}^0}$

Check $\beta_{n_p} \in [\beta_{n_p}^{\min}, \beta_{n_p}^{\max}]$ (If NO goto Step 2)

Compute $psm(\beta) = \|\beta\|_\infty$

If $psm(\beta) < psm$ then $psm = psm(\beta)$

Repeat Loop step

Algorithm end

The theory for the computation of smallest time delay that destabilises the closed-loop system has already been documented (Walton and Marshall, 1987; Marshall *et al.*, 1992) but it is useful to give the computational steps as the algorithm below.

Firstly, define some notation for the forward path:

$$G_{\text{fp}}(s) = \tilde{G}_{m\beta}(s, \beta_1, \ldots, \beta_{n_p-1}) e^{-q_{n_p} s} G_c(s)$$

which may be written as a ratio of two polynomial forms:

$$G_{\text{fp}}(s) = \frac{B(s, \tilde{\beta})}{A(s, \tilde{\beta})} e^{-q_{n_p} s}$$

where $\tilde{\beta} = [\beta_1, \ldots, \beta_{n_p-1}]^T \in \Re^{n_p-1}$.

Algorithm 2.2: Computation of destabilising time delay

Step 1 Initialisation
  Fix the values $\beta_1, \ldots, \beta_{n_p-1}$, thereby defining $\tilde{\beta} \in \Re^{n_p-1}$

Step 2 Solve frequency polynomial equation $p(\omega^2) = 0$
  Form $p(\omega^2) = A(j\omega, \tilde{\beta}) A(-j\omega, \tilde{\beta}) - B(j\omega, \tilde{\beta}) B(-j\omega, \tilde{\beta})$
  Solve $p(\omega^2) = 0$ for real positive values of ω
  Identify smallest positive frequency value $\omega_{\min}$

Step 3 Compute smallest destabilising time delay

$$\text{Compute } q_{n_p}^{\min} = \left(\frac{1}{\omega_{\min}}\right) \cos^{-1}\left(\text{Re}\left\{-\frac{A(j\omega, \tilde{\beta})}{B(j\omega, \tilde{\beta})}\right\}\right)$$

Algorithm end

Thus, the two algorithms together can be used as the basis for a program to compute the parametric stability margins for the two types of simple time delay process control models. The results for two examples follow.

Case Study 1: FOPDT and Ziegler–Nichols PI Tuning

The process model is defined by

$$G_p(s) = \frac{K e^{-T_D s}}{\tau s + 1}$$

The elements of the vector $q \in \Re^3$ are identified from the process model as $q_1 = K$, the static gain; $q_2 = \tau$, the time constant; and $q_3 = T_D$, the time delay. The nominal parameters are K^0, τ^0, T_D^0, from which the normalised uncertainty parameters are defined as $\beta_K = (K - K^0)/K^0$, $\beta_\tau = (\tau - \tau^0)/\tau^0$ and $\beta_{T_D} = (T_D - T_D^0)/T_D^0$ with $-1 < \beta_K < \infty, -1 < \beta_\tau < \infty$ and $-1 < \beta_{T_D} < \infty$. The nominal process model occurs at $\beta = 0 \in \Re^3$.

The PI controller is given by

$$G_c(s) = k_P\left(1 + \frac{1}{\tau_i s}\right)$$

The identification for polynomials $A(s, \tilde{\beta})$ and $B(s, \tilde{\beta})$ follows from the forward path transfer function

$$G_{\text{fp}}(s) = \frac{K e^{-T_D s}}{\tau s + 1} k_P\left(1 + \frac{1}{\tau_i s}\right) = \frac{B(s, \tilde{\beta})}{A(s, \tilde{\beta})} e^{-q_{n_p} s}$$

Plant Test

Nominal parameters: $K^0 = 1, \tau^0 = 1, T_D^0 = 0.25$

Ultimate data: $K_U = 6.94, P_U = 0.92$

Quarter Amplitude Decay PI tunings: $k_P = 0.45 \times K_U = 3.123, \tau_i = 0.833 \times P_U = 0.766$

PSM results: $\beta_K = 0.238, \beta_\tau = -0.239, \beta_{T_D} = 0.2385$

Parametric stability margin: $psm = 0.239$

Interpretation: the process plant can be incorrect in *all* its model parameters by up to almost ±24% before the closed-loop system with the Quarter Amplitude Decay PI tunings based on the nominal model becomes unstable.

Case Study 2: SOPDT and Ziegler–Nichols PI Tuning

The process model is defined by

$$G_p(s) = \frac{Ke^{-T_D s}}{(s^2/\omega_n^2) + (2\zeta s/\omega_n) + 1}$$

The elements of the vector $q \in \Re^4$ are identified from the process model as $q_1 = K$, the static gain; $q_2 = \zeta$, the damping; $q_3 = \omega_n$, the natural frequency; and $q_4 = T_D$, the time delay. The nominal parameters are $K^0, \zeta^0, \omega_n^0, T_D^0$, from which the normalised uncertainty parameters are defined as $\beta_K = (K - K^0)/K^0$, $\beta_\zeta = (\zeta - \zeta^0)/\zeta^0$, $\beta_{\omega_n} = (\omega_n - \omega_n^0)/\omega_n^0$, and $\beta_{T_D} = (T_D - T_D^0)/T_D^0$ with $-1 < \beta_K < \infty$, $-1 < \beta_\zeta < \infty$, $-1 < \beta_{\omega_n} < \infty$ and $-1 < \beta_{T_d} < \infty$. The nominal process model occurs at $\beta = 0 \in \Re^4$.

The PI controller is given by

$$G_c(s) = k_P \left(1 + \frac{1}{\tau_i s}\right)$$

The identification of the polynomials $A(s, \tilde{\beta})$ and $B(s, \tilde{\beta})$ follows from the forward path transfer function

$$G_{fp}(s) = \frac{Ke^{-T_D s}}{(s^2/\omega_n^2) + (2\zeta s/\omega_n) + 1} k_P\left(1 + \frac{1}{\tau_i s}\right) = \frac{B(s, \tilde{\beta})}{A(s, \tilde{\beta})} e^{-q_{n_p} s}$$

Plant Test

Nominal parameters: $K^0 = 1, \zeta^0 = 0.6, \omega_n = 0.75, T_D^0 = 0.5$

Ultimate data: $K_U = 3.5, P_U = 4.39$

Quarter Amplitude Decay PI tunings: $k_P = 0.45 \times K_U = 1.575, \tau_i = 0.833 \times P_U = 0.43$

PSM results: $\beta_K = 0.1472, \beta_\zeta = -0.1472, \beta_{\omega_n} = 0.1472, \beta_{T_D} = 0.1472$

Parametric stability margin: $psm = 0.1472$

Interpretation: the process plant can be incorrect in *all* its model parameters by up to almost ±15% before the closed-loop system with the given Quarter Amplitude Decay PI tunings based on the nominal model becomes unstable.

Extension to Delay-Free Process Models

An interesting issue is how to extend the above parametric stability margin procedure to systems which are delay free. The existing method depends on the calculation of the system time delay for which the closed-loop system lies on the stability–instability boundary. However, recall from the discussion on stability robustness margins the link between the delay margin *DM* and the phase margin *PM*; the formula was given as

$$DM = \frac{PM°}{180°} \frac{\pi}{\omega_1}$$

For a given phase margin, this formula was interpreted as the amount of system time delay that could be introduced in the forward path before the closed-loop system became unstable. But interpreting this in a reverse direction, if a calculated delay margin is zero then the phase margin is also zero and the particular forward path transfer function is such that the closed-loop control system is on the verge of instability. It is this reinterpretation that motivates the procedure for calculating the parametric stability margin for a delay-free system.

The method is to introduce a fictitious time delay term, denoted T_{DM}, into the system description and use the existing method to calculate the value of this time delay for which the closed-loop system lies on the stability–instability boundary. Thus when this computed time delay T_{DM} is zero, the set of system parameters $q \in \Re^{n_p}$ corresponding to this condition represents a perturbed delay-free system transfer function for which the equivalent closed-loop system is on the verge of instability. The cases arising are as follows:

1. Nominal parameter vector q^0 for which the system is closed-loop stable. In this case $T_{DM} > 0$ and there is room to introduce a time delay before the closed-loop system becomes unstable.

2. There will be sets of parameters $q \neq q^0$ for which the system is still closed-loop stable, and in this case the condition $T_{DM} > 0$ will be found.

3. There will be sets of parameters $q \neq q^0$ for which the time delay computation yields $T_{DM} = 0$. In this case the closed-loop system is on the verge of instability. But since $T_{DM} = 0$, the open loop system is delay-free and this is the condition on which a parametric stability margin computation can be based.

4. There will be sets of parameters $q \neq q^0$ for which the time delay computation yields $T_{DM} < 0$. In this case the closed-loop system is unstable.

To set up the appropriate algorithms, introduce the following notation. The system model is defined as the delay free transfer function:

$$G_m(s,q) = \tilde{G}_m(s, q_1, \ldots, q_{n_p})$$

from which the β-vector model is derived as

$$G_m(s,q) = G_{m\beta}(s, \beta)$$

with $\beta \in \Re^{n_p}$ and $-1 < \beta_i^{\min} \leq \beta_i \leq \beta_i^{\max} < \infty$, where

$$\beta_i^{\min} = \frac{q_i^{\min} - q_i^0}{q_i^0} \quad \text{and} \quad \beta_i^{\max} = \frac{q_i^{\max} - q_i^0}{q_i^0}, \quad \text{for } i = 1, 2, \ldots, n_p$$

Note that in this case, since the system is delay free, there is no special identification of the parameter q_{n_p} and the parameter variation β_{n_p} with a system time delay. The concepts above involve a test to see whether the computed destabilising time delay is zero; such a test can only realistically be done to within a small ε-tolerance of zero, and this will form part of the algorithm. A computational procedure for the *psm* calculation follows.

2.3 PID Control Performance

Algorithm 2.3: Calculation of *psm* for closed-loop systems with delay-free process models

Step 1 Initialisation

Set up step sizes for a grid in n_p space for the parameter variations $\beta_1,\ldots,\beta_{n_p}$
Determine an ε-tolerance for the computed time delay T_{DM}
Set *psm* = largevalue

Step 2 Loop step

Set value of $\beta_1,\ldots,\beta_{n_p}$
Find smallest time delay T_{DM} which destabilises the closed-loop system using the forward path, $G_{fp}(s) = G_{m\beta}(s,\beta)e^{-T_{DM}s}G_{PID}(s)$
If $|T_{DM}| < \varepsilon$-tolerance then compute $psm(\beta) = \|\beta\|_\infty$ and if $psm(\beta) < psm$ then $psm = psm(\beta)$
Repeat Loop step

Algorithm end

In the algorithm step to compute the destabilising time delay there is a minor notational modification to Algorithm 2.2 which should be noted. This occurs in the definition of the polynomials $A(s,\tilde{\beta})$ and $B(s,\tilde{\beta})$ that arises from the new form for the forward path. In the delay-free case, the tilde over the β vector can be omitted since it is not needed and the following definitions used:

$$G_{fp}(s) = G_{m\beta}(s,\beta)e^{-T_{DM}s}G_c(s) = \frac{B(s,\beta)}{A(s,\beta)}e^{-T_{DM}s}$$

where $\beta = [\beta_1,\ldots,\beta_{n_p}]^T \in \mathfrak{R}^{n_p}$.

Case Study 3: Delay-Free Second-Order System and Ziegler–Nichols PI Tuning

The process model is defined by

$$G_p(s) = \frac{K}{(s^2/\omega_n^2) + (2\zeta s/\omega_n) + 1}$$

The elements of the vector $q \in \mathfrak{R}^3$ are identified from the process model as $q_1 = K$, the static gain; $q_2 = \zeta$, the damping; and $q_3 = \omega_n$, the natural frequency. The nominal parameters are K^0, ζ^0, ω_n^0, from which the normalised uncertainty parameters are defined as $\beta_K = (K - K^0)/K^0$, $\beta_\zeta = (\zeta - \zeta^0)/\zeta^0$ and $\beta_{\omega_n} = (\omega_n - \omega_n^0)/\omega_n^0$, with $-1 < \beta_K < \infty, -1 < \beta_\zeta < \infty$ and $-1 < \beta_{\omega_n} < \infty$. The nominal process model occurs at $\beta = 0 \in \mathfrak{R}^3$.

The PI controller is given by

$$G_c(s) = k_P\left(1 + \frac{1}{\tau_i s}\right)$$

The identification for polynomials $A(s,\beta)$ and $C(s,\beta)$ follows from the forward path transfer function

$$G_{fp}(s) = \frac{K}{(s^2/\omega_n^2) + (2\zeta s/\omega_n) + 1}[e^{-T_{DM}s}]k_P\left(1 + \frac{1}{\tau_i s}\right) = \frac{B(s,\beta)}{A(s,\beta)}e^{-T_{DM}s}$$

Plant Test

Nominal parameters: $K^0 = 1.5, \zeta^0 = 0.6, \omega_n = 2.0$

Use a neglected parasitic high-frequency transfer

$$G_{para}(s) = \left[\frac{1}{0.1s + 1}\right]$$

in forward path to enable the sustained oscillation experiment.

Ultimate data: $K_U = 5.1, P_U = 1.2$

Quarter Amplitude Decay PI tunings: $k_P = 0.45 \times K_U = 2.295, \tau_i = 0.833 \times P_U = 1.00$

PSM results: $\beta_K = 0.4115, \beta_\zeta = -0.4115, \beta_{\omega_n} = -0.4115$

Parametric Stability Margin: $psm = 0.4115$

Interpretation: the process plant can be incorrect in *all* its model parameters by up to almost ±41% before the closed-loop system with the given Quarter Amplitude Decay PI tunings based on the nominal process model becomes unstable.

In summary, this method of computing parametric stability margins for the simple types of process models used in process control applications is straightforward to use. The parametric stability margins calculated give very useful robustness information about any control design and the permissible margin for error in the accuracy of the derived process model. These computational methods have been used by Wade and Johnson (2003) in an assessment of the robustness inculcated by different PID control law cost functions.

2.4 State Space Systems and PID Control

State space models are common in control engineering studies. The systematic features of state space equations are invaluable when creating a physically based model to investigate process behaviour. Further, many control software tools use the highly convenient state space form as a basis for establishing a control system simulation. A major advantage of the state space description is that it is generically multivariable, which subsumes the single input, single output case within the same formulation. In this section, some analysis for state space PID control is presented and discussed.

2.4.1 Linear Reference Error Feedback Control

The system is represented by a state space model which is assumed to have the same number of input and output variables m and a state vector with n state variables, namely

$$\dot{x}(t) = Ax(t) + Bu(t)$$
$$y(t) = Cx(t)$$

where state $x \in \Re^n$, output and control are y and $u \in \Re^m$, and the input (or driving) matrix, system matrix and output (or measurement) matrix are respectively $B \in \Re^{n \times m}, A \in \Re^{n \times n}, C \in \Re^{m \times n}$, with $t_0 \leq t < \infty$. It is assumed that this model is for an operating condition around steady (constant) conditions and thus there is no loss of generality in assuming that $x(t_0) = 0 \in \Re^n$.

Linear reference error feedback control is investigated and this uses a control signal given by

$$u(t) = K_P e(t)$$

with $e(t) = y(t) - r(t)$ and $K_P \in \Re^{m \times m}$.

The model and control setup are shown in Figure 2.22.

The closed-loop analysis follows by substituting for the control law:

$$\dot{x}(t) = Ax(t) + Bu(t)$$
$$= Ax(t) + BK_P e(t)$$
$$= Ax(t) + BK_P(r(t) - y(t))$$

Using $y(t) = Cx(t)$ yields

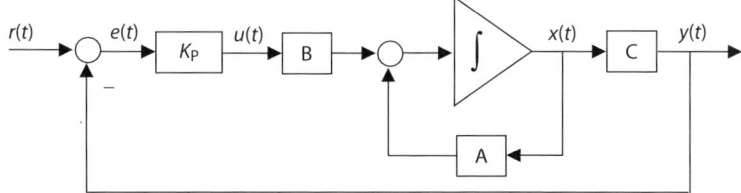

Figure 2.22 Linear reference error feedback control for a state space model.

$$\dot{x}(t) = [A - BK_P C]x(t) + BK_P r(t)$$
$$= [A_{CL}]x(t) + BK_P r(t)$$

where $A_{CL} = A - BK_P C \in \Re^{n \times n}$.

The system modes are the n eigenvalues of the system matrix $A \in \Re^{n \times n}$. Matrix $A_{CL} = A - BK_P C \in \Re^{n \times n}$ is the closed-loop system matrix and its n eigenvalues are the closed-loop system modes. The closed-loop system is stable if all the closed-loop system modes have values in the open left-half complex plane. The question as to the flexibility to assign the location of closed-loop system modes is answered by the following result.

Result 1

For a state space system model with $A \in \Re^{n \times n}$, $B \in \Re^{n \times m}$, $C \in \Re^{m \times n}$; if pair (A, B) is completely controllable and pair (C, A) is completely observable then m system modes of $A_{CL} = A - BK_P C \in \Re^{n \times n}$ are arbitrarily assignable.

An indication of the possible steady state closed-loop time domain performance is given by the following result.

Result 2

If feedback gain $K_P \in \Re^{m \times m}$ can be chosen so that closed-loop system matrix $A_{CL} = A - BK_P C \in \Re^{n \times n}$ is stable, then the steady state output y_∞ to a step reference signal of size r is given by

$$y_\infty = \lim_{t \to \infty} y(t) = [C(-A_{CL})^{-1} BK_P] r$$

The proof of Result 2 uses the final value theorem from Laplace transform theory. Observe that A_{CL} is itself a function of K_P, so it seems that it would be very fortuitous if the feedback gain K_P were chosen so that $[C(-A_{CL})^{-1} BK_P] = [I]$. Consequently it is highly likely that $y_\infty \neq r$ and there will be a non-zero offset error in steady state.

From the above analysis, the closed-loop control of the state space modelled system achieved by using linear reference error feedback control has the following practical advantage:

- The feedback mechanism uses the available measurable output vector $y \in \Re^m$.

The scheme also has the following theoretical disadvantages:

- Poor flexibility to assign the location of the n closed-loop system modes; at most only m can be assigned arbitrarily.
- There is a need for a method to select suitable closed-loop system modes, locations and place modes in the desired locations.
- Probable steady state offsets from desired output reference levels.

It should be noted that these are theoretical results and the applicability to the actual system will depend on the fidelity between the state space model and the actual system. However, to remove these disadvantages from the theoretical analysis a change in the structure of the feedback control law is proposed: a two degree of freedom control law is considered.

2.4.2 Two Degree of Freedom Feedback Control System

In this case the feedback law is amended to have a two degree of freedom structure comprising a reference control gain and a full state feedback control law. The new control law is defined as

$$u(t) = K_R r(t) - K_P x(t)$$

where the reference control gain is $K_R \in \Re^{m \times m}$ and the full state feedback gain is $K_P \in \Re^{m \times n}$. The configuration is shown in Figure 2.23.

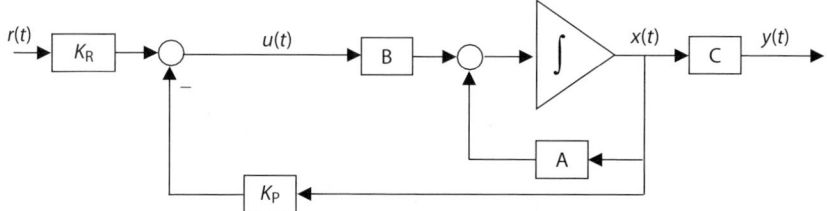

Figure 2.23 State space closed-loop system with two degree of freedom feedback.

The closed-loop analysis uses straightforward substitution for the control law, viz

$$\begin{aligned}\dot{x}(t) &= Ax(t) + Bu(t) \\ &= Ax(t) + B(K_R r(t) - K_P x(t)) \\ &= [A - BK_P]x(t) + BK_R r(t) \\ &= [A_{CL}]x(t) + BK_R r(t)\end{aligned}$$

where $A_{CL} = A - BK_P \in \Re^{n \times n}$.

The matrix $A_{CL} = A - BK_P \in \Re^{n \times n}$ is the closed-loop system matrix and its n eigenvalues are the closed-loop system modes. The flexibility to assign the location of closed-loop system modes is given by the following result.

Result 3

For a state space system model with $A \in \Re^{n \times n}$, $B \in \Re^{n \times m}$; if the pair (A, B) is completely controllable then $K_P \in \Re^{m \times n}$ can be chosen to arbitrarily assign the positions of all the modes the closed-loop system matrix $A_{CL} = A - BK_P \in \Re^{n \times n}$.

Some further analysis can be given to show how the reference controller gain $K_R \in \Re^{m \times m}$ can be chosen to improve the steady state performance of the control system. Assume that $K_P \in \Re^{m \times n}$ has been chosen so that $A_{CL} \in \Re^{n \times n}$ has only stable modes (this is permitted by Result 3); then in this case A_{CL} is invertible. Assume a step reference input of size r has been applied. In steady state conditions ($t \to \infty$) denote the constant state vector as $x_\infty \in \Re^m$ and the associated constant output as $y_\infty \in \Re^m$. The closed-loop state equation is

$$\dot{x}(t) = [A_{CL}]x(t) + BK_R r(t)$$

In steady state the condition $t \to \infty$ yields,

$$\begin{aligned}\dot{x}(\infty) = \dot{x}_\infty &= 0 \\ &= [A_{CL}]x_\infty + BK_R r\end{aligned}$$

Hence the invertibility of A_{CL} gives

$$x_\infty = -A_{CL}^{-1} B K_R r$$

The condition for the steady state output $y_\infty \in \Re^m$ to attain the step reference level r is given by $y_\infty = Cx_\infty = r$.

Substituting for $x_\infty \in \Re^n$ gives

$$[C(-A_{CL}^{-1})B]K_R r = [I]r$$

and, assuming the inverse exists, this may be solved for $K_R \in \Re^{m \times m}$ as

$$K_R = [C(-A_{CL}^{-1})B]^{-1}$$

Result 4

For a state space system model with $A \in \Re^{n \times n}, B \in \Re^{n \times m}, C \in \Re^{m \times n}$ which uses the two degree of freedom feedback law, $u(t) = K_R r(t) - K_P x(t)$. If the state feedback gain matrix is used the ensure that $A_{CL} = A - BK_P \in \Re^{n \times n}$ is closed-loop stable, then the step reference level r is attained if $K_R \in \Re^{m \times m}$ can be set to $K_R = [C(-A_{CL}^{-1})B]^{-1}$, assuming the outer inverse in the formula exists.

From the above analysis, the closed-loop control of the state space modelled system using a two degree of freedom control law has the following practical disadvantages:

- The feedback mechanism requires full state accessibility. This means that all the state variables of the system must be available in measured form. Since in many process control applications there are often numerous inaccessible state variables, this is a very unrealistic assumption.

- If model mismatch is present this will introduce errors into the calculation for the reference control gain and upset the ability of this control law to remove steady state offsets from desired constant reference levels.

However, the scheme now has some useful theoretical advantages:

- The n closed-loop modes can be arbitrarily assigned provided the system is completely controllable.

- The reference gain in the feedback law can be calculated to ensure that there are no steady state offsets from desired constant reference levels.

These state space analysis results have the useful role of describing precisely how the two degree of freedom control law will work. There may even be low state dimension cases where this control law is applicable in practice, but the applicability will depend on the fidelity between the state space model and the actual system. One of the difficulties is that model mismatch will introduce errors into the calculation for the reference control gain and this will prevent the control law from removing steady state offsets from the desired constant reference levels. What is needed is some integral action in the feedback law, and this is proposed next.

2.4.3 State Feedback With Integral Error Feedback Action

To overcome some of the steady state performance difficulties with the two degree of freedom control law, an integral state vector is introduced:

$$x_I(t) = \int^t e(\tau) d\tau$$

The integral state $x_I \in \Re^m$ satisfies a state space equation

$$\dot{x}_I(t) = e(t)$$
$$= r - y(t)$$

giving

$$\dot{x}_I(t) = r - Cx(t)$$

The system is now modelled using a composite state vector

$$\begin{bmatrix} x \\ x_I \end{bmatrix} \in \Re^{n+m}$$

for which the new composite state space model is derived as follows:

$$\begin{bmatrix} \dot{x} \\ \dot{x}_I \end{bmatrix} = \begin{bmatrix} Ax + Bu \\ r - Cx \end{bmatrix}$$

and

$$\begin{bmatrix} \dot{x} \\ \dot{x}_I \end{bmatrix} = \begin{bmatrix} A & 0 \\ -C & 0 \end{bmatrix} \begin{bmatrix} x \\ x_I \end{bmatrix} + \begin{bmatrix} B \\ 0 \end{bmatrix} u + \begin{bmatrix} 0 \\ I \end{bmatrix} r$$

The $(n + m)$ system modes are the eigenvalues of the composite system matrix:

$$\begin{bmatrix} A & 0 \\ -C & 0 \end{bmatrix}$$

and these are the n eigenvalues of system matrix A along with m zero eigenvalues. The flexibility to locate these $(n + m)$ system eigenvalues in closed loop using a composite state feedback law will depend on the pair

$$\left(\begin{bmatrix} A & 0 \\ -C & 0 \end{bmatrix}, \begin{bmatrix} B \\ 0 \end{bmatrix} \right)$$

being completely controllable; a result analogous to Result 3.

The state feedback law used with this new composite system model is

$$u(t) = -[K_P \quad K_I] \begin{bmatrix} x(t) \\ x_I(t) \end{bmatrix}$$

The closed-loop system diagram is constructed by using the following set of equations in order:

$$\dot{x}(t) = Ax(t) + Bu(t)$$
$$y(t) = Cx(t)$$
$$e(t) = r - y(t)$$
$$x_I(t) = \int^t e(\tau) d\tau$$

and then closing the loop with

$$u(t) = -K_P x(t) - K_I x_I(t)$$

Figure 2.24 shows the resulting diagram.

The analysis for the closed-loop system follows by substitution of the control law, viz

2.4 State Space Systems and PID Control

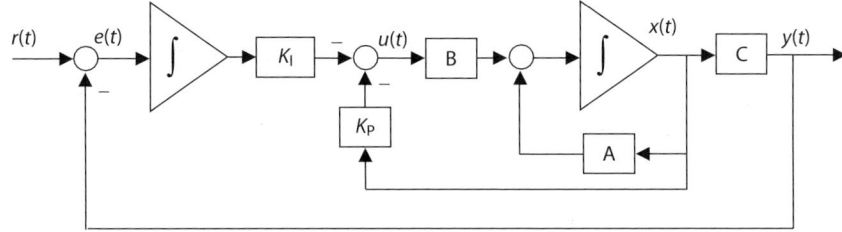

Figure 2.24 State and integral reference error feedback for a state space model system.

$$\begin{bmatrix} \dot{x} \\ \dot{x}_I \end{bmatrix} = \begin{bmatrix} A & 0 \\ -C & 0 \end{bmatrix} \begin{bmatrix} x \\ x_I \end{bmatrix} + \begin{bmatrix} B \\ 0 \end{bmatrix} [-K_P \quad -K_I] \begin{bmatrix} x \\ x_I \end{bmatrix} + \begin{bmatrix} 0 \\ I \end{bmatrix} r$$

giving the closed-loop system as

$$\begin{bmatrix} \dot{x} \\ \dot{x}_I \end{bmatrix} = \begin{bmatrix} A - BK_P & -BK_I \\ -C & 0 \end{bmatrix} \begin{bmatrix} x \\ x_I \end{bmatrix} + \begin{bmatrix} 0 \\ I \end{bmatrix} r$$

$$= [A_{CL}] \begin{bmatrix} x \\ x_I \end{bmatrix} + \begin{bmatrix} 0 \\ I \end{bmatrix} r$$

where

$$[A_{CL}] = \begin{bmatrix} A - BK_P & -BK_I \\ -C & 0 \end{bmatrix}$$

Note that if $K_I = 0$ so that there is no integral term, the complete controllability of pair (A, B) will allow K_P to be selected to attain arbitrary (stable) locations for the eigenvalues of $A - BK_P$. The introduction of integral action with a non-zero K_I gain will cause these eigenvalues to move to new (and possibly unstable) positions. The problem is then to find matrix valued gain matrices K_P and K_I to achieve satisfactory locations for the complete set of $(n + m)$ closed-loop system modes. However, the steady state analysis is examined under the assumption that the closed-loop system matrix $[A_{CL}]$ is stable.

Time Domain Analysis

For a step reference vector r and $t \to \infty$ introduce constant state and output vectors

$$x_\infty = \lim_{t \to \infty} x(t), \quad x_{I\infty} = \lim_{t \to \infty} x_I(t) \text{ and } y_\infty = \lim_{t \to \infty} y(t)$$

Then

$$\begin{bmatrix} \dot{x}_\infty \\ \dot{x}_{I\infty} \end{bmatrix} = \begin{bmatrix} 0 \\ 0 \end{bmatrix} = \begin{bmatrix} A - BK_P & -BK_I \\ -C & 0 \end{bmatrix} \begin{bmatrix} x_\infty \\ x_{I\infty} \end{bmatrix} + \begin{bmatrix} 0 \\ I \end{bmatrix} r$$

giving

(i) $[A - BK_P]x_\infty - BK_I x_{I\infty} = 0$

(ii) $-Cx_\infty + r = 0$ hence $y_\infty = Cx_\infty = r$

Result (ii) shows that there will be no steady offset and the output will attain the desired constant reference signal level.

s-Domain Analysis

Assuming that initial conditions are zero, the Laplace transforms of the closed-loop system equation

$$\begin{bmatrix} \dot{x} \\ \dot{x}_I \end{bmatrix} = \begin{bmatrix} A - BK_P & -BK_I \\ -C & 0 \end{bmatrix} \begin{bmatrix} x \\ x_I \end{bmatrix} + \begin{bmatrix} 0 \\ I \end{bmatrix} r$$

yield

$$sX(s) = (A - BK_P)X(s) - BK_I X_I(s)$$
$$sX_I(s) = -CX(s) + R(s)$$

Then

$$(sI - A)X(s) = -BK_P X(s) - BK_I X_I(s)$$

$$X_I(s) = \frac{-C}{s} X(s) + \frac{R(s)}{s}$$

Eliminating $X_I(s)$ between these equations yields

$$(sI - A)X(s) = -BK_P X(s) - BK_I \left(\frac{-C}{s} X(s) + \frac{R(s)}{s} \right)$$

and hence

$$X(s) = (sI - A)^{-1} \left\{ -BK_P X(s) - BK_I \left(\frac{-C}{s} X(s) + \frac{R(s)}{s} \right) \right\}$$

$$Y(s) = CX(s) = C(sI - A)^{-1} \left\{ -BK_P X(s) + B \frac{K_I}{s} CX(s) - B \frac{K_I}{s} R(s) \right\}$$

and setting $G_p(s) = C(sI - A)^{-1} B$ gives

$$Y(s) = -G_p(s) K_P X(s) + \frac{G_p(s) K_I}{s} Y(s) - \frac{G_p(s) K_I}{s} R(s)$$

This simplifies to

$$Y(s) = [sI - G_p(s) K_I]^{-1} \{ -sG_p(s) K_P X(s) - G_p(s) K_I R(s) \}$$

Assuming that $G_p(0) \in \Re^{m \times m}$ is finite and that the composite matrix $G_p(0) K_I \in \Re^{m \times m}$ is invertible, the final value theorem gives

$$y_\infty = \lim_{s \to 0} sY(s) = [-G_p(0) K_I]^{-1} \{ -G_p(0) K_P \lim_{s \to 0} s^2 X(s) - G_p(0) K_I \lim_{s \to 0} sR(s) \}$$

Then $y_\infty = [-G_p(0) K_I]^{-1} \{ G_p(0) K_I r \} = r$ and the set point vector r is achieved with the given feedback structure.

From the above closed-loop control analysis of the state space modelled system using state and integral reference error feedback the following comments can be made.

The method has the following practical advantage:

- The analysis given for the steady state behaviour of the feedback depends on the transfer function description, $G_p(s)$, whatever it might be. This means that the analysis defines a guaranteed property that the integral reference error feedback in the control law automatically removes steady state offsets from desired constant reference levels. This is the invaluable practical property of integral reference error feedback.

The method has the following practical disadvantage:

- The control is a full state feedback law and this requires the measured full state vector to be available; usually an unrealistic assumption.

However, the scheme now has the useful theoretical advantage:

- If the pair

$$\left(\begin{bmatrix} A & 0 \\ -C & 0 \end{bmatrix}, \begin{bmatrix} B \\ 0 \end{bmatrix} \right)$$

is completely controllable then the $(n + m)$ closed-loop modes will be arbitrarily assignable.

This is balanced by the theoretical disadvantage:

- A method to select suitable closed-loop system modes locations and place modes in the desired positions is required.

To achieve a comparison with the above state space control laws and analysis, a scheme with the structure of the classical PI is given next.

2.4.4 State Space Analysis for Classical PI Control Structure

Consider a system modelled by the usual state space form:

$$\dot{x}(t) = Ax(t) + Bu(t)$$
$$y(t) = Cx(t)$$

Denote the reference signal as $r(t) \in \Re^m$, a reference error as $e(t) \in \Re^m$ and the system output as $y(t) \in \Re^m$. The reference error expression is $e(t) = r - y(t)$.

Introduce the integral state vector $x_I \in \Re^m$ defined by

$$x_I(t) = \int^t e(\tau) d\tau$$

As before, the integral state $x_I \in \Re^m$ satisfies a state space equation:

$$\dot{x}_I(t) = e(t) = r - y(t)$$

giving

$$\dot{x}_I(t) = r - Cx(t)$$

Introduce a composite state vector:

$$\begin{bmatrix} x \\ x_I \end{bmatrix} \in \Re^{n+m}$$

Then the new composite state space model is given by

$$\begin{bmatrix} \dot{x} \\ \dot{x}_I \end{bmatrix} = \begin{bmatrix} Ax + Bu \\ r - Cx \end{bmatrix}$$

and

$$\begin{bmatrix} \dot{x} \\ \dot{x}_I \end{bmatrix} = \begin{bmatrix} A & 0 \\ -C & 0 \end{bmatrix} \begin{bmatrix} x \\ x_I \end{bmatrix} + \begin{bmatrix} B \\ 0 \end{bmatrix} u + \begin{bmatrix} 0 \\ I \end{bmatrix} r$$

The $(n + m)$ system modes are the eigenvalues of the composite system matrix

$$\begin{bmatrix} A & 0 \\ -C & 0 \end{bmatrix}$$

and these are then the n eigenvalues of system matrix A along with m zero eigenvalues. The flexibility to locate these $(n + m)$ system eigenvalues in closed loop using a composite state feedback law will depend on the pair

$$\left(\begin{bmatrix} A & 0 \\ -C & 0 \end{bmatrix}, \begin{bmatrix} B \\ 0 \end{bmatrix} \right)$$

being completely controllable – a result analogous to Result 3 above.

At this point, the classical PI control structure introduces a subtle difference, since the law is not a straightforward composite state feedback law. The classical PI control feedback law used with this new composite system model is

$$u(t) = K_P e(t) + K_I \int^t e(\tau) d\tau$$

A diagram can be constructed for the new closed-loop situation by using the following set of equations:

$$\dot{x}(t) = Ax(t) + Bu(t)$$
$$y(t) = Cx(t)$$
$$e(t) = r - y(t)$$
$$x_I(t) = \int^t e(\tau) d\tau$$

and closing the loop with

$$u(t) = K_P e(t) + K_I \int^t e(\tau) d\tau$$

The diagram is shown as Figure 2.25.

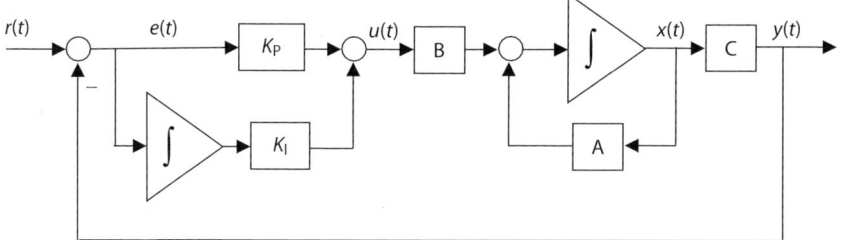

Figure 2.25 Control with a classical PI structure for a state space model.

Closed-loop analysis begins by expanding the control law:

$$u(t) = K_P e(t) + K_I \int^t e(\tau) d\tau$$
$$= K_P(r(t) - Cx(t)) + K_I x_I(t)$$

giving

$$u(t) = [-K_{\mathrm{P}}C \quad K_{\mathrm{I}}] \begin{bmatrix} x(t) \\ x_{\mathrm{I}}(t) \end{bmatrix} + [K_{\mathrm{P}}]r(t)$$

If this is compared with the state and integral reference error feedback of Section 2.4.3, namely

$$u(t) = -[K_{\mathrm{P}} \quad K_{\mathrm{I}}] \begin{bmatrix} x(t) \\ x_{\mathrm{I}}(t) \end{bmatrix}$$

it can be seen that in classical PI structure, the state feedback gain is not a completely free choice and this will restrict the arbitrary assignment of closed-loop modes over the state and integral reference error feedback of Section 2.4.3.

Continuing the analysis, the composite closed-loop system follows by direct substitution, namely

$$\begin{bmatrix} \dot{x} \\ \dot{x}_{\mathrm{I}} \end{bmatrix} = \begin{bmatrix} A & 0 \\ -C & 0 \end{bmatrix} \begin{bmatrix} x \\ x_{\mathrm{I}} \end{bmatrix} + \begin{bmatrix} B \\ 0 \end{bmatrix} \left\{ [-K_{\mathrm{P}}C \quad K_{\mathrm{I}}] \begin{bmatrix} x \\ x_{\mathrm{I}} \end{bmatrix} + [K_{\mathrm{P}}]r \right\} + \begin{bmatrix} 0 \\ I \end{bmatrix} r$$

$$= \begin{bmatrix} A - BK_{\mathrm{P}}C & BK_{\mathrm{I}} \\ -C & 0 \end{bmatrix} \begin{bmatrix} x \\ x_{\mathrm{I}} \end{bmatrix} + \begin{bmatrix} BK_{\mathrm{P}} \\ I \end{bmatrix} r$$

and hence

$$\begin{bmatrix} \dot{x} \\ \dot{x}_{\mathrm{I}} \end{bmatrix} = [A_{\mathrm{CL}}] \begin{bmatrix} x \\ x_{\mathrm{I}} \end{bmatrix} + \begin{bmatrix} BK_{\mathrm{P}} \\ I \end{bmatrix} r$$

where

$$A_{\mathrm{CL}} = \begin{bmatrix} A - BK_{\mathrm{P}}C & BK_{\mathrm{I}} \\ -C & 0 \end{bmatrix}$$

The closed-loop modes are the eigenvalues of $A_{\mathrm{CL}} \in \Re^{(n+m) \times (n+m)}$. If $K_{\mathrm{I}} = 0$ so that there is no integral term, the complete controllability of pair (A, B) and the complete observability of the pair (C, A) will allow K_{P} to be selected to attain arbitrary (stable) locations for only m eigenvalues of $A - BK_{\mathrm{P}}C$. The introduction of integral action with a non-zero K_{I} gain will cause these eigenvalues to move to new (and possibly unstable) positions. This gives rise to the problem of finding gain matrices K_{P} and K_{I} to achieve satisfactory locations for the $(n + m)$ closed-loop system modes. The classical PI control structure in state space form evidently has very restrictive mode placement flexibility. However, assuming that the closed loop is stable it is in the analysis of the steady state behaviour of the classical PI law that benefits are found.

Time Domain Analysis

For a step reference vector r and $t \to \infty$ introduce constant state and output vectors

$$x_{\infty} = \lim_{t \to \infty} x(t), \quad x_{\mathrm{I}\infty} = \lim_{t \to \infty} x_{\mathrm{I}}(t) \text{ and } y_{\infty} = \lim_{t \to \infty} y(t)$$

Then

$$\begin{bmatrix} \dot{x}_{\infty} \\ \dot{x}_{\mathrm{I}\infty} \end{bmatrix} = \begin{bmatrix} 0 \\ 0 \end{bmatrix} = \begin{bmatrix} A - BK_{\mathrm{P}}C & BK_{\mathrm{I}} \\ -C & 0 \end{bmatrix} \begin{bmatrix} x_{\infty} \\ x_{\mathrm{I}\infty} \end{bmatrix} + \begin{bmatrix} BK_{\mathrm{P}} \\ I \end{bmatrix} r$$

giving

(i) $[A - BK_{\mathrm{P}}C]x_{\infty} + BK_{\mathrm{I}}x_{\mathrm{I}\infty} + BK_{\mathrm{P}}r = 0$

(ii) $-Cx_{\infty} + r = 0$ hence $y_{\infty} = Cx_{\infty} = r$

Result (ii) shows that there will be no steady offset and the output will attain the desired constant reference signal level.

s-Domain Analysis

Assuming that initial conditions are zero, the Laplace transforms of the closed-loop system equation

$$\begin{bmatrix} \dot{x} \\ \dot{x}_I \end{bmatrix} = \begin{bmatrix} A - BK_PC & BK_I \\ -C & 0 \end{bmatrix} \begin{bmatrix} x \\ x_I \end{bmatrix} + \begin{bmatrix} BK_P \\ I \end{bmatrix} r$$

yield

$$sX(s) = (A - BK_PC)X(s) + BK_I X_I(s) + BK_P R(s)$$
$$sX_I(s) = -CX(s) + R(s)$$

and hence

$$(sI - A)X(s) = -BK_P CX(s) + BK_I X_I(s) + BK_P R(s)$$
$$X_I(s) = \frac{-C}{s} X(s) + \frac{R(s)}{s}$$

Eliminating $X_I(s)$ between these equations yields

$$(sI - A)X(s) = -BK_P CX(s) + BK_I\left(\frac{-C}{s} X(s) + \frac{R(s)}{s}\right) + BK_P R(s)$$

and hence

$$(sI - A)X(s) = -B\left(K_P + \frac{K_I}{s}\right)Y(s) + B\left(K_P + \frac{K_I}{s}\right)R(s)$$

so that

$$X(s) = -(sI - A)^{-1} B\left(\frac{K_P s + K_I}{s}\right) Y(s) + (sI - A)^{-1} B\left(\frac{K_P s + K_I}{s}\right) R(s)$$

and

$$Y(s) = CX(s) = C\left\{-(sI - A)^{-1} B\left(\frac{K_P s + K_I}{s}\right) Y(s) + (sI - A)^{-1} B\left(\frac{K_P s + K_I}{s}\right) R(s)\right\}$$

and setting $G_p(s) = C(sI - A)^{-1} B$ yields

$$Y(s) = -G_p(s)\left(\frac{K_P s + K_I}{s}\right) Y(s) + G_p(s)\left(\frac{K_P s + K_I}{s}\right) R(s)$$

This familiar expression simplifies to

$$Y(s) = [sI + G_p(s)(K_P s + K_I)]^{-1} G_p(s)(K_P s + K_I) R(s)$$

Assuming that $G_p(0) \in \Re^{m \times m}$ is finite and that the composite matrix $G_p(0) K_I \in \Re^{m \times m}$ is invertible, the final value theorem gives

$$y_\infty = \lim_{s \to 0} sY(s) = [G_p(0) K_I]^{-1} \{G_p(0) K_I\} r = r$$

and the set point vector r is achieved with the given feedback structure.

As in Section 2.4.3, this analysis for the steady state behaviour of the feedback depends on the transfer function description $G_p(s)$, whatever it might be. This means the analysis defines a guaranteed property that the integral reference error feedback automatically removes steady state offsets from desired constant reference levels. This analysis which transformed the state space description into a transfer function analysis is equivalent to that which can be obtained from the usual PI control diagram of Figure 2.26, where $G_P(s) = C(sI - A)^{-1}B$.

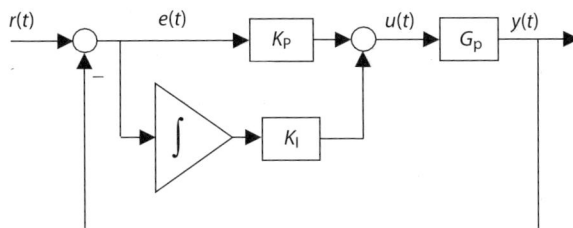

Figure 2.26 Classical PI control structure.

The closed-loop control analysis of the state space modelled system using error and integral reference error feedback enables the following comments to be made.

The method has the following practical advantage:

- The analysis given for the steady state behaviour of the feedback depends on the transfer function description $G_p(s)$, whatever it might be. This means that the analysis defines a guaranteed property that integral reference error feedback automatically removes steady state offsets from desired constant reference levels. This is the invaluable practical property of integral reference error feedback.

The method has the following theoretical disadvantage,

- The classical PI feedback structure is not a full state feedback solution and this leads to a loss of the arbitrary assignability of the $(n + m)$ closed-loop modes. Shaping the dynamics of the classical PI controller is likely to be problematic and restricted.

Concluding Remarks

This sequence of state space analyses has the benefit of providing precise statements about the properties of the various control laws. The disappointment is that the approach does not really provide many practical solutions for industrial use. It is frequently stated that the main difficulties of inaccessible state variables can be overcome by constructing an estimator or an observer. But this is to conveniently forget that (1) many industrial processes have very high order dynamics and consequently a state vector of large dimension, and (2) the performance of observers and the estimator generally depends on good process models – which in most cases are not available or highly uncertain. Therefore, the main value of the state space analysis method given in the preceding sections remains the insight gained into properties of the various control solutions.

2.5 Multivariable PID Control Systems

Common industrial practice is to use three-term control in structures of control loops. For simple problems, single loop systems prevail; for more complicated disturbance rejection problems cascade control is common. Beyond this, multiple control loops using three-term controllers are found along with hierarchies or layers of control loops. This preference for a layered structure of the control system has to do with trying to achieve transparency to the control objectives of the industrial system. Multivariable control *per se* is usually reserved for those situations where it is absolutely necessary. Some process units are difficult to control and often multivariable control is the only way of achieving the performance required. In this section, the emphasis is simply on introducing the structure and notation of the possible multivariable controller that might be found in later chapters of the book.

2.5.1 Multivariable Control

In some industrial processes the multivariable features of the system dominate, and to achieve satisfactory performance multivariable control methods must be used. This is a strong motivation to derive simple and effective methods for tuning the PID controllers being used within a multivariable controller. The most complicated multivariable controller is one which is fully cross-coupled. Consider the case for an m-square system: the mathematical expression for such a controller is given as follows:

1. *Matrix–vector expressions*

    System equation: $Y(s) = G_p(s)U(s)$, with output and control vectors satisfying $Y(s), U(s) \in \Re^m(s)$ and the m-square matrix of system transfer functions $G_p(s) \in \Re^{m \times m}(s)$.

    Error equation: $E(s) = R(s) - Y(s)$, with error, reference and output vectors satisfying $E(s), R(s), Y(s) \in \Re^m(s)$.

    Controller equation: $U(s) = G_c(s)E(s)$ with control and error vectors satisfying $U(s), E(s) \in \Re^m(s)$ and the m-square matrix of controller transfer functions $G_c(s) \in \Re^{m \times m}(s)$.

2. *Element-wise matrix vector definitions*

    The system equation $Y(s) = G_p(s)U(s)$ is given the element-wise description:

    $$\begin{bmatrix} y_1(s) \\ y_2(s) \\ \vdots \\ y_m(s) \end{bmatrix} = \begin{bmatrix} g_{11}(s) & g_{12}(s) & \cdots & g_{1m}(s) \\ g_{21}(s) & g_{22}(s) & \cdots & g_{2m}(s) \\ \vdots & \vdots & \ddots & \vdots \\ g_{m1}(s) & g_{m2}(s) & \cdots & g_{mm}(s) \end{bmatrix} \begin{bmatrix} u_1(s) \\ u_2(s) \\ \vdots \\ u_m(s) \end{bmatrix}$$

    where the output and control components are $y_i(s), u_i(s) \in \Re(s), i = 1, \ldots, m$ and the system matrix elements are $g_{ij}(s) \in \Re(s), i = 1, \ldots, m; j = 1, \ldots, m$.

    The error equation $E(s) = R(s) - Y(s)$ is given the element-wise description:

    $$\begin{bmatrix} e_1(s) \\ e_2(s) \\ \vdots \\ e_m(s) \end{bmatrix} = \begin{bmatrix} r_1(s) \\ r_2(s) \\ \vdots \\ r_m(s) \end{bmatrix} - \begin{bmatrix} y_1(s) \\ y_2(s) \\ \vdots \\ y_m(s) \end{bmatrix}$$

    with error, reference and output vector elements satisfying $e_i(s), r_i(s), y_i(s) \in \Re(s), i = 1, \ldots, m$.

    Finally, the controller equation $U(s) = G_c(s)E(s)$ is given the element-wise description:

    $$\begin{bmatrix} u_1(s) \\ u_2(s) \\ \vdots \\ u_m(s) \end{bmatrix} = \begin{bmatrix} g_{c11}(s) & g_{c12}(s) & \cdots & g_{c1m}(s) \\ g_{c21}(s) & g_{c22}(s) & \cdots & g_{c2m}(s) \\ \vdots & \vdots & \ddots & \vdots \\ g_{cm1}(s) & g_{cm2}(s) & \cdots & g_{cmm}(s) \end{bmatrix} \begin{bmatrix} e_1(s) \\ e_2(s) \\ \vdots \\ e_m(s) \end{bmatrix}$$

    where the error and control vector elements satisfy $e_i(s), u_i(s) \in \Re(s), i = 1, \ldots, m$ and the controller matrix elements are $g_{cij}(s) \in \Re(s), i = 1, \ldots, m; j = 1, \ldots, m$.

2.5 Multivariable PID Control Systems

The block diagram is the usual unity feedback control loop diagram except that matrix–vector rules apply. This is shown in Figure 2.27.

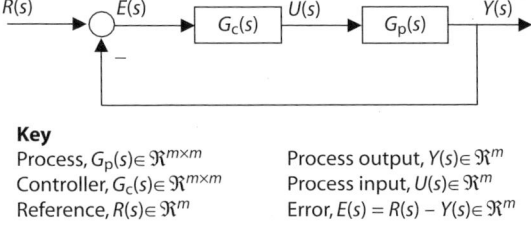

Key
Process, $G_p(s) \in \Re^{m \times m}$
Controller, $G_c(s) \in \Re^{m \times m}$
Reference, $R(s) \in \Re^m$

Process output, $Y(s) \in \Re^m$
Process input, $U(s) \in \Re^m$
Error, $E(s) = R(s) - Y(s) \in \Re^m$

Figure 2.27 Unity feedback structure – multivariable system description.

In general, the use of multivariable process control introduces two structural problems:

1. For a multivariable process, the input–output structure of the controller may have to be determined. Globally this is the problem of which m control inputs are to be used to control which m process outputs. Additionally, if the design procedure to synthesise the multivariable controller introduces either a diagonal or a banded structure into the controller matrix then this input–output pairing issue becomes far more critical since the success of the multivariable controller may depend on a good choice for the pairing. There has been much work on the pairing problem and the classic monograph is by McAvoy (1983); also good reviews can be found in the books by Skogestad and Postlethwaite (1996) and Albertos and Sala (2004).

2. After deciding the input–output structure of the multivariable control strategy, the type of PID controllers to be used in the individual controller matrix elements has to be determined. In the academic literature this is often cast as a so-called restricted structure controller problem. This arises because many multivariable synthesis procedures produce controllers whose order is at least the same as the process. Optimal methodologies often automatically produce a matrix of very high-order controller elements. Thus, restricted structure controllers are those whose structure has been fixed independently of the plant order. Even further, in some cases the controller parameters are fixed to a certain range of allowed values. In general, these restricted structure controllers are of a lower order than the plant they control. Typical examples of restricted structure controllers commonly employed in industry are phase lead, phase lag, phase lead-lag and PID controllers. In this book some chapters are devoted to this strategy of deducing a fixed structure controller from a full optimal control solution.

Decentralised Multivariable Control

Industrial engineers usually seek a physical understanding to the basis of multivariable controller designs and to reduce the amount of complexity existing in a fully cross-coupled multivariable controller, sparse decentralised controllers are often preferred. These decentralised controllers are also known as multi-loop controllers, since this is how they appear on a block diagram. They have the advantage of a simpler structure and, accordingly, require fewer tuning parameters to be found. The full range of structural simplifications that can be used include upper and lower triangular multivariable controllers, diagonally banded multivariable controllers or diagonal multivariable controllers *per se*; the definitions are given next.

1. *Upper and lower triangular multivariable controllers*
   The controller equation $U(s) = G_c(s)E(s)$ has the element-wise descriptions:
   (a) Upper triangular structure

$$\begin{bmatrix} u_1(s) \\ u_2(s) \\ \vdots \\ u_m(s) \end{bmatrix} = \begin{bmatrix} g_{c11}(s) & g_{c12}(s) & \cdots & g_{c1m}(s) \\ 0 & g_{c22}(s) & \cdots & g_{c2m}(s) \\ \vdots & \vdots & \ddots & \vdots \\ 0 & 0 & \cdots & g_{cmm}(s) \end{bmatrix} \begin{bmatrix} e_1(s) \\ e_2(s) \\ \vdots \\ e_m(s) \end{bmatrix}$$

(b) Lower triangular structure

$$\begin{bmatrix} u_1(s) \\ u_2(s) \\ \vdots \\ u_m(s) \end{bmatrix} = \begin{bmatrix} g_{c11}(s) & 0 & \cdots & 0 \\ g_{c21}(s) & g_{c22}(s) & \cdots & 0 \\ \vdots & \vdots & \ddots & \vdots \\ g_{cm1}(s) & g_{cm2}(s) & \cdots & g_{cmm}(s) \end{bmatrix} \begin{bmatrix} e_1(s) \\ e_2(s) \\ \vdots \\ e_m(s) \end{bmatrix}$$

2. *The diagonal band multivariable controller structure*
   The example shown is the controller equation $U(s) = G_c(s)E(s)$ for which the number of controls is $m = 5$ and which has a diagonal band which is three elements wide.

$$\begin{bmatrix} u_1(s) \\ u_2(s) \\ u_3(s) \\ u_4(s) \\ u_5(s) \end{bmatrix} = \begin{bmatrix} g_{c11}(s) & g_{c12}(s) & 0 & 0 & 0 \\ g_{c21}(s) & g_{c22}(s) & g_{c23}(s) & 0 & 0 \\ 0 & g_{c32}(s) & g_{c33}(s) & g_{c34}(s) & 0 \\ 0 & 0 & g_{c43}(s) & g_{c44}(s) & g_{c45}(s) \\ 0 & 0 & 0 & g_{c54}(s) & g_{c55}(s) \end{bmatrix} \begin{bmatrix} e_1(s) \\ e_2(s) \\ e_3(s) \\ e_4(s) \\ e_5(s) \end{bmatrix}$$

3. *The decentralised or diagonal multivariable controller*
   The controller equation $U(s) = G_c(s)E(s)$ is given a diagonal structure:

$$\begin{bmatrix} u_1(s) \\ u_2(s) \\ \vdots \\ u_m(s) \end{bmatrix} = \begin{bmatrix} g_{c11}(s) & 0 & \cdots & 0 \\ 0 & g_{c22}(s) & \cdots & 0 \\ \vdots & \vdots & \ddots & \vdots \\ 0 & 0 & \cdots & g_{cmm}(s) \end{bmatrix} \begin{bmatrix} e_1(s) \\ e_2(s) \\ \vdots \\ e_m(s) \end{bmatrix}$$

The diagonal structure is very popular in studies of decentralised multivariable control. In block diagram form diagonal decentralised multivariable control has a very simple loop structure. Consider an example where $m = 3$; then the details are as follows.

Process equation
$Y(s) = G_p(s)U(s)$, with output and control vectors satisfying $Y(s)$, $U(s) \in \Re^3(s)$ and system transfer functions matrix $G_p(s) \in \Re^{3\times 3}(s)$.

Error equation
$E(s) = R(s) - Y(s)$, with error, reference and output vectors $E(s), R(s), Y(s) \in \Re^3(s)$.

Controller equation
$U(s) = G_c(s)E(s)$, with control and error vectors $U(s), E(s) \in \Re^3(s)$ and controller transfer function matrix $G_c(s) \in \Re^{3\times 3}(s)$ having diagonal structure:

$$\begin{bmatrix} u_1(s) \\ u_2(s) \\ u_3(s) \end{bmatrix} = \begin{bmatrix} g_{c11}(s) & 0 & 0 \\ 0 & g_{c22}(s) & 0 \\ 0 & 0 & g_{c33}(s) \end{bmatrix} \begin{bmatrix} e_1(s) \\ e_2(s) \\ e_3(s) \end{bmatrix}$$

This naturally reduces to the following three controller equations:

$$u_1(s) = [g_{c11}(s)]e_1(s)$$
$$u_2(s) = [g_{c22}(s)]e_2(s)$$
$$u_3(s) = [g_{c33}(s)]e_3(s)$$

The multi-loop or decentralised structure is shown clearly in Figure 2.28.

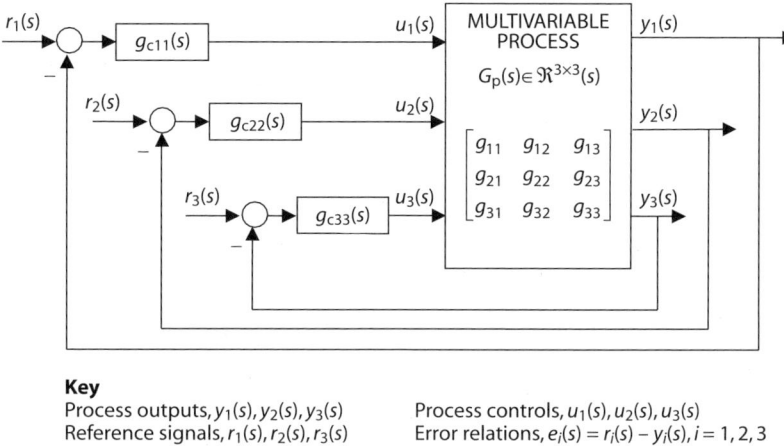

Key
Process outputs, $y_1(s), y_2(s), y_3(s)$ Process controls, $u_1(s), u_2(s), u_3(s)$
Reference signals, $r_1(s), r_2(s), r_3(s)$ Error relations, $e_i(s) = r_i(s) - y_i(s), i = 1, 2, 3$

Figure 2.28 Diagonal decentralised multivariable control where $m = 3$.

If a diagonal decentralised structure can be given an underlying physical justification then the control solution becomes very attractive to practitioners. Further, in the event of component failure in a loop, it is often relatively easy to stabilise the system manually, since only one loop is affected by the failure. A multivariable control design method which tries to inculcate a physical basis into the diagonal decentralised structure is the decoupling method. More on the decoupling concept can be found in later chapters in the book and by reference to the books by Skogestad and Postlethwaite (1996) and Albertos and Sala (2004), among others.

2.5.2 Cascade Control Systems

Many industrial processes are sequential in operation; one process output becomes a process input to the next stage in the process line. Sometimes the processes are connected continuously by a linking material input such as steam or a fluid feedstock. An important source of upset in sequential processes is a variation in the quality of the incoming material supply and a mismatch with the setup of the receiving unit. If an internal process measurement or a measurement between processes is available then a simple multi-loop control system often used in this situation is the cascade control system. Earlier in Chapter 1 there was a very good example of a cascade control system. The discussion of the industrial operator interfaces for a DCS provided by Siemens revealed a cascade control loop where an internal temperature measurement was being used to correct for any disturbances in a drying process. An important observation was that the DCS system had standard tools to facilitate the easy setup of a cascade control loop which is a highly structured form of multi-loop control.

To recap, cascade control arises from the control of two sequential processes where the output of the first or inner process supplies the second or outer process in the sequence. It is assumed that a measurement is available of the output of the inner process and that there is a measurement of the outer process variable. Thus there are two main objectives for cascade control:

1. Use the inner measure to attenuate the effect of supply disturbances or any internal process disturbance on the outer process in the sequence.
2. Use the outer process measurement to control the process final output quality.

Figure 2.29 shows the cascade control system structure.

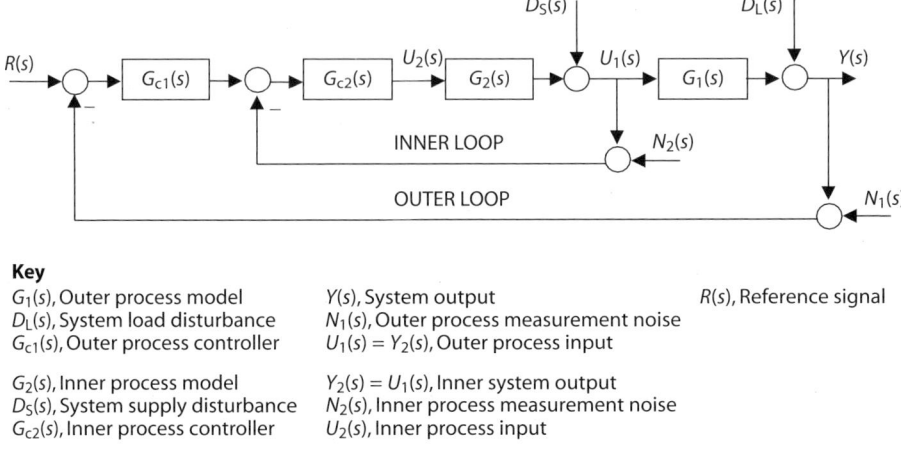

Key
$G_1(s)$, Outer process model
$D_L(s)$, System load disturbance
$G_{c1}(s)$, Outer process controller

$G_2(s)$, Inner process model
$D_S(s)$, System supply disturbance
$G_{c2}(s)$, Inner process controller

$Y(s)$, System output
$N_1(s)$, Outer process measurement noise
$U_1(s) = Y_2(s)$, Outer process input

$Y_2(s) = U_1(s)$, Inner system output
$N_2(s)$, Inner process measurement noise
$U_2(s)$, Inner process input

$R(s)$, Reference signal

Figure 2.29 Cascade control loop.

The Outer Loop

The outer loop is sometimes referred to as the Master or Primary loop. It contains the process output $Y(s)$, which is under primary control. The outer process is denoted $G_1(s)$ and the whole process is subject to load disturbances denoted $D_L(s)$. The outer loop equation is

$$Y(s) = G_1(s)U_1(s) + D_L(s)$$

where the important connecting relation is $U_1(s) = Y_2(s)$, so that the output from the inner process $Y_2(s)$ becomes the input $U_1(s)$ to the outer process. The outer process output is to be controlled to attain a given reference signal $R(s)$ and the measured output used in the comparator is corrupted by outer process measurement noise $N_1(s)$. Thus, the overall control objective is to make the outer process output, $Y(s)$ track the reference $R(s)$ in the presence of process load disturbance $D_L(s)$ and outer process measurement noise, $N_1(s)$. In the case of three-term control being used in the cascade loop, good reference tracking and load disturbance rejection will require integral action in the outer controller.

The Inner Loop

The inner loop is sometimes referred to as the Slave or Secondary loop. The loop contains the inner or supply process denoted $G_2(s)$. This inner process is subject to supply disturbances, denoted $D_S(s)$, and the inner process equation is

$$Y_2(s) = G_2(s)U_2(s) + D_S(s)$$

The output from the inner process becomes the input to the outer process, namely $U_1(s) = Y_2(s)$. The control of the inner process uses the inner loop and this comprises an inner comparator and an inner loop measurement of output $Y_2(s)$, which is corrupted by inner process measurement noise $N_2(s)$. The main inner loop control objective is to attenuate the effect of the supply disturbances $D_S(s)$. These usually represent variations in quality of the material supply (flow rate fluctuations, temperature variations, for example) to the outer process. A second objective for the inner loop is to limit the effect of actuator or inner process gain variations on the control system performance. Such gain variations might arise from changes in operating point due to set point changes or sustained disturbances. When three-term control is used in the cascade loop, fast supply load disturbance rejection will require a fast inner loop design, and this possibly means a high-gain proportional inner controller.

The global performance of the cascade system can be examined using the full closed-loop transfer function analysis along with a corresponding decomposed performance diagram as shown in Figure 2.30.

The closed-loop transfer function analysis, based on Figure 2.29, yields

$$Y(s) = \left[\frac{G_1(s)G_2(s)G_{c2}(s)G_{c1}(s)}{DEN(s)}\right](R(s) - N_1(s)) + \left[\frac{1 + G_2(s)G_{c2}(s)}{DEN(s)}\right]D_L(s)$$
$$+ \left[\frac{G_1(s)}{DEN(s)}\right]D_S(s) - \left[\frac{G_1(s)G_2(s)G_{c2}(s)}{DEN(s)}\right]N_2(s)$$

where $DEN(s) = 1 + G_2(s)G_{c2}(s) + G_1(s)G_2(s)G_{c2}(s)G_{c1}(s)$. This is given block diagram form in Figure 2.30.

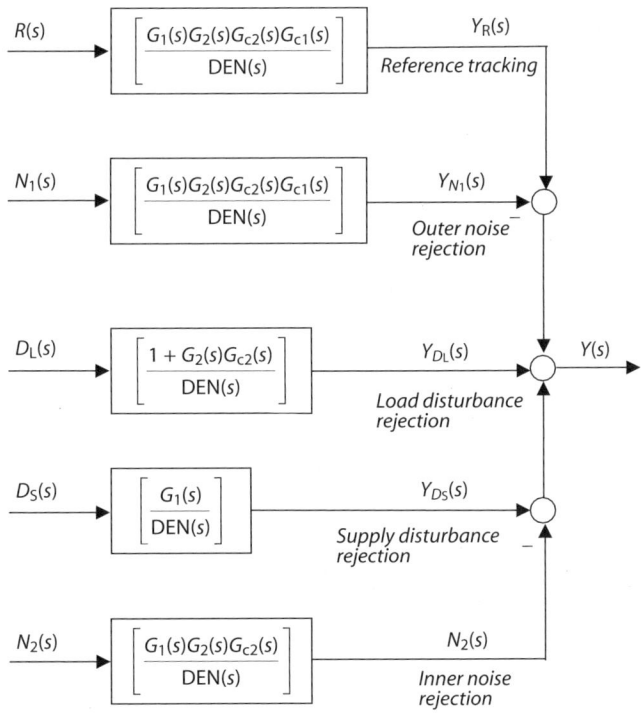

Key $DEN(s) = 1 + G_2(s)G_{c2}(s) + G_1(s)G_2(s)G_{c2}(s)G_{c1}(s)$

Figure 2.30 Cascade control objectives decomposed.

PID Cascade Control Performance

There are two controllers to be selected and tuned: the outer controller $G_{c1}(s)$ and the inner controller, $G_{c2}(s)$. The usual approach is that for good tracking of step reference signals then outer controller $G_{c1}(s)$

should be of PI form. The inner controller $G_{c2}(s)$ can be just P for speed of response, or if this is inadequate PI for the rejection of low-frequency supply disturbance signals. Thus common cascade control structures are often termed PI/P and PI/PI. The use of derivative action is usually avoided in the presence of significant measurement noise. It is quite useful to know qualitatively what can be achieved by cascade control for simple models, with simple disturbance types and for different three term control structures. The results of such an investigation are given in Table 2.4.

Table 2.4 Typical cascade control for simple process models.

Outer process	Outer controller forms	Inner process	Inner controller forms
$G_1(s) = \left[\dfrac{K_1}{\tau_1 s + 1}\right]$	$G_{c1}(s) = k_{P1} + \dfrac{k_{I1}}{s}$	$G_2(s) = \left[\dfrac{K_2}{\tau_2 s + 1}\right]$	$G_{c2}(s) = k_{P2} + \dfrac{k_{I2}}{s}$

Performance	Outer: $G_{c1}(s)$-P Inner: $G_{c2}(s)$-P	Outer: $G_{c1}(s)$-P Inner: $G_{c2}(s)$-PI	Outer: $G_{c1}(s)$-PI Inner: $G_{c2}(s)$-P	Outer: $G_{c1}(s)$-PI Inner: $G_{c2}(s)$-PI
Step ref. tracking	Offset exists	Offset exists	Offset eliminated	Offset eliminated
Outer noise rejection High frequency	−40 dB per decade	−20 dB per decade	−40 dB per decade	−40 dB per decade
Load disturbance rejection Step model	Offset exists	Offset exists	Offset eliminated	Offset eliminated
Supply disturbance rejection Step model	Offset exists	Offset eliminated	Offset eliminated	Offset eliminated
Inner noise rejection High frequency	−40 dB per decade	−40 dB per decade	−40 dB per decade	−40 dB per decade

Acknowledgements

The method of parametric stability margin reported in this chapter arose from research into wastewater system control and was developed by Michael Johnson and Matthew Wade (Industrial Control Centre, Glasgow). Dr L. Giovanini (Industrial Control Centre, Glasgow) and Jonas Balderud (Lund University, Sweden) kindly reviewed the state space section on PID control. The clarity of presentation in the section improved as a result. Dr L. Giovanini is also thanked for his review of the remainder of the chapter.

References

Albertos, P. and Sala, A. (2004) *Multivariable Control Systems*. Springer-Verlag, London.
Blevins, T.L., McMillan, G.K., Wojsznis, W.K. and Brown, M.W. (2003) *Advanced Control Unleashed*. ISA Publishers, Research Triangle Park, NC.

References

Gudaz, J. and Zhang, Y. (2001) Robustness based loop tuning. *ISA Conference*, Houston, USA, September.

Marshall, J.E., Gorecki, H., Korytowski, A. and Walton, K. (1992) *Time Delay Systems*. Ellis Horwood, Chichester.

McAvoy, T.J. (1983) *Interaction Analysis*. ISA Publishers, Research Triangle Park, NC.

Skogestad, S. and Postlethwaite, I. (1996) *Multivariable Feedback Control*. John Wiley, Chichester.

Wade, M.J. and Johnson, M.A. (2003) Towards automatic real-time controller tuning and robustness: an industrial application. *IEEE Industry Applications Society 38th Annual Meeting*, Grand America Hotel, Salt Lake City, Utah.

Walton, K. and Marshall, J.E. (1987) Direct method for TDS stability analysis. *Proc. IEE*, **134D**, 101–107.

Wilkie J., Johnson, M.A. and Katebi, M.R. (2002) *Control Engineering: An Introductory Course*. Palgrave, Basingstoke.

Ziegler, J.G. and Nichols, N.B. (1942) Optimum settings for automatic controllers. *Trans. ASME*, **64**, 759–768.

3 On-line Model-Free Methods

Learning Objectives

3.1 Introduction

3.2 Iterative Feedback Tuning

3.3 The Controller Parameter Cycling Tuning Method

3.4 Summary and Future Directions

Acknowledgements

Appendix 3.A

References

Learning Objectives

In a discipline dominated by model-based procedures the objective of producing a model-free control design method might seem remote from mainstream control engineering concerns. However, developing a model usually incurs a cost: a full physically based model will usually require expensive human input and a detailed empirical model may require many data collection trials from the process. The real question turns on the type of control problem being solved. Is high-performance control required? Is the problem multivariable? The more demanding the control problem and the performance specification, the stronger will be the need for detailed models to inform the solution.

For a significant class of industrial problems, straightforward PID control solutions suffice, and a prized feature of a technique like the relay experiment for PID control design lies in the reduced requirement for modelling input. So, even if it is only to learn how these methods work, it is interesting to investigate methods that do not use a model at all. Two different approaches are presented and the learning objectives for the chapter are to:

- Briefly review the model-free paradigm
- Understand the innovative construction of the Iterative Feedback Tuning method

- Examine the use of Iterative Feedback Tuning in an industrial example
- Learn about an alternative model-free method, the Controller Parameter Cycling method

3.1 Introduction

Much recent research in the control community has focused on finding viable approaches for using nonlinear control in industrial applications. Most of this work assumes that some form of mathematical model will be available, and therein is a real applications difficulty; nonlinear models are often hard to derive, can be expensive to develop and may be difficult to use in a routine control design exercise. If the nonlinearity of the process has to be captured and used in a control system, then linearisation at a number of operating points coupled with a bank of switched linear controllers, namely a gain or a controller schedule, is one common way forward. If a model is available, the use of nonlinear model-based predictive control is one method that is in particular vogue at present.

A different approach to generating routine controllers for industrial plant is to use only a little model information and rely on the robustness of the controller for success. The on-line relay experiment method of Åström and Hägglund (1985) was particularly successful in industrial applications. At the heart of the relay method is a nonparametric identification principle. The simplicity of this approach inspired extensions (Yu, 1999) and alternative nonparametric methods for automated industrial three-term controller design (Crowe and Johnson, 1998, 1999, 2000). Some of these methods are presented in later chapters of the book.

In 1994, Hjalmarsson *et al.* published the first in a series of papers on a *model-free* approach to restricted complexity controller design (Hjalmarsson *et al.*, 1994, Hjalmarsson *et al.*, 1998). The method was termed the Iterative Feedback Tuning (IFT) method of controller design. The principle behind this approach is to use a set of specific process experiments to produce data for use in a stochastic optimisation routine to optimise a simple controller, for example a PID controller. The immediate attraction of the procedure for the industrial control engineer is to remove the need for an explicit process model. However, before looking at the specific details of two model-free methods some basic principles of this paradigm are reviewed.

3.1.1 A Model-Free Control Design Paradigm

The model-free control design paradigm was pre-dated by procedures for on-line model-based optimisation of controllers, for example the methods due to Trulsson and Ljung (1985). Consequently, there are features of the earlier work that have been retained by the model-free approaches. For ease of discussion, these features are listed next.

Restricted Structure Controllers

Many optimal synthesis procedures automatically produce controllers whose order is at least the same as the process. The use of high-order controllers may be required in high-performance control systems such as those used in aerospace applications, but in more straightforward industrial applications, being able to understand how the controller operates and reliability in implementation may be more important concerns for the process engineer. In consequence, low-order approximations to high-order controllers are often sought. Restricted structure controllers, then, are those whose structure has been fixed independently of the plant order. Even further, in some cases the controller parameters are fixed to a certain range of allowed values. Typical examples of industrial restricted structure controllers are phase lead, phase lag, phase lead-lag and PID controllers. There is one more aspect to the restricted structure idea: having fixed the controller structure, the controller parameters are often calculated by

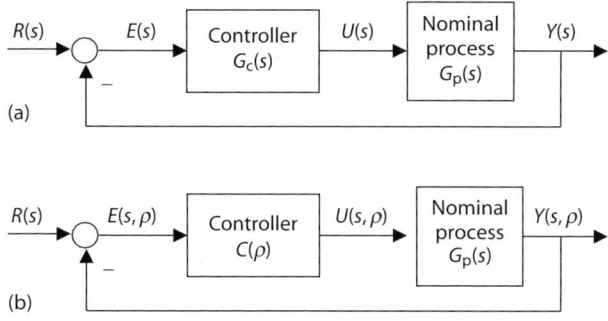

Figure 3.1 Unity feedback control loop.

selecting them to optimise the full design cost function or criterion. The cost function or criterion is discussed shortly, but first some restricted structure controller parameter notation is introduced.

The unity feedback control loop is used as shown in Figure 3.1(a) and the restricted structure controller is assumed to have n_c parameters denoted $\rho_i \in \Re, i = 1, \ldots, n_c$; clearly the controller parameters can be written in vector form as $\rho \in \Re^{n_c}$. The restricted structure controller can then be written as $G_c(s) = C(\rho)$. For example, given a decoupled PI control

$$G_c(s) = k_P + \frac{k_I}{s}$$

the following simple identification holds. Let $\rho_1 = k_P$, $\rho_2 = k_I$; thus $n_c = 2$ and $\rho \in \Re^2$, so that

$$G_c(s) = k_P + \frac{k_I}{s} = \rho_1 + \frac{\rho_2}{s} = C(\rho)$$

To indicate the dependence of time signals in the system on the controller parameter vector, the following notation for the time signals of error, control and process output respectively are given as $e(t,\rho), u(t,\rho)$ and $y(t,\rho)$. The arguments (t,ρ) denote the obvious dependence of these signals on time t and the dependence of these signals on the particular value of the controller parameter vector $\rho \in \Re^{n_c}$. A similar notation is introduced for the Laplace transforms of these signals: $E(s,\rho) = L\{e(t,\rho)\}$, $U(s,\rho) = L\{u(t,\rho)\}$ and $Y(s,\rho) = L\{y(t,\rho)\}$. In this case, the arguments (s,ρ) denote the obvious dependence of these signals on the Laplace variable s and the dependence of these signal transforms on the particular value of the controller parameter vector $\rho \in \Re^{n_c}$. Figure 3.1(b) shows the new notation used in the unity feedback system.

An Optimisation Criterion

Stochastic and deterministic optimal control methods try to capture desirable control design objectives in one cost function or criterion. The optimisation of the criterion then synthesises a controller with the requisite properties. The optimisation process itself can be an analytical step or the outcome of a numerical routine. This is a powerful idea that has dominated some academic control circles since the original LQ and LQG cost functions were investigated by Kalman in the early 1960s. The range of cost functions available for consideration has grown over the decades to include, for example, those that inculcate robustness properties. Some typical examples of relevance to this chapter are:

Stochastic cost functions
General cost function:

$$J(\rho) = E\{g(y(t,\rho), u(t,\rho))\}$$

Weighted sum of output and control variances:

$$J(\rho) = \tfrac{1}{2} E\{ y^2(t,\rho) + \alpha^2 u^2(t,\rho) \}$$

Deterministic cost functions
Linear quadratic cost function – time domain

$$J(\rho) = \frac{1}{2T_f} \int_0^{T_f} [\{Q_y e(t,\rho)\}^2 + \lambda^2 \{R_u u(t,\rho)\}^2] dt$$

Linear quadratic cost function – Laplace s domain

$$J(\rho) = \frac{1}{2\pi} \int_{-j\infty}^{j\infty} [E(-s,\rho)Q_y(-s)Q_y(s)E(s,\rho) + \lambda^2 U(-s,\rho)R_u(-s)R_u(s)U(s,\rho)] ds$$

An important feature of setting up a design cost function is the selection of the weighting functions. Useful material on this can be found in the book by Grimble (1994) and some material on this is also given in Chapter 12.

An Optimisation Routine

Assume that a suitable design LQ cost function has been selected:

$$J(\rho) = \frac{1}{2T_f} \int_0^{T_f} [\{Q_y e(t,\rho)\}^2 + \lambda^2 \{R_u u(t,\rho)\}^2] dt$$

and that a restricted structure controller has been specified, $G_c(s) = C(\rho)$. Then, for every value of the controller parameter $\rho \in \Re^{n_c}$ such that the closed loop is stable, the LQ cost function will return a finite value. Thus the optimal restricted structure controller can be found by solving the following optimisation problem:

$$\min_{\text{w.r.t. } \rho \in \Re^{n_c}} J(\rho)$$

subject to $C(\rho)$ stabilising the closed loop.

This is a fixed structure LQ optimal control problem with weighted error and control signals. Incorporating a limit process $T_f \to \infty$ will yield the steady state version of the optimisation problem. To simplify the subsequent analysis, the weighting filters are set to $Q_y = 1, R_u = 1$. The optimisation of the cost function follows a simple iterative Newton algorithm using the gradient direction. Firstly, the gradient of the cost function with respect to the controller parameter vector $\rho \in \Re^{n_c}$ is found using Leibniz's Theorem for the differentiation of an integral (Abramowitz and Stegun, 1972).

Lemma 3.1: Cost Gradient Formula
The gradient of the cost functional

$$J(\rho) = \frac{1}{2T_f} \int_0^{T_f} \{(e^2(t,\rho)) + \lambda^2(u^2(t,\rho))\} dt$$

with respect to controller parameter vector $\rho \in \Re^{n_c}$ is given by

$$\frac{\partial J}{\partial \rho} = \frac{1}{T_f} \int_0^{T_f} \left\{ e(t,\rho) \frac{\partial e(t,\rho)}{\partial \rho} + \lambda^2 u(t,\rho) \frac{\partial u(t,\rho)}{\partial \rho} \right\} dt$$

□

The basic Newton numerical procedure follows.

Algorithm 3.1: Basic controller optimisation

Step 1   *Initialisation*
- Choose cost weighting λ^2
- Choose costing time interval T_f
- Choose convergence tolerance ε
- Set loop counter $k = 0$
- Choose initial controller parameter vector $\rho(k)$

Step 2   *Gradient calculation*

Calculate gradient

$$\frac{\partial J}{\partial \rho}(k) = \frac{1}{T_f} \int_0^{T_f} \left\{ e(t,\rho(k)) \frac{\partial e(t,\rho(k))}{\partial \rho} + \lambda^2 u(t,\rho(k)) \frac{\partial u(t,\rho(k))}{\partial \rho} \right\} dt$$

If $\left\| \frac{\partial J}{\partial \rho}(k) \right\| < \varepsilon$ then stop

Step 3   *Update calculation*

Select or calculate the update parameters γ_k and R_k

Compute

$$\rho(k+1) = \rho(k) - \gamma_k R_k^{-1} \frac{\partial J}{\partial \rho}(k)$$

Update $k = k + 1$ and goto Step 2

Algorithm end

Remarks 3.1

(a) Setting $R = I$ gives an algorithm from the steepest descent family of optimisation routines.

(b) A selection for γ_k has to be made. In the case of a simple steepest descent algorithm, this can be a fixed step or line search step. A fixed step procedure can be computational costly. A line search procedure is usually based on finding a bracket around the minimum in the direction of search. This might be followed by fitting a quadratic function and computing the optimum quadratic minimising step. These ideas have their origin in work from the 1960s (Fletcher, 1987).

(c) Setting $R = H(\rho(k))$, where H is the Hessian matrix, produces a Newton iteration for the optimisation; in this case

$$H_{ij} = \frac{\partial^2 J}{\partial \rho_i \partial \rho_j}$$

Algorithm 3.1 serves as a fundamental framework for this chapter, since in the sequel two methods are described for obtaining the gradient. The first is that of the Iterative Feedback Tuning technique and the second is the method of controller parameter cycling.

3.2 Iterative Feedback Tuning

A deterministic version of the Iterative Feedback Tuning (IFT) method is given with the objective of understanding the key innovation introduced to enable model-free computation of the cost function gradient. A simulation study from the wastewater industry shows the application of the method. The section closes with a discussion of the issue of computing the Hessian, some remarks on the stochastic work published in the literature and the extension of the Iterative Feedback Tuning method to multivariable systems.

3.2.1 Generating the Cost Function Gradient

The system setup is the unity feedback system as shown in Figure 3.1. The template algorithm for on-line optimisation of the cost function is that of Algorithm 3.1. For a given value of the controller parameter vector $\rho(k) \in \mathfrak{R}^{n_c}$, the algorithm requires a computation of the gradient

$$\frac{\partial J}{\partial \rho}(k) = \frac{1}{T_f} \int_0^{T_f} \left\{ e(t,\rho(k)) \frac{\partial e(t,\rho(k))}{\partial \rho} + \lambda^2 u(t,\rho(k)) \frac{\partial u(t,\rho(k))}{\partial \rho} \right\} dt$$

The key innovation of the IFT method is to find this gradient from the closed-loop system signals without computing any intermediate system models. This means that the actual closed-loop system is used to produce the necessary data and compute the gradient function. The route to this begins from some standard closed-loop relationships given as Lemma 3.2.

Lemma 3.2: Closed-loop relationships

In the standard closed loop of Figure 3.1(b), the reference, reference error, controller output and the system output signals are denoted $R(s)$, $E(s,\rho)$, $U(s,\rho)$, $Y(s,\rho)$ respectively. Then

$$Y(s,\rho) = \left[\frac{G_p(s)C(\rho)}{1+G_p(s)C(\rho)} \right] R(s) = T(s,\rho)R(s)$$

$$E(s,\rho) = \left[\frac{1}{1+G_p(s)C(\rho)} \right] R(s) = S(s,\rho)R(s)$$

and

$$U(s,\rho) = \left[\frac{C(\rho)}{1+G_p(s)C(\rho)} \right] R(s) = S(s,\rho)C(\rho)R(s)$$

where the controller, process, sensitivity and complementary sensitivities have transfer function expressions denoted by $C(\rho)$, $G_p(s)$, $S(s,\rho)$, $T(s,\rho)$ respectively. ∎

Achieving the IFT model-free implementation begins by noting that to compute the gradient

$$\frac{\partial J}{\partial \rho}(k)$$

requires the four signals

$$e(t,\rho(k)), \frac{\partial e(t,\rho(k))}{\partial \rho} \text{ and } u(t,\rho(k)), \frac{\partial u(t,\rho(k))}{\partial \rho}$$

Lemma 3.2 shows that time domain signals $e(t,\rho(k)), u(t,\rho(k))$ can be computed directly from the closed-loop system. With the aid of some signal storage, computation of the remaining two signals

$$\frac{\partial e(t,\rho(k))}{\partial \rho}, \frac{\partial u(t,\rho(k))}{\partial \rho}$$

uses the results of Lemma 3.3.

Lemma 3.3: Gradient term transforms

The two terms in the gradient expression are given transform representation as follows. Recall

$$E(s,\rho) = \left[\frac{1}{1+G_p(s)C(\rho)}\right]R(s) = S(s,\rho)R(s)$$

Then

$$\frac{\partial E(s,\rho)}{\partial \rho} = \left[\left(\frac{-1}{C(\rho)}\right)\left(\frac{\partial C(\rho)}{\partial \rho}\right)\right]T(s,\rho)E(s,\rho)$$

Denote

$$U(s,\rho) = \left[\frac{C(\rho)}{1+G_p(s)C(\rho)}\right]R(s) = S(s,\rho)C(\rho)R(s)$$

Then

$$\frac{\partial U(s,\rho)}{\partial \rho} = \left[\left(\frac{\partial C(\rho)}{\partial \rho}\right)\right]S(s,\rho)E(s,\rho)$$

□

Motivated by this lemma, the error signal $E(s,\rho)$ can be used in a closed-loop identification step to generate the gradient expressions as follows.

1. *Computing the term $\partial E(s,\rho)/\partial \rho$*

   If the closed-loop system is fed the signal $E(s,\rho)$ at the reference input, the system output can be recorded as $Y^1(s,\rho) = T(s,\rho)E(s,\rho)$. Then, from Lemma 3.3,

   $$\frac{\partial E(s,\rho)}{\partial \rho} = \left[\left(\frac{-1}{C(\rho)}\right)\left(\frac{\partial C(\rho)}{\partial \rho}\right)\right]T(s,\rho)E(s,\rho) = \left[\left(\frac{-1}{C(\rho)}\right)\left(\frac{\partial C(\rho)}{\partial \rho}\right)\right]Y^1(s,\rho)$$

   Hence the gradient signal component $\partial E(s,\rho)/\partial \rho$ is found as

   $$\frac{\partial E(s,\rho)}{\partial \rho} = -[G_{grad}(s,\rho)]Y^1(s,\rho) \text{ where } G_{grad}(s,\rho) = \left[\left(\frac{1}{C(\rho)}\right)\left(\frac{\partial C(\rho)}{\partial \rho}\right)\right]$$

2. *Computing the term $\partial U(s,\rho)/\partial \rho$*

   If the closed-loop system is fed the signal $E(s,\rho)$ at the reference input, then the system controller output can be recorded as $U^1(s,\rho) = S(s,\rho)C(\rho)E(s,\rho)$. Then, from Lemma 3.3,

   $$\frac{\partial U(s,\rho)}{\partial \rho} = \left[\left(\frac{\partial C(\rho)}{\partial \rho}\right)\right]S(s,\rho)E(s,\rho) = \left[\left(\frac{1}{C(\rho)}\right)\left(\frac{\partial C(\rho)}{\partial \rho}\right)\right]S(s,\rho)C(\rho)E(s,\rho)$$

   and

$$\frac{\partial U(s,\rho)}{\partial \rho} = \left[\left(\frac{1}{C(\rho)}\right)\left(\frac{\partial C(\rho)}{\partial \rho}\right)\right] U^1(s,\rho)$$

Hence the gradient signal component $\partial U(s,\rho)/\partial \rho$ can be found as

$$\frac{\partial U(s,\rho)}{\partial \rho} = [G_{\text{grad}}(s,\rho)] U^1(s,\rho) \text{ where } G_{\text{grad}}(s,\rho) = \left[\left(\frac{1}{C(\rho)}\right)\left(\frac{\partial C(\rho)}{\partial \rho}\right)\right]$$

Although the consequences of Lemma 3.3 have been presented in the Laplace domain, the equivalent time domain interpretations show how the time domain gradient equation is to be calculated for a given value of the controller parameters $\rho(k) \in \Re^{n_c}$. This computation requires one set of system responses due to the reference $r(t)$ and a second set of system responses due to the signal $e(t,\rho(k))$ injected at the reference input port. Figure 3.2 illustrates this sequence of computations. The necessary processing of plant data to obtain the time domain signals

$$e(t,\rho(k)), \frac{\partial e(t,\rho(k))}{\partial \rho} \text{ and } u(t,\rho(k)), \frac{\partial u(t,\rho(k))}{\partial \rho}$$

uses the *actual* system and requires no models and hence the procedure is genuinely *model-free*. Storage of the signals $e(t,\rho(k)), u(t,\rho(k))$ is required and then the error signal $e(t,\rho(k))$ is fed into the reference input port and the new signals $y^1(t,\rho(k)), u^1(t,\rho(k))$ are collected and processed for entry into the gradient calculation. This processing will require a numerical procedure to effect the equivalent time domain operation of $G_{\text{grad}}(s,\rho)$ on the time domain equivalent of signals $-Y^1(s,\rho)$ and $U^1(s,\rho)$. For the purposes of simulation studies, Figure 3.2 shows how the various time signals needed to compute the gradient can be generated for a given vector of controller parameters $\rho(k) \in \Re^{n_c}$.

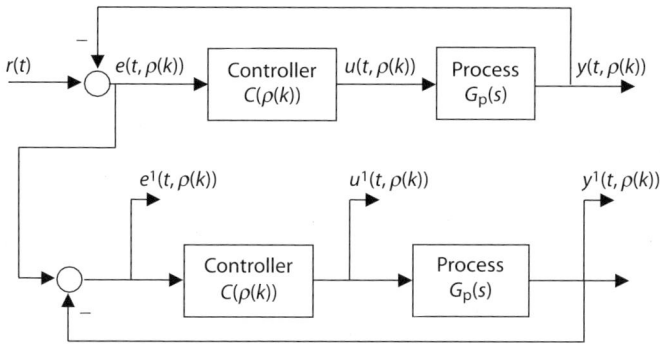

Figure 3.2 Signal generation for the gradient.

Algorithm 3.2 captures these findings for the generation of the kth gradient where a suitable time interval T_f has been chosen.

Algorithm 3.2: Gradient computation: kth step
Step 1   Initialise kth *step*
         Set up controller using $\rho(k)$.
Step 2   *Responses over time interval* T_f
         Run closed-loop system with reference input $r(t)$.
         Record the signals $e = e(t,\rho(k)), u = u(t,\rho(k))$.
         Run closed-loop system with $e = e(t,\rho(k))$ at the reference input.

Record the signals $y^1 = y^1(t,\rho(k)), u^1 = u^1(t,\rho(k))$.

Step 3 *Processing responses*

To produce the time signals

$$\frac{\partial e(t,\rho(k))}{\partial \rho} \text{ and } \frac{\partial u(t,\rho(k))}{\partial \rho}$$

process the recorded signals using time domain equivalents of the transfer function relations

$$\frac{\partial E(s,\rho)}{\partial \rho} = -[G_{\text{grad}}(s,\rho)]Y^1(s,\rho) \text{ and } \frac{\partial U(s,\rho)}{\partial \rho} = [G_{\text{grad}}(s,\rho)]U^1(s,\rho)$$

with

$$G_{\text{grad}}(s,\rho) = \left[\left(\frac{1}{C(\rho)}\right)\left(\frac{\partial C(\rho)}{\partial \rho}\right)\right]$$

Step 4 *Calculate gradient*

Use time domain formula to compute gradient

$$\frac{\partial J}{\partial \rho}(k) = \frac{1}{T_f}\int_0^{T_f}\left\{e(t,\rho(k))\frac{\partial e(t,\rho(k))}{\partial \rho} + \lambda^2 u(t,\rho(k))\frac{\partial u(t,\rho(k))}{\partial \rho}\right\}dt$$

Algorithm end

In practice, careful data collection procedures can automate the processing required to generate the gradient. This requires routines running in parallel with the data collection. Note that each gradient requires two response passes of the closed-loop system. Further, note that only system time responses are recorded and no knowledge of a system model is required. The processing of system responses requires knowledge of the controller $C(\rho(k))$.

3.2.2 Case Study – a Wastewater Process Example

A wastewater treatment plant reduces or removes potentially harmful components from an influent fluid and delivers a less harmful effluent to a receiving water body. Typically, the influent, which flows to the treatment plant through a sewer system network, contains household and industrial wastewater and rainwater. At the plant, the influent wastewater passes through several treatment stages before discharge to a receiving water body. These processing stages are termed preliminary, primary, secondary and tertiary treatment stages.

Preliminary treatment is the initial storage and coarse screening of the wastewater and may include grit removal, flow distribution and flow equalisation. The primary treatment phase is commonly a number of sedimentation tanks that partially remove suspended solids from the fluid mass. The secondary treatment phase is usually considered the most important part of the treatment process, as it is designed to remove the largest proportion of the biological pollution load. Secondary treatment can also be adjusted to remove specific components, for example phosphorus and nitrogen. In the first phase of secondary treatment, the process involves substrate degradation by selected micro-organisms in aerobic, anaerobic or anoxic conditions; this is the activated sludge process. The second phase of the secondary treatment process is final clarification, which is similar to primary sedimentation, but allows the balancing of solids and liquid within the secondary phase by returning activated sludge to designated sections of the plant, for example the biological tanks or primary clarifiers. The tertiary stage is used for polishing the effluent before discharge, but may also provide removal of specific pollutants based on the plant discharge requirements. Some common processes used are filters and carbon absorbers with chemical dosing.

To ensure that wastewater treatment plant efficiency is optimised, control technology has been developed to enable automatic and precise monitoring and control of many aspects of the process. One example is the control of dissolved oxygen in the bioreactors of the secondary treatment phase.

Wastewater Treatment Plant Model

The benchmark model for the development of application-specific models and controls for wastewater treatment plants is the Activated Sludge Model No. 1 (Henze et al., 1987). The model is often denoted ASM No. 1. The Activated Sludge Model No. 1 comprises 13 nonlinear equations describing component mass balances and sub-processes taking place in the biochemical reactions of the activated sludge process. An important control variable in the process is the amount of dissolved oxygen (DO) in the wastewater. In aerated systems, oxygen is mechanically introduced into the liquor and indirectly used to control some of the other system state variables that establish the effluent quality.

The Activated Sludge Model No. 1 involves nonlinear differential equations with more than 50 uncertain parameters. The model is numerically stiff. Although the activated sludge process operates on a time unit of *days*, the dissolved oxygen (DO) variable is manipulated on a *minutes* time-scale. For example, experimental and simulation results have shown that using relatively simple three-term control systems it is possible to reach DO setpoints within minutes. Lee et al. (1991) and Miklev et al. (1995) both show evidence of this behaviour. In this case study, a bioreactor with a volume of 1000 m^3 has been simulated using the Activated Sludge Model No. 1. The dissolved oxygen control loop framework is shown in Figure 3.3. The influent concentrations, stoichiometric and kinetic parameters for the model are assumed to remain constant during the simulation time. All the values have been taken from the COST Simulation Benchmark (Copp, 2001), and the values are archived in Appendix 3.A.

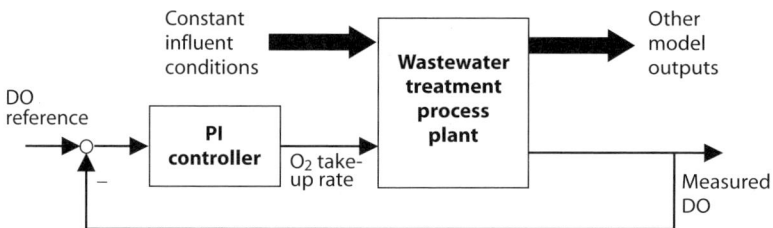

Figure 3.3 Dissolved oxygen control loop.

Dissolved Oxygen Loop Modelling Results

To investigate the expectation of highly nonlinear model behaviour, an assessment was performed on the dissolved oxygen loop to determine the characteristics of the plant model for a range of input oxygen take-up rates $99 \leq u_{Ox} \leq 121$. Open loop experiments were performed using MATLAB/SIMULINK software. For a range of input step responses, the first-order process model

$$G_p(s) = \left[\frac{K}{\tau s + 1}\right]$$

where the process gain is K and the time constant is τ, was fitted and the relationship between model gain K and time constant τ versus input signal size was plotted. The results, given in Figure 3.4, show that this particular loop exhibits only a mild degree of nonlinearity.

Consequently, an average first-order linear system model was chosen for the IFT tuning simulation runs and the final controller values were tested with the full nonlinear model. This has the advantage of

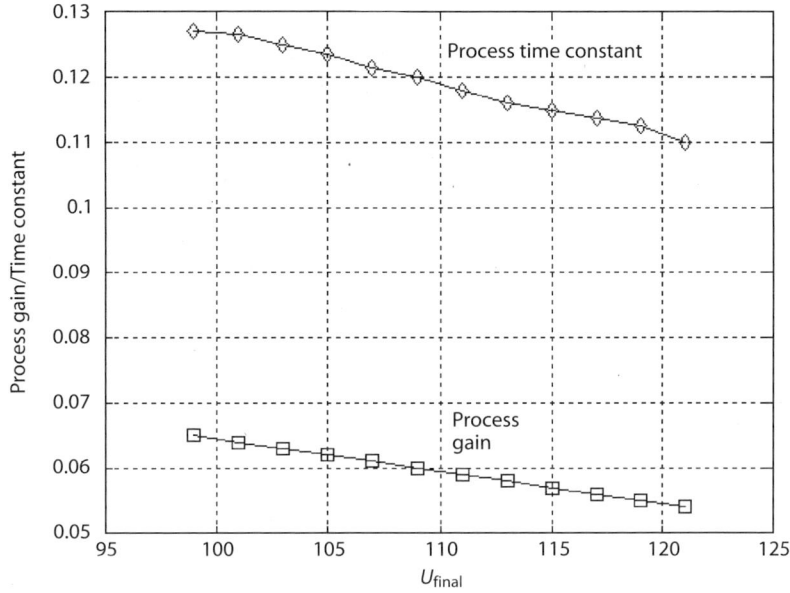

Figure 3.4 Plot of process gain and time constant for a range of input step sizes.

avoiding repeated runs of the full nonlinear model and reducing the simulation time needed in the studies. The model selected was

$$G_p(s) = \left[\frac{0.0585}{0.1186s + 1}\right]$$

where $\tau = 0.1186$ days. Thus, in the midpoint of the range $99 \leq u_{Ox} \leq 121$, say $u_{Ox} = 110$, the measured DO was found to be 6.43 mg/l and the DO reference was selected to be $DO_{ref} = 6.5$ mg/l.

Iterative Feedback Tuning PI Controller Results

The DO Control Loop Design Specification
The performance for the DO loop was set as a desired rise time of 0.021 days, which is approximately 30 minutes. Considering that the open loop system time constant is 171 minutes, this is quite a stringent design specification.

The Ziegler–Nichols Tuning Results
An initial PI controller tuning exercise used the standard Ziegler–Nichols reaction curve method (Ziegler and Nichols, 1942; Wilkie et al., 2001). The outcome was a PI controller with $k_P = 5.94$ and $k_I = 51.56$. The ZN tuning does not meet design specification.

Iterative Feedback Tuning – Specifying the Algorithmic Parameters
Table 3.1 shows the parameter values of the algorithm set up for a cost function of the form

$$J(\rho) = \frac{1}{2T_f} \int_0^{T_f} \{y^2(t) + \lambda^2 u^2(t)\} dt$$

where the output $y(t)$ is the measured DO and the control $u(t)$ is the O_2 rate. The weighting $\lambda^2 > 0$ is to be tuned to achieve the control specification.

Table 3.1 IFT algorithm parameters.

Parameter	Value	Reason
Time period T_f	1.5 days	Achieve steady state
Convergence tolerance ε	–	None set
Initial controller vector $\rho(0)$	$\rho_1(0) = k_P = 5.94$ $\rho_2(0) = k_I = 51.56$	Start from Z–N tuning
Setting the Hessian R	$R = I$	Steepest descent algorithm
Setting the cost function weighting λ^2	Case 1: $\lambda^2 = 10^{-3}$	Selected to draw contour maps
	Case 2: $\lambda^2 = 10^{-5}$	Selected to speed output response
Step size γ_k	Case 1: Fixed size	Investigate optimisation efficiency
	Case 2: Optimised	

IFT Tuning – Case 1: $\lambda^2 = 10^{-3}$

In this case, where $\lambda^2 = 10^{-3}$, the cost function contours in the gain space of the PI controller are spaced sufficiently to allow a contour plot to be drawn. Consequently, an interesting comparison of the progress of fixed step and optimised step line searches is shown in Figure 3.5. The fixed step routine gives an

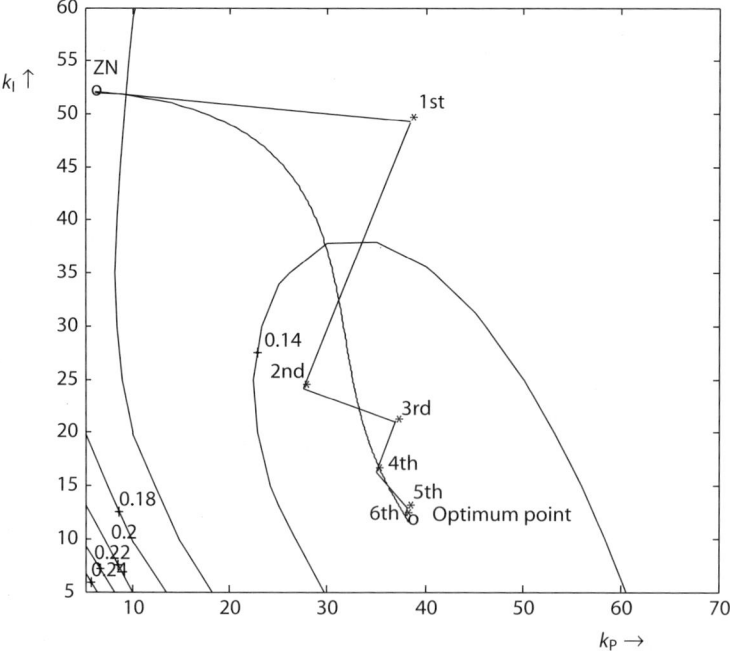

Figure 3.5 Contour map of optimisation progress, Case 1: $\lambda^2 = 10^{-3}$.

excessive and impractical number of gradient calculations. However, optimised step line searches give a much reduced and more realistic number of gradient and cost calculations; the details are in Table 3.2.

Table 3.2 Computations in the optimised line search routine.

Iteration cycle no.	Gradient calculations	Function calculations
1	1	7
2	1	6
3	1	5
4	1	4
5	1	4
6	1	3

IFT Tuning – Case 2: $\lambda^2 = 10^{-5}$

Step response tests with the IFT controller from Case 1 showed that the response was not meeting the desired rise time specification. Hence a new set of trials were performed to obtain a tuning with $\lambda^2 = 10^{-5}$. Conventional optimal control wisdom is that reducing the size of λ^2 will speed up the response. Table 3.3 shows the controller gains for the various cases and Figure 3.6 shows step responses for the various cases using the nonlinear model.

Table 3.3 Controller gains for different tuning cases.

Tuning details	k_P	k_I
Ziegler–Nichols	5.94	51.56
Case 1: $\lambda^2 = 10^{-3}$	38.27	11.63
Case 2: $\lambda^2 = 10^{-5}$	348.63	218.70

In Figure 3.6, the step response graph C arising from the IFT results of Case 1, $\lambda^2 = 10^{-3}$, is not acceptable since it does not meet the desired specification. Graph B arises from the ZN tuning and although it does not meet the design specification it looks more reasonable than Graph C. The graph A is the fast response that comes from the results of Case 2, $\lambda^2 = 10^{-5}$, but the PI gains as shown in Table 3.3 are unrealistically large and would undoubtedly lead to actuator saturation. The inescapable conclusion is that some refinement of the control performance specification is needed, with a possible concomitant change in the cost function parameters.

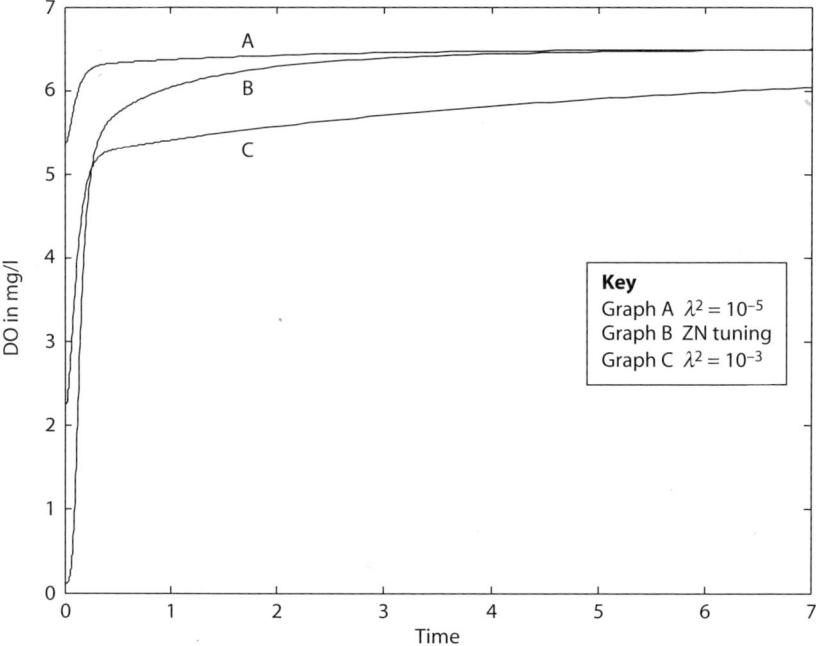

Figure 3.6 Step responses from the nonlinear model.

3.2.3 Some Remarks on Iterative Feedback Tuning

Iterative Feedback Tuning is a *model-free* control design method that has three features:

- *Restricted structure controller*
  The structure of the controller is specified by the designer. Typically, the controller will belong to the PID controller category, but the method generically extends to any specific constant-coefficient controller parameter set sought by the design engineer.

- *An optimisation problem to give a control design solution*
  The IFT method optimises a cost function that is selected to capture the desired control solution. Many different cost functions can be used, including those that do not belong to the well-established solution knowledge-base of optimal control. Despite the wealth of material available on the standard cost functions, such as the LQ and LQG cost functions, guidelines on how to turn a control specification into a cost function for use with the IFT method would be very useful. Some work in this direction is that due to Lequin *et al.* (1999).

- *An on-line method for generating cost function gradients*
  The Iterative Feedback Tuning method finds an optimal controller using an on-line numerical optimisation routine. This routine is configured as either a steepest descent or a Newton algorithm. The innovation introduced by Hjalmarsson *et al.* (1994, 1995) is a model-free method of generating the time signals for the calculation of the cost function gradient. The time signals depend only on closed-loop system responses and do not require system models, hence the model-free designation for the method. The simple case study reported in this chapter shows that there is always a need to look carefully at the implementation of the IFT optimisation routine to ensure that it is operating in an efficient manner. In an on-line situation, the method will use real plant trials and this makes such

an investigation imperative since it will be important to minimise the disturbance to the process operations.

The presentation of this chapter is a deterministic version of the Iterative Feedback Tuning method. This simple presentation was intended as an introduction of the main elements of the technique. The published literature of the method is now quite extensive and a number of references are given at the end of this chapter. It is, however, pertinent to make some observations on some of the other aspects of this technique.

Generating the Hessian
Theoretically (and hopefully practically), a significant advantage accrues if the Hessian data is available to turn the basic optimisation steps of Algorithm 3.1 from a steepest descent routine into a Newton procedure. However, if the analysis for the Hessian elements is derived for the deterministic formulation of this chapter it seems that a third set of system responses is required. The fact that the number of on-line system experiments is beginning to bloom leads to concern that the on-line generation of the Hessian will have to be conducted carefully to minimise both the number of experiments and disruption to the process.

Stochastic Versions of the Iterative Feedback Tuning Method
The seminal papers for the Iterative Feedback Tuning method due to Hjalmarsson *et al.* (1994, 1995) adopt a general formulation that incorporates the following features.

- A system description involving a stochastic process output disturbance
- A two degrees of freedom control law
- Use of a stochastic optimisation approach (Robbins and Munro, 1951).
- A restricted structure control law.

Consequently, the published literature already accommodates the presence of process and measurement noise in the closed-loop system behaviour. Many industrial systems suffer from the presence of these noise sources and it is useful to know that the general framework of Iterative Feedback Tuning has a stochastic process setting.

Iterative Feedback Tuning for Multivariable Systems
Some of the more recent Iterative Feedback Tuning publications have dealt with the application of the technique to multivariable systems (Hjalmarsson and Birkeland, 1998; Hjalmarsson, 1999). The introductory scalar deterministic formulation of this chapter can also be given multivariable system form to show how the multivariable version of Iterative Feedback Tuning works. The steps of the basic procedure are little changed but there are some differences; for example, the analysis is now a matrix–vector analysis. An interesting difference is that in the scalar system case, the second set of system responses needed for the gradient computation were generated by feeding a stored error signal into the system reference port; see Figure 3.2. However, in the multivariable case, the entry port for generating the responses needed for the gradient calculation moves to the controller signal port. The multivariable system analogy of Figure 3.2 is shown as Figure 3.7, where the new entry port is clearly seen. This will have some implications for the practical implementation of multivariable system Iterative Feedback Tuning.

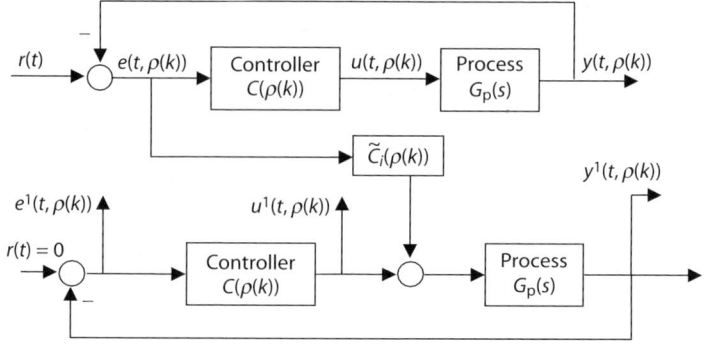

Key
Process, $G_p(s) \in \Re^{m \times m}$
Controller, $C(\rho(k)) \in \Re^{m \times m}$; Controller parameter vector, $\rho \in \Re^{n_c}$
Signal processor, $C_i(\rho(k)) = \dfrac{\partial C(\rho(k))}{\partial \rho_i} \in \Re^{m \times m}; i = 1,\ldots, n_c$

Figure 3.7 Multivariable IFT signal generation.

3.3 The Controller Parameter Cycling Tuning Method

The numerical study for the Iterative Feedback Tuning method showed the importance of using an efficient numerical routine for the search in the negative gradient direction optimisation step. The optimisation framework of a Newton algorithm quite naturally uses the Hessian information; however, a difficulty for the method of Iterative Feedback Tuning is the generation of the Hessian information from closed-loop response data. Thus, what is sought is a more efficient way of performing the gradient and Hessian generation step but retaining the novelty of using a model-free closed-loop system philosophy.

One possible route is to follow a classical gradient computation procedure. This would be based on perturbing the gain vector from $(\rho_1(k),\ldots, \rho_i(k),\ldots, \rho_{n_c}(k))$ to $(\rho_1(k),\ldots, \rho_i(k) + \Delta\rho_i,\ldots, \rho_{n_c}(k))$ and calculating the respective cost functions $J(\rho_1(k),\ldots, \rho_i(k),\ldots, \rho_{n_c}(k)) \in \Re$ and $J(\rho_1(k),\ldots, \rho_i(k) + \Delta\rho_i,\ldots, \rho_{n_c}(k)) \in \Re$, so that numerical differences could be used to calculate an expression for the ith element of the gradient vector as

$$\left\{\dfrac{\partial J}{\partial \rho(k)}\right\}_i = \dfrac{J(\rho_1(k),\ldots, \rho_i(k) + \Delta\rho_i,\ldots, \rho_{n_c}(k)) - J(\rho_1(k),\ldots, \rho_i(k),\ldots, \rho_{n_c}(k))}{\Delta\rho_i}$$

By repeating this procedure some n_c times the full gradient vector

$$\dfrac{\partial J}{\partial \rho(k)} \in \Re^{n_c}$$

can be approximated. Further numerical perturbations can be used to calculate the Hessian information. It would then be possible to use the Newton method to calculate updated controller parameters. However, this straightforward numerical method suffers from the problems of large numbers of gain perturbations and system response generations to calculate the gradient and Hessian information. In the sequel, a new and novel procedure is proposed as a coherent solution to this problem.

3.3.1 Generating the Gradient and Hessian – Some Theory

Consider the unity closed-loop feedback system as shown in Figure 3.1, where the controller is characterised by means of a parameter vector $\rho \in \Re^{n_c}$, for which n_c is the number of controller parameters. If this controller parameter vector is subject to a vector of perturbations $\Delta\rho(t_\Delta) \in \Re^{n_c}$ then Taylor's Theorem gives

$$J(\rho + \Delta\rho(t_\Delta)) = J(\rho) + \Delta\rho(t_\Delta)^T \frac{\partial J}{\partial \rho} + R_2(\xi)$$

and

$$J(\rho + \Delta\rho(t_\Delta)) = J(\rho) + \Delta\rho(t_\Delta)^T \frac{\partial J}{\partial \rho} + \frac{1}{2}\Delta\rho(t_\Delta)^T H(\rho)\Delta\rho(t_\Delta) + R_3(\xi)$$

where

$$\frac{\partial J}{\partial \rho} \in \Re^{n_c}, \quad R_2(\xi) = \tfrac{1}{2}\Delta\rho(t_\Delta)^T H(\xi)\Delta\rho(t_\Delta)$$

and $R_3(\xi)$ is a third-order residual, with

$$H_{ij} = \frac{\partial J}{\partial \rho_i \partial \rho_j}, H \in \Re^{n_c \times n_c} \text{ and } \rho_i < \xi_i < \rho_i + \Delta\rho_i(t_\Delta), i = 1, \ldots, n_c$$

In these expressions the perturbation of the controller parameters is time-varying and this property is used in these Taylor expansions to obtain gradient and Hessian extraction results. Two forms of time-varying controller parameter perturbation are possible:

1. Sine gain perturbations: $\Delta\rho_i(t_\Delta) = \delta_i \sin(n_i\omega_0 t_\Delta), \delta_i \in \Re, i = 1, \ldots, n_c$
2. Cosine gain perturbations: $\Delta\rho_i(t_\Delta) = \delta_i \cos(n_i\omega_0 t_\Delta), \delta_i \in \Re, i = 1, \ldots, n_c$

Trigonometric function orthogonality plays a key role in the results so that along with the perturbation is a gradient extraction signal, and these too can take two forms.

1. Sine gradient extraction signal: $g_{ext}(t) = \sin(n_i\omega_0 t_\Delta)$
2. Cosine gradient extraction signal: $g_{ext}(t) = \cos(n_i\omega_0 t_\Delta)$

Thus the possible combinations that can form the basis of the theoretical result are shown in the table below.

Case	Gain perturbation $\Delta\rho_i(t_\Delta)$	Gradient extraction signal
1	$\Delta\rho_i(t_\Delta) = \delta_i \sin n_i\omega_0 t_\Delta, \delta_i \in \Re$	$g_{ext}(t) = \sin(n_i\omega_0 t_\Delta)$
2	$\Delta\rho_i(t_\Delta) = \delta_i \sin n_i\omega_0 t_\Delta, \delta_i \in \Re$	$g_{ext}(t) = \cos(n_i\omega_0 t_\Delta)$
3	$\Delta\rho_i(t_\Delta) = \delta_i \cos n_i\omega_0 t_\Delta, \delta_i \in \Re$	$g_{ext}(t) = \cos(n_i\omega_0 t_\Delta)$
4	$\Delta\rho_i(t_\Delta) = \delta_i \cos n_i\omega_0 t_\Delta, \delta_i \in \Re$	$g_{ext}(t) = \sin(n_i\omega_0 t_\Delta)$

With the experience of hindsight, it turns out that only Cases 1 and 3 can form the basis of a theoretical result. Further, to achieve the orthogonality which plays a key role in the proof of the theorems it is necessary to define a set of integers that will specify multiples of a fundamental gain perturbation frequency ω_0. These sets of n_c integers are given the following notation:

1. For the sine gain perturbations and sine gradient extraction signal case: $\mathfrak{I}_G^s(n_c)$
2. For the cosine gain perturbations and the cosine gradient extraction signal case: $\mathfrak{I}_G^c(n_c)$

Theorem 3.1: Gradient extraction: sine gain perturbations – sine extraction signals
Consider a time-varying controller gain vector perturbation $\Delta\rho(t_\Delta) \in \mathfrak{R}^{n_c}$, where $\Delta\rho_i(t_\Delta) = \delta_i \sin(n_i \omega_0 t_\Delta)$, $\delta_i \in \mathfrak{R}$, $n_i \in \mathfrak{I}_G^s(n_c)$, $i = 1, \ldots, n_c$ and $T_0 = 2\pi/\omega_0$. Then

$$\int_0^{T_0} J(\rho + \Delta\rho(t_\Delta))\sin(n_i \omega_0 t_\Delta) dt_\Delta = \left(\frac{\delta_i \pi}{\omega_0}\right)\frac{\partial J}{\partial \rho_i} + O(\delta_{max}^2), i = 1, \ldots, n_c$$

where

$$\mathfrak{I}_G^s(n_c) = \{n_i; n_i \neq n_j; i,j \in [1, \ldots, n_c]\} \text{ and } \delta_{max} = \max_{i=1,\ldots,n_c}\{\delta_i\}$$

Proof
From Taylor's Theorem:

$$J(\rho + \Delta\rho(t_\Delta)) = J(\rho) + \Delta\rho(t_\Delta)^T \frac{\partial J}{\partial \rho} + \frac{1}{2}\Delta\rho(t_\Delta)^T H(\xi)\Delta\rho(t_\Delta)$$

where

$$H_{ij} = \frac{\partial J}{\partial \rho_i \partial \rho_j}, H \in \mathfrak{R}^{n_c \times n_c} \text{ and } \rho_i < \xi_i < \rho_i + \Delta\rho_i(t_\Delta), i = 1, \ldots, n_c$$

Thus if the perturbation is made time-varying,

$$J(\rho + \Delta\rho(t_\Delta)) = J(\rho) + \sum_{i=1}^{n_c}\Delta\rho_i(t_\Delta)\frac{\partial J}{\partial \rho_i} + \frac{1}{2}\sum_{i=1}^{n_c}\sum_{j=1}^{n_c}H_{ij}(\xi(t_\Delta))\Delta\rho_i(t_\Delta)\Delta\rho_j(t_\Delta)$$

For $i = 1, \ldots, n_c$, let the sine controller parameter perturbation be given by

$$\Delta\rho_i(t_\Delta) = \delta_i \sin(n_i \omega_0 t_\Delta), \delta_i \in \mathfrak{R}$$

Then

$$J(\rho + \Delta\rho(t_\Delta)) = J(\rho) + \sum_{i=1}^{n_c}\delta_i \frac{\partial J}{\partial \rho_i}\sin(n_i \omega_0 t_\Delta) + \frac{1}{2}\sum_{i=1}^{n_c}\sum_{j=1}^{n_c}\delta_i \delta_j H_{ij}(\xi(t_\Delta))\sin(n_i \omega_0 t_\Delta)\sin(n_j \omega_0 t_\Delta)$$

Using $T_0 = 2\pi/\omega_0$,

$$\int_0^{T_0} J(\rho + \Delta\rho(t_\Delta))\sin(n_i \omega_0 t_\Delta) dt_\Delta = J(\rho)\int_0^{T_0}\sin(n_i \omega_0 t_\Delta) dt_\Delta$$

$$+ \delta_i \frac{\partial J}{\partial \rho_i}\int_0^{T_0}\sin^2(n_i \omega_0 t_\Delta) dt_\Delta + \sum_{\substack{j=1 \\ j \neq i}}^{n_c}\delta_i \frac{\partial J}{\partial \rho_i}\int_0^{T_0}\sin(n_j \omega_0 t_\Delta)\sin(n_i \omega_0 t_\Delta) dt_\Delta$$

$$+ \frac{1}{2}\sum_{i=1}^{n_c}\sum_{j=1}^{n_c}\delta_i \delta_j \int_0^{T_0} H_{ij}(\xi(t_\Delta))\sin^2(n_i \omega_0 t_\Delta)\sin(n_j \omega_0 t_\Delta) dt_\Delta$$

3.3 The Controller Parameter Cycling Tuning Method

The individual terms compute as follows.

(a) $\int_0^{T_0} \sin(n_i\omega_0 t_\Delta)dt_\Delta = \left[\dfrac{-\cos(n_i\omega_0 t_\Delta)}{n_i\omega_0}\right]_0^{T_0} = \dfrac{-\cos(2n_i\pi)}{n_i\omega_0} + \dfrac{1}{n_i\omega_0} = 0$

(b) $\int_0^{T_0} \sin^2(n_i\omega_0 t_\Delta)dt_\Delta = \left[\dfrac{t_\Delta}{2} - \dfrac{\sin(2n_i\omega_0 t_\Delta)}{4n_i\omega_0}\right]_0^{T_0} = \dfrac{\pi}{\omega_0} - \dfrac{\sin(4n_i\pi)}{4n_i\omega_0} = \dfrac{\pi}{\omega_0}$

(c) $\int_0^{T_0} \sin(n_j\omega_0 t_\Delta)\sin(n_i\omega_0 t_\Delta)dt_\Delta = \left[\dfrac{\sin((n_j - n_i)\omega_0 t_\Delta)}{2(n_j - n_i)\omega_0} - \dfrac{\sin((n_j + n_i)\omega_0 t_\Delta)}{2(n_j + n_i)\omega_0}\right]_0^{T_0}$

$= \dfrac{\sin(2(n_j - n_i)\pi)}{2(n_j - n_i)\omega_0} - \dfrac{\sin(2(n_j + n_i)\pi)}{2(n_j + n_i)\omega_0} = 0$

with $n_j^2 \neq n_i^2$.

(d) $\delta_i \delta_j \int_0^{T_0} H_{ij}(\xi(t_\Delta))\sin^2(n_i\omega_0 t_\Delta)\sin(n_j\omega_0 t_\Delta)dt_\Delta < (\delta_{max})^2 B < \infty$

where B is a finite bound. Thus $R_2(\xi)$ is dominated by $(\delta_{max})^2$, where

$$\delta_{max} = \max_{i=1,\dots,n_c}\{\delta_i\}$$

and is written as $O(\delta_{max}^2)$.

Substitution of the above into the integral expression yields

$$\int_0^{T_0} J(\rho + \Delta\rho(t_\Delta))\sin(n_i\omega_0 t_\Delta)dt_\Delta = \left(\dfrac{\delta_i \pi}{\omega_0}\right)\dfrac{\partial J}{\partial \rho_i} + O(\delta_{max}^2)$$

❏

The theorem requires a suitable selection of the perturbation integration period T_0. Then the fundamental frequency is calculated as $\omega_0 = 2\pi/T_0$. The proof also leads to the definition of the integers $\mathfrak{J}_G^s(n_c)$ with $\mathfrak{J}_G^s(n_c) = \{n_i; n_i \neq n_j; i, j \in [1, \dots, n_c]\}$. As already stated, a second version of the theorem is possible with cosine controller parameter perturbations $\Delta\rho_i(t_\Delta) = \delta_i \cos n_i\omega_0 t_\Delta$, $\delta_i \in \mathfrak{R}$, $\mathfrak{J}_G^c(n_c)$, $i = 1, \dots, n_c$ and a cosine gradient extraction signal.

Trigonometric function orthogonality also plays a key role in the result for the extraction of the Hessian information. Hence, as with the gradient extraction result, it is possible to construct a table of possible extraction signal sets for the Hessian.

Cases	Gain perturbation	Gradient extraction signal	Sub-case	Hessian extraction signal
1	$\Delta\rho_i(t_\Delta) = \delta_i \sin n_i\omega_0 t_\Delta$, $\delta_i \in \mathfrak{R}$	$g_{ext}(t) = \sin n_i\omega_0 t_\Delta$	1A	$h_{ext}(t) = \sin n_i\omega_0 t_\Delta$
			1B	$h_{ext}(t) = \cos n_i\omega_0 t_\Delta$
2	$\Delta\rho_i(t_\Delta) = \delta_i \cos n_i\omega_0 t_\Delta$, $\delta_i \in \mathfrak{R}$	$g_{ext}(t) = \cos n_i\omega_0 t_\Delta$	2A	$h_{ext}(t) = \cos n_i\omega_0 t_\Delta$
			2B	$h_{ext}(t) = \sin n_i\omega_0 t_\Delta$

Once more, the benefit of hindsight shows that only Case 1 with Sub-case 1B and Case 2 with Sub-case 2A can form the basis of a theoretical result. To achieve the orthogonality which plays a key role in the proof of the Hessian extraction theorems it is necessary to define a second set of integers that will specify multiples of a fundamental gain perturbation frequency ω_0. The Hessian is symmetric and has $n_c(n_c-1)/2$ unknown elements; consequently introduce some notation for two cases of integer sets:

1. For an orthogonality based on a sine perturbation of the controller parameters and a cosine extraction of the Hessian, introduce a set of $n_c(n_c-1)/2$ integers, denoted $\mathfrak{J}_H^{s-c}(n_c)$.

2. For an orthogonality based on a cosine perturbation of the controller parameters and a cosine extraction of the Hessian, introduce a set of $n_c(n_c-1)/2$ integers, denoted $\mathfrak{J}_H^{c-c}(n_c)$.

Theorem 3.2: Hessian extraction – sine gain perturbation and cosine extraction
Consider a time-varying controller gain vector perturbation $\Delta \rho_i(t_\Delta) = \delta_i \sin(n_i \omega_0 t_\Delta)$ with $\delta_i \in \mathfrak{R}$, $i=1,\ldots,n_c$, and where the set of integers $\mathfrak{J}_G^s(n_c)$ have been used for gradient extraction with $T_0 = 2\pi/\omega_0$.
Define a set of $n_c(n_c-1)/2$ integers, denoted $\mathfrak{J}_H^{s-c}(n_c)$ such that

(a) $n_{ij} = n_i + n_j$ with $n_i, n_j \in \mathfrak{J}_G^s(n_c)$

(b) $n_{ij} \neq n_{i_1 j_1}$

(c) $n_{ij} \neq n_{i_1} - n_{j_1}$ for all pairs $(i,j), (i_1, j_1)$ where $i=1,\ldots,n_c$; $j=1,\ldots,n_c$ and $i_1 = 1,\ldots,n_c$; $j_1 = 1,\ldots,n_c$

Then

$$\int_0^{T_0} J(\rho + \Delta\rho(t_\Delta))\cos(n_{ij}\omega_0 t_\Delta) dt_\Delta = \left(-\frac{\delta_i \delta_j \pi f(i,j)}{4\omega_0}\right)\frac{\partial^2 J}{\partial \rho_i \partial \rho_j} + O(\delta_{max}^3)$$

where

$$f(i,j) = \begin{cases} 1 & i=j \\ 2 & \text{otherwise} \end{cases}$$

and

$$\delta_{max} = \max_{i=1,\ldots,n_c} \{\delta_i\}$$

Proof
From Taylor's Theorem

$$J(\rho + \Delta\rho(t_\Delta)) = J(\rho) + \Delta\rho(t_\Delta)^T \frac{\partial J}{\partial \rho} + \frac{1}{2}\Delta\rho(t_\Delta)^T H(\rho)\Delta\rho(t_\Delta) + R_3(\xi)$$

where $\rho_i < \xi_i < \rho_i + \Delta\rho_i(t_\Delta), i=1,\ldots,n_c$.
The time-varying controller gain vector sine perturbation $\Delta\rho(t_\Delta) \in \mathfrak{R}^{n_c}$ is given by

$$\Delta\rho_i(t_\Delta) = \delta_i \sin(n_i \omega_0 t_\Delta)$$

with $\delta_i \in \mathfrak{R}$ and $n_i \in \mathfrak{J}_G^s(n_c)$, $i=1,\ldots,n_c$. Then

$$J(\rho + \Delta\rho(t_\Delta)) = J(\rho) + \sum_{i=1}^{n_c} \delta_i \frac{\partial J}{\partial \rho_i}\sin(n_i\omega_0 t_\Delta) + \frac{1}{2}\sum_{i=1}^{n_c}\sum_{j=1}^{n_c}\delta_i\delta_j H_{ij}(\rho)\sin(n_i\omega_0 t_\Delta)\sin(n_j\omega_0 t_\Delta) + R_3(\xi)$$

3.3 The Controller Parameter Cycling Tuning Method

To find the (i_1, j_1)th element of $H_{ij}(\rho)$, introduce the frequency multiple $n_{i_1 j_1}$ where $i_1 = 1, \ldots, n_c$; $j_1 = 1, \ldots, n_c$. Hence using $T_0 = 2\pi/\omega_0$

$$\int_0^{T_0} J(\rho + \Delta\rho(t_\Delta))\cos n_{i_1 j_1} \omega_0 t_\Delta dt_\Delta = J(\rho) \int_0^{T_0} \cos(n_{i_1 j_1} \omega_0 t_\Delta) dt_\Delta$$

$$+ \sum_{i=1}^{n_c} \delta_i \frac{\partial J}{\partial \rho_i} \int_0^{T_0} \sin(n_i \omega_0 t_\Delta) \cos(n_{i_1 j_1} \omega_0 t_\Delta) dt_\Delta$$

$$+ \frac{1}{2} \sum_{i=1}^{n_c} \sum_{j=1}^{n_c} \delta_i \delta_j H_{ij}(\rho) \int_0^{T_0} \sin(n_i \omega_0 t_\Delta) \sin(n_j \omega_0 t_\Delta) \cos(n_{i_1 j_1} \omega_0 t_\Delta) dt_\Delta$$

$$+ \int_0^{T_0} R_3(\xi(t_\Delta)) \cos(n_{i_1 j_1} \omega_0 t_\Delta) dt_\Delta$$

The individual terms are resolved as follows,

(a) $\displaystyle\int_0^{T_0} \cos(n_{i_1 j_1} \omega_0 t_\Delta) dt_\Delta = \left[-\frac{\sin(n_{i_1 j_1} \omega_0 t_\Delta)}{n_{i_1 j_1} \omega_0} \right]_0^{T_0} = 0$

(b) $\displaystyle\int_0^{T_0} \sin(n_i \omega_0 t_\Delta) \cos(n_{i_1 j_1} \omega_0 t_\Delta) dt_\Delta = -\frac{1}{2} \left[\frac{\cos((n_i - n_{i_1 j_1})\omega_0 t_\Delta)}{(n_i - n_{i_1 j_1})\omega_0} + \frac{\cos((n_i + n_{i_1 j_1})\omega_0 t_\Delta)}{(n_i + n_{i_1 j_1})\omega_0} \right]_0^{T_0}$

when $n_i \neq n_{i_1 j_1}$

$$\int_0^{T_0} \sin(n_i \omega_0 t_\Delta) \cos(n_{i_1 j_1} \omega_0 t_\Delta) dt_\Delta = -\frac{1}{2} \left[\frac{\cos((n_i - n_{i_1 j_1})2\pi)}{(n_i - n_{i_1 j_1})\omega_0} + \frac{\cos((n_i + n_{i_1 j_1})2\pi)}{(n_i + n_{i_1 j_1})\omega_0} \right]$$

$$+ \frac{1}{2} \left[\frac{1}{(n_i - n_{i_1 j_1})\omega_0} + \frac{1}{(n_i + n_{i_1 j_1})\omega_0} \right] = 0$$

when $n_i = n_{i_1 j_1}$

$$\int_0^{T_0} \sin(n_i \omega_0 t_\Delta) \cos(n_{i_1 j_1} \omega_0 t_\Delta) dt_\Delta = \int_0^{T_0} \sin(n_i \omega_0 t_\Delta) \cos(n_i \omega_0 t_\Delta) dt_\Delta$$

$$= \frac{1}{2} \int_0^{T_0} \sin(2 n_i \omega_0 t_\Delta) dt_\Delta = -\frac{1}{2} \left[\frac{\cos(2 n_i \omega_0 t_\Delta)}{2 n_i \omega_0} \right]_0^{T_0}$$

$$= \frac{1}{4 n_i \omega_0} [-\cos 4 n_i \pi + \cos(0)] = 0$$

(c) $\displaystyle I_{ij} = \int_0^{T_0} \sin(n_i \omega_0 t_\Delta \delta_i) \sin(n_j \omega_0 t_\Delta) \cos(n_{i_1 j_1} \omega_0 t_\Delta) dt_\Delta$

$$= \int_0^{T_0} \frac{1}{2} [\cos((n_i - n_j)\omega_0 t_\Delta) - \cos((n_i + n_j)\omega_0 t_\Delta)] \cos(n_{i_1 j_1} \omega_0 t_\Delta) dt_\Delta$$

Assuming that $n_{i_1 j_1} \neq (n_i + n_j)$ and $n_{i_1 j_1} \neq (n_i - n_j)$, then

$$I_{ij} = \frac{1}{2}\left[\frac{\sin((n_i - n_j - n_{i_1 j_1})\omega_0 t_\Delta)}{2(n_i - n_j - n_{i_1 j_1})\omega_0} + \frac{\sin((n_i - n_j + n_{i_1 j_1})\omega_0 t_\Delta)}{2(n_i - n_j + n_{i_1 j_1})\omega_0}\right]_0^{T_0}$$

$$-\frac{1}{2}\left[\frac{\sin((n_i + n_j - n_{i_1 j_1})\omega_0 t_\Delta)}{2(n_i + n_j - n_{i_1 j_1})\omega_0} + \frac{\sin((n_i + n_j + n_{i_1 j_1})\omega_0 t_\Delta)}{2(n_i + n_j + n_{i_1 j_1})\omega_0}\right]_0^{T_0} = 0$$

This covers all of the non-selective cases.

Now choose $n_{i_1 j_1} = (n_i + n_j)$ with $n_{i_1 j_1} \neq (n_i - n_j)$:

$$I_{ij} = \frac{1}{2}\int_0^{T_0} \cos((n_i - n_j)\omega_0 t_\Delta)\cos(n_{i_1 j_1}\omega_0 t_\Delta)dt_\Delta - \frac{1}{2}\int_0^{T_0}\cos^2((n_i + n_j)\omega_0 t_\Delta)dt_\Delta$$

The first integral is zero since $n_{i_1 j_1} \neq (n_i - n_j)$; hence,

$$I_{ij} = -\frac{1}{2}\left[\frac{t_\Delta}{2} + \frac{\sin(2(n_i + n_j)\omega_0 t_\Delta)}{4(n_i + n_j)\omega_0}\right]_0^{T_0} = -\frac{\pi}{2\omega_0}$$

(d) $\int_0^{T_0} R_3(\xi(t_\Delta))\cos(n_{i_1 j_1}\omega_0 t_\Delta)dt_\Delta < B(\delta_{max})^3 < \infty$ where B is a finite bound.

Thus, $R_3(\xi)$ is dominated by $(\delta_{max})^3$ where

$$\delta_{max} = \max_{i=1,\ldots,n_c}\{\delta_i\}$$

and is written as $O(\delta_{max}^3)$.

Substitution of the above resolution for the terms of the integral expression gives

$$\int_0^{T_0} J(\rho + \Delta\rho(t_\Delta))\cos(n_{ij}\omega_0 t_\Delta)dt_\Delta = \left(-\frac{\delta_i \delta_j \pi f(i,j)}{4\omega_0}\right)\frac{\partial^2 J}{\partial \rho_i \partial \rho_j} + O(\delta_{max}^3)$$

where

$$f(i,j) = \begin{cases} 1 & i = j \\ 2 & \text{otherwise} \end{cases}$$

□

The above Hessian extraction theorem specifies the second orthogonality index set $\mathfrak{I}_H^{s\sim c}(n_c)$. This second set is constructed assuming that the gradient extraction used the orthogonality index set $\mathfrak{I}_G^s(n_c)$ and defines a set of $n_c(n_c - 1)/2$ integers, denoted $\mathfrak{I}_H^{s\sim c}(n_c)$ such that

(a) $n_{ij} \notin \mathfrak{I}_G^s(n_c)$

(b) $n_{ij} = n_i + n_j$ with $n_i, n_j \in \mathfrak{I}_G^s(n_c)$

(c) $n_{ij} \neq n_{i_1 j_1}$

(d) $n_{ij} \neq n_{i_1} - n_{j_1}$ for all pairs $(i,j), (i_1, j_1)$, where $i = 1, \ldots, n_c$; $j = 1, \ldots, n_c$ and $i_1 = 1, \ldots, n_c$; $j_1 = 1, \ldots, n_c$.

A second version of these results would use gradient extraction using cosine perturbations and the orthogonality integer set $\mathfrak{I}_G^c(n_c)$. This can be used with Hessian extraction based on cosines and requires an additional orthogonality set of $n_c(n_c-1)/2$ integers, denoted $\mathfrak{I}_H^{c \sim c}(n_c)$, where $n_{ij} \in \mathfrak{I}_H^{c \sim c}(n_c)$ if

(a) $n_{ij} \notin \mathfrak{I}_G^c(n_c)$

(b) $n_{ij} = n_i + n_j$

(c) $n_{ij} \neq n_{i_1 j_1}$

(d) $n_{ij} \neq n_{i_1} - n_{j_1}$ for all pairs (i,j), (i_1, j_1), where $i = 1, \ldots, n_c$; $j = 1, \ldots, n_c$ and $i_1 = 1, \ldots, n_c$; $j_1 = 1, \ldots, n_c$

Apart from these theoretical issues, other numerical considerations are necessary to convert this theory into a working algorithm The necessary discussion is given in the next section.

3.3.2 Issues for a Controller Parameter Cycling Algorithm

A comprehensive and unified theory was presented for cost function gradient and Hessian identification. The more serious task is to turn this theory into a working algorithm and to initiate the numerical investigations to make the algorithm efficient for on-line system use.

Numerical Selection of T_f, T_0, ω_0 for the Algorithm

In the calculation of the cost function, it is not practical to allow the on-line experiment to continue for an infinitely long time. Thus the integration period is fixed so that T_f is five times the dominant time constant in the process, where it is assumed that the system has settled to within 1% of the steady state value after this time has elapsed. From this, $T_0 = T_f$ and this enables the basic perturbation frequency to be determined as $\omega_0 = 2\pi/T_0$.

Selection of Gain Perturbation Amplitudes

The controller parameter perturbations are given by $\Delta \rho_i(t_\Delta) = \delta_i \sin(n_i \omega_0 t_\Delta)$, $\delta_i \in \mathfrak{I}$, $i = 1, \ldots, n_c$ and within this it is necessary to select the n_c perturbation amplitudes $\delta_i \in \mathfrak{R}$. These amplitudes will be system-dependent. However, there are several considerations here: firstly, the perturbation should not cause closed-loop instability, and secondly, the perturbation size is directly linked to the accuracy of gradient and Hessian extraction as given in the theorems. Finally, the amplitude size can be chosen to minimise the disturbance to the system outputs. This issue will be of particular concern to production personnel when conducting on-line experiments.

Construction of the Orthogonality Integer Sets $\mathfrak{I}_G^s(n_c)$, $\mathfrak{I}_H^{s \sim c}(n_c)$, $\mathfrak{I}_G^c(n_c)$, $\mathfrak{I}_H^{c \sim c}(n_c)$

An integer multiple of ω_0 must be assigned to each controller parameter perturbation $\Delta \rho_i(t_\Delta)$. The choice of frequency multiples depends on the choice of whether a sine or cosine controller parameter perturbation function is chosen. If only the gradient is to be extracted, then the rules for the construction of $\mathfrak{I}_G^s(n_c)$ or $\mathfrak{I}_G^c(n_c)$ are straightforward. If, however, the Hessian is also to be extracted then the construction of the pair $\mathfrak{I}_G^s(n_c)$, $\mathfrak{I}_H^{s \sim c}(n_c)$ or the pair $\mathfrak{I}_G^c(n_c)$, $\mathfrak{I}_H^{c \sim c}(n_c)$ is more involved.

Table 3.4 shows the feasible integer multiples of frequency ω_0 for the case of a single PID controller with three gains, so that $n_c = 3$. Thus, for a sine perturbation and a cosine extraction function for the Hessian, the duplication of a frequency between the $\mathfrak{I}_G^s(n_c)$ and the set $\mathfrak{I}_H^{s \sim c}(n_c)$ is permissible. This can be seen, for example, in Table 3.4, where $n_2 = n_{11} = 4$ in $\mathfrak{I}_H^{s \sim c}(n_c)$.

The choice of the controller perturbation frequency multiples cannot be made in isolation from the choice of the Hessian extraction frequency multiples. Algorithm 3.3 details the steps required in deriving a set of gain and Hessian extraction frequencies for a sine gain parameter perturbation, whilst Algorithm 3.4 gives similar details for a cosine gain perturbation.

Table 3.4 Feasible frequency multiples where $n_c = 3$.

$\Im_G^s(n_c)$	$n_1 = 2$	$n_2 = 4$	$n_3 = 5$
$\Im_H^{s\sim c}(n_c)$	$n_{11} = 4$	$n_{12} = 6$	$n_{13} = 7$
	–	$n_{22} = 8$	$n_{23} = 9$
	–	–	$n_{33} = 10$
$\Im_G^c(n_c)$	$n_1 = 1$	$n_2 = 5$	$n_3 = 8$
$\Im_H^{c\sim c}(n_c)$	$n_{11} = 2$	$n_{12} = 6$	$n_{13} = 9$
	–	$n_{22} = 10$	$n_{23} = 13$
	–	–	$n_{33} = 16$

Algorithm 3.3: Selection of frequencies for Hessian element extraction with sine controller parameter perturbation

Step 1 *Initialisation*

Choose a set of gain perturbation integers

$$\Im_G(n_c) = \{I_1, I_2, \ldots, I_{n_c}\} \text{ such that } I_1 < I_2 < \ldots < I_{n_c}$$

and n_c is the number of controller parameters.
Set S_Flag = 0, D_Flag = 0

Step 2 *Generation of sum terms*

Construct the $n_c \times n_c$ matrix, S, such that $S(i,j) = I_i + I_j$
where $i = 1, \ldots, n_c, j = 1, \ldots, n_c$

Step 3 *Generation of difference terms*

Construct the $n_c \times n_c$ matrix, D, such that $D(i,j) = |I_i - I_j|$
where $i = 1, \ldots, n_c, j = 1, \ldots, n_c$

Step 4 *Test for extraction frequency conflict*

Step 4a

For each of the leading diagonal elements of S,
    If $S(i,i) = S(i,j)$ then S_Flag = 1
    where $i = 1, \ldots, n_c, j = 1, \ldots, n_c$ and $i \ne j$
    If $S(i,i) = D(i,j)$ then S_Flag = 1
    where $i = 1, \ldots, n_c, j = 1, \ldots, n_c$ and $i \ne j$
If S_Flag = 1 then goto Step 1
Else record the frequency given by $S(i,i)$

Step 4b

For each of the purely upper triangular elements of S,
    If $S(i,i) = S(i_1, j_1)$
    Then If $D(i,j) \ne S(i_1, j_1)$
    Then D_Flag = 1
    Else (S_Flag = 1 and D_Flag = 0)
    where $i = 1, \ldots, n_c, j = 1, \ldots, n_c, i_1 = 1, \ldots, n_c, j_1 = 1, \ldots, n_c$

If S_Flag = 1 then go to Step 1
Else record the frequency given by $S(i,j)$
If D_Flag = 1 then record the frequency given by $S(i,j)$
Algorithm end

Remarks 3.2
1. The generation of the controller perturbation frequency multiples can be performed using nested For loops in a coded version of the algorithm.

2. The algorithm can generate different sets of gain perturbation and Hessian extraction frequencies.

3. It can be seen from Algorithm 3.3 that the selection of the gain perturbation frequencies and the frequencies that are used to extract the Hessian elements cannot be considered in isolation.

Algorithm 3.4: Selection of frequencies for Hessian element extraction with cosine controller parameter perturbation

Step 1 *Initialisation*

Choose a set of gain perturbation integers

$$\mathfrak{I}_G(n_c) = \{I_1, I_2, \ldots, I_{n_c}\} \text{ such that } I_1 < I_2 < \ldots < I_{n_c}$$

and n_c is the number of controller parameters.
Set D_Flag = 0, P_Flag = 0, S_Flag = 0

Step 2 *Generation of sum terms*

Construct the $n_c \times n_c$ matrix, S, such that $S(i,j) = I_i + I_j$
where $i = 1, \ldots, n_c, j = 1, \ldots, n_c$

Step 3 *Generation of difference terms*

Construct the $n_c \times n_c$ matrix, D, such that $D(i,j) = I_i + I_j$
where $i = 1, \ldots, n_c, j = 1, \ldots, n_c$

Step 4 *Test for extraction frequency conflict*

For each of the gain perturbation frequencies $\mathfrak{I}_G(n_c) = \{I_1, I_2, \ldots, I_{n_c}\}$
  If $I_i = S(i,j)$ or $I_i = D(i,j)$ then P_Flag = 1
    where $i = 1, \ldots, n_c, j = 1, \ldots, n_c$
If P_Flag = 1 then got to Step 1

Step 5 *Test for extraction frequency conflict*

Step 5a

For each of the leading diagonal elements of S,
  If $S(i,i) = S(i,j)$ then S_Flag = 1
    where $i = 1, \ldots, n_c, j = 1, \ldots, n_c$ and $i \neq j$
  If $S(i,i) = D(i,j)$ then S_Flag = 1
    where $i = 1, \ldots, n_c, j = 1, \ldots, n_c$ and $i \neq j$
If S_Flag = 1 then goto Step 1
Else record the frequency given by $S(i,i)$

Step 5b

For each of the purely upper triangular elements of S,
  If $S(i,i) = S(i_1, j_1)$
    Then If $D(i,j) \neq S(i_1, j_1)$
      Then D_Flag = 1
      Else (S_Flag = 1 and D_Flag = 0)
  where $i = 1, \ldots, n_c, j = 1, \ldots, n_c, i_1 = 1, \ldots, n_c, j_1 = 1, \ldots, n_c$

If S_Flag = 1 then go to Step 1
Else record the frequency given by $S(i,j)$
If D_Flag = 1 then record the frequency given by $S(i,j)$
Algorithm end

Selection of Integration Formula

The time-scales over which the cost function is calculated and the parameter perturbation signals evolve are separate. The cost function is calculated in real time, whereas the parameter perturbation signal uses the time-domain, t_Δ. Standard numerical integration formulae are used to compute the two extraction integrals:

$$\int_0^{T_0} J(\rho + \Delta\rho(t_\Delta))\sin(n_i\omega_0 t_\Delta)dt_\Delta \quad \text{and} \quad \int_0^{T_0} J(\rho + \Delta\rho(t_\Delta))\cos(n_i\omega_0 t_\Delta)dt_\Delta$$

These use the discrete time where $t\Delta = kT$, $k = 0, 1, ..., N$ and T is the integration step size or sampling interval. Clearly, step size T has to be chosen to achieve maximal accuracy from the integration method and provide sufficient resolution for the maximum frequency of the perturbation and extraction signals.

Exploitation of Symmetry Properties to Reduce Computation Requirements

From the proof of Theorem 3.1 recall that with a time-varying parameter perturbation, $\Delta\rho(t_\Delta) = \delta_i \sin(n_i\omega_0 t_\Delta)$, the perturbed cost function is given by

$$J(\rho + \Delta\rho(t_\Delta)) = J(\rho) + \sum_{i=1}^{n_c} \delta_i \frac{\partial J}{\partial \rho_i}\sin(n_i\omega_0 t_\Delta) + \frac{1}{2}\sum_{i=1}^{n_c}\sum_{j=1}^{n_c}\delta_i\delta_j H_{ij}(\xi(t_\Delta))\sin(n_i\omega_0 t_\Delta)\sin(n_j\omega_0 t_\Delta)$$

Thus, it can be seen that the cost is composed of a sum of constant terms and sinusoidal terms. Since a sinusoidal function possesses odd symmetry, the cost function $J(\rho + \Delta\rho(t_\Delta))$ will also have odd symmetry. If the cost is determined over half of the time period T_0, then by exploiting the odd symmetry property the entire time-varying cost can be constructed. Similarly, if the parameter perturbation is a cosine function then it can be shown that the resulting cost function will possess even symmetry. Thus only half the time-varying cost function is required to be determined for either sine or cosine controller parameter perturbation. As an illustration of this principle, the time-varying cost function relating to the third iteration of the case study example discussed later in the chapter is shown in Figure 3.8. It can be seen in Figure 3.8 that the axis of symmetry is about a vertical line drawn at the centre of the x-axis, thus, only the first 10 s of the time-varying cost function is required to be determined since symmetry can be employed to generate the remaining part of the cost function.

Reaching the Region of Convergence

It is known that Newton type algorithms are not globally convergent and if the algorithm is initialised outside the region of convergence the controller parameter updates are chosen to move in the direction of the convergence region. Similarly, if the Hessian estimate is negative definite then this cannot be used and additional steps are used for the parameter updates until a value is reached where the Hessian estimate becomes positive definite. This has been discussed by Ljung (1987) and involves replacing the Hessian by

$$H_{LM} = \left(\frac{\partial J}{\partial \rho}\right)\left(\frac{\partial J}{\partial \rho}\right)^T + \alpha I \quad \text{where} \quad \alpha > 0$$

This is known as the Levenberg–Marquardt procedure. When the Hessian becomes positive definite, the Newton method reverts to using the Hessian estimate in the parameter updates.

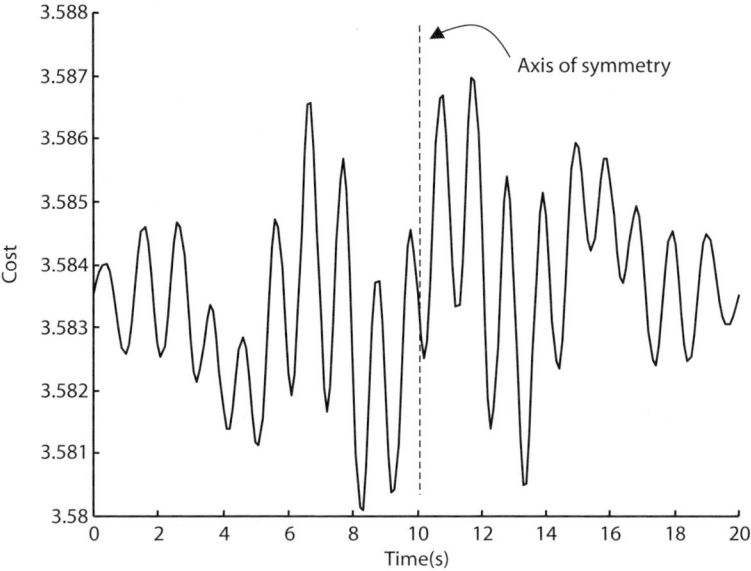

Figure 3.8 Time-varying cost function showing odd symmetry.

3.3.3 The Controller Parameter Cycling Algorithm

To create a new controller parameter cycling method, the framework of the optimisation Algorithm 3.1 is used with the estimates of the cost function gradient and Hessian from Theorems 3.1 and 3.2. In an application, the controller structure will be fixed *a priori*, and can be used to define the integer sets $\Im_G^s(n_c)$, $\Im_H^{s\sim c}(n_c)$ or $\Im_G^c(n_c)$, $\Im_H^{c\sim c}(n_c)$. The algorithm finds estimates of both the gradient and the Hessian and uses these to determine the next set of controller parameters whilst minimising the cost function. The algorithm is given as follows.

Algorithm 3.5: Optimisation by controller parameter cycling

Step 1   Initialisation and setup

Choose $\rho \in \Re^{n_c}$

Choose cost weighting $\lambda > 0$ and cost time interval T_f

Set $T_0 = T_f$ and compute $\omega_0 = 2\pi / T_0$

Choose perturbation sizes $\{\delta_i, i = 1, \ldots, n_c\}$

Find sets $\Im_G^s(n_c)$, $\Im_H^{s\sim c}(n_c)$, $\Im_G^c(n_c)$, $\Im_H^{c\sim c}(n_c)$ as appropriate

Set value of N; set $T = T_0 / N$

Choose convergence tolerance ε

Set loop counter $k = 0$; choose $\rho(k) \in \Re^{n_c}$

Step 2   Gradient and Hessian calculation

Calculate gradient $\dfrac{\partial J}{\partial \rho}(k) \in \Re^{n_c}$ (Theorem 3.1)

Calculate Hessian $H_{ij} = \dfrac{\partial^2 J}{\partial \rho_i \partial \rho_j}(k)$, $H \in \Re^{n_c \times n_c}$ (Theorem 3.2)

If $\left\| \dfrac{\partial J}{\partial \rho}(k) \right\| < \varepsilon$ and $H(k) > 0 \in \Re^{n_c \times n_c}$ then stop

Step 3 Update calculation
Select or calculate the update step size γ_k

If $H(k) > 0 \in \mathfrak{R}^{n_c \times n_c}$ compute $\rho(k+1) = \rho(k) - \gamma_k [H(k)]^{-1} \dfrac{\partial J}{\partial \rho}(k)$

Else compute $\rho(k+1) = \rho(k) - \gamma_k \left[\left(\dfrac{\partial J}{\partial \rho}\right)\left(\dfrac{\partial J}{\partial \rho}\right)^T + \alpha_k I \right]^{-1} \dfrac{\partial J}{\partial \rho}(k)$

Set $k = k + 1$ and goto Step 2
Algorithm end

3.3.4 Case Study – Multivariable Decentralised Control

An application of the method to a two-input, two-output system controlled by a decentralised PID controller system is reported in the following.

Multivariable Process and Decentralised Control

The system to be controlled is considered typical of those found in process industries and the system transfer function matrix is (Zhuang and Atherton, 1994)

$$G_p(s) = \dfrac{1}{d(s)} \begin{bmatrix} 1.5s + 1 & 0.15s + 0.2 \\ 0.45s + 0.6 & 0.96s + 0.8 \end{bmatrix}$$

where $d(s) = 2s^4 + 8s^3 + 10.5s^2 + 5.5s + 1$.

Two PID controllers in a decentralised structure are to be tuned. The decentralised structure is shown in Figure 3.9. The PID controllers are of the parallel type and are given by

$$G_{cii}(s) = k_{Pii} + \dfrac{k_{Iii}}{s} + k_{Dii}s, \; i = 1, 2$$

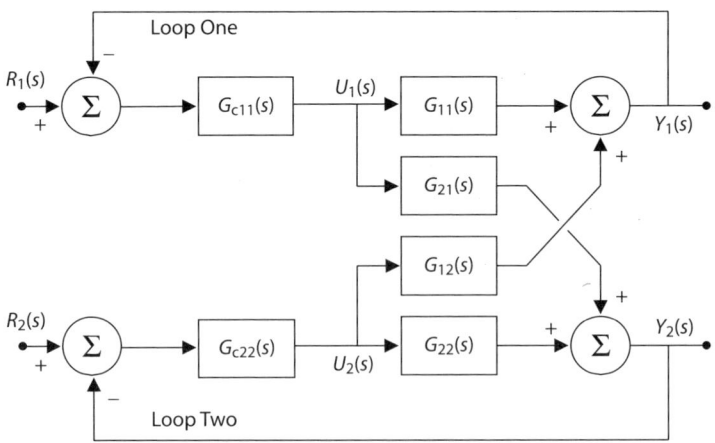

Figure 3.9 Decentralised control structure with two PID controllers.

Thus the number of controller parameters to be tuned is six and hence $n_c = 6$. The fixed structure controllers $G_{cii}(s)$, $i = 1, 2$ are to be tuned such that the LQ cost function to be minimised is given by

$$J = \int_0^{T_f} \{e^T(t)e(t) + u^T(t)\Lambda u(t)\}\,dt$$

Algorithm Setup

In this case, unit weightings are given to the error terms, and equal weighting is given to each of the control terms using $\Lambda = \text{diag}\{0.01, 0.01\}$. The initial PID controllers were derived using a relay experiment and Ziegler–Nichols rules. During the runs of the tuning Algorithm 3.5, the controller parameter perturbation sizes were set as $\delta_i = 0.001$; $i = 1, \ldots, 6$. The cost function integration period was chosen to be $T_f = 20$; hence $T_0 = T_f = 20$ and $\omega_0 = 0.1\pi$ rad s^{-1}.

Algorithm 3.3 was used to develop the Hessian extraction frequencies for a sinusoidal controller parameter perturbation using the integer set $\mathfrak{S}_G^s(n_c) = \{1,4,5,16,19,20\}$. The results from the use of the algorithm are given in Table 3.5.

Table 3.5 Hessian extraction frequency multiples $n_c = 6$.

$\mathfrak{S}_G^s(n_c)$	$n_1 = 1$	$n_2 = 4$	$n_3 = 5$	$n_4 = 16$	$n_5 = 19$	$n_6 = 20$
$\mathfrak{S}_H^{s \sim c}(n_c)$	$n_{11} = 2$	$n_{12} = 5$	$n_{13} = 6$	$n_{14} = 17$	$n_{15} = 18$	$n_{16} = 19$
	–	$n_{22} = 8$	$n_{23} = 9$	$n_{24} = 12$	$n_{25} = 23$	$n_{26} = 16$
	–	–	$n_{33} = 10$	$n_{34} = 11$	$n_{35} = 14$	$n_{36} = 25$
	–	–	–	$n_{44} = 32$	$n_{45} = 35$	$n_{46} = 36$
	–	–	–	–	$n_{55} = 38$	$n_{56} = 39$
	–	–	–	–	–	$n_{66} = 40$

Algorithm Performance

The simulation was carried out using MATLAB/SIMULINK. The gain perturbation and the calculation of the cost function was implemented using SIMULINK, while the estimation of the gradient, Hessian and updated controller parameters was carried out using MATLAB. Sinusoidal perturbation of the controller parameters was used in this example.

Cost Function Evolution

The evolution of the cost function $J(\rho + \Delta\rho(t_\Delta))$ over the period $t_\Delta = T_0$ is shown in Figure 3.10. This presents more evidence that the cost exhibits odd symmetry as predicted since sinusoidal excitation of the controller parameters was used. Figure 3.10 was generated from the third iteration of Algorithm 3.5. The graphs of the cost function $J(\rho + \Delta\rho(t_\Delta))$ for the subsequent iterations of the algorithm all have a similar form and show odd symmetry.

Algorithm Iteration Progress

A plot of $\|\partial J/\partial \rho\|_2$ versus algorithm iteration index is shown in Figures 3.11 whilst the plot of the cost function values $J(\rho(k))$ for the algorithm iterations is shown in Figure 3.12. As can be seen from these plots, progress towards convergence was ultimately successful. However, for the first three iterations of the algorithm the Hessian estimate was negative definite and the Hessian was replaced by the Levenberg–Marquardt procedure with $\alpha I = 0.01 I_6$. At the fourth iteration it was found that the

138 On-line Model-Free Methods

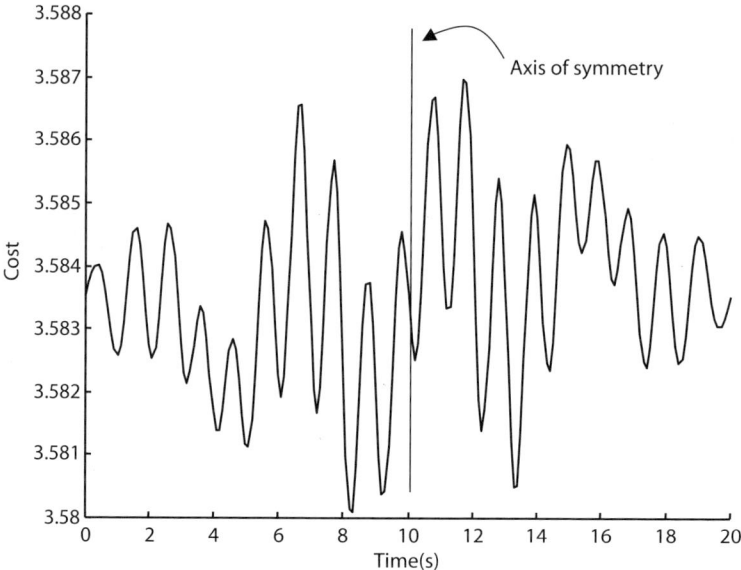

Figure 3.10 Evolution of the cost function $J(\rho + \Delta\rho(t_\Delta))$.

algorithm returned negative values for the updated integral gain parameters for both controllers. Consequently the algorithm was adjusted using $\alpha I = 0.1 I_6$. After two further iterations the Hessian estimate was positive definite and was used in the Newton update. Since the Hessian estimate was corrupted by noise, its use had to be conservative. To do this the controller parameter update size was limited using $\gamma_k \in (0,1)$. For the next two iterations, $\gamma_k = 0.1$ and the values of $\|\partial J/\partial \rho\|_2$ and cost function value decreased. For iterations 10 and 11, $\gamma_k = 1.0$. After iteration 11, the cost function value and $\|\partial J/\partial \rho\|_2$ increased. The step size was reset at $\gamma_k = 0.1$ and a new iteration 11 performed where the cost function value and $\|\partial J/\partial \rho\|_2$ decreased; the step size $\gamma_k = 0.1$ was retained for the remaining iterations. From Figure 3.11 the evolution of $\|\partial J/\partial \rho\|_2$ shows that the minimum of the cost function is found after approximately 15 iterations of the algorithm when $\|\partial J/\partial \rho\|_2 < 10^{-3}$. Thus the controller parameters at iteration 15 give the minimum value of the cost function. In order to verify that further iterations of the algorithm do attain a minimum, a large number of additional iterations were performed. It can be seen from Figures 3.11 and 3.12 that the cost function values tend to an asymptotic value and that $\|\partial J/\partial \rho\|_2$ limits at zero.

Convergence of the Controller Parameters
The evolution of the controller parameters is shown in Figures 3.13–3.15 inclusive.

Convergence of the proportional controller parameters k_{P11}, k_{P22}
As can be seen from Figure 3.13, updates of the controller parameter k_{P22} tend to wander during the first ten or so iterations of the algorithm. However, this did not cause any problems with the stability of the closed-loop system and as can be seen, after approximately 10 iterations of the algorithm the movement in value of k_{P22} became less pronounced.

Convergence of the integral controller parameters k_{I11}, k_{I22}
From Figure 3.14 it can be seen that for both controllers the evolution of the integral gains k_{I11} k_{I22} was relatively smooth. After an initial rapid reduction in the value of the integral gain for both controllers, the rate of change of gains k_{I11} k_{I22} fell to a low value approximately after iteration 10 of the algorithm.

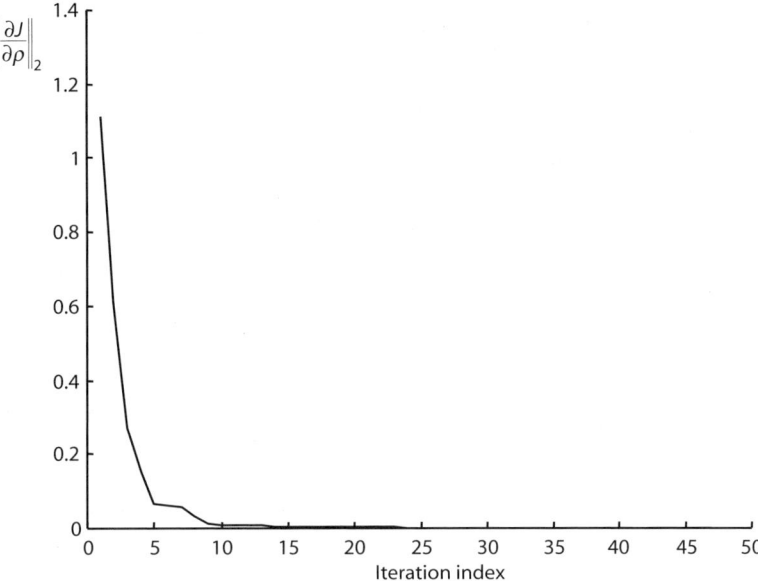

Figure 3.11 Plot of $\left\|\partial J/\partial \rho\right\|_2$ versus algorithm iteration index.

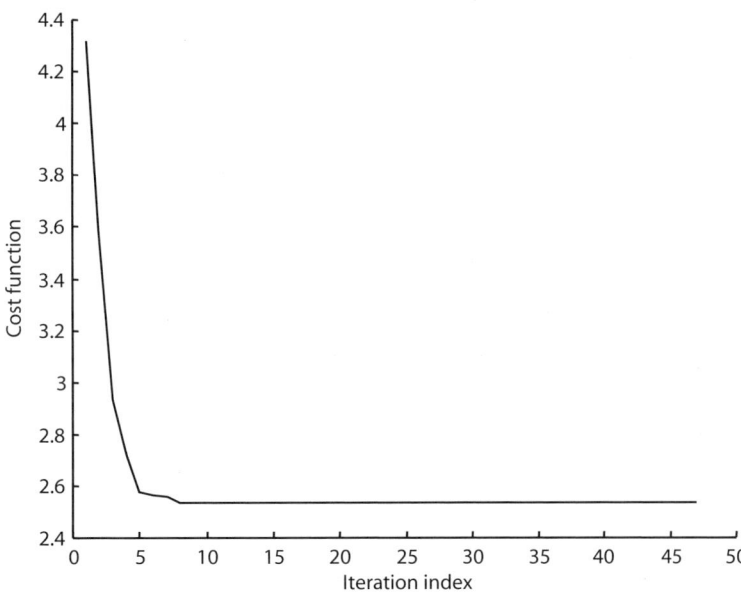

Figure 3.12 Plot of the cost function values $J(\rho(k))$ versus algorithm iteration index.

Convergence of the derivative controller parameters k_{D11}, k_{D22}

From Figure 3.15 it can be seen there is an initial rapid change in the values of the controller derivative gains. The controller parameter k_{D22} has a larger variation in its value than k_{D11}; however the stability of the closed-loop system was unaffected by the variations in the controller parameters during the course of the tuning algorithm. As can be seen from Figure 3.15, after approximately ten iterations of the algorithm the rate of change of the controller derivative gain parameters fell to a low value.

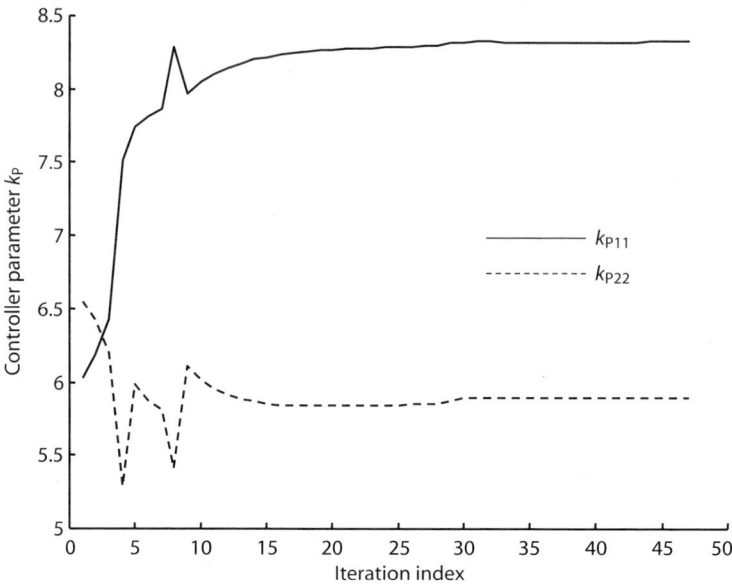

Figure 3.13 Plot of the proportional controller parameters k_{P11}, k_{P22} versus iteration index.

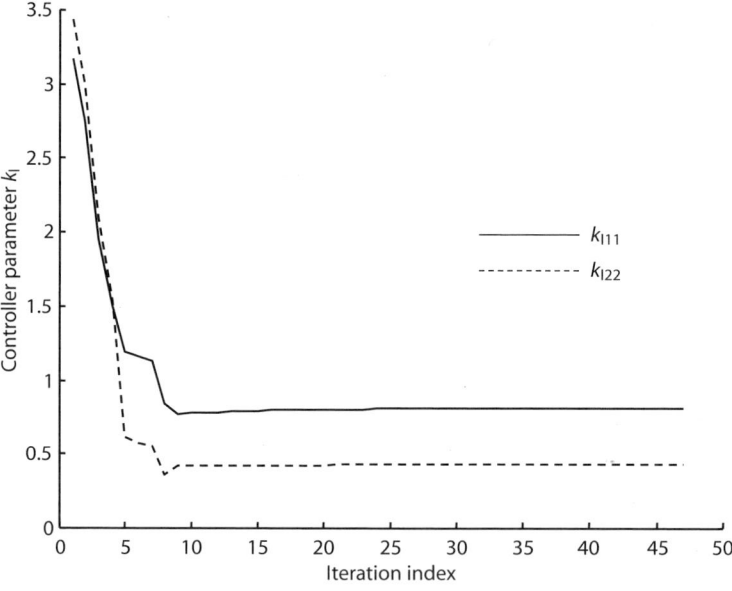

Figure 3.14 Plot of the integral controller parameters k_{I11} k_{I22} versus iteration index.

Optimised Decentralised Control Performance

From the plot of $\|\partial J/\partial \rho\|_2$ versus algorithm iteration index (Figure 3.11) and the plot of the cost function values $J(\rho(k))$ versus algorithm iteration index (Figure 3.12) it can be seen that the gradient norm tends to zero and that the value of the cost function tends to an asymptote as the number of algorithm iterations increases. However, the time response of the final version of the decentralised controller

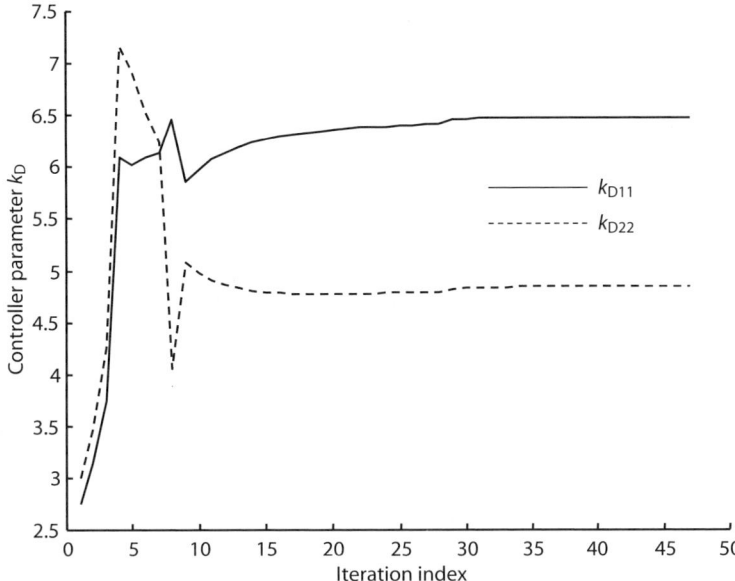

Figure 3.15 Plot of the derivative controller parameters k_{D11} k_{D22} versus iteration index.

cannot be inferred from the cost function or gradient values; for this control system responses are required.

The block diagram of the decentralised control system used in the assessment of the initial and final control system performance was shown in Figure 3.9. This showed two loops, and the designations *Loop One* and *Loop Two* will be used in the descriptions of the controller performance assessment.

A step input was applied to the reference input of *Loop One* at time $t = 0$ s and a step input was applied to the reference input of *Loop Two* at time $t = 100$ s. Figure 3.16 compares the closed-loop step response for the output of *Loop One* of the system with the controller parameters set to the initial Z–N values and the final Controller Parameter Cycling method tuned values. As can be seen from Figure 3.16 the Ziegler–Nichols tuned controller gives a peak percentage overshoot of approximately 50% compared with no overshoot from the Controller Parameter Cycling method tuned controller to the initial step input. Both controllers achieve similar rise times. The settling times achieved for the controllers are, based on a ±2% criterion, 12.5 s for the Z–N controller and 4.5 s for the Controller Parameter Cycling tuned controller. The Integral of the Square reference Error (ISE) was calculated during the simulation for both the Z–N tuned controllers and the Controller Parameter Cycling (CPC) tuned controllers. For the initial step input applied to the reference input of *Loop One* the Z–N tuned controllers had a value $ISE = 1.251$. The corresponding ISE value for the system using the CPC tuned controllers was $ISE = 0.9012$ for *Loop One*.

When the step input is applied to *Loop Two* the effect on *Loop One* is much reduced when the CPC tuned controllers are used compared with the response when the Z–N tuned controllers are used. The Z–N tuned controllers give a percentage overshoot of 13% and a percentage undershoot of 12%. This can be compared with a percentage overshoot of 5% and no undershoot for the CPC tuned controllers. The settling time for the Z–N tuned controllers is approximately 11 s compared with 3.5 s for the CPC tuned controllers. The settling time was based on a ±2% criterion. The ISE figure for *Loop One* when a step input is applied to *Loop Two* is $ISE = 0.05$ for the Z–N tuned controllers compared with a value $ISE = 0.05$ when the CPC tuned controllers are used.

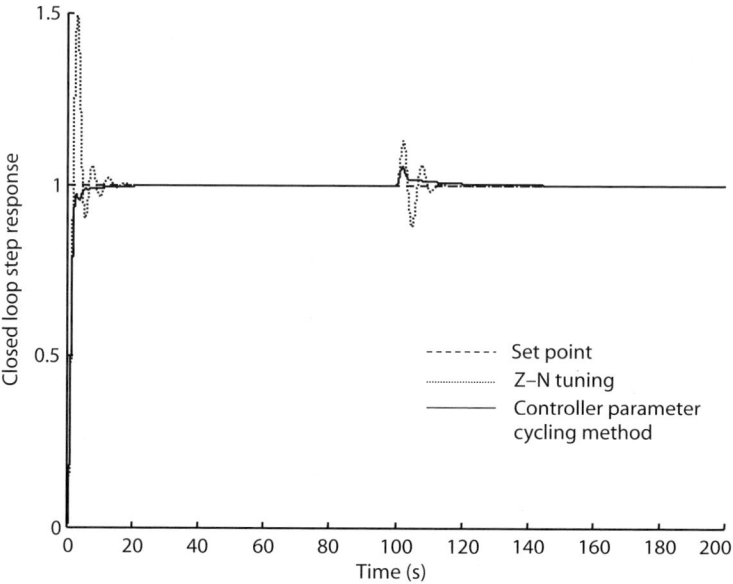

Figure 3.16 Closed-loop response for *Loop One*.

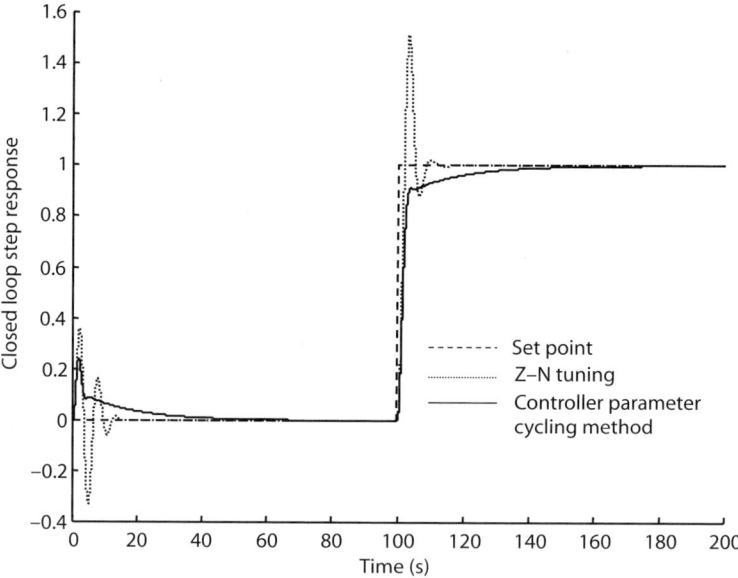

Figure 3.17 Closed-loop response for *Loop Two*.

Figure 3.17 shows the step response for the closed-loop output of *Loop Two* when a step input is applied to the reference input of *Loop One* at time $t = 0$ s and a step input is applied to the reference input of *Loop Two* at time $t = 100$ s. From Figure 3.17 it can be seen that when a step input is applied to the reference input of *Loop One* that the interaction into *Loop Two* has a peak overshoot of 35.7% and a peak undershoot of 33.5% for the Z–N tuned controllers, compared with a peak percentage overshoot of 24% for the CPC tuned controllers.

The settling time for the interaction disturbance is 12 s for the Z–N tuned controllers and 28 s for the CPC-based controllers. The *ISE* calculations for the Z–N and CPC tuned controllers are respectively *ISE* = 0.3794 and *ISE* = 0.1663. Thus it can be seen that although the Z–N tuned controllers give a faster response than the CPC tuned controllers the disturbance rejection properties of the CPC tuned controllers are improved over the original Z–N tuned controllers. When the step input is applied to the reference input of *Loop Two*, then from Figure 3.17 it can be seen that the peak percentage overshoot of the output of *Loop Two* using the Z–N tuned controller is approximately 51% compared with 24% for the corresponding figure using the CPC tuned controllers. The settling time for the closed-loop system using the Z–N tuned controllers is 5 s compared with 30 s for the system using the CPC tuned controllers. The settling time was calculated using a ±2% criterion. The rise time of the closed-loop output of *Loop Two* when the Z–N controllers are used is 2 s compared with 3 s for the system when the CPC tuned controllers are used. The rise time was based on the time taken for the output of *Loop Two* to go from 10% to 90% of its final value. The ISE for the step input applied at time $t = 100$ s is $ISE = 1.4686$ for the system using the Z–N tuned controllers and $ISE = 1.3447$ for the system using the CPC based controllers. Thus although the Z–N tuned controllers give a faster response than is achieved using the CPC-based controllers, the CPC-based controllers give an improved control performance as regards overshoot and ISE error reduction.

3.4 Summary and Future Directions

Iterative Feedback Tuning Method

The method of Iterative Feedback Tuning uses an on-line optimisation approach and is able to exploit the extensive existing knowledge on optimal control design. The major innovation introduced by the method was to generate the cost function gradient without using a system model. In fact, the actual system replaces the model and the method is genuinely *model-free*. The basic framework for the optimisation routine can be configured as a steepest descent method or as a Newton algorithm. The production of the Hessian matrix appears to involve quite a heavy computational load with three layers of system responses needed. The technique was originally developed in a stochastic framework and recently the multivariable system versions have been published. Further investigation of the numerical and implementation aspects seems a useful future research direction for the method.

Controller Parameter Cycling Method

The concept of a model-free control technique is an intriguing one and the innovative construction of the method of Iterative Feedback Tuning is inspirational. One of the shortcomings of IFT is that the generation of Hessian information is not a simple operation. This motivated the development of the Controller Parameter Cycling method, where the gradient and second-order Hessian information arise from one unified theoretical framework.

The details of a first version of the new procedure were reported in this chapter. Whilst the procedure was successful in tuning a multivariable decentralised PID controller, certain issues require further investigation. Experience with several examples has shown that many of the cost functions of fixed structure controllers are often very flat and it is necessary to ensure that sensible parameter updates are used in the routine. In the version of the algorithm described here, the Levenberg–Marquardt procedure is used to try to ensure that successive parameter updates continue in the cost minimising direction; it would be useful to examine other procedures. Even in the Controller Parameter Cycling method, the implementation of the gradient and Hessian extraction involves a heavy computational burden in terms of system runs and minimising iteration steps. Symmetry can be exploited to reduce the computational load and an intelligent strategy can be used to ensure the algorithm uses computed data to maximum effect in the sequence of iteration steps. Further development is envisaged in these areas.

Acknowledgements

The first part of this chapter on the Iterative Feedback Tuning method was written by Michael Johnson. Doctoral students at the Industrial Control Centre, Alberto Sanchez and Matthew Wade, introduced and wrote the material on a wastewater process application. MSc graduate Kasetr Mahathanakiet (Thailand) performed the computations for application of Iterative Feedback Tuning to the wastewater process.

The second part of the chapter was devoted to the new method of Controller Parameter Cycling. James Crowe and Michael Johnson jointly wrote this presentation of the technique. The method of controller parameter cycling is original to James Crowe.

Appendix 3.A

Table 3.6 Activated Sludge Model Parameters (Copp, 2001).

Stoichiometric parameters

Parameter	Symbol	Units	Value
Autotrophic yield	Y_A	G XBA COD formed (g N utilised)$^{-1}$	0.24
Heterotrophic yield	Y_H	G XBH COD formed (g COD utilised)$^{-1}$	0.67
Fraction of biomass to particulate products	f_p	dimensionless	0.08
Fraction nitrogen in biomass	i_{XB}	G N (g COD)$^{-1}$ in biomass	0.08
Fraction nitrogen in biomass	i_{XP}	G N (g COD)$^{-1}$ in X_p	0.06

Kinetic parameters

Parameter	Symbol	Units	Value
Maximum heterotrophic growth rate	μ_{mH}	day^{-1}	4.0
Half-saturation (hetero. growth)	K_S	G COD m^{-3}	10.0
Half-saturation (hetero. oxygen)	K_{OH}	G O$_2$ m^{-3}	0.2
Half-saturation (hetero. nitrate)	K_{NO}	G NO$_3$ N m^{-3}	0.5
Heterotrophic decay rate	b_H	day^{-1}	0.3
Anoxic growth rate correction factor	η_G	dimensionless	0.8
Anoxic hydrolysis rate correction factor	η_h	dimensionless	0.8

Table 3.6 (continued)

Maximum specific hydrolysis rate	k_h	G X_s (g X_{BH} COD day)$^{-1}$	3.0
Half-saturation (hydrolysis)	K_X	G X_s (g X_{BH} COD)$^{-1}$	0.1
Maximum autotrophic growth rate	μ_{mA}	day^{-1}	0.5
Half-saturation (auto.growth)	K_{NH}	G NH_3–N m^{-3}	1.0
Autotrophic decay	b_A	day^{-1}	0.05
Half-saturation (auto.oxygen)	$K_{O,A}$	G O_2 m^{-3}	0.4
Ammonification rate	k_a	M^3 (g COD day)$^{-1}$	0.05

References

Abramowitz, M. and Stegun, I.A. (eds.) (1972) *Handbook of Mathematical Functions*. Dover Publications, New York.

Åström, K.J. and Hägglund, T. (1985) Method and an apparatus in tuning a PID regulator. US Patent No. 4549123.

Copp, J. (2001) The COST simulation benchmark: description and simulator manual. *Technical Report from the COST Action 624 and 682*.

Crowe, J. and Johnson, M.A. (1998) New approaches to nonparametric identification for control applications. *Preprints IFAC Workshop on Adaptive Systems in Control and Signal Processing*, Glasgow, 26–28 August, pp. 309–314.

Crowe, J. and Johnson, M.A. (1999) A new nonparametric identification procedure for online controller tuning. *Proceedings American Control Conference*, San Diego, 2–4 June, pp. 3337–3341.

Crowe, J. and Johnson, M.A. (2000) Automated PI controller tuning using a phase locked loop identifier module. *Proceedings IECON 2000*, IEEE International Conference on Industrial Electronics, Control and Instrumentation, Nagoya, Japan, 22–28 October.

Crowe, J., Johnson, M.A. and Grimble, M.J. (2003) PID parameter cycling to tune industrial controllers – a new model-free approach. *Proceedings 13th IFAC Symposium on System Identification*, Rotterdam, The Netherlands, 27–29 August.

Fletcher, R. (1987) *Practical Methods of Optimisation*, 2nd edn. John Wiley & Sons, Chichester.

Gevers, M. (2000) A decade of progress in iterative process control design: from theory to practice. *Proceedings ADCHem 2000*, Pisa, pp. 677–688.

Grimble, M.J. (1994) *Robust Industrial Control*. Prentice Hall, Hemel Hempstead.

Henze, M., Grady, C.P.L., Gujer, W., Marais, G.v.R. and Matsuo, T. (1987) Activated Sludge Model No. 1. *IAWQ Scientific and Technical Report No.1*, IAWQ.

Hjalmarsson, H. (1995) Model free tuning of controllers: experience with time varying linear systems. *Proceedings European Control Conference*, Rome, September, pp. 2869–2874.

Hjalmarsson, H. (1998) Iterative feedback tuning. *Preprints of IFAC Workshop on Adaptive Systems in Control and Signal Processing*, Glasgow, August.

Hjalmarsson, H. (1999) Efficient tuning of linear multivariable controllers using iterative feedback tuning. *Int. J. Adapt. Control Signal Process.*, **13**, 553–572.

Hjalmarsson, H. and Gevers, M. (1997) Frequency domain expressions of the accuracy of a model free control design scheme. *IFAC System Identification*, Kitakyushu, Fukuoka, Japan.

Hjalmarsson, H. and Birkeland, T. (1998) Iterative feedback tuning of linear MIMO systems. *Proceedings Conference on Decision and Control*, Tampa, December, pp. 3893–3898.

Hjalmarsson, H., Gunnarsson, S. and Gevers, M. (1994) A convergent iterative restricted complexity control design scheme. *Proceedings Conference on Decision and Control*, Lake Buena Vista, Florida, December, pp. 1735–1740.

Hjalmarsson, H., Gunnarsson, S. and Gevers, M. (1995) Optimality and sub-optimality of iterative identification and control design schemes. *Proceedings American Control Conference*, Seattle, pp. 2559–2563.

Hjalmarsson, H., Gevers, M., Gunnarsson, S. and Lequin, O. (1998) Iterative feedback tuning: theory and applications. *IEEE Control System Society Magazine*, August, pp. 26–41.

Lee, S.C., Hwang, Y.B., Chang, H.N. and Chang, Y.K. (1991) Adaptive control of dissolved oxygen concentration in a bioreactor. *Biotechnology and Bioengineering*, **37**, 597–607.

Lequin, O., Gevers, M. and Triest, L. (1999) Optimising the settle-time with iterative feedback tuning. *IFAC Triennial*, Beijing, Paper I-36-08-3, pp. 433–437.

Mahathanakiet, K., Johnson, M.A., Sanchez, A. and Wade, M. (2002) Iterative feedback tuning and an application to wastewater treatment plant. *Proceedings Asian Control Conference*, Singapore, September, pp. 256–261.

Miklev, R., Frumer, W., Doblhoff-Dier, O. and Bayer, K. (1995) Strategies for optimal dissolved oxygen (DO) control, Preprints 6th IFAC Conference on Computer Applications in Biotechnology, Garmisch, Germany, pp 315 – 318.

Robbins, H. and Munro, S. (1951) A stochastic approximation method. *Ann. Math. Stat.*, **22**, 400–407.

Trulsson, E. and Ljung, L. (1985) Adaptive control based on explicit criterion minimisation. *Automatica*, **21**(4), 385–399.

Wilkie, J., Johnson, M.A. and Katebi, M.R. (2002) *Control Engineering: An Introduction*. Palgrave, Basingstoke.

Yu, C.C. (1999) *Autotuning of PID Controllers: Relay Feedback Approach*. Springer-Verlag, London.

Zhuang, M. and Atherton, D.P. (1994) PID controller design for a TITO system. *IEE Proc. CTA*, **141**(2), 1994

Ziegler, J.G. and Nichols, N.B. (1942) Optimum settings for automatic controllers. *Trans. ASME*, **64**, 759–768.

4 Automatic PID Controller Tuning – the Nonparametric Approach

Learning Objectives

4.1 Introduction

4.2 Overview of Nonparametric Identification Methods

4.3 Frequency Response Identification with Relay Feedback

4.4 Sensitivity Assessment Using Relay Feedback

4.5 Conversion to Parametric Models

4.6 Case Studies

References

Learning Objectives

For many processes, PID control can provide the performance required by the industrial engineer and the production plant operator. However, the issues of industrial concern will relate to providing simple procedures for achieving satisfactory performance, for making sure that the control loop is optimally tuned despite changes in the process and materials and for finding efficient ways to tune and maintain the many hundreds of PID control loops that are found in large-scale industrial production systems. The relay experiment due to Åström and Hägglund was a very successful solution to many of the problems posed above by the industrial control engineer. Although valid for a restricted but nonetheless wide group of practical system applications, the original relay methodology depends on a describing function approximation and uses simplistic PID tuning rules. In this chapter, extensions are described for both these aspects of the relay technique – identification accuracy and more advanced PID tuning.

The learning objectives for this chapter are to:

- Introduce the idea of nonparametric identification and reviewing the basic methods available

- Understand how to modify the relay experiment to enhance the estimation accuracy of Nyquist curve points
- Use the relay experiment to find other controller design points such as the maximum sensitivity point and the gain and phase margins
- Learn how to extend the relay method to identify simple parametric models

4.1 Introduction

PID controllers have undergone many changes over the last few decades in terms of the underlying technology implementing the control function and the interface to the sensors and actuators of processes. They have evolved from the pneumatic configurations of the 1940s through the electrical analogue devices of the 1960s to the current modern microprocessor-based controller technology that has been available since the 1980s.

While controller implementation has changed tremendously, the basic functions of the controller and the PID algorithm have not changed and probably will not do so in the future. However, new possibilities and functionalities have become possible with a microprocessor-driven PID controller. Modern process controllers often contain much more than just the basic PID algorithm. Fault diagnosis, alarm handling, signal scaling, choice of type of output signal, filtering, and simple logical and arithmetic operations are becoming common functions to be expected in modern PID controllers. The physical size of the controller has shrunk significantly compared to the analogue predecessors, and yet the functions and performance have greatly increased. Furthermore, riding on the advances in adaptive control and techniques, modern PID controllers are becoming intelligent. Many high-end controllers appearing in the market are equipped with autotuning and self-tuning features. No longer is tedious manual tuning an inevitable part of process control. The role of operators in PID tuning has now been very much reduced to initial specifications; thereafter the controllers can essentially take charge to yield respectable performance.

Different systematic methods for tuning of PID controllers are available, but irrespective of the particular design method preferred, the following three phases are applicable:

1. The process is disturbed with specific control inputs or control inputs automatically generated in the closed loop.

2. The response to the disturbance is analysed, yielding a model of the process that may be in a nonparametric or parametric form.

3. Based on this model and certain operation specifications, the control parameters are determined.

Automatic tuning of controllers means quite simply that the above procedures are automated so that the disturbances, model calculation and choice of controller parameters all occur within the same controller. In this way, the work of the operator is made simpler, so that instead of having to calculate suitable controller parameters, the operator only needs to initiate the tuning process. The operator may have to give the controller some information about the process and the expected performance before the tuning is performed, but this information will be considerably simpler to specify than the controller parameters.

This chapter will focus on the process modelling aspects of the overall control autotuning procedures. First in the chapter is a brief review of common nonparametric identification methods that are suitable for facilitating automatic control tuning. The nonparametric methods described in the review aim to access or determine the process time or frequency functions directly without first selecting a confined set of possible models. These methods do not directly employ a finite-dimensional parameter vector in the search for a best description, although it is possible to derive parametric models from the nonparametric information obtained. After this brief review, the chapter focuses on the relay feedback method of identification, and illustrates the basic configuration along with variants of the method. Case studies of successful applications of the different methods are reported to highlight the principles and flexibility that are obtainable in the basic version and the variants of the relay experiment.

4.2 Overview of Nonparametric Identification Methods

This section provides a short description of several nonparametric identification methods in both the time and frequency domains. These methods can be used in conjunction with many control tuning strategies.

4.2.1 Transient Response Methods

Many methods for automatic tuning of controllers are based on deliberately disturbing the process via a specific change in the control signal and analysing the process response in the time domain. These types of methods are often referred to as transient response methods, since the information required to tune the controller is derived from the transient response of the process following the change in input. Transient response methods provide a quick and easy insight into the cause and effect relationships existing in a process and are probably still the most widely used method of identification in industrial practice. A drawback is that the information obtained can be rather limited, given the practical limits in the choice of the input signals. Two common inputs used to trigger a transient response are the impulse and step signals.

The impulse response method injects an impulse signal into the process. An impulse is essentially a pulse input of very large amplitude. The response to the impulse can then be recorded. The impulse response allows the prediction of the process response to other inputs, and thus it serves as a useful nonparametric model for the purpose of tuning a controller. The basic weakness is that many real processes do not allow the use of a pulse of arbitrary amplitude due to safety considerations and the need to avoid exciting the nonlinear dynamics of the process. On the other hand, if the amplitude of the pulse is too small, the effect of measurement noise becomes significant relative to the dynamic response, and this usually adversely affects the accuracy of the identification.

The step response method is based on the use of a step input signal. When the operator wishes to tune the controller, the control signal is changed to the form of a step, and the controller records how the process variable reacts to the step input, thereby producing the process step response. The step response is another nonparametric model in the time domain, which can be used for control tuning. It gives approximate estimates of the delay, gain and time constant of the process. Given these process parameter estimates, many tuning methods (including variations of the Ziegler–Nichols methods) can be used to derive the final control parameters. As with the impulse response method, one difficulty in carrying out the tuning process automatically is in selecting the size of the step input. The user will want the step size to be as small as possible so that the process is not disturbed more than necessary. On the other hand, it will be easier to determine the process model parameters if the disturbance is large, since the step response of the process variable must be distinguished from the prevailing noise. A trade-off between the two conflicting requirements is usually necessary.

More on the use of transient response methods can be found in Chapter 8.

4.2.2 Relay Feedback Methods

An increasing number of PID tuning methods are based on the frequency response information of the process, particularly for the frequency range around the ultimate frequency, that is, where the process has a phase shift of $-\pi$ rad. The popular stability limit method due to Ziegler and Nichols (1942) for PID control tuning involves generating oscillations at this frequency by iteratively increasing the proportional gain in the loop until the closed loop reaches the verge of instability. Then the amplitude and frequency of the oscillations will relate to the ultimate gain and frequency of the process, and they can be used to tune controllers directly. Unfortunately, this method is not easy to automate, as it means operating to the stability limit. In addition, the method may result in the amplitude of the process variable becoming too large to be acceptable in practice.

The relay feedback method is a non-iterative alternative, first pioneered by Åström and co-workers in the early 1980s, which can automatically derive the same information without incurring the problems of the stability limit method. The configuration of the relay feedback approach is shown in Figure 4.1. Under this setup, the ultimate frequency ω_π and gain k_π of the process can be determined automatically and efficiently, and with the process under closed loop regulation.

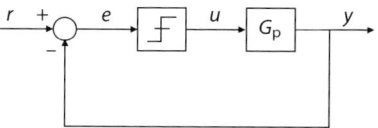

Figure 4.1 Relay feedback system.

4.2.3 Fourier Methods

The basic relay feedback method employs only stationary relay oscillations for the frequency response estimation. The advantages are that the resultant estimations are simple and robust to noise, non-zero initial conditions and disturbances. However, the relay transients in process input and output are not used; that is, the multiple frequencies in the signals are not adequately analysed to yield more information of the process. This limits the identifiability of the process dynamics, since this process dynamic information is mainly contained in the process transient response.

Fourier analysis will be useful for nonparametric modelling of processes with multiple frequencies in the input and output signals. Consider a process denoted $G_p(s)$. If the input has finite energy, the frequency function $G_p(j\omega)$ can be computed from the Fourier transforms of the input and the output:

$$G_p(j\omega) = \frac{Y(j\omega)}{U(j\omega)}$$

Normally, $u(t)$ and $y(t)$ are only available over a finite time interval $0 \leq t \leq T_f$, from which the following approximations may then be made:

$$Y(j\omega) = \int_0^{T_f} y(t) e^{-j\omega t} dt$$

and

$$U(j\omega) = \int_0^{T_f} u(t) e^{-j\omega t} dt$$

If the input contains pure sinusoids, the frequency function can be estimated with arbitrary accuracy as the time interval tends to infinity. For inputs that do not contain pure sinusoids, the estimation error grows with the noise to signal ratio, even for large T_f. The Fourier method can be used on deliberately generated signals to the process in the open loop. It can also be applied to transients generated during a setpoint change or a load disturbance with the process under closed-loop control.

The Fourier method can also be used for identifying multiple points on process frequency response from a single relay test with appropriate processing of the relay transient, if the oscillations are stationary. If there are 2^n (where n is an integer) samples of sustained and periodic oscillations are available, the Fourier Transform can be efficiently computed using the Fast Fourier Transform (FFT) algorithm. However, this is only an approximation since the actual oscillation input and output signals are neither absolutely integrable nor strictly periodic. It is possible to introduce a decay exponential to the process input and output so that the modified input and output will approximately decay to zero in a finite time interval. FFT can then be employed to obtain the process frequency response (Tan et al., 1999). This method uses the same relay experiment as in the basic method and it can yield multiple frequency response points.

4.2.4 Phase-Locked Loop Methods

The Phase-Locked Loop (PLL) method is an alternative nonparametric identification method to the relay method (Crowe and Johnson, 1998) which aims to produce a more accurate estimate of the system frequency response. It can be used with the process in open loop or in closed loop. In comparison with the relay experiment, it is claimed to offer more flexibility along with increased estimation accuracy.

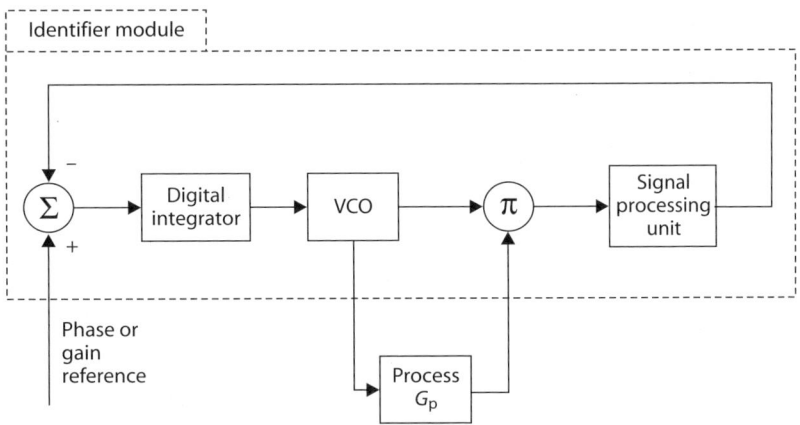

Figure 4.2 Phase-locked loop identifier module.

The configuration of the PLL identifier module, applied directly to the process $G_p(s)$, is shown in Figure 4.2 (Crowe and Johnson, 2000). The identifier has a feedback structure that uses either a phase or a gain reference value as an input to a comparator. A digital model of a controlled oscillator (a voltage-controlled oscillator) is used to provide the process excitation as well as a sinusoidal reference path to the signal processing unit. The signal processing unit is used to extract frequency response information (either the phase or gain data) and feed this to the comparator. The integrator will ensure that the system converges to the required input phase or magnitude reference value. In Chapter 6, Crowe and Johnson give a more detailed description of the PLL identifier, subsequently demonstrating its use in closed loop automated PID controller tuning in Chapter 7.

The PLL identifier can also be used with the process under closed-loop control, as shown in Figure 4.3. The phase crossover point of the process can be found using a phase reference $\phi_{ref} = -\pi$. The gain crossover point can be found by setting the gain reference to $G_{ref} = 1$. Once the gain or phase response information of the closed loop at the required frequency is obtained, and if the controller is known, then the gain and phase response of $G_p(j\omega)$ can be correspondingly obtained from

$$|G_p(j\omega)| = \frac{|G_{yr}(j\omega)|}{|1-G_{yr}(j\omega)|\,|G_c(j\omega)|}$$

and

$$\arg\{G_p(j\omega)\} = \arg\{G_{yr}(j\omega)\} - \arg\{1-G_{yr}(j\omega)\} - \arg\{G_c(j\omega)\}$$

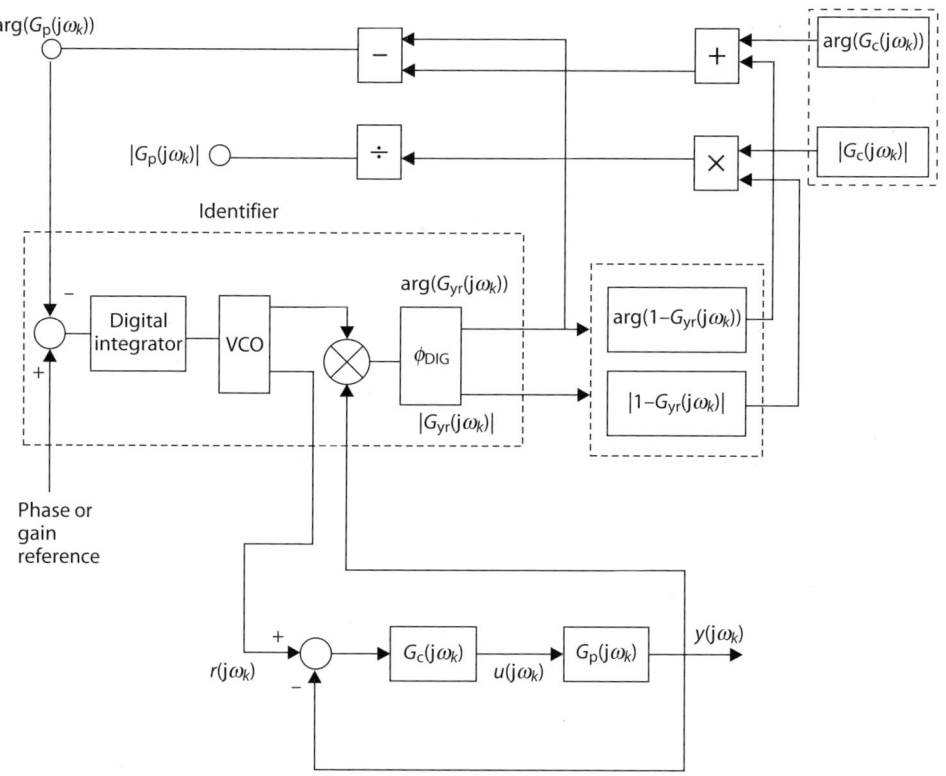

Figure 4.3 Closed-loop identification using PLL identifier module.

4.3 Frequency Response Identification with Relay Feedback

The relay feedback method has been the subject of much interest in recent years and it has been field tested in a wide range of applications. There are many attractive features associated with the relay feedback technique. First, for most industrial processes, the arrangement automatically results in a sustained oscillation approximately at the ultimate frequency of the process. From the oscillation amplitude, the ultimate gain of the process can be estimated. This alleviates the task of input specification from the user and therefore is in delightful contrast to other frequency-domain based methods requiring the frequency characteristics of the input signal to be specified. This particular feature of the relay feedback technique greatly facilitates automatic tuning procedures, since the arrangement will automatically give an important point of the process frequency response. Secondly, the relay feedback method is a closed-loop test and the process variable is maintained around the setpoint value. This keeps the process in the linear region where the frequency response is of interest and this works well on highly nonlinear processes; the process is never far away from the steady state conditions. Current identification techniques relying on transient analysis such as impulse or step tests do not possess this property, and they are therefore ineffective for processes with nonlinear dynamics. Thirdly, the relay feedback technique does not require prior information of the system time constants for a careful choice of the sampling period. The choice of the sampling period is often problematic for traditional parameter estimation techniques. If the sampling interval is too long, the dynamics of the process will not be adequately captured in the data, and consequently the accuracy of the model obtained will be poor. While a conservative safety-first approach towards this decision may be to select the smallest sampling period supported by the data acquisition equipment, this would result in too much data collection with inconsequential information. A corrective action then is data decimating in the post-treatment phase that for real-time parameter estimation may not be tolerable. Spared of these cumbersome and difficult decisions, the relay feedback method is therefore an attractive method to consider in autotuning applications.

Relay systems can be traced back to their classical configurations. In the 1950s, relays were mainly used as amplifiers, but owing to the development of electronic technology, such applications are now obsolete. In the 1960s, relay feedback was applied to adaptive control. One prominent example of such applications is the self-oscillating adaptive controller developed by Minneapolis Honeywell that uses relay feedback to attain a desired amplitude margin. This system was tested extensively for flight control systems, and it has been used in several missiles. It was in the 1980s that Åström successfully applied the relay feedback method to autotune PID controllers for process control, and initiated a resurgence of interest in relay methods, including extensions of the method to more complex systems. Åström and Hägglund (1995) have given a recent survey of relay methods.

This section focuses on nonparametric frequency response identification using a relay feedback. A review of the basic relay method is first provided, followed by variants of the basic method that expand its applicability to a larger class of processes and other scenarios which may require better accuracy or faster control tuning time.

4.3.1 Basic Idea

The ultimate frequency ω_π of a process, where the phase lag is $-\pi$ rad, can be determined automatically from an experiment with relay feedback, as shown in Figure 4.1.

The usual method employed to analyse such systems is the describing function method that replaces the relay with an "equivalent" linear time-invariant system. For estimation of the critical point (ultimate gain and ultimate frequency), the self-oscillation of the overall feedback system is of interest. Here, for the describing function analysis, a sinusoidal relay input $e(t) = a\sin(\omega t)$ is considered, and the resulting signals in the overall system are analysed. The relay output $u(t)$ in response to $e(t)$ would be a square-wave having a frequency ω and an amplitude equal to the relay output level μ. Using a Fourier series expansion, the periodic output $u(t)$ is written as

$$u(t) = \frac{4\mu}{\pi} \sum_{k=1}^{\infty} \frac{\sin(2k-1)\omega t}{2k-1}$$

The describing function (DF) of the relay $N(a)$ is simply the complex ratio of the fundamental component of $u(t)$ to the input sinusoid, namely

$$N(a) = \frac{4\mu}{\pi a}$$

Since the describing function analysis ignores harmonics beyond the fundamental component, define here the residual ρ as the entire sinusoid-forced relay output minus the fundamental component (that is, the part of the output that is ignored in the describing function development) so that

$$\rho(t) = \frac{4\mu}{\pi} \sum_{k=2}^{\infty} \frac{\sin(2k-1)\omega t}{2k-1}$$

In the describing function analysis of the relay feedback system, the relay is replaced with its quasi-linear equivalent DF, and a self-sustained oscillation of amplitude a and frequency ω_{osc} is assumed. Then, if $G_p(s)$ denotes the transfer function of the process, the variables in the loop must satisfy the following loop equations:

$$E = -Y$$
$$U = N(a)E$$
$$Y = G_p(j\omega_{osc})U$$

From these relations, it must follow that

$$G_p(j\omega_{osc}) = -\frac{1}{N(a)}$$

Relay feedback estimation of the critical point for process control is thus based on the key observation that the intersection of the Nyquist curve of $G_p(j\omega)$ and $-1/N(a)$ in the complex plane gives the critical point of the linear process. Hence, if there is a sustained oscillation in the system of Figure 4.1, then in the steady state the oscillation must be at the ultimate frequency:

$$\omega_\pi = \omega_{osc}$$

and the amplitude of the oscillation is related to the ultimate gain k_π by

$$k_\pi = \frac{4\mu}{\pi a}$$

It may be advantageous to use a relay with hysteresis as shown in Figure 4.4, so that the resultant system is less sensitive to measurement noise.

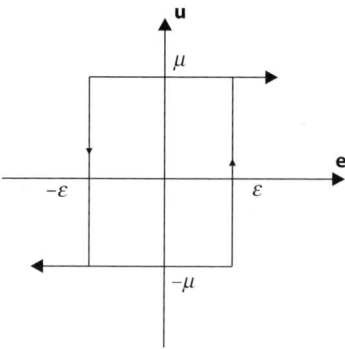

Figure 4.4 Relay with hysteresis.

The inverse negative describing function of the relay with hysteresis is given by

$$-\frac{1}{N(a)} = \frac{\pi}{4\mu}(\sqrt{a^2 - \varepsilon^2} + j\varepsilon)$$

In this case, the oscillation corresponds to the point where the negative inverse describing function of the relay crosses the Nyquist curve of the process as shown in Figure 4.5. With hysteresis present, there is an additional parameter ε, which can, however, be set automatically based on a pre-determination of the measurement noise level.

Relay tuning is an attractively simple method for extracting the critical point of a process. Accompanying the method are three main limitations. First, the accuracy of the estimation could be poor for certain processes. Secondly, the relay experiment yields only one point of the process frequency response that may not be adequate in many applications other than basic PID tuning. Thirdly, the method is essentially an off-line method, which is sensitive to the presence of disturbances and generally requires the stationary condition to be known. To overcome these limitations, some modifications to the basic method are necessary; these are covered in the next sections.

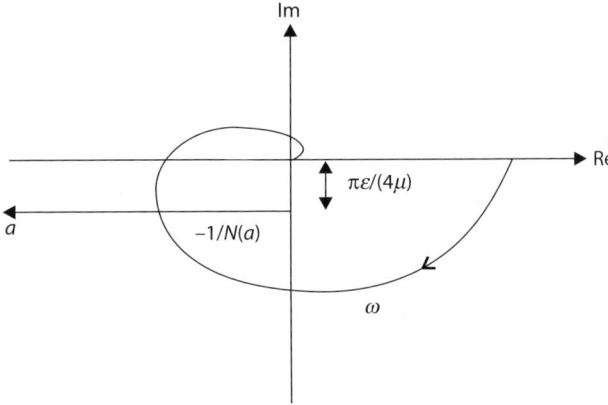

Figure 4.5 Negative inverse describing function of the hysteresis relay.

4.3.2 Improved Estimation Accuracy

While the relay feedback experiment and controller design will yield sufficiently accurate results for many of the processes encountered in the process control industry, there are some potential problems associated with such techniques. These arise as a result of the approximations used in the development of the procedures for estimating the critical point, namely the ultimate frequency and ultimate gain. In particular, the basis of most existing relay-based procedures for critical point estimation is the describing function method. This method is approximate in nature, and under certain circumstances, the existing relay-based procedures could result in estimates of the critical point that are significantly different from their real values.

The accuracy of the relay feedback estimation depends on the residual $\rho(t)$, which determines whether, and to what degree, the estimation of the critical point will be successful. For the relay,

$$\rho(t) = \frac{4\mu}{\pi} \sum_{k=2}^{\infty} \frac{\sin(2k-1)\omega t}{2k-1}$$

consists of all the harmonics in the relay output. The amplitudes of the third and fifth harmonics are about 30% and 20% that of the fundamental component and are not negligible if fairly accurate analysis results are desirable; therefore they limit the class of processes for which describing function analysis is adequate, because for accuracy the process must attenuate these signals sufficiently. This is the fundamental assumption of the describing function method, which is also known as the *filtering hypothesis*. Mathematically, the filtering hypothesis requires that the process $G_p(s)$ must satisfy the following two equations:

$$|G_p(jk\omega_\pi)| \ll |G_p(j\omega_\pi)|, k = 3, 5, 7, \ldots$$

and

$$|G_p(jk\omega_\pi)| \to 0 \text{ as } k \to \infty$$

Note that this equation pair requires the process to be not simply low-pass, but rather low-pass at the ultimate frequency. This is essential, as the delay-free portion of the process may be low-pass, but the delay may still introduce higher harmonics within the bandwidth. Typical processes that fail the filtering hypothesis are processes with a long time delay and processes with frequency response resonant peaks so that the undesirable frequencies are boosted instead of being attenuated. This provides one explanation for the poor results associated with these processes.

Having observed the accuracy problems associated with conventional relay feedback estimation, the design of a modified relay feedback that addresses the issue of improved estimation accuracy is considered next. Thus consider the modified relay feedback system of Figure 4.6, where the process is assumed to have the transfer function $G_p(s)$.

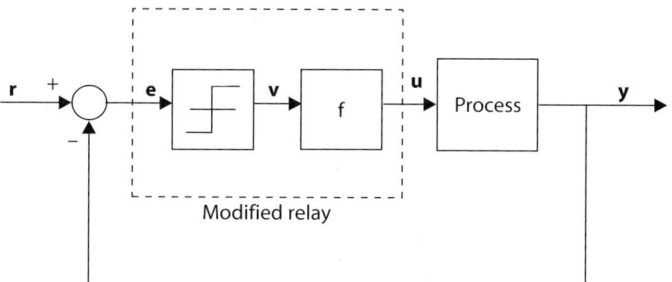

Figure 4.6 Modified relay feedback system.

Define the mapping function $f: \Re \to \Re$ such that

$$u(t) = f(v(t)) = v_1(t) = \frac{4\mu}{\pi} \sin \omega t$$

where $v_1(t)$ is the fundamental harmonic of $v(t)$ and μ is the amplitude of the relay element. For this modified relay feedback, it turns out that the following property may be stated.

Proposition 4.1

Consider the use of the modified relay feedback defined by the system of Figure 4.6 and the mapping function $f: \Re \to \Re$ such that

$$u(t) = f(v(t)) = v_1(t) = \frac{4\mu}{\pi} \sin \omega t$$

and where the process is assumed to have the transfer function $G_p(s)$. The set of signals

4.3 Frequency Response Identification with Relay Feedback

$$u(t) = \frac{4\mu}{\pi}\sin(\omega^* t)$$

$$y(t) = \frac{4\mu}{\pi}A^*\sin(\omega^* t + \phi^*)$$

$$v(t) = N_r\left(-\frac{4\mu}{\pi}A^*\sin(\omega^* t + \phi^*)\right)$$

where $N_r(\cdot)$ denotes the relay function, describes an invariant set of the dynamical system defined by Figure 4.6 with $\phi^* = \arg\{G_p(j\omega^*)\} = -\pi$ and $A^* = |G_p(j\omega^*)|$.

Proof

Assume that for some ω^*, the setup of Figure 4.6 admits a solution of the form

$$u(t) = \frac{4\mu}{\pi}\sin(\omega^* t)$$

It will be shown that this solution is consistent with the definitions of the other signals in the loop so that this solution describes an invariant set of the dynamical system defined by Figure 4.6. With input

$$u(t) = \frac{4\mu}{\pi}\sin(\omega^* t)$$

it follows that the output is given by

$$y(t) = \frac{4\mu}{\pi}A^*\sin(\omega^* t + \phi^*)$$

where $A^* = |G_p(j\omega^*)|$ and $\phi^* = \arg G_p(j\omega^*)$.

Then, since $e(t) = -y(t)$, it follows that

$$v(t) = N_r(e(t)) = N_r\left(-\frac{4\mu}{\pi}A^*\sin(\omega^* t + \phi^*)\right)$$

where $N_r(\cdot)$ denotes the relay function. Since this is the case, clearly $v_1(t)$, the fundamental harmonic of $v(t)$, is given by

$$v_1(t) = -\frac{4\mu}{\pi}\sin(\omega^* t + \phi^*)$$

This in turn implies that

$$u(t) = f(v(t)) = -\frac{4\mu}{\pi}\sin(\omega^* t + \phi^*)$$

Thus, having satisfied the loop equations, the two relations

$$u(t) = \frac{4\mu}{\pi}\sin(\omega^* t)$$

$$u(t) = f(v(t)) = -\frac{4\mu}{\pi}\sin(\omega^* t + \phi^*)$$

are consistent for

$$\phi = -(2n+1)\pi$$

where n is a non-negative integer, and this characterises a class of admissible solutions.

Therefore, for $n = 0$, the set of signals:

$$u(t) = \frac{4\mu}{\pi}\sin(\omega^* t)$$

$$y(t) = \frac{4\mu}{\pi} A^* \sin(\omega^* t + \phi^*)$$

$$v(t) = N_r\left(-\frac{4\mu}{\pi} A^* \sin(\omega^* t + \phi^*)\right)$$

describes an invariant set of the dynamical system defined by Figure 4.6 with $\phi = -\pi$, as claimed. Note that the above invariant equation set is clearly periodic in t.

∎

Remark 4.1 The equations for the invariant set established in Proposition 4.1 provide a suitable basis for estimation of the critical point for process control with improved accuracy. This is because the analysis used to prove Proposition 4.1 does not depend on approximations, so that theoretically it is possible to calculate the critical point exactly from observations of the invariant set.

To estimate the critical point using the arrangement of Figure 4.6 and the result of Proposition 4.1, assume that an oscillation is observed which corresponds to the (admissible) solution $\phi^* = \arg\{G_p(j\omega^*)\} = -\pi$. This implies that $\omega^* = \omega_{osc}$. Thus, the ultimate frequency is obtained directly from the measurement of the frequency of the oscillation observed. For the critical point, the remaining parameter to be estimated is the ultimate gain, and from the measurements this may be obtained from the ratio

$$k_\pi = \frac{a_u}{a_y} = \frac{4\mu}{\pi a_y}$$

where a_u and a_y are the observed amplitudes of the oscillations in $u(t)$ and $y(t)$, respectively. As in conventional relay feedback, the relay magnitude μ may be used as a design parameter to appropriately size the magnitude of the oscillations for situations with different levels of noise.

Remark 4.2 The analysis to prove Proposition 4.1 is essentially a time-domain analysis, and no approximations were involved in establishing the existence of the invariant set equations and the frequency ω^* that characterised it. This may be considered as an improvement over existing methods, which are based on describing function analysis and which consequently yield estimation procedures for the critical point that involve approximations. The existing methods have the advantage of simplicity, and under certain circumstances, the improved procedure here only yields small gains in accuracy. However, there are other circumstances (underdamped processes and processes with significant time delay) where the gains in accuracy are significant.

It is interesting to note that the improved accuracy is also evident from the same describing function analysis used for the existing methods. Thus a describing function analysis applied to the system of Figure 4.4 shows that, corresponding to the input $e(t) = a\sin(\omega t)$, the output of the nonlinearity $u(t)$ would also be a sinusoid described by

$$u(t) = \frac{4\mu}{\pi}\sin(\omega t)$$

The describing function of the nonlinear system can then be obtained as

$$N(a) = \frac{4\mu}{\pi a}$$

and it is quite straightforward to check that the residual $\rho(t) = 0$ so that the filtering hypothesis of the describing function analysis is satisfied strictly. It then follows that analysis using the different tools of describing functions also indicates that improved accuracy will be obtained from the procedure here.

Implementation Procedures

The system described by Figure 4.6 and the mapping equation,

$$u(t) = f(v(t)) = v_1(t) = \frac{4\mu}{\pi} \sin \omega t$$

defines the basic elements required for the improved accuracy technique. For the construction of implementation procedures, the key point to note is that the module realising the mapping function $f: \mathfrak{R} \to \mathfrak{R}$ should be designed to extract the fundamental harmonic and apply it as the signal $u(t)$. Thus there are various practical procedures and variations that may be used.

For a simple practical implementation, an initial approximate estimate of the ultimate frequency can be obtained from two or three switches of the relay under normal relay feedback, and then function $f: \mathfrak{R} \to \mathfrak{R}$ can be turned on using this initial value. The function f is then updated iteratively using the observed signals. The steps in such a procedure are as follows.

Algorithm 4.1: Improved relay feedback experiment.

Step 1 Put the process under relay feedback; that is set $f(v(t)) = v(t)$, to obtain an approximate estimate of its ultimate frequency $\tilde{\omega}_\pi$ from m oscillations, say with $m = 2$.

Step 2 Denote T_n as the time corresponding to the nth switch of the relay to $v = \mu$, and $\tilde{\omega}_{\pi,n-1}$ as the ultimate frequency estimate just prior to $t = T_n$. For $n > m$, update $f(\cdot)$ to output

$$u(t) = f(v(t)) = \frac{4\mu}{\pi} \sin \omega_{\pi,n-1}(t - T_n)$$

Step 3 Obtain a new estimate of the ultimate frequency $\tilde{\omega}_{\pi,n}$ from the resultant oscillation.

Step 4 Repeat Step 2 and Step 3 until successive estimates of the ultimate frequency show a satisfactory convergence.

Algorithm end

Estimation Accuracy

The use of the modified relay feedback for critical point estimation was investigated using simulations, and the results are tabulated and compared with critical point estimation using basic relay feedback in Tables 4.1–4.4.

Table 4.1 shows the results for an overdamped process with different values for the dead time. Tables 4.2 and 4.3 show the respective results for an underdamped process and an overdamped process. Each of these processes has a (stable) process zero but have different values for the dead time. Finally, the results for a non-minimum phase process with different values for the dead time are shown in Table 4.4. From the tables, it can be seen that critical point estimation using the proposed modified relay feedback consistently yields improved accuracy over the basic relay feedback.

The problem of improving the accuracy of the results from the basic relay experiment has been of interest to many control engineers. For example, Saeki (2002) recently proposed a new adaptive method for the identification of the ultimate gain that can also yield improved accuracy. Instead of using relay feedback, a saturation nonlinearity and an adaptive gain are inserted in the loop, where the gain adapts to the difference between the input and output of the saturation function. The gain decreases monotonically and converges to a constant, which can be very close to the ultimate gain.

Table 4.1 Process $G_p(s) = \left[\dfrac{e^{-sL}}{s+1}\right]$.

	Real process		Basic relay				Modified relay			
L	k_π	ω_π	$\hat{k}_\pi$	PE	$\hat{\omega}_\pi$	PE	$\hat{k}_\pi$	PE	$\hat{\omega}_\pi$	PE
0.5	3.81	3.67	3.21	15.7	3.74	1.8	3.80	0.3	3.66	0.3
2.0	1.52	1.14	1.46	4.2	1.16	1.7	1.51	0.7	1.14	0.0
5.0	1.13	0.53	1.28	13.2	0.55	3.7	1.13	0.0	0.53	0.0
10.0	1.04	0.29	1.27	22.4	0.29	2.3	1.04	0.0	0.29	0.0

PE: Percentage error

Table 4.2 Process $G_p(s) = \left[\dfrac{(s+0.2)e^{-sL}}{s^2+s+1}\right]$.

	Real process		Basic relay				Modified relay			
L	k_π	ω_π	$\hat{k}_\pi$	PE	$\hat{\omega}_\pi$	PE	$\hat{k}_\pi$	PE	$\hat{\omega}_\pi$	PE
0.5	3.48	3.61	2.97	14.7	3.70	2.5	3.47	0.3	3.61	0.0
2.0	1.09	1.27	1.20	10.1	1.28	0.8	1.08	0.9	1.26	0.8
5.0	1.19	0.70	1.10	7.7	0.62	11.4	1.19	0.0	0.70	0.0
10.0	2.17	0.38	1.16	46.5	0.31	18.4	2.17	0.0	0.38	0.0

PE: Percentage error

Table 4.3 Process $G_p(s) = \left[\dfrac{(s+0.2)e^{-sL}}{(s+1)^2}\right]$.

	Real process		Basic relay				Modified relay			
L	k_π	ω_π	$\hat{k}_\pi$	PE	$\hat{\omega}_\pi$	PE	$\hat{k}_\pi$	PE	$\hat{\omega}_\pi$	PE
0.5	4.26	4.02	3.81	10.6	4.16	3.5	4.25	0.2	4.02	0.0
2.0	2.07	1.35	2.37	15.0	1.38	2.2	2.07	0.0	1.35	0.0
5.0	2.09	0.65	1.98	5.3	0.61	6.2	2.09	0.0	0.65	0.0
10.0	2.78	0.35	1.93	30.6	0.31	11.4	2.76	0.7	0.35	0.0

PE: Percentage error

Table 4.4 Process $G_p(s) = \left[\dfrac{(-s+0.2)e^{-sL}}{(s+1)^2} \right]$.

	Real process		Basic relay				Modified relay			
L	k_π	ω_π	$\hat{k}_\pi$	PE	$\hat{\omega}_\pi$	PE	$\hat{k}_\pi$	PE	$\hat{\omega}_\pi$	PE
0.5	1.97	0.83	1.64	17.0	0.72	13.3	1.96	0.5	0.83	0.0
2.0	2.31	0.51	1.54	33.0	0.53	3.9	2.31	0.0	0.51	0.0
5.0	2.97	0.31	1.51	49.2	0.35	13.0	2.98	0.3	0.31	0.0
10.0	3.69	0.20	1.51	59.1	0.23	15.0	3.67	0.5	0.20	0.0

PE: Percentage error

4.3.3 Estimation of a General Point

With the basic relay feedback approach, only one point on the process Nyquist curve is determined. It is possible, for example, to cascade a known linear dynamical system to the system in Figure 4.1 to obtain a frequency other than the ultimate frequency. For example, an integrator can be cascaded to obtain the point where the process Nyquist curve crosses the negative imaginary axis. Similarly, to obtain a point with a frequency below the ultimate frequency, a first-order lag can be designed and inserted in cascade with the relay. However, with these modifications, the frequency of interest cannot be pre-specified; it is fixed by the particular choice of cascaded linear element. This technique also has the problem that the introduction of the linear system affects the amplitude response of the original process, and in the case of a gain reduction, a smaller signal-to-noise ratio (SNR) could affect the estimation accuracy adversely.

Variations of the basic relay configuration are possible to give a negative inverse describing function that is a ray through the origin in the third or fourth quadrant of the complex plane. With these variations, it is possible to obtain a point on the process frequency response at an arbitrarily specified phase lag. Such a point is needed, for example, in the general Ziegler–Nichols frequency response method, or in the automatic tuning of controllers for servo systems (Tan *et al.*, 2001).

Adaptive Delay

An alternative approach is based on the use of an adaptive delay. With the adaptive delay, the frequency at a specified process lag of interest, $-\pi + \phi$ with $\phi \in [0,\pi]$, can be obtained without affecting the amplitude response of the original process. The modified arrangement is similar to Figure 4.6 but with $f: \Re \rightarrow \Re$ defined such that

$$u(t) = f(v(t)) = v(t - L_a(\omega))$$

where $L_a(\omega) = \phi/\omega$ is an adaptive time delay function and ω is the oscillating frequency of $v(t)$.

The feature will be investigated using a describing function analysis. Consider the modified relay in the dashed box of Figure 4.6, which consists of a relay cascaded to the delay function $f: \Re \rightarrow \Re$. Corresponding to the reference input $e(t) = a\sin(\omega t)$, the output of the modified relay can be expanded using the Fourier series and shown to be

$$u(t) = \frac{4\mu}{\pi} \sum_{k=1}^{\infty} \frac{\sin((2k-1)\omega t - \phi)}{2k-1}$$

Hence the describing function of the modified relay can be computed as

$$N(a) = \frac{4\mu}{\pi a} e^{-j\phi}$$

The negative inverse describing function

$$-\frac{1}{N(a)} = \frac{\pi a}{4\mu} e^{j(-\pi+\phi)}$$

is thus a straight line segment through the origin, as shown in Figure 4.7.

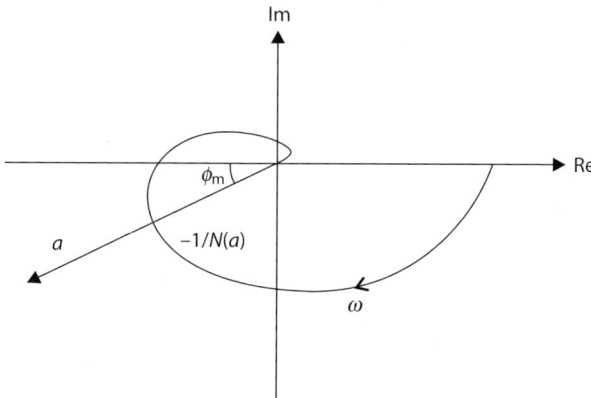

Figure 4.7 Negative inverse describing function of the modified relay.

The resultant amplitude and frequency of oscillation thus correspond to the intersection between $-1/N(a)$ and the process Nyquist curve. Hence, at the specified phase lag $(-\pi + \phi)$ rad, the inverse gain k_ϕ and the frequency of the process ω_ϕ can be obtained from the output amplitude a and the frequency ω_{osc} as

$$k_\phi = \frac{4\mu}{\pi a}$$

and

$$\omega_\phi = \omega_{osc}$$

It is thus possible to automatically track a frequency at the specified phase lag $(-\pi + \phi)$ rad for $\phi \in [0, \pi)$ without affecting the amplitude response of the original process. In this respect, the configuration can be considered as a generalised relay experiment where the conventional relay experiment then appears as a special case with $\phi = 0$. As with the basic relay feedback method, this technique facilitates single-button tuning; a feature which is invaluable for autonomous and intelligent control applications.

In the following, a practical implementation of procedure is presented to obtain two points on the process Nyquist curve: $G_p(j\omega_\pi)$ at the phase lag $-\pi$ rad and $G_p(j\omega_\phi)$ at the specified phase lag $(-\pi + \phi)$ rad.

Algorithm 4.2: Identifying two process Nyquist curve points.
Step 1 Estimate the process ultimate gain

$$k_\pi = \frac{1}{|G_p(j\omega_\pi)|}$$

and frequency ω_π with the normal relay feedback.

Step 2 With these estimates, an initial guess for the frequency $\widetilde{\omega}_\phi$ at the specified phase of $(-\pi + \phi)$ rad is

$$\widetilde{\omega}_\phi = \frac{\pi - \phi}{\pi}\omega_\pi$$

and L_a is initialised as $L_a = \phi / \widetilde{\omega}_\phi$.

Step 3 Continue the relay experiment, and adapt the delay function $f: \Re \rightarrow \Re$ to the oscillating frequency ω_{osc}, namely $L_a = \phi/\omega_{osc}$.

Upon convergence, $\omega_\phi = \omega_{osc}$, and

$$k_\phi = \frac{1}{|G_p(j\omega_\phi)|} = \frac{4\mu}{\pi a}$$

where a is the amplitude of process output oscillation

Algorithm end

Two-Channel Relay

Another method to yield a general point uses a two-channel relay configuration as proposed by Friman and Waller (1997). The basic construction is shown in Figure 4.8. A describing function with a phase lag may be decomposed into two orthogonal components. These components may be conveniently chosen to be along the real and imaginary axes. In this method, an additional relay, operating on the integral of the error, is added in parallel to the basic relay. With this method, the desired phase lag ϕ can be specified by selecting proper design parameters h_1 and h_2, as

$$\tan\phi = \frac{h_2}{h_1}$$

If $h_1 = h_2$, then $\phi = \pi/4$. It is clear that different combinations of h_1 and h_2 may result in the same ratio, and therefore the same phase angle. However, the absolute values of h_1 and h_2 must be chosen such that the resultant input $u(t)$ to the process does not saturate the actuator.

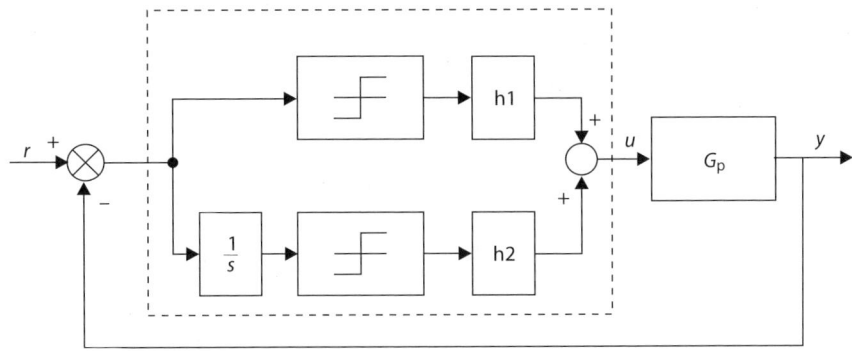

Figure 4.8 Setup of the two-channel relay tuning.

4.3.4 Estimation of Multiple Points

While the methods presented in Section 4.3.3 can give a general frequency of interest, tuning time is increased proportionally when more frequency estimations are required, especially if high accuracy is desirable. This is particularly true in the case of a process with a long time delay, where tuning time is considerably longer. A further extension of the procedure is possible which allows multiple frequency estimations in one single relay experiment. The arrangement and implementation is similar to Figure 4.6, but with the mapping function $f: \Re \to \Re$ defined such that

$$u(t) = f(v(t)) = \sum_{k=1}^{k_m} a_k \sin k\omega t, \quad k_m \in Z^+$$

where $v_1(t) = a_1 \sin(\omega t)$ is the fundamental frequency of the input $v(t)$, $a_k = 4\mu/k\pi$ and $k_m \omega$ is the upper bound of the frequencies injected. Assuming the oscillations are stationary and periodic, the multiple points on the frequency response can then be estimated as

$$G_p(jk\omega_\pi) = \frac{\int_{-T_\pi/2}^{T_\pi/2} y(t) e^{-jk\omega_\pi t} dt}{\int_{-T_\pi/2}^{T_\pi/2} u(t) e^{-jk\omega_\pi t} dt}, \quad k = 1, \ldots, k_m$$

where $T_\pi = 2\pi/\omega_\pi$.

The steps in a practical implementation of the procedure would be as follows. For good estimation accuracy from this procedure, the estimation procedure is restricted to two frequencies, so that $k_m = 2$.

Algorithm 4.3: Dual relay procedure

Step 1 Put the process under relay feedback, that is set $f(v(t)) = v(t)$ to obtain an initial estimate of its ultimate frequency $\tilde{\omega}_\pi$ from m oscillations, say with $m = 2$.

Step 2 Denote T_n as the time corresponding to the nth switch of the relay to $v = \mu$, and $f(\cdot)$ as the ultimate frequency estimate just prior to $t = T_n$. For $n > m$, update $f(\cdot)$ to output

$$u(t) = f(v(t)) = \frac{4\mu}{\pi} \sin \tilde{\omega}_{\pi,n-1}(t - T_n) + \frac{2\mu}{\pi} \sin 2\tilde{\omega}_{\pi,n-1}(t - T_n)$$

Step 3 Obtain a new estimate of the ultimate frequency $\tilde{\omega}_{\pi,n}$ from the resultant oscillation.

Step 4 Repeat Steps 2 and Step 3 until successive estimates of the ultimate frequency show a satisfactory convergence. At convergence, the amplitude and phase of the resultant oscillation can be obtained from

$$G_p(jk\omega_\pi) = \frac{\int_{-T_\pi/2}^{T_\pi/2} y(t) e^{-jk\omega_\pi t} dt}{\int_{-T_\pi/2}^{T_\pi/2} u(t) e^{-jk\omega_\pi t} dt}$$

Algorithm end

4.3.5 On-line relay tuning

One of the main features of the relay autotuning method, which probably accounts for its success more than any other associated features, is that it is a closed-loop method, and therefore on–off regulation of the process may be maintained even when the relay experiment is being conducted. However, the approach has several important practical constraints related to the structure that have remained largely unresolved to date.

1. First, it has a sensitivity problem in the presence of disturbance signals, which may be real process perturbation signals or equivalent ones arising from the varying process dynamics, nonlinearities and uncertainties present in the process. For small and constant disturbances, given that stationary conditions are known, an iterative solution has been proposed, essentially by adjusting the relay bias until symmetrical limit cycle oscillations are recovered. However, for general disturbance signals, there has been no effective solution to date.

2. Secondly, relating partly to the first problem, relay tuning may only begin after stationary conditions are attained in the input and output signals, so that the relay switching levels may be determined with respect to these conditions and the static gain of the process. In practice, under open-loop conditions it is difficult to determine when these conditions are satisfied and therefore when the relay experiment may be initiated.

3. Thirdly, the relay autotuning method is not applicable to certain classes of process that are not relay-stabilisable, such as the double integrator and runaway processes. For these processes, relay feedback is not effective in inducing stable limit cycle oscillations.

4. Finally, the basic relay method is an off-line tuning method. Some information on the process is first extracted with the process under relay feedback. This information is subsequently used to commission the controller. Off-line tuning has associated implications in the tuning control transfer, affecting operational process regulation, which may not be acceptable for certain critical applications. Indeed, in certain key process control areas (for example, vacuum control and environment control problems) directly affecting downstream processes, it may be just too expensive or dangerous for the control loop to be broken for tuning purposes, and tuning under tight continuous closed-loop control (not the on–off type) is necessary.

In this subsection, an on-line configuration of the relay method is presented which can be effective against some of the above-mentioned constraints of the basic relay autotuning method whilst retaining the simplicity of the original method. The configuration of the on-line method is given in Figure 4.9. Schei *et al.* (1992) proposed a similar configuration for iterative automatic tuning of PID controllers.

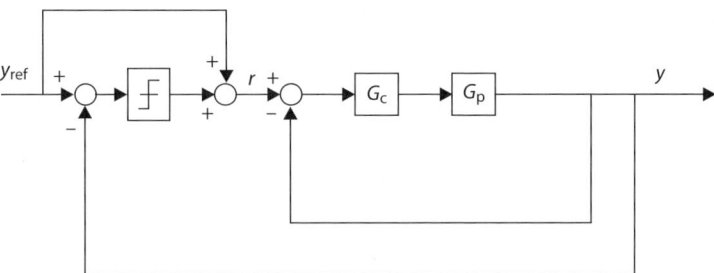

Figure 4.9 Configuration for on-line tuning of a closed-loop system.

In the on-line configuration, the relay is applied to an inner loop comprising a controller-stabilised process in the usual manner. The initial settings of the controller can be conservative and intended primarily to stabilise the process. The settings may be based on simple prior information about the process or default settings may be used. Indeed, practical applications of controller automatic tuning methods have been mainly to derive better values of current or default control settings. Thus the proposed configuration does not strictly pose additional and stringent prerequisites for its usage, but rather it uses information on the system that is already available in many cases.

In this configuration, if the system settles in a steady state limit cycle oscillation at frequency ω_π (ultimate frequency of the closed loop), then $G_{yr}(j\omega_\pi)$ can be obtained in the usual manner. It then follows that one point of the frequency response of $G_p(j\omega)$ may be obtained at $\omega = \omega_\pi$ as

$$G_p(j\omega_\pi) = \frac{G_{yr}(j\omega_\pi)}{G_{c,0}(j\omega_\pi)(1 - G_{yr}(j\omega_\pi))}$$

where $G_{c0}(s)$ denotes the initial PID controller during the tuning process.

4.4 Sensitivity Assessment Using Relay Feedback

Apart from control tuning, the relay feedback approach can also be used for control performance assessment purposes. In this section, one such application of the relay feedback method towards assessment of sensitivity will be illustrated. The method is based on deriving sensitivity parameters from the nonparametric frequency response of the compensated system.

4.4.1 Control Robustness

It has always been an important design objective to achieve adequate robustness for control systems functioning under harsh practical conditions. A good design for a control system is expected to be sufficiently robust to unmodelled dynamics as well as giving a good response to extraneous signals that arise from time to time during the system operations, including noise and load disturbances. Control robustness is also commonly used as an indication of how well the controller has been tuned, and whether retuning should be initiated. In the frequency domain, the maximum sensitivity and the stability margins provide an assessment of the robustness of a compensated system.

The maximum sensitivity, denoted M_S, fulfils the main requirements of a good design parameter for robustness (Åström et al., 1998). Robustness can usually be guaranteed by imposing a bound on the maximum sensitivity, typically in the range from 1.3 to 2.0. Lower M_S values give better robustness at the expense of a slower reaction (Åström and Hägglund, 1995). Several PID tuning rules have been established, where the maximum sensitivity is used as a design parameter (Åström et al., 1998; Panagopoulos et al., 1999). For a quick assessment of robustness, the stability margins (gain and phase margins) are the classical performance indicators, which have been in existence for a long time, and have found an affinity with practising control engineers. They have also been widely used as design specifications for the tuning of PID controllers (Åström and Hägglund, 1984; O'Dwyer et al., 1999).

The assessment of maximum sensitivity and stability margins of a control system usually requires a lengthy, nonparametric frequency response identification procedure (Ljung, 1999). Motivated by the relay feedback method pioneered by Åström and Hägglund (1995) to identify key process parameters efficiently for the tuning of PID controllers, this section will illustrate the use of a relay-type procedure and apparatus to automatically identify these robustness indicators from a control system. The experiment is a more elaborate one than the basic relay experiment to identify one critical point for PID tuning, since more information is clearly necessary for such an assessment. The apparatus uses a relay in series with a time delay element. The procedure involves sweeping the time delay through a range (which can be determined) and generating a series of sustained oscillations. Based on the amplitude and frequency of the oscillations, a chart (to be explained) of the proximity (to the critical point) versus phase can be systematically plotted. The maximum sensitivity and stability margins can be directly identified from the chart.

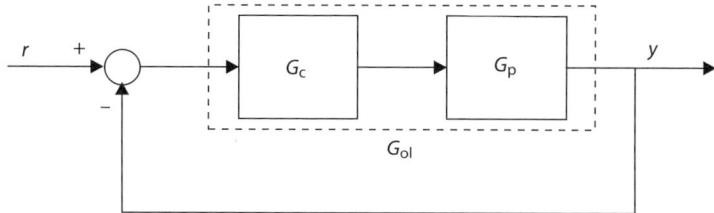

Figure 4.10 Feedback control system.

4.4.2 Maximum Sensitivity

A feedback control system comprising of a controller $G_c(s)$ and the plant $G_p(s)$ is shown in Figure 4.10. The compensated system, otherwise known as the loop transfer function, is defined as $G_{ol}(s) = G_p(s)G_c(s)$. The sensitivity function, which has many useful physical interpretations (Morari and Zafiriou, 1989; Davison *et al.*, 1999), is denoted $S(s)$ and is defined as

$$S(s) = \left[\frac{1}{1+G_p(s)G_c(s)}\right] = \left[\frac{1}{1+G_{ol}(s)}\right]$$

The maximum sensitivity M_S is defined as the maximum value of the sensitivity function over frequency, namely

$$M_S = \max_{0 \le \omega \le \infty} \left|\frac{1}{1+G_{ol}(j\omega)}\right| = \max_{0 \le \omega \le \infty} |S(j\omega)|$$

Figure 4.11 shows M_S graphically and, equivalently, as the inverse of the shortest distance from the locus of $G_{ol}(j\omega)$ to the critical point $s = -1 + j0$. The distance from other points on $G_{ol}(j\omega)$ to the critical point is thus always larger than $1/M_S$.

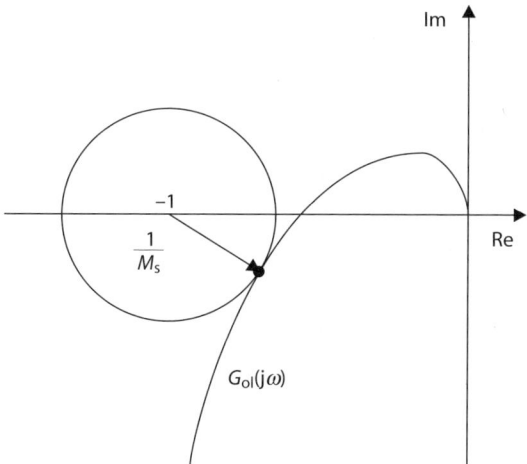

Figure 4.11 Definition of M_S.

Figure 4.12 shows two points on $G_{ol}(j\omega)$, $i = 1, 2$ at the respective phase lags of $-\pi + \phi(\omega_i)$, $i = 1, 2$ so that $\phi(\omega_i) = \pi + \arg G_{ol}(j\omega_i)$, $i = 1, 2$.

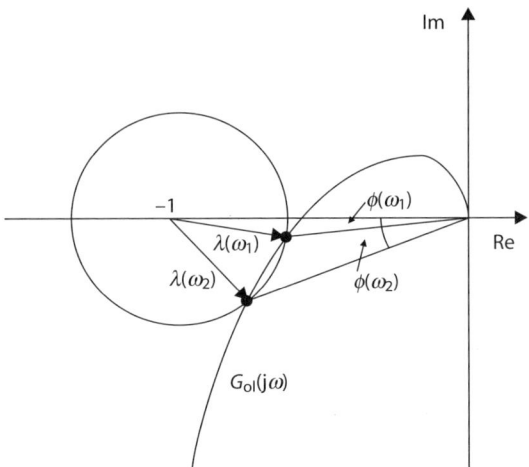

Figure 4.12 Relationship between $G_{ol}(j\omega_i)$ and $\lambda(\omega_i)$.

Both points are at a distance of $\lambda(\omega_1) = \lambda(\omega_2)$ from the critical point $s = -1 + j0$. Thus the points can be viewed as the intersection points of a circle centred at $s = -1 + j0$ with a radius of $\lambda(\omega_1)$. Direct vector manipulation gives

$$\lambda(\omega) = |G_{ol}(j\omega) + 1|$$

Consider a series of circles all centred at $s = -1$ with different radii λ. Following the definition of M_S, the circle at $s = -1 + j0$ with radius $\lambda(\omega^*) = 1/M_S$ will intersect the frequency response $G_{ol}(j\omega)$ at only one point, $\omega = \omega^*$, (Figure 4.11). If the radius is larger than $1/M_S$, so that $\lambda > 1/M_S$, intersection will occur at more points, typically two. On the other hand, if the radius is smaller than $1/M_S$, so that $\lambda < 1/M_S$, there will be no intersection. In this observation, lies the main idea for the identification of the maximum sensitivity.

Assume that $G_{ol}(j\omega)$ is available. A plot of $\lambda(\omega) = |G_{ol}(j\omega) + 1|$ versus $\phi(\omega)$ will typically exhibit the characteristics as shown in Figure 4.13.

From this plot, the turning point, where there is a one-to-one mapping from $\lambda(\omega)$ to $\phi(\omega)$, can be located. This is the minimum point on the plot in Figure 4.13, which corresponds to $\omega = \omega^*$, where there is only one intersection. Denoting this specific value of $\lambda(\omega^*)$ as λ^*, the maximum sensitivity can thus be obtained as $M_S = 1/\lambda^*$.

In the next subsection, it is explained how this $\lambda - \phi$ chart can be generated automatically and efficiently via a modified relay experiment.

4.4.3 Construction of the $\lambda - \phi$ Chart

Consider a modified relay feedback configuration as shown in Figure 4.14.

The modified relay comprises a normal relay in series with a time delay element that simply delays the action of the relay by L time units. Under this configuration, sustained oscillation at frequency ω_L can be usually attained from which the point at a phase lag of $-\pi + \omega_L L$ rad can be obtained. Then, by sweeping the time delay over a suitable range, values of $G_{ol}(j\omega)$ in the third and fourth quadrants of the complex plane can be obtained.

Assume that for a specific time delay L and a relay amplitude d, a sustained oscillation with amplitude a and frequency ω_L is obtained. Suppose $G_{ol}(j\omega_L) = \alpha_L + j\beta_L$; then α_L and β_L can be computed from the oscillations using the following equations that obtain from a describing function analysis:

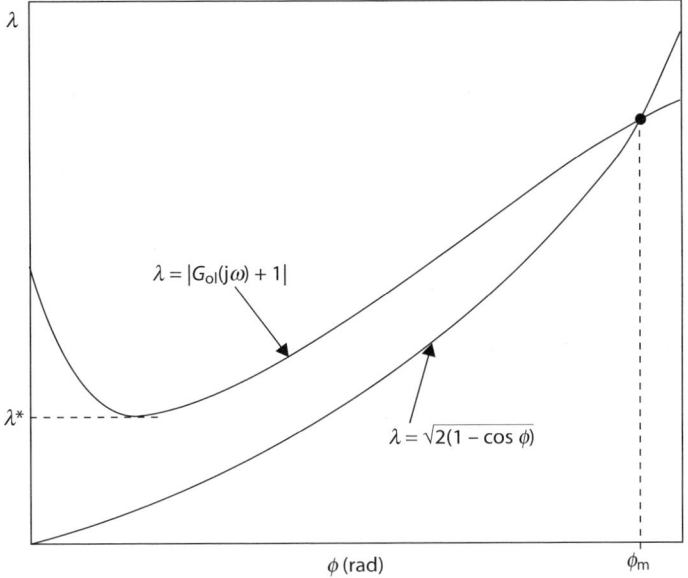

Figure 4.13 Typical plot of $\lambda(\omega)$ versus $\phi(\omega)$.

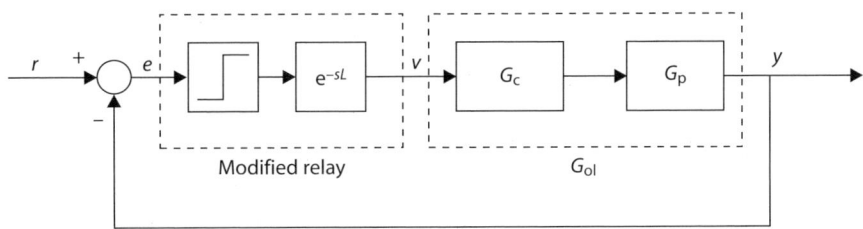

Figure 4.14 Relay with a pure delay.

$$\alpha_L = \frac{\pi a \cos(\omega_L L - \pi)}{4d}$$

$$\beta_L = \frac{\pi a \sin(\omega_L L - \pi)}{4d}$$

The describing function analysis is an approximation method; thus there is some inevitable error expected with these estimates. Alternatively, assuming the oscillations are periodic, Fourier or spectral analysis applied to the signals v and y yields an efficient estimate of $G_{ol}(j\omega_L)$ at $\omega = \omega_L$ as

$$G_{ol}(jk\omega_L) = \frac{\int_0^{T_L} y(t) e^{-jk\omega_L t} dt}{\int_0^{T_L} v(t) e^{-jk\omega_L t} dt}$$

In this way, $G_{ol}(j\omega_L)$ is obtained and thus the corresponding $\lambda(\omega_L)$ can be computed. By sweeping L over a range, the plot of $\lambda - \phi$ can be generated using describing function analysis or Fourier analysis method.

Potential users of the approach will be concerned to learn what range of L should be used, since this range will determine the duration of the identification experiment. To this end, it may be noted that the parameters to be determined can typically be derived from $G_{ol}(j\omega_L)$ in the third quadrant of the complex plane. The experiment can thus be terminated when $\lambda(\omega_L)$ is close to $\pi/2$, and $\gamma(\pi/2) \leq L\omega_L$, where based on empirical experience, the user selectable parameter γ can lie in the range $0.8 \leq \gamma \leq 1.0$. Alternatively, the experiment can also terminate once all the required parameters (maximum sensitivity and stability margins) are found. It is also possible, in certain cases, to pre-determine an upper bound L_{upp} for L by identifying the frequency $\tilde{\omega}$, where $G_{ol}(j\tilde{\omega})$ lies on the negative imaginary axis. This can be done by adding an integrator to the compensated system $G_{ol}(s)$ (Åström and Hägglund, 1995), in addition to the modified relay. Then it is straightforward to pre-determine the upper bound L_{upp} as $L_{upp} = \pi/\tilde{\omega}$.

4.4.4 Stability Margins Assessment

The gain and phase margins of a compensated system are the classical stability indicators that are familiar to most control engineers. In fact, many PID design rules have been formulated to achieve desired stability margins that are specified by the users.

The gain margin G_m and phase margin ϕ_m of a compensated system are defined as follows

$$G_m = \frac{1}{|G_{ol}(j\omega_u)|}$$
$$\phi_m = \phi(\omega_g) - \arg G_{ol}(j\omega_u)$$

where ω_u is the phase-crossover frequency ($\arg G_{ol}(j\omega_u) = -\pi$) and ω_g is the gain-crossover frequency ($|G_{ol}(j\omega_g)| = 1$, as in Figure 4.15). Typical desired values of G_m and ϕ_m lie in the ranges $2 \leq G_m \leq 5$ and $\pi/6 \leq \phi_m \leq \pi/3$ respectively. Following Åström and Hägglund (1995), it may be noted that the following relations hold:

$$G_m > \frac{M_s}{M_s - 1}$$

and

$$\phi_m > 2\sin^{-1}\left(\frac{1}{M_s}\right)$$

This implies that typically $\omega_g < \omega < \omega_u$. The gain and phase margins can be simultaneously identified from the same $\lambda - \phi$ chart generated earlier without additional experimentation. Referring to Figure 4.15, based on the margin definitions, it can be shown that the gain margin of the compensated system can be obtained as

$$G_m = \frac{1}{1 - \lambda(\omega_u)}$$

where $\lambda(\omega_u)$ corresponds to $\phi = 0$.

The procedure to identify the phase margin ϕ_m is slightly more elaborate. It requires the equivalent point on the chart to be located where $|G_{ol}(j\omega)| = 1$. This can be done by substituting $|G_{ol}(j\omega)| = 1$ into equation for $\lambda(\omega) = |G_{ol}(j\omega) + 1|$, thus obtaining the locus of $|G_{ol}(j\omega)| = 1$, which can be shown to be described by

$$\lambda(\omega) = \sqrt{2(1 - \cos\phi)}$$

The intersection between this locus and the earlier plotted $\lambda - \phi$ curve will yield ϕ_m (Figure 4.16).

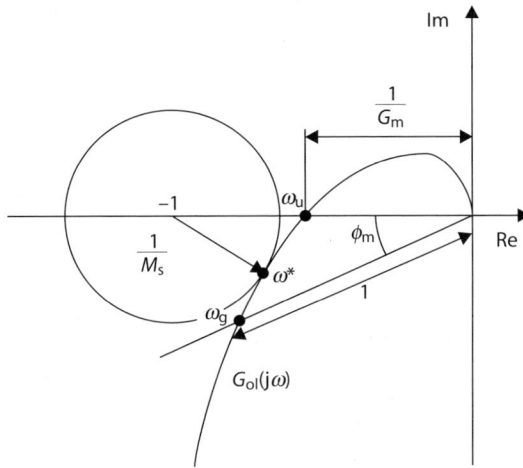

Figure 4.15 Definition of gain margin G_m and phase margin ϕ_m.

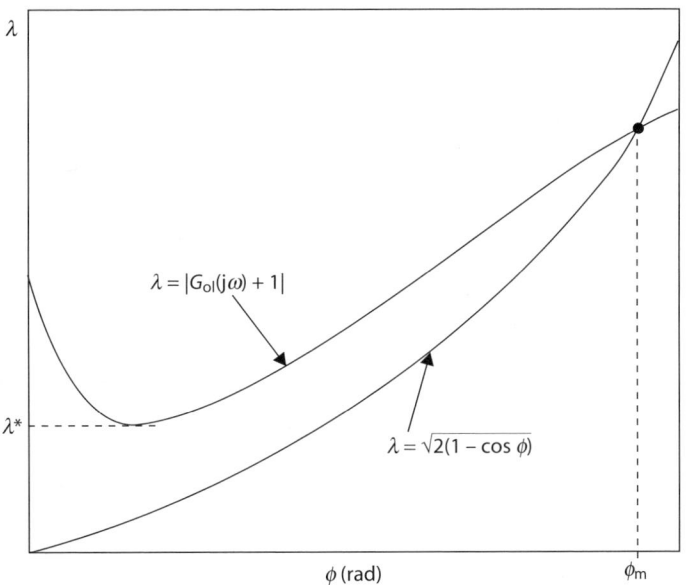

Figure 4.16 Identification of G_m and ϕ_m from the $\lambda - \phi$ plot.

4.5 Conversion to Parametric Models

The identification methods described in the previous sections may yield the frequency response of the process at certain frequencies. However, for simulation or certain control purposes, transfer function models fitted to the frequency response estimation are more directly useful. Pintelon *et al.* (1994) give a good survey of parameter identification of transfer functions without dead time in the frequency domain. With dead time, the parameter identification is usually a nonlinear problem (Palmor and Blau, 1994) and difficult to solve. While the focus of the chapter is on nonparametric models, a short section on the derivation of simple parametric models from the nonparametric information obtained will be useful. In this section, the identification of common low-order transfer function models from the frequency response estimation will be illustrated. Such models have proven to be relevant and adequate particularly in the process control industry. Algorithms are presented to calculate the parameters of these approximate transfer function models from the frequency response information acquired using the techniques discussed in the earlier sections. Such an identification technique is readily automated, and, as such, it is useful for autotuning applications.

4.5.1 Single and Multiple Lag Processes

It is well known that many processes in the process industry are of low-order dynamics, and they can be adequately modelled by a rational transfer function of the First-Order Plus Dead Time (FOPDT) model:

$$G_p(s) = \left[\frac{Ke^{-sL}}{Ts+1}\right]$$

where K, T and L are real parameters to be estimated. It is desirable to distinguish this class of processes from higher-order ones since controller design and other supervisory tasks are often much simplified with the use of the model.

Two points on the process Nyquist curve are sufficient to determine the three parameters of the FOPDT process model. Assume that the following Nyquist points of the process $G_p(s)$ have been obtained: $G_p(0)$ and $G_p(j\omega_1)$, $\omega_1 \neq 0$. Then the process static gain is

$$K = G_p(0)$$

Equating the process and model at $\omega = \omega_1$ and using the relations to define quantities k_1, ϕ_1 yields

$$\frac{1}{k_1} = |G_p(j\omega_1)| = \frac{K}{\sqrt{1+T^2\omega_1^2}}$$

$$\phi_1 = \arg G_p(j\omega_1) = -\arctan T\omega_1 - L\omega_1$$

It follows that

$$T = \frac{1}{\omega_1}\sqrt{k_1^2 K^2 - 1} \quad \text{and} \quad L = -\frac{1}{\omega_1}(\phi_1 + \arctan \omega_1 T)$$

A more general model adequate for processes with a monotonic open-loop step response is of the following form

$$G_p(s) = \left[\frac{Ke^{-sL}}{(Ts+1)^n}\right]$$

where K, T and L are real parameters to be estimated, and $n \in Z^+$ is an order estimate of the process.

Lundh (1991) proposed a method for order estimation based on this model type. Lundh noted that the maximal slope of the frequency response magnitude is a measure of the process complexity, and the slope near the critical frequency was then used to estimate the relative degree of the process, choosing one of three possible models to represent the process. In Lundh's method, FFTs are performed on the input and output of the process, from which the amplitudes of the first and third harmonics of the frequency spectrum are used to compute the amplitude gains at these two frequencies. From these frequencies, the slope of the frequency response magnitude at the geometrical mean of the harmonics is then calculated. A simpler method for order estimation is proposed here that does not require a Fourier transform on the input and output signals of the process.

4.5 Conversion to Parametric Models

Assume that the following Nyquist points of the process $G_p(s)$ have been obtained: $G_p(j\omega_1)$ and $G_p(j\omega_2)$, where neither ω_1 nor ω_2 are zero in value and $\omega_1 \neq \omega_2$. Equating the gain of the process and model at ω_1 and ω_2, and defining quantities k_1 and k_2 gives

$$\frac{1}{k_1} = |G_p(j\omega_1)| = \frac{K}{\left(\sqrt{1+T^2\omega_1^2}\right)^n}$$

and

$$\frac{1}{k_2} = |G_p(j\omega_2)| = \frac{K}{\left(\sqrt{1+T^2\omega_2^2}\right)^n}$$

Eliminating the unknown variable T between these two equations yields

$$\left[1+\left(\frac{\omega_2}{\omega_1}\right)^2 ((Kk_1)^{2/n}-1)\right]^{n/2} = Kk_2$$

Equating the phase of the process and model at ω_1 and ω_2, and defining quantities ϕ_1 and ϕ_2 gives

$$-\phi_1 = \arg G_p(j\omega_1) = -n\arctan T\omega_1 - L\omega_1$$

and

$$-\varphi_2 = \arg G_p(j\omega_2) = -n\arctan T\omega_2 - L\omega_2$$

The first of these phase equations is used to find model parameter L. And, as a means of finding the best fit for the order $n \in Z^+$ where n lies in a user-specified range, $1 \leq n \leq n_{max}$, the second phase equation is recast as a cost function defined by

$$J(n) = |-\phi_2 + n\arctan T\omega_2 + L\omega_2|$$

The algorithm will prescribe model parameters that give the minimum value of the cost, namely the algorithm selects $n = n_{min}$, where

$$J(n_{min}) = \min_n \{J(n)\}$$

Thus a simple algorithm to obtain the parameters K, T and L and order $n \in Z^+$ for the process model is outlined below, where a specified upper bound on the order of the process is denoted n_{max}.

Algorithm 4.4: Model parameter computation

Step 1 Loop step: From $n = 1$ to $n = n_{max}$

Compute K from $\left[1+\left(\frac{\omega_2}{\omega_1}\right)^2 ((Kk_1)^{2/n}-1)\right]^{n/2} = Kk_2$

Compute T from $\frac{1}{k_1} = |G_p(j\omega_1)| = \frac{K}{\left(\sqrt{1+T^2\omega_1^2}\right)^n}$ or $\frac{1}{k_2} = |G_p(j\omega_2)| = \frac{K}{\left(\sqrt{1+T^2\omega_2^2}\right)^n}$

Compute L from $-\phi_1 = \arg G_p(j\omega_1) = -n\arctan T\omega_1 - L\omega_1$

Use the values of n, K, T and L to compute the cost function:

$$J(n) = |-\phi_2 + n\arctan T\omega_2 + L\omega_2|$$

Step 2 At the end of the loop step, select the set of model parameters n, K, T and L corresponding to
$n = n_{\min}$, where $J(n_{\min}) = \min_{n}\{J(n)\}$.
Algorithm end

4.5.2 Second-Order Modelling

Second-Order Plus Dead Time (SOPDT) models can be used to represent both monotonic and oscillatory processes. In this subsection, a non-iterative method is proposed for model identification.

The stable second-order plus dead time model is

$$G_p(s) = \left[\frac{e^{-sL}}{(as^2 + bs + c)}\right]$$

Assume that the process frequency response data is available at the points $G_p(j\omega_i)$, $i = 1, ..., M$. It is required to be fitted to $G_p(s)$ of the stable second-order plus dead time model form such that

$$G_p(j\omega_i) = \left[\frac{e^{-\omega_i L}}{a(j\omega_i)^2 + b(j\omega_i) + c}\right] \text{ where } i = 1, ..., M$$

The determination of the model parameters a, b, c and L in the stable second-order plus dead time model seems to be a nonlinear problem. One way of solving this problem is to find the optimal a, b and c given L, and then iteratively determine L using a suitable search algorithm. To avoid the iterative process modulus and phase equations are used.

The modulus of the SOPDT model equation is computed to give

$$[\omega_i^4 \quad \omega_i^2 \quad 1]\theta = \frac{1}{|G_p(j\omega_i)|^2}$$

where $\theta = [a^2 \quad (b^2 - 2ac) \quad c^2]^T \in \mathfrak{R}^3$ and $i = 1, ..., M$.

This equation is independent of L and forms a system of M linear equations in $\theta \in \mathfrak{R}^3$ which can be solved for vector θ using the linear least squares method. The model parameters are recovered as

$$[a \quad b \quad c] = [\sqrt{\theta_1} \quad \sqrt{\theta_2 + 2\sqrt{\theta_1 \theta_3}} \quad \sqrt{\theta_3}]$$

In a similar way, the phase relation for the stable SOPDT model equation gives

$$\omega_i L = -\arg G_p(j\omega_i) - \tan^{-1}\left(\frac{b\omega_i}{c - a\omega_i^2}\right), \quad i = 1, ..., M$$

Hence parameter L can also be obtained with the least squares method.

4.6 Case Studies

In this section, case studies involving the applications of the relay feedback methods, presented in earlier sections, to the identification of nonparametric models will be presented. Where control results are presented which are based on the models obtained, references to specific literature on the control designs used will be given.

A common coupled-tanks apparatus is used for the case studies, although it should be noted that the results of the case studies are obtained at different times and from different coupled tanks (though of the same type). The schematic of the coupled-tanks apparatus is shown in Figure 4.17, and a photograph of the apparatus is shown in Figure 4.18. The pilot scale process consists of two square tanks, Tank 1 and

4.6 Case Studies   175

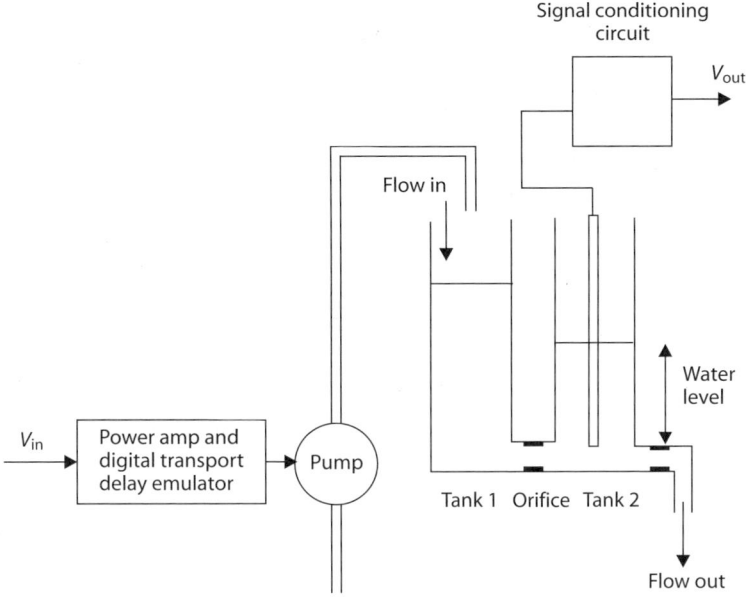

Figure 4.17 Schematic of the coupled-tanks system.

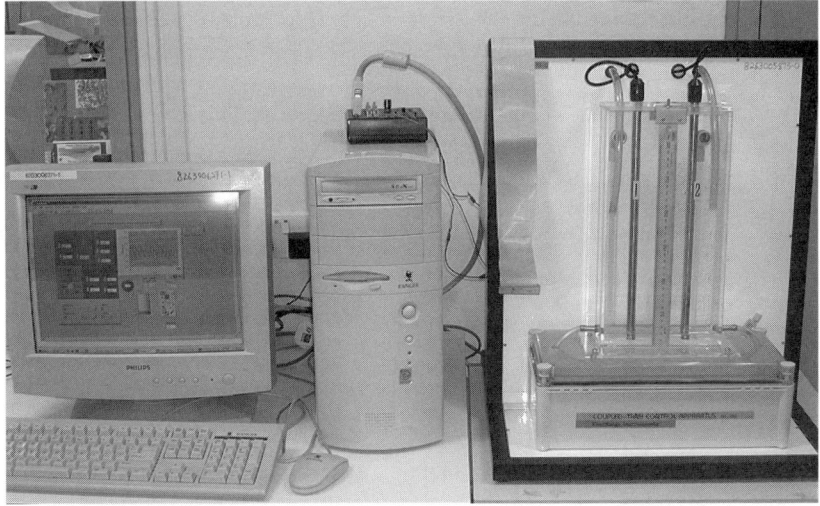

Figure 4.18 Photograph of the coupled-tanks system.

Tank 2, coupled to each other through an orifice at the bottom of the tank wall. The effect of transport delay has been intentionally incorporated in the experiment setup by means of additional digital electronic hardware that cascades a time delay into the open-loop process. (Such transport delays might arise, for example, when there is a long vented and gravity-fed leg.) The inflow (control input) is supplied by a variable speed pump that takes water from a reservoir and pumps it into Tank 1 though a long tube. The orifice between Tank 1 and Tank 2 allows the water to flow into Tank 2. In the case studies, it is an objective to identify the frequency response of the process and to control it with the voltage to drive the pump as input and the water level in Tank 2 as process output. The coupled-tanks apparatus is connected to a PC via an A/D and D/A board. LabVIEW 6.0 from National Instruments is used as the control development platform. This coupled-tanks pilot process has process dynamics that are representative of many fluid-level control problems found in the process control industry.

Case Study 4.1: Improved Estimation Accuracy for the Relay Experiment

In this case study, the modified relay feedback procedure to obtain improved estimation accuracy of the critical point, described in Section 4.3.2, will be illustrated. In the real-time experiments, both the basic relay feedback procedure and the modified relay feedback procedure are used to estimate the critical point of the coupled-tanks process. The estimates obtained are shown in Table 4.5.

Table 4.5 Relay feedback experiment results.

Critical parameter	Basic relay feedback	Modified relay feedback	Spectral analysis
Ultimate gain k_π	6.88	5.90	5.95
Ultimate frequency ω_π	0.0747	0.0571	0.058

These estimates are then used to autotune PID controllers for the process and the performance in closed-loop setpoint tracking are compared. The results are shown in Figure 4.19 and Figure 4.20 for PID controllers tuned using the Ziegler–Nichols formula. Visual inspection of Figure 4.19 and Figure 4.20 shows that the improved accuracy in critical point estimation has yielded improved closed-loop control performance.

The trade-off is that a longer tuning time is required. This longer tuning time is to be expected, as the basic relay feedback procedure essentially settles to limit cycle behaviour in approximately two periods while the proposed modified relay feedback technique works by generating iterative updates for the

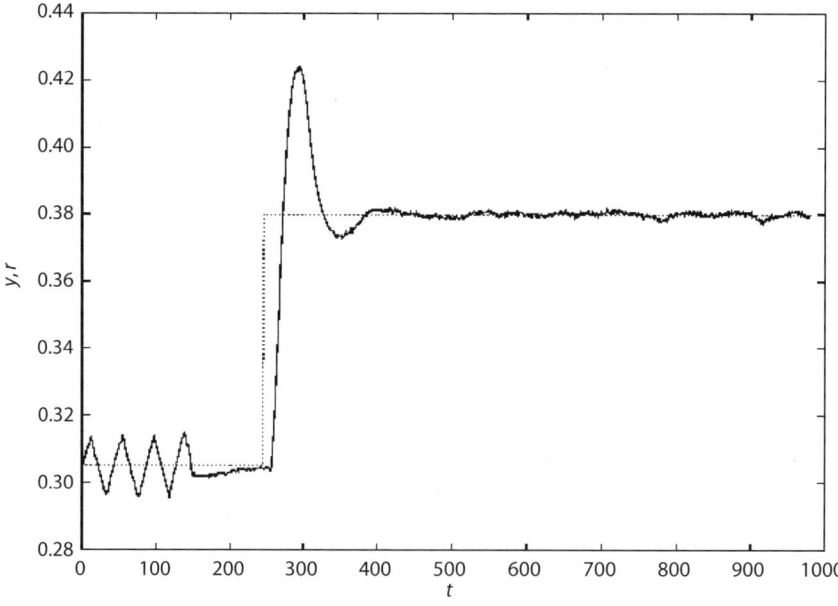

Figure 4.19 Basic relay tuning and closed-loop performance using the Z–N formula.

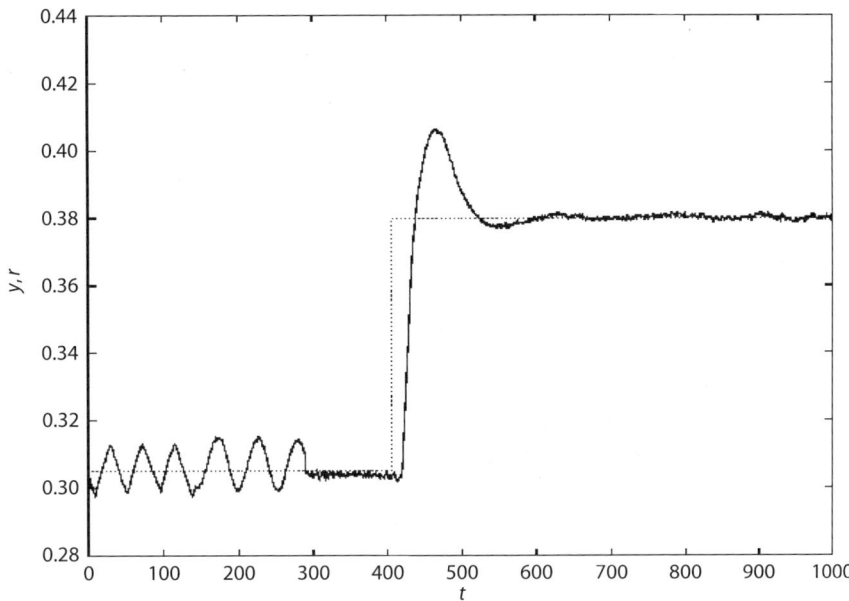

Figure 4.20 Modified relay tuning and closed-loop performance using the Z–N formula.

function $f: \mathfrak{R} \to \mathfrak{R}$. The modified technique thus requires a few more oscillation periods before attaining convergence.

Case Study 4.2: Estimation of a General Point

This case study will apply the procedures to identify a general point as described in Section 4.3.3. It is desired to obtain both the critical point and the point with a phase lag of $\phi = \pi/4$. Thereafter, it is desired to tune a PID controller to attain desired phase and gain margins following the work by Tan *et al.* (1996).

The results of the real-time experiments are shown in Figure 4.21. The default margin specifications of $G_m = 2$ and $\phi_m = \pi/4$ are chosen. From $t = 0$ to $t = 750$, the relay-based frequency response estimation is run and the two points of the process frequency response are obtained as $(k_m, \omega_\pi) = (7.33, 0.03)$ and $(k_\phi, \omega_\phi) = (3.16, 0.017)$. The PID control design procedure as reported by Tan *et al.* (1996) is used and the autotuning is completed. In the simulation test, a setpoint change is made at $t = 900$. In addition, the experiments also included a 10% load disturbance in the process at $t = 1600$. It can be seen from Figure 4.21 that the PID control thus tuned attains a tight setpoint response and shows good closed-loop regulation of load disturbances.

Case Study 4.3: Estimation of Multiple Points

In this case study, multiple points are estimated simultaneously using the procedures described in Section 4.3.4. An additional time delay of 500 is added digitally to simulate transport time delay on the coupled-tanks apparatus. The estimation procedures are invoked from $t = 0$ to $t = 6000$, after which two points of the frequency response are obtained. Following the transfer function fitting procedures in Section 4.5, a transfer function model is fitted as

$$G_p(s) = \left[\frac{0.013 e^{-503s}}{(s^2 + 2.089s + 0.012)}\right]$$

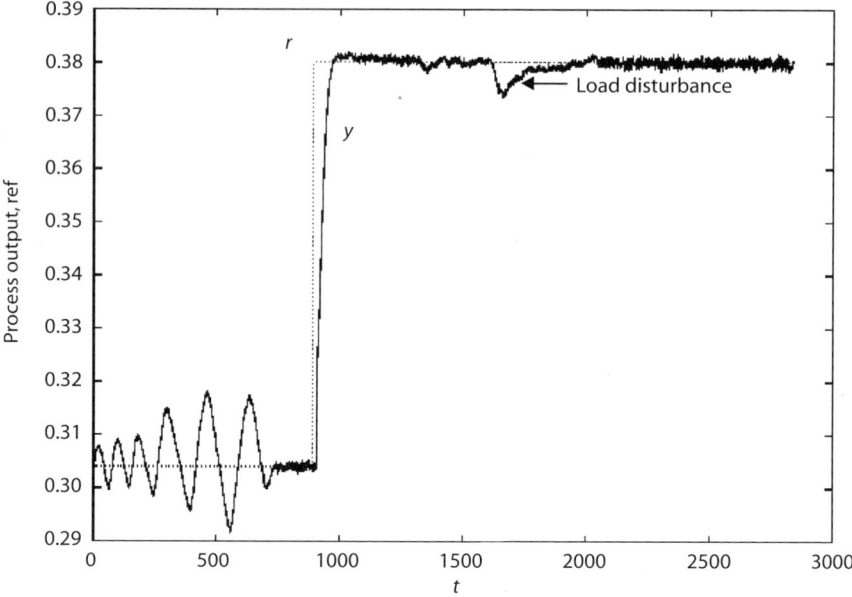

Figure 4.21 Closed-loop performance of the PID control based on the estimation of general points.

A Smith predictor controller is tuned following the recommendations of Lee *et al.* (1995). From Figure 4.22, it can be seen that good closed-loop setpoint tracking and load disturbance regulation responses are achieved.

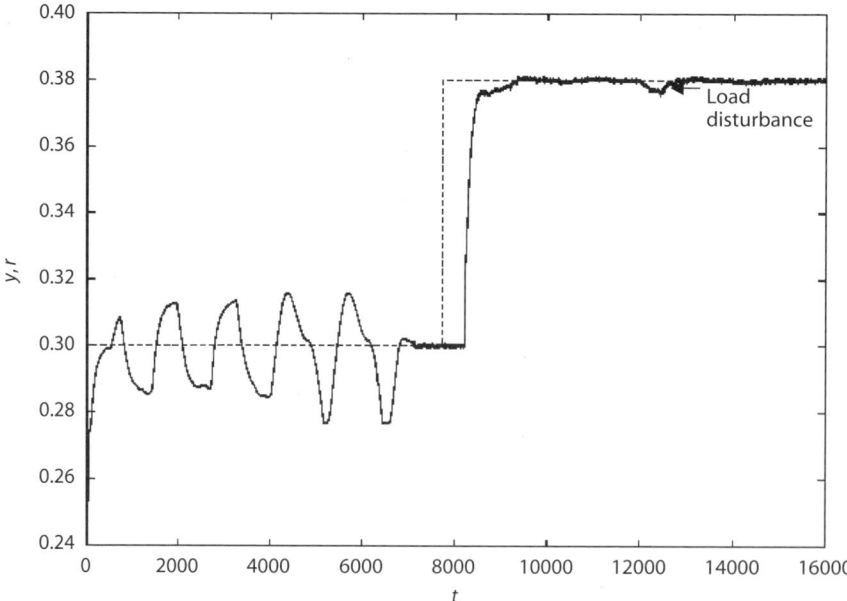

Figure 4.22 Closed-loop performance of the controller based on multiple-points estimation.

Case Study 4.4: On-line Relay Tuning

In this case study, the on-line relay tuning procedures described in Section 4.3.5 will be illustrated. In the initial part of the experiment from $t = 0$ to $t = 70$, the on-line relay tuning method is conducted on a closed-loop system with a poorly tuned controller. It yields stable limit cycle oscillations as shown in Figure 4.23. From the oscillations, the controller is then tuned using a direct control tuning method (Tan et al., 2000). A setpoint change occurs at $t = 130$. The good tracking performance of the control system is shown in Figure 4.23.

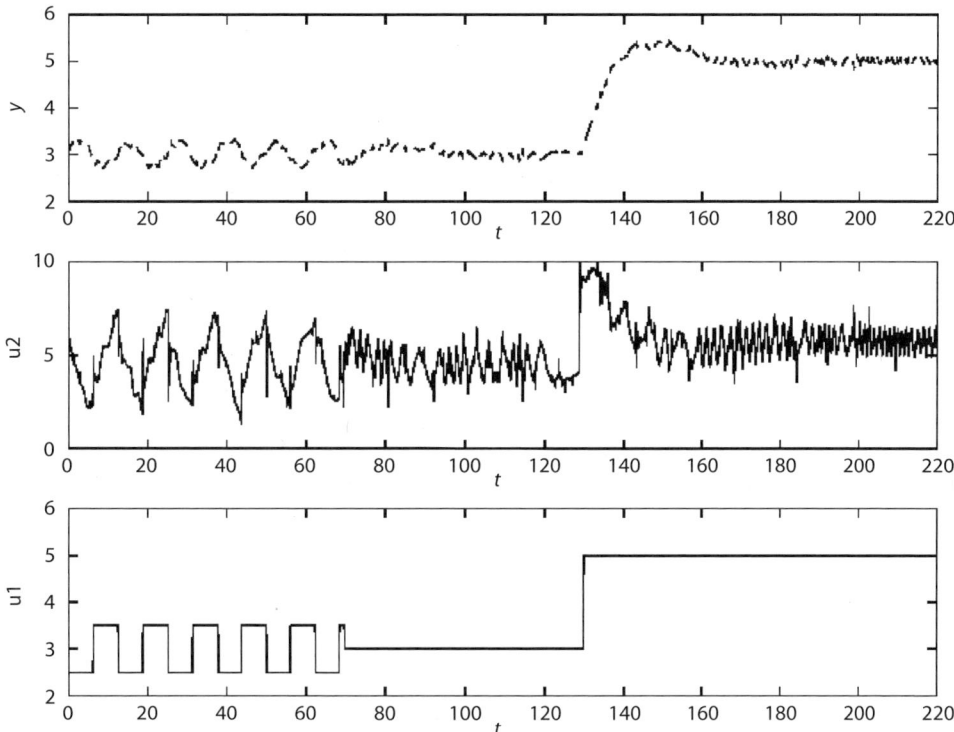

Figure 4.23 Real-time experiment results.

Case Study 4.5: Sensitivity Assessment

In this case study, the performance of a PI control loop for the coupled-tanks apparatus will be assessed using the procedures described in Section 3.4. The following PI controller is used:

$$G_c(s) = 25 + \frac{0.5}{s}$$

Following the procedures described in Section 4.4.3, L is varied from 0.5 to 2.5. Table 4.6 shows the variation of λ and ϕ with L.

Figure 4.24 shows the $\lambda - \phi$ plot, from which the following can be identified:

1. maximum sensitivity $M_S = 1.47$

2. gain margin, $G_m = 7.142$

3. phase margin, $\phi_m = 0.75$

Table 4.6 The variation of λ and ϕ with L.

L	0.50	1.00	1.10	1.20	1.30	1.40	1.50	2.00	2.20	2.50
ϕ	0.12	0.35	0.38	0.42	0.45	0.49	0.55	0.78	1.10	1.31
λ	0.84	0.75	0.72	0.69	0.69	0.68	0.70	0.73	0.96	1.15

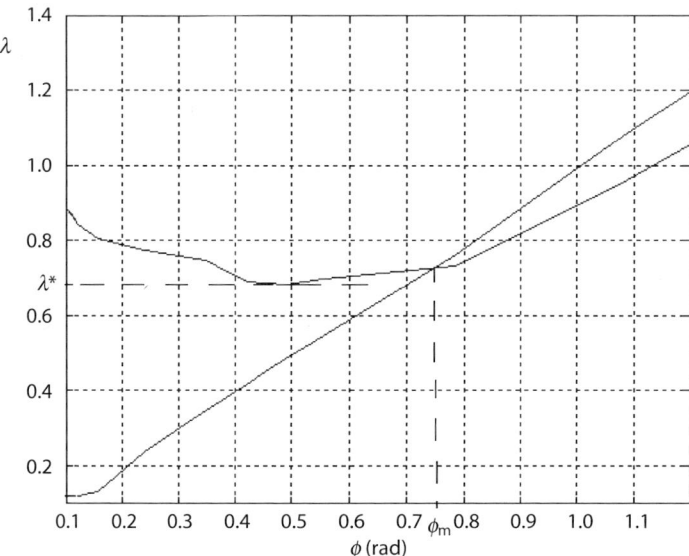

Figure 4.24 Identification of M_S and the stability margins from the real-time experiment.

References

Åström, K.J. and Hägglund, T. (1984) Automatic tuning of simple regulators with specifications on phase and amplitude margins. *Automatica*, **20**(5), 645–651.

Åström, K.J. and Hägglund, T. (1995) *PID Controllers: Theory, Design and Tuning*. Instrument Society of America, Research Triangle Park, NC.

Åström, K.J., Panagopoulos, H. and Hägglund, T. (1998) Design of PI controllers based on non-convex optimisation. *Automatica*, **35**(5), 585–601.

Crowe, J. and Johnson, M.A. (1998) *GB Patent No. 9802358.3: Phase Locked Loop Identification*.

Crowe, J. and Johnson, M.A. (2000) A process identifier and its application to industrial control. *IEE Proc. Control Theory Appl.*, **147**(2), 196–204.

Davison, D.E., Kabamba, P.T. and Meerkov, S.M. (1999) Limitations of disturbance rejection in feedback system with finite bandwidth. *IEEE Trans. Automatic Control*, **44**(6), 1132–1144.

Friman, M. and Waller, K. (1994) Auto-tuning of multi-loop control system. *Ind. Eng. Chem. Res.*, **33**, 1708.

Lee, T.H., Wang, Q.G. and Tan, K.K. (1995) A modified relay-based technique for improved critical point estimation in process control. *IEEE Control Syst. Technol.*, **3**(3), 330–337.

Ljung, L. (1999) *System Identification: Theory for the User*, 2nd edn. Prentice Hall, Upper Saddle River, NJ.

Lundh, M. (1991) Robust adaptive control. PhD Thesis, Lund Institute of Technology, Lund, Sweden.

Morari, M. and Zafiriou, E. (1989) *Robust Process Control*. Prentice Hall, Englewood Cliffs, NJ.

O'Dwyer, A. (1999) *Proceedings of the 2nd Wismarer Automatisierungssymposium*, Wismar, Germany, September.

Palmor, Z.J. and Blau, M. (1994) An auto-tuner for Smith dead-time compensators. *Int. J. Control*, **60**(1), 117–135.

Panagopoulos, H., Åström, K.J. and Hägglund, T. (1999) Design of PID controllers based on constrained optimisation (Invited Paper). *Proc. 1999 American Control Conference (ACC'99)*, San Diego, CA.

Pintelon, R., Guillaume, P., Rolain, Y., Schoukens, J. and Van Hamme, H. (1994) Parametric identification of transfer functions in frequency domain – a survey. *IEEE Trans. Automatic Control*, **39**(11), 2245–2259.

Saeki, M. (2002) A new adaptive identification method of critical gain for multivariable plants. *Asian Journal of Control* (Special issue on Advances in PID Control), **4**(4), 464–471.

Schei, D.E., Edgar, T.F. and Mellichamp, D.A. (1992) A method for closed-loop automatic tuning of PID controllers. *Automatica*, **28**(3), 587–591.

Tan, K.K., Wang, Q.G. and Lee, T.H. (1996) Enhanced automatic tuning procedure for PI/PID control for process control. *AIChE Journal*, **42**(9), 2555–2562.

Tan, K.K., Wang, Q.G. and Hang, C.C. with T. Hägglund (1999) *Advances in PID Control*. Springer-Verlag, London.

Tan, K.K., Lee, T.H., Dou, H. and Huang, S. (2001) *Precision Motion Control*. Springer-Verlag, London.

Tan, K.K., Lee, T.H. and Jiang, X. (2001) Online relay identification, assessment and tuning of PID controller. *J. Process Control*, **11**, 483–496.

Ziegler, J.G. and Nichols, N.B. (1942) Optimum settings for automatic controllers. *Trans. ASME*, **64**, 759–768.

5 Relay Experiments for Multivariable Systems

Learning Objectives

5.1 Introduction

5.2 Critical Points of a System

5.3 Decentralised Relay Experiments for Multivariable Systems

5.4 A Decentralised Multi-Input, Multi-Output PID Control System Relay-Based Procedure

5.5 PID Control Design at Bandwidth Frequency

5.6 Case Studies

5.7 Summary

References

Further Reading

Learning Objectives

The application of the relay experiment to single-input, single-output systems is widely accepted and the experiment has been used in commercial PID control hardware units and PID control software packages. However, there is a body of work on the relay experiments in multi-input, multi-output systems, although it is difficult to gauge whether industrial application has been attempted, and still less whether successful results have been obtained. This chapter is motivated by the results that can be obtained by simultaneous relays operating in a decentralised control structure for control of a multivariable system.

The learning objectives for this chapter are to:

- Understand the concept of the critical point for multivariable systems.
- Be able to setup and use the decentralised relay experiment method for two-input, two-output systems.
- Extend the relay experiment to a general decentralised multivariable PID control system.
- Understand the method and value of PID control design at bandwidth frequency.
- Investigate the performance of the methods using some case study material of simulated multivariable industrial system models.

5.1 Introduction

A fundamental problem associated with relays in feedback control systems is the determination of limit cycle frequencies and the associated system gains. In recent years, the existence of various automatic tuning techniques for industrial SISO controllers has led to renewed interest in relay control and particularly in the limit cycles associated with such control (Åström and Hägglund, 1988; Palmor and Blau, 1994). Relays in these applications are used mainly for the identification of points on the process Nyquist curve or at the point of system closed-loop instability from which essential information for tuning industrial process controllers, such as the PID controller, can be obtained. When the relay technique is extended to MIMO systems there are three possible schemes (Wang et al., 1997):

1. *Independent single relay feedback* (IRF): in this method, only one loop at a time is closed subject to relay feedback, while all the other loops are kept open. The main disadvantage is that independent single relay feedback does not excite any multivariable interaction directly, and it is thus difficult to tune a fully cross-coupled multivariable PID controller.

2. *Sequential relay feedback* (SRF): the main idea of sequential relay tuning is to tune the multivariable system loop-by-loop, closing each loop once it is tuned, until all the loops are tuned and closed. To tune each loop, a relay feedback configuration is set up to determine the ultimate gain and frequency of the corresponding loop. The PI/PID controller coefficients are then computed on the basis of this information and the loop closed. Loh *et al.* (1993) have presented an autotuning method for multi-loop PID controllers using a combination of relay autotuning and sequential loop closing. The method, which requires very little process knowledge, has the advantage of guaranteed closed-loop stability.

3. *Decentralised relay feedback* (DRF): this method uses a simultaneous set of completely closed-loop relay tests in all loops while the independent single relay feedback and sequential relay feedback methods only have partial closed-loop relay tests. Closed-loop testing is preferred to open-loop tests since a closed-loop test keeps outputs close to the setpoints and causes less perturbation to the process. In addition, the decentralised relay feedback excites and identifies multivariable process interaction, whilst the independent single relay feedback and sequential relay feedback methods may not excite any multivariable process interaction directly. Palmor *et al.* (1995a) have proposed an algorithm for automatic tuning of decentralised control for two-input and two-output (TITO) plants. The main objective of the autotuner is to identify the desired critical point (DCP).

Using the results from relay experiments, either a fully cross-coupled or a multi-loop controller can be selected and tuned use with MIMO processes (Wang *et al.*, 1997). Decentralised controllers, sometimes known as multi-loop controllers, have a simpler structure and, accordingly, less tuning parameters than the fully cross-coupled ones. In addition, in the event of component failure, it is relatively easy to stabilise the system manually, since only one loop is directly affected by the failure (Skogestad and Morari, 1989).

This chapter opens with a review of the critical points for MIMO processes and the calculation of process gains at specific system frequencies. Special mention is made of the zero frequency and the bandwidth frequency points for gain identification. An algorithm for the automatic tuning of decentralised controllers for two-input and two-output plants has been devised (Palmor *et al.*, 1995a). The main objective of the algorithm is to use decentralised relay feedback to identify a desired critical point. The advantage of the method due to Palmor *et al.* is that it identifies the desired critical point with almost no *a priori* information about the process. The desired critical point finding routine due to Palmor *et al.* (1995a) is considered in depth. This two-input, two-output system method is followed by the presentation of an extension to an L loop decentralised relay scheme. The next section of the chapter considers the selection of the desired critical point to achieve control system robustness and suggests that the bandwidth point should be used. Some case studies that show results for the Palmor *et al.* tuning method and the robust tuning method close the chapter.

5.2 Critical Points of a System

Knowledge of the desired critical point is essential to the tuning procedure. Critical points of a system depend on the specification and the complexity of the system. The relay experiment is one method that can be used to identify the critical points of processes. During the identification phase with the relay experiment all the controllers are replaced by relays, thus generating limit cycle oscillations. Many PID tuning methods then use the critical point information to tune the PID controller. For this, the ultimate gain and ultimate period of process are measured. The critical points of different systems are explained next.

5.2.1 Critical Points for Two-Input, Two-Output Systems

A block diagram of the decentralised control system for a two-input, two-output process is shown in Figure 5.1, where $y(t) \in \Re^2$ denotes the vector of process outputs, $u(t) \in \Re^2$ is the vector of manipulated variables or control signals and $r(t) \in \Re^2$ is the vector of the reference signals. Similarly, the vector of loop errors is given as $e(t) \in \Re^2$.

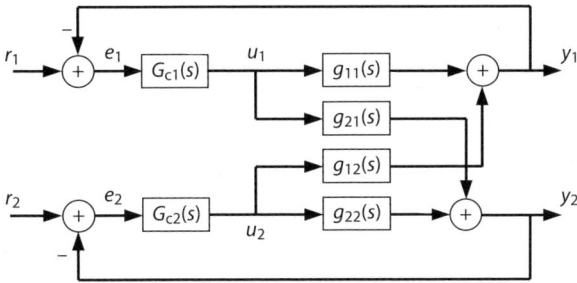

Figure 5.1 Two-input, two-output system with decentralised controller.

If the case of proportional control is considered, then in a SISO system there are usually only a finite number of critical gains that bring the system to the verge of instability, and in many cases there is only one such gain. In the decentralised system, the two controllers are both set to proportional gains of the form

$$G_{c1}(s) = k_{P_1}$$
$$G_{c2}(s) = k_{P_2}$$

The critical point is found by increasing the proportional terms until the outputs produce a stable oscillation and the system is on the verge of instability. Unlike SISO plants, there are many critical points for two-input, two-output systems (Palmor et al., 1995a). Thus there are many pairs of proportional gains (k_{P1}, k_{P2}) which can be used in the controllers $G_{c1}(s)$ and $G_{c2}(s)$ and which lead to a condition of neutral stability. The pure oscillations produced are evidence of poles on the imaginary axis. These pairs of critical gains are denoted, (k_{P1cr}, k_{P2cr}), and the set of these gains is called the stability limit set of the system. A typical example is shown in Figure 5.2.

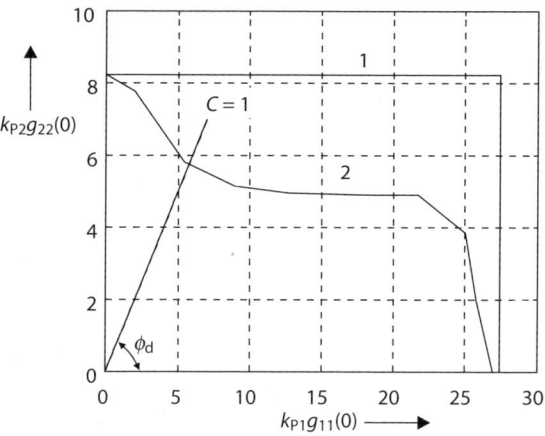

Figure 5.2 Typical example of a stability limit set of a two-input, two-output system (after Palmor et al., 1995a).

In Figure 5.2, each point on one of the curves corresponds to a pair of critical gains (k_{P1cr}, k_{P2cr}) and a critical frequency, which together are referred to as a critical point. The points on the axes of the figure represent the situation where one loop is open (either $k_{P1} = 0$ or $k_{P2} = 0$). If the system does not have full interaction, a situation where either $g_{12}(s), g_{21}(s)$ or both are zero, the stability limit set takes the rectangular form of curve 1 in Figure 5.2. In this case the two critical gains are independent of each other, and the system becomes unstable when either one of the gains exceeds the SISO critical value. The curve 2 in Figure 5.2 shows the stability limit set for a system with interaction terms present. The deviation from the rectangular shape of curve 1 may be regarded as a measure of the interaction present in the system. In choosing a critical point on which to base a controller tuning procedure, different critical points will lead to different controller settings, and hence different overall control performance. Therefore it is necessary to specify on which critical point the tuning should be based. This critical point will be referred to as the *desired critical point*. The choice of the desired critical point depends on the relative importance of loops, which is commonly expressed through a weighting factor (Luyben, 1986). The weighting factor can depend on different criteria and system properties such as the relationship between loop gains in steady state (Palmor et al., 1995a) or the relationship between loop gains at bandwidth frequency.

5.2.2 Critical Points for MIMO Systems

Generalising the discussion from a two-input, two-output system to an $(L \times L)$ multivariable system is fairly straightforward. Figure 5.3 shows a block diagram of a decentralised control system for a MIMO process with L loops.

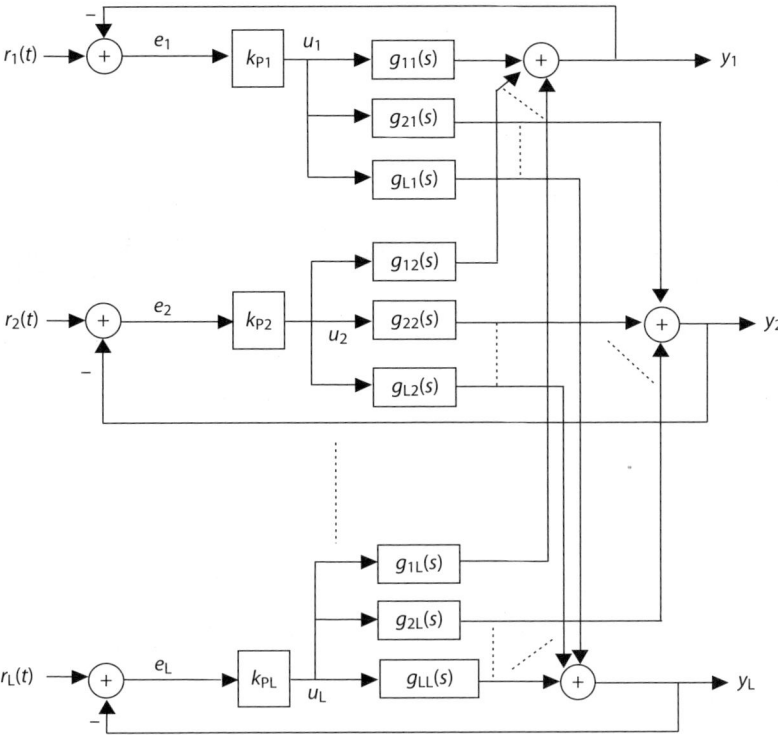

Figure 5.3 L-loop decentralised control system with proportional control.

The process outputs, the control signals, the reference input, and the loop error vectors are given respectively as $y(t), u(t), r(t), e(t) \in \Re^L$. As would be expected in the structure of decentralised control, the control signal u_i in the ith loop, where $i = 1, \ldots, L$, depends only on the error $e_i(t)$ in the same loop. For an $(L \times L)$ MIMO system, the stability limits form a set in $\Re^L$ so that the desired critical point can be selected using $(L-1)$ weighting factors; for example, the weighting factor of the $(i+1)$th loop relative to the ith loop, where $i = 1, \ldots, L-1$. As an example, consider a three-input, three-output system. In this case the stability limits do not form a point as in a SISO system or a curve as in a two-input, two-output system, but a surface, as shown in Figure 5.4. Since $L = 3$, the desired critical point can be selected using two weighting factors, for example weighting the second loop relative to the first and weighting the third loop relative to second loop.

5.3 Decentralised Relay Experiments for Multivariable Systems

The single-input, single-output system relay experiment can be generalised to decentralised control of multivariable systems. As with the single-input, single-output case, the proportional terms in the diagonal controller structure of Figure 5.3 are replaced with individual relays. This setup is shown in Figure 5.5.

On closing the loop, for a significant set of industrially relevant systems a stable oscillatory output occurs with the multivariable closed-loop system on the verge of instability. It is then possible to use some simple data collection and processing to identify the critical point found. The critical point comprises the critical gains and critical frequencies. To exploit this property in PID controller tuning it

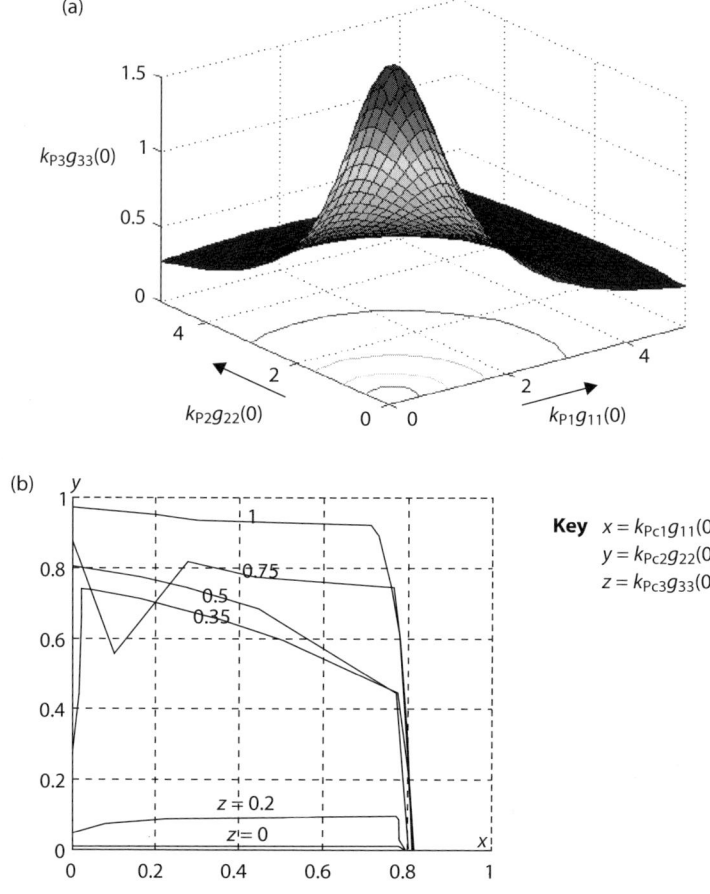

Figure 5.4 (a) 3D plot for the critical points of a three-input, three-output system; (b) Graphs for the critical points of a three-input, three-output system.

is necessary to be able to target and reach a *desired* critical point and then use the critical point data in the PID tuning routine. This is the key advance which is at the core of the development to be presented in this section. However, the section proceeds by first showing how particular system gain values can be determined from the relay experiment output. This is then followed by the key result characterising the critical points occurring within the decentralised control relay experiment. A discussion of the two-input, two-output system case is given as a prelude to the general L-loop multivariable system case to be covered later in the chapter.

5.3.1 Finding System Gains at Particular Frequencies

One of the steps needed in the full decentralised control system relay autotune procedure is to determine the system gains at zero frequency and the critical frequency. The outcomes on activating the relays simultaneously of the experiment setup in Figure 5.5 are:

1. The outputs $y_1(t), \ldots, y_L(t)$ exhibit a steady oscillation with a common period, P_{cr}.

2. The oscillations in the outputs $y_1(t), \ldots, y_L(t)$ have different amplitudes $a_1(t), \ldots, a_L(t)$ respectively.

3. The cycles in the L-loops exhibit time shifts and are not generally synchronised.

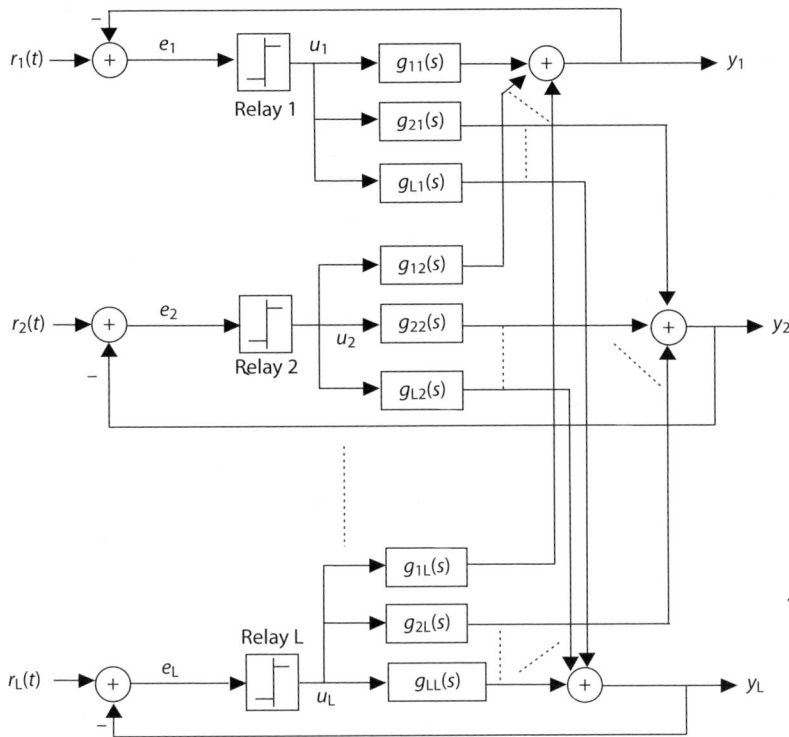

Figure 5.5 *L*-loop decentralised control system with relay control.

The periodicity of the process inputs $u_i(t)$ and outputs $y_i(t)$ can be used in a time-domain Fourier analysis to extract process gain information. Let the process transfer function matrix be defined as $G_p(s) \in \Re^{L \times L}(s)$, let the common frequency of oscillation be ω_{cr} and define the common period as $P_{cr} = 2\pi/\omega_{cr}$, then the system relationship at ω is

$$y(j\omega) = G_p(j\omega)u(j\omega)$$

with

$$y(j\omega) = \begin{bmatrix} y_1(j\omega) \\ y_2(j\omega) \\ \vdots \\ y_L(j\omega) \end{bmatrix} = \begin{bmatrix} \dfrac{1}{P_{cr}} \int_0^{P_{cr}} y_1(t) e^{-j\omega t} dt \\ \dfrac{1}{P_{cr}} \int_0^{P_{cr}} y_2(t) e^{-j\omega t} dt \\ \vdots \\ \dfrac{1}{P_{cr}} \int_0^{P_{cr}} y_L(t) e^{-j\omega t} dt \end{bmatrix}$$

and

$$u(j\omega) = \begin{bmatrix} u_1(j\omega) \\ u_2(j\omega) \\ \vdots \\ u_L(j\omega) \end{bmatrix} = \begin{bmatrix} \dfrac{1}{P_{cr}} \int_0^{P_{cr}} u_1(t) e^{-j\omega t} dt \\ \dfrac{1}{P_{cr}} \int_0^{P_{cr}} u_2(t) e^{-j\omega t} dt \\ \vdots \\ \dfrac{1}{P_{cr}} \int_0^{P_{cr}} u_L(t) e^{-j\omega t} dt \end{bmatrix}$$

Since the frequency domain input–output equation is a vector equation, it is not sufficient to determine $G_p(j\omega)$. Consequently, the relay experiments must be repeated L times. So the relay amplitude of the dominant loop is slightly increased or the relay amplitude of other loops is decreased to ensure that the oscillation frequencies for each experiment remain close to that of the previous experiment. Now, the frequency domain input–output equation can be written for L relay experiments as a matrix equation:

$$Y(j\omega) = G_p(j\omega)U(j\omega)$$

with process transfer function matrix $G_p(s) \in \Re^{L \times L}(s)$ and where

$$G_p(j\omega) = \begin{bmatrix} g_{11}(j\omega) & g_{12}(j\omega) & \cdots & g_{1L}(j\omega) \\ \vdots & \vdots & \ddots & \vdots \\ g_{L1}(j\omega) & g_{L2}(j\omega) & \cdots & g_{LL}(j\omega) \end{bmatrix}$$

$$Y(j\omega) = [Y^1(j\omega) \quad Y^2(j\omega) \quad \cdots \quad Y^L(j\omega)]$$

$$Y^k(j\omega) = [y_1^k(j\omega) \quad y_2^k(j\omega) \quad \cdots \quad y_L^k(j\omega)]^T \text{ for } k = 1, \ldots, L$$

and

$$U(j\omega) = [U^1(j\omega) \quad U^2(j\omega) \quad \cdots \quad U^L(j\omega)]$$

$$U^k(j\omega) = [u_1^k(j\omega) \quad u_2^k(j\omega) \quad \cdots \quad u_L^k(j\omega)]^T \text{ for } k = 1, \ldots, L$$

The superscript k indicates the kth experiment. Thus a solution can be found for the system gains $G_p(j\omega)$ at $s = j\omega$ as

$$G_p(j\omega) = Y(j\omega)U^{-1}(j\omega)$$

where $G_p(j\omega), Y(j\omega), U(j\omega) \in C^{L \times L}$.

This means that from inputs and outputs signal of the relay experiment, the value of outputs and inputs at a specified frequency are extracted in each experiment. After L relay experiments these values are inserted into the equation $G_p(j\omega) = Y(j\omega)U^{-1}(j\omega)$ to determine the process gains at the specified frequency.

One specific set of system gains used in later routines is the set of steady state system gains. To calculate these, set $\omega = 0$ in the above equations and the result is

$$G_p(0) = Y(0)U^{-1}(0)$$

In this case, it is possible to use a set of L different independent reference inputs to achieve the result. A second set of system gains of interest are those for the critical frequency ω_{cr}. In this case, the interest lies in the fact that the critical frequency may be taken as an approximation for the bandwidth. To use the above routine for estimating the system gain at critical frequency ω_{cr}, the computations use $\omega = \omega_{cr}$ in the above equations to give

$$G_p(j\omega_{cr}) = Y(j\omega_{cr})U^{-1}(j\omega_{cr})$$

5.3.2 Decentralised Relay Control Systems – Some Theory

The decentralised relay feedback system of Figure 5.5 has yielded a result in which the relay elements can be quite general, incorporating relay, hysteresis and dead zone. The implications of this general result provide the concepts needed to create a procedure for decentralised PID control design using the simultaneous application of relays in the L-loop system.

Theorem 5.1: General decentralised relay feedback system result

Consider the decentralised relay feedback system of Figure 5.5. If a limit cycle exists then its period P_{cr} and the time shifts h remain the same for all m_i, b_i and Δ_i satisfying

$$m_1 : m_2 : \ldots : m_L = \delta_1 : \delta_2 : \ldots : \delta_L$$
$$m_i : b_i : \Delta_i = 1 : \beta_i : \gamma_i, \quad i = 1, \ldots, L$$

where δ_i, β_i and γ_i are nonnegative constants, b_i is the hysteresis of the ith relay, Δ_i is the dead zone of the ith relay and m_i is the height of the ith relay.

Proof (Palmor *et al.*, 1995b) ☐

From Theorem 5.1 it follows that in the relay setup, the critical points depend only on the ratio m_i/m_{i+1}. Although these results are theoretically independent of the absolute magnitudes of m_i and m_{i+1}, these relay sizes do have practical significance. To reduce the effect of noise and to cause measurable changes in output, $y \in \Re^L$, the relay sizes m_i, $i = 1, \ldots, L$ must be sufficiently large. Hence, there is lower bound on m_i, denoted by $\underline{m_i}$. Conversely, there is an upper bound on m_i, denoted by $\overline{m_i}$. This is to ensure that the relay size is not so large that it cannot be used due to actuator saturation constraints or because the allowed change in $y \in \Re^L$ is restricted possibly for production process reasons. Hence Theorem 5.1 and the preceding discussion indicate that:

(a) By using L simultaneous relays loops, a critical point for L-inputs and L-outputs system is identified.

(b) In practice, due to possible constraints on the size of the controls and the output variables, the amplitude of the ith relay is restricted to the range $\underline{m_i} \leq m_i \leq \overline{m_i}$.

(c) All sets $(m_1, m_2, \ldots, m_L)$ such that ratio m_i/m_{i+1} is constant correspond to one critical point.

(d) It is possible to change the critical point by changing the relay height ratio m_i/m_{i+1}.

(e) Changing the critical point implies that it is possible to "move" along the stability boundary by varying the ratios m_i/m_{i+1}.

From parts (c), (d) and (e) above, for all sets $(m_1, m_2, \ldots, m_L)$ such that $m_1 : m_2 : \ldots : m_L = \delta_1 : \delta_2 : \ldots : \delta_L$ leads to limit cycle with the same period P_{cr} and same time shift and same ratio of relay input amplitudes $a_1 : a_2 : \ldots : a_L = v_1 : v_2 : \ldots : v_L$, where a_i is the input to the ith relay and v_i is a constant. This presents the potential to establish a sequence of relay experiments to find the data of a desired critical point. In this sequence introduce the iteration counter as k. Thus, the relationship between ratio of the ith and $(i + 1)$th relay input amplitudes and the set of ratios of relay heights in the kth relay experiment may be postulated as

$$\frac{a_i(k)}{a_{i+1}(k)} = f_k\left(\frac{m_1(k)}{m_2(k)}, \frac{m_2(k)}{m_3(k)}, \ldots, \frac{m_{L-1}(k)}{m_L(k)}\right)$$

This functional relationship can be used to iteratively find the system critical point at an indirectly specified desired location for PID controller tuning. These concepts are all collated to develop a full algorithm for a two-input, two-output system algorithm in the next section.

5.3.3 A Decentralised Two-Input, Two-Output PID Control System Relay-Based Procedure

A special case of Theorem 5.1 has been used to develop a new automatic tuning algorithm for a decentralised two-input, two-output PID control system (Krasney, 1991; Palmor *et al.*, 1995b). The feedback

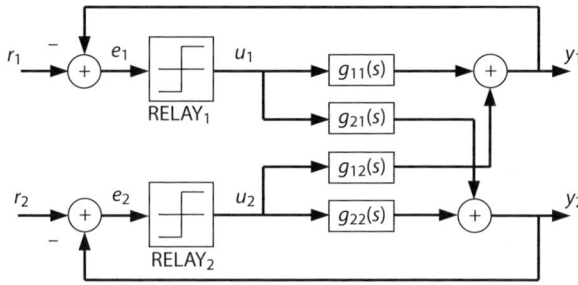

Figure 5.6 Two-loop decentralised control system with relay control.

system configuration is shown in Figure 5.6. In the kth relay experiment the two relay heights are denoted for relay 1 as $m_1(k)$ and for relay 2 as $m_2(k)$.

Corollary 5.1: Two-loop decentralised control system with relay control
In the kth relay experiment in the two-loop decentralised control system denote for loops 1 and 2, respectively, the relay heights as $m_1(k), m_2(k)$ and the relay input amplitudes as $a_1(k), a_2(k)$. Then all pairs $m_1(k), m_2(k)$ such that the ratio

$$M_{12}(k) = \frac{m_1(k)}{m_2(k)}$$

is constant leads to limit cycles with the same period P_{cr}, the same time shift and the same value for the relay input amplitude ratio

$$A_{12}(k) = \frac{a_1(k)}{a_2(k)}$$

Proof (Krasney, 1991) ☐

To construct an algorithm it is necessary to consider the following subroutines.

Choosing a Desired Critical Point
Select a value for the criterion C_d which specifies a desired critical point:

$$\tan \phi_d = \frac{C_d}{1} = \frac{k_{P_{2cr}} g_{22}(0)}{k_{P_{1cr}} g_{11}(0)}$$

The desired critical point criterion C_d is a weighting factor of the second loop relative to the first loop. Selecting $C_d > 1$ means that loop 2 is required to be under tighter control relative to loop 1 in steady state. The relative weighting can be alternatively expressed as the angle $\phi_d = \tan^{-1} C_d$ defined in the loop-gain plane. The angle ϕ connects the origin to a desired critical point in the loop-gain plane as shown in the sketch of Figure 5.7. The subscript "d" denotes a *desired* critical point giving a desired angle ϕ_d in the loop-gain plane.

Finding the Steady State Gains
Consider the two-input, two-output system shown in Figure 5.6 with at least one input with non-zero mean; then the error signals $e_i(t), i = 1, 2,$ and the control signals $u_i(t), i = 1, 2,$ are also non-zero mean. Thus a d.c. balance or, alternatively, comparing the constant terms of the Fourier series of outputs and controls signal in the ith experiment and using the equation $G_p(0) = Y(0)U^{-1}(0)$ provides a computational method to find the steady state gains. To perform this calculation the following quantities are required:

5.3 Decentralised Relay Experiments for Multivariable Systems

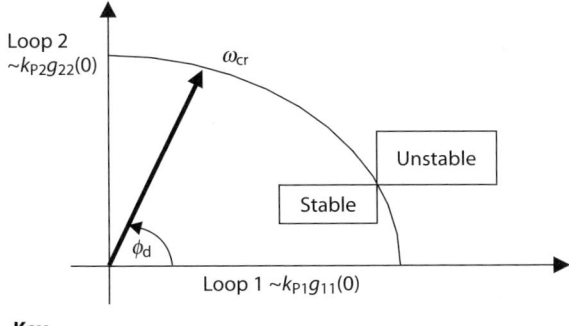

Key
$\phi_d = \tan^{-1} C_d$
ϕ_d small – loop 1 tighter
ϕ_d large – loop 2 tighter

Figure 5.7 Desired critical point selection.

$$Y(0) = \begin{bmatrix} y_1^1 & y_1^2 \\ y_2^1 & y_2^2 \end{bmatrix}, \quad U(0) = \begin{bmatrix} u_1^1 & u_1^2 \\ u_2^1 & u_2^2 \end{bmatrix}$$

with

$$y^k = \begin{bmatrix} \frac{1}{P_{cr}} \int_0^{P_{cr}} y_1^k(t)\,dt \\ \frac{1}{P_{cr}} \int_0^{P_{cr}} y_2^k(t)\,dt \end{bmatrix}, \quad u^k = \begin{bmatrix} \frac{1}{P_{cr}} \int_0^{P_{cr}} u_1^k(t)\,dt \\ \frac{1}{P_{cr}} \int_0^{P_{cr}} u_2^k(t)\,dt \end{bmatrix}, \quad k=1,2$$

The d.c. process gains recovered from the computation are

$$G_p(0) = \begin{bmatrix} g_{11}(0) & g_{12}(0) \\ g_{21}(0) & g_{22}(0) \end{bmatrix} = Y(0) U^{-1}(0) \in \Re^{2 \times 2}$$

Reference inputs r_1 and r_2 are chosen to ensure that the matrix $U(0)$ is non-singular; typically the following might be selected: $r_1^1(t) \neq 0, r_2^1(t) = 0$ and $r_1^2(t) = 0, r_2^2(t) \neq 0$. The computation requires the results from just two relay experiments and then the steady state gains can be identified.

Initialisation of First Two Relay Experiments

After just two limit cycle experiments in two-input, two-output systems, all the steady state gains are identified and two critical points can be found. In the first two experiments, it is recommended that the largest allowed relay ratio spreads (namely $M_{12}(1) = \overline{m}_1 / \underline{m}_2$ and $M_{12}(2) = \underline{m}_1 / \overline{m}_2$) are used respectively. These relay ratios will lead to two critical points which provide two good starting points for the next step. A good approximation for the critical point gains is given by

$$k_{P_i cr} = 4m_i / \pi a_i, \quad i=1,2 \text{ and } \omega_{cr} = 2\pi / P_{cr}$$

Establishing an Update Relationship

Having identified two critical points, the question now is how to determine the relay ratio for the next experiments in a systematic fashion that will lead to a rapid convergence to the desired critical point. Ultimately, this procedure is generalised into an iterative procedure. The data available is:

Iteration $k = 1$

Apply $M_{12}(1) = \dfrac{\overline{m_1}}{\underline{m_1}}$; yields $A_{12}(1) = \dfrac{a_1(1)}{a_2(1)}$

Iteration $k = 2$

Apply $M_{12}(2) = \dfrac{\overline{m_1}}{m_2}$; yields $A_{12}(2) = \dfrac{a_1(2)}{a_2(2)}$

Relationship 1: The design constraint
The desired critical point criterion relationship is rearranged as

$$\frac{k_{P1cr}}{k_{P2cr}} = \frac{g_{22}(0)}{C_d g_{11}(0)}$$

The approximation for the critical point gains $k_{Picr} = 4m_i/\pi a_i$, $i = 1,2$ leads to

$$\frac{k_{P1cr}}{k_{P2cr}} = \left(\frac{4m_1}{\pi a_1}\right) / \left(\frac{4m_2}{\pi a_2}\right) = \frac{m_1 a_2}{m_2 a_1} = \left(\frac{m_1}{m_2}\right) / \left(\frac{a_1}{a_2}\right) = \frac{M_{12}}{A_{12}}$$

Hence eliminating the quantity k_{P1cr}/k_{P2cr} between these two equations leads to the desired relationship between relay height ratio $M_{12} = m_1/m_2$ and relay input amplitude ratio $A_{12} = a_1/a_2$ as

$$\frac{M_{12}}{A_{12}} = \frac{g_{22}(0)}{C_d g_{11}(0)}$$

and

$$M_{12} = \frac{m_1}{m_2} = A_{12} \frac{1}{C_d} \left|\frac{g_{22}(0)}{g_{11}(0)}\right| = \frac{a_1}{a_2} \frac{1}{C_d} \left|\frac{g_{22}(0)}{g_{11}(0)}\right|$$

This relationship is made dependent on iteration k by simply indexing the computed ratios as follows:

$$M_{12}(k) = A_{12}(k) \times \frac{1}{C_d} \left|\frac{g_{22}(0)}{g_{11}(0)}\right|$$

Relationship 2: The stability boundary interpolation
The stability boundary is tracked by introducing a linear relationship for the functional dependence suggested by Theorem 5.1. Hence, write the functional relationship

$$\frac{a_1}{a_2} = f\left(\frac{M_1}{M_2}\right)$$

as

$$A_{12}(k) = f(M_{12}(k)) = b_1 \times M_{12}(k) + b_2$$

Then assuming data from two iterations $k-1, k$ is available, the simultaneous equations to be solved are

$$A_{12}(k-1) = b_1 \times M_{12}(k-1) + b_2$$
$$A_{12}(k) = b_1 \times M_{12}(k) + b_2$$

from which the constants b_1 and b_2 are found as

$$b_1 = \frac{A_{12}(k-1) - A_{12}(k)}{M_{12}(k-1) - M_{12}(k)}$$
$$b_2 = A_{12}(k-1) - b_1 M_{12}(k-1)$$

Eliminating between Relationships 1 and 2

To predict the relay ratio $M_{12}(k+1)$ required for the next iteration of the algorithm, set the iteration counter to $k+1$ and eliminate the future and unknown relay input amplitude ratio $A_{12}(k+1)$ between the simultaneous equations

$$M_{12}(k+1) = A_{12}(k+1) \times \frac{1}{C_d} \left| \frac{g_{22}(0)}{g_{11}(0)} \right|$$
$$A_{12}(k+1) = b_1 \times M_{12}(k+1) + b_2$$

This yields

$$\left[1 - b_1 \frac{1}{C_d} \left| \frac{g_{22}(0)}{g_{11}(0)} \right| \right] M_{12}(k+1) = b_2 \frac{1}{C_d} \left| \frac{g_{22}(0)}{g_{11}(0)} \right|$$

and hence

$$M_{12}(k+1) = \frac{b_2 \frac{1}{C_d} \left| \frac{g_{22}(0)}{g_{11}(0)} \right|}{\left[1 - b_1 \frac{1}{C_d} \left| \frac{g_{22}(0)}{g_{11}(0)} \right| \right]} = \frac{b_2}{\left[C_d \left| \frac{g_{11}(0)}{g_{22}(0)} \right| - b_1 \right]}$$

Introducing a Convergence Test

If the linear function approximation $A_{12}(k) = b_1 \times M_{12}(k) + b_2$ is perfectly accurate then the prediction of the next relay ratio $M_{12}(k+1)$ leads exactly to the desired critical point. In most cases several iterations will be required and the usual convergence tolerance test is introduced. At the kth iteration, compute the current value of the angle $\phi(k)$ from

$$\tan\phi(k) = \frac{k_{P_{2cr}} g_{22}(0)}{k_{P_{1cr}} g_{11}(0)} = \left(\frac{4m_2(k)}{\pi a_2(k)} \right) \left(\frac{\pi a_1(k)}{4m_1(k)} \right) \frac{g_{22}(0)}{g_{11}(0)} = \frac{A_{12}(k) g_{22}(0)}{M_{12}(k) g_{11}(0)}$$

A tolerance ε is defined, and algorithm stops when $|\phi(k) - \phi_d| \leq \varepsilon$. Otherwise, the procedure continues in the same fashion where the best two experiments, namely, those that are closer to ϕ_d play the role of experiments $k, k-1$ in the above equation steps.

Algorithm 5.1: Two-input, two-output system decentralised relay experiment

Step 1 Data input

    Input the value selected for the criterion C_d
    Input the relay height bounds $\underline{m}_1 \leq m_1 \leq \overline{m}_1, \underline{m}_2 \leq m_2 \leq \overline{m}_2$
    Input the convergence tolerance ε
    Input the maximum iteration count k_{max}

Step 2 Initial two-relay experiments

    Record input and output data from relay experiments $k = 1, 2$

$$y_{data}^k = \begin{bmatrix} y_1^k(t) \\ y_2^k(t) \end{bmatrix}, \quad u_{data}^k = \begin{bmatrix} u_1^k(t) \\ u_2^k(t) \end{bmatrix}, \quad t_0 \leq t \leq t_0 + P_{cr}$$

Step 3 *Computation of initial gains and other data*

Form matrices $Y(0) = \begin{bmatrix} y_1^1 & y_1^2 \\ y_2^1 & y_2^2 \end{bmatrix}$, $U(0) = \begin{bmatrix} u_1^1 & u_1^2 \\ u_2^1 & u_2^2 \end{bmatrix}$

$$y^k = \begin{bmatrix} \frac{1}{P_{cr}} \int_0^{P_{cr}} y_1^k(t) dt \\ \frac{1}{P_{cr}} \int_0^{P_{cr}} y_2^k(t) dt \end{bmatrix}, \quad u^k = \begin{bmatrix} \frac{1}{P_{cr}} \int_0^{P_{cr}} u_1^k(t) dt \\ \frac{1}{P_{cr}} \int_0^{P_{cr}} u_2^k(t) dt \end{bmatrix}, \quad k = 1, 2$$

Compute $G_p(0) = \begin{bmatrix} g_{11}(0) & g_{12}(0) \\ g_{21}(0) & g_{22}(0) \end{bmatrix} = Y(0) U^{-1}(0)$

Compute $\phi_d = \tan^{-1} C_d$

Compute $M_{12}(1) = \dfrac{\overline{m_1}}{\underline{m_1}}$ and $A_{12}(1) = \dfrac{a_1(1)}{a_2(1)}$

Compute $M_{12}(2) = \dfrac{\overline{m_1}}{\underline{m_2}}$ and $A_{12}(2) = \dfrac{a_1(2)}{a_2(2)}$

$k = 2$

Step 4 *Convergence test*

Compute $\phi(k) = \tan^{-1} \dfrac{A_{12}(k) g_{22}(0)}{M_{12}(k) g_{11}(0)}$

If $|\phi(k) - \phi_d| \leq \varepsilon$ then $k_{Picr} = 4 m_i / \pi a_i$, $i = 1, 2$, $\omega_{cr} = 2\pi / P_{cr}$ and STOP
If $k > k_{max}$ then STOP with maximum iteration count exceeded
If $|\phi(k) - \phi_d| > \varepsilon$ then CONTINUE

Step 5 *Update computation*

Compute $b_1 = \dfrac{A_{12}(k-1) - A_{12}(k)}{M_{12}(k-1) - M_{12}(k)}$

$b_2 = A_{12}(k-1) - b_1 M_{12}(k-1)$

Compute $M_{12}(k+1) = \left[\dfrac{b_2}{C_d |g_{11}(0) / g_{22}(0)| - b_1} \right]$

Step 6 *Next relay experiment*

Use $M_{12}(k+1)$ to update relay sizes $m_1(k+1)$, $m_2(k+1)$
Run simultaneous relay experiment

Compute $A_{12}(k+1) = \dfrac{a_1(k+1)}{a_2(k+1)}$

Update iteration counter k to $k+1$
Goto Step 4

Algorithm end

PID Tuning

This simple algorithm shows excellent convergence properties, and once the desired critical point has been found the PID controller coefficients can be determined in a straightforward manner. A simple choice is to use the Ziegler and Nichols (1942) rules.

5.4 A Decentralised Multi-Input, Multi-Output PID Control System Relay-Based Procedure

This is an extension of the concepts and notation of the two-input, two-output system procedure to an L-loop decentralised relay control scheme. The presentation follows the steps of the previous section and concludes with a complete algorithm where the loop counter is denoted i and the iteration counter is denoted k.

Choosing the Desired Critical Point

To choose the critical point use the following criteria:

$$\tan \phi_{di} = \frac{C_{di}}{1} = \frac{k_{P_{i+1\mathrm{cr}}} g_{i+1,i+1}(0)}{k_{P_{i\mathrm{cr}}} g_{ii}(0)}, \quad i=1,\ldots,L-1$$

where $\tan \phi_{di}$, or equivalently C_{di} is the weighting factor of the $(i+1)$th loop relative to the ith loop. If C_{di} is selected so that $C_{di} > 1$ then this implies that the $(i+1)$th loop is required to be under tighter control relative to the ith loop.

Computing the Steady State Gains

The initial set of L relay experiments is used to identify the steady state gains of process simultaneously with the first L critical points needed to start the linear interpolation procedure. The set of reference inputs $r_i(t)$, $i=1,\ldots,L$ are chosen to ensure that the matrix $U(0)$ is non-singular. Typically, the following might be selected for the kth experiment: $r_i^k(t) = \mu_i \delta_{ik}$; $i, k=1,\ldots,L$, where $\{\mu_i; i=1,\ldots,L\}$ is a set of constants selected to size the step reference changes and $\{\delta_{ij}; i,j=1,\ldots,L\}$ is the usual Kronecker delta. For each of the kth experiments the following quantities should be computed for $k = 1, 2, \ldots, L$:

$$y^k = \begin{bmatrix} \frac{1}{P_{\mathrm{cr}}} \int_0^{P_{\mathrm{cr}}} y_1^k(t)\,dt \\ \vdots \\ \frac{1}{P_{\mathrm{cr}}} \int_0^{P_{\mathrm{cr}}} y_L^k(t)\,dt \end{bmatrix} \quad \text{and} \quad u^k = \begin{bmatrix} \frac{1}{P_{\mathrm{cr}}} \int_0^{P_{\mathrm{cr}}} u_1^k(t)\,dt \\ \vdots \\ \frac{1}{P_{\mathrm{cr}}} \int_0^{P_{\mathrm{cr}}} u_L^k(t)\,dt \end{bmatrix}$$

Subsequently, the following composite matrices are formed:

$$Y(0) = \begin{bmatrix} y_1^1 & \cdots & y_1^L \\ \vdots & \ddots & \vdots \\ y_L^1 & \cdots & y_L^L \end{bmatrix} \quad \text{and} \quad U(0) = \begin{bmatrix} u_1^1 & \cdots & u_1^L \\ \vdots & \ddots & \vdots \\ u_L^1 & \cdots & u_L^L \end{bmatrix}$$

Finally, the steady state gains are recovered from the computation

$$G_{\mathrm{p}}(0) = \begin{bmatrix} g_{11}(0) & \cdots & g_{1L}(0) \\ \vdots & \ddots & \vdots \\ g_{L1}(0) & \cdots & g_{LL}(0) \end{bmatrix} = Y(0) U^{-1}(0)$$

Initialisation of the First L Relay Experiments

To identify the first L critical points use is made of the relay height bounds. These are selected to ensure that sufficient movement occurs in the output for measurement purposes without imperilling the experiments with actuator saturation or violations. The relay height bounds are given as $\underline{m}_i \leq m_i \leq \overline{m}_i$, where $i=1,\ldots,L$. It is recommended to use following ratios for the first L relay experiments:

$$M_i(k) = \frac{m_i(k)}{m_{i+1}(k)}$$

where $i=1,\ldots,L-1$; $k=1,\ldots,L$ and

$$m_i(k) = \begin{cases} \overline{m}_i & i=k \\ \underline{m}_i & i \neq k \end{cases}$$

Establishing an Update Relationship

From L relay experiments, L critical points have been identified and the information available is as follows:

kth relay experiment, where $k=1,\ldots,L$

Relay heights	$m_1(k), m_2(k), \ldots, m_L(k)$
Relay input amplitudes	$a_1(k), a_2(k), \ldots, a_L(k)$
Critical control gains	$k_{P_{icr}}(k) = 4m_i(k)/\pi a_i(k),\ i=1,2,\ldots,L$
Critical oscillation frequency	$\omega_{cr}(k) = 2\pi/P_{cr}(k)$

From this the following ratios are computed:

Relay height ratios	$M_{i,i+1}(k) = \dfrac{m_i(k)}{m_{i+1}(k)},\ i=1,\ldots,L-1$
Relay input amplitude ratios	$A_{i,i+1}(k) = \dfrac{a_i(k)}{a_{i+1}(k)},\ i=1,\ldots,L-1$

Relationship 1: The design constraint

For the kth relay experiment the design condition is rearranged as

$$\frac{k_{P_{icr}}}{k_{P_{(i+1)cr}}} = \frac{g_{i+1,i+1}(0)}{C_{di}g_{ii}(0)}, \quad \text{for } i=1,\ldots,L-1$$

From the critical gain formulae

$$\frac{k_{P_{icr}}}{k_{P_{(i+1)cr}}} = \frac{4m_i(k)}{\pi a_i(k)} \frac{\pi a_{i+1}(k)}{4m_{i+1}(k)} = \frac{m_i(k)}{m_{i+1}(k)} \frac{a_{i+1}(k)}{a_i(k)} = \frac{M_{i,i+1}(k)}{A_{i,i+1}(k)}, \quad \text{for } i=1,\ldots,L-1$$

Hence eliminating between the above equations

$$\frac{k_{P_{icr}}}{k_{P_{(i+1)cr}}} = \frac{g_{i+1,i+1}(0)}{C_{di}g_{ii}(0)} = \frac{M_{i,i+1}(k)}{A_{i,i+1}(k)} \quad \text{for } i=1,\ldots,L-1 \text{ and } k=1,\ldots,L$$

The desired relationship between relay height ratio

$$M_{i,i+1}(k) = \frac{m_i(k)}{m_{i+1}(k)}$$

5.4 A Decentralised Multi-Input, Multi-Output PID Control System Relay-Based Procedure

and relay input amplitude ratio

$$A_{i,i+1}(k) = \frac{a_i(k)}{a_{i+1}(k)}$$

is

$$M_{i,i+1}(k) = A_{i,i+1}(k)\frac{g_{i+1,i+1}(0)}{C_{di}g_{ii}(0)} \text{ for } i = 1, \ldots, L-1 \text{ and } k = 1, \ldots, L$$

Relationship 2: The stability boundary interpolation

The stability boundary is tracked by introducing a linear relationship for the functional dependence suggested by Theorem 5.1. This dependence may be written as

$$A_{1,2}(k) = b_{11}M_{1,2}(k) + b_{12}M_{2,3}(k) + \ldots + b_{1,L-1}M_{L-1,L}(k) + b_{1L}$$
$$A_{2,3}(k) = b_{21}M_{1,2}(k) + b_{22}M_{2,3}(k) + \ldots + b_{2,L-1}M_{L-1,L}(k) + b_{2L}$$
$$\vdots$$
$$A_{L-1,L}(k) = b_{L-11}M_{1,2}(k) + b_{L-12}M_{2,3}(k) + \ldots + b_{L-1,L-1}M_{L-1,L}(k) + b_{L-1,L}$$

This linear system may be represented in matrix–vector form as

$$A_R(k) = B\begin{bmatrix} M_R(k) \\ 1 \end{bmatrix}$$

where $A_R(k) \in \mathfrak{R}^{L-1}, M_R(k) \in \mathfrak{R}^{L-1}$ and $B \in \mathfrak{R}^{L-1 \times L}$.

However, the data from the L relay experiments, $k = 1, \ldots, L$, leads to

$$A_R = [A_R(1) \quad A_R(2) \quad \ldots \quad A_R(L)]$$
$$= \begin{bmatrix} B\begin{bmatrix} M_R(1) \\ 1 \end{bmatrix} & B\begin{bmatrix} M_R(2) \\ 1 \end{bmatrix} & \ldots & B\begin{bmatrix} M_R(L) \\ 1 \end{bmatrix} \end{bmatrix}$$
$$= B\begin{bmatrix} M_R(1) & M_R(2) & \ldots & M_R(L) \\ 1 & 1 & & 1 \end{bmatrix}$$

and

$$A_R = B\widetilde{M}_R$$

where $A_R(k) \in \mathfrak{R}^{L-1 \times L}, \widetilde{M}_R(k) \in \mathfrak{R}^{L \times L}$ and $B \in \mathfrak{R}^{L-1 \times L}$.

This linear equation above may be solved for the matrix $B \in \mathfrak{R}^{L-1 \times L}$ as

$$B = A_R \widetilde{M}_R^{-1}$$

Eliminating Between Relationships 1 and 2

To predict the vector of relay ratios $M_R(k+1)$ required for the next iteration of the algorithm, the situation at iteration count $k+1$ must be considered. Firstly, the design constraint is investigated, where for $i = 1, \ldots, L-1$

$$\frac{m_i(k+1)}{m_{i+1}(k+1)} = M_{i,i+1}(k+1) = A_{i,i+1}(k+1)\frac{g_{i+1,i+1}(0)}{C_{di}g_{ii}(0)}$$

This may be rearranged and written in matrix form as

$$A_R(k+1) = \Lambda_d M_R(k+1)$$

where

$$\Lambda_d = \text{diag}\left\{\frac{C_{di}g_{ii}(0)}{g_{i+1,i+1}(0)}\right\} \in \Re^{L-1 \times L-1} \text{ and } A_R(k+1), M_R(k+1) \in \Re^{L-1}$$

Then the functional relationship for the stability boundary at iteration count $k+1$ is recalled as

$$A_R(k+1) = B\begin{bmatrix} M_R(k+1) \\ 1 \end{bmatrix}$$

This matrix–vector equation may be partitioned as

$$A_R(k+1) = B_1 M_R(k+1) + B_2$$

where $A_R(k+1), M_R(k+1), B_2 \in \Re^{L-1}$,

$$B_1 = \begin{bmatrix} b_{1,1} & \cdots & b_{1,L-1} \\ \vdots & \ddots & \vdots \\ b_{L-1,1} & \cdots & b_{L-1,L-1} \end{bmatrix} \in \Re^{L-1 \times L-1} \text{ and } B_2 = \begin{bmatrix} b_{1,L} \\ \vdots \\ b_{L-1,L} \end{bmatrix} \in \Re^{L-1}$$

The simultaneous equations to be solved are therefore

$$A_R(k+1) = \Lambda_d M_R(k+1)$$
$$A_R(k+1) = B_1 M_R(k+1) + B_2$$

where $A_R(k+1), M_R(k+1), B_2 \in \Re^{L-1}$ and $\Lambda_d, B_1 \in \Re^{L-1 \times L-1}$.

Performing the elimination gives

$$B_1 M_R(k+1) + B_2 = \Lambda_d M_R(k+1)$$
$$[\Lambda_d - B_1] M_R(k+1) = B_2$$

and the new $(k+1)$th relay ratio vector is obtained as

$$M_R(k+1) = [\Lambda_d - B_1]^{-1} B_2$$

Introducing a Convergence Test

In most cases of L-loop problems several iterations will be required and the usual norm convergence tolerance test is introduced. For $i = 1, \ldots, L-1$, the desired critical point yields the following quantities for attainment: $\phi_{di} = \tan^{-1} C_{di}, i = 1, \ldots, L-1$.

Define a vector version of these target values: $\Phi_d \in \Re^{L-1}$ where $\Phi_d(i) = \phi_{di}; i = 1, \ldots, L-1$.

At the kth iteration, compute the current values of the set of angles $\phi_i(k)$ from

$$\tan \phi_i(k) = \frac{k_{P_{i+1cr}}(k) g_{i+1,i+1}(0)}{k_{P_{i1cr}}(k) g_{ii}(0)} = \frac{A_{i,i+1}(k) g_{i+1,i+1}(0)}{M_{i,i+1}(k) g_{ii}(0)}$$

where $i = 1, \ldots, L-1$.

Define a vector of the current values of the set of angles as

$$\Phi_k \in \Re^{L-1}, \text{ where } \Phi_k(i) = \phi_i(k) = \tan^{-1}\left(\frac{A_{i,i+1}(k) g_{i+1,i+1}(0)}{M_{i,i+1}(k) g_{ii}(0)}\right), i = 1, \ldots, L-1$$

A tolerance ε is defined, and the algorithm is considered converged when $\|\Phi_k - \Phi_d\| \leq \varepsilon$; otherwise the procedure is continued until the convergence tolerance or a maximum iteration count is exceeded.

5.4 A Decentralised Multi-Input, Multi-Output PID Control System Relay-Based Procedure

Algorithm 5.2: L-loop multi-input, multi-output system decentralised relay experiment

Step 1 Data input

Input the selected values for the criterion set $\{C_{di}, i=1,\ldots,L-1\}$
Input the set of relay height bounds $\{\underline{m}_i \leq m_i \leq \overline{m}_i, i=1,\ldots,L\}$
Input the convergence tolerance ε
Input the maximum iteration count $k_{\max}$

Step 2 Initial L relay experiments

Record input and output data from relay experiments $k=1,2,\ldots,L$

$$y_{\text{data}}^k = \begin{bmatrix} y_1^k(t) \\ \vdots \\ y_L^k(t) \end{bmatrix}, \quad u_{\text{data}}^k = \begin{bmatrix} u_1^k(t) \\ \vdots \\ u_L^k(t) \end{bmatrix}, \quad t_0 \leq t \leq t_0 + P_{\text{cr}}$$

Step 3 Computation of initial gains and other data

Form matrices $Y(0) = \begin{bmatrix} y_1^1 & \cdots & y_1^L \\ \vdots & \ddots & \vdots \\ y_L^1 & \cdots & y_L^L \end{bmatrix}$, $U(0) = \begin{bmatrix} u_1^1 & \cdots & u_1^L \\ \vdots & \ddots & \vdots \\ u_L^1 & \cdots & u_L^L \end{bmatrix}$

$$y^k = \begin{bmatrix} \frac{1}{P_{\text{cr}}} \int_0^{P_{\text{cr}}} y_1^k(t)\,dt \\ \vdots \\ \frac{1}{P_{\text{cr}}} \int_0^{P_{\text{cr}}} y_L^k(t)\,dt \end{bmatrix}, \quad u^k = \begin{bmatrix} \frac{1}{P_{\text{cr}}} \int_0^{P_{\text{cr}}} u_1^k(t)\,dt \\ \vdots \\ \frac{1}{P_{\text{cr}}} \int_0^{P_{\text{cr}}} u_L^k(t)\,dt \end{bmatrix}, \quad k = 1,2,\ldots,L$$

Compute $G_p(0) = \begin{bmatrix} g_{11}(0) & \cdots & g_{1L}(0) \\ \vdots & \ddots & \vdots \\ g_{L1}(0) & \cdots & g_{LL}(0) \end{bmatrix} = Y(0)U^{-1}(0)$

Compute $\phi_{di} = \tan^{-1} C_{di}$, $i = 1,\ldots,L-1$

Define vector $\Phi_d \in \Re^{L-1}$ where $\Phi_d(i) = \phi_{di}$, $i = 1,\ldots,L-1$

Compute relay height ratios

$$M_i(k) = \frac{m_i(k)}{m_{i+1}(k)}, \quad i = 1,\ldots,L-1 : k = 1,\ldots,L$$

where $m_i(k) = \begin{cases} \overline{m}_i & i = k \\ \underline{m}_i & i \neq k \end{cases}$

Compute relay input amplitude ratios $A_{i,i+1}(k) = \dfrac{a_i(k)}{a_{i+1}(k)}$, $i = 1,\ldots,L-1$

Compute matrix $\Lambda_d = \text{diag}\left\{\dfrac{C_{di}g_{ii}(0)}{g_{i+1,i+1}(0)}\right\} \in \Re^{L-1 \times L-1}$

Set iteration counter $k = L$

Step 4 Convergence test

Compute vector $\Phi_k \in \Re^{L-1}$ of current values of the angle set where $\Phi_k(i) = \phi_i(k) = \tan^{-1}\left(\dfrac{A_{i,i+1}(k)g_{i+1,i+1}(0)}{M_{i,i+1}(k)g_{ii}(0)}\right)$, $i = 1,\ldots,L-1$

If $\|\Phi_k - \Phi_d\| \le \varepsilon$ then $K_{icr} = 4m_i/\pi a_i$, $i=1,2,\ldots,L$, $\omega_{cr} = 2\pi/P_{cr}$ and STOP
If $k > k_{max}$ then STOP with maximum iteration count exceeded
If $\|\Phi_k - \Phi_d\| > \varepsilon$ then CONTINUE

Step 5 Update computation

Form $A_R = [A_R(k-L+1) \quad \ldots \quad A_R(k-1) \quad A_R(k)]$

Form $\widetilde{M}_R = \begin{bmatrix} M_R(k-L+1) & \ldots & M_R(k-1) & M_R(k) \\ 1 & & 1 & 1 \end{bmatrix}$

Compute $B = A_R \widetilde{M}_R^{-1}$
Partition matrix B to form

$B_1 = \begin{bmatrix} b_{1,1} & \ldots & b_{1,L-1} \\ \vdots & \ddots & \vdots \\ b_{L-1,1} & \ldots & b_{L-1,L-1} \end{bmatrix} \in \Re^{L-1 \times L-1}$ and $B_2 = \begin{bmatrix} b_{1,L} \\ \vdots \\ b_{L-1,L} \end{bmatrix} \in \Re^{L-1}$

Compute $M_R(k+1) = [\Lambda_d - B_1]^{-1} B_2$

Step 6 Next relay experiment

Use $M_R(k+1)$ to update relay sizes $\{m_i(k+1), i=1,\ldots,L\}$
Run simultaneous relay experiment

Compute $(k+1)$th relay input amplitude ratios $A_{i,i+1}(k+1) = \dfrac{a_i(k+1)}{a_{i+1}(k+1)}$, $i=1,\ldots,L-1$

Update iteration counter to $k+1$
Goto Step 4

Algorithm end

In the methods of Algorithms 5.1 and 5.2 the relative importance of the loops has been determined using a critical point dependent only on the diagonal elements of the steady state process gain matrix. Consequently this does not consider the interaction between loops represented by the off-diagonal gains in the process gain matrix. It has previously been shown that for PID tuning, decoupling the system around bandwidth frequency has the best robustness (Katebi et al., 2000a) and in the next section an extension of the L-loop procedure is described in which new weighting factors are introduced using the process gains at bandwidth frequency.

5.5 PID Control Design at Bandwidth Frequency

The usual unity feedback configuration is shown in Figure 5.8. For such a single-input, single-output system description, the closed-loop bandwidth frequency ω_{bw} is defined as the lowest frequency such that $|T(j\omega_{bw})| = |T(0)|/\sqrt{2}$, where the complementary sensitivity function is defined as

$$T(s) = \dfrac{G_p(s)G_c(s)}{1 + G_p(s)G_c(s)}$$

In the single-input, single-output system case, the loop bandwidth ω_{bw} is usually very close to the gain crossover frequency ω_{gco}, where $|G_p(j\omega_{gco})G_c(j\omega_{gco})| = 1$. It has been demonstrated that for

5.5 PID Control Design at Bandwidth Frequency

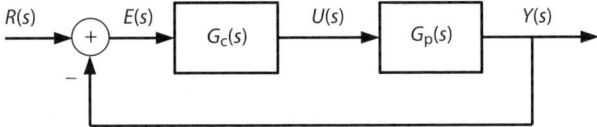

Figure 5.8 Unity feedback block diagram of process and controller.

typically acceptable designs the loop bandwidth can be bounded in terms of the gain crossover frequency by $\omega_{gco} \leq \omega_{bw} \leq 2\omega_{gco}$ (Maciejowski, 1989). However, the situation to be investigated is the L-loop decentralised control of an $(L \times L)$ multivariable process.

Selecting the Range of Oscillation Frequency

If an $(L \times L)$ process is controlled by decentralised relay feedback, its outputs usually oscillate in the form of limit cycles after an initial transient. Theorem 5.1 indicates that for a typical coupled multivariable process, the L outputs normally have the same oscillation frequency but different phases. This common frequency of oscillation is denoted ω_{cr} and satisfies $\omega_{cr} = \omega_i, i = 1, \ldots, L$. The describing function method was extended by Loh and Vasnani (1994) to analyse multivariable oscillations under decentralised relay feedback. For this analysis, it is assumed that the $(L \times L)$ process transfer function matrix has low pass transfer functions in each of its elements and that at least one of its characteristic loci has at least 180° phase lag. In the decentralised relay feedback, let the relay heights be $\{m_i; i = 1, \ldots, L\}$, the input signals to the relays have amplitudes $\{a_i; i = 1, \ldots, L\}$, and let the describing function matrix of the decentralised relay controller be defined as $K(a,m) = \text{diag}\{k_i(a_i, m_i)\}$ where $k_i(a_i, m_i) = 4m_i / \pi a_i; i = 1, \ldots, L$.

Theorem 5.2: Common oscillation frequency of decentralised relay feedback system

If a decentralised relay feedback system oscillates at a common frequency then at least one of the characteristic loci of $G_p(j\omega)K(a,m)$ passes through the critical point $s = -1 + j0$ on the complex plane, and the oscillation frequency corresponds to the frequency at which the crossing through point $s = -1 + j0$ occurs. Further, if the process is stable then the limit cycle oscillation is stable, the outermost characteristic locus of $G_p(j\omega)K(a,m)$ passes through the point $-1 + 0j$ and the process critical frequency is the same as the critical frequency of the outermost characteristic locus.

Proof (Loh and Vasnani, 1994) ◻

It should be noted that the crossing condition and the oscillation frequency in Theorem 5.2 are related to matrix $K(a, m)$, and this cannot be calculated until the oscillations are observed and the amplitudes $\{a_i; i = 1, \ldots, L\}$ of relay inputs measured from the oscillation waveforms. It would be useful if the frequency were given in terms of process information only and independently of the relay controller. Therefore, consider the situation of each of the Gershgorin bands of the $L \times L$ multivariable process $G_p(s)$. For each band, the region in the vicinity of the crossover on the negative real axis is of interest as shown in Figure 5.9.

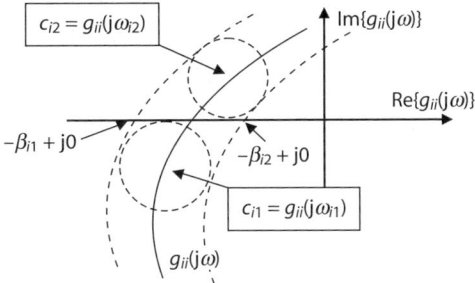

Figure 5.9 Negative real axis crossover of a Gershgorin band for process $G_p(s)$.

Let $c_{i1} = g_{ii}(j\omega_{i1})$ and $c_{i2} = g_{ii}(j\omega_{i2})$ be the centres of circles that are tangential to the negative real axis, and let $s_{i1} = -\beta_{i1} + j0$ and $s_{i2} = -\beta_{i2} + j0$ be the points at which the outer curve and inner curve of the ith Gershgorin band respectively intersect the negative real axis. If the ith Gershgorin band does not intersect the negative real axis, $[\omega_{i1},\omega_{i2}]$ is defined to be empty. The following result gives an estimate of ω_{cr} in terms of ω_{i1} and ω_{i2}.

Theorem 5.3: Bounds for common oscillation frequency

If the decentralised relay feedback system oscillates at a common frequency ω_{cr}, then there exists $i \in [1,2,...,L]$ such that $\omega_{cr} \in [\omega_{i1},\omega_{i2}]$.

Proof (Wang *et al.*, 1997) ❑

In view of Theorems 5.2 and 5.3, the oscillation frequency ω_{cr} for a stable process depends on which characteristic locus of $G_p(s)$ is moved outermost by the multiplication of the corresponding relay element describing function $k_i(a_i,m_i) = 4m_i/\pi a_i$. In general, the gain $k_i(a_i,m_i) = 4m_i/\pi a_i$ can be enlarged by increasing the ratios of the relay amplitudes in the ith loop with respect to the other loops. This will create a desired outmost loop which is called the *dominant* loop. It should be noted that the dominant loop remains dominant and the critical frequency varies little with a fairly large change of relay amplitude ratios unless an inner characteristic locus moves to become a new outermost one.

Theorem 5.4: Relay size condition for common oscillation frequency

If the decentralised relay feedback system for a stable process oscillates at a common frequency and for some i

$$k_i(a_i,m_i) > \frac{k_j(a_j,m_j)\beta_{j1}}{\beta_{i2}}, \quad j=1,2,...,L; \; j \neq i$$

then only the ith characteristic locus of $G_p(j\omega)K(a,m)$ passes through the point $-1+j0$ and the oscillation frequency satisfies $\omega_{cr} \in [\omega_{i1},\omega_{i2}]$.

Proof (Wang *et al.*, 1997) ❑

By Theorem 5.4, if the relay amplitudes are varied such that the resulting describing function gain matrix $K'(a,m)$ still satisfies

$$k'_i(a_i,m_i) > \frac{k'_j(a_j,m_j)\beta_{j1}}{\beta_{i2}}, \quad j=1,2,...,L; \; j \neq i$$

then the resulting limit-cycle oscillation frequency is expected to be in the range $[\omega_{i1},\omega_{i2}]$, and thus close to the previous value if the interval $[\omega_{i1},\omega_{i2}]$ is small. In general the conditions

$$k_i(a_i,m_i) > \frac{k_j(a_j,m_j)\beta_{j1}}{\beta_{i2}}, \quad j=1,2,...,L; \; j \neq i$$

remain true if the amplitude of the dominant loop increases or decreases one or more relay amplitudes among the other loops. This discussion indicates that by increasing the amplitude of the dominant loop or decreasing one or more relay amplitudes among the other loops, the limit-cycle oscillation frequency is expected to remain constant. This means that choosing the *dominant* loop has the effect of choosing the range of frequency $[\omega_{i1},\omega_{i2}]$ and bounding ω_{cr}.

Selecting the Desired Critical Point Criteria

Since a significant performance parameter is the loop gain at bandwidth frequency, the critical point criteria are chosen as follows:

$$\tan\phi_{di} = \frac{C_{di}}{1} = \frac{k_{P_{i+1cr}}\tilde{g}_{i+1,i+1}(\omega_{bw})}{k_{P_{icr}}\tilde{g}_{ii}(\omega_{bw})}, \quad i=1,\ldots,L-1$$

where ϕ_{dk}, or equivalently C_{di} is the weighting factor of the $(i+1)$th loop relative to the ith loop. If C_{di} is selected so that $C_{di} > 1$ then this implies that the $(i+1)$th loop is required to be under tighter control relative to the ith loop. In these formulae the gain quantities $\tilde{g}_{ii}(\omega_{bw})$; $i=1,\ldots,L$ are derived as a constant real approximation to $g_{ii}(s)$; $i=1,\ldots,L$ at the particular point $s=j\omega_{bw}$ using the ALIGN algorithm (Kouvaritakis, 1974), where

$$\begin{bmatrix} \tilde{g}_{11}(\omega_{bw}) & \cdots & \tilde{g}_{1L}(\omega_{bw}) \\ \vdots & \ddots & \vdots \\ \tilde{g}_{L1}(\omega_{bw}) & \cdots & \tilde{g}_{LL}(\omega_{bw}) \end{bmatrix} = \begin{bmatrix} g_{11}(j\omega_{bw}) & \cdots & g_{1L}(j\omega_{bw}) \\ \vdots & \ddots & \vdots \\ g_{L1}(j\omega_{bw}) & \cdots & g_{LL}(j\omega_{bw}) \end{bmatrix} \in \Re^{L\times L}$$

Thus the normalisation of the critical gain uses the loop gain at bandwidth frequency instead of the loop gain at zero frequency.

Finding the Gains at Bandwidth Frequency

To identify the process frequency response $G_p(j\omega)$ at the critical oscillation frequency $\omega_{cr} \approx \omega_{bw}$, the procedure outlined in Section 5.3.1 can be used.

Algorithm 5.3: *L-loop multi-input multi-output system decentralised relay experiment with bandwidth gain weighting*

Step 1 Data input

Input the selected values for the criterion set $\{C_{di}, i=1,\ldots,L-1\}$
Input the set of relay height bounds $\{\underline{m}_i \leq m_i \leq \overline{m}_i, i=1,\ldots,L\}$
Input the convergence tolerance ε
Input the maximum iteration count k_{max}

Step 2 Initial L-relay experiments

Record input and output data from relay experiments $k=1,2,\ldots,L$

$$y_{data}^k = \begin{bmatrix} y_1^k(t) \\ \vdots \\ y_L^k(t) \end{bmatrix}, \quad u_{data}^k = \begin{bmatrix} u_1^k(t) \\ \vdots \\ u_L^k(t) \end{bmatrix}, \quad t_0 \leq t \leq t_0 + P_{cr}$$

Record common frequency of oscillation ω_{osc}

Step 3 Computation of initial gains and other data

$$\text{Form matrices } Y(j\omega_{osc}) = \begin{bmatrix} y_1^1 & \cdots & y_1^L \\ \vdots & \ddots & \vdots \\ y_L^1 & \cdots & y_L^L \end{bmatrix}, \quad U(j\omega_{osc}) = \begin{bmatrix} u_1^1 & \cdots & u_1^L \\ \vdots & \ddots & \vdots \\ u_L^1 & \cdots & u_L^L \end{bmatrix}$$

$$y^k = \begin{bmatrix} \frac{1}{P_{cr}}\int_0^{P_{cr}} y_1^k(t)e^{-j\omega_{osc}t}dt \\ \vdots \\ \frac{1}{P_{cr}}\int_0^{P_{cr}} y_L^k(t)e^{-j\omega_{osc}t}dt \end{bmatrix}, \quad u^k = \begin{bmatrix} \frac{1}{P_{cr}}\int_0^{P_{cr}} u_1^k(t)e^{-j\omega_{osc}t}dt \\ \vdots \\ \frac{1}{P_{cr}}\int_0^{P_{cr}} u_L^k(t)e^{-j\omega_{osc}t}dt \end{bmatrix}, \quad k=1,2,\ldots,L$$

$$\text{Compute } G_p(j\omega_{osc}) = \begin{bmatrix} g_{11}(j\omega_{osc}) & \cdots & g_{1L}(j\omega_{osc}) \\ \vdots & \ddots & \vdots \\ g_{L1}(j\omega_{osc}) & \cdots & g_{LL}(j\omega_{osc}) \end{bmatrix} = Y(j\omega_{osc})U^{-1}(j\omega_{osc})$$

Compute $\begin{bmatrix} \tilde{g}_{11}(\omega_{\text{bw}}) & \cdots & \tilde{g}_{1L}(\omega_{\text{bw}}) \\ \vdots & \ddots & \vdots \\ \tilde{g}_{L1}(\omega_{\text{bw}}) & \cdots & \tilde{g}_{LL}(\omega_{\text{bw}}) \end{bmatrix} = \text{ALIGN} \begin{bmatrix} g_{11}(j\omega_{\text{bw}}) & \cdots & g_{1L}(j\omega_{\text{bw}}) \\ \vdots & \ddots & \vdots \\ g_{L1}(j\omega_{\text{bw}}) & \cdots & g_{LL}(j\omega_{\text{bw}}) \end{bmatrix}$

Compute $\phi_{di} = \tan^{-1} C_{di}$, $i = 1, \ldots, L-1$
Define vector $\Phi_d \in \Re^{L-1}$ where $\Phi_d(i) = \phi_{di}$, $i = 1, \ldots, L-1$
Compute relay height ratios

$$M_i(k) = \frac{m_i(k)}{m_{i+1}(k)}, \quad i = 1, \ldots, L-1;\ k = 1, \ldots, L$$

where $m_i(k) = \begin{cases} \overline{m}_i & i = k \\ \underline{m}_i & i \neq k \end{cases}$

Compute relay input amplitude ratios $A_{i,i+1}(k) = \dfrac{a_i(k)}{a_{i+1}(k)}$, $i = 1, \ldots, L-1$

Compute matrix $\Lambda_d = \text{diag}\left\{\dfrac{C_{di}\tilde{g}_{ii}(\omega_{\text{osc}})}{\tilde{g}_{i+1,i+1}(\omega_{\text{osc}})}\right\} \in \Re^{L-1 \times L-1}$

Set iteration counter $k = L$

Step 4 *Convergence test*
Compute the vector $\Phi_k \in \Re^{L-1}$ of current values of angle set, where

$$\Phi_k(i) = \phi_i(k) = \tan^{-1}\left(\frac{A_{i,i+1}(k)\tilde{g}_{i+1,i+1}(\omega_{\text{osc}})}{M_{i,i+1}(k)\tilde{g}_{ii}(\omega_{\text{osc}})}\right), \quad i = 1, \ldots, L-1$$

If $\|\Phi_k - \Phi_d\| \leq \varepsilon$ then $K_{icr} = 4m_i/\pi a_i$, $i = 1, 2, \ldots, L$, $\omega_{cr} = 2\pi/P_{cr}$ and STOP
If $k > k_{\max}$ then STOP with maximum iteration count exceeded
If $\|\Phi_k - \Phi_d\| > \varepsilon$ then CONTINUE

Step 5 *Update computation*

Form $A_R = [A_R(k-L+1) \quad \cdots \quad A_R(k-1) \quad A_R(k)]$

Form $\tilde{M}_R = \begin{bmatrix} M_R(k-L+1) & \cdots & M_R(k-1) & M_R(k) \\ 1 & & 1 & 1 \end{bmatrix}$

Compute $B = A_R \tilde{M}_R^{-1}$
Partition matrix B to form

$$B_1 = \begin{bmatrix} b_{11} & \cdots & b_{1,L-1} \\ \vdots & \ddots & \vdots \\ b_{L-1,1} & \cdots & b_{L-1,L-1} \end{bmatrix} \in \Re^{L-1 \times L-1} \text{ and } B_2 = \begin{bmatrix} b_{1L} \\ \vdots \\ b_{L-1,L} \end{bmatrix} \in \Re^{L-1}$$

Compute $M_R(k+1) = [\Lambda_d - B_1]^{-1} B_2$

Step 6 *Next relay experiment*
Use $M_R(k+1)$ to update relay sizes $\{m_i(k+1), i = 1, \ldots, L\}$
Run simultaneous relay experiment

Compute $(k+1)$th relay input amplitude ratios $A_{i,i+1}(k+1) = \dfrac{a_i(k+1)}{a_{i+1}(k+1)}$, $i = 1, \ldots, L-1$

                Update iteration counter to $k + 1$
                Goto Step 4
        Algorithm end

5.6 Case Studies

In this section two case study systems are considered. The two versions of the algorithm are used and results compared in the first example. In the second study the convergence of the algorithm is the main point of interest. For ease of discussion, the following labelling is used.

- Method 1 – Implements Algorithm 5.2 with desired critical point scaled by the zero frequency gain $G_p(0) \in \Re^{L \times L}$.

- Method 2 – Implements Algorithm 5.3 with desired critical point scaled by the bandwidth frequency gain $\widetilde{G}_p(\omega_{bw}) \in \Re^{L \times L}$.

Once the desired critical point has been found, the PID controller parameters are obtained from simple application of the Ziegler–Nichols rules. For the purposes of comparing the results, indices are computed for nominal performance, robustness stability and robustness performance of closed-loop systems.

For nominal performance
The H_∞ norm of the weighted sensitivity function is computed. The desirable condition is $WS = \left\| W_1(s)[I + G_p(s)G_c(s)]^{-1} \right\|_\infty < 1$, and the smaller the norm the better is the disturbance rejection.

For robust stability
The H_∞ norm of the weighted complementary sensitivity function is computed. The desirable condition is $WT = \left\| W_2(s)G_p(s)G_c(s)[I + G_p(s)G_c(s)]^{-1} \right\|_\infty < 1$.

For robust performance
Robust performance is guaranteed if $\mu(M) < 1$ (Maciejowski, 1989). Based on the criterion proposed by Gagnon *et al.* (1999) an interesting extension which uses the control sensitivity to penalise the actuator variations, which are often important in terms of control energy cost and actuator saturation, wear and tear and hence plant availability, is given as the robust performance index

$$J_{RP} = \min_{G_c(s)} \left\{ \sum_{i=1}^{m} W_\rho \left| \mu_{\Delta G(s)}[M(s)] - 1 \right|^2 + W_u \mu_{\Delta G(s)}[C(s)] \right\}$$

where an extended process description is used, with

$$\widetilde{Y}(s) = [M(s)]\widetilde{U}(s)$$
$$\widetilde{Y}(s) = \begin{bmatrix} y(s) \\ e(s) \end{bmatrix}, \widetilde{U}(s) = \begin{bmatrix} u(s) \\ r(s) \end{bmatrix}$$

$\widetilde{G}_p(s) = [I + W_1(s)\Delta G(s)]G_p(s)$ is the plant model with uncertainty $\Delta G(s), C(s) = G_c(s)[I + G_p(s)G_c(s)]^{-1}$, W_ρ and W_u are appropriate weightings, $\mu[\cdot]$ is the structured singular value and m represents the number of frequency points at which the criterion is evaluated.

5.6.1 Case Study 1: The Wood and Berry Process System Model

The process model due to Wood and Berry (1973) is given by

$$G_p(s) = \begin{bmatrix} \dfrac{12.8e^{-s}}{16.7s+1} & \dfrac{-18.9e^{-3s}}{21.0s+1} \\ \dfrac{6.6e^{-7s}}{10.9s+1} & \dfrac{-19.6e^{-3s}}{14.4s+1} \end{bmatrix}$$

The data collected for Method 1 and Method 2 are summarised in Table 5.1. Both methods converged in four iterations. The desired critical points found for Methods 1 and 2 are, respectively, $k_{P1cr} = 0.402$, $k_{P2cr} = -0.261$ and $k_{P1cr} = 0.1682$, $k_{P2cr} = -0.3223$.

Table 5.1 Data of algorithm progress.

Iteration count and method		Relay height		Relay amplitude		Critical gain		Crit. freq.	Tolerance
		m_1	m_2	a_1	a_2	k_{P1cr}	k_{P2cr}	ω_{cr}	$|\phi - \phi_d|$
1	Method 1	0.3	0.03	0.224	0.154	1.7	−0.248	1.56	32.5
	Method 2	0.3	0.03	0.224	0.154	1.7	−0.248	1.56	41
2	Method 1	0.02	0.08	0.212	0.307	0.12	−0.331	0.529	31.5
	Method 2	0.02	0.08	0.212	0.307	0.12	−0.331	0.529	12
3	Method 1	0.092	0.08	0.138	0.441	0.853	−0.23	0.498	22.7
	Method 2	0.0312	0.08	0.217	0.3445	0.1831	−0.2957	0.5216	3.0
4	Method 1	0.059	0.08	0.186	0.389	0.402	−0.261	0.491	0.44
	Method 2	0.0286	0.08	0.2165	0.3165	0.1682	−0.3223	0.5271	1.7

The desired critical point achieved by the two methods is seen in Figure 5.10(a). Figure 5.10(b) shows that both methods have produced a controlled system which has achieved nominal performance.

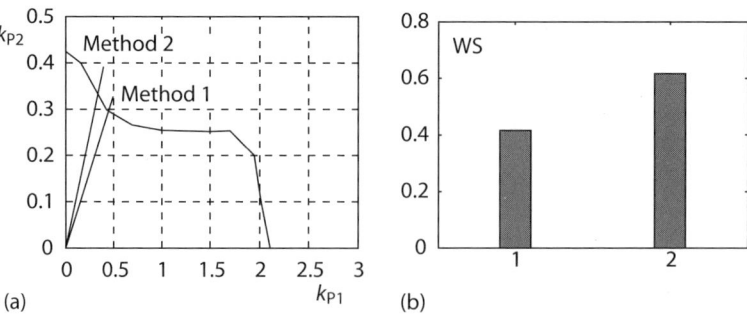

Figure 5.10 (a) Critical point location; (b) nominal performance index.

A comparison of robust stability and robust performance indices of the control schemes resulting from the two methods shows Method 2 having marginally better index values than Method 1, as shown in Figures 5.11(a) and (b).

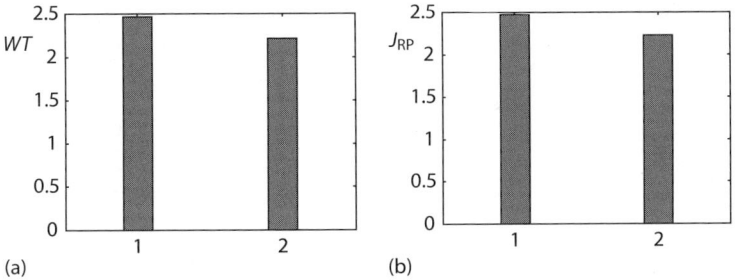

Figure 5.11 (a) Robust stability index; (b) robust performance index.

The step responses (Figures 5.12 and 5.13) show very little difference between the two methods.

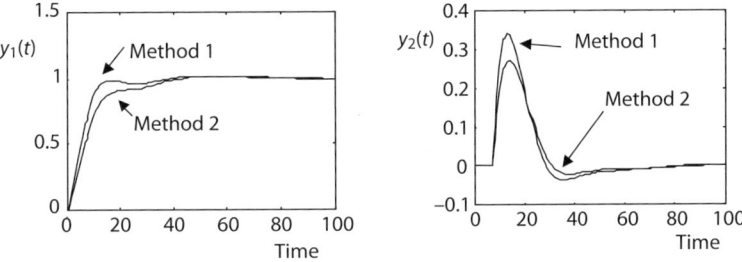

Figure 5.12 Step responses $r_1(t) = 1; r_2(t) = 0$.

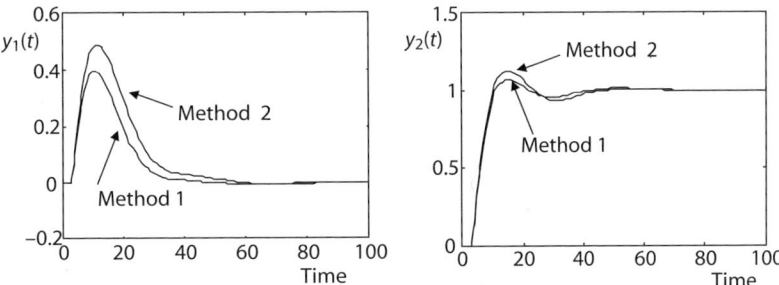

Figure 5.13 Step responses $r_1(t) = 0; r_2(t) = 1$.

5.6.2 Case Study 2: A Three-Input, Three-Output Process System

The process transfer function matrix is

$$G_p(s) = \begin{bmatrix} \dfrac{-1e^{-s}}{6s+1} & \dfrac{1.5e^{-s}}{15s+1} & \dfrac{0.5e^{-s}}{10s+1} \\ \dfrac{0.5e^{-2s}}{s^2+4s+1} & \dfrac{0.5e^{-3s}}{s^2+4s+1} & \dfrac{0.513e^{-s}}{s+1} \\ \dfrac{0.375e^{-3s}}{10s+1} & \dfrac{-2.0e^{-2s}}{10s+1} & \dfrac{-2.0e^{-3s}}{3s+1} \end{bmatrix}$$

The data collected for Method 1 is summarised in Table 5.2. The results show that the algorithm of Method 1 for a three-input, three-output system ($L = 3$) converges to desired critical point in seven iterations.

Table 5.2 Data of algorithm progress for three-input, three-output system.

Iter. No.	Relay height			Relay amplitude			Critical gain			Crit. freq.	Angle	
	m_1	m_2	m_3	a_1	a_2	a_3	k_{P1cr}	k_{P2cr}	k_{P3cr}	ω_{cr}	ϕ_1	ϕ_2
1	−7	0.7	−0.7	1.04	0.656	0.24	8.57	1.36	3.71	1.73	4.53	85
2	−1.25	2.5	−0.7	0.498	1.254	1.787	3.2	2.54	0.5	0.683	21.7	38.2
3	−0.446	2.5	0.738	0.583	1.165	1.787	0.974	2.73	0.526	0.712	54.5	37.6
4	−0.944	2.5	−0.862	0.55	1.21	1.86	2.19	2.63	0.59	0.7	31	41.9
5	−0.633	2.5	−1.024	0.585	1.338	1.993	1.38	2.38	0.654	0.728	40.8	47.7
6	−0.553	2.5	−0.936	0.59	1.293	1.969	1.193	2.46	0.605	0.725	45.9	44.5
7	−0.564	2.5	−0.948	0.594	1.31	1.95	1.21	2.43	0.619	0.726	45.1	45.2

5.7 Summary

The chapter opened with a short review and discussion of the importance of the critical point specification for multi-loop relay experiments in multivariable systems. The well-known Palmor decentralised relay identification method for PID tuning for two-input, two-output systems was presented in some detail to expose the structure and derivation of the method. This method was extended from two-input, two-output systems to square L-input, L-output systems.

In the extended algorithm, the tuning procedure consists of two stages. In the first stage, the desired critical point, which consists of the steady state gains of L-loops and a critical frequency, is identified. In the second stage, the data of the desired critical point is used to tune the PID controller by the Ziegler–Nichols rule or its modifications.

Focusing on the selection of the desired critical point, a new criterion for tuning of decentralised PID controller for MIMO plants was introduced. In the proposed method the bandwidth frequency of the system was used to calculate the desired critical point. The process gains at the crossover frequency, which are required for the appropriate desired critical point choice of the system, are determined using

L relay experiments. The robustness and performance of the method were compared with the Palmor method. The limited examples presented and other extensive practical experience indicate that the new method is more robust and converges faster than Palmor method.

References

Åström, K.J. and Hägglund, T. (1988) *Automatic Tuning of PID Controllers*. Instrument Society of America, Research Triangle Park, NC.

Katebi, M.R., Moradi, M.H. and Johnson, M.A. (2000) Comparison of stability and performance robustness of multivariable PID tuning methods. *IFAC Workshop on Digital Control*, Terrassa, Spain.

Kouvaritakis, B. (1974) Theory and practice of the characteristic locus design method. *Proc. IEE*, **126**, 542–548.

Loh, A.P. and Vasnani, V.U. (1994) Describing function matrix for multivariable systems and its use in multi-loop PI design. *J. Process Control*, **4**(4), 115–120.

Loh, A.P., Hang, C.C., Quek, C.K. and Vasnani, V.U. (1993) Auto-tuning of multi-loop PI controllers using relay feedback. *Ind. Eng. Chem. Res.*, **322**, 1102–1107.

Maciejowski, J.M. (1989) *Multivariable Feedback Design*. Addison-Wesley, Wokingham.

Palmor, Z.J. and Blau, M. (1994) An auto-tuner for Smith dead time compensator. *Int. J. Control*, **60**, 117–135.

Palmor, Z.J., Halevi, Y. and Krasney, N. (1995) Automatic tuning of decentralised PID controllers for TITO processes. *Automatica*, **31**(7), 1001–1010.

Skogestad, S. and Morari, M. (1989) Robust performance of decentralised control systems by independent design. *Automatica*, **25**, 119–125.

Wang, Q., Zou, B., Lee, T.H., Bi, Q., Queek, C.K. and Vasnani, V.U. (1997) Auto-tuning of multivariable PID controllers from decentralised relay feedback. *Automatica*, **33**(3), 319–330.

6 Phase-Locked Loop Methods

Learning Objectives

6.1 Introduction

6.2 Some Constructive Numerical Solution Methods

6.3 Phase-Locked Loop Identifier Module – Basic Theory

6.4 Summary and Discussion

References

Learning Objectives

The sustained oscillations experiment devised by Ziegler and Nichols proved to be of enduring value in PID controller tuning for a significant class of industrial systems. The experiment involves maintaining the process in closed-loop proportional control and increasing the proportional gain until the closed-loop system reaches the verge of instability. At this point estimates of the ultimate period and gain data are made for use in rule-based PID controller design formulae. The relay experiment is the modern incarnation of this experiment as used in much current process control technology. An alternative route to the same data is the recently introduced phase-locked loop identifier concept. Over the next two chapters, the basic theory for the phase-locked loop is presented and the flexibility inherent in the identifier explored. This chapter opens with some background on the relay experiments that motivated the search for an alternative identifier method. A constructive numerical solution is presented before the phase-locked loop method is formulated. The remainder of the chapter is devoted to presenting the basic theory for the phase-locked loop identifier module.

The learning objectives for this chapter are to:

- Learn more about the relay experiment and its related implementation aspects.

- Investigate a possible constructive numerical solution method for ultimate data based on Nyquist geometry.

- Draw some conclusions about the relative merits of the bisection and prediction methods.
- Understand the construction and basic theory of the phase-locked loop identifier module.
- Learning about some noise management techniques for the phase-locked loop identifier.
- Understanding some disturbance management techniques for the phase-locked loop identifier.

6.1 Introduction

The engineer in the process industries often finds that there is no system model available on which to base the design of the system controller. Faced with this problem there are two possible ways forward to produce a model of the system, either physical modelling or system identification is followed. In physical modelling the equations relating to the physics or chemistry of the system are used to produce a model from first principles. In system identification, test signals are input to the system and then the input and output data are analysed to produce the system model.

The possible benefits to be realised from constructing a physical model of the system include a greater understanding of the underlying operation of the system and improved interaction between the interdisciplinary staff team involved in producing the model. However, the drawbacks of developing a physical model are often more influential in determining whether this route is followed. Physical model determination can be costly in the time and expert personnel needed for success. There is a risk that the system may be so complex that it is not possible to produce a model from first principles. If the system is highly complex the model may contain a number of parameters which are unknown and hence may require either additional experiments or other known system data for accurate model parameter calibration. The system may contain nonlinearity, so the model may be unsuitable at certain operating points.

System identification techniques are not a panacea for the production of system models. They do, however, provide the engineer with a number of benefits. Simple model fits may give an indication of the best model form for accurate system representation. The control engineer is likely to learn about the range of frequencies over which the model will be valid. Simple models are often relatively easy to produce from measured system input and output data. The low cost of producing identified models can be attractive and is often a decisive factor with management in selecting the system identification route. As with physical modelling there are also a number of disadvantages. The identified models produced may only be valid for a certain type of operating point, input signal or system.

Nonparametric System Identification

System identification techniques can be categorised into two main groups: parametric techniques and nonparametric techniques. Parametric system identification is perhaps the more widely known method (Eykhoff, 1974; Ljung, 1987). This involves proposing a model which is parameterised by a finite dimensional vector of parameters. The next step is to choose the model parameters such that the error between observed data and model data is minimised with respect to the model parameters (Ljung and Glover, 1981).

The nonparametric system identification method (Wellstead, 1981; Ljung, 1987; Söderström and Stoica, 1989) is perhaps not as well known as the parametric methods. These techniques do not usually produce the system model in an analytical form. For example, the model may take the form of a data table, a graph or information about one or two important data points from an inherent analytical model. Typical graphical representations of the model are Bode or Nyquist plots containing the frequency domain data for the system transfer function over a certain range of frequencies. Of particular interest in PID controller design is the location of the system phase and gain crossover points. The method of sustained oscillation formulated by Ziegler and Nichols (1942) is precisely an example of nonparametric identification for the phase crossover point. Ziegler and Nichols used the sustained oscillation data in a set of rules (formulae) which defined the PID controller coefficients directly.

This direct approach to PID control design is successful for a large number of simple processes and the method became extremely popular within the process industries. Eventually a number of competing rule-based methods emerged, and the control engineer should always verify that the performance that will result from using any particular rule-based method is both desirable and satisfactory. However, it should not be considered that the tuning of a large number of control loops is a trivial task. Even if the computation of the parameters of a PID controller is relatively simple, the Ziegler and Nichols procedure still required the process loop to be brought to the verge if instability and there is evidence which shows that engineers still have difficulty in choosing an appropriate set of tuning parameters (Hersh and Johnson, 1997).

To ease the problems of performing the sometimes difficult on-line nonparametric identification experiment, of correctly selecting the PID controller parameters and of tuning the sheer number of loops found in any large scale industrial process, the procedure of PID controller tuning was automated in process controller units which entered the market from the 1980s onward. A very interesting procedure to automate the Ziegler and Nichols method of determining the phase crossover point was proposed by Åström and Hägglund (1985). In this method, a simple relay replaced the PID controller in the control loop and for a usefully significant class of industrial systems a stable limit cycle at the phase crossover frequency is automatically established with the system in closed loop.

6.1.1 The Relay Experiment

Consider the closed-loop system block diagrams shown in Figure 6.1. In Figure 6.1(a), the typical unity closed-loop feedback system is shown with a PID controller in the forward path. In Figure 6.1(b) the configuration of the relay experiment is shown where the controller has been replaced by a simple on–off relay unit of height $\pm h$.

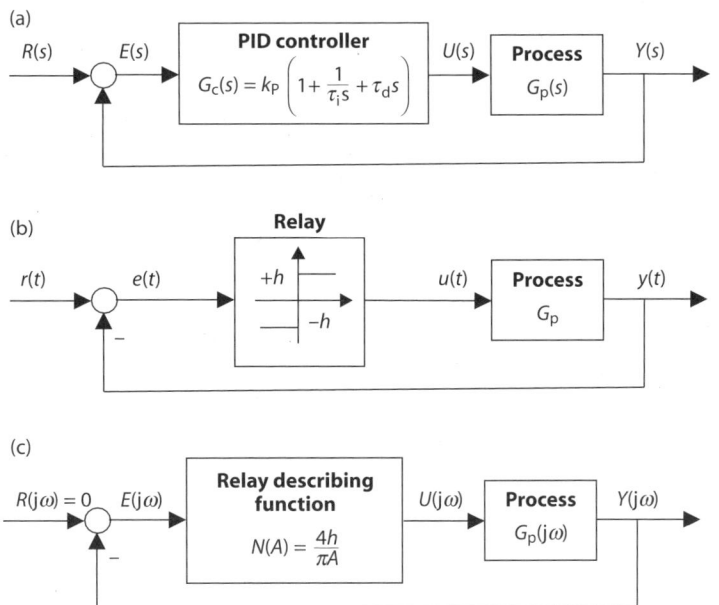

Figure 6.1 (a) PID controller in a unity feedback system; (b) relay experiment setup; (c) analysis setting with relay describing function $N(A)$.

In Figure 6.1(c) the setup for an analysis of the relay experiment using a describing function is shown. In this figure, the process is represented by the transfer function $G_p(j\omega)$ and the relay by the frequency independent describing function $N(A)$ with $N(A) = 4h/\pi A$, where the relay height is $\pm h$, and the signal at the input to the relay has amplitude A. Let the signal at the input to the relay be $E(j\omega) = A\exp(j\omega t)$; then the input to the process is $U(j\omega) = N(A)E(j\omega)$ and the process output is $Y(j\omega) = N(A)G_p(j\omega)E(j\omega)$. In steady state operation, there is no loss in generality in setting the reference input to zero, so that $R(j\omega) = 0$ and $E(j\omega) = -Y(j\omega)$.

Thus, going round the loop yields

$$E(j\omega) = A\exp(j\omega t) = -N(A)G_p(j\omega)A\exp(j\omega t)$$

Hence

$$[1 - N(A)G_p(j\omega)]A\exp(j\omega t) = 0$$

and the describing function condition for sustained oscillation is

$$G_p(j\omega) = \frac{-1}{N(A)}$$

This complex relation can be expressed as two equivalent conditions:

$$|G_p(j\omega)| = \frac{1}{N(A)} \text{ and } \arg(G_p(j\omega)) = -\pi$$

The frequency of this oscillation will be the frequency at which the phase crossover occurs, so that the critical frequency is denoted $\omega_{-\pi}$. From the system output response at the phase crossover point and the relay characteristics, the value of the ultimate gain and period can be determined and used in the tuning formula for PID controller design. This is the basic principle behind the relay experiment. The significant advantages of the experiment are:

1. The experiment is successful for a significant group of industrial systems.

2. The experiment is very simple to set up and implement.

3. The system is in closed loop at all times.

4. The limit cycle established is a stable oscillation.

5. The limit cycle automatically occurs at the critical frequency.

6. The closed-loop system does not approach the verge of instability as in the sustained oscillation experiment.

6.1.2 Implementation Issues for the Relay Experiment

For many situations, the relay experiment provides a sufficiently accurate estimate of the ultimate data. However, there are circumstances where inaccurate estimates are obtained and there can be degradation in the performance of the control loop from a controller designed on the basis of the inaccurate values. The potential for poor estimation performance can come from the method itself or the process experimental conditions. In the relay experiment there are three main sources for error when estimating the phase crossover point:

1. The describing function method of linearising the relay characteristic.

2. The presence of load disturbances.

3. The presence of measurement noise in the system output response.

6.1 Introduction

The relay experiment literature is quite mature, and various methods have been proposed to eliminate these effects. Some of these solution methods, particularly to overcome the shortcomings of the accuracy of the describing function method were presented in Chapter 4. The next few sections describe these aspects of the relay method and the implementation issues arising.

Harmonic Generation Reduction Techniques

The describing function method assumes that the relay is excited by a single sinusoidal signal and the relay output is then described by a Fourier series. The relay describing function is then defined as the ratio of the first harmonic of the relay output signal to the magnitude of the excitation. If the sinusoidal input to the relay has frequency ω_0 then the Fourier series representation of the relay output can be written as

$$y(t) = \sum_{n=-\infty}^{\infty} \frac{2h}{n\pi} \sin^2\left(\frac{n\pi}{2}\right) \exp\left[j\left(n\omega_0 t - \left(\pi n - \frac{\pi}{2}\right)\right)\right]$$

The sequence of magnitudes of the fundamental and the first few harmonics is tabulated below:

Frequency	$1 \times \omega_0$	$2 \times \omega_0$	$3 \times \omega_0$	$4 \times \omega_0$	$5 \times \omega_0$	$6 \times \omega_0$	...
Coefficients	$M_1 = \frac{4h}{\pi}$	0	$\frac{4h}{3\pi} = \frac{1}{3}M_1$	0	$\frac{4h}{5\pi} = \frac{1}{5}M_1$	0	...

For the describing function method to work well it is inherently assumed that the process satisfies $|G(jn\omega_{-\pi})| \ll |G(j\omega_{-\pi})|, n = 3, 5, 7, \ldots$ and $|G(jn\omega_{-\pi})| \to 0$ as $n \to \infty$.

Processes that do not satisfy this assumption will have considerable power present in the harmonics of the system response and this will give rise to errors in the estimation of the phase crossover point using the standard technique. The techniques developed to overcome this problem fall into two main categories.

Methods that Modify the Relay Characteristic

In this class of techniques the relay is replaced by an alternative nonlinear device. As an example, Shen *et al.* (1996a) showed that the harmonic content of the relay output is reduced by adopting a relay with a saturation characteristic, as shown in Figure 6.2.

The analysis of the method shows that the choice of the saturation relay gain to give exact identification of the phase crossover point is system-dependent and cannot be determined prior to the relay

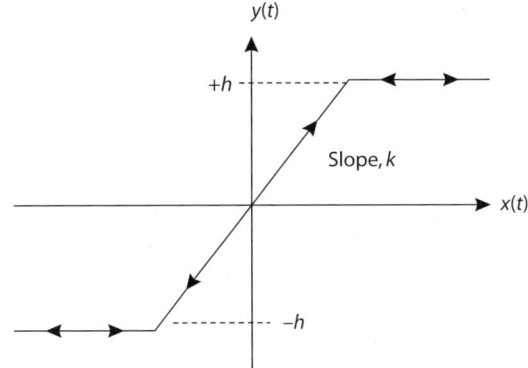

Figure 6.2 Saturation relay characteristic.

experiment being performed. There is also a further problem in that the saturation relay gain must be greater than the minimum value given by

$$k_{min} = \frac{1}{|G(j\omega_{-\pi})|}$$

if a stable limit cycle is to be established by the experiment. The minimum value of gain is shown to be equal to the ultimate gain of the process. Shen *et al.* (1996a) give a value of $k = 1.4k_{min}$ as a starting point for the saturation relay gain that will allow limit cycling of the process with an improved estimate of the phase crossover point. The initial estimate of the minimum gain is derived by carrying out a relay test with a standard relay; after the estimate of the ultimate gain is found this value can be used to set the gain for the saturation relay experiment to improve the estimate of the phase crossover point.

Methods that Filter the Relay Output

Lee *et al.* (1995) propose that the standard relay be modified such that there is no generation of harmonics in the relay output other than the fundamental frequency of the relay excitation. This method differs from that of Shen *et al.* (1996a) in that a nonlinear system is added to the output of the standard relay. The nonlinear system is used to provide a mapping such that the excitation signal fed to the process is only the fundamental component of the standard relay. A full description of this technique and its extensions has been given previously in Chapter 4.

Reducing Load Disturbance-Induced Errors

Load disturbances occur during normal process operation and are usually caused by a change in demand of the process. The effect of a load disturbance on the relay experiment is to produce an output from the standard relay which has an unequal mark to space ratio.

Example 6.1

Consider the case of a process modelled by

$$G_p(s) = \frac{0.6}{(1+s)^3}$$

In a simulation test, a load disturbance of 0.5 is added at time 20 seconds to the output of the system, which is in closed loop with a standard relay. The resulting system responses are shown in Figure 6.3.

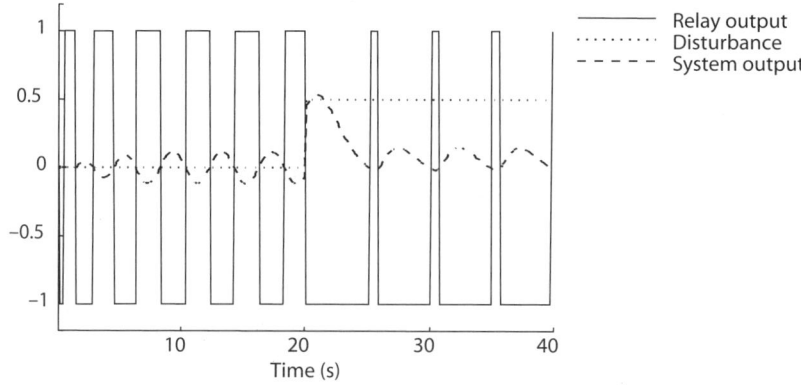

Figure 6.3 Relay experiment with load disturbance.

It can be seen from Figure 6.3 that after the application of the load disturbance, the mark to space ratio of the relay output is changed. There is a further problem, since if the magnitude of the load disturbance is greater than or equal to the steady state magnitude response of the system, then the relay fails to produce any oscillations at all; this can be seen in Figure 6.4.

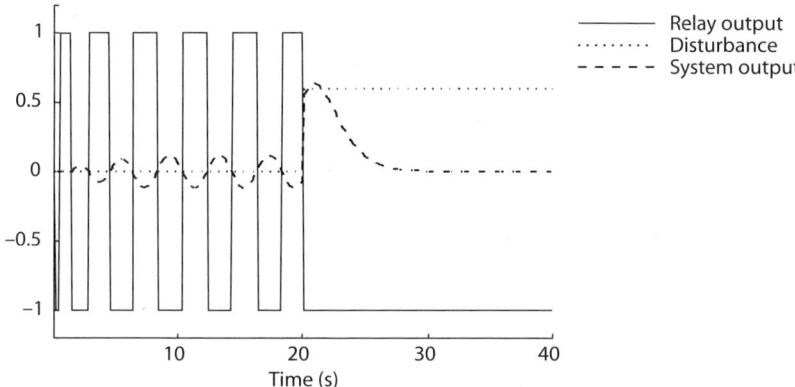

Figure 6.4 Relay experiment with load disturbance equal to system steady state magnitude.

Clearly, under load disturbance conditions there will be substantial error in the estimation if the relay output signal is used in the calculation of the phase crossover point. Shen *et al.* (1996b) proposed the use of a biased relay characteristic (Figure 6.5) to overcome the effect of a step load disturbance.

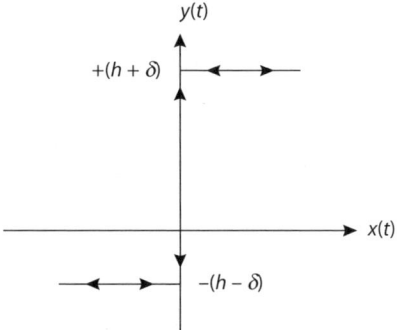

Figure 6.5 Biased relay characteristic.

This has the effect of introducing an additional bias term in the Fourier series of the relay output. The magnitude of the bias must be chosen correctly in order to obtain exact cancellation of the load disturbance term. Shen *et al.* (1996b) give the bias δ required for the biased relay unit as $\delta = -h(\Delta a/a)$, where h is the relay height, a is half the peak to peak amplitude of the system output and Δa is the bias which has been added to the system output due to the load disturbance effect. Consequently, the bias required for the relay can be calculated from the available process data on-line and it is possible to negate the effects of the load disturbance on the estimation of the phase crossover point.

Improving the Experiment Accuracy in the Presence of Noise

The accuracy of the relay experiment is adversely affected if there is measurement noise present in the system output. Spurious switching of the relay can be experienced near the zero crossing of the system output due to the presence of the measurement noise. Åström and Hägglund (1995) addressed this

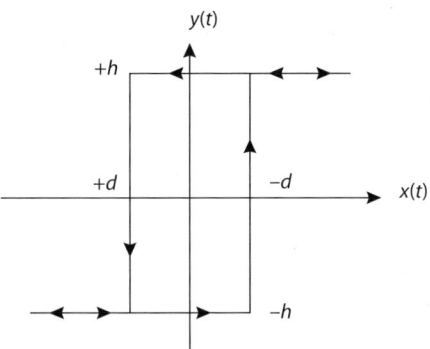

Figure 6.6 Relay characteristic with hysteresis.

problem by using a relay characteristic with hysteresis in place of the standard relay. Figure 6.6 shows a relay characteristic with hysteresis.

The value of the relay hysteresis $\pm d$ is set such that it is greater than the peak value of the noise signal. The use of hysteresis to prevent noise-induced switching of the relay output introduces imprecision into the identification of the $\omega_{-\pi}$ point on the frequency response curve. The frequency-independent describing function of a relay with hysteresis is

$$N(A) = \frac{4h}{\pi A}\left[\sqrt{1-\left(\frac{d}{A}\right)^2} - j\frac{d}{A}\right]$$

where h is the relay height and d is the hysteresis value. Consequently, the stable limit cycle is established where

$$G_p(j\omega) = \frac{-1}{N(A)}$$

and in this case this will occur when

$$G_p(j\omega) = \frac{-\pi}{4h}[\sqrt{A^2 - d^2} + jd]$$

Thus the intersection between the system frequency response and the describing function takes place away from the negative real axis and this does not identify the $\omega_{-\pi}$ point, but a point possibly in its neighbourhood. Hence although the spurious switching of the relay at the zero crossing point of the system response may have been remedied, the estimation accuracy of the $\omega_{-\pi}$ point has been compromised.

6.1.3 Summary Conclusions on the Relay Experiment

The simplicity of the relay experiment has ensured that it has become the preferred method for implementing the Ziegler–Nichols sustained oscillation procedure. It is thought that the relay experiment has been adopted in many commercially produced autotuning controllers and has certainly influenced progress towards the push-button autotune PID culture. The relay experiment and its application are still current research topics for academics around the world (for example Yu, 1998; Tan et al., 1999; Wang et al., 2003). However, as shown in the preceding sections, there are some shortcomings to the relay experimental procedure. These have motivated the academic community to propose various *ad hoc* remedies for these failings in the method. In practice, it is likely that control vendors have installed jacketing procedures in any commercial implementation to protect and ensure a successful outcome to the autotuner application.

Commercial relay-based autotuner technology has garnered around twenty years of development experience and it seems instrumental to initiate a search for a new methodology which might eventually replace the relay basis of current autotuner technology. It was this search which motivated the developments reported in this chapter. The investigation eventually led to the use of an identification module based on a phase-locked loop concept. The route taken by the research was not a direct one and the next section features geometrically based solution methods. The difficult problem of convincing industrial engineers to use these rather academic geometric numerical procedures led to the more accessible phase-locked loop concepts. The phase-locked loop ideas are considered in the remainder of the chapter, with the subsequent PID control applications appearing in Chapter 7.

6.2 Some Constructive Numerical Solution Methods

The sine wave experiment is a nonparametric identification technique that can be used to identify any point on the frequency response of a linear system. The method critically depends on the assumption of system linearity and involves the injection of a sinusoid of known amplitude, frequency and phase and measuring the system output response after any transients have decayed. The magnitude and phase of the frequency response at the excitation frequency can be obtained from the output response and the known input signal data. If it is desired to identify the frequency response phase crossover point for an unknown process using the sine wave experiment, then a trial and error approach might use the following algorithm.

Algorithm 6.1: Trial and error sine wave experiment

Step 1   Initialisation
  Specify initial sinusoidal frequency ω_0
  Set input signal amplitude A
  Set phase convergence tolerance *eps*
  Set iteration counter $k = 0$

Step 2   Loop step
  Excite the system with input $u_k(t) = A\sin(\omega_k t)$
  Allow transients to decay and a steady oscillating output to emerge
  Form output, $y_k(t) = B\sin(\omega_k t + \phi_k)$
  Calculate system phase $\phi_k(\omega_k)$
  Convergence test: If $|(-\pi) - \phi_k(\omega_k)| < eps$ goto Step 3
  Increase or decrease ω_k as appropriate to give ω_{k+1}
  Update k to $k + 1$ and goto Step 2

Step 3   Output results
  Critical frequency $\omega_{-\pi} = \omega_k$
  System magnitude $|G_p(j\omega_{-\pi})| = B/A$

Algorithm end

Thus it can be seen that the time required to identify the process phase crossover point could be very long. The weakness of the algorithm lies in the loop step, where a systematic way of increasing or decreasing ω_k to give ω_{k+1} such that $\phi_{k+1}(\omega_{k+1})$ is closer to the target phase of $(-\pi)$ than the current estimate $\phi_k(\omega_k)$ is needed. Initially two methods were investigated to achieve this aim:

1. A bisection method
2. A prediction method

6.2.1 Bisection Method

This method relies on defining two frequencies such that the phase of the system frequency response at those two frequencies gives a bound on the critical phase of $\phi_{-\pi} = -\pi$ rad. The next test frequency is then determined from the data derived from the first two sine wave tests using a bisection formula. The new sine wave test is carried out at this frequency. From the data obtained between the three sine wave tests, a new test frequency is calculated from the two frequencies which give the smallest bound on the frequency response phase of $\phi_{-\pi} = -\pi$ rad. The geometrical construction of the method is shown in Figure 6.7. The assumption made in this method is that the frequency along the Nyquist plot is linearly related to its length.

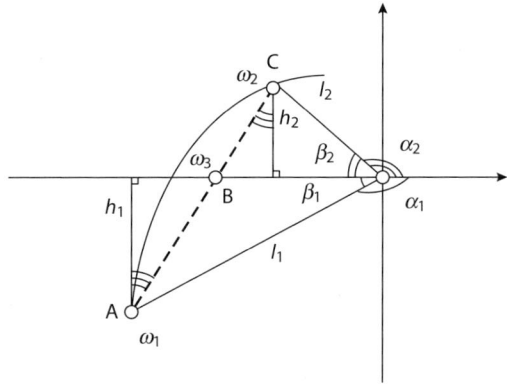

Figure 6.7 Bisection method: geometrical construction.

Using Figure 6.7, the following quantities are defined:

$l_1 = |G_p(j\omega_1)|,\ \alpha_1 = \arg(G_p(j\omega_1)),\ -\pi < \alpha_1 < 0,\ \beta_1 = \pi + \alpha_1,\ 0 < \beta_1$

$l_2 = |G_p(j\omega_2)|,\ \alpha_2 = 2\pi + \arg(G_p(j\omega_2)),\ 0 < \alpha_2 < \pi,\ \beta_2 = \pi - \alpha_2,\ 0 < \beta_2$

The assumption of a linear frequency relationship gives the following ratio:

$AB:(\omega_3 - \omega_1) :: AC:(\omega_2 - \omega_1)$

and using Figure 6.7, the ratio accrues:

$AB:h_1 :: AC:(h_1 + h_2)$

From the two ratios, it can be shown that

$$\omega_3 = \frac{\omega_1 l_2 \sin \beta_2 + \omega_2 l_1 \sin \beta_1}{l_1 \sin \beta_1 + l_2 \sin \beta_2}$$

This relationship performs the role of an update equation in Algorithm 6.2.

Algorithm 6.2: Bisection method

Step 1 Initialisation

Choose two initial frequencies such that $\phi_2 = \arg(G_p(j\omega_2)) < -\pi < \arg(G_p(j\omega_1)) = \phi_1$

Set input signal amplitude A
For ω_1 case
    Excite the system with input $u_1(t) = A\sin(\omega_1 t)$
    Allow transients to decay; steady oscillating output $y_1(t) = B\sin(\omega_1 t + \phi_1)$ emerges
    Calculate magnitude $l_1 = |G_p(j\omega_1)| = B/A$
    Calculate phase $\alpha_1 = \arg(G_p(j\omega_1)) = \phi_1, \beta_1 = \pi + \alpha_1, 0 < \beta_1$
For ω_2 case
    Excite the system with input $u_2(t) = A\sin(\omega_2 t)$
    Allow transients to decay; steady oscillating output $y_2(t) = B\sin(\omega_2 t + \phi_2)$ emerges
    Calculate magnitude $l_2 = |G_p(j\omega_2)| = B/A$
    Calculate phase $\phi_2 = \arg(G_p(j\omega_2)), \phi_2 < -\pi, \beta_2 = -\pi - \phi_2, 0 < \beta_2$
Set phase convergence tolerance eps

Step 2   Loop step

$$\text{Compute } \omega_3 = \frac{\omega_1 l_2 \sin\beta_2 + \omega_2 l_1 \sin\beta_1}{l_1 \sin\beta_1 + l_2 \sin\beta_2}$$

    Excite the system with input $u_3(t) = A\sin(\omega_3 t)$
    Allow transients to decay; steady oscillating output $y_3(t) = B\sin(\omega_3 t + \phi_3)$ emerges
    Compute system magnitude $l_3 = |G_p(j\omega_3)| = B/A$
    Compute phase $\phi_3 = \arg(G_p(j\omega_3)), \phi_3 < 0$
    Convergence test If $|(-\pi) - \phi_3(\omega_3)| < eps$ goto Step 3
    If $(-\pi) < \phi_3(\omega_3) < 0$ then $l_1 := l_3$ and $\beta_1 := \pi + \phi_3$
    If $\phi_3(\omega_3) < -\pi$ then $l_2 := l_3$ and $\beta_2 := -\pi - \phi_3$
    Goto Step 2

Step 3   Output results
    Critical frequency $\omega_{-\pi} = \omega_3$
    System magnitude $|G_p(j\omega_{-\pi})| = l_3$
Algorithm end

The method generates a sequence of frequencies at which a sine wave experiment has to be performed. The frequencies used to generate the test data are selected so that the process phase crossover point is always bounded. This is the outcome of the two *if* statements in Step 2 of the algorithm. The method is considered to have converged when the condition $|(-\pi) - \phi_3(\omega_3)| < eps$ is satisfied, thus numerical accuracy of the final result (assuming convergence) is dependent on the size of the convergence tolerance, eps.

Bisection Method Case Study Results

The bisection method was used for determining the critical frequency of the all pole process

$$G_p(s) = \frac{10}{(1 + sT_1)(1 + sT_2)(1 + sT_3)}$$

Five sets of pole time constants were chosen for the test. Table 6.1 lists the pole time constants and the associated theoretical values of critical frequency and magnitude.

Case 1: A Slowly Converging Case

Case 1 has the pole time constants: $T_1 = 5$ s, $T_2 = 2$ s, $T_3 = 10$ s. The frequencies used to initiate the routine were $\omega_1 = 0.2$ rad s^{-1} and $\omega_2 = 0.6$ rad s^{-1}. The routine converged to $\omega_{-\pi} = 0.4123$ rad s^{-1} after 44 iterations. However, a less severe convergence tolerance reduces this number of iterations significantly. For

Table 6.1 Test process pole time constants, critical frequencies and magnitudes.

Case no.	Pole time constants (units of seconds)			Critical frequency	Critical magnitude
	T_1	T_2	T_3	$\omega_{-\pi}$ rad s^{-1}	$|G(j\omega_{-\pi})|$
1	5	2	10	0.4123	0.7937
2	1	1	3	1.2910	0.9375
3	3	1	6	0.7454	0.7142
4	0.2	1	0.7	3.6840	0.7625
5	0.5	1.3	0.1	5.4065	0.4299

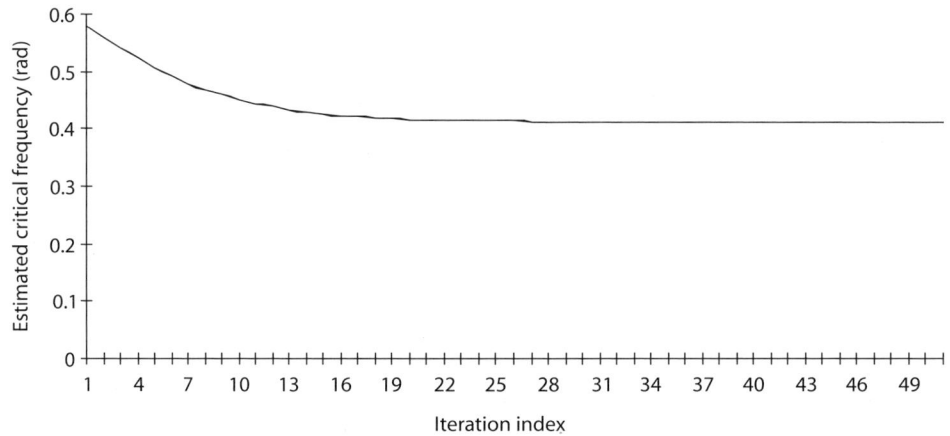

Figure 6.8 Estimated critical frequency for Case 1: $T_1 = 5$, $T_2 = 2$, $T_3 = 10$.

example, the convergence condition $|(-\pi) - \phi(\omega_k)| < 0.1$ required only 21 iterations for convergence. Figure 6.8 shows the progress of the algorithm.

Case 2: A Rapidly Converging Case

Case 2 has the process pole time constants $T_1 = 1$ s, $T_2 = 1$ s, $T_3 = 3$ s. The bisection method converged to $\omega_{-\pi} = 1.2910$ rad s^{-1} after nine iterations. If the convergence tolerance was reduced to $|(-\pi) - \phi(\omega_k)| < 0.1$, then convergence occurred in four iterations. The two starting frequencies for the routine were $\omega_1 = 1.07$ rad s^{-1} and $\omega_2 = 1.47$ rad s^{-1}. Figure 6.9 shows how the estimate for the phase crossover frequency progressed for the process pole time constants of Case 2.

6.2.2 Prediction Method

The prediction method seeks to approach the critical frequency from one direction only rather than creating a closing bound on the critical frequency, as in the bisection method. Thus this method relies on choosing two frequencies which will give points on the system frequency response such that

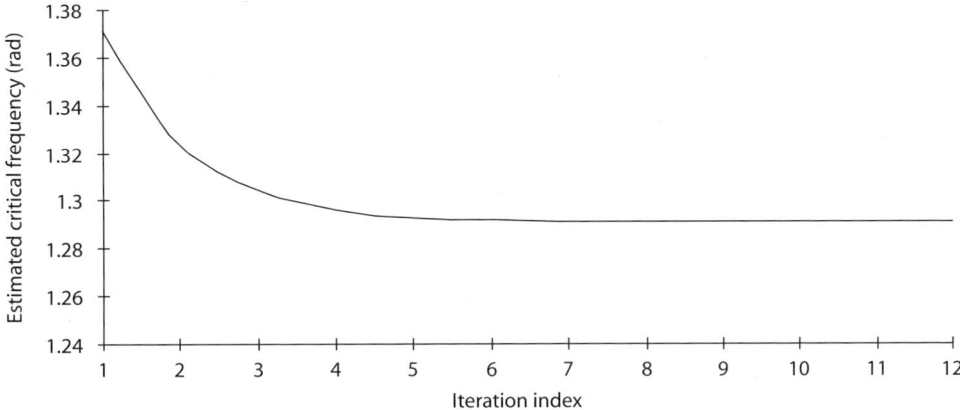

Figure 6.9 Estimated critical frequency for Case 2: $T_1 = 1, T_2 = 1, T_3 = 3$.

$-\pi < \arg G(j\omega_2) < \arg(G(j\omega_1))$. A prediction formula is then used to determine the next test frequency such that $-\pi < \arg(G(j\omega_3)) < \arg(G(j\omega_2))$. The geometrical construction of the method is shown in Figure 6.10. As with the bisection method there is an assumption that the frequency along the Nyquist curve is linearly related to length.

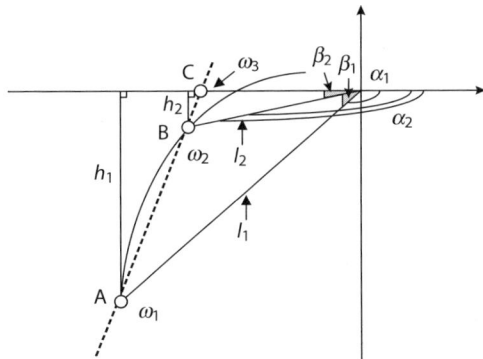

Figure 6.10 Prediction method geometrical construction.

From Figure 6.10:

$l_1 = |G(j\omega_1)|, \quad \alpha_1 = \arg(G(j\omega_1)), \quad \beta_1 = \pi + \alpha_1$
$l_2 = |G(j\omega_2)|, \quad \alpha_2 = \arg(G(j\omega_2)), \quad \beta_2 = \pi + \alpha_2$

The assumption on the linear frequency relationship gives the following ratio:

$CB:CA::(\omega_3 - \omega_2):(\omega_3 - \omega_1)$

This gives rise to the geometrical relationship:

$CB:h_2::CA:h_1$

These two ratios can be used to give the prediction relationship

$$\omega_3 = \frac{\omega_2 l_1 \sin \beta_1 - \omega_1 l_2 \sin \beta_2}{l_1 \sin \beta_1 - l_2 \sin \beta_2}$$

Algorithm 6.3: Prediction method

Step 1 Initialisation

Choose two initial frequencies such that $-\pi < \arg(G_p(j\omega_2)) < \arg G_p(j\omega_1)$
Set input signal amplitude A
For $k = 1, 2$ case
  Excite the system with input $u_k(t) = A\sin(\omega_k t)$
  Allow transients to decay; steady oscillating output $y_k(t) = B\sin(\omega_k t + \phi_k)$ emerges
  Calculate magnitude $l_k = |G_p(j\omega_k)| = B/A$
  Calculate phase $\alpha_k = \arg(G_p(j\omega_k)) = \phi_k, \beta_k = \pi + \alpha_k$
Set phase convergence tolerance eps

Step 2 Loop step

$$\text{Compute } \omega_3 = \frac{\omega_2 l_1 \sin \beta_1 - \omega_1 l_2 \sin \beta_2}{l_1 \sin \beta_1 - l_2 \sin \beta_2}$$

Excite the system with input $u_3(t) = A\sin(\omega_3 t)$
Allow transients to decay; steady oscillating output $y_3(t) = B\sin(\omega_3 t + \phi_3)$ emerges
Compute system magnitude $l_3 = |G_p(j\omega_3)| = B/A$
Compute phase $\alpha_3 = \arg(G_p(j\omega_3)) = \phi_3, \phi_3 < 0, \beta_3 = \pi + \alpha_3$
Convergence test: If $|(-\pi) - \phi_3(\omega_3)| < eps$ goto Step 3
Update $l_1 := l_2$ and $\beta_1 := \beta_2$
Update $l_2 := l_3$ and $\beta_2 := \beta_3$
Goto Step 2

Step 3 Output results

Critical frequency $\omega_{-\pi} = \omega_3$
System magnitude $|G_p(j\omega_{-\pi})| = l_3$

Algorithm end

This method generates the frequencies at which the sine wave test is to be performed and approaches the phase crossover point of the process from one direction only. The method is considered to have converged when the error in the phase angle of the system frequency response at the new frequency is less than a tolerance value of *eps*.

Prediction Method Case Study Results

To enable a comparison between the bisection and prediction methods to be made, the same processes of Table 6.1 were used in the application of the prediction method.

Case 1 Results

The frequencies chosen to initialise the method were $\omega_1 = 0.1$ rad s^{-1} and $\omega_2 = 0.2$ rad s^{-1}. The method converged to $\omega_{-\pi} = 0.4123$ rad s^{-1} after eight iterations. However, if the convergence tolerance was relaxed to $|(-\pi) - \phi(\omega_k)| < 0.1$, then convergence occurred in six iterations. Figure 6.11 shows the evolution of the estimate of the critical frequency during the progress of the algorithm.

Case 2 Results

For the process of case 2, the method converged to $\omega_{-\pi} = 1.2910$ after eight iterations. After six iterations, the method had produced a phase crossover frequency estimate satisfying $|(-\pi) - \phi(\omega_k)| < 0.1$. The starting point frequencies for the initialisation of the routine were $\omega_1 = 0.5$ rad s^{-1} and $\omega_2 = 0.6$ rad s^{-1}. Figure 6.12 shows how the phase crossover point frequency point estimate evolved for a process with pole time constants $T_1 = 1$ s, $T_2 = 1$ s, $T_3 = 3$ s.

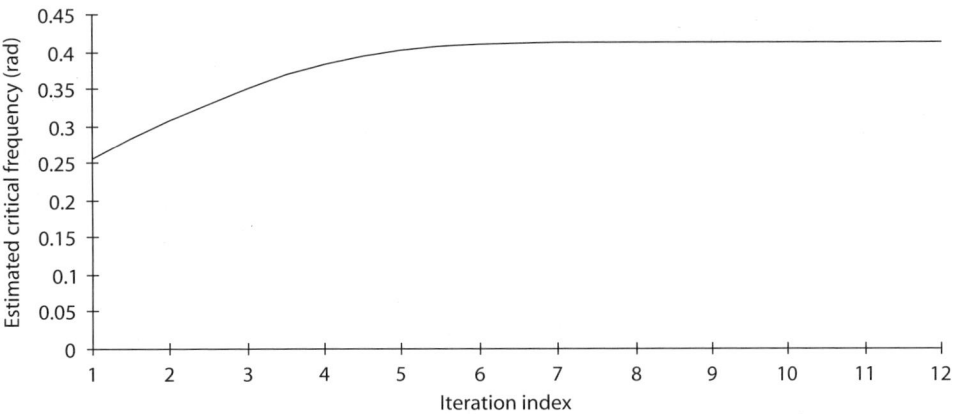

Figure 6.11 Estimated critical frequency for Case 1: $T_1 = 5, T_2 = 2, T_3 = 10$.

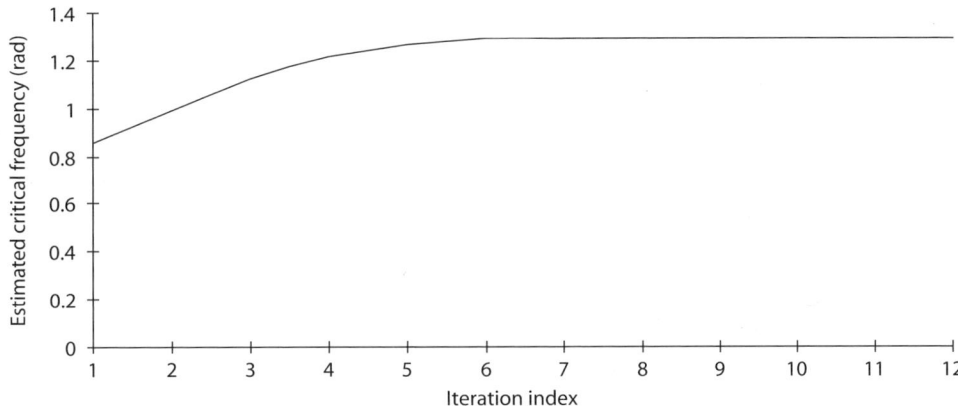

Figure 6.12 Estimated critical frequency for Case 2: $T_1 = 1, T_2 = 1, T_3 = 3$.

6.2.3 Bisection and Prediction Method – a Comparison and Assessment

Only limited trials were performed for the two methods on the same set of five processes as given in Table 6.1. The experience gained was sufficient to make a preliminary assessment of comparative performance and to identify practical implementation difficulties. The detail of the convergence rates for the two methods on the five all real pole processes can be seen in Table 6.2.

The Bisection Method

- *Initialisation*: the bisection method requires two starting frequencies which bound the target critical frequency $\omega_{-\pi}$. As would be expected in any practical situation, the system to be identified is unknown and it is difficult to select two initial bounding frequencies *a priori*. It may prove very time-consuming to determine these two frequencies to start the routine.

- *Convergence characteristics*: once the bisection method has been initiated the routine maintains a bound on the critical frequency and does not require jacketing software or logic guards to ensure convergence. Convergence is a linear process and the speed of convergence of the method is strongly related to the choice of initial bounding frequencies.

Table 6.2 Comparison of bisection and prediction method convergence rates.

Case no.	Pole time constants			Bisection method		Prediction method	
	T_1	T_2	T_3	$|\text{error}| \leq 10^{-3}$	$|\text{error}| \leq 10^{-1}$	$|\text{error}| \leq 10^{-3}$	$|\text{error}| \leq 10^{-1}$
1	5	2	10	44	21	8	6
2	1	1	3	9	4	8	6
3	3	1	6	19	8	8	6
4	0.2	1	0.7	14	6	9	7
5	0.5	1.3	0.5	10	4	10	8

- *System type*: time delay systems have a spiral frequency response. For such systems it is possible for the bisection method to converge to the wrong frequency, namely one which gives a system frequency response phase of $-\pi(1+2n), n = 1, 2, \ldots$.

The Prediction Method

- *Initialisation*: for the prediction method, the choice of starting frequencies is easier than that of the bisection method, since the prediction method will converge given any two starting frequencies as long as they do not give rise to a system phase angle that is in the fourth quadrant. To minimise the number of iterations, and hence the number of sine wave experiments necessary, the selection of the initial starting frequencies needs some care. This remains a difficult problem if the system is unknown.

- *Speed of convergence*: generally, the convergence rate for the prediction method was faster than that of the bisection method and hence fewer iterations and sine wave experiments were required. This can be seen from the data of Table 6.2.

- *System type*: if the system contains a time delay then it is possible to converge to the wrong point on the process frequency response – one in the fourth quadrant for example. The prediction algorithm was found to require more jacketing software and logic safeguards than the bisection routine to ensure convergence.

The methods have some flexibility in the problems that can be solved. For example, both methods have the potential to target a different phase point on the frequency response. It is very easy to adjust the geometry of both methods to introduce a phase ray with a pre-specified phase angle. An interesting problem would be to investigate whether it is also possible to develop a similar geometric framework to target the frequency at a pre-specified system *gain* value. Thus theoretically the methods have some potential for further development.

Both the bisection and prediction methods have linear convergence. They have similar problems in implementation, although once initialised the bisection routine seems to be robust whilst the prediction routine seems to have the advantage of faster convergence. It might be possible to combine the best features of both algorithms to develop a hybrid bisection–prediction algorithm. Convergence speed might be improved if sequential process model fitting to the data generated occurred as the algorithm progressed; alternatively, super-linear convergence might be achieved using Newton techniques. However, these pragmatic suggestions are not supported by an elegant unified principle as exists in the

6.3 Phase-Locked Loop Identifier Module – Basic Theory

relay experiment; consequently other routes were sought. The phase-locked loop concept emerged as a possible alternative approach and the remainder of the chapter is devoted to this methodology.

Phase-locked loops have been used in a number of applications ranging from speed control loops to the detection of FM signals for radio receivers (Best, 1987). The analysis of phase-locked loop systems used in the above applications has been covered in detail by a number of authors (Rugh, 1981; Best, 1987). In the application of a phase-locked loop to system identification, the loop is used to extract the phase information from the process under test and provide the appropriate excitation for the system. The phase-locked loop system which was first devised was based on a number of ideas implemented as an analogue circuit, as shown in Figure 6.13.

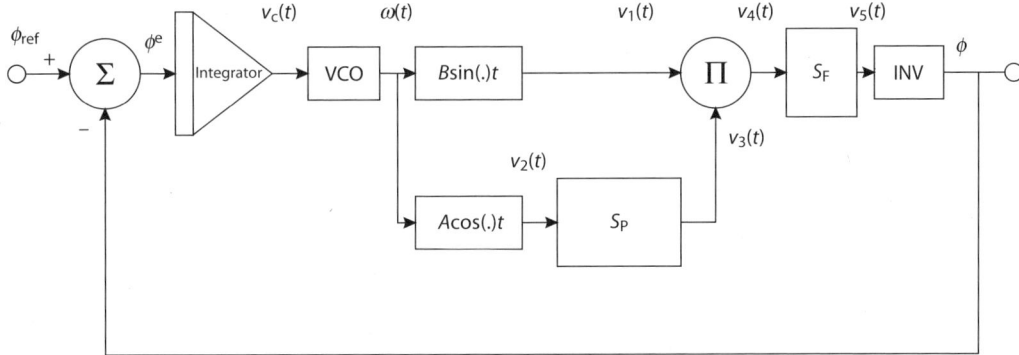

Figure 6.13 Phase-locked loop identifier module – analogue prototype.

The identifier module of Figure 6.13 was conceived as a feedback circuit around a voltage-controlled oscillator (VCO). The reference input to the loop was the desired process phase angle, so to drive the loop towards the critical frequency, the phase reference is $\phi^{\text{ref}} = -\pi$. Thus at the comparator, the phase error signal input to the integrator is given by $\phi^e(t) = \phi^{\text{ref}}(t) - \phi(t)$. The integrator in the loop is designed to ensure that the constant phase reference signal is attained in steady state conditions. The integrator equation is straightforwardly

$$v_c(t) = \int_0^t \phi^e(\tau) d\tau$$

A voltage-controlled oscillator produces the sinusoidal excitation signals of appropriate frequency. This frequency is proportional to the integral of the phase reference error $\omega(t) = K_{\text{VCO}} v_c(t)$, where the gain of the voltage controlled oscillator is denoted K_{VCO}. The voltage-controlled oscillator generates two sinusoidal signals, $v_1(t) = B\sin \omega t$ and $v_2(t) = A\cos \omega t$. The signal $v_2(t)$ is used to excite the process and, after transients have decayed, yields the steady state output signal

$$v_3^{SS}(t) = S_p(v_2(t)) = A|G_p(j\omega)|\cos(\omega t + \phi(\omega))$$

Across the multiplier, the steady state output is

$$v_4^{SS}(t) = v_1(t) v_3^{SS}(t) = \frac{AB}{2}|G_p(j\omega)|[\sin(2\omega t + \phi(\omega)) - \sin \phi(\omega)]$$

This signal is used in the nonlinear block to extract and identify the phase achieved by the current frequency value. This current value of the phase is then fed to the comparator to complete the loop. The amplitude A of the process excitation signal is user-selected to minimise disruption to the process outputs. The equation for $v_4^{SS}(t)$ shows that a convenient choice for amplitude B is to use $B = 2/A$. In this way, the dependence of the signal $v_4^{SS}(t)$ on signal amplitudes A and B through the multiplier $AB/2$ is removed. The concepts in this analogue prototype were used to construct a phase-locked loop identifier module that retained the feature of continuous-time process excitation and used digital processing in the data extraction components of the outer loop.

6.3.1 The Digital Identifier Structure

The analogue concepts in Figure 6.13 led to an identifier that is based on a digital phase-locked loop principle as shown in Figure 6.14. The main components of the identification module have the following functions:

1. A feedback structure which uses a phase or gain reference at an input comparator.

2. A digital model of a voltage-controlled oscillator. This provides a frequency value and generates the system sinusoidal excitation signal and a sinusoidal reference signal.

3. A digital signal processing unit that extracts the actual measured system phase or gain value. This value passes to the comparator to close the loop.

4. A digital integrator unit to ensure the identifier unit converges to the given constant system phase or gain reference.

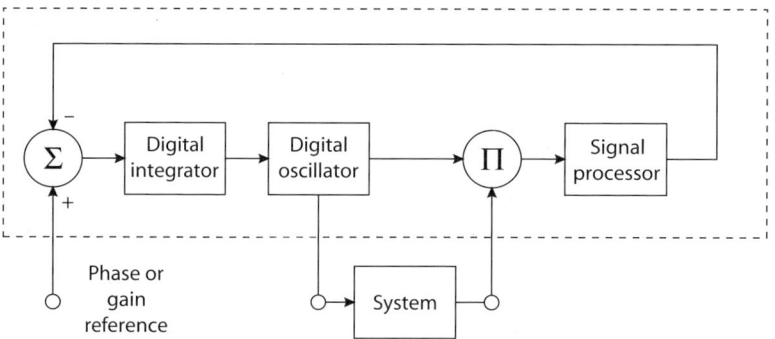

Figure 6.14 Digital identifier module.

The operation of the process analyser in digital form, as shown in Figure 6.14, comprises two basic processes. One process is the excitation of the system being identified and the operation of the signal processing unit to extract phase and/or gain values from the signal at the output of the multiplier. This is an operation on continuous time signals.

The second process is the operation of the overall digital loop and the convergence of the loop. This can be considered either as a closed-loop stability problem or as a convergence problem for a numerical routine. These two processes are discussed in the next two sections.

Phase and Gain Data Extraction in a Digital Identification Module

The digital identifier module, as shown in Figure 6.14, implements the main components of the analogue precursor using digital equivalents. The voltage-controlled oscillator block is modelled in digital form. The loop filter is replaced by a digital processor, so that the duty of extracting information from the continuous time signal $v_4^{SS}(t)$ (Figure 6.13) is performed by the signal processor block of Figure 6.14. This block implements a digitally based method for phase/gain data extraction for which the main details are specified in Theorem 6.1.

Theorem 6.1: Phase and gain data extraction
If the signal at the multiplier output (Figure 6.14), which takes the form

$$v(t) = |G_p(j\omega)|[\sin(2\omega t + \phi(\omega)) - \sin(\phi(\omega))]$$

is sampled at its maximum and minimum values, then

$$|G_p(j\omega)| = \frac{v(t_{max}) - v(t_{min})}{2}$$

and

$$\phi(\omega) = \sin^{-1}\left[-\frac{1}{2|G_p(j\omega)|}(v(t_{max}) + v(t_{min}))\right]$$

where t_{max} and t_{min} are the times at which a maximum and a minimum occurs in the signal.

Proof

$$v(t) = |G_p(j\omega)|[\sin(2\omega t + \phi(\omega)) - \sin(\phi(\omega))]$$

$$\frac{dv(t)}{dt} = 2\omega|G_p(j\omega)|\cos(2\omega t + \phi(\omega))$$

The derivative function yields turning points at

$$2\omega t + \phi(\omega) = \frac{\pi}{2}(1 + 4n) \text{ and } 2\omega t + \phi(\omega) = \frac{\pi}{2}(3 + 4n), n = 0, 1, 2, \ldots$$

The second derivative function is

$$\frac{d^2v(t)}{dt^2} = -4\omega^2|G_p(j\omega)|\sin(2\omega t + \phi(\omega))$$

Then maxima occur at

$$2\omega t + \phi(\omega) = \frac{\pi}{2}(1 + 4n)$$

and minima at

$$2\omega t + \phi(\omega) = \frac{\pi}{2}(3 + 4n)$$

since

$$\text{sgn}\left(-4\omega^2|G_p(j\omega)|\sin\left(\frac{\pi}{2}(1 + 4n)\right)\right) = -\text{ve}$$

and

$$\text{sgn}\left(-4\omega^2|G_p(j\omega)|\sin\left(\frac{\pi}{2}(3 + 4n)\right)\right) = +\text{ve}$$

Hence maxima occur in $v(t)$ when

$$t_{max} = \frac{(\pi/2)(1+4n) - \phi(\omega)}{2\omega}$$

and minima will occur in $v(t)$ when

$$t_{min} = \frac{(\pi/2)(3+4n) - \phi(\omega)}{2\omega}$$

Substitution in the above equations yields

$$v(t_{max}) = |G_p(j\omega)|[1 - \sin(\phi(\omega))] \text{ and } v(t_{min}) = -|G_p(j\omega)|[1 + \sin(\phi(\omega))]$$

These two equations can be rearranged to give the results of the theorem.

❑

To implement the theorem results as a digital processing algorithm, a convention is defined regarding the times at which maxima and minima occur. Let the times at which the ith maximum and minimum occur be defined as $t_{max}(i)$ and $t_{min}(i)$ respectively; then the operation of the digital processing block is given by Algorithm 6.4.

Algorithm 6.4: Phase and gain data extraction

Step 1 *Initialisation*

Specify closeness tolerances $Tol1$, $Tol2$, $Tol3$

Step 2 Detect the first maximum; store the values $t_{max}(i)$ and $v(t_{max}(i))$

Step 3 Detect the first minimum; store the values $t_{min}(i)$ and $v(t_{min}(i))$

Step 4 *Calculation step*

$$\text{Compute } |G_p(j\omega(i))| = \frac{v(t_{max}(i)) - v(t_{min}(i))}{2}$$

$$\text{Compute } \phi(\omega(i)) = \sin^{-1}\left[-\frac{1}{2|G_p(j\omega(i))|}(v(t_{max}(i)) + v(t_{min}(i)))\right]$$

Determine $\omega(i)$

Step 5 *Test convergence*

If $||G_p(j\omega(i))| - |G_p(j\omega(i-1))|| \leq Tol1$
And $|\phi(\omega(i)) - \phi(\omega(i-1))| \leq Tol2$
And $|\omega(i) - \omega(i-1)| \leq Tol3$
Then converged and STOP
Goto Step 2

Algorithm end

This algorithm accomplishes the extraction of phase and gain data efficiently within each iteration cycle. Next, the convergence of the sequence of loop cycles is investigated.

Identifier Closed-Loop Convergence Theory

Within the phase-locked loop identifier module (Figure 6.14) there is a sine wave test. It is known that if a sinusoidal signal is applied to a linear and stable system then after the transient has decayed the output response will be sinusoidal, changed in amplitude and shifted in phase. Thus the analysis of closed-loop convergence of the phase-locked loop is a quasi-steady state analysis which assumes that the inner sine wave test has passed the transient stage and is yielding increasingly steady results. The digital framework for the analysis of the convergence of the phase-locked loop module, as shown in Figure 6.15, incorporates this assumption. For the kth sine wave experiment let the signal processing unit be producing a data stream ϕ_{kt} that is the tth estimate of system phase ϕ_k. If the experiment was allowed to go to completion then $\phi_k = \lim_{t \to \infty} \phi_{kt}$, but since this is not practical it is usual to truncate the estimation process at some $\tilde{t} < \infty$ with an associated estimate $\phi_{k\tilde{t}}$. For this reason, Figure 6.15 does not include the system test loops and shows the system Φ_{DIG} with output $\phi_{k\tilde{t}}$. The other components in the digital block diagram of Figure 6.15 are straightforward, comprising a comparator, a digital integrator unit, a gain K_{VCO} to represent the action of the voltage-controlled oscillator, and finally the signal processing block represented by Φ_{DIG}.

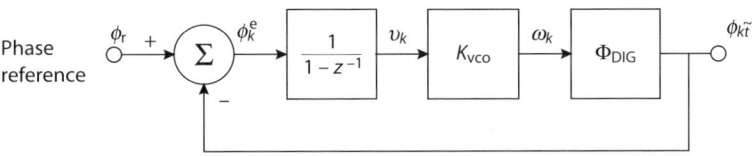

Figure 6.15 Digital phase-locked loop identifier module – steady state representation.

Some key results which provide the basis for the convergence theory now follow. From Figure 6.15 obtain

Error equation $\qquad \phi_k^e = \phi_r - \phi_{k\tilde{t}}$

Integrator equation $\qquad v_{k+1} = v_k + \phi_k^e$

Oscillator equation $\qquad \omega_k = K_{\text{VCO}} v_k$

Signal processing unit output $\qquad \phi_{k\tilde{t}} = \Phi_{\text{DIG}}(\omega_k)$

At the $(k+1)$th iteration it follows that

$$\omega_{k+1} = K_{\text{VCO}} v_{k+1}$$

and

$$\omega_{k+1} = K_{\text{VCO}}(v_k + \phi_k^e) = K_{\text{VCO}} v_k + K_{\text{VCO}} \phi_k^e$$

giving

$$\omega_{k+1} = \omega_k + K_{\text{VCO}}(\phi_r - \phi_{k\tilde{t}})$$

This recursive relation for ω_k is an iterative scheme of the classical variety (Schwartz, 1989), namely $\omega_{k+1} = f(\omega_k)$, for which there is a fixed-point lemma.

Lemma 6.1: Fixed-point lemma

If ω_* is a fixed point of the recursive scheme

$$\omega_{k+1} = \omega_k + K_{\text{VCO}}(\phi_r - \phi_{k\tilde{t}})$$

then

$$\phi(\omega_*) = \phi_r$$

Proof

The recursive scheme is $\omega_{k+1} = \omega_k + K_{\text{VCO}}(\phi_r - \phi_{k\tilde{t}})$.

Indicate the dependence of phase estimates on frequency ω_k and let the sine wave experiment go to completion; then $t \to \infty$ and

$$\phi_k(\omega_k) = \lim_{t \to \infty} \phi_{kt}(\omega_k)$$

Hence $\omega_{k+1} = \omega_k + K_{VCO}(\phi_r - \phi_k(\omega_k))$. Let $\omega_{k+1} = \omega_k = \omega_*$; then with $\phi(\omega_*) = \phi_k(\omega_*)$, $\omega_* = \omega_* + K_{VCO}(\phi_r - \phi(\omega_*))$ and the result follows.

□

The fixed-point result shows that the recurrent relation is satisfied by a frequency solution ω_* having the desired property that $\phi(\omega_*) = \phi_r$. It is also possible to state an analogous result for a system gain reference. This leads the way to a similar phase-locked loop module which has gain references rather than phase references. Convergence properties are given by Theorem 6.2.

Theorem 6.2: Sufficient conditions for convergence

The algorithmic relationship of the phase-locked loop module (Figure 6.15) given by $\omega_{k+1} = \omega_k + K_{VCO}(\phi_r - \phi_{k\tilde{t}})$, where K_{VCO} is the voltage-controlled oscillator gain and ϕ_r is the phase reference value, extends to the iterative relation $(\omega_* - \omega_{k+1}) = (\omega_* - \omega_k) + K_{VCO}\eta_{k\tilde{t}}(\phi_r - \phi_k)$, where ω_* is a fixed point of the recurrent relation and

$$\eta_{k\tilde{t}} = \left[1 - \left(\frac{\phi_{k\tilde{t}} - \phi_k}{\phi_r - \phi_k}\right)\right]$$

The above iterative relation can be given as

$$|(\omega_* - \omega_{k+1})| \leq \left[\prod_{j=0}^{k} L_j\right] |(\omega_* - \omega_0)|$$

where

$$L_j = |1 - K_{VCO}\eta_{j\tilde{t}}\phi'(\omega_{*j})|, \omega_j \leq \omega_{*j} \leq \omega_* \text{ and } \phi'(\omega_{*j}) = \frac{d\phi(\omega_{*j})}{d\omega}$$

Sufficient conditions for convergence are:

(i) If there exists $\tilde{k}$ such that for all $k > \tilde{k}, |L_k| < 1$ then $|\omega_* - \omega_{k+1}| \to 0$ as $k \to \infty$.

(ii) If $L_\infty = \max_j \{|L_j|\}_{j=0}^\infty$ and $L_\infty < 1$, then $|\omega_* - \omega_{k+1}| \to 0$ as $k \to \infty$.

Proof

From $\omega_{k+1} = \omega_k + K_{VCO}(\phi_r - \phi_{k\tilde{t}})$ we obtain

$$\omega_* - \omega_{k+1} = \omega_* - \omega_k - K_{VCO}(\phi_r - \phi_{k\tilde{t}} + \phi_k - \phi_k)$$
$$= \omega_* - \omega_k - K_{VCO}(\phi_r - \phi_k - (\phi_{k\tilde{t}} - \phi_k))$$

and

$$\omega_* - \omega_{k+1} = \omega_* - \omega_k - K_{VCO}\left(1 - \frac{(\phi_{k\tilde{t}} - \phi_k)}{(\phi_r - \phi_k)}\right)(\phi_r - \phi_k)$$

Hence

$$(\omega_* - \omega_{k+1}) = (\omega_* - \omega_k) - K_{VCO}\eta_{k\tilde{t}}(\phi_r - \phi_k), \text{ where } \eta_{k\tilde{t}} = \left[1 - \frac{\phi_{k\tilde{t}} - \phi_k}{\phi_r - \phi_k}\right]$$

If ω_* is a fixed point of the scheme, then from Lemma 6.1 $\phi(\omega_*) = \phi_r$ and

$$(\omega_* - \omega_{k+1}) = (\omega_* - \omega_k) - K_{VCO}\eta_{k\tilde{t}}(\phi(\omega_*) - \phi_k)$$

From the Mean Value Theorem there exists ω_{*k} such that

$$\phi'(\omega_{*k}) = \frac{\phi(\omega_*) - \phi(\omega_k)}{(\omega_* - \omega_k)}, \text{ where } \omega_k \leq \omega_{*k} \leq \omega_*.$$

Hence

$$(\omega_* - \omega_{k+1}) = (\omega_* - \omega_k) - K_{VCO}\eta_{k\tilde{t}}\phi'(\omega_{*k})(\omega_* - \omega_k)$$
$$= (1 - K_{VCO}\eta_{k\tilde{t}}\phi'(\omega_{*k}))(\omega_* - \omega_k)$$

Repeated application of this recurrent relation obtains the key result of the theorem:

$$(\omega_* - \omega_{k+1}) = \prod_{j=0}^{k}(1 - K_{VCO}\eta_{j\tilde{t}}\phi'(\omega_{*j}))(\omega_* - \omega_0)$$

Define $L_j = |1 - K_{VCO}\eta_{j\tilde{t}}\phi'(\omega_{*j})|$; then the sufficient conditions for convergence emerge from the relation

$$|\omega_* - \omega_{k+1}| \leq \left[\prod_{j=0}^{k} L_j\right]|\omega_* - \omega_0|$$

□

Remarks 6.1
1. In the case where $\tilde{t} \to \infty$ then $\phi_{k\tilde{t}} = \phi_k$ and $\eta_{k\tilde{t}} = 1$.
2. The key to convergence lies in the appropriate selection of $\tilde{t}$ (the cut-off point for the accuracy within Φ_{DIG}) and the selection of K_{VCO}.
3. The convergence analysis of the system gives the bounds over which the digital controlled oscillator gain may be varied and convergence maintained. To improve the convergence rate of the identifier module an adaptive K_{VCO} gain can be considered.

Adaptive Gain for Digital Controlled Oscillator
At the $(k+1)$th step of the routine

$$(\omega_* - \omega_{k+1}) = (1 - K_{VCO}\eta_{k\tilde{t}}\phi'(\omega_{*k}))(\omega_* - \omega_k)$$

and, for global convergence, the condition is

$$(\omega_* - \omega_{k+1}) = \prod_{j=0}^{k}(1 - K_{VCO}\eta_{j\tilde{t}}\phi'(\omega_{*j}))(\omega_* - \omega_0)$$

The sufficient conditions for convergence are

$$L_j = |1 - K_{VCO}\eta_{j\tilde{t}}\phi'(\omega_{*j})| < 1, \text{ for all } j = 0, 1, \ldots$$

These equations have two implications:

Bounds for K_{VCO}
The sufficient conditions for convergence give

$$0 < K_{VCO}\eta_{j\tilde{t}}\phi'(\omega_{*j}) < 2$$

Optimum values for K_{VCO}
The $(k+1)$th step of the routine would give a solution in one step if

$$(\omega_* - \omega_{k+1}) = (1 - K_{VCO}\eta_{k\tilde{t}}\phi'(\omega_{*k}))(\omega_* - \omega_k) = 0$$

This requires the condition

$$K_{VCO} = \frac{1}{\eta_{k\tilde{t}}\phi'(\omega_{*k})}$$

If the gain K_{VCO} were updated at every iteration then convergence could be increased. Since it is not possible to calculate exact values for $\eta_{j\tilde{t}}$ and $\phi'(\omega_{*j})$ estimates must be used. Such estimates are given by

$$\hat{\eta}_{k\tilde{t}} = 1 - \frac{\Delta\phi}{\phi_r - (\phi_{j\tilde{t}} - \Delta\phi)}$$

and

$$\hat{\phi}'(\omega_{*j}) = \frac{\phi_j - \phi_{j-1}}{\omega_j - \omega_{j-1}}$$

where $\Delta\phi$ is the tolerance in the phase values given in Algorithm 6.4, viz $\Delta\phi = Tol2$.

Case Studies

It was important to use industry standard software for the research programme being pursued for the phase-locked loop identifier. A key factor here was the likelihood that the step to PID controller development would be taken and that the end product would be demonstrated to industrial engineers. Consequently it was decided to implement the phase-locked loop identifier module using the LabVIEW (Student edition version 3.1) software. The relay experiment was implemented in MATLAB (Student edition V.4). Both techniques were found to be relatively easy to implement and the LabVIEW software enabled a very professional test-bed module to be constructed, as shown in Figure 6.16.

A comparison for the performance of the relay experiment and the phase-locked loop identifier module was performed on the basis of the following assessment criteria:

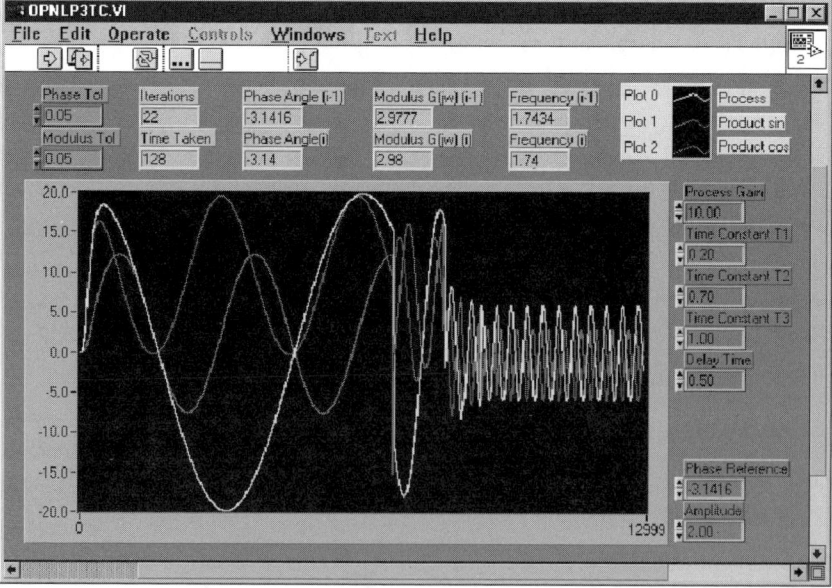

Figure 6.16 Digital phase-locked loop identifier module – LabVIEW implementation.

1. The accuracy of the estimated phase crossover point
2. The time taken to determine the estimate

The describing function analysis of the relay experiment relies on the assumption that the system to be identified has a low-pass characteristic at the phase crossover point frequency. This assumption is violated by systems which have large time delays and by systems which are underdamped. The test schedule selected a set of processes which included time delay and underdamped systems, both of which are commonly encountered in the process industries. In all the tests the phase crossover frequency was estimated and subsequently the system magnitude was evaluated according to the rules of the method and the describing function for the relay experiment, and directly from the waveforms in the phase-locked loop.

Systems With Time Delays

The objective of the following three cases is to demonstrate how the identification accuracy of the relay method of system identification suffers when the low-pass assumption is not met at the phase crossover frequency. First the system characteristics are given.

System 1: (Shen et al., 1996a) $G_1(s) = \dfrac{12.8e^{-s}}{1+16.8s}$

This process model is a first-order, type zero system with a relatively small time delay. The theoretical values of magnitude and frequency of the phase crossover point for the process given are $|G_1(j\omega_{-\pi})| = 0.4736$, $\omega_{-\pi} = 1.6078$ rad s^{-1}.

System 2: (Shen et al., 1996a) $G_2(s) = \dfrac{37.7e^{-10s}}{(1+2s)(1+7200s)}$

This process model is a second-order, type zero system with a large time delay. The theoretical values of the magnitude and frequency of the phase crossover point for the system are $|G_2(j\omega_{-\pi})| = 0.0385$, $\omega_{-\pi} = 0.1315$ rad s^{-1}.

System 3: $G_3(s) = \dfrac{(10s-1)e^{-s}}{(1+2s)(1+4s)}$

This process model has a non-minimum phase characteristic and a relative order of one. The theoretical values of the magnitude and frequency of the process phase crossover point for the system are $|G_3(j\omega_{-\pi})| = 1.7369$, $\omega_{-\pi} = 0.3363$ rad s^{-1}. The progress of a typical relay experiment and a phase-locked loop experiment is shown for System 3.

System 3 Relay Experiment

Figure 6.17 shows the output response of System 3 during the relay experiment for a relay height of unity. As can be seen, the output response is not sinusoidal in nature. The estimated values for the phase crossover point from the relay experiment are $|G_3(j\hat{\omega}_{-\pi})| = 0.671$, $\hat{\omega}_{-\pi} = 1.696$ rad s^{-1}. In this case, the output response contains a large contribution from the odd harmonics of the fundamental frequency, consequently, erroneous results are expected from the describing function-based analysis. The errors in the estimated data for the phase crossover point of $G_3(j\omega)$ are −61% and 400% for the magnitude and frequency respectively. The time to determine the estimates was approximately 20 s.

System 3 Phase-Locked Loop Identifier Module Experiment

The phase-locked loop method of estimating the phase crossover point for System 3 is shown in Figures 6.18 and 6.19. The phase crossover point estimates from the phase-locked loop identifier are $|G_3(j\hat{\omega}_{-\pi})| = 1.7369$, $\hat{\omega}_{-\pi} = 0.3363$ rad s^{-1}. The estimated data are exact for all practical purposes. The time taken to achieve the estimates was 117 s, which is slower than the relay experiment, but the accuracy of the phase-locked loop method is seen to be superior to that of the relay experiment.

238 Phase-Locked Loop Methods

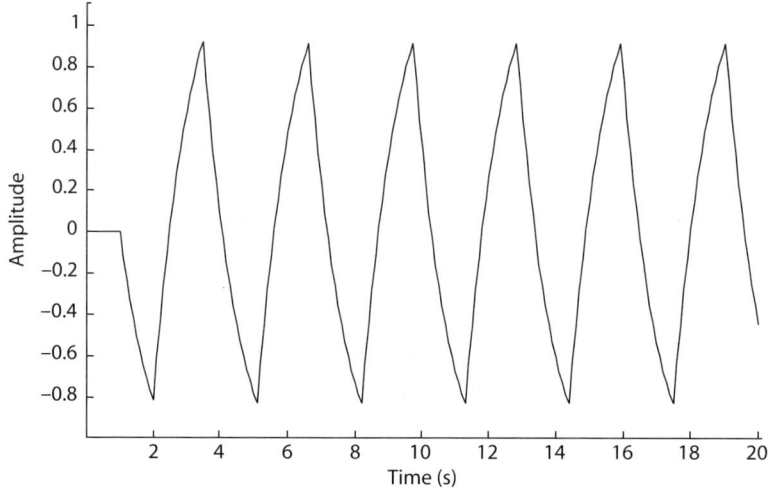

Figure 6.17 Relay experiment output for System 3.

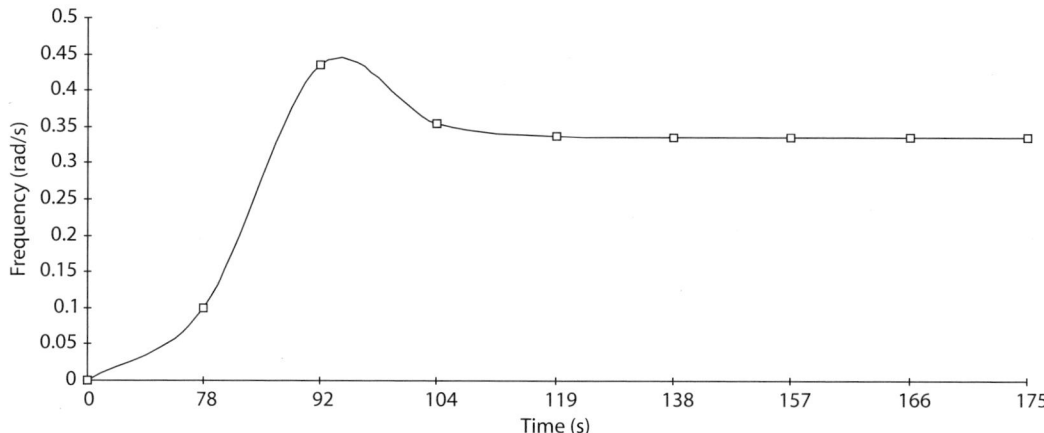

Figure 6.18 Phase-locked loop progress toward the phase crossover frequency for System 3: $\omega_{-\pi} = 0.3363$ rad s^{-1}.

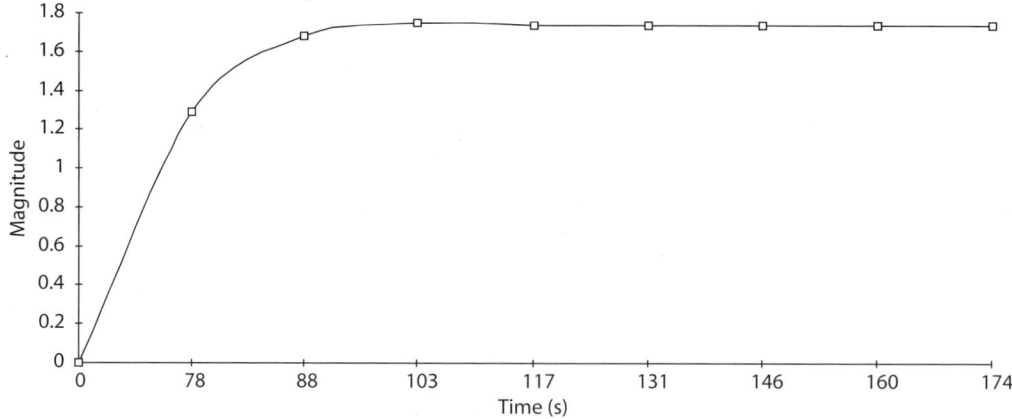

Figure 6.19 Phase-locked loop progress toward the phase crossover magnitude for System 3: $|G_3(j\omega_{-\pi})| = 1.7369$.

Table 6.3 Systems 0 to 3: experimental comparison.

Units = rad s−1; Time = seconds

System			Theoretical	Relay experiment	Phase-locked loop
System 0		Magnitude	0.5308	0.5315	0.5308
$G_0(s) = \dfrac{1}{(s+1)^6}$		Frequency	0.4142	0.4147	0.4142
		Time to estimate	N/A	45	173
System 1		Magnitude	0.4736	0.611	0.4737
$G_1(s) = \dfrac{12.8 e^{-s}}{(16.8s+1)}$		Frequency	1.6078	1.549	1.6074
		Time to estimate	N/A	14	166
System 2		Magnitude	0.0385	0.0462	0.0384
$G_2(s) = \dfrac{37.7 e^{-10s}}{(2s+1)(7200s+1)}$		Frequency	0.1315	0.127	0.1317
		Time to estimate	N/A	150	633
System 3		Magnitude	1.7369	0.671	1.7369
$G_3(s) = \dfrac{(10s-1) e^{-s}}{(2s+1)(4s+1)}$		Frequency	0.3363	1.696	0.3363
		Time to estimate	N/A	20	117

A summary table of the results for the three test systems is given in Table 6.3. The processes chosen did not have a low-pass characteristic at the process phase crossover point and thus the estimates obtained from the relay experiment were not expected nor found to be accurate. As a *control* experiment, results for System 0 are included in Table 6.3, since this is an example of a system which *does* satisfy the low-pass assumptions at phase crossover frequency. Here it can be seen that the relay experiment gives satisfactory results. The phase-locked loop method of system identification was shown to give accurate estimates of the process phase crossover point for all four systems. A comparison of the time taken to obtain an estimate of the process phase crossover point shows the phase-locked loop method is approximately four to five times slower than the relay method. In these experiments, the phase-locked loop accuracy tolerances have been selected to be stringent, typically an error of less than 10^{-4} was specified. This naturally prolongs the experiment time. Consequently an important consideration is to choose phase-locked loop error tolerances which match the requirements of the problem. Other ways to reduce the experimentation time include using an adaptive gain or having a random initialisation of the VCO initial excitation frequency.

Underdamped Systems with Time Delays

The second set of tests considers systems that are second-order underdamped with time delays. The details for the two systems chosen are given next.

System 4: $G_4(s) = \dfrac{5.625e^{-2.5s}}{s^2 + 0.6s + 2.25}$

This is an underdamped second-order system with a time delay of 2.5 s and gain 2.5. The damping factor and natural frequency of the system are respectively 0.2 and 1.5 rad s^{-1}. The theoretical values of the magnitude and frequency of the phase crossover point of the process are $|G_4(j\omega_{-\pi})| = 4.3169$ and $\omega_{-\pi} = 1.0539$ rad s^{-1}.

System 5: $G_5(s) = \dfrac{1.078e^{-10s}}{s^2 + 0.14s + 0.49}$

This is also an underdamped second-order system with a time delay of 10 s and a damping factor and natural frequency of 0.1 and 0.7 rad s^{-1} respectively, and a gain of 2.2. This has quite a low value of damping factor and exhibits oscillatory behaviour. The theoretical phase crossover point data values for magnitude and frequency for the system are $|G_5(j\omega_{-\pi})| = 2.6940$ and $\omega_{-\pi} = 0.3035$ rad s^{-1}. Typical experimental details for System 5 are exhibited next.

System 5 Relay Experiment
Figure 6.20 shows the system output response for a relay excitation, where the relay height was chosen as unity. As can be seen from the figure, the system takes a long time for the transients to decay, although

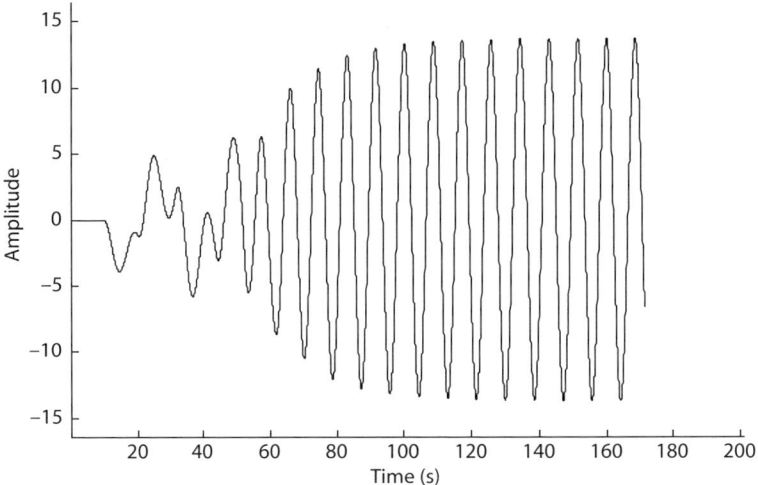

Figure 6.20 Relay experiment output for System 5.

once the transients have decayed the response is relatively sinusoidal in appearance. This should help to reduce the error in the phase crossover point data estimates produced by the additional harmonics present in the output signal. The estimated phase crossover point data is $|G_5(j\hat{\omega}_{-\pi})| = 1.0780$, $\hat{\omega}_{-\pi} = 0.750$ rad s^{-1}. Hence the errors in the estimation of the phase crossover point data are −60% for the magnitude and 147% for the frequency. The time taken to achieve the estimate was approximately 140 s.

System 5 Phase-Locked Loop Experiment
The Figures 6.21 and 6.22 show how the data for the phase-locked loop experiment for frequency and magnitude estimates evolved. The estimated values of the phase crossover point data for system 5 are $|G_5(j\hat{\omega}_{-\pi})| = 2.6940$ and $\hat{\omega}_{-\pi} = 0.3035$ rad s^{-1}. The estimate of the phase crossover point data is

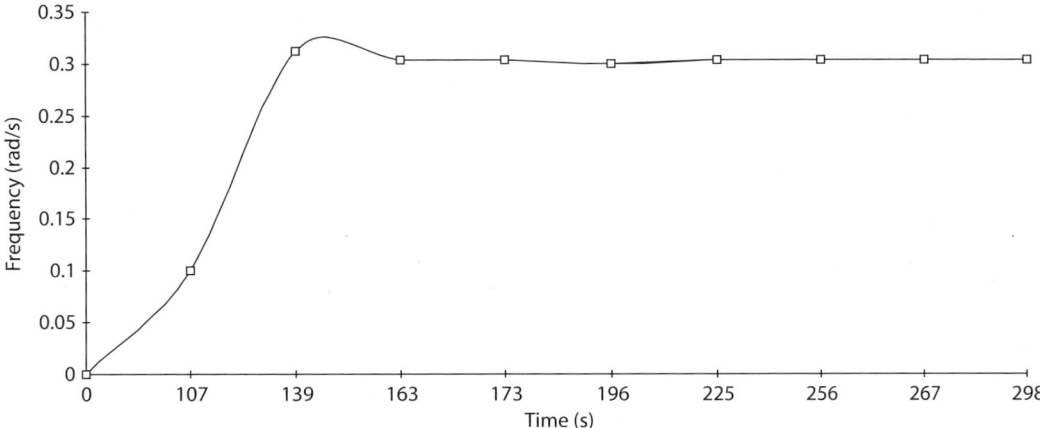

Figure 6.21 Phase-locked loop progress toward the phase crossover frequency for System 5: $\omega_{-\pi} = 0.3035$ rad s^{-1}.

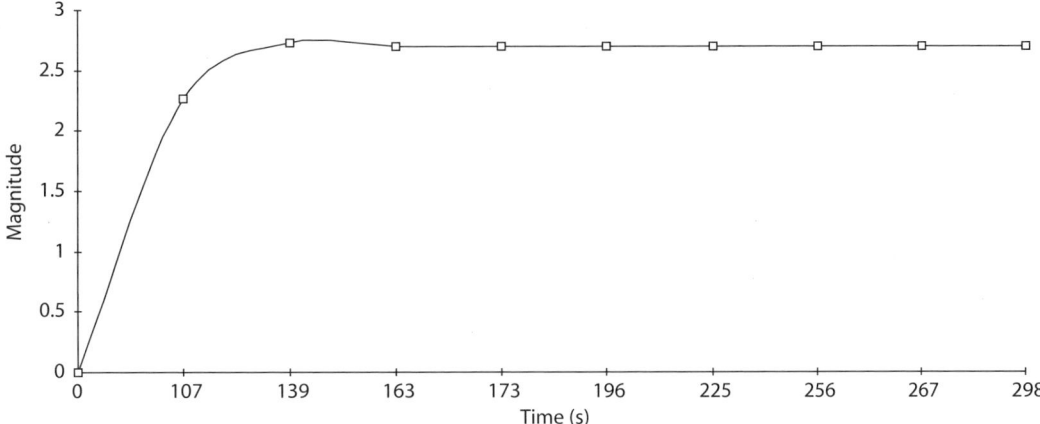

Figure 6.22 Phase-locked loop progress toward the phase crossover magnitude for System 5: $|G_5(j\omega_{-\pi})| = 2.6940$ rad s^{-1}.

practically exact, with the estimate being achieved in 256 s. Thus the phase-locked loop method is much more accurate than the relay experiment, but requires more time to carry out the estimation.

Summary results for the last two experiments are found in Table 6.4. These results are similar to those of the outcomes in Table 6.3. It can be seen that the relay experiment obtains an estimate of the phase crossover point more quickly than the phase-locked loop method. However, the accuracy of the phase-locked loop method is very much better than that of the relay experiment with practically zero error in its estimates compared to the large errors of the relay experiment.

If the estimates of the phase crossover point are to be used in a rule-based design for a PID controller then the phase-locked loop method offers results of higher accuracy from a wider range of systems over those obtained from the relay experiment. However, the experiments also show that the phase-locked loop method is approximately four to five times slower than the relay method. This extended identification time can be reduced by careful selection of the accuracy tolerances used in the phase-locked loop method, by using an adaptive gain or by having a random initialisation of the VCO initial excitation frequency. In an industrial situation, where it may be necessary to tune controllers on a number of similar loops, the results coming from the first experiments can be used to initialise the subsequent experiments.

Table 6.4 Systems 0, 4, 5: experimental comparison.

Frequency units = rad s^{-1}; Time = seconds

System		Theoretical	Relay experiment	Phase-locked loop
System 0	Magnitude	0.5308	0.5315	0.5308
$G_0(s) = \dfrac{1}{(s+1)^6}$	Frequency	0.4142	0.4147	0.4142
	Time to estimate	N/A	45	173
System 4	Magnitude	4.3169	0.733	4.3162
$G_4(s) = \dfrac{5.625e^{-2.5s}}{(s^2 + 0.6s + 2.25)}$	Frequency	1.0539	1.0414	1.0537
	Time to estimate	N/A	25	122
System 5	Magnitude	2.6940	1.0780	2.6940
$G_5(s) = \dfrac{1.078e^{-10s}}{(s^2 + 0.14s + 0.49)}$	Frequency	0.3035	0.750	0.3035
	Time to estimate	N/A	140	256

6.3.2 Noise Management Techniques

The phase-locked loop method of nonparametric identification must be capable of producing accurate estimates of specific points of the system frequency response in the presence of noise. To gain an insight to the effect of noise on the convergence of the discrete update process an analysis extending the equations of Theorem 6.2 is performed. This analysis shows that if the phase-locked loop module converges the noise in the sequence of estimates is attenuated. To ensure the minimum potential for disruption to the process of convergence a Kalman filter routine is proposed on the outcomes of the sine wave experiment. A sequence of phase estimates is taken from the transient output of the system under test and the Kalman filter smooths this sequence. This aids the determination of when the phase estimates have converged sufficiently to allow the output of the signal processing block and hence the feedback loop to be updated.

Convergence in the Presence of Process Noise

The case where the external reference signal of the phase-locked loop identifier is a desired phase ϕ_r is modelled by allowing additive noise to enter the convergence process via corrupted estimates of the current phase value $\phi_{k\tilde{t}}$. The system is as shown in Figure 6.23, where white noise is added at the exit from the signal processing unit.

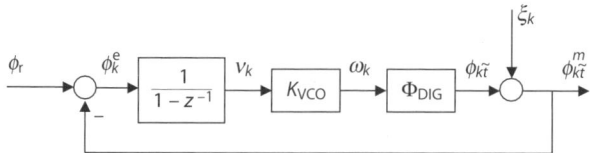

Figure 6.23 Phase-locked loop module with additive noise.

The noise process is assumed to have zero mean value and a variance of σ^2, and the noise samples are uncorrelated with each other. The noise statistics are given as

$$E\{\xi_k\} = 0, \, k = 0, 1, \ldots$$

$$E\{\xi_k \xi_j\} = \begin{cases} \sigma^2 & k = j \\ 0 & k \neq j \end{cases}$$

where $E\{\cdot\}$ denotes expectation.

The digital system equations from Figure 6.23 are

Error equation	$\phi_k^e = \phi_r - \phi_{k\tilde{t}}^m$
Integrator equation	$v_{k+1} = v_k + \phi_k^e$
Oscillator equation	$\omega_k = K_{VCO} v_k$
Signal processing unit output	$\phi_{k\tilde{t}} = \Phi_{DIG}(\omega_k)$
Noise model	$\phi_{k\tilde{t}}^m = \phi_{k\tilde{t}} + \xi_k$

The composite recurrent relationship that can be derived from these equations is

$$\omega_{k+1} = \omega_k + K_{VCO}(\phi_r - \phi_{k\tilde{t}}) + K_{VCO}\xi_k$$

Then using

$$\eta_{k\tilde{t}} = \left[1 - \left(\frac{\phi_{k\tilde{t}} - \phi_k}{\phi_r - \phi_k}\right)\right]$$

yields

$$\omega_{k+1} = \omega_k + K_{VCO}\eta_{k\tilde{t}}(\phi_r - \phi_k) + K_{VCO}\xi_k$$

Application of the fixed-point result gives

$$\omega_* - \omega_{k+1} = \omega_* - \omega_k - K_{VCO}\eta_{k\tilde{t}}(\phi_r - \phi_k) - K_{VCO}\xi_k$$

and use of the mean value theorem gives

$$e_{k+1} = L_k e_k - K_{VCO}\xi_k$$

where $e_{k+1} = \omega_* - \omega_{k+1}, e_k = \omega_* - \omega_k$ and $L_k = 1 - K_{VCO}\eta_{k\tilde{t}}\phi'(\omega_{*k})$.

The noise analysis begins with a mean result:

$$E\{e_{k+1}\} = L_k E\{e_k\} - K_{VCO} E\{\xi_k\}$$

giving $\bar{e}_{k+1} = L_k \bar{e}_k$ where $\bar{e}_{k+1} = E\{e_{k+1}\}, \bar{e}_k = E\{e_k\}$.

A variance analysis follows as

$$\begin{aligned} Var(e_{k+1}) &= E\{(e_{k+1} - \bar{e}_{k+1})^2\} \\ &= E\{(L_k e_k - K_{VCO}\xi_k - L_k \bar{e}_k)^2\} \\ &= L_k^2 E\{(e_k - \bar{e}_k)^2\} - 2L_k K_{VCO} E\{(e_k - \bar{e}_k)\xi_k\} + K_{VCO}^2 E\{\xi_k^2\} \end{aligned}$$

and

$$Var(e_{k+1}) = L_k^2 Var(e_k) + K_{VCO}^2 \sigma^2$$

This recurrent relation for the convergence error variance shows two effects:

1. If the routine is converging then $|L_k|<1$ and the convergence error variance at step k is further attenuated by a factor of L_k^2, where $L_k^2 < |L_k| < 1$.

2. This must be balanced against the second effect arising from the term $K_{VCO}^2 E\{\xi_k^2\} = K_{VCO}^2 \sigma^2$. For convergence, K_{VCO} must be chosen so that $|L_k|<1$ and further, if $|K_{VCO}|<1$ then this will attenuate the introduction of the noise term at the kth step; otherwise much depends on the size of the noise variance, for if σ^2 is very small then the convergence process will be virtually deterministic in operation. Conversely, if σ^2 is large then it may be necessary to carry out some form of processing, for example Kalman filtering on the estimates in the signal processing unit.

The above analysis and discussion shows that if the noise effect from the process sine wave experiment on the estimation procedure can be reduced, then the effect of noise on the loop convergence will be negligible.

Derivation of a Kalman Filter for the Phase-Locked Loop Identifier

If the system on which the sine wave test is being performed is subject to process and measurement noise this will have the effect of perturbing the estimates of the current phase, a value that is used by the outer estimate update process. It was shown in the previous section that if the noise from the process reaching the digital signal process is significantly reduced then the operation of the estimate updating loop is closer to being deterministic. Thus if process noise is a problem in the sine wave tests then measures must be taken to ensure that this has a minimal effect on the estimate updating loop. To achieve this, a Kalman filter is used in the signal processing unit.

Consider the case where the frequency corresponding to a given phase reference is to be found. The digital signal processing unit is then producing a series of phase estimates. If the system under test exhibits noise in the output then this will disturb the phase estimates. The output of the signal processing unit may be written as $\phi_k = \Phi_{DIG}(\omega_k)$. A key property is that when the sine wave experiment transients have decayed, then ϕ_k is essentially a constant. Thus, for the purposes of constructing a Kalman filter, the problem can be considered as the estimation of a constant from a measurement contaminated by noise. Consequently, the situation is modelled as a series of estimates of phase disturbed by additive noise being passed through a Kalman filter. To derive the Kalman filter equations the following notation is defined:

1. Let k_i be the counter for the sequence of estimates within the signal processing unit.

2. Let the measured value of the phase within the digital signal processing unit for a frequency of ω_k and when the sequence counter is k_i be denoted $\Phi_{DIG}(\omega_k, k_i)$.

3. Let the actual value of the phase within the digital signal processing unit for a frequency of ω_k and when the sequence counter is k_i be denoted $\phi_k(k_i)$.

4. Let the additive noise be denoted $\xi(k_i)$ with statistics $E\{\xi(k_i)\} = 0$ and $E\{\xi(k_i)\xi(k_j)\} = R(k_i)\delta_{ij}$.

Figure 6.24 shows this setup and notation.

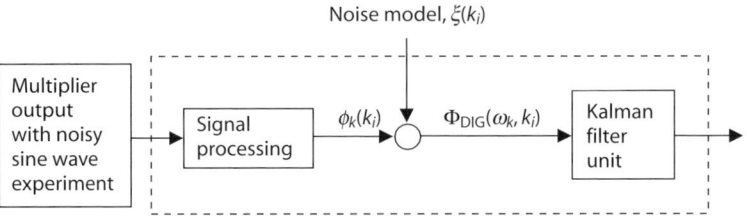

Figure 6.24 Digital signal processing unit and Kalman filter.

6.3 Phase-Locked Loop Identifier Module – Basic Theory

Table 6.5 Discrete Kalman filter equations.

Process equation	$X(k+1) = A_k X(k) + B_k U(k) + n_p(k)$
Observation equation	$Z(k) = C_k X(k) + n_m(k)$
Process noise $n_p(k)$	$E\{n_p(i)\} = 0, E\{n_p(i)n_p(j)^T\} = Q_i \delta_{ij}$
Measurement noise $n_m(k)$	$E\{n_m(i)\} = 0, E\{n_m(i)n_m(j)^T\} = R_i \delta_{ij}$
Prediction equation	$\hat{X}(k+1 \mid k) = A_k \hat{X}(k \mid k) + B_k U(k)$
Prediction covariance	$P(k+1 \mid k) = A_k P(k \mid k) A_k^T + Q_k$
Kalman gain	$K_{k+1} = P(k+1 \mid k) C_{k+1}^T (C_{k+1} P(k+1 \mid k) C_{k+1}^T + R_{k+1})^{-1}$
Correction equation	$\hat{X}(k+1 \mid k+1) = \hat{X}(k+1 \mid k) + K_{k+1}(Z(k+1) - C_{k+1}\hat{X}(k+1 \mid k))$
Correction covariance	$P(k+1 \mid k+1) = P(k+1 \mid k) - K_{k+1} C_{k+1} P(k+1 \mid k)$

State estimate at time instant k given information up to and including time instant j denoted $\hat{X}(k \mid j)$.

The Kalman filter equations used to implement the discrete form of the Kalman filter for a process described by the state space triple $S(A_k, B_k, C_k)$ are given in Table 6.5 (Grewal and Andrews, 1993).

The process and observation models and the related noise statistics are now defined. The process equation is $\phi_k(k_i + 1) = [1]\phi_k(k_i)$, giving $A_{k_i} = [1]$, $B_{k_i} = [0]$ with process noise statistics $Q_{k_i} = [0]$, $k_i = 0, \ldots$. The observation equation is $\Phi_{DIG}(\omega_k, k_i) = \phi_k(k_i) + \xi(k_i)$, giving $C_{k_i} = [1]$ with measurement noise statistics $R_{k_i} = R(k_i)$, $k_i = 0, \ldots$. If these parameter values are inserted into the equations of Table 6.5 then the equations of Table 6.6 are obtained.

Table 6.6 Discrete Kalman filter equations for signal-processing module.

Process equation	$\phi_k(k_i + 1) = [1]\phi_k(k_i)$
Observation equation	$\Phi_{DIG}(\omega_k, k_i) = \phi_k(k_i) + \xi(k_i)$
Process noise	$E\{n_p(i)\} = 0, Q_{k_i} = [0], k_i = 0, \ldots$
Measurement noise $\xi(k_i)$	$E\{\xi(k_i)\} = 0, E\{\xi(k_i)\xi(k_j)^T\} = R(k_i)\delta_{ij}, k_i = 0, \ldots$
Prediction equation	$\hat{\phi}_k(k_i + 1 \mid k_i) = \hat{\phi}_k(k_i \mid k_i)$ simplifies to $\hat{\phi}_k(k_i + 1) = \hat{\phi}_k(k_i)$
Prediction covariance	$P(k_i + 1 \mid k_i) = P(k_i \mid k_i)$ simplifies to $P(k_i + 1) = P(k_i)$
Kalman gain	$K_{k_i+1} = P(k_i \mid k_i)(P(k_i \mid k_i) + R(k_i))^{-1}$ simplifies to $K_{k_i+1} = P(k_i)/(P(k_i) + R(k_i))$
Correction equation	$\hat{\phi}_k(k_i + 1 \mid k_i + 1) = \hat{\phi}_k(k_i \mid k_i) + K_{k_i+1}(\Phi_{DIG}(\omega_k, k_i + 1) - \hat{\phi}_k(k_i \mid k_i))$ simplifies to $\hat{\phi}_k(k_i + 1) = \hat{\phi}_k(k_i) + K_{k_i+1}(\Phi_{DIG}(\omega_k, k_i + 1) - \hat{\phi}_k(k_i))$
Correction covariance	$P(k_i + 1 \mid k_i + 1) = P(k_i \mid k_i) - P(k_i \mid k_i)[P(k_i \mid k_i) + R(k_i \mid k_i)]^{-1} P(k_i \mid k_i)$ simplifies to $P(k_i + 1) = P(k_i)R(k_{i+1})/[P(k_i) + R(k_{i+1})]$

The simplified equations in Table 6.6 are implemented in the phase-locked loop identifier module using the following algorithm.

Algorithm 6.5: Kalman filter implementation

Step 1 Initialisation
Initialise $k_i = 0, \hat{\phi}(k_i)$ and $P(k_i) = 10{,}000$
Set value for $R(k_i)$, $k_i = 0, 1, \ldots$

Step 2 Loop step
Calculate the gain term $K_{k_i+1} = P(k_i)/(P(k_i) + R(k_i))$
Receive phase measurement $\phi_{\text{DIG}}(\omega_k, k_i + 1)$
Update the phase estimate:

$$\hat{\phi}_k(k_i + 1) = \hat{\phi}_k(k_i) + K_{k_i+1}(\Phi_{\text{DIG}}(\omega_k, k_i + 1) - \hat{\phi}_k(k_i))$$

Update the variance estimate:

$$P(k_i + 1) = \frac{P(k_i)R(k_i + 1)}{(P(k_i) + R(k_i + 1))}$$

Test whether the phase estimate has converged; if it has EXIT digital signal processing unit with updated phase estimate
Update the iteration index $k_i = k_i + 1$; goto step 2
Algorithm end

When noise is present in the process, the use of a Kalman filter in the inner process will smooth the convergence of the phase-locked loop estimates. There are two tuning parameters:

1. The variance of the error between the ϕ_k parameter and its estimate $\hat{\phi}_k(k_i)$ is denoted $P(k_i)$. When this is initialised it gives an indication of the uncertainty associated with the estimate $\hat{\phi}_k(0)$.

2. The variance of the measurement noise ξ_k, is denoted $R(k_i)$. If the responses from the sine wave experiments exhibit significant noise then $R(k_i)$ will be selected to be large, and conversely if there is little process measurement noise it will be small.

Case Study Application of Kalman Filter Algorithm

The simulation results for two experiments are presented for a third-order system with time delay present. The system for which the phase crossover frequency is to be identified in the presence of noise is given by

$$G_p(s) = \frac{10e^{-2s}}{(1 + 0.2s)(1 + 0.7s)(1 + s)}$$

The theoretical values for the phase crossover point are $|G_p(j\omega_{-\pi})| = 6.4033$ and $\omega_{-\pi} = 0.8597$ rad s^{-1}. Noise was added to the output of the Φ_{DIG} block for both the phase angle and magnitude estimates. Kalman filters were then used to remove the noise from the corrupted estimates. The phase-locked loop method was tested with two noise signals, the first had unit variance, the second had a variance of 0.5 and both had zero mean.

Case Study 1: Additive Noise with a Large Standard Deviation
For this particular case the noise had unit standard deviation and was zero mean. To find the phase crossover frequency the reference phase is $\phi_r = -\pi$; thus the additive noise is quite significant when compared to the reference phase magnitude. The progress of the phase-locked loop identifier towards the phase crossover frequency $\omega_{-\pi} = 0.8597$ can be seen in Figure 6.25 and progress towards the phase

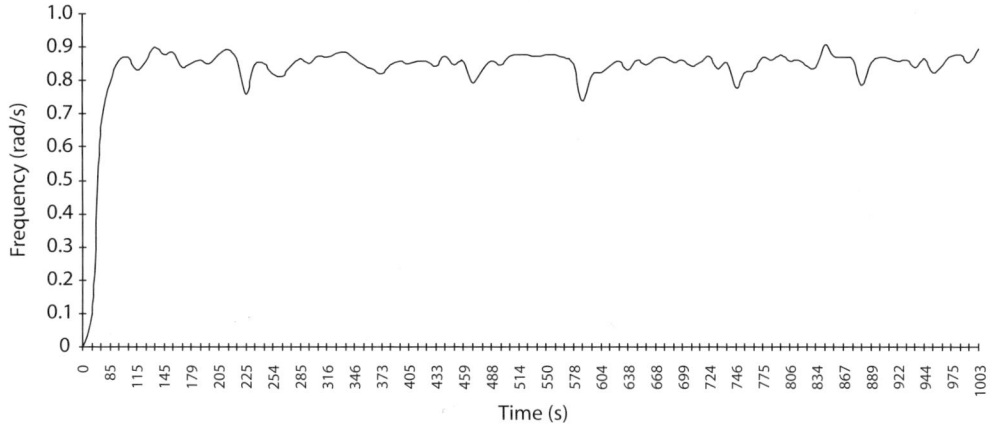

Figure 6.25 Progress towards the phase crossover frequency $\omega_{-\pi} = 0.8597$ rad s^{-1}.

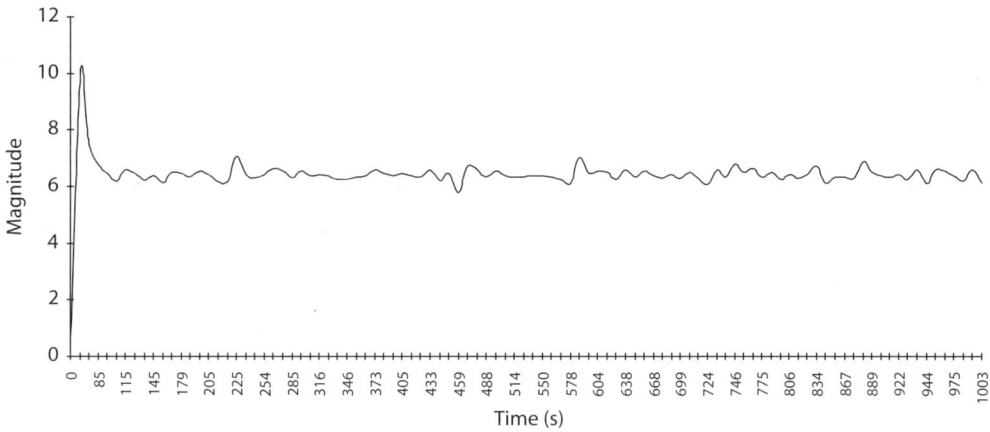

Figure 6.26 Progress toward the magnitude $|G_p(j\omega_{-\pi})| = 6.4033$.

crossover magnitude $|G_p(j\omega_{-\pi})| = 6.4033$ is in Figure 6.26. The figures show that the Kalman filter has substantially reduced the effects of the noise on the estimated parameters and demonstrated improved estimation of the phase crossover point data from a highly corrupted measurement signal.

Case Study 2: Additive Noise with a Lower Standard Deviation
The experiment was repeated but reducing the standard deviation from unity to 0.707 (the variance changes from unity to 0.5). As might be expected the graphs showing progress of the Kalman filter values are very much smoother. The progress of the phase-locked loop identifier towards the phase crossover frequency $\omega_{-\pi} = 0.8597$ rad s^{-1} can be seen in Figure 6.27 and progress towards the phase crossover magnitude $|G_p(j\omega_{-\pi})| = 6.4033$ is in Figure 6.28.

The main conclusion from these limited trials is that the estimation time required by the phase-locked loop method can be relatively long when the measurement signal is noise corrupted. One important reason for this is that the Kalman filter covariance is reset at each outer loop update to allow the tracking of the new phase angle estimate. However, it should be noted that even with a highly corrupted signal the phase-locked loop method did converge to a value that can be used to give an estimate of the phase crossover point; hence the Kalman filter is an effective method of noise management

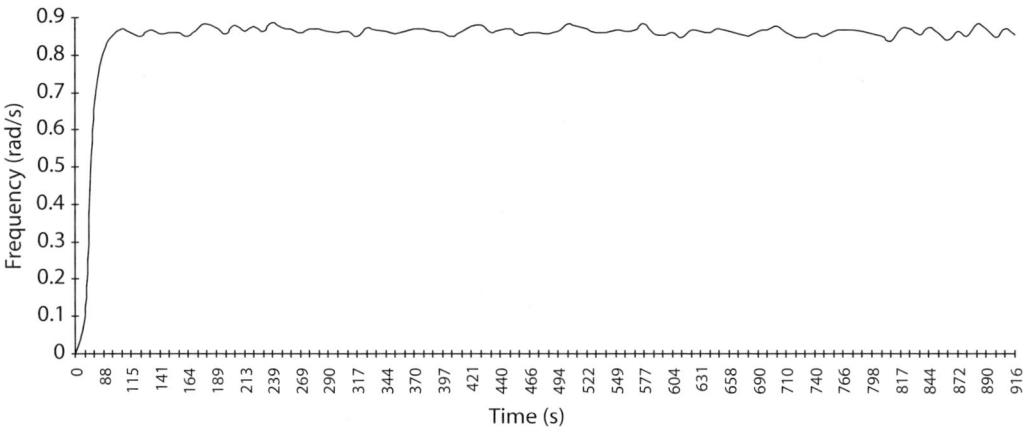

Figure 6.27 Progress towards the phase crossover frequency $\omega_{-\pi} = 0.8597$ rad s^{-1}.

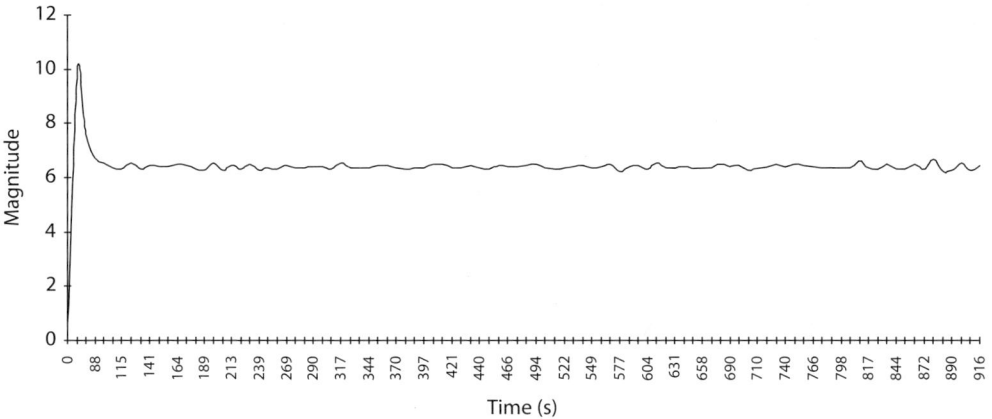

Figure 6.28 Progress toward the magnitude $|G_p(j\omega_{-\pi})| = 6.4033$.

for use in the identifier module. The second case study experiment showed, as might be expected, that as the noise variance decreases the quality of the estimate increases.

6.3.3 Disturbance Management Techniques

Disturbances occur in the normal operation of process plants. The types of disturbance observed are often related to changes in the supply of materials and energy to the process or in the loading of the process. In the analysis of the effects of the disturbances on the performance of the phase-locked loop method, two cases are considered: step disturbance models and sinusoidal disturbance models. The Figure 6.29 shows the model of how the disturbance term enters the identification module.

It is assumed that the detection signal from the VCO is the sine wave $vco(t) = \sin(\omega_1 t)$, whilst the linear system model $G_p(s)$ has been excited with the signal $exc(t) = \cos(\omega_1 t)$ so that the steady state system response is $y_{ss}(t) = |G_p(j\omega_1)|\cos(\omega_1 t + \phi(\omega_1))$. Two methods are considered for the removal of the effects of the disturbance term in the signal processing unit. These are time averaging and Fourier analysis of the multiplier output and are investigated for two different types of disturbance signal model.

6.3 Phase-Locked Loop Identifier Module – Basic Theory

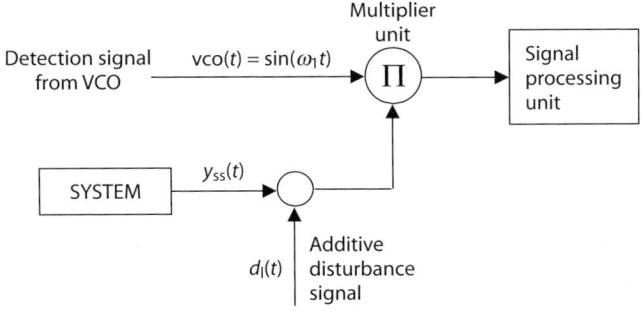

Figure 6.29 Model of load disturbance entry to signal processing unit.

Constant Load Disturbance
The disturbance signal $d_l(t)$ is assumed to be a step signal of the form

$$d_l(t) = \begin{cases} d & t \geq 0 \\ 0 & t < 0 \end{cases}$$

The steady state multiplier output will be the product

$$y(t) = vco(t) \times (d_l(t) + y_{ss}(t))$$

and hence

$$y(t) = \sin(\omega_1 t)(d + |G_p(j\omega_1)|\cos(\omega_1 t + \phi(\omega_1)))$$

This output is partitioned into two parts:

$$y(t) = y_1(t) + y_2(t)$$

where

$$y_1(t) = d\sin(\omega_1 t)$$
$$y_2(t) = \frac{|G_p(j\omega_1)|}{2}\sin(2\omega_1 t + \phi(\omega_1)) - \frac{|G_p(j\omega_1)|}{2}\sin(\phi(\omega_1))$$

and the output can be treated analytically in two different ways.

Time Averaging the Multiplier Output
If the output is averaged over a period or over multiple periods $nT, n = 1, \ldots$ then

$$\bar{y} = \bar{y}_1 + \bar{y}_2$$

where

$$\bar{y}_1 = \frac{1}{T}\int_{t_0}^{t_0+T} y_1(t)dt = \frac{1}{T}\int_{t_0}^{t_0+T} d\sin(\omega_1 t)dt = 0$$

and

$$\bar{y}_2 = \frac{1}{T}\int_{t_0}^{t_0+T} y_2(t)dt = \frac{1}{T}\int_{t_0}^{t_0+T}\left(\frac{|G_p(j\omega_1)|}{2}\sin(2\omega_1 t + \phi(\omega_1)) - \frac{|G_p(j\omega_1)|}{2}\sin(\phi(\omega_1))\right)dt$$

Hence

$$\bar{y}_2 = -\frac{|G_p(j\omega_1)|}{2}\sin(\phi(\omega_1))$$

with t_0 as the integration start time.

Clearly the effect of the step disturbance signal can be easily removed from the multiplier output by averaging the signal over an integer multiple of T periods.

Fourier Transform of the Multiplier Output

If multiplier output $y(t)$ is Fourier transformed then

$$Y(j\omega) = jd\pi(\delta(\omega+\omega_1)-\delta(\omega-\omega_1)) + \frac{|G_p(j\omega_1)|}{2}\cos\phi(\omega_1)j\pi(\delta(\omega+2\omega_1)-\delta(\omega-2\omega_1))$$
$$+ \frac{|G_p(j\omega_1)|}{2}\sin\phi(\omega_1)\pi(\delta(\omega-2\omega_1)+\delta(\omega+2\omega_1)) - |G_p(j\omega_1)|\sin\phi(\omega_1)\pi\delta(\omega)$$

Thus it can be seen that the signal energy is distributed over the zero, fundamental and second harmonic frequencies. The disturbance term is contained in the fundamental component of the multiplier output signal; hence this analysis shows that it is possible to determine the magnitude of the disturbance term. The magnitude and phase response data can then be recovered from the zero frequency and second harmonic components.

Sinusoidal Disturbance Signal

It is assumed that the detection signal from the VCO is the sine wave $vco(t) = \sin(\omega_1 t)$ and the disturbance term is a sinusoidal signal given by $d_l(t) = d\sin(\omega_0 t)$. The steady state multiplier output is given by

$$y(t) = vco(t) \times (d_l(t) + y_{ss}(t))$$

and hence

$$y(t) = \sin(\omega_1 t)(d\sin(\omega_0 t) + |G_p(j\omega_1)|\cos(\omega_1 t + \phi(\omega_1)))$$

This output can be partitioned as

$$y(t) = y_d(t) + \tilde{y}(t)$$

where the additional component due to the disturbance term is given by $y_d(t) = d\sin(\omega_0 t)\sin(\omega_1 t)$.

Time-Averaging the Multiplier Output

If the disturbance component $y_d(t)$ is averaged over multiple periods $nT, n = 1, \ldots$, where $T = 2\pi/\omega_1$, then for $n = 1, 2, \ldots$

$$\bar{y}_d(t_0) = \frac{1}{T}\int_{t_0}^{t_0+nT} y_d(t)dt$$

and

$$\bar{y}_d(t_0) = \frac{d\omega_0\omega_1}{2\pi n(\omega_0^2+\omega_1^2)}\left[\sin(\omega_0 t_0)\sin\left(\frac{2n\pi\omega_0}{\omega_1}\right) - \cos(\omega_0 t_0)\left(\cos\left(\frac{2n\pi\omega_0}{\omega_1}\right) - 1\right)\right]\sin(\omega_1 t_0)$$
$$-\frac{d\omega_0\omega_1}{2\pi n(\omega_0^2+\omega_1^2)}\left[\cos(\omega_0 t_0)\sin\left(\frac{2n\pi\omega_0}{\omega_1}\right) - \sin(\omega_0 t_0)\left(\cos\left(\frac{2n\pi\omega_0}{\omega_1}\right) - 1\right)\right]\cos(\omega_1 t_0)$$

where t_0 is the start time for the integration operation.

Thus, three cases are investigated for the disturbance frequency ω_0:

1. $\omega_0 = k\omega_1$
   If $\omega_0 = k\omega_1, k=1,2,\ldots,$ then the component $\bar{y}_d(t_0)$ is zero. Hence if the disturbance frequency is an integer multiple of the excitation frequency it can easily be removed from the multiplier output signal.

2. $\omega_0 \ll \omega_1$
   The equation for $\bar{y}_d(t_0)$ is approximated by

   $$\bar{y}_d(t_0) \cong \frac{d\omega_0}{\omega_1}\left[\frac{\omega_0^2 t_0}{\omega_1}\sin(\omega_1 t_0) - \cos(\omega_1 t_0)\right]$$

   Thus, the condition $\omega_0 \ll \omega_1$ implies $(\omega_0/\omega_1) \ll 1$, and it can be seen that if the disturbance frequency is very much lower than the excitation frequency the effect of the disturbance on the estimation will be reduced. It can also be seen that the point t_0 at which the integration starts has an effect on the size of the disturbance term.

3. $\omega_0 \gg \omega_1$
   For the condition where $\omega_0 \gg \omega_1$, the equation for $\bar{y}_d(t_0)$ will tend to zero, and hence if there is a high frequency disturbance present this will have little effect on the accuracy of the estimation.

Fourier Transform of the Multiplier Output

If the Fourier transform is applied to the multiplier output $y(t)$ then

$$Y(j\omega) = -\frac{d\pi}{2}[\delta(\omega+(\omega_1+\omega_0)) + \delta(\omega-(\omega_1+\omega_0))] + \frac{\delta\pi}{2}[\delta(\omega+(\omega_1-\omega_0)) + \delta(\omega-(\omega_1-\omega_0))]$$

$$+ \frac{|G_p(j\omega_1)|}{2}\cos\phi(\omega_1)j\pi[\delta(\omega+2\omega_1) - \delta(\omega-2\omega_1)]$$

$$+ \frac{|G_p(j\omega_1)|}{2}\sin(\phi(\omega_1))\pi[\delta(\omega-2\omega_1) + \delta(\omega+2\omega_1)]$$

$$- \frac{|G_p(j\omega_1)|}{2}\sin(\phi(\omega_1))\pi\delta(\omega)$$

Thus it can be seen that the required phase and magnitude information is found in the zero frequency and the $2\times\omega_1$ frequency terms whilst the disturbance term is found to be the $\pm(\omega_1+\omega_0)$ and $\pm(\omega_1-\omega_0)$ frequency terms. As with the time averaging case, three sub-cases are investigated.

1. $\omega_0 = k\omega_1$
   If the disturbance frequency is $\omega_0 = k\omega_1, k=1,2,\ldots$ then the sum and difference frequencies occur at multiples of the fundamental frequency and can be removed. This would only create a problem for multipliers of the fundamental frequency of 1 and 3. This is because the sum and difference frequencies of the disturbance would contaminate the information-bearing frequencies of zero and twice the fundamental.

2. $\omega_0 \ll \omega_1$
   If the disturbance frequency is very much less than the excitation or fundamental frequency then it can be seen that the components of the multiplier output, due to the disturbance term, will occur at approximately the fundamental frequency ω_1. Thus it will be possible to recover the phase and magnitude data from the components of the multiplier output which are at zero frequency and twice the fundamental frequency.

3. $\omega_0 \gg \omega_1$

   When the disturbance frequency is very much greater than the excitation or fundamental frequency then it can be seen that the disturbance component of the multiplier output signal will be at the disturbance frequency. It is therefore possible to recover the process magnitude and phase data from the zero frequency and twice the fundamental frequency directly from the Fourier transformed multiplier output signal.

Implementation for Load Disturbance Rejection and Case Study Results

The method used to model a load disturbance was to add a step signal to the process output and determine if the phase-locked loop identifier module was still able to recover the estimated magnitude and phase angle data for the disturbed system. The analysis demonstrated that time averaging or use of the Fourier transform was able to reject the disturbance terms in the multiplier signal. Thus the method used to recover the phase angle and magnitude data from the system for the demonstration in this section has been changed from the Peak and Trough method (Algorithm 6.4) to one based on Fourier analysis.

Using Fourier analysis to extract phase and magnitude information requires an analysis of the multiplier output signal, which is a complex waveform containing a number of harmonics of the system excitation frequency. To accomplish this, the Discrete Fourier Transform (DFT) was implemented using the Fast Fourier Transform (FFT) algorithm (Cooley and Tukey, 1965). This reduces the number of multiplication and addition operations from $O(n^2)$ for the DFT to $O(n \log_2(n))$. At the kth step the excitation frequency has reached ω_k, and the multiplier output signal with step disturbance signal addition can be shown to have a Fourier transform of

$$Y(j\omega) = jd\pi(\delta(\omega+\omega_k) - \delta(\omega-\omega_k)) + \frac{|G_p(j\omega_k)|}{2}\cos\phi(\omega_k)j\pi(\delta(\omega+2\omega_k) - \delta(\omega-2\omega_k))$$
$$+ \frac{|G_p(j\omega_k)|}{2}\sin\phi(\omega_k)\pi(\delta(\omega-2\omega_k) + \delta(\omega+2\omega_k)) - |G_p(j\omega_k)|\sin\phi(\omega_k)\pi\delta(\omega)$$

Within this transform, the disturbance term can be found as the component at the excitation frequency ω_k. The zero frequency term can be shown to be $c_{0S} = -|G_p(j\omega_k)|\sin(\phi(\omega_k))$, where the subscript S refers to the sine wave detection signal of the multiplier. The component at $2\omega_k$ is given by $c_{2S} = |G_p(j\omega_k)|\sin(2\omega_k t + \phi(\omega_k))$. Thus, it is possible to extract the magnitude and phase angle information from c_{2S}. In practice this is difficult to achieve due to the fact that if the waveform sampling is not initiated at a positive-going zero crossing there will be a shift in the phase angle value corresponding to the time difference in the starting point of the sampling of the waveform and the zero crossing point. To overcome this difficulty the zero frequency term was used. As has been stated previously, the LabVIEW implementation of the phase-locked loop method utilises two identification channels, one using a sine wave and the other a cosine wave, for the detection signals of the multiplier and conveniently presents the results in magnitude and phase form. If the FFT of the cosine channel is analysed then the zero frequency component is $c_{0C} = |G_p(j\omega_k)|\cos(\phi(\omega_k))$. The $2\omega_k$ component of the cosine channel is given by $c_{2C} = |G_p(j\omega_k)|\cos(2\omega_k t + \phi(\omega_k))$. Hence from these equations for c_{0S} and c_{0C} the phase angle of the system frequency response at ω_k can be found as

$$\phi(\omega_k) = -\tan^{-1}\left(\frac{c_{0S}}{c_{0C}}\right)$$

Similarly the magnitude data can be found from the equations for c_{2S} and c_{2C} as $|G_p(j\omega_k)| = \sqrt{c_{2S}^2 + c_{2C}^2}$.

Algorithm 6.6: Fourier extraction method (8-point DFT)

Step 1 Initialisation

   Specify closeness tolerances, *Tol1*, *Tol2*, *Tol3*

Set sample counter $n = 0$

Step 2 **Gather sample data**

Read the excitation frequency ω_k

Set sample time $T = \dfrac{\pi}{4\omega_k}$

Set timer to T
Store sine channel multiplier output sample n
Store cosine channel multiplier output sample n
Set $n = n + 1$
If $n = 7$ goto Step 4

Step 3 **Time delay**

Wait until timer has expired, then go to Step 2

Step 4 **DFT calculation**

Calculate the DFT for both sine and cosine channels using the samples of the multiplier output responses

Step 5 **Output results**

From the zero frequency components c_{0C}, c_{0S} of the DFT obtain

$$\phi(\omega_k) = -\tan^{-1}\left(\dfrac{c_{0S}}{c_{0C}}\right)$$

From the second harmonic components c_{2C}, c_{2S} of the DFT obtain

$$|G_p(j\omega_k)| = \sqrt{c_{2S}^2 + c_{2C}^2}$$

Step 6 **Convergence test**

If $|\phi(\omega_k) - \phi(\omega_{k-1})| \leq Tol1$ and
If $||G_p(j\omega_k)| - |G_p(j\omega_{k-1})|| \leq Tol2$ and
If $|\omega_k - \omega_{k-1}| \leq Tol3$ then STOP
Goto Step 2

Algorithm end

The phase-locked loop method of system identification was now tested to determine whether it could derive an accurate estimate of the process phase crossover point data after the application of a step disturbance.

Case Study Results

The phase crossover point is to be identified for the system given by

$$G_p(s) = \dfrac{10e^{-0.2s}}{(1+2s)(1+3s)(1+5s)}$$

The theoretical values of magnitude and frequency of the phase crossover point are given as $|G_p(j\omega_{-\pi})| = 1.2862$ and $\omega_{-\pi} = 0.5307$ rad s^{-1}. A step disturbance of magnitude 4 was added to the process output at time 140 s, the disturbance was removed at time 410 s. This represents a disturbance of approximately 310% when compared with the peak magnitude of the process at the phase crossover point frequency, when the excitation signal magnitude has a peak value of one. The evolution of the critical frequency estimate and the critical system magnitude estimate during the on–off step disturbance test is shown in Figures 6.30 and 6.31 respectively. The values of phase angle and magnitude which were estimated by the phase-locked loop method using Fourier data extraction during the application of the

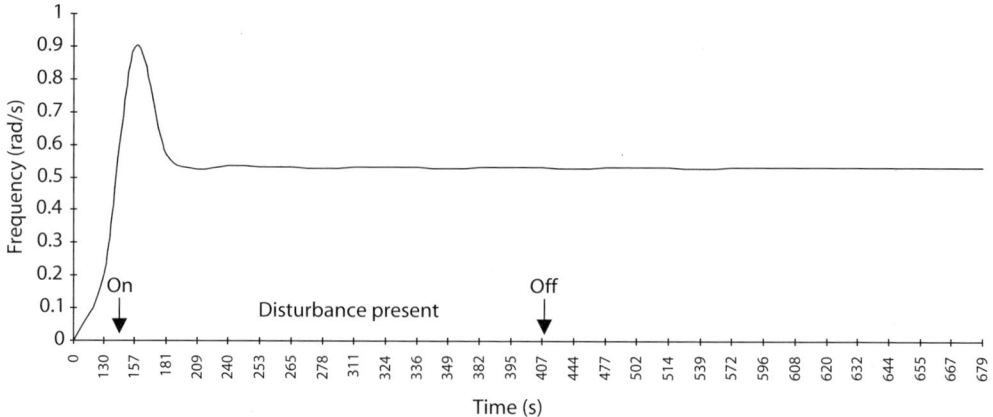

Figure 6.30 Progress toward critical frequency: on–off step disturbance test.

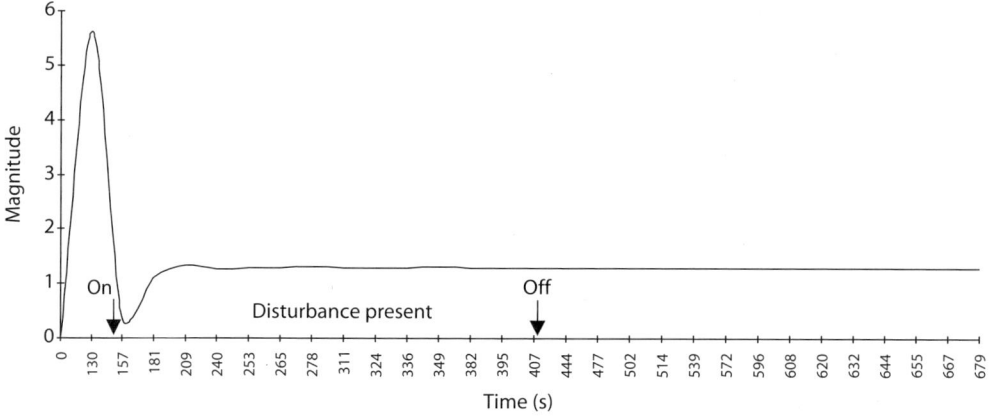

Figure 6.31 Progress toward critical magnitude: on–off step disturbance test.

disturbance were $|G_p(j\hat{\omega}_{-\pi})| = 1.2932$ and $\hat{\omega}_{-\pi} = 0.5312$ rad s^{-1} and were found to be $|G_p(j\hat{\omega}_{-\pi})| = 1.2864$ and $\hat{\omega}_{-\pi} = 0.5309$ rad s^{-1} when the disturbance was removed. Thus the estimates of the process phase crossover point can be seen to be practically equal to the theoretical values.

The phase-locked loop method, when coupled to Fourier analysis data extraction, is seen to be immune to the effects of a step load disturbance. Thus the phase-locked loop method can be considered to be robust in operation for step type load disturbances.

To use the phase-locked loop method of system identification under load disturbance conditions it is necessary to equip the signal processing module with either time average or Fourier transform routines on the multiplier output signal. Unfortunately, the method of using the peak and trough data values for data extraction is not feasible when a disturbance term is present in the process output signal.

If the load disturbance is a step type then the accuracy of the phase-locked loop method is not compromised. If the disturbance is a sinusoid then the accuracy of the method does suffer. The degree to which the accuracy of the method is compromised depends on the frequency of the sinusoidal disturbance term relative to that of the process excitation signal. If the disturbance term is an integer multiple of the excitation signal then it can be removed either by time averaging or by using Fourier transform techniques. It is possible for the Fourier transform method of disturbance rejection to fail. This failure is for the case where the disturbance frequency is 1× or 3× that of the excitation frequency. If the disturbance frequency is either very much greater or very much less than the process excitation frequency then both averaging and Fourier transform methods can be used to extract the phase and magnitude data from the multiplier output. Thus with the appropriate software in the signal processing unit, the phase-locked loop method of system identification can be used in situations where there is likely to be a load disturbance of the process output.

6.4 Summary and Discussion

As evidenced by the large body of publications produced on the subject, the relay experiment for system ultimate data estimation has been a fertile source of research. Over a number of years since *circa* 1985, the shortcomings of the relay approach have slowly become apparent and the desire to extend the application has grown. Various modifications to the original relay experiment have been published and pursued. These have been modifications to eliminate the error due to the use of the describing function method and to improve the accuracy of the relay experiment in the presence of measurement noise and process disturbances. The use of the relay experiment has been extended from single loop identification applications to the tuning of cascade and multi-input, multi-output systems. Other extensions have considered the estimation of several frequency response points, and the determination of the parameters of first- and second-order plus dead time process models along with analytical methods to determine the output of a closed-loop relay system for higher order systems.

From this extensive development arises the question "What potential future development can be achieved with the relay experiment?". It would appear that research and development in the relay experiment paradigm has moved from obtaining accurate *nonparametric* model data to using the relay to obtain an accurate *parametric* model of the process. However, it can be confidently asserted that in the absence of an elegant competing device with many more advantages than the relay experiment, the relay experiment will continue to be used in proprietary autotune controllers. After all, the information supplied by the relay experiment is sufficiently accurate to allow a rule-based design to provide a satisfactory level of system performance from a PID controller for a significant class of industrial systems. This immediately gives rise to the question, "Is there an elegant competing device with many more advantages than the relay experiment waiting to be discovered?". This question motivated the phase-locked loop development reported upon in this chapter.

It was the intention of the system identification research that led to the phase-locked loop method to provide an identification method with all of the benefits of the relay experiment but also to provide the added flexibility of being able to identify any point on the frequency response curve of a system. The phase-locked loop method does represent a clear departure from the relay experiment (Crowe and Johnson, 1998a,b,c; 1999). The method has been shown to provide a high degree of identification accuracy and is able to give more accurate results than the relay experiment. However, in all of the simulation examples performed, the relay experiment takes less time to achieve the identification albeit to a lower degree of accuracy. The main advantage that the phase-locked loop method has over the relay experiment is that there is no restriction as to the point on the frequency response curve to be identified. The user can easily specify any phase angle or any magnitude value as a reference input to the identifier module. A further advantage of the phase-locked loop method is that the intermediate points that are identified, as the error between the reference value and the actual value reduces, are accurate and can be used as estimates of the frequency response of the system being identified.

It is useful to note that there have been some recent accounts of other applications of the phase-locked loop concept published in the literature. In Balestrino *et al.* (2000), the term Sinusoidal AutoTune Variation (SATV) is used to describe a method involving phase-locked loop concepts. The identifier

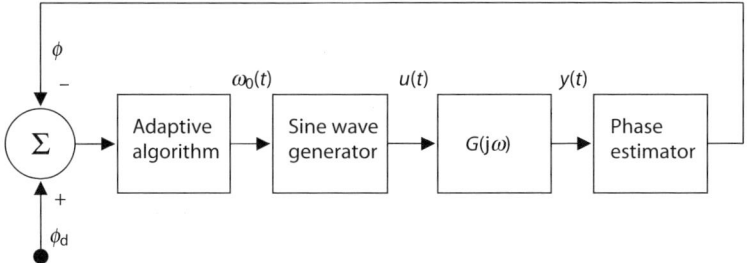

Figure 6.32 Conceptual diagram of a phase/frequency estimator (Clarke and Park, 2003).

described is a mixture of relay experiment and phase-locked loop ideas. Balestrino *et al.* used the method of synchronous detection to extract the phase information; this technique is known to have good noise rejection properties (Söderström and Stoica, 1989).

In another contribution to the use of phase-locked loop concepts in control systems, Clarke and Park (2003) have devised a *phase/frequency estimator* for system frequency response identification with the structure shown in Figure 6.32.

The adaptive algorithm block in Figure 6.32 implements the frequency update using the rule

$$\frac{d\omega_0(t)}{dt} = -Ke(t) = K(\arg(G(j\omega_0)) - \phi_d)$$

Since the output of the adaptive algorithm is the next excitation frequency $\omega_0(t)$ for input to the process, this is equivalent to the effect of using an integrator with gain K. Although there are many points of similarity between the phase-locked loop identifier described in this chapter and the Clarke and Park structure of Figure 6.32 there is one key difference. The phase-locked loop identifier described here operates in quasi-discrete time whereas that of Clarke and Park operates in continuous time. The Clarke and Park publication also gave a very thorough and useful investigation of routines for phase detection. Their work showed that the Hilbert transform phase detector gave an improved performance with regard to noise rejection than did zero-crossing detection methods. It is hoped that more implementation knowledge on phase-locked loop systems will be published in the literature in future years.

One significant advantage of the phase-locked loop identifier is its inherent identification flexibility. These aspects of the phase-locked loop module will be described in the next chapter and will include phase margin identification, peak resonance search and frequency response data for (a) a closed-loop unknown process with unknown controller and (b) a closed-loop unknown process with known controller. All of these operational modes have interesting uses, but the goal is automated PI control design algorithms. In Chapter 7, routines for automated PI control design for (a) gain margin and phase margin and (b) maximum sensitivity and phase margin specifications are also given.

References

Åström, K.J. and Hägglund, T. (1985) *US Patent 4,549,123: Method and an apparatus in tuning a PID regulator.*
Åström, K.J. and Hägglund, T. (1995) *PID Controllers: Theory, Design and Tuning.* Instrument Society of America, Research Triangle Park, NC.
Balestrino, A., Landi, A. and Scani, L. (2000) ATV techniques: troubles and remedies. *ADCHem 2000, IFAC International Symposium on Advanced Control of Chemical Processes*, Pisa, Italy, 14–16 June.
Best, R.E. (1997) *Phase-locked Loops: Design Simulation, and Applications.* McGraw-Hill, New York.
Cooley, J.W. and Tukey, J.W. (1965) An algorithm for the machine calculation of Fourier series. *Math. Comput.*, **19**, 297–301.

Clarke, D.W. and Park, J.W. (2003) Phase-locked loops for plant tuning and monitoring. *IEE Proceedings Control Theory and Applications*, **150**(1), 155–169.

Crowe, J. and Johnson, M.A. (1998a) *UK Patent Application GB Patent No. 9802358.3, Process and system analyser.*

Crowe, J. and Johnson, M.A. (1998b) New approaches to nonparametric identification for control applications. *Preprints IFAC Workshop on Adaptive Systems in Control and Signal Processing* (309–314), Glasgow, 26–28 August.

Crowe, J. and Johnson, M.A. (1998c) A phase-lock loop identifier module and its application. *IChemE Conference, Advances in Process Control*, Swansea, September.

Crowe, J. and Johnson, M.A. (1999) A new nonparametric identification procedure for online controller tuning. *Proc. American Control Conference*, pp. 3337–3341, San Diego, CA.

Eykhoff, P. (1974) *System Identification: Parameter and State Estimation.* John Wiley & Sons, Chichester.

Grewal, M.S. and Andrews, A.P. (1993) *Kalman Filtering.* Prentice Hall, Englewood Cliffs, NJ.

Hersh, M. and Johnson, M.A. (1997) A study of advanced control systems in the workplace. *Control Eng. Practice*, **5**(6), 771–778.

Lee, T.H., Wang, Q.G. and Tan, K.K. (1995) A modified relay based technique for improved critical point estimation in process control. *IEEE Trans. Control Systems Technology*, **3**(3), 330–337.

Ljung, L. (1987) *System Identification: Theory for the User.* Prentice Hall, Englewood Cliffs, NJ.

Ljung, L. and Glover, K. (1981) Frequency domain versus time domain methods in system identification. *Automatica*, **17**(1), 71–86.

Rugh, W.J. (1981) *Nonlinear System Theory.* Johns Hopkins University Press, Baltimore, MD.

Schwartz, H.R. (1989) *Numerical Analysis.* John Wiley & Sons, Chichester.

Shen, S.-H., Yu, H.-D. and Yu, C.-C. (1996a) Use of saturation-relay feedback for autotune identification. *Chem. Engrg. Sci.*, **51**(8), 1187–1198.

Shen, S.-H., Yu, H.-D. and Yu, C.-C. (1996b) Use of biased-relay feedback for system identification. *AIChE Journal*, **42**(4), 1174–1180.

Söderström T. and Stoica, P. (1989) *System Identification.* Prentice Hall International, Hemel Hampstead.

Tan, K.K., Wang, Q.G., Hang, C.C. and Hägglund, T.J. (1999) *Advances in PID Control.* Springer-Verlag, London.

Wang, Q.G., Lee, T.H. and Lin, C. (2003) *Relay Feedback.* Springer-Verlag, London.

Wellstead, P.E. (1981) Nonparametric methods of system identification. *Automatica*, **17**(1), 55–69.

Yu, C.-C. (1998) *Autotuning of PID Controllers.* Springer-Verlag, London.

Ziegler, J.G. and Nichols, N.B. (1942) Optimum settings for automatic controllers. *Trans. ASME*, **64**, 759–768.

7 Phase-Locked Loop Methods and PID Control

Learning Objectives

7.1 Introduction – Flexibility and Applications

7.2 Estimation of the Phase Margin

7.3 Estimation of the Parameters of a Second-Order Underdamped System

7.4 Identification of Systems in Closed Loop

7.5 Automated PI Control Design

7.6 Conclusions

References

Learning Objectives

The phase-locked loop identifier module is essentially a feedback loop able to lock on to a system frequency that satisfies an auxiliary condition such as a system phase or gain specification. This structure gives the module significant flexibility that can be used in a wide range of applications. The applications fall into two categories. Firstly, there are the single-task applications such as the classical task of finding the phase crossover frequency. An example of a more demanding single-task application is to find the peak resonance frequency in a system with dominant second-order model characteristics. In the second category of applications, the identifier module is part of a larger scheme to realise a more complex set of conditions. Good examples in this category are the two automated PI control design algorithms that have complex specifications of (a) desired gain and phase margin values and (b) desired maximum sensitivity and phase margin values. Both these algorithms are presented in this chapter.

The learning objectives for this chapter are to:

- Understand the application flexibility of the phase-locked loop identifier module.

- Study several single-task applications including estimation of phase margin and the system peak resonance frequency.
- Learn how to use the module to identify systems in closed loop.
- Use the identifier module to automate PID control design to attain desired classical robustness specifications.

7.1 Introduction – Flexibility and Applications

A major objective in developing an alternative approach to the relay experiment in PID tuning is to devise a method with additional flexibility in various types of different applications. Now that the basic theoretical framework for the phase-locked loop identifier module has been presented some of the applications flexibility inherent in the procedure can be examined. The basic theory was presented for phase angle searches, but an immediate consequence of the framework is that it can also be applied to system modulus reference specifications. For example, the identifier can be used to find a system phase margin where the phase-locked loop reference will instead be $|G_p|_{ref} = 1$. This captures the property that the system modulus satisfies $|G_p(j\omega_1)| = 1$ at the phase margin frequency ω_1. This application is presented in Section 7.2. Exploiting the ability of the phase-locked loop to seek specific gain values, a procedure that uses a peak magnitude search to find the peak resonance frequency is described in Section 7.3. It is a short step from the data of the peak resonance point to the parameters of an underdamped second-order system. In Section 7.4, procedures for the nonparametric identification of systems in closed loop are presented. The two cases identified depend on the available knowledge of the controller. These important procedures lead the way to use the phase-locked loop identifier to tune PID controllers on-line with classical frequency-domain control specifications.

The relay experiment of Åström and Hägglund (1985) was an elegant way to implement PID autotuner facilities in a process controller and from the 1980s onward push-button on-line PID controller tuning became a standard feature in hardware controllers and software-based process control supervisor systems. In Section 7.5, the phase-locked loop identifier module is used to automatically tune a PI controller with designs that additionally satisfy various combinations of the classical robustness measures of pre-specified gain and phase margin and maximum sensitivity values. The utility of these classical control performance specifications was discussed briefly in Chapter 2 and can be found in most introductory control engineering course textbooks (for example Wilkie et al., 2002). Thus, the presentation in this section concentrates on how the phase-locked loop is used in autotune mode for these more closely specified three-term control designs. In both of the methods described it is assumed that the process to be controlled is unknown, that the existing controller is known and that the identifications required are performed by a phase-locked loop identifier module.

7.2 Estimation of the Phase Margin

The phase margin of a system occurs at the gain crossover frequency ω_1 and is defined as the interior angle from the phase angle ray of a system when the modulus of the gain is unity to the negative real axis. Hence for a system $G_p(s)$, the phase margin in radians will be

$$\phi_m = \arg(G_p(j\omega_1)) + \pi$$

where ω_1 is the gain crossover frequency at which the condition $|G_p(j\omega_1)| = 1$ holds. In classical control theory, the phase margin serves as a robustness measure and is related to the damping of the system

(D'Azzo and Houpis, 1975). Schei (1994) gives a method whereby the relay experiment can be modified to allow it to identify the gain crossover point. As has been shown in Chapter 6, the accuracy of the relay method is system-dependent and hence accurate results cannot be guaranteed for every system type. However, the phase-locked loop method of system identification can use either phase angle or gain reference values. In gain reference mode, the gain estimates are output by the signal processing unit of the module, and thus the determination of the phase margin, which requires reference value $|G|_{ref} = 1$, can be trivially accomplished. This means that the outer loop of the basic phase-locked loop will be a feedback loop in system gain with the gain reference $|G|_{ref} = 1$ at the comparator.

Case Study

The test system is

$$G_p(s) = \frac{3.3}{(1+0.1s)(1+0.2s)(1+0.7s)}$$

The theoretical values for the system frequency response at the gain crossover point are:

Phase margin frequency: $\omega_1 = 3.4021 \text{ rad s}^{-1}$

System phase: $\arg(G(j\omega_1)) = -2.0986$

Hence the phase margin is $\phi_m = 1.0430$ rad (equivalently, $\phi_m = 59.76°$).

The progress of the phase-locked loop identifier module to convergence is shown in Figures 7.1 and 7.2. The outcomes of the phase-locked loop procedure are shown in Table 7.1.

It can be seen that the method gives an almost exact estimate for the gain crossover point. The time taken to obtain the estimate was approximately 140 s. Thus from this example and experience with other examples not reported here, the phase-locked loop method can be used to identify the phase margin point successfully.

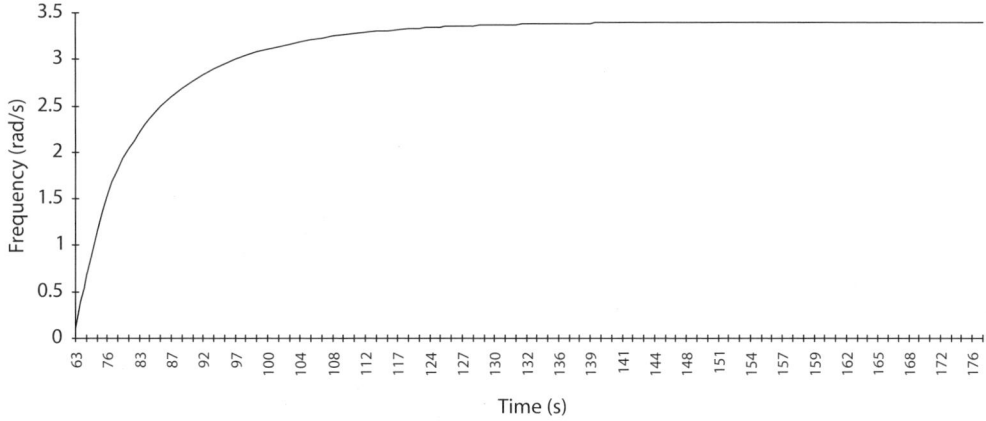

Figure 7.1 Phase-locked loop progress towards gain crossover frequency $\omega_1 = 3.4021 \text{ rad s}^{-1}$.

7.3 Estimation of the Parameters of a Second-Order Underdamped System

In this application, the ability of the phase-locked loop procedure to use a gain reference is exploited in a search routine to find the parameters of an underdamped second-order system. Let the system be represented by

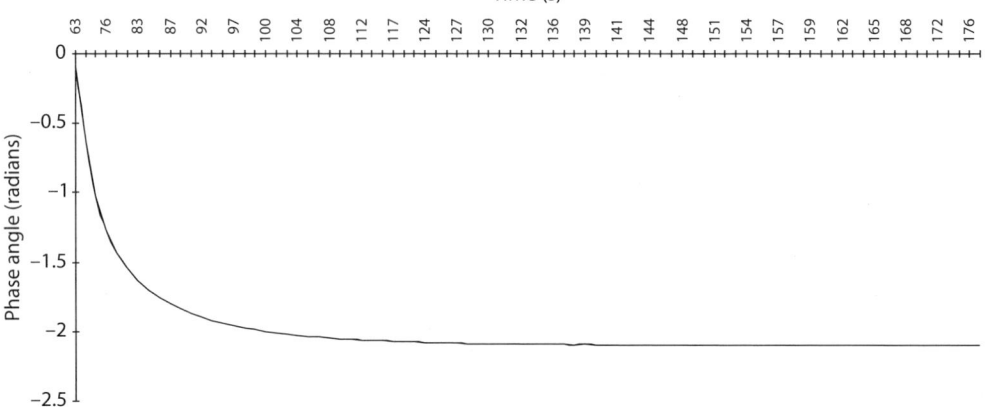

Figure 7.2 Phase-locked loop progress towards system phase at gain crossover frequency $\arg(G(j\omega_1)) = -2.0986$ rad.

Table 7.1 System gain crossover estimation results.

	Theoretical values	Numerical results
ω_1 rad s^{-1}	3.4021	3.3945
$\arg\{G(j\omega_1)\}$ rad	−2.0986	−2.0965
ϕ_{PM} rad	1.0430	1.0451

$$G_p(s) = \frac{K\omega_n^2}{s^2 + 2\zeta\omega_n s + \omega_n^2}$$

where ω_n is the natural frequency, ζ is the damping factor and K is the zero frequency gain.

Identification Method 1
The parameters of the system can be determined by identifying the two points on the frequency response curve at which the system phase angles are $\phi_{-\pi/2} = -(\pi/2)$ and $\phi_{-\pi/4} = -(\pi/4)$. If the frequencies at which these two points occur are denoted by $\omega_{-\pi/2}$ and $\omega_{-\pi/4}$ respectively then it can be shown that $\zeta = (\omega_{-\pi/2}^2 - \omega_{-\pi/4}^2) / (2\omega_{-\pi/2}\omega_{-\pi/4})$ and $K = 2\zeta |G(j\omega_{-\pi/2})|$.

The phase-locked loop method can be used to find the two frequencies required to determine the parameters of the system by applying the appropriate phase angle reference values $\phi_{-\pi/2}$ and $\phi_{-\pi/4}$, and conducting two phase-locked loop experiments.

Identification Method 2
A second method that can be used to estimate the parameters of a second-order underdamped system is to determine the point at which the peak resonance ω_p of the system occurs (Crowe and Johnson, 1999). If the peak resonance point is identified, it can be shown that

$$\zeta = \frac{1}{\sqrt{\tan^2(-\phi(\omega_p)) + 2}}, \quad k = 2|G_p(j\omega_p)|\zeta\sqrt{1-\zeta^2} \text{ and } \omega_n = \frac{\omega_p}{\sqrt{1-2\zeta^2}}$$

where $\phi(\omega_p)$ and $|G_p(j\omega_p)|$ are respectively the phase angle and the gain of the system at the resonance peak. The method used to determine the peak resonance is to determine a bound on the peak resonance point and then use a search by golden section (Fletcher, 1987) to narrow the bound at each iteration until the bound has reduced to within a sufficient tolerance. The method used by the phase-locked loop system is shown in the following algorithm.

Algorithm 7.1: Peak resonance search routine

Step 1 Use phase angle reference $\phi_1 = -(\pi/6)$ to find ω_1 and $|G_p(j\omega_1)|$
Step 2 Use phase angle reference $\phi_2 = -(\pi/6) + 0.4$ to find ω_2 and $|G_p(j\omega_2)|$
Step 3 Use phase angle reference $\phi_3 = -(\pi/2)$ to find ω_3 and $|G_p(j\omega_3)|$
Step 4 Use phase angle reference $\phi_4 = \phi_2 + (\phi_3 - \phi_2)\left(\dfrac{3-\sqrt{5}}{2}\right)$ to find ω_4 and $|G_p(j\omega_4)|$

Step 5 If $|G_p(j\omega_2)| > |G_p(j\omega_4)|$
Then $\phi_3 \leftarrow \phi_4, |G_p(j\omega_3)| \leftarrow |G_p(j\omega_4)|$ and $\omega_3 \leftarrow \omega_4$
Else $\phi_1 \leftarrow \phi_2, |G_p(j\omega_1)| \leftarrow |G_p(j\omega_2)|$ and $\omega_1 \leftarrow \omega_2$
$\phi_2 \leftarrow \phi_4, |G_p(j\omega_2)| \leftarrow |G_p(j\omega_4)|$ and $\omega_2 \leftarrow \omega_4$

Step 6 If $|\phi_2 - \phi_1| > |\phi_3 - \phi_2|$

Then $\phi_4 = \phi_2 + \left(\dfrac{3-\sqrt{5}}{2}\right)(\phi_2 - \phi_1)$

Else $\phi_4 = \phi_2 + \left(\dfrac{3-\sqrt{5}}{2}\right)(\phi_3 - \phi_2)$

Step 7 If $|G_p(j\omega_3)| - |G_p(j\omega_1)| <$ tolerance STOP
Step 8 Use phase angle reference ϕ_4 to find ω_4 and $|G_p(j\omega_4)|$
Step 9 Goto Step 5
Algorithm end

Case Study

The test system is

$$G_p(s) = \frac{2.16}{s^2 + 0.96s + 1.44}$$

This system has the following parameters: a gain of $K = 1.5$, a natural frequency of $\omega_n = 1.2$ rad s^{-1} and a damping factor of $\zeta = 0.4$. The results of the parameter estimation are shown in Figures 7.3 to 7.5.

From the results of the trial the estimated parameter values were

$$\hat{\omega}_n = 1.20 \text{ rad s}^{-1}, \hat{K} = 1.50 \text{ and } \hat{\zeta} = 0.40$$

The percentage error in the estimates of the natural frequency, gain factor and damping were found to be 0.1%, 0.6% and 0.8% respectively. The form which the graphs take requires an explanation. The search by golden section requires that three points are initially identified; the next point to be identified is then calculated from the previous three identifications as described in Algorithm 7.1, Steps 1 to 4. The calculation of the estimated parameters of the second-order system uses the mid-point data from the bounds of the resonance peak; thus until the first two points have been identified there is no estimate available for the calculation of the parameters of the second-order system. Referring to Figures 7.3 to 7.5, at time $t = 97$ s two points have been estimated and the first estimate of the parameters of the second-order system are calculated. During time $t = 97$ s to $t = 161$ s the third bound and the next bound estimate, corresponding to Steps 3 and 4 of Algorithm 7.1, are being estimated and hence there is no

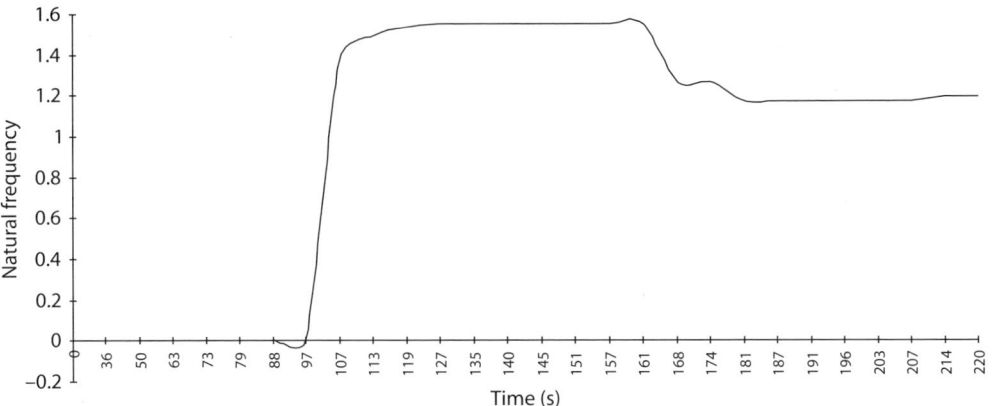

Figure 7.3 Phase-locked loop progress towards natural frequency $\omega_n = 1.2\,\text{rad s}^{-1}$.

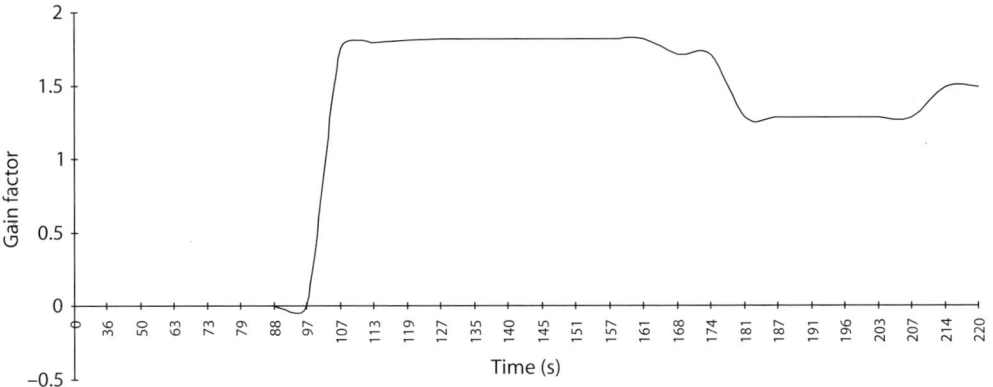

Figure 7.4 Phase-locked loop progress towards system gain $K = 1.5$.

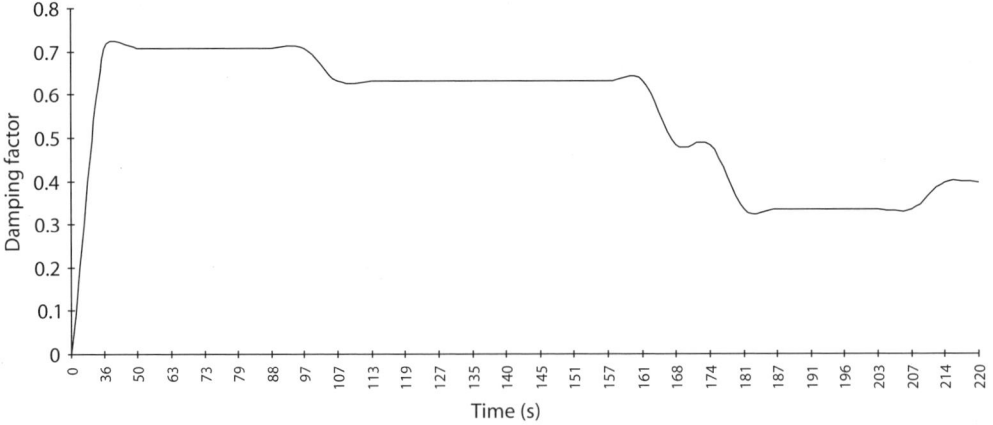

Figure 7.5 Phase-locked loop progress towards damping $\zeta = 0.4$.

change in the mid-point bound estimate. After time $t = 161$ s the search by golden section algorithm is narrowing the bound on the peak of the resonance curve and hence the parameter estimates of the second-order system are improved as the bound narrows. This example shows the versatility of the phase-locked loop method with its ability to easily find any point on the frequency response curve of a system.

7.4 Identification of Systems in Closed Loop

In the applications given so far, the processes have been identified as open loop systems (Crowe and Johnson, 2000a). However, there are circumstances where it is not possible or desirable to perform open-loop identification. For example, a process may be open-loop unstable and hence requires to be identified in closed loop. Another common circumstance occurs when production personnel require a PID controller to be retuned but are not willing to allow the process to go into open-loop operation in case the product quality is adversely affected. In these cases closed-loop identification and tuning are necessary. In the analysis to follow, two situations are discussed (Crowe and Johnson, 2000b):

1. Closed-loop identification without transfer function knowledge of either the controller or the process.

2. Closed-loop identification with transfer function knowledge of the controller, but where the process transfer function is unknown.

The first case is a complete black box identification scheme where there is no prior knowledge of the process or of the controller. In the second and more usual situation, the controller transfer function is known and the process has to be identified. Both of these cases will be discussed and analysed using the unity feedback configuration of Figure 7.6 where the process is represented by $G_p(s)$, the controller by $G_c(s)$ and the forward path transfer function by $G_{fp}(s) = G_p(s)G_c(s)$.

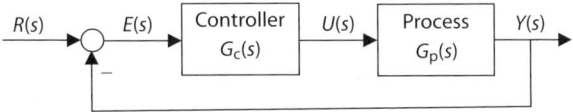

Figure 7.6 Unity feedback system.

7.4.1 Identification of an Unknown System in Closed Loop with an Unknown Controller

Consider the unity feedback system of Figure 7.6 where $G_p(s)$ represents the process and $G_c(s)$ represents the controller. The closed-loop transfer function relationships are

$$\frac{Y(s)}{R(s)} = \left[\frac{G_c(s)G_p(s)}{1+G_c(s)G_p(s)}\right] \text{ and } \frac{U(s)}{R(s)} = \left[\frac{G_c(s)}{1+G_c(s)G_p(s)}\right]$$

Division of these relationships gives

$$\left(\frac{Y(s)}{R(s)}\right) / \left(\frac{U(s)}{R(s)}\right) = \left[\frac{G_c(s)G_p(s)}{1+G_c(s)G_p(s)}\right] / \left[\frac{G_c(s)}{1+G_c(s)G_p(s)}\right] = G_p(s)$$

Using the identity $s = j\omega$, it is readily shown that

$$\arg\left\{\frac{Y(j\omega)}{R(j\omega)}\right\} - \arg\left\{\frac{U(j\omega)}{R(j\omega)}\right\} = \arg G_p(j\omega)$$

and

$$\left|\frac{Y(j\omega)}{R(j\omega)}\right| \bigg/ \left|\frac{U(j\omega)}{R(j\omega)}\right| = |G_p(j\omega)|$$

These equations motivate two identifications, one between the reference input $R(s)$ and the controller output $U(s)$ and a second between the reference input $R(s)$ and the closed-loop output $Y(s)$. From these it is possible to identify the process $G_p(s)$ in magnitude (gain) and phase. Most importantly, without having any information about the process or the controller, the process can still be identified when it is connected in a closed-loop configuration. The setup for the identification is shown in Figure 7.7.

In the scheme of Figure 7.7, Identifier 1 is used to estimate the magnitude and phase data of

$$\frac{U(j\omega_k)}{R(j\omega_k)} = G_1(j\omega_k)$$

and also provides the system excitation signal. Identifier 2 utilises the same detection signal for its multiplier as does Identifier 1 and is used to estimate the phase and magnitude data for

$$\frac{Y(j\omega_k)}{R(j\omega_k)} = G_2(j\omega_k)$$

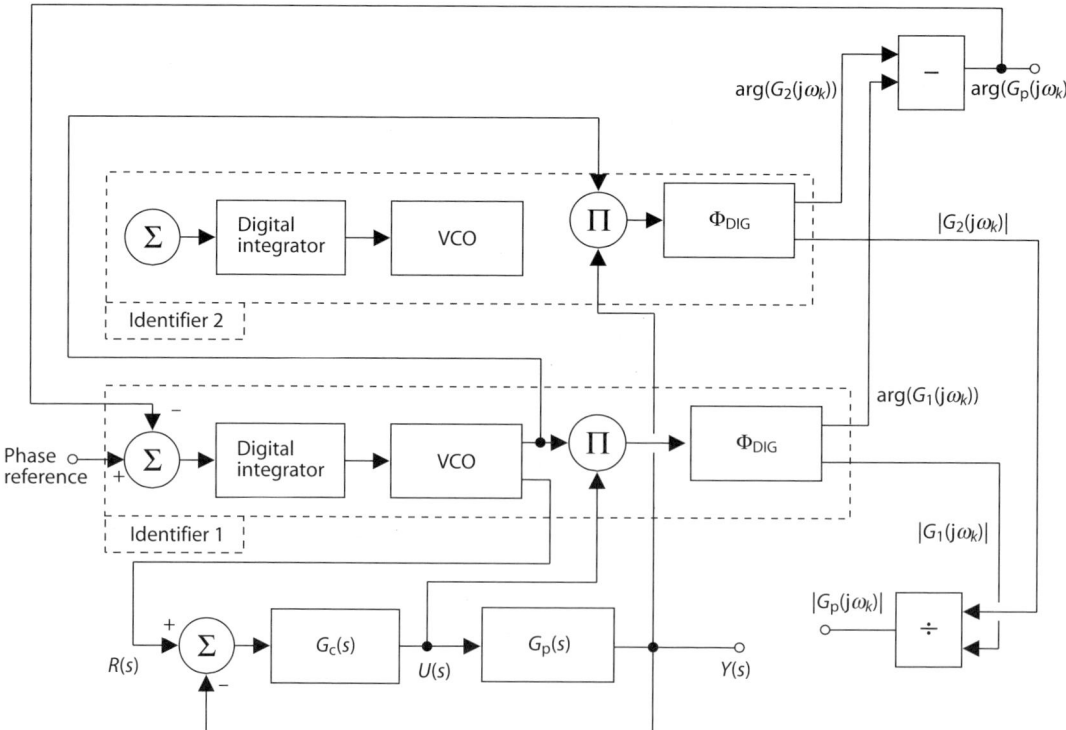

Figure 7.7 Identifier setup for closed-loop identification of a process without controller knowledge.

7.4 Identification of Systems in Closed Loop

The phase angle feedback to Identifier 1 is derived from subtracting the phase output of Identifier 1 from that of Identifier 2. Hence it can be seen that if the phase reference is given as $\phi_r = -\pi$ rad then the combination of Identifier One and Two will determine the phase crossover point of the process $G_p(s)$.

Case Study

The test process is

$$G_p(s) = \frac{10}{(1+0.2s)(1+0.7s)(1+s)}$$

and the closed-loop PID controller is

$$G_c(s) = 0.79\left(1 + \frac{1}{0.85s} + 0.21s\right)$$

The purpose of the identification was to find the system parameters at the phase crossover point. The theoretical values and those estimated from the phase-locked loop procedure for the frequency and magnitude of the process at the phase crossover point are given in Table 7.2. The progress of the phase-locked loop identifiers towards critical frequency and the associated gain value is shown in Figures 7.8 and 7.9.

Table 7.2 Unknown controller, unknown process identification results.

	Theoretical values	Numerical results
$\omega_{-\pi}$	3.6841 rad s^{-1}	3.6405 rad s^{-1}
$\lvert G_p(j\omega_{-\pi})\rvert$	0.7625	0.78920

The time taken to converge to the estimates was approximately 250 s. The percentage errors in the estimates are −1.18% and 3.51% for the frequency and magnitude estimates respectively. The accuracy of the identification suffers from the problem that the estimates are calculated from the results of two separate but simultaneous identifications. Thus the identified phase crossover point estimation error will be a combination of the errors of the individual estimates. In later LabVIEW software implementations for this mode of operation these errors were eliminated.

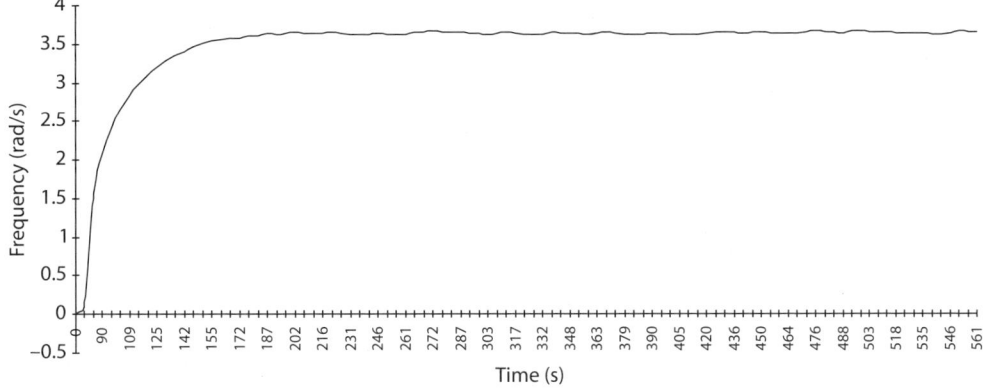

Figure 7.8 Phase-locked loop progress towards critical frequency $\omega_{-\pi} = 3.6841$ rad s^{-1}.

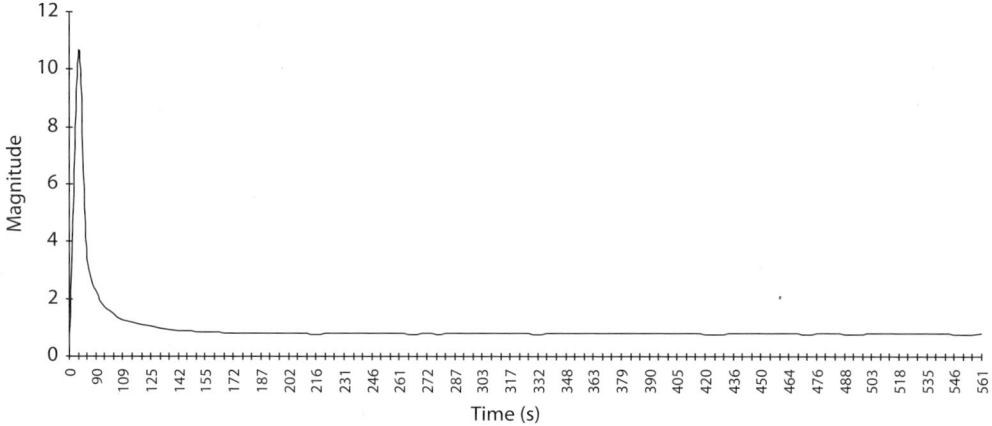

Figure 7.9 Phase-locked loop progress towards the critical magnitude $|G(j\omega_{-\pi})| = 0.7625$.

7.4.2 Identification of an Unknown System in Closed Loop with a Known Controller

The method again uses the unity feedback structure of Figure 7.6, and the closed relationship is given as

$$\frac{Y(s)}{R(s)} = \left[\frac{G_c(s)G_p(s)}{1 + G_c(s)G_p(s)}\right] = G_{CL}(s)$$

Since the controller transfer function $G_c(s)$ is assumed to be known, this closed-loop expression can be rearranged as

$$G_c(s)G_p(s) = G_{CL}(s)(1 + G_c(s)G_p(s))$$

and hence

$$G_p(s) = \left[\frac{G_{CL}(s)}{G_c(s)[1 - G_{CL}(s)]}\right]$$

Setting $s = j\omega$, the frequency response relationship for the process is

$$G_p(j\omega) = \frac{G_{CL}(j\omega)}{(1 - G_{CL}(j\omega))G_c(j\omega)}$$

Assume that the closed-loop transfer function $G_{CL}(j\omega)$ is identified and written as

$$G_{CL}(j\omega) = |G_{CL}(j\omega)|(\cos(\phi_{CL}(\omega)) + j\sin(\phi_{CL}(\omega)))$$

where $\phi_{CL}(\omega) = \arg(G_{CL}(j\omega))$. Then

$$|1 - G_{CL}(j\omega)| = \sqrt{(1 - |G_{CL}(j\omega)|\cos(\phi_{CL}(\omega)))^2 + (|G_{CL}(j\omega)|\sin(\phi_{CL}(\omega)))^2}$$

and

$$\arg(1 - G_{CL}(j\omega)) = \tan^{-1}\left[\frac{|G_{CL}(j\omega)|\sin(\phi_{CL}(\omega))}{1 - |G_{CL}(j\omega)|\cos(\phi_{CL}(\omega))}\right]$$

If these results are computed from the closed-loop identification and the controller transfer function is known then the process can be identified. The process magnitude and phase angle follow respectively as

$$|G_p(j\omega)| = \frac{|G_{CL}(j\omega)|}{|1-G_{CL}(j\omega)||G_c(j\omega)|}$$

and

$$\arg(G_p(j\omega)) = \arg(G_{CL}(j\omega)) - \arg(1-G_{CL}(j\omega)) - \arg(G_c(j\omega))$$

Figure 7.10 shows how the identifier is configured to perform the identification.

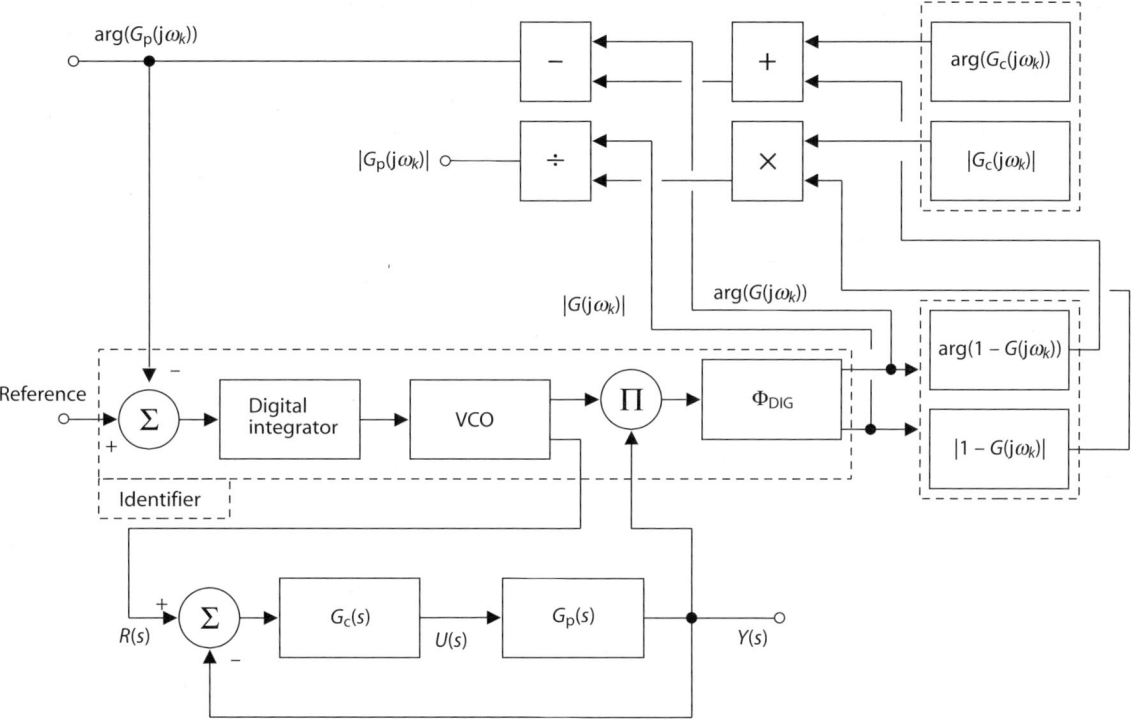

Figure 7.10 Identifier setup for known controller, unknown process closed-loop identification.

The identifier supplies the excitation for the system and identifies the closed-loop frequency response points:

$$\frac{Y(j\omega_k)}{R(j\omega_k)} = G_{CL}(j\omega_k)$$

The closed-loop identification data is further processed to derive the process frequency response for the frequency ω_k. Since the phase angle of the process is used to close the feedback loop around the identifier, the identifier will converge to the frequency at which the process $G_p(s)$ has a phase angle equal to the reference phase angle. Hence for the case of identifying the phase crossover point of the process, the phase-locked loop converges to a frequency of excitation that produces a phase shift $-\pi$ rad across the process and *not* to an excitation frequency that produces a phase shift of $-\pi$ rad across the closed-loop system.

Case Study

This test process was

$$G_p(s) = \frac{10}{(1+0.2s)(1+0.7s)(1+s)}$$

in closed loop with the controller:

$$G_c(s) = 0.79\left(1 + \frac{1}{0.85s} + 0.21s\right)$$

The process $G_p(s)$ was to be identified at its phase crossover point. The theoretical values and the values obtained from the phase-locked loop identification for the frequency and magnitude of the process frequency response at the phase crossover point are shown in Table 7.3.

Table 7.3 Known controller, unknown process identification results.

	Theoretical values	Numerical results
$\omega_{-\pi}$	3.6841 rad s^{-1}	3.6851 rad s^{-1}
$\lvert G_p(j\omega_{-\pi})\rvert$	0.7625	0.7574

The performance of the phase-locked loop identifier module for the known controller, unknown process configuration is shown in Figures 7.11 and 7.12.

Hence it can be seen that the error in the estimated data is minimal. The accuracy of this method is improved over that of Section 7.4.1, since the controller frequency values can be calculated exactly. The overall error in the estimation is therefore that of the closed-loop estimation and the phase-locked loop method can be used to identify a system in closed loop when the transfer function of the controller is known.

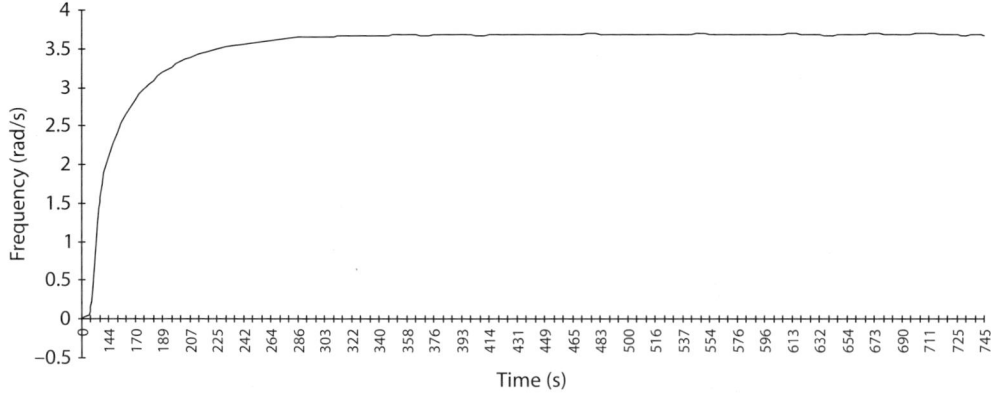

Figure 7.11 Phase-locked loop progress towards critical frequency $\omega_{-\pi} = 3.6841$ rad s^{-1}.

7.5 Automated PI Control Design

A major objective of the development of the phase-locked loop method of system identification was to exploit the method's flexibility to automate PID controller tuning for a more demanding desired

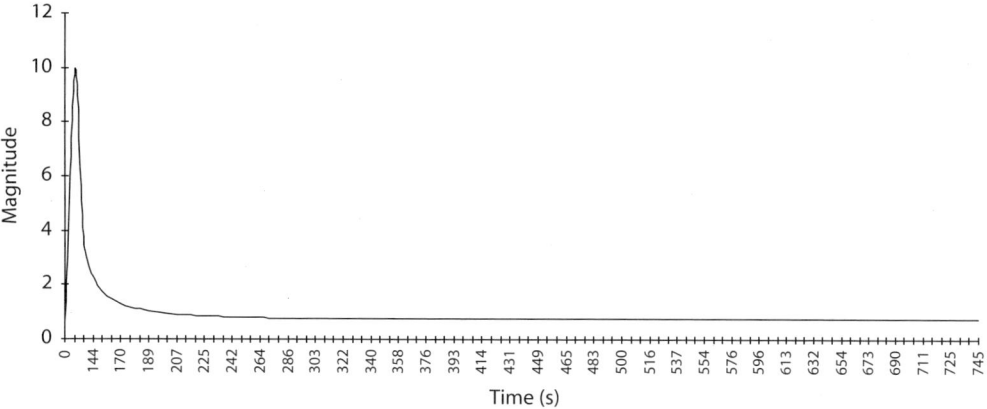

Figure 7.12 Phase-locked loop progress towards the critical magnitude $|G(j\omega_{-\pi})|=0.7625$.

controller specification. In this category of applications, the identifier module becomes a building block as part of a larger scheme to realise a more complex set of conditions. The two main autotune routines presented in this section are an automated PI design to satisfy desired gain and phase margin values (Section 7.5.2) and an automated PI design with a maximum sensitivity and phase margin specification pair (Section 7.5.3). However, the chapter opens with a generic look at the phase-locked loop identifier issues for automated PID control design. This is a summary of the different identifier modes required and an introduction to an all-purpose phase-locked loop identifier module.

7.5.1 Identification Aspects for Automated PID Control Design

In the automated PI controller design methods to follow, gain margin, phase margin and maximum sensitivity design specifications will be given and data sought for the unknown process at various specific frequencies. One all-purpose phase-locked loop identifier module structure is used and this incorporates switching to instigate the appropriate identification needed. Figure 7.13 shows this all-purpose phase-locked loop identifier module.

The notation used is as follows: the unknown process is $G_p(s)$, the new controller being computed is $G_{nc}(s)$, the existing *known* controller is $G_c(s)$ and the closed-loop system is denoted $G_{CL}(s)$. The identification in the algorithms to follow *always* retains the known existing PID controller in the loop, and uses a parallel virtual computation to place specifications on and to find the required frequencies for the new compensated forward path, $G_{fp}(s) = G_p(s)G_{nc}(s)$. The phase-locked loop structure of Figure 7.13 was implemented using LabVIEW software as a prototype autotune device; a typical LabVIEW front panel is shown in Figure 7.14.

Phase Crossover Frequency Identification

In the automated PI design procedures, a sequence of new controllers is generated for which the phase crossover frequencies are required as if each new controller is installed in the forward path. The information available is the phase-locked loop identification of the closed-loop process $G_{CL}(j\omega)$ and the known controller transfer function which can be used to compute $G_c(j\omega)$. These two quantities are used to compute the unknown process phase at frequency ω from the relation

$$\arg(G_p(j\omega)) = \arg(G_{CL}(j\omega)) - \arg(1 - G_{CL}(j\omega)) - \arg(G_c(j\omega))$$

The condition for the phase crossover frequency $\omega_{-\pi}$ of the forward path transfer with a new controller installed is $\arg\{G_{fp}(j\omega_{-\pi})\} = \arg\{G_p(j\omega_{-\pi})G_{nc}(j\omega_{-\pi})\} = -\pi$ rad. Thus to perform the appropriate

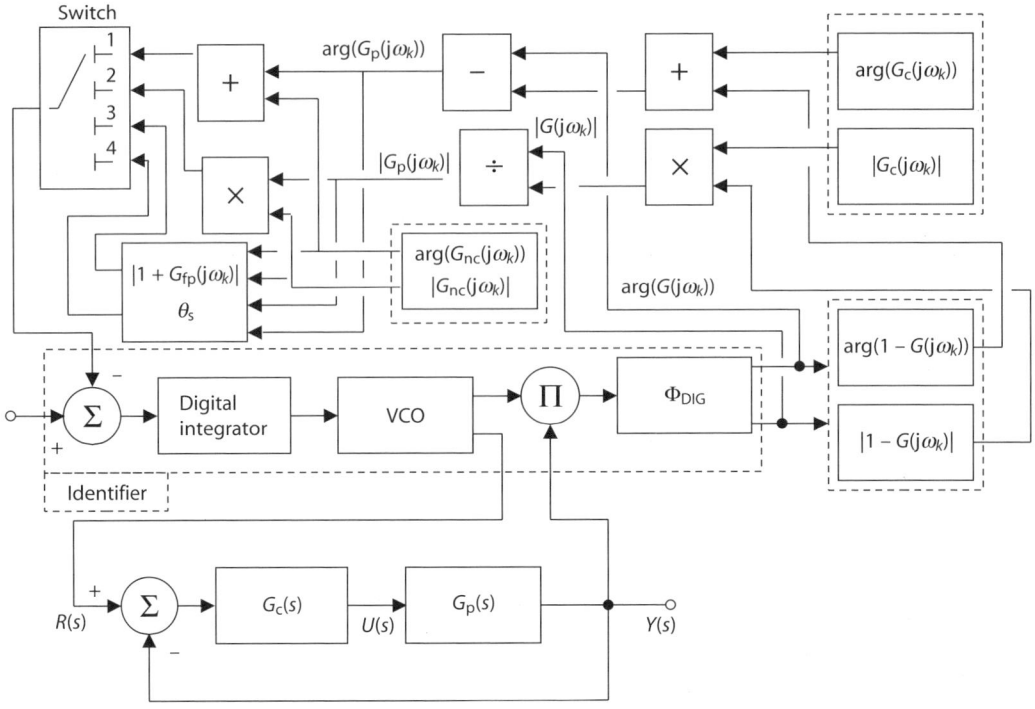

Figure 7.13 All-purpose phase-locked loop identifier module for autotune control.

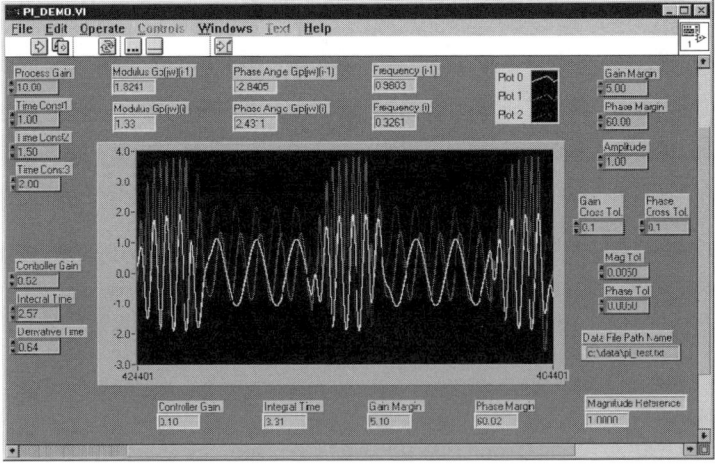

Figure 7.14 PI Autotune front panel view.

phase-locked loop identification, the switch in Figure 7.13 is set to position 1 and the phase-locked loop is driven with a phase reference input set to $\phi_r = -\pi$ rad. In the kth iteration, the computation of $\arg\{G_p(j\omega_k)\}$ is then used to form a phase reference error equation, $e_k = (-\pi) - (\arg\{G_p(j\omega_k)\} + \arg\{G_{nc}(j\omega_k)\})$. When this error is zero, the frequency $\omega_{-\pi}$ for the virtual forward path transfer $G_{fp}(s) = G_p(s)G_{nc}(s)$ is found.

7.5 Automated PI Control Design

Gain Crossover Frequency Identification.

A similar procedure to that described above for the phase crossover frequency is followed for the computation of the gain crossover frequency, denoted ω_1. As in the above case, the information available is the phase-locked loop identification of the closed-loop process $G_{CL}(j\omega)$ and the known controller transfer function which can be used to compute $G_c(j\omega)$. These two quantities are used to compute the unknown process magnitude at frequency ω from the relation

$$|G_p(j\omega)| = \frac{|G_{CL}(j\omega)|}{|1-G_{CL}(j\omega)||G_c(j\omega)|}$$

The condition for ω_1, the gain crossover frequency of the forward path transfer with a new controller installed, is $|G_{fp}(j\omega_1)| = |G_p(j\omega_1)| \times |G_{nc}(j\omega_1)| = 1$. Thus to perform the appropriate phase-locked loop identification, the switch in Figure 7.13 is set to position 2 and the phase-locked loop is driven with a gain reference input set to $G_r = 1$. In the kth iteration, the computation of $|G_p(j\omega_k)|$ is then used to form a gain reference error equation $e_k = 1 - |G_p(j\omega_k)| \times |G_{nc}(j\omega_k)|$. When this error is zero, the frequency ω_1 for the virtual forward path transfer $G_{fp}(s) = G_p(s)G_{nc}(s)$ is found.

Maximum Sensitivity Frequency Identification

In the maximum sensitivity frequency identification, the objective is to use the phase-locked loop identifier to find the maximum sensitivity value for a given forward path system. Consequently, this identification uses properties from the Nyquist geometry for the definition of maximum sensitivity, as shown in Figure 7.15. The maximum sensitivity circle is the circle of largest radius centred at $(-1, 0)$, which is tangential to the forward path frequency response $G_{fp}(j\omega)$. This circle has radius $r = 1/M_S$, and from this radius the maximum sensitivity value is computed. However, the point to be identified by the phase-locked loop is the point $G_{fp}(j\omega_S)$ where the tangential circle touches the forward path frequency response. The frequency at which this occurs is the maximum sensitivity frequency ω_S. There are two geometrical properties which are important to the identification procedure:

1. The first property is the relationship between the angles θ_S and ψ_S. From Figure 7.15, the right-angled triangle shown leads to the identity $(\pi/2) + \theta_S + \psi_S = \pi$, and this is easily rearranged to give $\psi_S = (\pi/2) - \theta_S$. It is this relation which provides a route for estimating the *a priori* unknown angle θ_S needed by the second maximum sensitivity property.

2. The second geometric property is concerned with creating a condition which will be used to supply a reference input to drive the phase-locked loop identification to the maximum sensitivity point $G_{fp}(j\omega_S)$ at which the frequency is ω_S. The property is the simple expression for $\tan\theta_S$, which is easily derived from Figure 7.15 as

$$\tan\theta_S = \frac{|\text{Im}(G_{fp}(j\omega_S))|}{1-|\text{Re}(G_{fp}(j\omega_S))|}$$

The way in which these two properties are used is described next. The identification of maximum sensitivity is determined from a sequence of pairs of points on the forward path frequency response. It is assumed that the forward path transfer $G_{fp}(j\omega)$ does not satisfy the desired maximum sensitivity constraint and intersects the $1/M_S$ circle at two points with frequencies ω_A and ω_B. These frequencies form the important bound $\omega_A \leq \omega_S \leq \omega_B$. This is shown in Figure 7.16, where it can be seen that these intersections occur at the condition

$$1/M_S = |1 + G_{fp}(j\omega_A)| = |1 + G_{fp}(j\omega_B)|$$

From these two points on the frequency response $G_{fp}(j\omega)$, the angles θ_A and θ_B can be found as

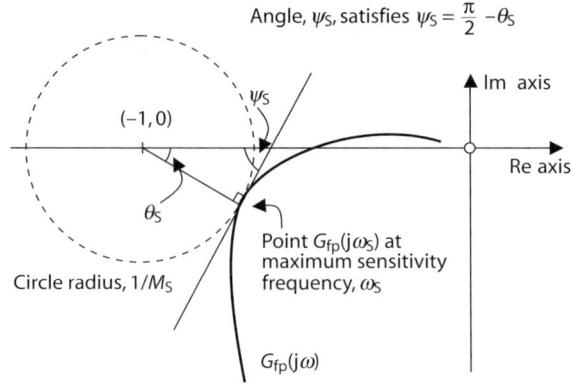

Figure 7.15 Tangency condition for maximum sensitivity frequency point $G_{fp}(j\omega_S)$.

$$\theta_A = \tan^{-1}\left\{\frac{|\text{Im}(G_{fp}(j\omega_A))|}{1-|\text{Re}(G_{fp}(j\omega_A))|}\right\} \text{ and } \theta_B = \tan^{-1}\left\{\frac{|\text{Im}(G_{fp}(j\omega_B))|}{1-|\text{Re}(G_{fp}(j\omega_B))|}\right\}$$

The values of θ_A and θ_B are then used to calculate the angles ψ_A and ψ_B as $\psi_A = (\pi/2) - \theta_A$ and $\psi_B = (\pi/2) - \theta_B$. The geometric mean of angles ψ_A and ψ_B is used to form an estimate of the angle ψ_S as $\hat{\psi}_S = \sqrt{\psi_A \psi_B}$. Hence an estimate for θ_S follows as $\hat{\theta}_S = (\pi/2) - \hat{\psi}_S$. It should be noted that the geometric mean estimator for $\hat{\psi}_S$ favours the smaller of the two angles ψ_A and ψ_B. This theory leads to the two identification steps needed to find frequencies and system data to design for a maximum sensitivity specification.

Step 1: Finding the Bounding Frequencies ω_A and ω_B

The forward path data at the two frequencies ω_A and ω_B are needed such that $\omega_A \leq \omega_S \leq \omega_B$. It has already been seen that these are the intersection points of the Nyquist plot of $G_{fp}(j\omega)$ and the $1/M_S$ circle (see Figure 7.16), and these points occur at the condition

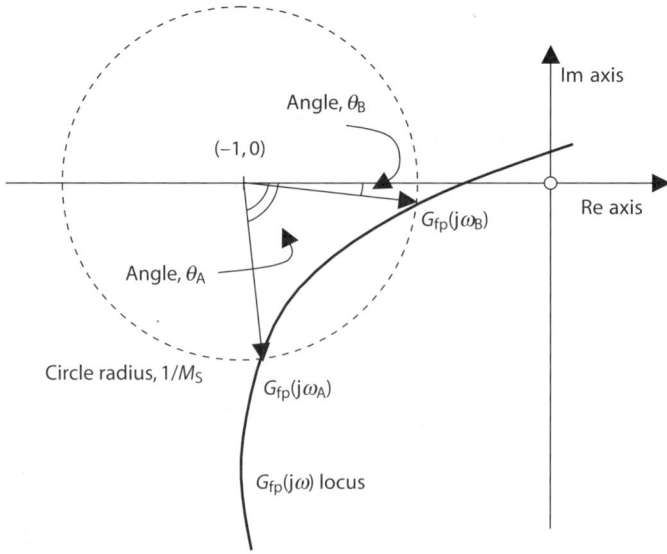

Figure 7.16 Intersection of the $G_{fp}(j\omega)$ locus and the $1/M_S$ circle.

$$1/M_S = |1 + G_{fp}(j\omega_A)| = |1 + G_{fp}(j\omega_B)|$$

and hence

$$1/M_S = |1 + G_p(j\omega_A)G_{nc}(j\omega_A)| = |1 + G_p(j\omega_B)G_{nc}(j\omega_B)|$$

Thus, with the switch in Figure 7.13 in position 3 and a reference input of $1/M_S$, the bounding frequencies are found using the phase-locked loop identifier module and a reference error given by

$$e_k = \left(\frac{1}{M_S}\right) - |1 + G_p(j\omega_k)G_{nc}(j\omega_k)|$$

From the reference error equation it is obvious that when the phase-locked loop identifier drives the reference error to zero it will not automatically seek a second frequency at which the error is zero. To ensure that both frequencies are found, when finding ω_A the phase-locked loop identifier starts from a frequency ω_A^0 such that $\omega_A^0 < \omega_A$ and when finding ω_B the identifier starts from frequency ω_B^0 such that $\omega_B^0 > \omega_B$. In the examples given in the section, to find ω_A the initial frequency was chosen at the low value of $\omega_A^0 = 0.1$ rad s^{-1} and to find ω_B the initial frequency was set at $\omega_B = \omega_{-\pi}$. Using this reference error and the suitable starting frequencies ω_A^0 and ω_B^0, the phase-locked loop identifier automatically produces values of $|G_{fp}(j\omega_A)|, \arg\{G_{fp}(j\omega_A)\}$ and $|G_{fp}(j\omega_B)|, \arg\{G_{fp}(j\omega_B)\}$.

Step 2: Finding the Maximum Sensitivity Frequency ω_S

From the two points $|G_{fp}(j\omega_A)|, \arg\{G_{fp}(j\omega_A)\}$ and $|G_{fp}(j\omega_B)|, \arg\{G_{fp}(j\omega_B)\}$ the angles θ_A and θ_B can be found as

$$\theta_A = \tan^{-1}\left\{\frac{|\text{Im}(G_{fp}(j\omega_A))|}{1 - |\text{Re}(G_{fp}(j\omega_A))|}\right\} \text{ and } \theta_B = \tan^{-1}\left\{\frac{|\text{Im}(G_{fp}(j\omega_B))|}{1 - |\text{Re}(G_{fp}(j\omega_B))|}\right\}$$

This is followed by computing $\psi_A = (\pi/2) - \theta_A, \psi_B = (\pi/2) - \theta_B$ and then $\hat{\psi}_S = \sqrt{\psi_A \psi_B}$, followed by the estimate for θ_S as $\hat{\theta}_S = (\pi/2) - \hat{\psi}_S$. Using the estimate of θ_S, the frequency ω_S can be identified using the condition

$$\theta_S = \tan^{-1}\left\{\frac{|\text{Im}(G_p(j\omega) \times G_{nc}(j\omega))|}{1 - |\text{Re}(G_p(j\omega) \times G_{nc}(j\omega))|}\right\}$$

With the switch in Figure 7.13 in position 4 and a reference input $\hat{\theta}_S$, the phase-locked loop identifier module will be driven to find ω_S by using a reference error equation

$$e_k = (\hat{\theta}_S) - \tan^{-1}\left\{\frac{|\text{Im}(G_p(j\omega_k) \times G_{nc}(j\omega_k))|}{1 - |\text{Re}(G_p(j\omega_k) \times G_{nc}(j\omega_k))|}\right\}$$

From these three identification cases it can be seen that it is the extensive flexibility of the phase-locked loop module that enables these various frequency points to be found.

7.5.2 PI Control with Automated Gain and Phase Margin Design

The objective for the design is to find a PI controller that attains pre-specified values of phase margin and gain margin (Crowe and Johnson, 2001; 2002a). The new PI controller is treated as a virtual controller, since at all times during the on-line design phase the original controller remains in the loop. Only on an instruction from the process engineer would the original controller be replaced or updated at the end of the design phase. The gain and phase margin constraints are stability robustness measures and typical design ranges are (Åström and Hägglund, 1995):

Gain margin $2 < GM < 5$

Phase margin in degrees: $30° < \phi_{PM} < 60°$

in radians: $\pi/6 < \phi_{PM} < \pi/3$

Gain Margin and Phase Margin Design Equations

The design is concerned with finding a new PI controller, denoted by $G_{nc}(s)$, in cascade with an unknown process $G_p(s)$ such that the new forward path satisfies both the gain margin and phase margin specifications. The new PI controller is given in transfer function form as

$$G_{nc}(s) = k_P + \frac{k_I}{s}$$

and in frequency domain form as

$$G_{nc}(j\omega) = k_P + \frac{k_I}{j\omega}$$

The unknown process $G_p(s)$ is given the Cartesian frequency domain form $G_p(j\omega) = G_{pR}(\omega) + jG_{pI}(\omega)$. The new compensated forward path transfer function is given by $G_{fp}(s) = G_p(s)G_{nc}(s)$. Using these identities and setting $s = j\omega$, the frequency domain equations follow as

$$G_{fp}(j\omega) = G_p(j\omega)G_{nc}(j\omega)$$
$$= (G_{pR}(\omega) + jG_{pI}(\omega))\left(k_P + \frac{k_I}{j\omega}\right)$$

and

$$G_{fp}(j\omega) = \left(G_{pR}(\omega)k_P + \frac{G_{pI}(\omega)k_I}{\omega}\right) + j\left(G_{pI}(\omega)k_P - \frac{G_{pR}(\omega)k_I}{\omega}\right)$$

The desired gain margin is denoted by GM, and thus the specified phase crossover point is given as

$$s_{-\pi} = -\frac{1}{GM} + j0$$

The specified phase crossover point occurs at the forward path transfer function frequency of $\omega_{-\pi}$; consequently, the gain margin design equation is

$$G_{fp}(j\omega_{-\pi}) = s_{-\pi}$$

Hence

$$\left(G_{pR}(\omega_{-\pi})k_P + \frac{G_{pI}(\omega_{-\pi})k_I}{\omega_{-\pi}}\right) + j\left(G_{pI}(\omega_{-\pi})k_P - \frac{G_{pR}(\omega_{-\pi})k_I}{\omega_{-\pi}}\right) = -\frac{1}{GM} + j0$$

Equating the real and imaginary parts of the above equation gives the matrix–vector gain margin design equation:

$$\begin{bmatrix} G_{pR}(\omega_{-\pi}) & \dfrac{G_{pI}(\omega_{-\pi})}{\omega_{-\pi}} \\ G_{pI}(\omega_{-\pi}) & -\dfrac{G_{pR}(\omega_{-\pi})}{\omega_{-\pi}} \end{bmatrix} \begin{bmatrix} k_P \\ k_I \end{bmatrix} = \begin{bmatrix} -\dfrac{1}{GM} \\ 0 \end{bmatrix}$$

The desired phase margin is denoted by ϕ_{PM} rad, and this occurs at the forward path transfer function frequency ω_1 where $|G_{fp}(j\omega_1)|=1$. The phase angle (radians) at this point is $\phi(\omega_1)=-\pi+\phi_{PM}$; thus the specified gain crossover point is $s_1 = e^{-j(\pi-\phi_{PM})} = -e^{j\phi_{PM}} = -\cos\phi_{PM} - j\sin\phi_{PM}$ and using the forward path equation the phase margin design equation is

$$G_{fp}(j\omega_1) = s_1$$

Hence

$$\left(G_{pR}(\omega_1)k_P + \frac{G_{pI}(\omega_1)k_I}{\omega_1}\right) + j\left(G_{pI}(\omega_1)k_P - \frac{G_{pR}(\omega_1)k_I}{\omega_1}\right) = -\cos\phi_{PM} - j\sin\phi_{PM}$$

Equating the real and imaginary parts of the above identity yields the matrix–vector phase margin design equation:

$$\begin{bmatrix} G_{pR}(\omega_1) & \dfrac{G_{pI}(\omega_1)}{\omega_1} \\ G_{pI}(\omega_1) & -\dfrac{G_{pR}(\omega_1)}{\omega_1} \end{bmatrix} \begin{bmatrix} k_P \\ k_I \end{bmatrix} = \begin{bmatrix} -\cos\phi_{PM} \\ -\sin\phi_{PM} \end{bmatrix}$$

The gain and phase margin equations are needed to specify the new PI controller. Although these equations are linear in the PI controller gains k_P and k_I, the equations are nonlinear, being dependent on the unknown process transfer function $G_p(s)$ and the unknown frequencies $\omega_{-\pi}$ and ω_1.

Automated Gain Margin and Phase Margin PI Controller Design Algorithm

The phase-locked loop module is used in the automated PI control design procedure only to perform the identification steps of the routine. The identification steps are performed in closed loop with the existing controller remaining *in situ*. A joint gain and phase margin specification uses the composite equation suite for the new PI compensator gains k_P, k_I as

$$\begin{bmatrix} G_{pR}(\omega_{-\pi}) & \dfrac{G_{pI}(\omega_{-\pi})}{\omega_{-\pi}} \\ G_{pI}(\omega_{-\pi}) & -\dfrac{G_{pR}(\omega_{-\pi})}{\omega_{-\pi}} \\ G_{pR}(\omega_1) & \dfrac{G_{pI}(\omega_1)}{\omega_1} \\ G_{pI}(\omega_1) & -\dfrac{G_{pR}(\omega_1)}{\omega_1} \end{bmatrix} \begin{bmatrix} k_P \\ k_I \end{bmatrix} = \begin{bmatrix} -\dfrac{1}{GM} \\ 0 \\ -\cos\phi_{PM} \\ -\sin\phi_{PM} \end{bmatrix}$$

This can be written in a more compact matrix–vector form as

$$[X]K = Y^D \quad \text{with } K = \begin{bmatrix} k_P \\ k_I \end{bmatrix} \text{ and } Y^D = \begin{bmatrix} -\dfrac{1}{GM} & 0 & -\cos\phi_{PM} & -\sin\phi_{PM} \end{bmatrix}^T$$

Using this matrix–vector equation and the phase-locked loop identifier module an automated PI controller design algorithm can be constructed.

Algorithm 7.2: Automated PI controller design for desired gain and phase margin values

Step 1 Design specification
  Select the desired gain margin GM and phase margin ϕ_{PM}
  Select the desired convergence tolerance *tol*

$$\text{Compute } Y^D = \begin{bmatrix} -\dfrac{1}{GM} & 0 & -\cos\phi_{PM} & -\sin\phi_{PM} \end{bmatrix}^T$$

Step 2 *Initialisation step*

Choose initial PI controller gains (can be those already used with the closed-loop system):
$K_0 = [k_P(0) \quad k_I(0)]^T$
Initialise counter $n = 0$

Step 3 *Identification step*

Step 3a: Phase crossover frequency identification

Use phase-locked loop to find $\omega_{-\pi}(n)$ for the forward path $G_{fp}(s) = G_p(s)G_{nc}(s)$
Use known $K_n = [k_P(n) \quad k_I(n)]^T$ to solve for $G_{pR}(\omega_{-\pi}), G_{pI}(\omega_{-\pi})$

Step 3b: Gain crossover frequency identification

Use phase-locked loop to find $\omega_1(n)$ for the forward path $G_{fp}(s) = G_p(s)G_{nc}(s)$
Use known $K_n = [k_P(n) \quad k_I(n)]^T$ to solve for $G_{pR}(\omega_{-\pi}), G_{pI}(\omega_{-\pi})$

Step 3c: Convergence test step

$$\text{Compute } Y_n = \begin{bmatrix} -\dfrac{1}{GM(n)} & 0(n) & -\cos\phi_{PM}(n) & -\sin\phi_{PM}(n) \end{bmatrix}^T$$

If $\|Y^D - Y_n\| < tol$, then STOP

Step 4 *Controller update calculation*

Use $\omega_{-\pi}, \omega_1, G_{pR}(\omega_{-\pi}), G_{pI}(\omega_{-\pi}), G_{pR}(\omega_1), G_{pI}(\omega_1)$
Form $[X_n]$, coefficient matrix of $[X_n]K_{n+1} = Y^D$
Solve $K_{n+1} = [X_n^T X_n]^{-1} X_n^T Y^D$
Update $n := n + 1$
Goto Step 3

Algorithm end

To establish a convergence proof for the algorithm requires a more formal presentation of the steps being followed. The forward path transfer function of the compensated system is given by $G_{fp}(s) = G_p(s)G_{nc}(s)$ where the new controller is

$$G_{nc}(s) = k_P + \frac{k_I}{s}$$

From this it can be seen that the independent variables of the problem are the controller gains k_P and k_I. Recall that in the nth iteration the controller coefficient vector is

$$K_n = \begin{bmatrix} k_P(n) \\ k_I(n) \end{bmatrix}$$

Then for a given $K_n \in \Re^2$ the phase-locked loop identifier is used to find $\omega_{-\pi}(K_n) \in \Re_+$ and $\omega_1(K_n) \in \Re_+$. This identification operation may be written

$$\begin{bmatrix} \omega_{-\pi} \\ \omega_1 \end{bmatrix} = \omega_v(K_n)$$

where $\omega_v(K): \Re^2 \to \Re_+^2$ and $\omega_v(K)$ is nonlinear. The next step is to use $\omega_v(K_n)$ to calculate an updated value of K using the least squares formula

$$K_{n+1} = [X^T(\omega_v)X(\omega_v)]^{-1} X^T(\omega_v) Y^D$$

where $X(\omega_v): \Re_+^2 \to \Re^2$. Formally this may be written

$$K_{n+1} = f(\omega_v) = f(\omega_v(K_n)) = g(K_n)$$

where $f: \Re_+^2 \to \Re^2$, $\omega_v: \Re^2 \to \Re_+^2$ and $g: \Re^2 \to \Re^2$. The composite function $g: \Re^2 \to \Re^2$ features in the following convergence theorem.

Theorem 7.1: Sufficient conditions for the convergence of the joint gain and phase margin algorithm

The independent variables of the joint gain and phase margin algorithm are defined as $K = [k_p \ \ k_I]^T$, where $K \in \Re^2$. The operation of identification and recursive solution of the least squares equations can be written as $K_{n+1} = g(K_n)$, where $g: \Re^2 \to \Re^2$ and the iteration counter is $n = 0, 1, \ldots$. Assuming that a problem solution exists which satisfies the desired gain and phase margin specification and that this is given by the fixed point solution, $K_* = g(K_*)$ then sufficient conditions for convergence follow from

$$\|K_* - K_{n+1}\| \leq \left(\prod_{j=0}^{n} \mu_j\right) \|K_* - K_0\|$$

where

$$\mu_j = \sup_{0 < \alpha_j < 1} \|g'(K_j + \alpha_j(K_* - K_j))\|$$

The sufficient conditions for convergence are

Case (a): $\exists j_1 \geq 0$ such that $\forall j \geq j_1, \mu_j$ satisfies $|\eta_j| < 1$

Case (b): $|\eta_j| < 1$ for all $j = 0, 1, \ldots$

Proof

(i) $K_* = g(K_*)$ and $K_{n+1} = g(K_n)$

(ii) $K_* - K_{n+1} = g(K_*) - g(K_n)$ and hence $\|K_* - K_{n+1}\| = \|g(K_*) - g(K_n)\|$

Applying Luenberger's mean value inequality result (Luenberger, 1969) gives

$$\|K_* - K_{n+1}\| \leq \mu_n \|K_* - K_n\|$$

where

$$\mu_n = \sup_{0 < \alpha < 1} \|g'(K_n + \alpha(K_* - K_n))\|$$

(iii) Repeated application of this inequality yields

$$\|K_* - K_{n+1}\| \leq \left(\prod_{j=0}^{n} \mu_j\right) \|K_* - K_0\|$$

where

$$\mu_j = \sup_{0 < \alpha_j < 1} \|g'(K_j + \alpha_j(K_* - K_j))\|$$

(iv) Proof of Case (a)

Write

$$\prod_{j=0}^{n} \mu_j = \left(\prod_{j=j_1+1}^{n} \mu_j \right) \left(\prod_{j=0}^{j_1} \mu_j \right)$$

$$= k_{j_1} \left(\prod_{j=j_1+1}^{n} \mu_j \right)$$

Thus if $|\mu_j| < 1$ for all $j \geq j_1$ then as $n \to \infty$, $\Pi_{j=0}^{n} \mu_j \to 0$ and convergence follows.

(v) Case (b) is a special case of case (a).

☐

This theorem is a classic contraction mapping result. Satisfaction of the result depends on a number of properties; the existence of a closed region, denoted $\Theta \subseteq \Re^2$, through which the solution iterates $\{K_j, j = 0, 1, ...\}$ travel, and conditions on the Jacobian

$$\left[\frac{\partial g}{\partial K} \right] = g'(K) : \Re^2 \to \Re^2$$

over this closed region which includes K_*. Schwarz (1989) links the contraction property to the necessary and sufficient condition that the spectral radius of the Jacobian, $g'(K)$, must satisfy $\rho(g'(K)) < 1$ over the closed region Θ. However, the particular difficulties associated with finding analytical conditions for the convergence of the autonomous PI algorithm include the function of a function implicit in the relation $g(K)$, where the Jacobian can be expressed as

$$g'(K) = \left[\frac{\partial g}{\partial K} \right] = \left[\frac{\partial f}{\partial \omega_v} \right] \times \left[\frac{\partial \omega_v}{\partial K} \right]$$

the complexity of the function $f : \Re_+^2 \to \Re^2$, and the indirect computational access to the function, $\omega_v : \Re^2 \to \Re_+^2$

Case Study System Results

Several processes were chosen to demonstrate the ability of the tuning algorithm to achieve a joint gain and phase margin specification. These processes are representative of those found in the process industries. Moreover, as will be seen in the case studies, the sufficient conditions emerging from Theorem 7.1 give insight into possible convergence mechanisms.

Case Study 1

The system $G_1(s)$ is a high-order non-oscillatory process represented by the transfer function

$$G_1(s) = \frac{1}{(s+1)^6}$$

The required design specification for the compensated forward path was chosen as a gain margin of 3 and a phase margin of 60°. The results of the algorithm in terms of the controller parameters together with the gain margin, phase margin, phase crossover frequency and gain crossover frequency for the initial and final values are shown in Table 7.4.

The initial PI controller parameters were derived using a relay experiment followed by the application of Ziegler–Nichols tuning rules for a PI controller. These initial controller coefficients were considered to be significantly removed from those required to achieve the desired gain and phase margin values. The choice of gain margin and phase margin values was made so that the resulting compensated system would have good stability robustness and good robustness to process parameter variations.

7.5 Automated PI Control Design

Table 7.4 Initial and final values for the process $G_1(s)$.

Initial values for $G_1(s)G_c(s)$ Frequency units: rad s^{-1}

k_p	k_i	Gain margin	Phase margin (degrees)	Phase crossover frequency	Gain crossover frequency
0.9372	0.1077	2.2	85.5	0.5098	0.2256

Final values for $G_1(s)G_{nc}(s)$

0.4265	0.1545	3.0	60.0	0.4308	0.1561

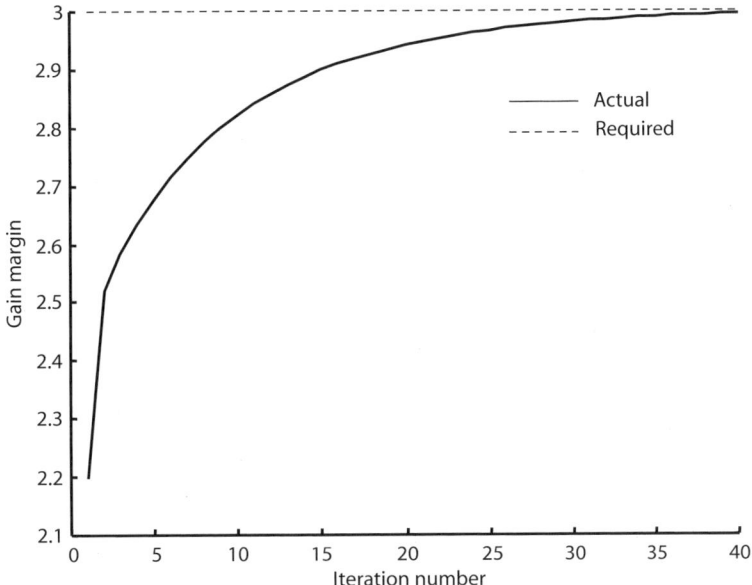

Figure 7.17 Evolution of gain margin for the process $G_1(s)$.

Figure 7.17 shows how the gain margin of the newly compensated forward path of the process $G_1(s)G_{nc}(s)$ evolves as the iterations of the algorithm are completed. After approximately 12 iterations of the algorithm, the gain margin is within 5% of the desired value and is within 2% of the desired value after approximately 23 iterations.

Figure 7.18 shows how the phase margin evolves as the PI controller gain margin and phase margin tuning algorithm progresses. From the figure it can be seen that the phase margin is practically attained after five iterations of the algorithm. This type of behaviour – fast attainment of the desired phase margin – was a feature of all of the simulations performed on representative process models from the process industries. Correspondingly, the attainment of the desired gain margin was always observed to converge much more slowly.

In Theorem 7.1, it was stated that sufficient conditions for convergence are:

(i) $\exists j_1 \geq 0$ such that $\forall j \geq j_1$ μ_j satisfies $|\mu_j| < 1$.

(ii) $|\mu_j| < 1$ for all $j = 0, 1, \ldots, \infty$.

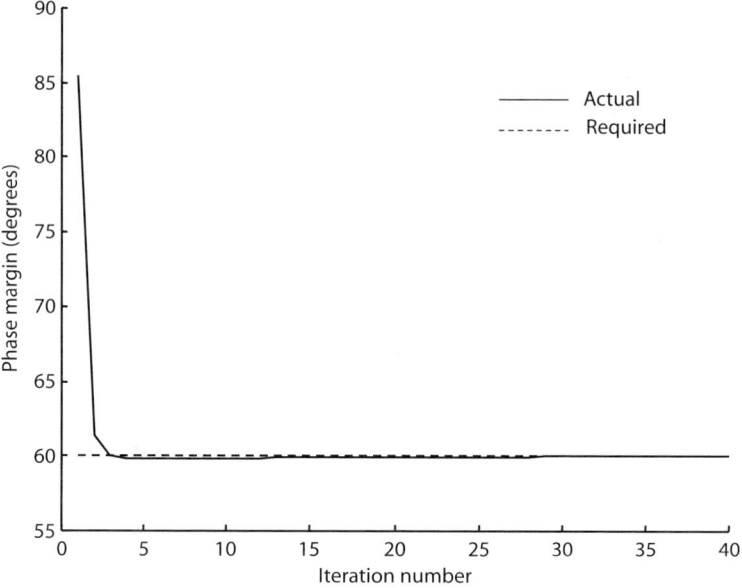

Figure 7.18 Evolution of phase margin for the process $G_1(s)$.

The results of the simulation can be used to construct an estimate of μ_n for each iteration of the algorithm; the estimate for μ_n is given by

$$\hat{\mu}_n = \frac{\|K_* - K_{n+1}\|}{\|K_* - K_n\|} \leq \mu_n$$

where n is the iteration count and the value of K_* is taken to be the converged value of the sequence $K_n, n = 0, 1, \ldots$. Figure 7.19 shows the progress of μ_n as the algorithm evolved. From the figure it can be seen that the estimates of μ_n satisfy both sufficient condition cases (i) and (ii) of Theorem 7.1 for each iteration of the algorithm. To ensure that the behaviour shown in Figure 7.19 was a truly representative estimate of the sequence $\mu_n, n = 0, 1, \ldots$ many more iterations were performed than with the computation of the final controller parameters and the gain and phase margin graphs.

Since the results are available, it is useful to compare the performance of the two PI control designs. Figure 7.20 shows the closed-loop step response and disturbance rejection properties of $G_1(s)$ in closed loop with the initial Ziegler–Nichols controller

$$G_c(s) = 0.9372 + \frac{0.1077}{s}$$

and the final controller

$$G_{nc}(s) = 0.4265 + \frac{0.1545}{s}$$

A unit step is applied at time $t = 0$ s and a disturbance of magnitude 0.15 is applied at time $t = 100$ s. It can be seen that the initial controller gives a sluggish response with no overshoot whereas the final controller gives an underdamped response with an overshoot to the initial step of approximately 8%. The new controller response has a shorter settling time and removes the disturbance perturbation more quickly than the initial poorly tuned controller.

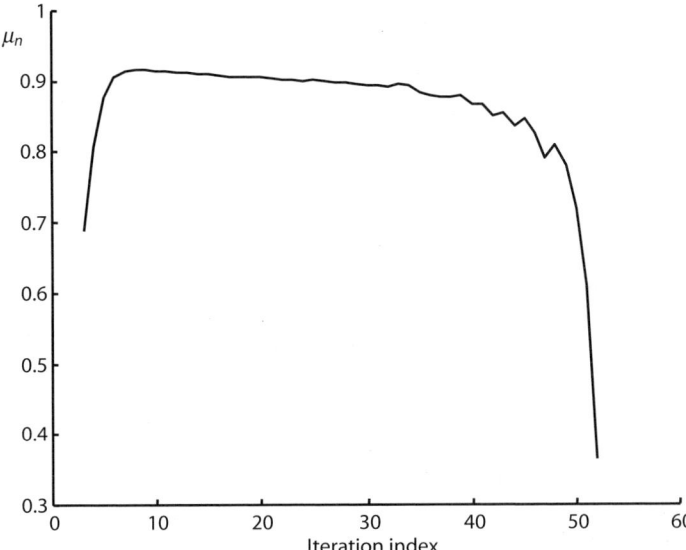

Figure 7.19 Evolution of μ_n for the process $G_1(s)$.

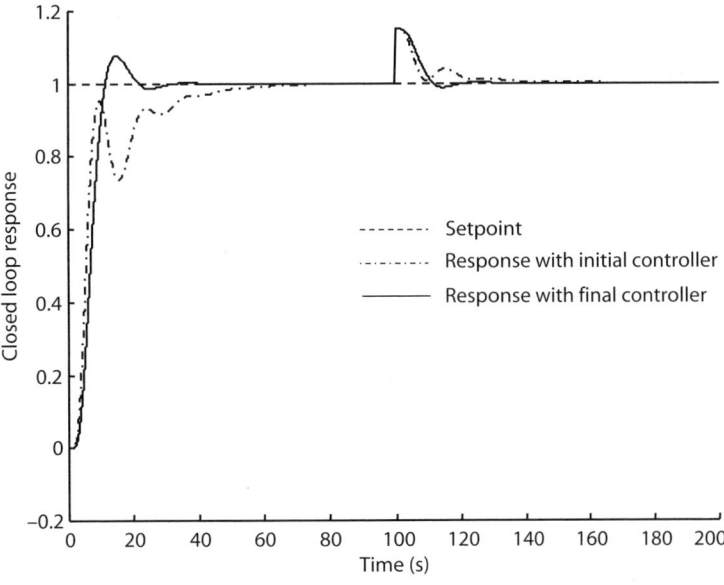

Figure 7.20 Closed-loop response with initial and final PI controllers for the process $G_1(s)$.

Case Study 2
The system $G_2(s)$ is a third-order process with a time delay and is represented by the transfer function

$$G_2(s) = \frac{e^{-s}}{(s+1)(s+3)^2}$$

The initial controller settings were based on Ziegler–Nichols tuning parameters for a PI controller with the data for the tuning rules supplied from a relay experiment. The initial and final PI controller tuning

Table 7.5 Initial and final values for the process $G_2(s)$.

Initial values for $G_2(s)G_c(s)$ Frequency units: rad s^{-1}

k_p	k_i	Gain margin	Phase margin (degrees)	Phase crossover frequency	Gain crossover frequency
7.064	1.938	2.33	92.0	1.2443	0.3046

Achieved values for $G_2(s)G_{nc}(s)$

| 3.5687 | 3.0777 | 3.0 | 60.0 | 1.0 | 0.3436 |

parameters are given in Table 7.5 along with the values of gain and phase margin and phase and gain crossover frequencies resulting from the use of the Algorithm 7.2. The design specification was a desired gain margin of 3 and a desired phase margin of 60°.

The progress of the algorithm towards the desired gain margin is shown in Figure 7.21 and towards the phase margin in Figure 7.22. From Figure 7.21 it can be seen that the gain margin is within 5% of the desired value within approximately 10 iterations of the algorithm and within 2% after 25 iterations. This compares with the results in Figure 7.22 which show that the phase margin is practically achieved after approximately five iterations.

As with Case Study 1, the estimate of the convergence parameter μ_n was constructed and the sequence of estimates obtained is shown in Figure 7.23. It is interesting to observe from the figure that the estimate of μ_n evolves to case (i) of Theorem 7.1. The figure shows that for the first five iterations $\hat{\mu}_n > 1$ and then after iteration five it can be seen that $\hat{\mu}_n < 1$. Extra iterations of the algorithm are shown in Figure 7.23 to demonstrate that the trend in the behaviour of $\hat{\mu}_n$ continued.

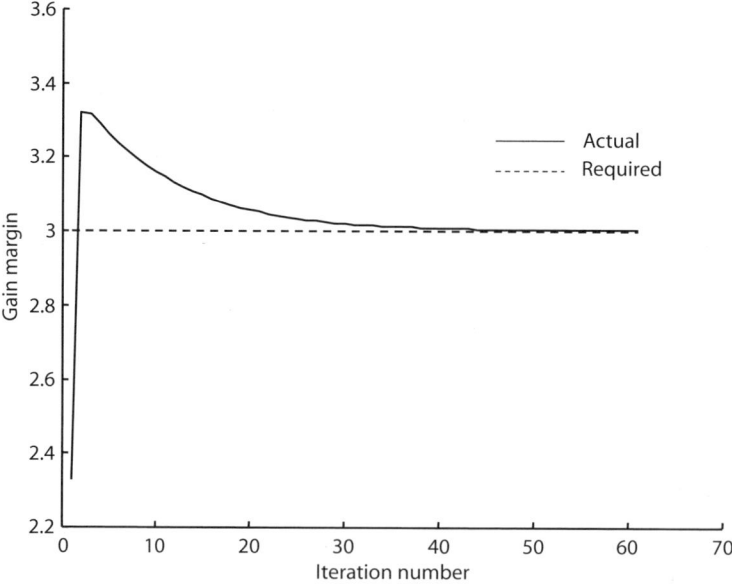

Figure 7.21 Evolution of the gain margin for the process $G_2(s)$.

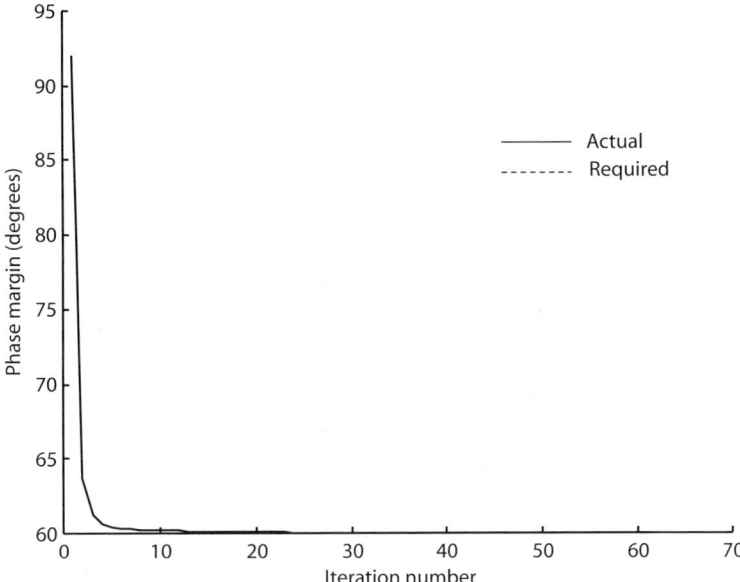

Figure 7.22 Evolution of the phase margin for the process $G_2(s)$.

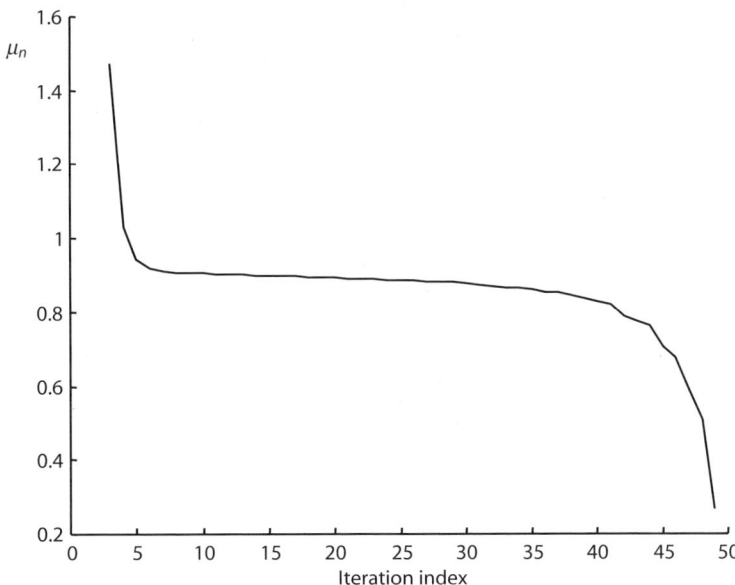

Figure 7.23 Evolution of μ_n for process $G_2(s)$.

By referring to Figure 7.24, it can be seen that the step response of the closed-loop system with the final controller is an improvement on that obtained with the initial Ziegler–Nichols controller, as is the disturbance rejection property.

However, in both of these simulation studies it is difficult to compare the time domain responses of the initial and final closed-loop controllers since the design specifications are for a robust controller

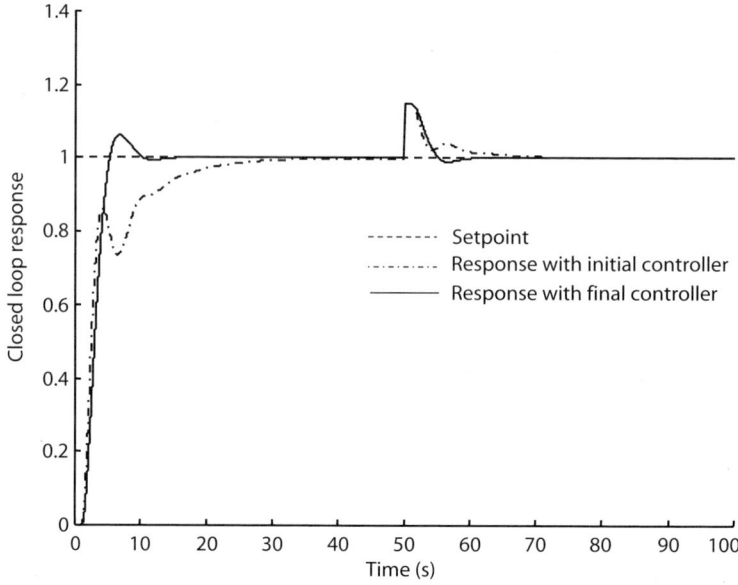

Figure 7.24 Closed-loop response with initial and final PI controllers for process $G_2(s)$.

rather than enhanced closed-loop time domain step or disturbance rejection properties. As can be seen from Tables 7.4 and 7.5, in both cases there is an improvement in the respective gain and phase margin of the new compensated system compared with the values obtained with the initial Ziegler–Nichols PI controller.

7.5.3 PI Control with Automated Maximum Sensitivity and Phase Margin Design

In the process control industries it is common to find that the process setpoint is changed relatively infrequently. Another common and difficult feature of process industry control is the presence of unknown process disturbances. For these reasons, the knowledgeable process control engineer will seek to tune a loop controller to give good disturbance rejection properties rather than good reference tracking performance. The specification of a desired maximum sensitivity value is often linked to this need to improve the disturbance rejection performance of a loop controller. The phase margin specification remains a stability robustness requirement which complements rather than replicates the performance objectives of a maximum sensitivity specification. Thus the second of the automated PI control design methods combines a desired maximum sensitivity condition with a desired phase margin value (Crowe and Johnson, 2002b). The maximum sensitivity and phase margin constraints have typical design ranges (Åström and Hägglund, 1995):

Maximum sensitivity $1.3 < M_S < 2$

Phase margin   In degrees: $30° < \phi_{PM} < 60°$

In radians: $\pi/6 < \phi_{PM} < \pi/3$

Maximum Sensitivity and Phase Margin Design Equations

This design procedure is concerned with finding a new PI controller, denoted by $G_{nc}(s)$, in cascade with an unknown process, denoted $G_p(s)$, such that the new forward path jointly satisfies desired maximum sensitivity and phase margin specifications. The new PI controller is given in transfer function form as

7.5 Automated PI Control Design

$$G_{nc}(s) = k_P + \frac{k_I}{s}$$

and in frequency domain form as

$$G_{nc}(j\omega) = k_P + \frac{k_I}{j\omega}$$

The unknown process $G_p(s)$ is given the Cartesian frequency domain form $G_p(j\omega) = G_{pR}(\omega) + jG_{pI}(\omega)$ and the new compensated forward path transfer function is given by $G_{fp}(s) = G_p(s)G_{nc}(s)$.

Using the above identities and setting $s = j\omega$, the frequency domain equations follow as

$$G_{fp}(j\omega) = G_p(j\omega)G_{nc}(j\omega) = (G_{pR}(\omega) + jG_{pI}(\omega))\left(k_P + \frac{k_I}{j\omega}\right)$$

and

$$G_{fp}(j\omega) = \left(G_{pR}(\omega)k_P + \frac{G_{pI}(\omega)k_I}{\omega}\right) + j\left(G_{pI}(\omega)k_P - \frac{G_{pR}(\omega)k_I}{\omega}\right)$$

The desired maximum sensitivity value is denoted by M_S; thus the geometry of the Nyquist plane as shown in Figure 7.15 shows that the desired maximum sensitivity point is given as

$$s_{M_s} = -\left(1 - \frac{1}{M_S}\cos\theta_S\right) - j\frac{1}{M_S}\sin\theta_S$$

The specified maximum sensitivity point occurs at the forward path transfer function frequency of ω_S; consequently the maximum sensitivity design equation is $G_{fp}(j\omega_S) = s_{M_s}$ and hence

$$\left(G_{pR}(\omega_S)k_P + \frac{G_{pI}(\omega_S)k_I}{\omega_S}\right) + j\left(G_{pI}(\omega_S)k_P - \frac{G_{pR}(\omega_S)k_I}{\omega_S}\right) = -\left(1 - \frac{1}{M_S}\cos\theta_S\right) - j\frac{1}{M_S}\sin\theta_S$$

Equating the real and imaginary parts of the above equation gives the maximum sensitivity design matrix–vector equation:

$$\begin{bmatrix} G_{pR}(\omega_S) & \dfrac{G_{pI}(\omega_S)}{\omega_S} \\ G_{pI}(\omega_S) & -\dfrac{G_{pR}(\omega_S)}{\omega_S} \end{bmatrix} \begin{bmatrix} k_P \\ k_I \end{bmatrix} = \begin{bmatrix} -\left(1 - \dfrac{1}{M_S}\cos\theta_S\right) \\ -\dfrac{1}{M_S}\sin\theta_S \end{bmatrix}$$

where ω_S is the frequency at which the *new* compensated forward path is tangent to the $1/M_S$ circle and θ_S is the angle between the negative real axis, the $(-1,0)$ point and the point where the new compensated forward path locus is tangent to the $1/M_S$ circle.

The derivation of the phase margin design equations was given fully in Section 7.5.2; however, salient points are given here for completeness. The desired phase margin is denoted by ϕ_{PM} rad and this occurs at the forward path transfer function frequency ω_1 where $|G_{fp}(j\omega_1)| = 1$. The phase angle in radians at this point is $\phi(\omega_1) = -\pi + \phi_{PM}$; thus the specified gain crossover point is given as $s_1 = e^{-j(\pi - \phi_{PM})} = -e^{j\phi_{PM}} = -\cos\phi_{PM} - j\sin\phi_{PM}$. Using the forward path equation the phase margin design equation is $G_{fp}(j\omega_1) = s_1$ and this relation yields the matrix–vector phase margin design equation

$$\begin{bmatrix} G_{pR}(\omega_1) & \dfrac{G_{pI}(\omega_1)}{\omega_1} \\ G_{pI}(\omega_1) & -\dfrac{G_{pR}(\omega_1)}{\omega_1} \end{bmatrix} \begin{bmatrix} k_P \\ k_I \end{bmatrix} = \begin{bmatrix} -\cos\phi_{PM} \\ -\sin\phi_{PM} \end{bmatrix}$$

Automated Maximum Sensitivity and Phase Margin PI Controller Design Algorithm

The maximum sensitivity and phase margin equation pairs are the key equations needed in the specification of the new PI controller. Although these equations are linear in the PI controller gains k_P and k_I, the equations are nonlinear being dependent on the unknown process transfer function $G_p(s)$ and the frequencies ω_S and ω_1. For a joint maximum sensitivity and phase margin specification, the respective equation pairs are combined to form the matrix–vector design suite:

$$\begin{bmatrix} G_{pR}(\omega_S) & \dfrac{G_{pI}(\omega_S)}{\omega_S} \\ G_{pI}(\omega_S) & -\dfrac{G_{pR}(\omega_S)}{\omega_S} \\ G_{pR}(\omega_1) & \dfrac{G_{pI}(\omega_1)}{\omega_1} \\ G_{pI}(\omega_1) & -\dfrac{G_{pR}(\omega_1)}{\omega_1} \end{bmatrix} \begin{bmatrix} k_P \\ k_I \end{bmatrix} = \begin{bmatrix} -\left(1 - \dfrac{1}{M_S}\cos\theta_S\right) \\ -\dfrac{1}{M_S}\sin\theta_S \\ -\cos\phi_{PM} \\ -\sin\phi_{PM} \end{bmatrix}$$

This set of equations can be written in compact matrix–vector form as

$$[X]K = Y^D(\theta_S) \text{ with } K = \begin{bmatrix} k_P \\ k_I \end{bmatrix}$$

and

$$Y^D(\theta_S) = \begin{bmatrix} -\left(1 - \dfrac{1}{M_S}\cos\theta_S\right) & -\dfrac{1}{M_S}\sin\theta_S & -\cos\phi_{PM} & -\sin\phi_{PM} \end{bmatrix}^T$$

Unlike the joint gain and phase margin algorithm the maximum sensitivity and phase margin procedure is far more complicated. From experience with running and testing a prototype algorithm, steps have to be taken to ensure that a viable sequence of iterates is obtained. In general there is a finite range of values for the angle θ_S and first experimental trials with the automated algorithm led to a realisation that it was essential to generate a sequence of points that bounded at least a subset of the possible values for θ_S. To ensure that this occurred, it was necessary to limit the size of updates that could be taken by the algorithm. Consequently, the algorithm was reconstructed so that the locus of $G_{fp}(j\omega)$ intersected the $1/M_S$ circle at two points, as in Figure 7.16, and the term $1/M_S$ in the design equation suite was replaced with a reduced value, denoted d given by

$$d = \kappa_T \sqrt{\dfrac{1}{M_S}}$$

where

$$\kappa_T = \sqrt{(1 - |G_{fp}(j\omega_S)|\cos(\pi + \phi))^2 + (|G_{fp}(j\omega_S)|\sin(\pi + \phi))^2} \text{ and } \phi = \arg(G_p(j\omega_k))$$

when

$$\hat{\theta}_S = \tan^{-1}\left(\dfrac{|\text{Im}(G_{fp}(j\omega_k))|}{1 - |\text{Re}(G_{fp}(j\omega_k))|}\right)$$

and ω_S is the frequency identified for the location of $\hat{\theta}_S$. The geometric mean for d is taken although the geometric mean could be replaced by a weighting factor on $1/M_S$. Using estimates for θ_S and d the design equation set was reformulated as:

$$\begin{bmatrix} G_{pR}(\omega_S) & \dfrac{G_{pI}(\omega_S)}{\omega_S} \\ G_{pI}(\omega_S) & -\dfrac{G_{pR}(\omega_S)}{\omega_S} \\ G_{pR}(\omega_1) & \dfrac{G_{pI}(\omega_1)}{\omega_1} \\ G_{pI}(\omega_1) & -\dfrac{G_{pR}(\omega_1)}{\omega_1} \end{bmatrix} \begin{bmatrix} k_P \\ k_I \end{bmatrix} = \begin{bmatrix} -(1-d\cos\hat{\theta}_S) \\ -d\sin\hat{\theta}_S \\ -\cos\phi_{PM} \\ -\sin\phi_{PM} \end{bmatrix}$$

With obvious identification this set of equations can be written in the matrix–vector form as

$$[X]K = Y(n) \text{ with } K = \begin{bmatrix} k_P \\ k_I \end{bmatrix}$$

where n is the iteration counter and

$$Y(n) = [-(1-d\cos\hat{\theta}_S) \quad -d\sin\hat{\theta}_S \quad -\cos\phi_{PM} \quad -\sin\phi_{PM}]^T$$

Using this constructional device, the full algorithm description can now be given.

Algorithm 7.3: Automated PI controller design for desired maximum sensitivity and phase margin specifications

Step 1 *Design specification*
Select the required maximum sensitivity M_S and phase margin θ_{PM} for the controller design.

Step 2 *Initialisation step*
Choose the initial PI controller gains $K_0 = [k_P(0) \quad k_I(0)]^T$
Set the convergence tolerance *tol*
Initialise counter $n = 0$

Step 3 *Identification step*

  Step 3a: *Gain crossover frequency identification*
  Use the phase-locked loop to find $\omega_1(n)$ for the forward path $G_{fp}(s) = G_p(s)G_{nc}(s)$
  Use $K_n = [k_P(n) \quad k_I(n)]^T$ to solve for $G_{pR}(\omega_1)$ and $G_{pI}(\omega_1)$

  Step 3b: *Maximum sensitivity frequency identification*
  Use the phase-locked loop to identify $\omega_A(n)$ and $\omega_B(n)$ for the forward path $G_{fp}(s) = G_p(s)G_{nc}(s)$
  Use $K_n = [k_P(n) \quad k_I(n)]^T$ to solve for $G_{pR}(\omega_A), G_{pI}(\omega_A), G_{pR}(\omega_B)$ and $G_{pI}(\omega_B)$
  Calculate $\theta_A(\omega_A), \theta_B(\omega_B)$ and $\hat{\theta}_S(\theta_A, \theta_B)$
  Use phase-locked loop with $\hat{\theta}_S(\theta_A, \theta_B)$ to identify $\omega_S(n)$
  Use $\omega_S(n)$ to calculate $\phi = \arg(G_p(j\omega_k))$
  Compute

  $$\kappa_T = \sqrt{(1-|G_{fp}(j\omega_S)|\cos(\pi+\phi))^2 + (|G_{fp}(j\omega_S)|\sin(\pi+\phi))^2}$$

  Compute $d = \kappa_T \sqrt{\dfrac{1}{M_S}}$

  Step 3c: *Convergence test step*
  Compute $Y(n), Y^D$ with

  $$Y(n) = [-(1-d\cos\hat{\theta}_S(n)), -d\sin\hat{\theta}_S(n), -\cos\phi_{PM}(n), -\sin\phi_{PM}(n)]^T$$

$$Y^D(\hat{\theta}_S) = [-(1-(1/M_S)\cos\hat{\theta}_S(n)), -(1/M_S)\sin\hat{\theta}_S(n), -\cos\phi_{PM}, -\sin\phi_{PM}]^T$$

If $\|Y^D(\hat{\theta}_S) - Y(n)\| < tol$ then STOP

Step 4 Controller update calculation
Use $\omega_S, \omega_1, G_{pR}(\omega_S), G_{pI}(\omega_S), G_{pR}(\omega_1)$ and $G_{pI}(\omega_1)$
Form $[X_n]$, coefficient matrix of $[X_n]K_{n+1} = Y^D$
Solve $K_{n+1} = [X_n^T X_n]^{-1} X_n^T Y^D$
Update $n := n+1$
Goto Step 3

Algorithm end

In this case, a prototype autotuner was implemented using MATLAB. Experience showed that the controller initialisation step could sometimes use the existing closed-loop controller parameters. However, the main complexity of the algorithm resides in the inner iterative routines being run to find the nonparametric data for the process. These routines are subject to inner convergence tolerances. If these tolerances are too stringent, then the phase-locked loop module is slow to converge. Experience has shown that low values of convergence accuracy give satisfactory solutions. A theoretical result similar to Theorem 7.1 can be given, but it is omitted due to space constraints.

Case Study System Results

Some case study examples of the performance of the algorithm are given next.

Case Study 3
The process to be used in this study is given by

$$G_1(s) = \frac{1}{(s+1)^6}$$

The desired phase margin and maximum sensitivity design values are 60° and 1.7 respectively. The initial controller parameters were derived from the Ziegler–Nichols rules for tuning a PI controller and the data required by the method was obtained from the results of a relay experiment. Table 7.6 details the values of the controller tuning parameters, the maximum sensitivity and phase margin and tangency angle for the initial and final values of the design.

Figure 7.25 shows how the maximum sensitivity of the new compensated forward path evolves during the progress of the algorithm. It can be seen that the design maximum sensitivity is within 5% of the desired value after 28 iterations and within 2% after 46 iterations of the algorithm. However, as can be seen from Figure 7.25, it takes a large number of further iterations to achieve the *exact* value required.

Table 7.6 Initial and converged controller results for the process $G_1(s)$.

k_P	k_i	Maximum sensitivity	Phase margin (degrees)	Tangency angle θ_S (degrees)
Initial values for $G_1(s)G_c(s)$				
0.687	0.2268	2.29	45.7	–
Achieved values for $G_1(s)G_{nc}(s)$				
0.4671	0.1591	1.70	60.0	18.0

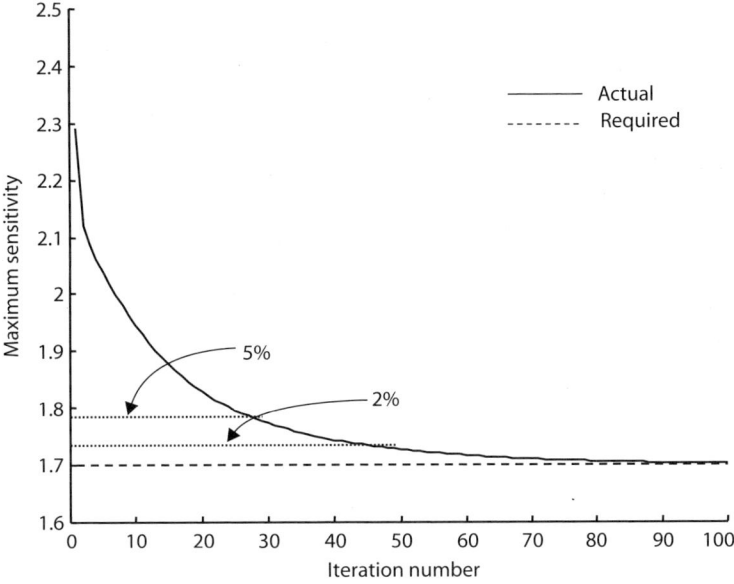

Figure 7.25 Evolution of the maximum sensitivity for the process $G_1(s)$.

Figure 7.26 shows how the phase margin of the compensated forward path evolves during the progress of the algorithm. It can be seen that the required phase margin is *exactly* attained after approximately 70 iterations of the algorithm. This is a recurring feature of the results obtained using this algorithm, in that the required phase margin is *achieved* more rapidly than is the maximum sensitivity.

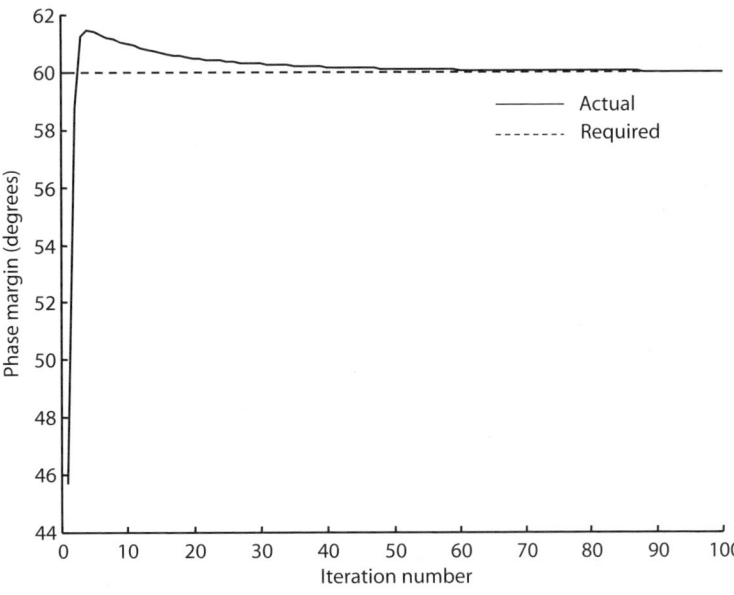

Figure 7.26 Evolution of the phase margin for the process $G_1(s)$.

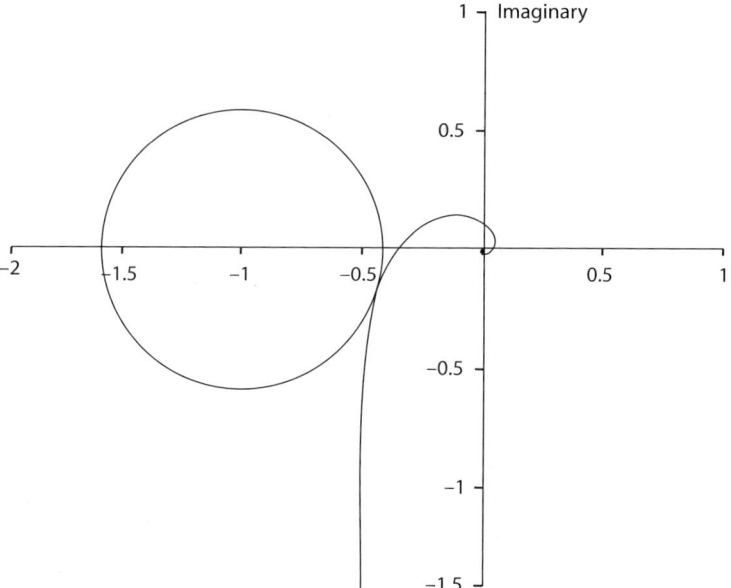

Figure 7.27 Nyquist diagram showing the compensated forward path of the process $G_1(s)$.

By reference to Figure 7.27 it can be seen that the desired design phase margin and maximum sensitivity have been achieved. The frequency range shown for the forward path of the system is 0.1 to 10 rad s^{-1}.

Case Study 4
The process used for this case study is given by

$$G_2(s) = \frac{e^{-s}}{(s+1)(s+3)^2}$$

The desired maximum sensitivity specification and phase margin for this design are 1.7 and 60° respectively. Table 7.7 details the initial and final values of the controller parameters and the phase margins and maximum sensitivities for the design.

The evolution of the phase margin and maximum sensitivity are shown in Figures 7.28 and 7.29 respectively. From Figure 7.28 it can be seen that the phase margin design specification is *exactly*

Table 7.7 Initial and converged controller results for the process $G_2(s)$.

k_p	k_i	Maximum sensitivity	Phase margin (degrees)	Tangency angle θ_s (degrees)
Initial values for $G_2(s)G_c(s)$				
7.064	1.938	1.82	92.0	–
Achieved values for $G_2(s)G_{nc}(s)$				
4.1478	3.2371	1.70	60.0	16.9

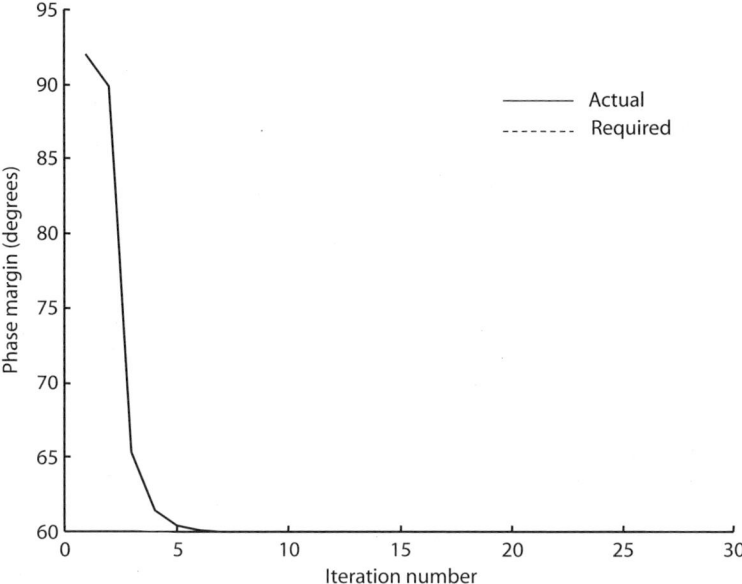

Figure 7.28 Evolution of the phase margin for the process $G_2(s)$.

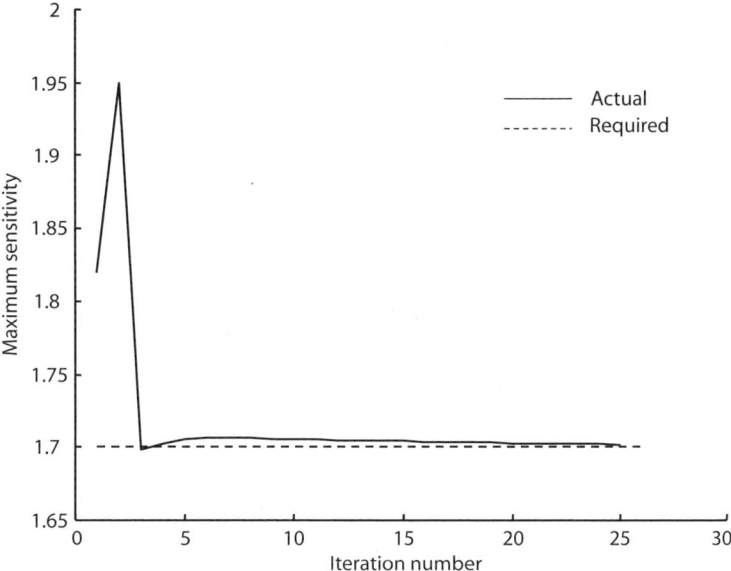

Figure 7.29 Evolution of the maximum sensitivity for the process $G_2(s)$.

attained after a very few iterations of the algorithm. The evolution of the maximum sensitivity is shown in Figure 7.29. It can be seen that after an initial large change in the sensitivity that there is a long slow movement towards the *exact* required value.

7.6 Conclusions

A considerable body of experience was reported in the chapter on using the phase-locked loop identifier module in identification and PID tuning tasks. An initial step was to package the identifier in an all-purpose shell that could have particular identification targets switched on and off at will. This enabled the module to be used in a number of more ambitious tasks than just identifying simple single points from the process response. This ability to use the identifier module as part of a larger scheme to realise a more complex set of conditions is seen as a particular strength of the phase-locked loop approach. A full assessment of the possible uses awaits input from other control engineers and practitioners.

In this chapter, the simple identification tasks of estimating the phase margin of an unknown process was added to that of finding the phase crossover data already demonstrated in Chapter 6. Two different possible ways of estimating the parameters of a second-order underdamped system were proposed. The route using a search technique to find the process peak resonance point was demonstrated successfully. This was a novel use of the identifier unit involving a golden section search routine. It was also an illustration of the potential use of the identifier unit as a block in a larger algorithmic scheme.

The identification of unknown systems in closed loop is an important step towards automated PID control design. A significant advantage of the relay experiment is the property of finding system data whilst the process loop remains in closed loop. In the case of the phase-locked loop module two different situations were investigated:

1. Identification of an unknown system in closed loop with an unknown controller.

2. Identification of an unknown system in closed loop with a known controller.

In both cases, the phase-locked loop module could be used to give satisfactory results. This opened the way to devising algorithms for automated PI controller design.

The final part of the chapter reported on the theme of using the phase-locked loop identifier in automated PID control design routines. A principle exploited in these procedures was the use of a virtual compensated forward path of the process for identifying the PI controller tuning needed to attain the desired controller performance specification. This enabled the process loop to remain closed and under the control of the existing process controller. As with the relay experiment, acceptance and installation of the new PI controller tuning remained an operator decision. Given that more understanding of classical robustness specifications was likely amongst industrial control engineers, the automated tuning used several of these robustness specifications as the target for the controller. In particular two automatic PI controller design algorithms were presented. These were set to obtain PI controller tunings to achieve:

1. A joint gain margin and phase margin specification.

2. A joint maximum sensitivity and phase margin specification.

In the case of the joint gain margin and phase margin specification, the algorithm performed tolerably well, with the phase margin specification being achieved within an iteration count in single figures. The gain margin specification took longer. The joint maximum sensitivity and phase margin PI design algorithm was altogether a more demanding task. In the version given here, considerable care was required to ensure that the algorithm progressed through a sequence of convergent iterations.

The potential of the phase-locked loop as a controller tuning tool is quite extensive. A number of development issues remain to be resolved:

1. *Implementation aspects*: the procedures presented in Chapters 6 and 7 are just one version of the phase-locked loop method and comparative studies of this and other algorithmic proposals (for example, the procedures due to Clarke and Park (2003)) would enable an implemental knowledge base to be established. For industrial use modular all-purpose phase-locked loop identifier units would be required.

2. *Industrial best practice*: if it is assumed that efficient, fast algorithms are developed in a modular form for industrial application then there is the whole area of best practice for industrial application of the phase-locked loop methodology to investigate. A user knowledge base is required to enable industrial control engineers to understand the different ways in which the identification tools can be used.

3. *Improved autotune three-term controller algorithms*: a start has been made on developing automated tuning algorithms based on the phase-locked loop module. The basic viability of two algorithms has been established but there is more research and development to be accomplished in this area.

4. *Cascade tuning algorithms*: the work of Hang *et al.* (1994) in using the relay experiment to tune classical cascade process control loops is well known. A similar procedure for using the phase-locked loop to tune cascade systems has also been published (Crowe *et al.*, 2003), but further research is needed for this important and common class of industrial control systems.

5. *Decentralised control structure tuning algorithms*: although there are many research contributions published on the decentralised PID control structure as a solution to industrial multivariable control problems, it would be very instructive to know the actual extent of these systems in industrial use. If there were a demand then tuning procedures using phase-locked loop methods would be required; some initial research in this area is due to Crowe (2003).

References

Åström, K.J. and Hägglund, T. (1985) *US Patent 4,549,123: Method and an apparatus in tuning a PID regulator.*

Åström, K.J. and Hägglund, T. (1995) *PID Controllers: Theory, Design and Tuning.* Instrument Society of America, Research Triangle Park, NC.

Clarke, D.W. and Park, J.W. (2003) Phase-locked loops for plant tuning and monitoring. *IEE Proc. Control Theory and Applications*, **150**(1), 155–169.

Crowe, J. (2003) *PhD Thesis.* Industrial Control Centre, University of Strathclyde, Glasgow.

Crowe, J. and Johnson, M.A. (1999) A new nonparametric identification procedure for online controller tuning. *Proc. American Control Conference*, San Diego, CA, pp. 3337-3341.

Crowe, J. and Johnson, M.A. (2000a) Process identifier and its application to industrial control. *IEE Proc. Control Theory and Applications*, **147**(2), 196–204.

Crowe, J. and Johnson, M.A. (2000b) Open and closed loop process identification by a phase locked loop identifier module. *ADCHem 2000*, IFAC International Symposium on Advanced Control of Chemical Processes, Pisa, 14–16 June.

Crowe, J. and Johnson, M.A. (2001) Automated PI control tuning to meet classical performance specifications using a phase locked loop identifier. *American Control Conference*, Arlington, VA, 25–27 June.

Crowe, J. and Johnson, M.A. (2002a) Towards autonomous PI control satisfying classical robustness specifications. *IEE Proc. Control Theory and Applications*, **149**(1), 26–31.

Crowe, J. and Johnson, M.A. (2002b) Automated maximum sensitivity and phase margin specification attainment in PI control. *Asian Journal of Control*, **4**(4).

Crowe, J., Johnson, M.A. and Grimble, M.J. (2003) On the closed loop identification of systems within cascade connected control strategies. *European Control Conference*, University of Cambridge, 1–4 September.

D'Azzo, J.J. and Houpis, C.H. (1975) *Linear Control System Analysis and Design, Conventional and Modern.* McGraw-Hill International, New York.

Fletcher, R. (1987) *Practical Methods of Optimisation*, 2nd edn. John Wiley & Sons, Chichester.

Hang, C.-C., Loh, A.P. and Vasnani, V.U. (1994) Relay feedback auto-tuning of cascade controllers. *IEEE Trans. Control Systems Technology*, 2(1), 42–45.

Luenberger, D.G. (1969) *Optimisation by Vector Space Methods*. John Wiley & Sons, New York.

Schei, T.S. (1994) Automatic tuning of PID controllers based on transfer function estimation. *Automatica*, 30(12), 1983–1989.

Schwartz, H.R. (1989) *Numerical Analysis*. John Wiley & Sons, Chichester.

Wilkie, J., Johnson, M.A. and Katebi, M.R. (2002) *Control Engineering: An Introductory Course*, Palgrave, Basingstoke.

8 Process Reaction Curve and Relay Methods Identification and PID Tuning

Learning Objectives

8.1 Introduction

8.2 Developing Simple Models from the Process Reaction Curve

8.3 Developing Simple Models from a Relay Feedback Experiment

8.4 An Inverse Process Model-Based Design Procedure for PID Control

8.5 Assessment of PI/PID Control Performance

References

Learning Objectives

The original methods arising from the process reaction curve approach avoided the explicit production of a process model and moved directly from some particular response measurements to a set of rules or formulae from which to compute the coefficients of the three terms in the PID controller. In this chapter, however, the PID controller tuning is based on the use of simple parametric models. Thus, first-order plus dead time (FOPDT) models or second-order plus dead time (SOPDT) models are identified using some characteristic measurements obtained from a transient step response test or from a relay feedback experiment. The design for the PI/PID controllers is then derived using these simple models and an internal model control (IMC) based methodology. This control design method uses an inverse process model and has degrees of freedom available which allow further control performance improvement. To complete the approach, the best achievable Integral of Absolute Error (IAE) cost function value and rise time of the system can be computed and used as a benchmark against which to assess the performance of the installed controller. Finally these methods all have the potential to be incorporated into a process controller unit to enhance the task of autotuning for the PID controllers.

The learning objectives for this chapter are to:

- Identify low-order process models from simple process reaction curve data.
- Use a relay feedback experiment to adaptively tune simple process models.
- Tune a PID controller using an inverse process model-based technique.
- Understand and use the idea of controller assessment based on classical performance indices.

8.1 Introduction

In model-based controller design, simple models are used to characterise the dynamics of a given process. Based on the simple dynamic model obtained, the PID controller parameters are then computed. For example, in the parametric Ziegler–Nichols tuning method, a process model of the following form was assumed:

$$G_p(s) = \left[\frac{Ke^{-\theta s}}{\tau s + 1}\right] \approx \left[\frac{Ke^{-\theta s}}{\tau s}\right] = \left(\frac{K\theta}{\tau}\right)\left[\frac{e^{-\theta s}}{\theta s}\right]$$

where the process gain, the time constant and the delay time are denoted by K, τ and θ respectively. Identification of this simple model structure used the step response of the process, as shown in Figure 8.1. In process engineering this type of system response is termed a process reaction curve or an S-curve, and this is the origin of the name for the family of tuning methods.

The process reaction curve is recorded and various response measurements made. As can be seen on Figure 8.1, the process gain K can be measured and the intersection of the tangent line with the baseline provides an estimate of the dead time θ. The slope of the tangent at the inflection point on the S-curve

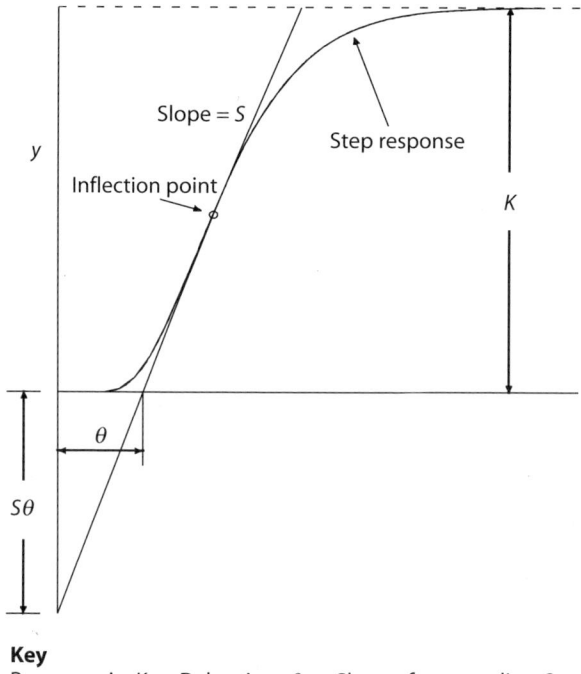

Key
Process gain, K Delay time, θ Slope of tangent line, S

Figure 8.1 Process reaction curve of a system.

can also be measured. This measured slope has the theoretical formula $S = K\theta/\tau$. Although the PID tuning is based on an inherent process model with parameters K, τ, θ, the PID controller coefficients are computed from formula given in terms of the measured values of K, S, θ. Modelling for controller design in this way is only approximate due to the need for a tangent line at the inflection point of the S-curve.

There are also quite a few methods in the literature which propose to model the S-shape step response curve with the so called first-order plus dead time (FOPDT) models or alternatively, overdamped second-order plus dead time (SOPDT) models. Parametric methods to tune the PID controllers have been presented in many different ways for the FOPDT model given by

$$G_p(s) = \left[\frac{Ke^{-\theta s}}{\tau s + 1}\right]$$

For example, tuning rules to minimise an integral performance index based on the FOPDT dynamics have been given in the forms

$$P = a\left(\frac{\theta}{\tau}\right)^b \text{ and } P = a + b\left(\frac{\theta}{\tau}\right)^c$$

where P represents a controller parameter (such as k_p, τ_i and τ_d) to be tuned for a PID controller, and a, b and c are regression constants. Furthermore, since the 1980s, Internal Model Control (IMC) theory has been used to formulate PID controllers, and as a result the need for simple models is even more important.

There are a few methods reported in the literature for identifying simple SOPDT models. In an early work of Oldenbourg and Sartorius (1948), the model being identified is

$$G_p(s) = \left[\frac{Ke^{-\theta s}}{(\tau_1 s + 1)(\tau_2 s + 1)}\right]$$

where the process gain, the two process time constants and the delay time are denoted by $K, \tau_1, \tau_2, \theta$, respectively. From the response as shown in Figure 8.2, the initial delay time θ and the lengths of segments l_1 and l_2 are measured.

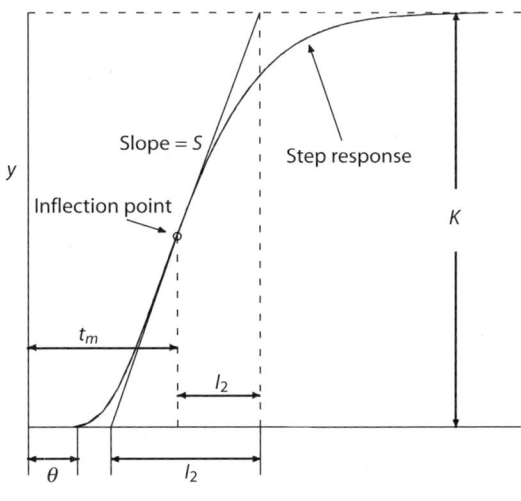

Key
Process gain, K Delay time, θ Slope of tangent line, S

Figure 8.2 Process step response of a system.

Mathematically, the value of τ_1/I_1 is related to τ_2/I_2 by the following two equations:

$$\tau_1 I_1 = \left(\frac{\tau_2}{\tau_1}\right)^{\tau_2/(\tau_1-\tau_2)} = x^{x/(1-x)} \text{ and } \frac{\tau_1}{I_1}x = \frac{\tau_2}{I_1}$$

where $x = \tau_2/\tau_1$.

On the other hand, the point $(\tau_1/I_1, \tau_2/I_1)$ also satisfies the relation

$$\frac{\tau_1}{I_1} + \frac{\tau_2}{I_2} = \frac{I_2}{I_1}$$

Consequently, based on the above equations and the measured segments, the two time constants can be found.

In an alternative method, Sundaresan et al. (1978) derived the following relation:

$$\lambda = (t_m - m)S = \chi e^{-\chi}, \text{ where } \chi = \frac{\ln \eta}{\eta - 1} \text{ and } \eta = \frac{\tau_2}{\tau_1} \leq 1.$$

The value of t_m is taken as the time at the inflection point of the response, and the quantity m is computed from the following integral:

$$m = \int_0^\infty [y(t_\infty) - y(t)]dt$$

As the input is a unit step change, $y(t_\infty)$ equals K (see Figure 8.2). A plot of η versus λ can thus be prepared. Hence by calculating the value of λ from the response, the value of η can be calculated. Then the two time constants and the dead time can be obtained from the following formulae:

$$\tau_1 = \eta^{1(1-\eta)}/S, \tau_2 = \eta \cdot \tau_1 \text{ and } \theta = m - \tau_1 - \tau_2$$

where, as previously mentioned, S is the slope of the tangent at the inflection point of the response.

Another method for identifying simple overdamped SOPDT models is found in an early textbook due to Harriott (1964). Harriott's method is based on two observations from a class of step responses of the process represented by the transfer function

$$\frac{1}{(\tau_1 s + 1)(\eta \tau_1 s + 1)}, \quad \eta \in [0,1]$$

First, it is found that almost all the step responses mentioned reach 73% of the final steady state approximately at a time of 1.3 times $\tau_1 + \tau_2$. For easy reference, this response position and time are designated as y_{73} and t_{73}, respectively, as shown in Figure 8.3(a). Second, at a time of $0.5(\tau_1 + \tau_2)$, all the step responses separate from each other most widely. Again, for easy reference, the time at $0.5(\tau_1 + \tau_2)$ is designated as t_*, and the corresponding response position is denoted as y^*. It is found that y^* is a function of $\tau_1/(\tau_1 + \tau_2)$ as shown in Figure 8.3(b). In other words, from the experimental response, the position of y_{73} is identified and the corresponding time t_{73} is recorded. Upon having a value for t_{73}, the value of t_* can be calculated. From this value of t_*, the position of y^* can be read from the step response again. When y^* is found, the value of $\tau_1/(\tau_1 + \tau_2)$ can be obtained by making use of Figure 8.3(b), and the two time constants can be computed.

The method as given above was originally formulated for systems without delay time, but it can be easily extended to include the estimation of delay time by using the relationship for the area m given as

$$m = \tau_1 + \tau_2 + \theta$$

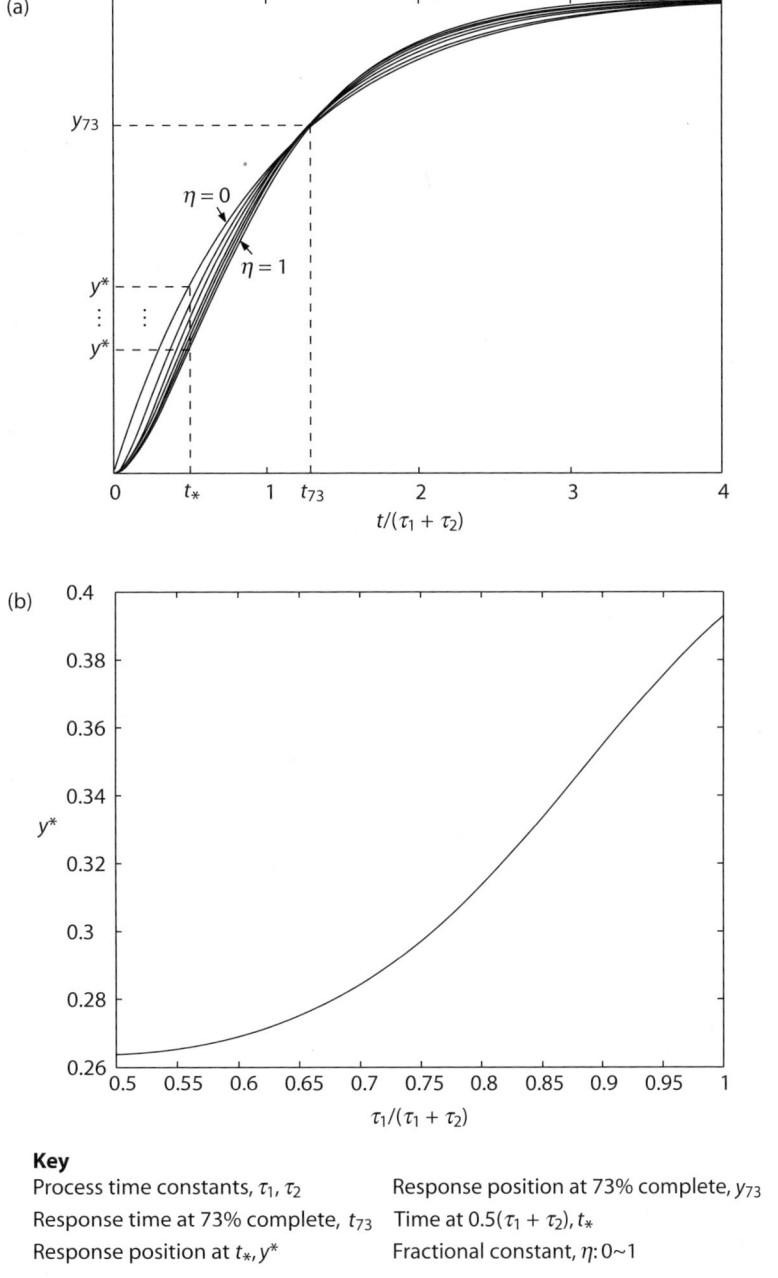

Figure 8.3 (a) Step responses of the processes; (b) relation between y^* and $\tau_1/(\tau_1 + \tau_2)$.

Thus, the model parameters can be estimated from the following equation set:

$$\theta = m - \frac{t_{73}}{1.3}$$

$$\tau_1 = \left(\frac{\tau_1}{\tau_1 + \tau_2}\right)_{\text{at } y^*} \times \left(\frac{t_{73} - t_0}{1.3}\right)$$

$$\tau_2 = \left\{1 - \left(\frac{\tau_1}{\tau_1 + \tau_2}\right)_{\text{at } y^*}\right\} \times \left(\frac{t_{73} - t_0}{1.3}\right)$$

where t_0 is the time at which the process begins to respond initially.

All of the above methods for developing SOPDT models are characterised by the use of data from the step responses instead of using whole sets of time series data of both input and output as might typically occur for classical parameter estimation procedures. In this chapter, the focus will be on developing a systematic procedure to identify these simple models from transient step response curves or from relay feedback experiments. Following this, PID controllers will be synthesised using these simple models and a proposed inverse-based method. This methodology will not resort to rule-based methods based on optimisation procedures.

8.2 Developing Simple Models from the Process Reaction Curve

In this section, the identification of simple dynamic models using a process reaction curve is presented. A process reaction curve for identification is generated by disturbing the process with a manual step input, and recording the process variable (PV) output as shown in Figure 8.4. Thus the term "process reaction curve" is usually considered as being synonymous with the transient step response of an open-loop process, excluding the controller in a control loop. However, on a real plant the process controller would be put into manual mode and the step change generated from the controller unit. The step change signal would be a change, say 10% of setpoint, added to the constant process input needed to hold the process at the desired production output level. In this way the disturbance to the production unit operation is minimised and controlled. The perceived effect is to record the transient step response of an open-loop process with the feedback loop broken, as shown in Figure 8.4.

The simple SOPDT models used for identification include the following forms

$$\frac{Y(s)}{U(s)} = G_p(s) = \begin{cases} \dfrac{K(1+as)e^{-\theta s}}{\tau^2 s^2 + 2\tau\zeta s + 1} & 0 < \zeta < 1 \quad \text{(model I)} \\ \dfrac{K(1+as)e^{-\theta s}}{(\tau s + 1)(\eta \tau s + 1)} & 0 < \eta \leq 1 \quad \text{(model II)} \end{cases}$$

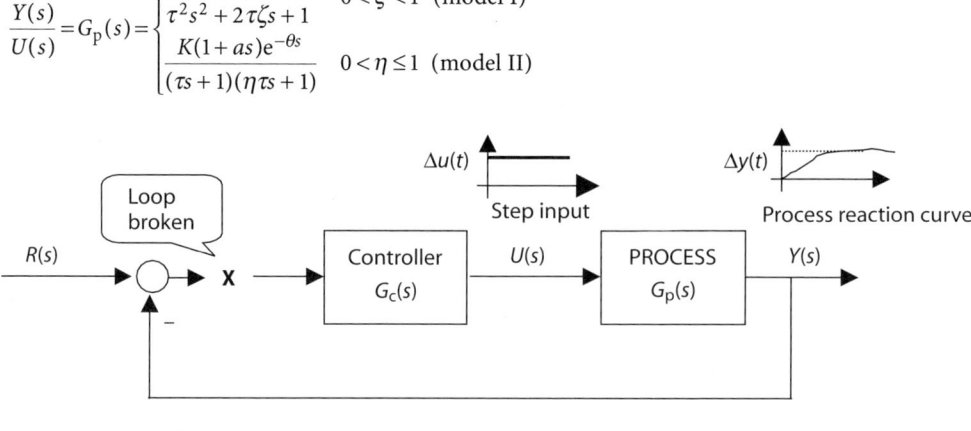

Figure 8.4 Generating the process reaction curve.

where model I is used for the underdamped second-order case with $\zeta < 1$ and model II is used for the overdamped second-order case with $\zeta \geq 1$.

In general, it is assumed that a process with high-order dynamics can be represented by the above models. For model-based controller design, models used are usually confined to the minimum phase case for which $a \geq 0$. Since the steady state gain K can be obtained easily by computing the ratio of changes of output Δy to the changes of input Δu, the identification will focus on the estimation of other parameters in the model.

The above simple SOPDT models for $G_p(s)$ can be changed into the dimensionless form by using the relations $\bar{s} = \tau s$ and $\bar{\theta} = \theta/\tau$. The dimensionless forms are given as

$$\frac{Y(s)}{KU(s)} = \frac{\bar{Y}(\bar{s})}{\bar{U}(\bar{s})} = \bar{G}_p(\bar{s}) = \begin{cases} \dfrac{(1+\bar{a}\,\bar{s})e^{-\bar{\theta}\bar{s}}}{\bar{s}^2 + 2\zeta\,\bar{s} + 1} & 0 < \zeta < 1 \text{ (model I)} \\[2ex] \dfrac{(1+\bar{a}\,\bar{s})e^{-\bar{\theta}\bar{s}}}{(\bar{s}+1)(\eta\,\bar{s}+1)} & 0 < \eta \leq 1 \text{ (model II)} \end{cases}$$

where $\bar{Y} = Y/K$ and $\bar{a} = a/\tau$. The unit step response resulting from the dimensionless model above is given by the following equations.

Model I

$$\bar{y}(\bar{t}) = 1 - \left[\frac{\zeta - \bar{a}}{\sqrt{1-\zeta^2}} \sin(\sqrt{1-\zeta^2}\,\bar{t}) + \cos(\sqrt{1-\zeta^2}\,\bar{t}) \right] e^{-\zeta \bar{t}}$$

Model II

$$\bar{y}(\bar{t}) = 1 - \left(\frac{1-\bar{a}}{1-\eta} \right) e^{-\bar{t}} - \left(\frac{\eta - \bar{a}}{1-\eta} \right) e^{-\bar{t}/\eta}$$

where $\bar{t} = (t - \theta)/\tau$.

In the following, a method of identification will be presented based on the work of Huang et al. (2001).

8.2.1 Identification Algorithm for Oscillatory Step Responses

When the reaction curve is oscillatory (as shown in Figure 8.5), the dimensionless equation for model I will be adopted.

To distinguish between process models that do not have a process zero (where $\bar{a} = 0$) and those which do have a zero (where $\bar{a} \neq 0$) the following lemma is needed.

Lemma 8.1

Consider a system of type Model I having $\zeta < 1$ and $\bar{a} \in (-\infty, \infty)$. Let $t_{p,i}$ and $t_{m,i}$ be the time instants when the output $y(t)$ reaches its ith peak and ith valley, respectively. Then, if and only if $\bar{a} = 0$, the following relationship holds:

$$t_{p,1} - \theta = t_{p,i+1} - t_{m,i} = t_{m,i} - t_{p,i} = \frac{P}{2} = \frac{\pi \tau}{\sqrt{1-\zeta^2}} \quad \forall i \geq 1$$

where P represents the oscillating period of the response, that is,

$$P = t_{p,i+1} - t_{p,i}, \quad \forall i \geq 1$$

Proof (Huang et al., 2001) ∎

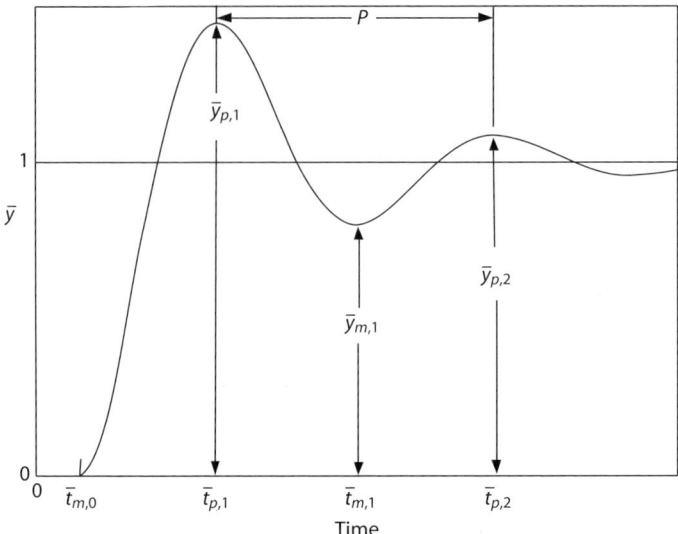

Figure 8.5 Step response for underdamped system with $\bar{a} \geq 0$.

Notice that in Figure 8.5 the time interval $t_{p,1} - \theta$ is denoted as $t_{p,1} - t_{m,0}$. Thus, the criterion to determine whether $\bar{a} = 0$ is simply to determine whether the time interval $t_{p,1} - t_{m,0}$ is equal to the time interval $t_{p,2} - t_{m,1}$.

Consequently, model parameters for model I processes can be estimated using the following algorithm.

Algorithm 8.1: SOPDT models for oscillatory step responses

Step 1 Complete the step response test, Measure P (see Figure 8.5).
Step 2 Compute the time constant τ using

$$\tau = \frac{P\sqrt{1-\zeta^2}}{2\pi} = \frac{P}{\sqrt{4\pi^2 + P^2 x^2}}$$

where

$$x = \begin{cases} \dfrac{1}{t_{p,1} - t_{p,2}} \ln\left[\dfrac{\bar{y}_{p,2} - 1}{\bar{y}_{p,1} - 1}\right] & \forall \bar{a} \geq 0 \\[2ex] \dfrac{1}{t_{m,1} - t_{p,1}} \ln\left[\dfrac{\bar{y}_{p,1} - 1}{1 - \bar{y}_{m,1}}\right] & \forall \bar{a} < 0 \end{cases}$$

and

$$\bar{y}_{p,i} = \bar{y}(\bar{t}_{p,i}) \quad i = 1, 2, \ldots$$

Step 3 Compute the damping ratio ζ using

$$\zeta = \begin{cases} \sqrt{\dfrac{\ln^2(\bar{y}_{p,1} - 1)}{\pi^2 + \ln^2(\bar{y}_{p,1} - 1)}} & \text{for } \bar{a} = 0 \\[2ex] \dfrac{Px}{\sqrt{4\pi^2 + P^2 x^2}} & \forall \bar{a} \neq 0 \end{cases}$$

Step 4 Compute $\bar{a}$ and process delay time θ using
(a) for $\bar{a} = 0$, compute $\theta = t_{p,1} + (\tau/\zeta)\ln(\bar{y}_{p,1} - 1)$
(b) for $\bar{a} \neq 0$, $\bar{a}$ and θ are computed using

$$\bar{a} = \begin{cases} \zeta + \sqrt{\zeta^2 + \left[1 - \left(\dfrac{\bar{y}_{p,1} - 1}{e^{-\zeta \bar{t}_{p,1}}}\right)^2\right]} & \forall \bar{a} > 0 \\[2ex] \zeta - \sqrt{\zeta^2 + \left[1 - \left(\dfrac{1 - \bar{y}_{m,1}}{e^{-\zeta \bar{t}_{m,1}}}\right)^2\right]} & \forall \bar{a} < 0 \end{cases}$$

and

$$\theta = \begin{cases} t_{p,1} - \dfrac{P}{2\pi}\left[\pi - \tan^{-1}\dfrac{\bar{a}\sqrt{1-\zeta^2}}{1-\bar{a}\zeta}\right] & \forall \bar{a} > 0 \\[2ex] t_{m,1} + \dfrac{P}{2\pi}\left[\tan^{-1}\dfrac{\bar{a}\sqrt{1-\zeta^2}}{1-\bar{a}\zeta}\right] & \forall \bar{a} < 0 \end{cases}$$

Algorithm end

Example 8.1

Consider the following two processes:

Process (a) $G_p(s) = \dfrac{e^{-s}}{(4s^2 + 2s + 1)(s^2 + s + 1)}$

Process (b) $G_p(s) = \dfrac{(2s + 1)e^{-2s}}{(4s^2 + 2s + 1)(s^2 + s + 1)}$

The step responses of these processes are as shown in Figure 8.6.

Both responses are oscillatory. It is found that only the response of process (a) has equal values for the time intervals $t_{p,2} - t_{m,1}$ and $t_{p,1} - t_{m,0}$. The identification results based on these observations are given in Table 8.1 and the unit step responses of the resulting identified models are also given in Figure 8.6.

8.2.2 Identification Algorithm for Non-Oscillatory Responses Without Overshoot

In this section the identification is presented for a process which has all the characteristics of an over-damped SOPDT model where the response will be non-oscillatory and without overshoot. As shown previously, the process response of a Model II type is

$$\bar{y}(\bar{t}) = 1 - \left(\dfrac{1-\bar{a}}{1-\eta}\right)e^{-\bar{t}} - \left(\dfrac{\eta - \bar{a}}{1-\eta}\right)e^{-\bar{t}/\eta}, \text{ where } \bar{t} = (t-\theta)/\tau$$

If it is assumed that $\bar{a} = 0$ and the model has no process zero, then the $\bar{t}_x$ value at any given x is a function of ζ or η. This value can be easily obtained by solving the above response equation with a single MATLAB instruction. For later use, the values of $\bar{t}_x$ for $x = 0.3, 0.5, 0.7, 0.9$ have been calculated and represented as polynomial functions of η as follows:

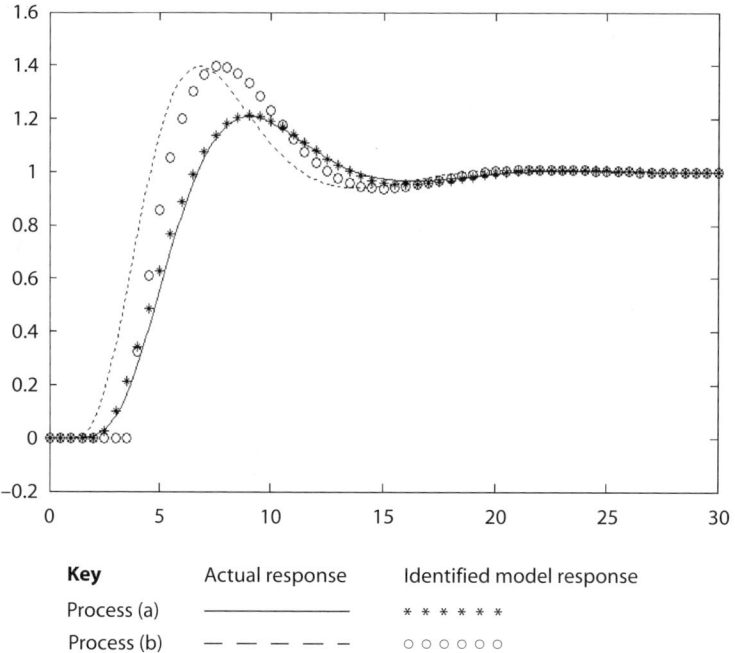

Figure 8.6 Step responses of processes (a) and (b) in Example 8.1.

Table 8.1 Identification results for the processes of Example 8.1.

Quantities computed		Process (a)		Process (b)	
		$\dfrac{e^{-s}}{(4s^2 + 2s + 1)(s^2 + s + 1)}$		$\dfrac{(2s+1)e^{-2s}}{(4s^2 + 2s + 1)(s^2 + s + 1)}$	
$t_{p,1}$	$y_{p,1}$	9.03	1.41	6.79	1.40
$t_{p,2}$	$y_{p,2}$	23.54	1.06	21.12	1.01
$t_{m,0}$	$t_{p,1} - t_{m,0}$	1.75	7.28	2.21	4.58
P		14.51		14.33	
χ		0.25		0.26	
τ		2.00		1.97	
ζ		0.44		0.51	
$\bar{a}$		0		1.33	
θ		2.03		3.50	
Identified models		$\dfrac{e^{-2.03s}}{(4.00s^2 + 1.76s + 1)}$		$\dfrac{(2.62s + 1)e^{-3.50s}}{(3.88s^2 + 2.01s + 1)}$	

$$\bar{t}_{0.3} = 0.3548 + 1.1211\eta - 0.5914\eta^2 + 0.2145\eta^3$$
$$\bar{t}_{0.5} = 0.6862 + 1.1682\eta - 0.1704\eta^2 - 0.0079\eta^3$$
$$\bar{t}_{0.7} = 1.1988 + 1.0818\eta + 0.4043\eta^2 - 0.2501\eta^3$$
$$\bar{t}_{0.9} = 2.3063 + 0.9017\eta + 1.0214\eta^2 + 0.3401\eta^3$$

These forms for the function representation of $\bar{t}_x(\eta)$ are obtained from polynomial regressions of the graphs shown in Figure 8.7.

For identifying SOPDT processes, a ratio that characterises the response in time domain, designated as $R(x)$, is defined as follows:

$$R(x) = \frac{M_\infty - t_{(x-0.2)}}{t_x - t_{(x-0.2)}}, \text{ where } x > 0.2$$

In the above equation, the function $M(t)$ denotes the integration of $e(t) = y_\infty - y(t)$ with respect to time t, namely

$$M(t) \equiv \int_0^t e(t')dt' = \int_0^t (y_\infty - y(t'))dt' = K\int_0^t (1 - \bar{y}(t'))dt'$$

The normalised value of $M(t)$, denoted $\overline{M}(\bar{t})$, is defined as

$$\overline{M}(\bar{t}) \equiv \int_0^{\bar{t}} (1 - \bar{y}(t'))dt'$$

Recalling that the time delay is denoted θ, variable $M(t)$ is related to $\overline{M}(\bar{t})$ by the following analysis:

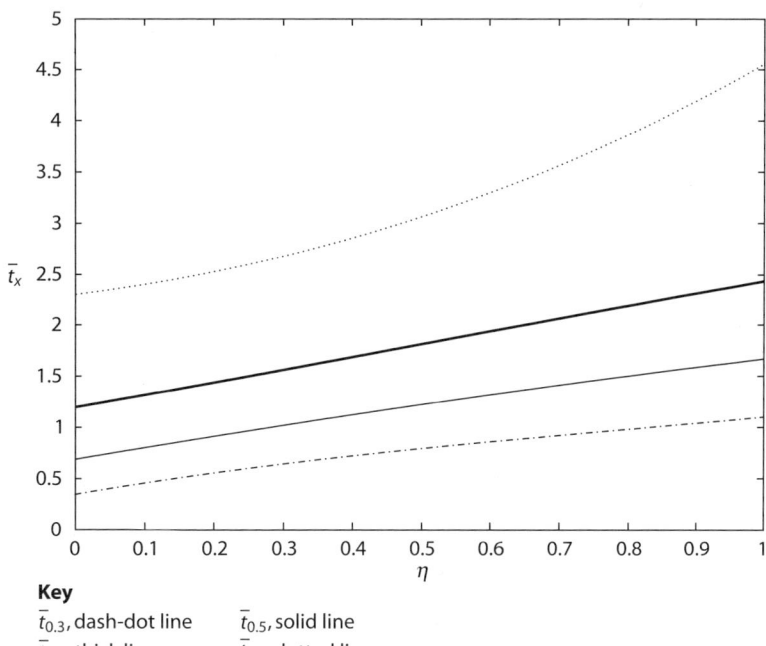

Key

$\bar{t}_{0.3}$, dash-dot line $\bar{t}_{0.5}$, solid line
$\bar{t}_{0.7}$, thick line $\bar{t}_{0.9}$, dotted line

Figure 8.7 Graphs for $\bar{t}_x$ for an overdamped process with $\bar{a} = 0$.

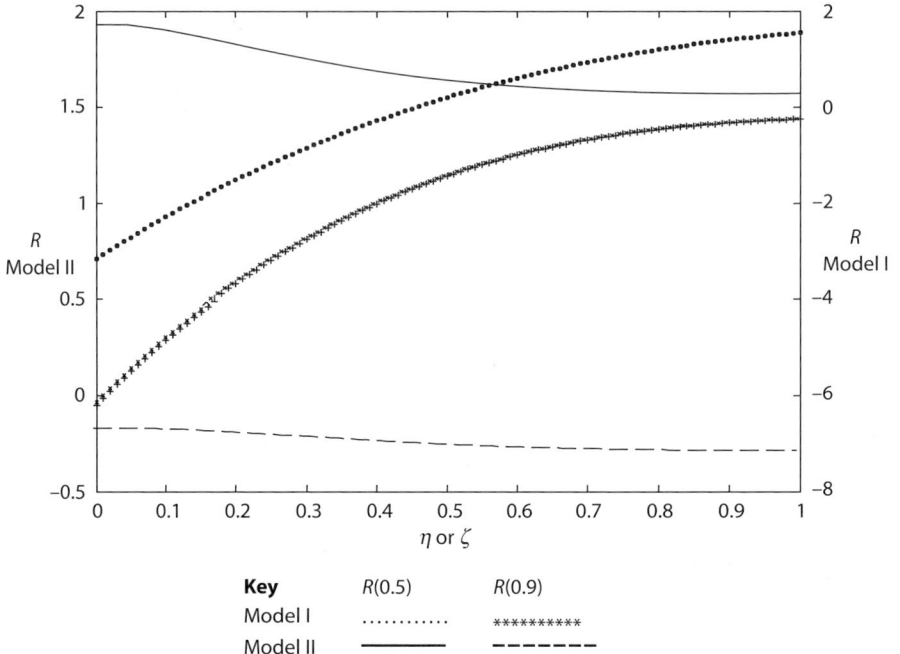

Figure 8.8 $R(x)$ for model I and model II with $\bar{a} = 0$.

$$M(t) = \int_0^t (y_\infty - y(t'))dt' = \int_0^\theta y_\infty dt' + \int_\theta^t (y_\infty - y(t'))dt'$$

$$= K\theta + K\int_\theta^t (1 - \bar{y}(t'))dt' = K\theta + K\int_\theta^{\bar{t}} (1 - \bar{y}(\bar{t}))\tau d\bar{t}$$

and

$$M(t) = K[\theta + \overline{M}(\bar{t})\tau]$$

Values of $R(x)$ at $x = 0.5$ and at $x = 0.9$ are plotted in Figure 8.8.

It is found that for each value of $R(x)$ at a given x there is only one value of η. As a result, an average of these two values can be taken as the estimate for η. For later use, the values of $R(x)$ at $x = 0.5$ and $x = 0.9$ have been calculated and regression methods used to give functional representations in polynomials of η as follows:

$$R(0.5) = 1.9108 + 0.2275\eta - 5.5504\eta^2 + 12.8123\eta^3 - 11.8164\eta^4 + 3.9735\eta^5$$
$$R(0.9) = -0.1871 + 0.0736\eta - 1.2329\eta^2 + 2.1814\eta^3 - 1.5317\eta^4 + 0.3937\eta^5$$

Thus, by using the polynomial equations and the experimental values of R, the value of η can be obtained. With this estimated value of η, the values of $\bar{t}_x$ at $x = 0.3, 0.5, 0.7, 0.9$ can be calculated enabling the time constant value τ to be estimated by the following equation:

$$\tau = \frac{t_{x,i} - t_{x,j}}{\bar{t}_{x,i} - \bar{t}_{x,j}}$$

where i and j are any integers.

Consequently, model parameters for Model II processes can be estimated using the following algorithm.

Algorithm 8.2: SOPDT models for non-oscillatory overdamped step responses

Step 1 Complete the step response test

Step 2 Finding η

Calculate $R(0.5)$ and $R(0.9)$ from the experimental step response.
Determine η from Figure 8.8 or the equation set

$$R(0.5) = 1.9108 + 0.2275\eta - 5.5504\eta^2 + 12.8123\eta^3 - 11.8164\eta^4 + 3.9735\eta^5$$

$$R(0.9) = -0.1871 + 0.0736\eta - 1.2329\eta^2 + 2.1814\eta^3 - 1.5317\eta^4 + 0.3937\eta^5$$

Take average value $\eta = \dfrac{\eta_{0.5} + \eta_{0.9}}{2}$

Step 3 Finding time constant τ

Use η estimate in the equation set

$$\bar{t}_{0.3} = 0.3548 + 1.1211\eta - 0.5914\eta^2 + 0.2145\eta^3$$
$$\bar{t}_{0.5} = 0.6862 + 1.1682\eta - 0.1704\eta^2 - 0.0079\eta^3$$
$$\bar{t}_{0.7} = 1.1988 + 1.0818\eta + 0.4043\eta^2 - 0.2501\eta^3$$
$$\bar{t}_{0.9} = 2.3063 + 0.9017\eta + 1.0214\eta^2 + 0.3401\eta^3$$

Use $\bar{t}_{0.3}, \bar{t}_{0.5}, \bar{t}_{0.7}$ and $\bar{t}_{0.9}$ to compute

$$\tau = \frac{1}{3}\left[\frac{t_{0.9} - t_{0.7}}{\bar{t}_{0.9} - \bar{t}_{0.7}} + \frac{t_{0.7} - t_{0.5}}{\bar{t}_{0.7} - \bar{t}_{0.5}} + \frac{t_{0.5} - t_{0.3}}{\bar{t}_{0.5} - \bar{t}_{0.3}}\right]$$

Step 4 Finding process delay time θ

$$\text{Compute } \theta = \frac{t_{0.9} + t_{0.7} + t_{0.5} + t_{0.3}}{4} - \left(\frac{\bar{t}_{0.9} + \bar{t}_{0.7} + \bar{t}_{0.5} + \bar{t}_{0.3}}{4}\right)\tau$$

Algorithm end

A demonstration of Algorithm 8.2 is given in Example 8.2.

Example 8.2

Consider the following three processes:

Process (a) $G_p(s) = \dfrac{1}{(s+1)^5}$

Process (b) $G_p(s) = \dfrac{e^{-0.5s}}{(2s+1)(s+1)(0.5s+1)}$

Process (c) $G_p(s) = \dfrac{e^{-0.2s}}{(s+1)^3}$

The actual step responses, as shown in Figure 8.9, are non-oscillatory and exhibit neither response overshoots nor inverse response features.

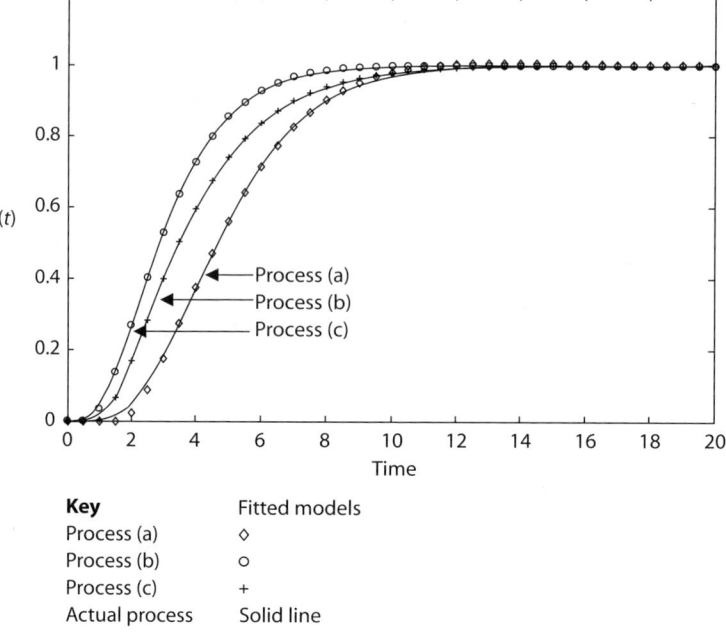

Figure 8.9 Actual and fitted step response graphs.

Table 8.2 Identification results for the processes of Example 8.2.

Quantities computed		Process (a)		Process (b)		Process (c)	
		$\dfrac{1}{(s+1)^5}$		$\dfrac{e^{-0.5s}}{(2s+1)(s+1)(0.5s+1)}$		$\dfrac{e^{-0.2s}}{(s+1)^3}$	
$t_{0.3}$	$t_{0.5}$	3.63	4.67	2.75	3.48	2.11	2.87
$t_{0.7}$	$t_{0.9}$	3.89	7.99	4.67	7.01	3.82	5.52
M_∞		5.00		4.00		3.20	
$R(0.5)$	$R(0.9)$	1.32	−0.42	1.57	−0.29	1.43	−0.36
ζ or η		0.85		0.74		0.92	
τ		2.02		1.80		1.41	
θ		1.53		0.86		0.61	
Identified models		$\dfrac{e^{-1.53s}}{(4.08s^2 + 3.43s + 1)}$		$\dfrac{e^{-0.86s}}{(1.80s+1)(1.33s+1)}$		$\dfrac{e^{-0.61s}}{(1.99s^2 + 2.59s + 1)}$	

The procedure of Algorithm 8.2 was applied and some important values relating to the steps of the identification are given in Table 8.2 along with the final fitted model parameters. The predicted step responses for each of the fitted models compared with those from the real process are also shown in Figure 8.9.

Summary Remarks

In this section, two algorithms were presented for estimating simple dynamic models using data that characterises the step response. These methods assumed that there is no right half-plane zero present in the process. A more general procedure for identifying simple models that include a right half-plane zero can be found in the work of Huang *et al.* (2001). The derivation and use of a criterion for determining model structure is not discussed here but results and more details can be also found in Huang *et al.* (2001).

Additional similar methods for identifying models of specific second-order transfer function form have also been reported by Huang and co-workers (1993; 1994). Alternative methods that identify simple second-order transfer function models have been reported by Wang *et al.* (2001), but these have not been described here since these methods use the extensive response time sequence data of the step response.

8.3 Developing Simple Models from a Relay Feedback Experiment

Conventionally, a relay experiment is used to estimate the ultimate gain and ultimate frequency of a given process so that rule-based PID control design may be applied. In this section, a different type of relay experiment is introduced as an identification method that aims to produce simple models for tuning PID controllers. The general structure of this new relay feedback experiment is shown in Figure 8.10.

The identification is conducted by compensating the output from the relay feedback loop with a simple compensator. The objective is to make the compensated output behave like the one from an integral plus dead time (IPDT) process. An on-line mechanism to adjust the compensator is presented which will lead to the identification of the parameters of the desired simple model.

To obtain simple and reduced order models, the use of a relay feedback experiment is very attractive due to several advantages. These advantages include (Yu, 1999):

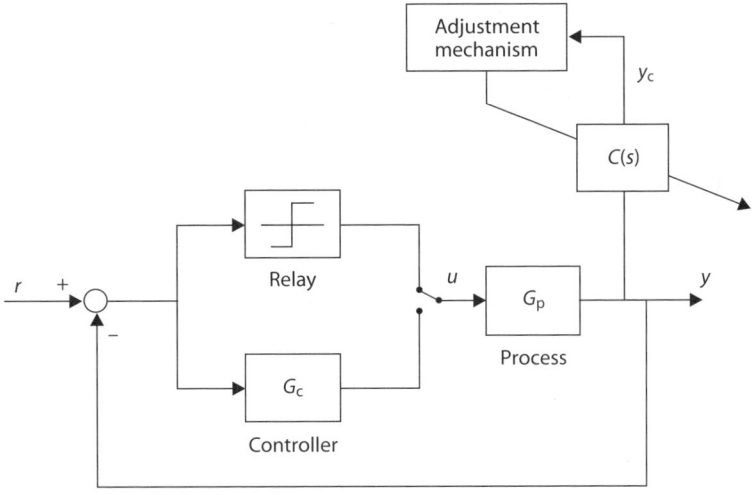

Figure 8.10 General structure of on-line identification scheme using relay feedback.

1. The process is operated in closed loop so that the output does not drift away from its desired target.
2. Relay feedback is a very time-efficient method for on-line testing procedures.
3. Although unknown disturbances and process noise can cause difficulties for the relay experiment, there are a number of simple remedies and modifications available.

Conventionally, the relay feedback test is used to implement the method of sustained oscillations and estimate the ultimate gain and ultimate frequency of a given process. However, the classical relay experiment has been extended in many works to identify parametric models (Li et al., 1991; Sung et al., 1996; Huang et al., 2000; Luyben, 2001). Hang et al. (2002) have recently published a comprehensive review of this area. In general, a relay experiment provides data to estimate one single point on the Nyquist plot, for example the data point corresponding to phase crossover frequency. Although there are modifications which allow the identification of a limited number of other points, this is often insufficient for identifying simple FOPDT and SOPDT process models.

8.3.1 On-line Identification of FOPDT Models

In this section, the simple compensated relay feedback structure of Figure 8.10 is to be used to identify the usual first-order plus dead time model given as

$$G_p(s) = \frac{Ke^{-\theta s}}{\tau s + 1}$$

In order to identify the FOPDT model, a dynamic compensator $C(s)$ is appended outside the loop (see Figure 8.10) to generate an instrumental output $y_c(t)$. During the experiment, the loop controller $G_c(s)$ is switched out and the loop is under relay feedback control. The compensator $C(s)$ is chosen to produce integral plus dead time output. Thus, consider a compensator of the form

$$C(s) = \frac{\hat{\tau} s + 1}{s}$$

where $\hat{\tau}$ is an increasingly improving estimate of the process time constant τ.

The forward path output is given as

$$Y_c(s) = G_{fp}(s)U(s) = G_p(s)C(s)U(s)$$

Consequently, the forward path transfer function becomes

$$G_{fp}(s) = G_p(s)C(s) = K\left(\frac{1}{s} + \frac{\hat{\tau} - \tau}{\tau s + 1}\right)e^{-\theta s}$$

and the output can be written in terms of two components, each driven by the same input:

$$Y_c(s) = \left[K\left(\frac{1}{s}\right)e^{-\theta s}\right]U(s) + \left[K\left(\frac{\hat{\tau} - \tau}{\tau s + 1}\right)e^{-\theta s}\right]U(s)$$

If the value of $\hat{\tau}$ is properly adjusted to converge to the actual process time constant τ, the process output reduces to the residual output:

$$Y_c(s) = \left[K\left(\frac{1}{s}\right)e^{-\theta s}\right]U(s)$$

Thus the output curve $y_c(t)$ will be the same as one generated by an IPDT process driven by the same input sequence. When this is achieved, the time constant term of the FOPDT process has been exactly

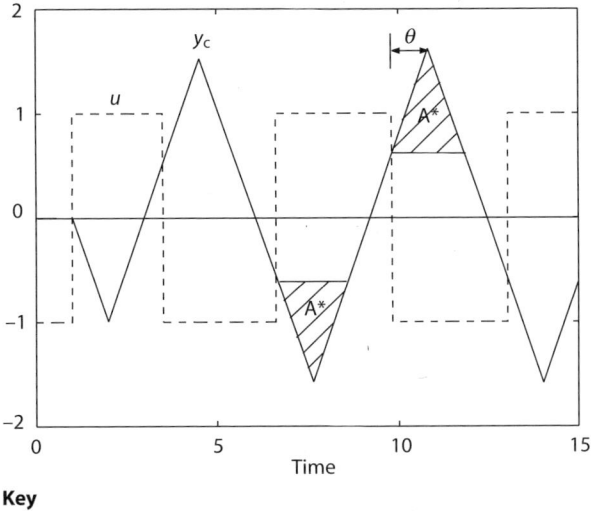

Figure 8.11 Input and compensated output waveforms for the IPDT process in an ideal relay feedback loop.

cancelled by the compensator $C(s)$ and the value of $\hat{\tau}$ will be equal to the actual value of the time constant τ.

The mechanisms which can be used to provide the adjustment scheme are derived from the knowledge that the desired output should be the output of an IPDT process driven by the relay input. Figure 8.11 shows this output and some measurable quantities which are available from this waveform.

An Estimate of Process Gain K

To estimate the process gain K, a biased relay with two levels (that is, $+h_1$ and $-h_2$) is used in the relay feedback loop. Then the process gain can be obtained as

$$K = \frac{\int_t^{t+P} y(t)\,dt}{\int_t^{t+P} u(t)\,dt}$$

where P is the oscillation period of output $y(t)$.

An Adjustment Mechanism for $\hat{\tau}$

The output shown in Figure 8.11 will be attained if the time constant estimate $\hat{\tau}$ converges to the unknown actual time constant τ. In the figure, the shaded triangular area A^* can be computed and used as the target value for a recursive algorithm to update the value of $\hat{\tau}$ to make it approach τ. When a biased-relay as mentioned is used to generate the response, the target value of A^* can be expressed as

$$A^* = \frac{1}{2} K h_1 \left(1 + \frac{h_1}{h_2}\right) \theta^2$$

Define E as the difference between the actual and target values of A^*, that is

$$E = A - A^*$$

The actual value of A^*, namely A, can be computed from a numerical integration of $y_c(t)$. Then an update equation can be given as

$$\hat{\tau}^{i+1} = \hat{\tau}^i - k_a \frac{dE}{d\hat{\tau}} E$$

where k_a is an adjustable gain for updating $\hat{\tau}$. The initial value of $\hat{\tau}$, namely $\hat{\tau}^0$, can be computed as

$$\hat{\tau}^0 = \frac{Kh_1}{\dot{y}_s}$$

where $\dot{y}_s$ is the positive slope of output as it crosses the setpoint value, h_1 is the magnitude of the positive relay output, and K is the process gain. The convergence of this algorithm can be guaranteed if the value of k_a is properly selected (Huang and Jeng, 2003).

An Estimate of Process Time Delay θ

The value of the process time delay θ can be also read from the output time waveform $y_c(t)$. Using Figure 8.11, the process time delay is found as the time that the output $y_c(t)$ takes to go from the base of the shaded triangle to the waveform peak.

In a true FOPDT process, the time delay can be taken as the time, denoted as d, that $y(t)$ takes to go from the setpoint value to the waveform peak in a relay feedback test. Thus a simple means to justify whether the FOPDT model is adequate for representing the process is to check whether the obtained time delay is close enough to d using the following criterion:

$$\Delta = \left|1 - \frac{\theta}{d}\right| \leq \varepsilon$$

where ε is an arbitrarily small value assigned to provide a tolerance of estimation errors.

8.3.2 On-line Identification of SOPDT Models

In cases where the unknown process does not exhibit oscillatory characteristics and the FOPDT model is found inadequate to represent processes of high-order dynamics, an SOPDT model should be identified instead. The SOPDT model is given as

$$G_p(s) = \frac{Ke^{-\theta s}}{(\tau_1 s + 1)(\tau_2 s + 1)}$$

The identification scheme retains the basic adaptive structure shown in Figure 8.10, and the theory for the method follows the same outline steps as used for the on-line identification of the FOPDT model above. However, in the SOPDT case the dynamic compensator $C(s)$ is now given as

$$C(s) = \frac{(\hat{\tau}_1 s + 1)(\hat{\tau}_2 s + 1)}{s(\tau_f s + 1)}$$

where $\hat{\tau}_1$, $\hat{\tau}_2$ are the estimated values for the process time constants, and τ_f is a small constant to make compensator $C(s)$ realisable.

As in the previous case, the forward path output is given as

$$Y_c(s) = G_{fp}(s)U(s) = G_p(s)C(s)U(s)$$

Substituting for the process and the compensator transfer functions (and neglecting τ_f in $C(s)$) enables the output to be written in terms of two components, each driven by the same input:

$$Y_c(s) = \left[K\left(\frac{1}{s}\right)e^{-\theta s}\right]U(s) + \left[K\left(\frac{Bs + C}{(\tau_1 s + 1)(\tau_2 s + 1)}\right)e^{-\theta s}\right]U(s)$$

where $B = \hat{\tau}_1 \hat{\tau}_2 - \tau_1 \tau_2$ and $C = (\hat{\tau}_1 + \hat{\tau}_2) - (\tau_1 + \tau_2)$.

Clearly, if $\hat{\tau}_1, \hat{\tau}_2$ converge to the process time constants τ_1, τ_2, then the process output converges to the residual output, that is

$$Y_c(s) = \left[K\left(\frac{1}{s}\right) e^{-\theta s} \right] U(s)$$

Consequently, the same principles used for identifying the FOPDT model can be applied to the case of SOPDT models, including the method of estimating the steady state gain.

An Adjustment Mechanism for $\hat{\tau}_1$ and $\hat{\tau}_2$

In the case of the SOPDT models, the output shown in Figure 8.11 will also be attained if the time constants are estimated correctly. In the waveform, the shaded triangular area A^* can be computed and used as the target value for a recursive update algorithm as follows.

As in the FOPDT case, define the area error E as

$$E = A - A^*$$

Then two update equations are defined as

$$\hat{\tau}_1^{i+1} = \hat{\tau}_1^i - k_{a1} \frac{\partial E}{\partial \hat{\tau}_1} E$$

and

$$\hat{\tau}_2^{i+1} = \hat{\tau}_2^i - k_{a2} \frac{\partial E}{\partial \hat{\tau}_2} E$$

where k_{a1} and k_{a2} are adjustable gains for updating $\hat{\tau}_1$ and $\hat{\tau}_2$, respectively. The vector

$$\nabla E = \left[\frac{\partial E}{\partial \hat{\tau}_1} \quad \frac{\partial E}{\partial \hat{\tau}_2} \right]^T$$

can be considered as the gradient direction of area E with respect to the two process time constants. This gradient has to be estimated and can be considered to give the corresponding direction of change in each iteration. The initial value of $\hat{\tau}_1$ can be computed as

$$\hat{\tau}_1^0 = \frac{Kh_1}{\dot{y}_s}$$

as defined above and the initial value of $\hat{\tau}_2$ is set as 0.8 times $\hat{\tau}_1^0$.

An Estimate of the Process Time Delay θ

When the result converges, the process time delay is found as the time that the output $y_c(t)$ takes to go from the base of the shaded triangle to the waveform peak.

8.3.3 Examples for the On-line Relay Feedback Procedure

In order to illustrate the proposed on-line identification method, a second-order plus dead time process and a high-order process were used in two separate examples.

Example 8.3: Identifying an FOPDT model using the on-line procedure

Consider the problem of identifying an FOPDT model for an unknown second-order plus dead time process. The unknown process is taken to be represented by

$$G_p(s) = \frac{e^{-2s}}{(5s+1)(s+1)}$$

Some initial features of the FOPDT model for this process are obtained using a biased relay ($h_1 = 1.2$ and $-h_2 = -1.0$) experiment. According to the procedures, the slope $\dot{y}_s$ is 0.205 and the initialising response time d is 2.45. The estimated process time constant $\hat{\tau}$ is updated according to the recursive procedure of Section 8.3.1:

$$\hat{\tau}^{i+1} = \hat{\tau}^i - k_a \frac{dE}{d\hat{\tau}} E$$

where the adjustable gain k_a is taken as unity, the target waveform area A^* is set to be $1.32\theta^2$, the value of θ is recorded from the current waveform of $y_c(t)$, and the initial value of $\hat{\tau}$ is set as 5.85. The compensator used in the on-line scheme is

$$C(s) = \frac{\hat{\tau}s + 1}{s}$$

As a result, the estimated time constant $\hat{\tau}$ converges after three iterations where the FOPDT model obtained is

$$\hat{G}_p(s) = \frac{e^{-2.66s}}{5.28s + 1}$$

To verify this obtained model, the calculated Δ is found to be less than 0.1 and, thus, it is considered that the resulting FOPDT model is adequate for representing this second-order process.

Example 8.4: Identifying an SOPDT model using the on-line procedure

Consider the following high-order plus dead time process:

$$G_p(s) = \frac{e^{-0.8s}}{(2s+1)^3(s+1)^2}$$

As in the previous example, an FOPDT model was first identified. However, the calculated Δ was 0.4, which indicated that the FOPDT model was not adequate for representing such high-order dynamics. Therefore the identification of an SOPDT model was pursued instead. The identification scheme used a compensator of the form

$$C(s) = \frac{(\hat{\tau}_1 s + 1)(\hat{\tau}_2 s + 1)}{s(0.01s + 1)}$$

where $\hat{\tau}_1, \hat{\tau}_2$ are estimates for the process time constants τ_1, τ_2. The time constant update used the pair of equations:

$$\hat{\tau}_1^{i+1} = \hat{\tau}_1^i - k_{a1} \frac{\partial E}{\partial \hat{\tau}_1} E$$

and

$$\hat{\tau}_2^{i+1} = \hat{\tau}_2^i - k_{a2} \frac{\partial E}{\partial \hat{\tau}_2} E$$

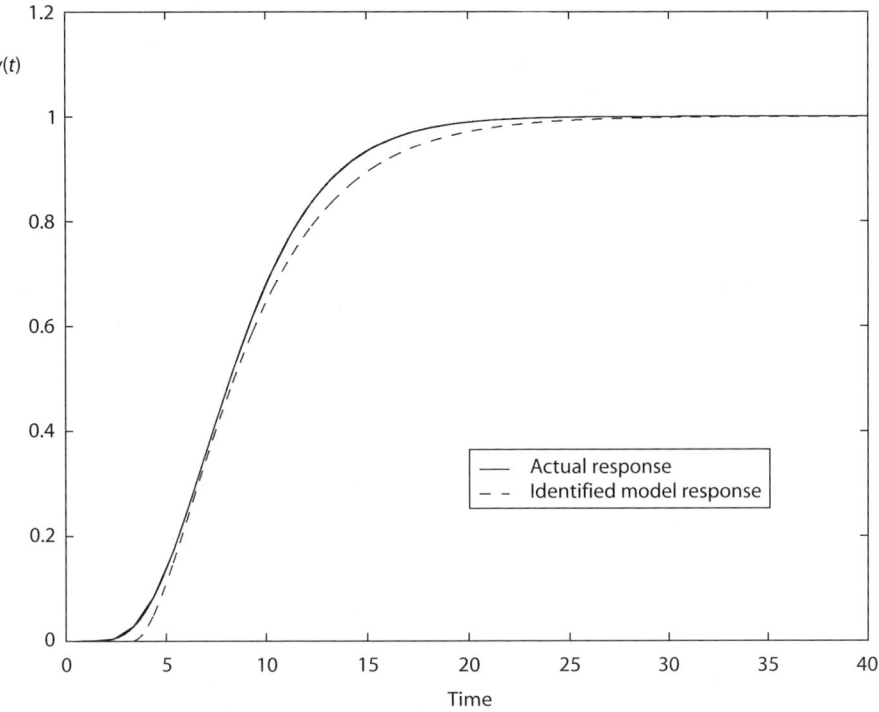

Figure 8.12 Step responses of actual process and identified model in Example 8.4.

where the adjustable gains were given values of $k_{a1} = k_{a2} = 0.5$. The estimation procedure converged after six iterations and the resulting SOPDT model was

$$\hat{G}_p(s) = \frac{e^{-3.30s}}{(3.60s + 1)(2.51s + 1)}$$

To show the result of the identification, a comparison of the responses is given in Figure 8.12.

8.3.4 Off-line Identification

In Sections 8.3.1 and 8.3.2, the identification procedure is conducted on-line with the relay feedback loop. This routine may be found to be time-consuming due to the iterative procedures and the slow process response. An off-line method is presented below which uses only the amplitude and the period of constant cycles.

If it is assumed that there is no process zero, the simple SOPDT models take the following two forms:

$$\frac{Y(s)}{U(s)} = G_p(s) = \begin{cases} \dfrac{Ke^{-\theta s}}{\tau^2 s^2 + 2\tau\zeta s + 1} & 0 < \zeta < 1 \quad (\text{model I}) \\[2ex] \dfrac{Ke^{-\theta s}}{(\tau s + 1)(\eta\tau s + 1)} & 0 < \eta \leq 1 \quad (\text{model II}) \end{cases}$$

where model I is used for the underdamped second-order case with $\zeta < 1$ and model II is used for the overdamped second-order case with $\zeta \geq 1$.

These simple SOPDT models for $G_p(s)$ can be changed into dimensionless forms by using the relations $\bar{s} = \theta s$ and $\bar{\tau} = \tau/\theta$ to yield

$$\frac{Y(s)}{KU(s)} = \frac{\overline{Y}(\overline{s})}{U(\overline{s})} = \overline{G}_p(\overline{s}) = \begin{cases} \dfrac{e^{-\overline{s}}}{\overline{\tau}^2 \overline{s}^2 + 2\overline{\tau}\zeta\overline{s} + 1} & 0 < \zeta < 1 \quad (\text{model I}) \\ \dfrac{e^{-\overline{s}}}{(\overline{\tau}\,\overline{s} + 1)(\eta\,\overline{\tau}\,\overline{s} + 1)} & 0 < \eta \leq 1 \quad (\text{model II}) \end{cases}$$

where $\overline{Y} = Y/K$.

A conventional relay feedback loop of which an ideal relay that has output magnitude h is then simulated. By disturbing the output from its equilibrium point with a pulse, the system will then be activated and the output $y(t)$ has an amplitude designated as a and a cycling period designated as P. Thus, by performing relay experiments numerically on model I and model II above, the responses in terms of normalised amplitude and normalised period are plotted as in Figure 8.13.

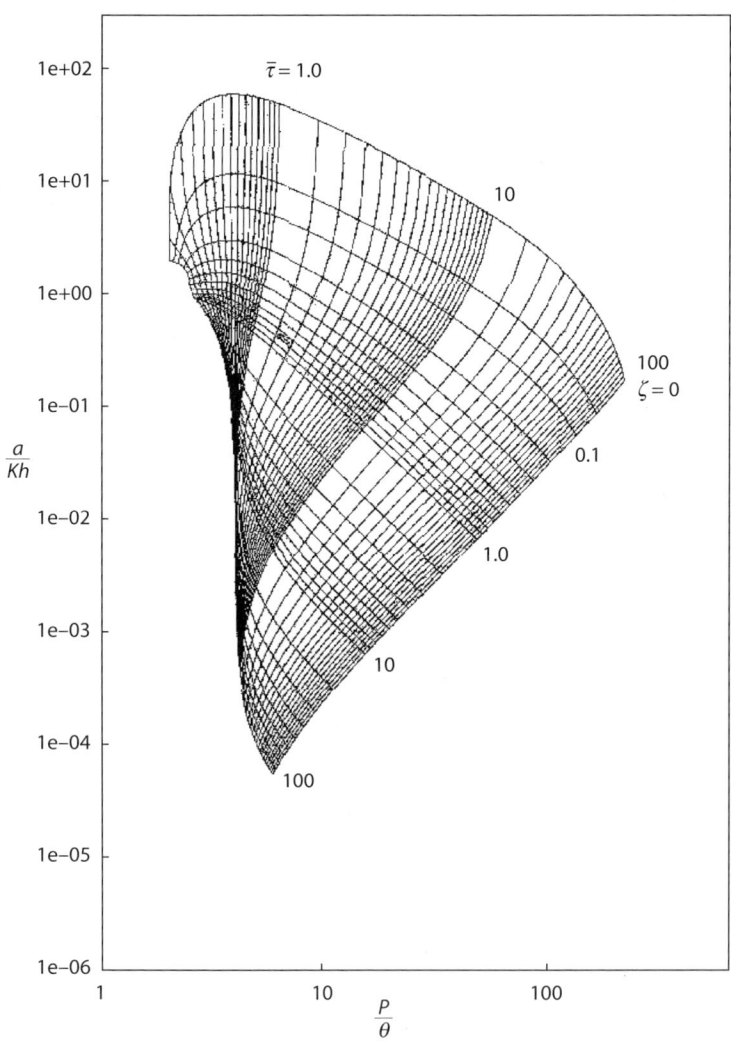

Figure 8.13 Normalised reaction curve $\overline{\tau}$ and ζ of relay feedback test.

8.3 Developing Simple Models from a Relay Feedback Experiment

It is then considered that in a dynamic system that can be represented by either of the model I or model II SOPDT model types, there will be a point in Figure 8.13 corresponding to the experimental magnitude and cycling period. However, there is one free parameter, namely the apparent dead time θ, to be determined. To fix this free parameter, one more equality relation has to be established. For control design purposes, it is desirable that the model and the actual process should have almost the same ultimate frequency ω_u. This design consideration becomes the additional equality, namely

$$\arg\{\hat{G}_p(j\omega_u)\} = \arg\left\{\hat{G}_p\left(j\frac{2\pi}{P}\right)\right\} = -\pi \text{ rad}$$

Based on the description given above, an iterative procedure for identification is given by the following algorithm.

Algorithm 8.3: Off-line relay feedback identification routine
Step 1 Compute or provide an initial value for the apparent dead time $\hat{\theta}$.
Step 2 Read new values of $\bar{\tau}$ and ζ from Figure 8.13 with the amplitude and cycling period obtained from the experiment.
Step 3 Use the resulting τ and ζ, together with $\hat{\theta}$, to check if $\arg\{\hat{G}_p(j\omega_u)\}$ equals $-\pi$. If not, go back to *Step 1* and repeat the procedures with a new guess of θ according to the one-dimensional search method until $\arg\{\hat{G}_p(j\omega_u)\}$ equals $-\pi$.
Algorithm end

Example 8.5

The results of this off-line procedure for a number of processes are given.

Case 1: Second-order plus dead time process
The process is

$$G_p(s) = \frac{e^{-2s}}{4s^2 + 3s + 1}$$

The identified SOPDT model is

$$\hat{G}_p(s) = \frac{e^{-2s}}{4s^2 + 3.2s + 1}$$

Case 2: High-order plus dead time process
The fifth-order process is

$$G_p(s) = \frac{e^{-0.8s}}{(2s+1)^3(s+1)^2}$$

which is the same one as in Example 8.4.
The identified SOPDT model is

$$\hat{G}_p(s) = \frac{e^{-3.31s}}{10.9561s^2 + 6.62s + 1}$$

Note that the identified model is similar to the result obtained in Example 8.4. The apparent time delay, that is 3.31, is larger than the true time delay of 0.8. This is because the reduction in dynamic order results in some extra time delay in the model to account for the missing process lags.

8.4 An Inverse Process Model-Based Design Procedure for PID Control

Three-term PID controllers have been widely used in chemical plants for process control. There is a wide variety of methods with different complexities that can be used to determine the parameters of PID controllers to meet given performance specifications. Through a very long period of development, PID controller design for SISO systems seems to be reaching a status of considered maturity. The book of Åström and Hägglund (1995) contains a good collection of PID control research papers along with a bibliography for the topic. There are also books (Åström and Hägglund, 1988, 1995; Hang et al. 1993; McMillan, 1994) that focus on autotuning principles and PID controller design. PID control tuning formulae based on the internal model control (IMC) principle for simple transfer function models have been given by Chien and Fruehauf (1990) and Hang et al. (1991). In this section, a contribution is made to PID autotuning with the derivation of tuning formulae for one degree of freedom (1 dof) PID controllers using an inverse process model-based approach.

8.4.1 Inverse Process Model-Based Controller Principles

Theoretically, non-minimum phase (nmp) elements in a process cannot be eliminated by any controller in a simple closed-loop system. On the other hand, in order to eliminate offsets, an integrator must be a part of the loop transfer function. As a result, the target forward path loop transfer function should consist of at least one integrator and the non-minimum phase elements of the process such as any right half-plane (RHP) zeros and the pure dead time if it is present. This combination of the integrator and the non-minimum phase elements constitute the basic loop transfer function of a control loop. To achieve this basic loop transfer function, many design methods try to use the controller in the loop as an explicit or implicit inverse process model. According to IMC theory, the nominal loop transfer function of a control system that has an inverse process model-based controller will be of the following form:

$$G_{lp}(s) = \left(\frac{G_{p+}(s)}{s}\right) F_{lp}(s)$$

where $F_{lp}(s)$ serves as a loop filter in the control system and the $G_{p+}(s)$ contains the non-minimum phase elements of the process and represents the non-invertible part of process $G_p(s)$.

Examples Due to Chien and Fruehauf (1990)

If the process is represented by a model of FOPDT transfer function

$$G_p(s) = \frac{Ke^{-\theta s}}{\tau s + 1}$$

then Chien and Fruehauf (1990) show that the resulting loop transfer function $G_{lp}(s)$ (namely $G_p(s)G_c(s)$) will be

$$G_{lp}(s) = \left(\frac{e^{-\theta s}}{s}\right)\left(\frac{1}{\lambda + \theta}\right)$$

or

$$G_{lp}(s) = \left(\frac{e^{-\theta s}}{s}\right)\left(\frac{1 + 0.5\theta s}{(\lambda + \theta)(\tau_f s + 1)}\right)$$

Note that in the above loop transfer function the loop filter $F_{lp}(s)$ has lead–lag form.
Similarly, for a process model of SOPDT type given by

$$G_p(s) = \frac{Ke^{-\theta s}}{\tau^2 s^2 + 2\tau\zeta s + 1}$$

compensation by a controller yields a the loop transfer function

$$G_{lp}(s) = \left(\frac{e^{-\theta s}}{s}\right)\left(\frac{1}{\lambda + \theta}\right)$$

A Methodology Based on Inverse Model-Based Approach

The inverse model-based approach presented here differs from the IMC method in the way the controller is synthesised. In the IMC approach, a target overall transfer function is assigned and is embedded in the so-called IMC filter. On the other hand, the inverse model-based approach is to assign a target loop transfer function, designated as $G_{lp}(s)$ and synthesise the controller according to the following equation:

$$G_c(s) = G_p^{-1}(s) G_{lp}(s)$$

In this way, it is more direct without encountering the mathematical approximation to the process time delay in the derivation of controllers. Also, the gain margin and phase margin can be clearly defined from the assigned $G_{lp}(s)$. Recently, a study of Huang and Jeng (2002) showed that it is possible to obtain optimal performance and reasonable robustness using a system as shown in Figure 8.14 with a loop transfer function of the following form:

$$G_{lp}(s) = \frac{k_0(1 + \alpha\theta s)e^{-\theta s}}{s}$$

This closed-loop system has an implicit assumption that a controller can be designed to achieve the target loop transfer function. In the following, the properties related to the robustness issue of the system in Figure 8.14 will be discussed.

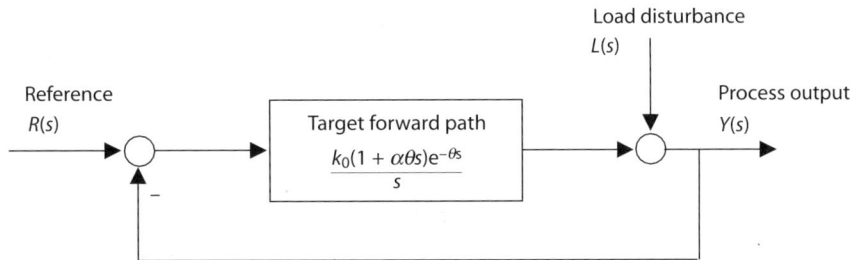

Figure 8.14 Equivalent inverse process model-based control system.

Introduce the normalising relationships $K_0 = k_0\theta$ and $\bar{s} = \theta s$; then the above target loop transfer becomes

$$G_{lp}(s) = \frac{k_0(1 + \alpha\theta s)e^{-\theta s}}{s} = \frac{K_0(1 + \alpha\bar{s})e^{-\bar{s}}}{\bar{s}}$$

The lead element $(1 + \alpha\theta s)$ in the above loop transfer function provides an extra degree of freedom for controller design. With the aid of this normalised loop transfer function, it is obvious that the phase crossover frequency of the loop ω_{pco} is only a function of parameter α. Hence, for any $\theta \neq 0$ the phase crossover frequency satisfies

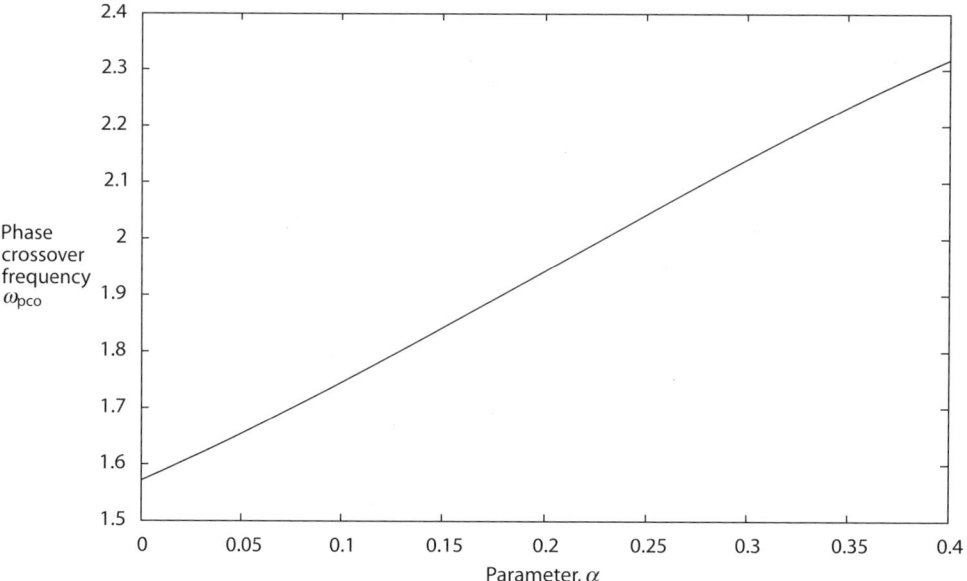

Figure 8.15 Graph of ω_{pco} versus α for the inverse process-based system.

$$-\omega_{pco}\alpha = \cot(\omega_{pco})$$

A graph of the phase crossover frequency ω_{pco} versus α is given in Figure 8.15, where it can be seen that the slope of ω_{pco} to the change of α is about 1.93. This graph leads to the following linear relationship: $\omega_{pco} = 1.93\alpha + 1.56$.

It is then easy to find the gain and phase margins of this equivalent system (Figure 8.14) as, respectively,

$$GM = \frac{\omega_{pco}}{K_0\sqrt{(1+\alpha^2\omega_{pco}^2)}}$$

and

$$\phi_{PM} = \frac{\pi}{2} - \frac{K_0}{\sqrt{(1+\alpha^2 K_0^2)}} + \tan^{-1}\frac{\alpha K_0}{\sqrt{(1-\alpha^2 K_0^2)}} \text{ rad}$$

Thus the gain and phase margins are functions of K_0 and α; both these equations are plotted in Figure 8.16 to show the relationship between the phase margin and gain margin over the range of K_0 and α. Note that the phase margin has been plotted in degrees in the Figure 8.16.

From Figure 8.16, it is found that the closed-loop system may have gain margin 2.1 and phase margin 60° if K_0 is set equal to 0.8 and α is set equal to 0.4. Simulation studies with the system structure of Figure 8.14 reveal that the value $\alpha = 0.4$ yields good performance and robustness properties and is selected as the proper choice for this parameter. If parameter α is chosen to have the value $\alpha = 0.4$, then the following relation is obtained:

$$\omega_{pco} = 1.93\alpha + 1.56 = 2.332$$

and the gain margin equation can be rearranged to give

8.4 An Inverse Process Model-Based Design Procedure for PID Control

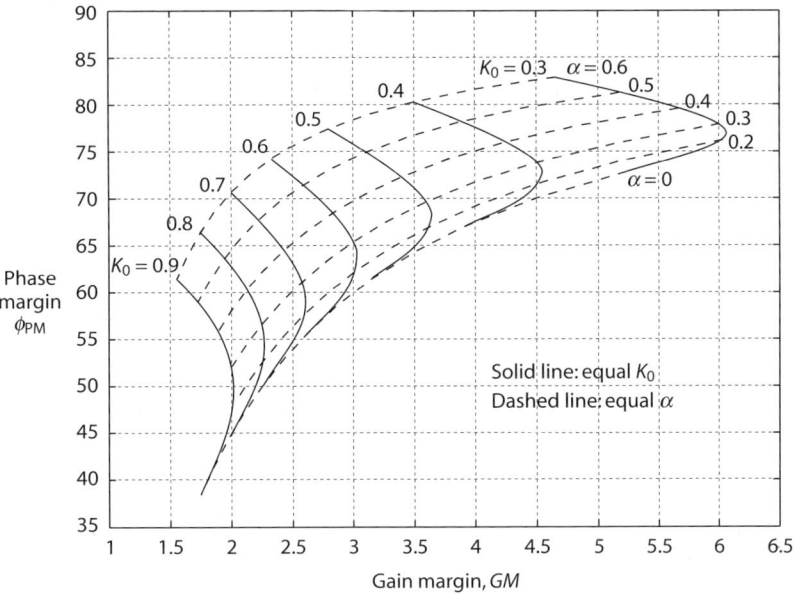

Figure 8.16 Gain margin versus phase margin (degrees) for range of K_0 and α.

$$K_0 = \frac{\omega_{pco}}{GM\sqrt{(1+\alpha^2\omega_{pco}^2)}} = \frac{1.7}{GM}$$

Thus, by specifying a gain margin, the overall normalised loop gain K_0 can be obtained, and with this loop gain the corresponding phase margin is also determined.

Finally, for practical synthesis it is necessary to introduce a low-pass filter into the target loop transfer function as follows:

$$G_{lp}(s) = \frac{K_0(1+0.4\bar{s})e^{-\bar{s}}}{\bar{s}(1+0.01\bar{s})}$$

Note that the time constant of this additional filter can be set to an arbitrarily small value, which is chosen as 0.01 here. It is this practical target loop transfer function which plays an important role in the synthesis of the three-term controllers described in the next section.

8.4.2 PI/PID Controller Synthesis

Based on the desired loop transfer function given above, a tuning formula for the PID controller can be easily derived for different processes. The key relationship is the forward path transfer equality:

$$G_p(s)G_c(s) = G_{lp}(s) = \frac{K_0(1+0.4\bar{s})e^{-\bar{s}}}{\bar{s}(1+0.01\bar{s})}$$

Into the forward path transfer $G_{lp}(s) = G_p(s)G_c(s)$ can be inserted different process models for $G_p(s)$ and different three-term controller structures, and the equality solved to synthesise the controller.

Case 1: FOPDT Process Model and PID Controller
The process model is

$$G_p(s) = \left(\frac{Ke^{-\theta s}}{\tau s + 1}\right)$$

Let the PID controller be

$$G_c(s) = k_p\left(1 + \frac{1}{\tau_i s}\right)\left(\frac{\tau_d s + 1}{\tau_f s + 1}\right)$$

The forward path relationship is

$$G_p(s)G_c(s) = \left(\frac{Ke^{-\theta s}}{\tau s + 1}\right)k_p\left(\frac{\tau_i s + 1}{\tau_i s}\right)\left(\frac{\tau_d s + 1}{\tau_f s + 1}\right)$$

$$= G_{lp}(s) = \frac{K_0(1 + 0.4\bar{s})e^{-\bar{s}}}{\bar{s}(1 + 0.01\bar{s})}$$

from which the controller parameters are determined as

$$k_p = \frac{0.65\tau_i}{K\theta},\ \tau_i = \tau,\ \tau_d = 0.4\theta \text{ and } \tau_f = 0.01\theta$$

Case 2: FOPDT Process Model and PI Controller

Let the PI controller be

$$G_c(s) = k_p\left(1 + \frac{1}{\tau_i s}\right)$$

The forward path relationship is

$$G_p(s)G_c(s) = \left(\frac{Ke^{-\theta s}}{\tau s + 1}\right)k_p\left(\frac{\tau_i s + 1}{\tau_i s}\right)$$

$$= G_{lp}(s) \approx \frac{K_0(1 + 0.4\bar{s})e^{-\bar{s}}}{\bar{s}(1 + 0.01\bar{s})}$$

The synthesis of the PI controller is not as direct, since the target loop transfer function cannot be exactly achieved. As an approximation, the controller parameters are determined as

$$k_p = \frac{0.55\tau_i}{K\theta} \text{ and } \tau_i = 0.4\theta + 0.9\tau$$

Case 3: Overdamped SOPDT Process Model and PID Controller

Consider an overdamped SOPDT process of the form

$$G_p(s) = \frac{Ke^{-\theta s}}{(\tau s + 1)(\eta \tau s + 1)}$$

Let the PID controller be

$$G_c(s) = k_p\left(1 + \frac{1}{\tau_i s}\right)\left(\frac{\tau_d s + 1}{\tau_f s + 1}\right)$$

The forward path relationship is

8.4 An Inverse Process Model-Based Design Procedure for PID Control

$$G_p(s)G_c(s) = \left(\frac{Ke^{-\theta s}}{(\tau s+1)(\eta \tau s+1)}\right)k_p\left(\frac{\tau_i s+1}{\tau_i s}\right)\left(\frac{\tau_d s+1}{\tau_f s+1}\right)$$

$$= G_{lp}(s) \approx \frac{K_0(1+0.4\bar{s})e^{-\bar{s}}}{\bar{s}(1+0.01\bar{s})}$$

This case is very similar to Case 2; hence, if the derivative time constant τ_d is set equal to τ, the controller parameters are determined as

$$k_p = \frac{0.55\tau_i}{K\theta}, \tau_i = 0.4\theta + 0.9\eta\tau, \tau_d = \tau \text{ and } \tau_f = 0.01\theta$$

Case 4: Underdamped SOPDT Process Model and PID Controller

Consider the underdamped SOPDT process model

$$G_p(s) = \frac{Ke^{-\theta s}}{(\tau^2 s^2 + 2\tau\zeta s + 1)}$$

Let the PID controller have an extra filter, namely

$$G_c(s) = k_p\left(1 + \frac{1}{\tau_i s} + \tau_d s\right)\left(\frac{1}{\tau_f s + 1}\right)$$

Then the forward path relationship is

$$G_p(s)G_c(s) = \left(\frac{Ke^{-\theta s}}{(\tau^2 s^2 + 2\tau\zeta s + 1)}\right)k_p\left(\frac{\tau_i \tau_d s^2 + \tau_i s + 1}{\tau_i s}\right)\left(\frac{1}{1+\tau_f s}\right)$$

$$= G_{lp}(s) = \frac{K_0 e^{-\bar{s}}}{\bar{s}(1+0.01\bar{s})}$$

Notice that in this case, the target loop transfer function used in the previous cases is not achievable. Based on the given loop transfer function, the controller parameters are determined as

$$k_p = \frac{0.55\tau_i}{K\theta}, \tau_i = 2\tau\zeta, \tau_d = \frac{\tau}{2\zeta} \text{ and } \tau_f = 0.01\theta$$

8.4.3 Autotuning of PID Controllers

It remains to incorporate the identification of simple models with the tuning formula described in Section 8.4.2 to illustrate the autotuning of PI/PID controllers. The autotuning procedure requires a step response from the open-loop process or from a relay feedback experiment. In this section the results from the example processes have been identified in Sections 8.2 and 8.3. By applying the resulting models and the tuning formula, autotuning of PID controllers is demonstrated and assessed.

Example 8.6: Non-oscillatory high-order process – autotuning using models from reaction curve method

A demonstration of Algorithm 8.2 was given in Example 8.2 for a non-oscillatory process given by

Process (b) $G_p(s) = \dfrac{e^{-0.5s}}{(2s+1)(s+1)(0.5s+1)}$

The identification algorithm used a reaction curve approach and yielded the model

$$\hat{G}_p(s) = \frac{e^{-0.86s}}{(1.8s+1)(1.33s+1)}$$

The PID controller was given in the series form

$$G_c(s) = k_p\left(1 + \frac{1}{\tau_i s}\right)\left(\frac{1+\tau_d s}{1+\tau_f s}\right)$$

According to the tuning formulae the controller parameters were found as

$k_p = 0.98$, $\tau_i = 1.54$, $\tau_d = 1.80$ and $\tau_f = 0.0086$

The responses resulting from the use of this and other controllers with the full process transfer function are shown in Figure 8.17. The performance of inverse-based controller tuning is satisfactory and similar that from IMC tuning. On the other hand, ZN tuning results in excessive overshoot and an over-oscillatory response.

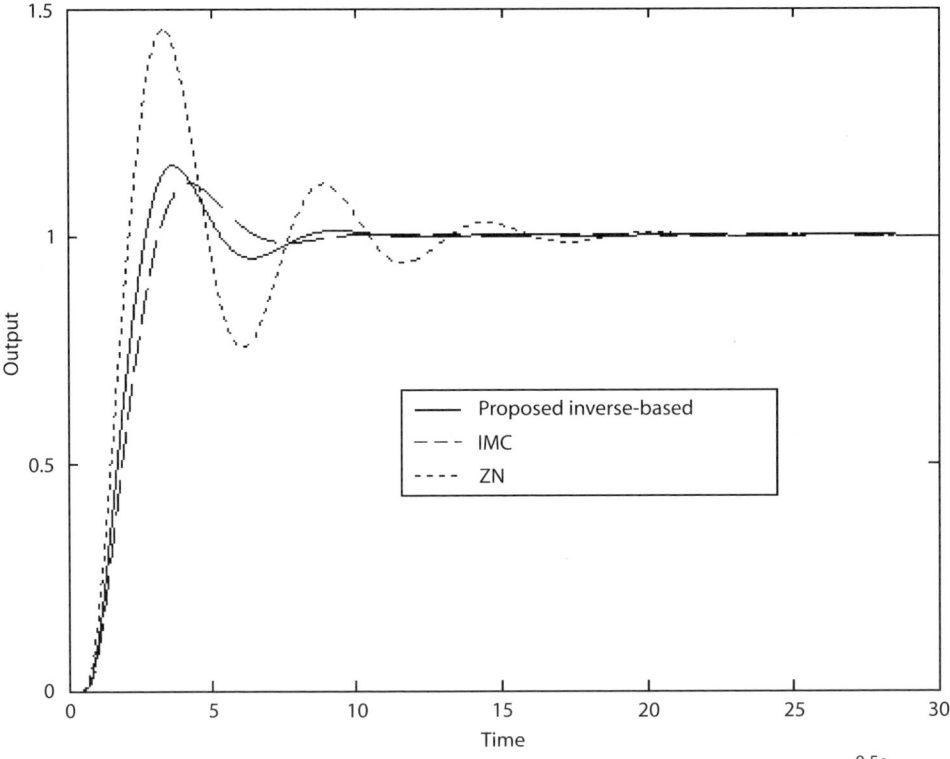

Figure 8.17 Comparison of step reference responses for the process $G_p(s) = \dfrac{e^{-0.5s}}{(2s+1)(s+1)(0.5s+1)}$.

Example 8.7: FOPDT model for a SOPDT process – autotuning using models from relay feedback test

In Example 8.3 the on-line relay feedback procedure was used to identify a FOPDT model for an unknown second-order plus dead time process. The unknown process is taken to be represented by

$$G_p(s) = \frac{e^{-2s}}{(5s+1)(s+1)}$$

8.4 An Inverse Process Model-Based Design Procedure for PID Control

The identified FOPDT model was

$$\hat{G}_p(s) = \frac{e^{-2.66s}}{5.28s + 1}$$

The PID controller was given in the series form as

$$G_c(s) = k_p \left(1 + \frac{1}{\tau_i s}\right)\left(\frac{1 + \tau_d s}{1 + \tau_f s}\right)$$

According to the tuning formulae the controller parameters were found as

$k_p = 1.29$, $\tau_i = 5.28$, $\tau_d = 1.06$ and $\tau_f = 0.0266$

For PI control, the controller has the form

$$G_c(s) = k_p \left(1 + \frac{1}{\tau_i s}\right)$$

and the PI controller parameters are

$k_p = 1.09$, $\tau_i = 5.82$

The output responses of the system to a step setpoint change are given in Figure 8.18 for the PID controller and Figure 8.19 for the PI controller. The responses resulting from both inverse-based PID and PI controllers are similar to that of IMC, and are much better than those of ZN tuned systems.

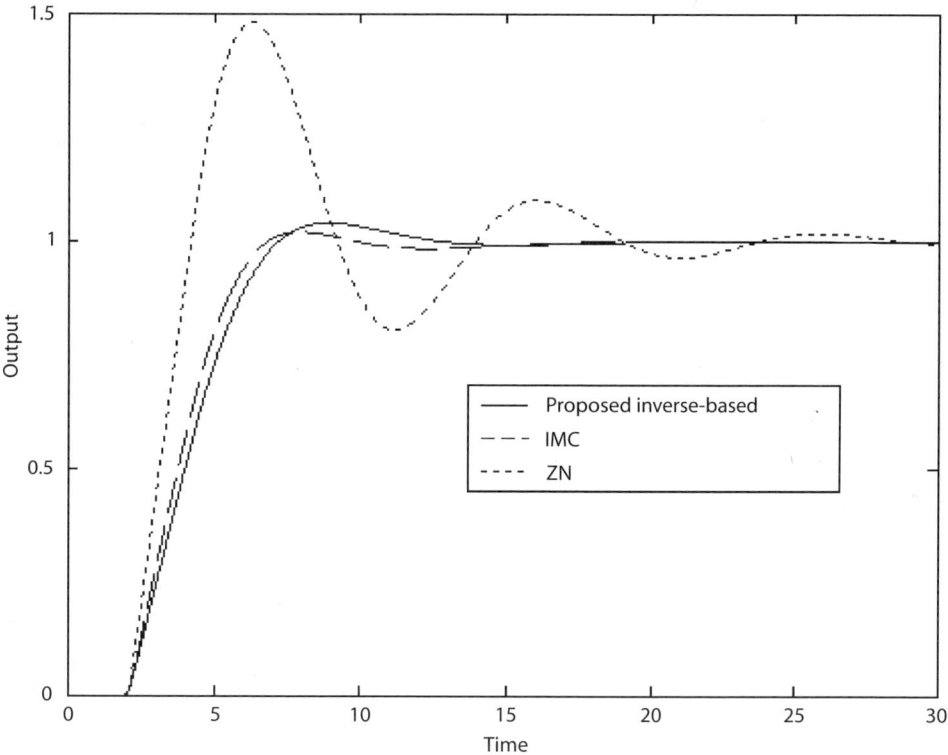

Figure 8.18 Comparison of PID control results for $G_p(s) = \dfrac{e^{-2s}}{(5s + 1)(s + 1)}$.

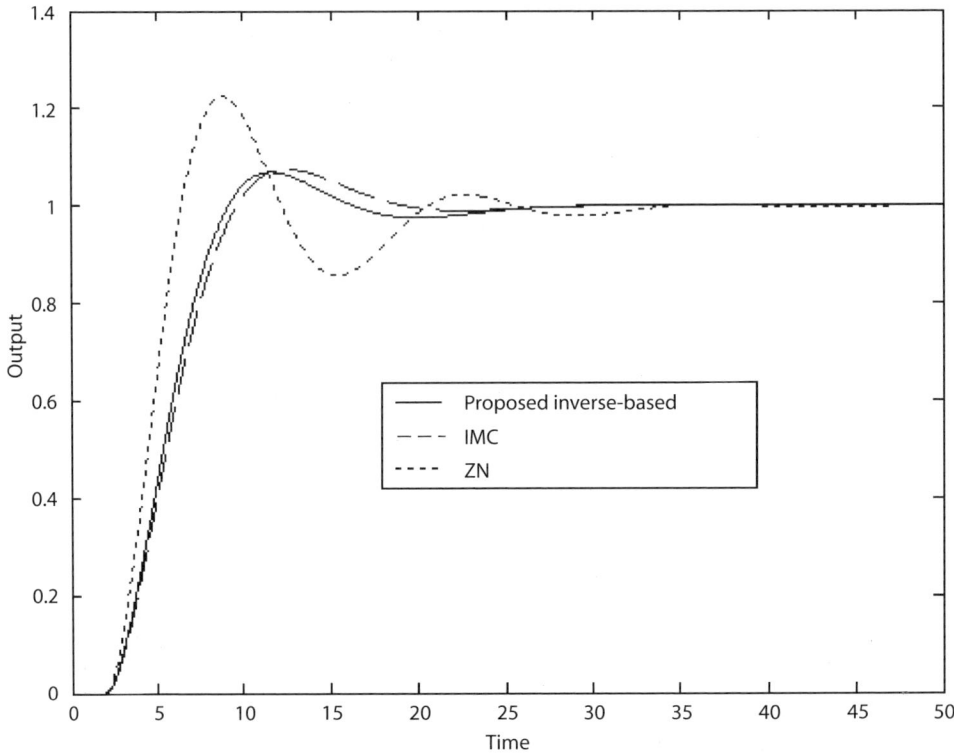

Figure 8.19 Comparison of PI control results for $G_p(s) = \dfrac{e^{-2s}}{(5s+1)(s+1)}$.

Example 8.8: Non-oscillatory high-order process – autotuning using models from relay feedback test

In Example 8.4 the on-line procedure was used to identify an SOPDT model for the following high-order plus dead time process:

$$G_p(s) = \frac{e^{-0.8s}}{(2s+1)^3(s+1)^2}$$

The outcome was the identified SOPDT model

$$\hat{G}_p(s) = \frac{e^{-3.3s}}{(3.6s+1)(2.51s+1)}$$

The PID controller took the series form

$$G_c(s) = k_p\left(1 + \frac{1}{\tau_i s}\right)\left(\frac{1+\tau_d s}{1+\tau_f s}\right)$$

and the controller parameters were found as

$k_p = 0.60$, $\tau_i = 3.58$, $\tau_d = 3.60$ and $\tau_f = 0.033$

The resulting output response to a step setpoint change is shown in Figure 8.20. The results from the other tuning rules are also given for comparison. The performance of inverse-based PID controller is satisfactory compared with those of IMC or ZN PID controllers.

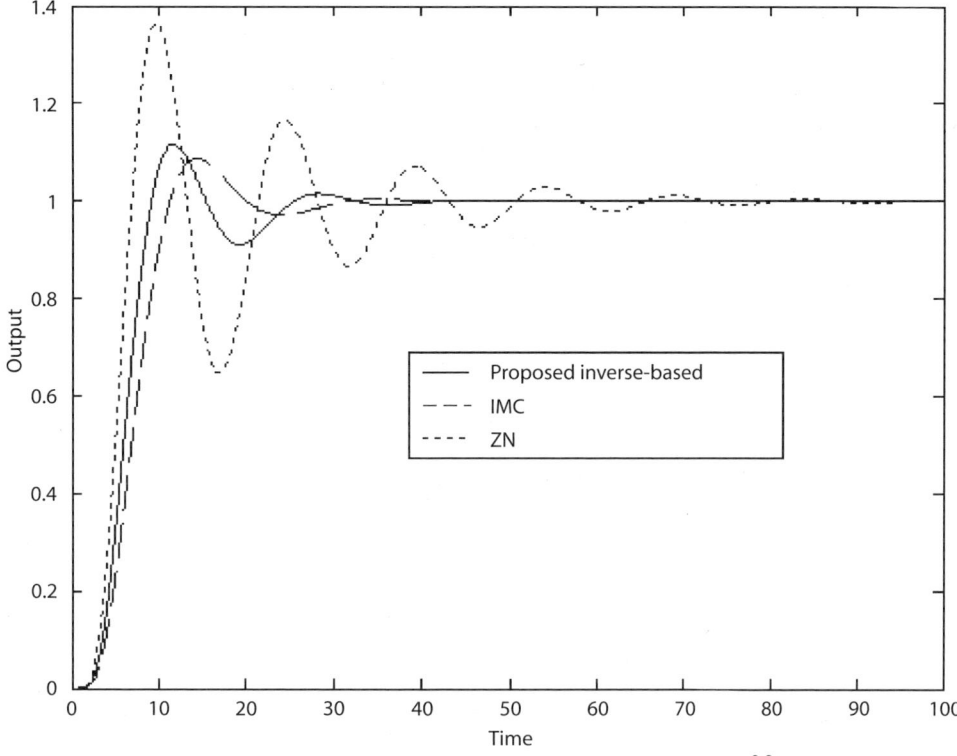

Figure 8.20 Comparison of control results for $G_p(s) = \dfrac{e^{-0.8s}}{(2s+1)^3(s+1)^2}$.

8.5 Assessment of PI/PID Control Performance

One of the common process control specifications is to provide good performance in tracking setpoint changes and rejecting output disturbances. Consequently, in real applications it would be invaluable to know how well the installed PID control system is performing against a benchmark of what is actually achievable from a particular loop. Quantifiable measures which can be used to perform an assessment of the performance of a conventional PID control system will be discussed in this section. In the spirit of the previous work in this chapter, the open-loop process in this control loop is assumed to be represented by simple models such as first-order or second-order plus dead time models. The benchmarks to be studied are the achievable optimal integral of absolute error (the so called IAE performance criterion, denoted J with optimal value denoted J^*) and the corresponding rise time t_r^* resulting from the closed-loop system tracking response. The optimal performance of the simple feedback systems is computed and represented with graphs of J^* and t_r^*. These graphs enable the performance of PID controllers to be assessed.

8.5.1 Achievable Minimal IAE Cost and Rise Time

The achievable dynamic performance of a system is important for the assessment of a current control system in operation. As with the results from the previous sections, this performance depends upon the open-loop process being controlled and the structure and tuning of the controller selected. One of the performance measures of a control system adopted for assessment in this section is the IAE value. The IAE performance criterion is defined as

$$J = \text{IAE} = \int_0^\infty |e(t)| dt$$

where $e(t)$ is the error or difference between the controlled variable and its setpoint. Typically, IAE is a measure of performance since the size and length of the error in either direction is proportional to lost revenue (Shinskey, 1990).

Based on the open-loop process (FOPDT or SOPDT) and type of controller (PI or PID) used, the IAE value subject to an unit step setpoint change can be minimised by adjusting the controller parameters (denoted P) in the corresponding loop transfer function. This is equivalent to solving the following optimisation problem:

$$J = \min_P \text{IAE} = \min_P \int_0^\infty |e(t)| dt$$

Huang and Jeng (2002) solved this optimisation problem numerically and the results were correlated into functions of process parameters. In Table 8.3 and Table 8.4, the optimal IAE values for PI and PID controllers are given as $\bar{J}^*_{pi}$ and $\bar{J}^*_{pid}$, which are normalised by time delay (thus $\bar{J}^*_{pi}$ and $\bar{J}^*_{pid}$ are optimal IAE values when time delay of the process is unity). Therefore these $\bar{J}^*_{pi}$ and $\bar{J}^*_{pid}$ values have to be multiplied by the time delay θ to become the actual optimal J^*_{pi} and J^*_{pid} values, that is, $J^*_{pi} = \bar{J}^*_{pi}\theta$ and $J^*_{pid} = \bar{J}^*_{pid}\theta$.

Table 8.3 Optimal PI control loop and IAE cost value $\bar{J}^*_{pi}$.

Process	$G_p(s)$	$G_{lp}(s) = G_p(s)G_c(s)$
FOPDT $\bar{\tau} = \tau/\theta \leq 5$	$\dfrac{Ke^{-\bar{s}}}{\bar{\tau}\bar{s}+1}$	$\dfrac{K_0(B\bar{s}+1)}{\bar{\tau}\bar{s}+1}\dfrac{e^{-\bar{s}}}{\bar{s}}$
	$\bar{J}^*_{pi} = 2.1038 - 0.6023\, e^{-1.0695\bar{\tau}}$	
FOPDT $\bar{\tau} = \tau/\theta \leq 5$	$\dfrac{Ke^{-\bar{s}}}{\bar{\tau}\bar{s}+1}$	$\dfrac{0.59e^{-\bar{s}}}{\bar{s}}$
	$\bar{J}^*_{pi} = 2.1038$	
SOPDT $\zeta \leq 2.0$	$\dfrac{Ke^{-\bar{s}}}{\bar{\tau}^2\bar{s}^2 + 2\bar{\tau}\zeta\bar{s}+1}$	$\dfrac{K_0(B\bar{s}+1)}{(\bar{\tau}^2\bar{s}^2+2\bar{\tau}\zeta\bar{s}+1)}\dfrac{e^{-\bar{s}}}{\bar{s}}$
	$\bar{J}^*_{pi} = \alpha(\zeta)\bar{\tau}^2 + \beta(\zeta)\bar{\tau} + \gamma(\zeta)$ $\alpha(\zeta) = 0.7444\zeta^3 - 1.4975\zeta^2 + 1.0202\zeta - 0.2525$ for $\zeta \leq 0.7$ $\alpha(\zeta) = 0.0064\zeta - 0.0203$ for $0.7 < \zeta < 2.0$ $\beta(\zeta) = 1.1193\zeta^{-0.9339}$ for $\zeta \leq 2.0$ $\gamma(\zeta) = -1.84675\zeta^2 + 17.9592\zeta - 2.7222$ for $\zeta \leq 0.5$ $\gamma(\zeta) = -0.0995\zeta^2 + 0.4893\zeta + 1.4712$ for $0.5 < \zeta \leq 2.0$	
SOPDT $\zeta > 2.0$	$\dfrac{Ke^{-\bar{s}}}{(\bar{\tau}_1\bar{s}+1)(\bar{\tau}_2\bar{s}+1)};\ \bar{\tau}_1 \geq \bar{\tau}_2$	$\dfrac{K_0(B\bar{s}+1)}{(\bar{\tau}_1\bar{s}+1)(\bar{\tau}_2\bar{s}+1)}\dfrac{e^{-\bar{s}}}{\bar{s}}$
	$\bar{J}^*_{pi} = -0.0173\bar{\tau}_2^2 + 1.7749\bar{\tau}_2 + 2.3514$	

8.5 Assessment of PI/PID Control Performance

Table 8.4 Optimal PID control loop and IAE cost value $\bar{J}_{pid}^*$.

Process	$G_p(s)$	$G_{lp}(s) = G_p(s)G_c(s)$
FOPDT $\bar{\tau} = \tau/\theta \leq 3$	$\dfrac{Ke^{-\bar{s}}}{\bar{\tau}\bar{s}+1}$	$\dfrac{K_0(A\bar{s}^2 + B\bar{s} + 1)}{\bar{\tau}\bar{s}+1} \dfrac{e^{-\bar{s}}}{\bar{s}}$
	$\bar{J}_{pid}^* = 1.38 - 0.1134\,e^{-1.5541\bar{\tau}}$	
FOPDT $\bar{\tau} = \tau/\theta > 3$	$\dfrac{Ke^{-\bar{s}}}{\bar{\tau}\bar{s}+1}$	$\dfrac{0.76(0.47\bar{s}+1)e^{-\bar{s}}}{\bar{s}}$
	$\bar{J}_{pid}^* = 1.38$	
SOPDT $\zeta \leq 1.1$	$\dfrac{Ke^{-\bar{s}}}{\bar{\tau}^2\bar{s}^2 + 2\bar{\tau}\zeta\bar{s} + 1}$	$\dfrac{K_0(A\bar{s}^2 + B\bar{s} + 1)}{(\bar{\tau}^2\bar{s}^2 + 2\bar{\tau}\zeta\bar{s} + 1)} \dfrac{e^{-\bar{s}}}{\bar{s}}$
	$\bar{J}_{pid}^* = 2.1038 - \lambda(\zeta)e^{-\mu(\zeta)\bar{\tau}}$ $\lambda(\zeta) = 0.4480\zeta^2 - 1.0095\zeta + 1.2904$ $\mu(\zeta) = 6.1998e^{-3.8888\zeta} + 0.6708$	
SOPDT $\zeta > 1.1$	$\dfrac{Ke^{-\bar{s}}}{(\bar{\tau}_1\bar{s}+1)(\bar{\tau}_2\bar{s}+1)};\ \bar{\tau}_1 \geq \bar{\tau}_2$	$\dfrac{K_0(B\bar{s}+1)}{(\bar{\tau}_2\bar{s}+1)} \dfrac{e^{-\bar{s}}}{\bar{s}}$
	$\bar{J}_{pid}^* = 2.1038 - 0.6728e^{-1.2024\bar{\tau}_2}$	

IAE Cost Values and Rise Times for FOPDT Process Models

For FOPDT processes with PI control (see Table 8.3), the value of the optimal IAE criterion is a function of the ratio of τ to θ. Notice that the value of the optimal IAE criterion remains approximately constant at 2.1θ when $\bar{\tau} > 5$, and for $\bar{\tau} \leq 5$ the value of the optimal IAE criterion lies in the range $1.7\theta \leq J_{pi}^* \leq 2.1\theta$. For those cases where $J_{pi}^* \approx 2.1\theta$, the reset time is approximately equal to the process time constant. In the case of FOPDT processes with PID control, a first-order filter is needed to fulfil the requirement for realisability. As a result, the value of the achievable optimal IAE criterion is a function not only of the ratio of τ to θ, but also of the time constant of the low pass filter. If this filter time constant is taken very small, the achievable IAE criterion will approach the value 1.38θ.

For both PI and PID control, the optimal rise time is obtained from simulation based on the optimal loop transfer function and is plotted as shown in Figure 8.21. For optimal PID control loops of FOPDT processes, it is found that the rise time is independent of the value of τ/θ. On the other hand, in the case of the optimal PI control loops, the rise time depends on the value of τ/θ when $\tau/\theta \leq 5$, as shown in Figure 8.21.

IAE Cost Values and Rise Times for SOPDT Process Models

Similarly, for SOPDT processes with PI/PID control, the results of optimal IAE cost values are functions of parameters of open-loop processes. Notice that for the PI control loop the function of the optimal IAE cost value is divided into two parts depending on the value of ζ (see Table 8.3). The first part applies to

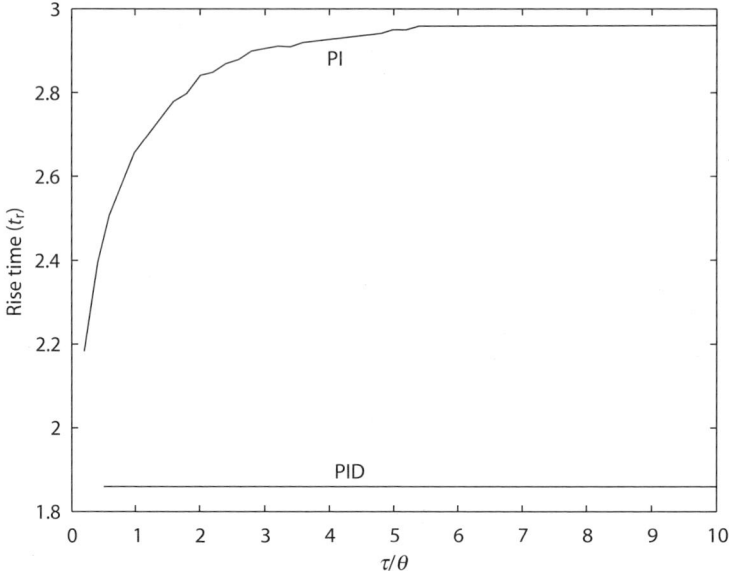

Figure 8.21 Optimal rise time t_r^* of a PI/PID control system for the FOPDT process model.

those systems which have $\zeta \leq 2.0$, where $\bar{J}_{pi}^*$ is a function of $\bar{\tau}$ and ζ. The second part applies to those $\zeta > 2.0$, where $\bar{J}_{pi}^*$ is only a function of the ratio of minor time constant to time delay $\bar{\tau}_2$. The IAE cost function values for the PID controller are also found to be given in two parts (see Table 8.4), that is for $\zeta \leq 1.1$ and for $\zeta > 1.1$. The rise times of these optimal systems are also obtained from simulation based on corresponding optimal loop transfer functions, and the results are plotted in Figure 8.22.

In fact, to justify which type of controller fits better to control a given process the system dynamic is not the only issue of concern. From the dynamical point of view, the actual achievable minimum IAE criterion value resulting from two designs based on different dynamic models can be compared. Although both FOPDT and SOPDT models can be used to model the same open-loop dynamics, the apparent dead time of the former is usually larger than that of the latter. Thus by applying the formulae in Table 8.3 and Table 8.4, two achievable minimal IAE values can be obtained and compared.

8.5.2 Assessment of PI/PID Controllers

PI/PID control systems have been widely used in the industrial process control and the performance of these control systems is dependent on the dynamics of the open-loop process. Huang and Jeng (2002) indicate that if the controller in a simple feedback loop has a general form, not restricted to the PI/PID controller, the achievable minimum IAE criterion value is found to be 1.38θ. To understand how well a PI/PID control system can perform referring to the optimal system with general controllers, an efficiency factor of the optimal PI/PID control systems (designated as η^*) for FOPDT and SOPDT processes is computed according to the formula

$$\eta^* = \frac{1.38\theta}{J_{pi}^* (\text{or } J_{pid}^*)}$$

It is found that for open-loop processes with second-order plus delay time dynamics the efficiency of PID control is mostly around 65% only, but for FOPDT processes the efficiency is very close to 100%. Thus, compared with the optimal simple feedback system, the PID controller is not as efficient for

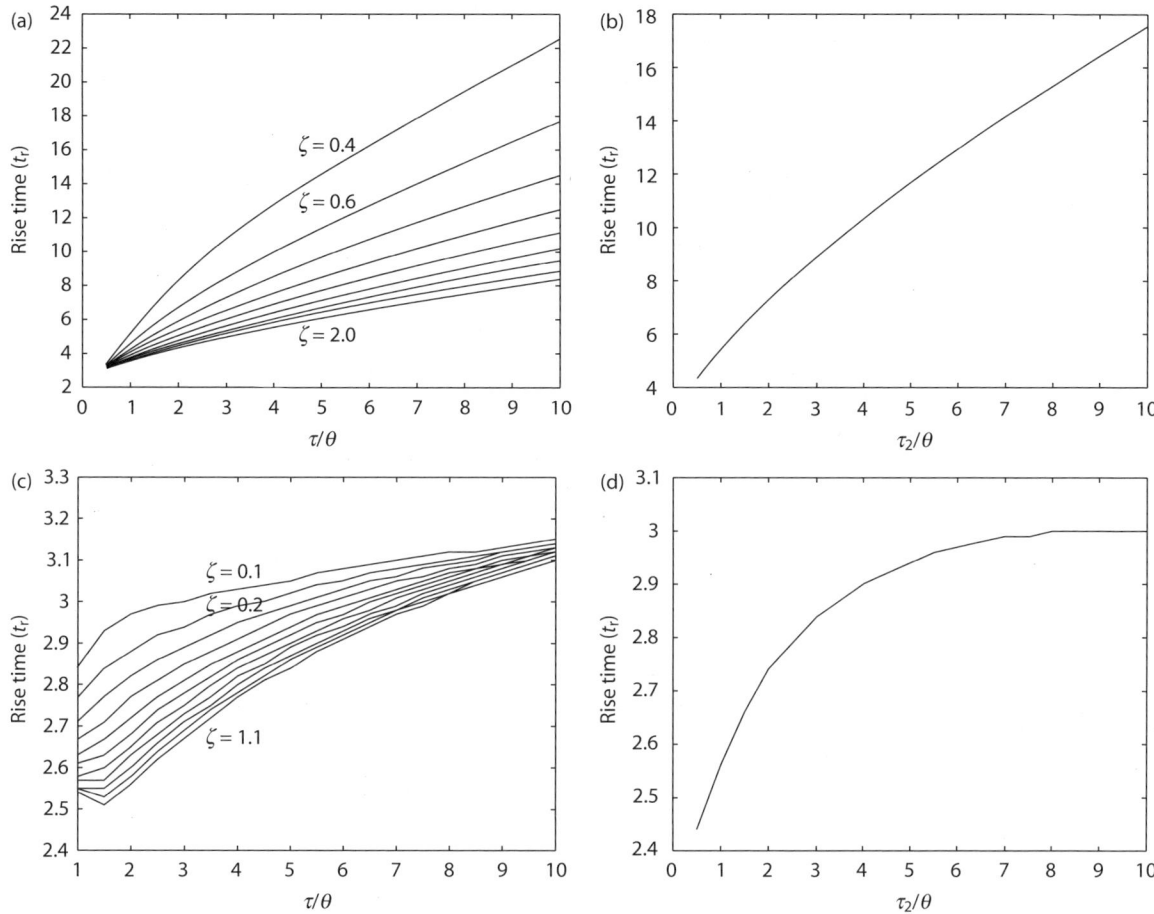

Figure 8.22 Optimal rise time t_r^* of PI/PID control systems for SOPDT processes. (a) Process G_p, $\zeta \leq 2.0$; Controller G_c, PI. (b) Process G_p, $\zeta > 2.0$; Controller G_c, PI. (c) Process G_p, $\zeta \leq 1.1$; Controller G_c, PID. (d) Process G_p, $\zeta > 1.1$; Controller G_c, PID.

SOPDT processes. On the other hand, the PID controller has the best efficiency for controlling FOPDT processes.

The graphs of $\bar{J}_{pi}^*$ and $\bar{J}_{pid}^*$ versus t_r^* as shown in Figures 8.23 and 8.24 can be used to highlight the performance of PI/PID loops. The status of the performance of a particular PI/PID control loop can be ascertained by computing the actual value for the loop integrated absolute error criterion $\bar{J}$, recording the actual loop rise time t_r and then locating the point $(\bar{J}, t_r)$ on the appropriate figure. If it is found that the point $(\bar{J}, t_r)$ is far away from the optimal region then the system is not performing well and re-tuning should be considered. Further, the actual location of the point will give an indication of the weakness of the control system. For example, if the point lies beneath the optimal region, it means that the tracking error is due to the response being too fast. It might be thought that the assessment requires knowledge of the open-loop dynamics, but from Figures 8.23 and 8.24 it can be seen that the optimal regions form fairly narrow bands. Consequently, the performance indicators should not be so sensitive to the actual process parameters; thus as long as the point $(\bar{J}, t_r)$ is located on the bands of the corresponding controller type, the performance of the system is close to optimal.

334 Process Reaction Curve and Relay Methods Identification and PID Tuning

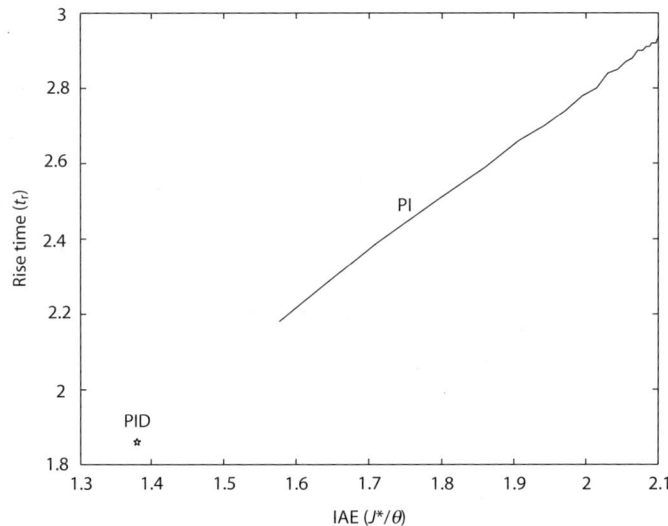

Figure 8.23 Graph of t_r^* vs. J^*/θ for optimal PI/PID control system of FOPDT processes.

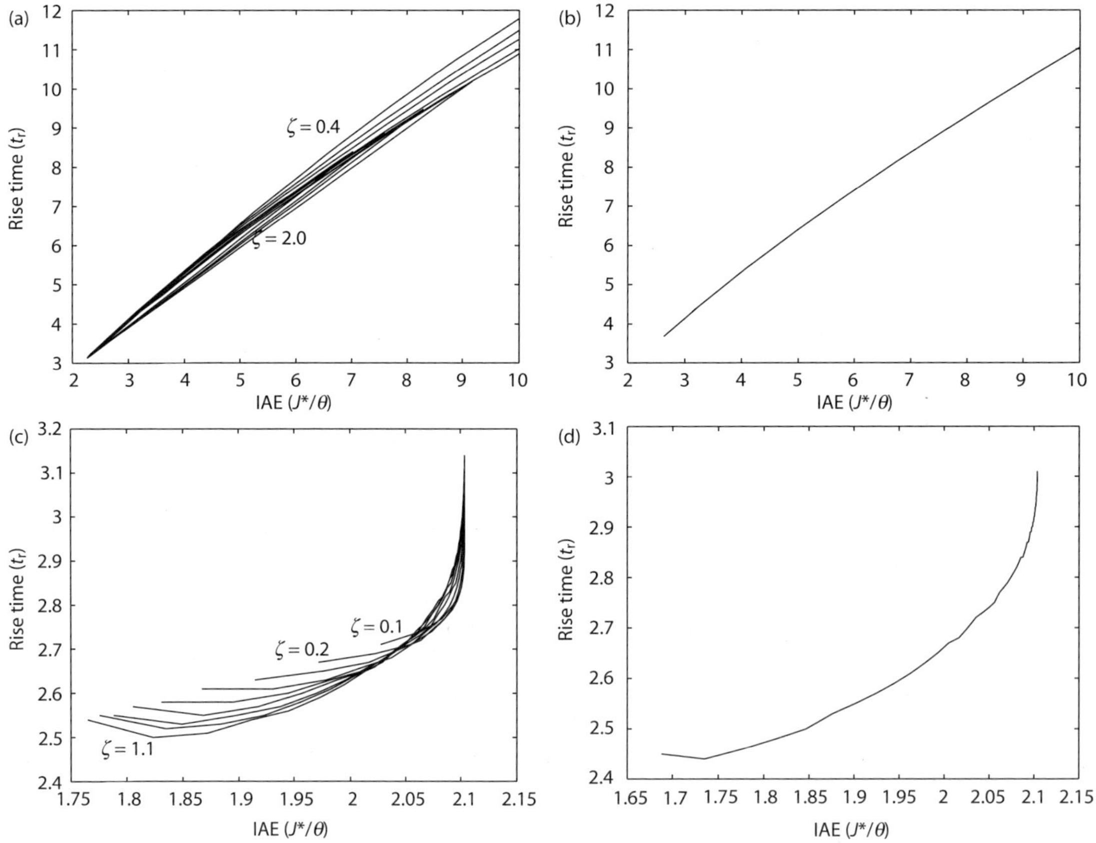

Figure 8.24 Graphs of t_r^* vs. J^*/θ for optimal PI/PID control system of SOPDT processes. (a) Process G_p, $\zeta \leq 2.0$; controller G_c, PI. (b) Process G_p, $\zeta > 2.0$; controller G_c, PI. (c) Process G_p, $\zeta \leq 1.1$; controller G_c, PID. (d) Process G_p, $\zeta > 1.1$; controller G_c, PID.

Example 8.9: Assessment of PID controller

Consider a PID control system for a process with the following model:

$$G_p(s) = \frac{e^{-s}}{4s^2 + 3.2s + 1}$$

The system starts with a PID controller with the following parameters:

$$k_p = 3.0, \ \tau_i = 3.0, \ \tau_d = 1.0$$

Both $\bar{J}$ and t_r are computed from the formula given in Table 8.5, and then the point $(\bar{J}, t_r)$ is located using the graph of Case (c) in Figure 8.24. It is found that the point is far off the optimal region, as shown in Figure 8.25 (point 1). Moreover, its location indicates that the large *IAE* value is due to the response being too fast; namely, the rise time is too small. For this reason, the controller is re-tuned to be more conservative to the second setting by decreasing k_p and increasing τ_i as given in Table 8.5. Although the *IAE* value is smaller than the first one, this setting is still not optimal due to its sluggish response (point 2 in Figure 8.25). In this manner, the parameters of PID controller are changed to the third and fourth sets as given in Table 8.5. The location of $(\bar{J}, t_r)$ as shown in Figure 8.25 indeed implies the status of the performance at each stage. Since the point $(\bar{J}, t_r)$ of the fourth setting falls into the optimal region, it is concluded that the current control system is close to optimal.

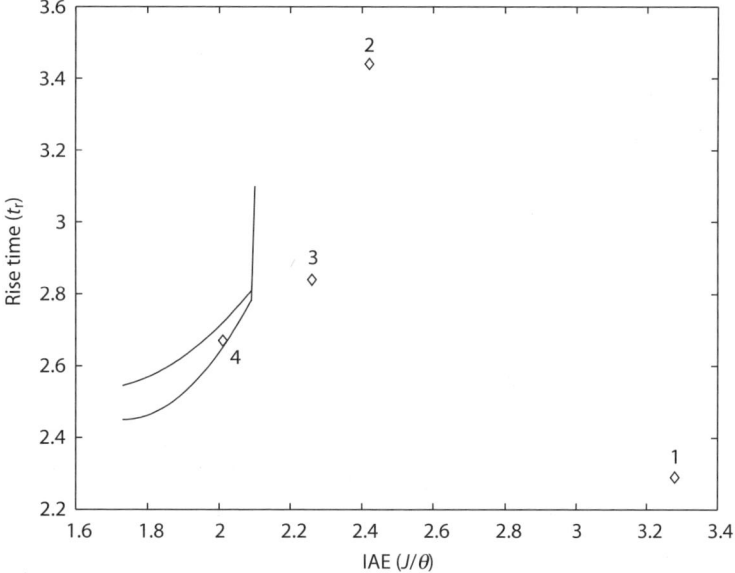

Figure 8.25 Assessment of PID control performance in Example 8.9.

Table 8.5 Parameters of PID controller and control performance in Example 8.9.

Setting	k_p	τ_i	τ_d	$\bar{J}$	t_r
Initial (1)	3.0	3.0	1.0	3.16	2.23
Second (2)	1.8	4.0	1.0	2.50	3.44
Third (3)	2.0	3.5	1.2	2.19	2.86
Fourth (4)	2.0	3.2	1.5	2.03	2.60

References

Åström, K.J. and Hägglund, T. (1988) *Automatic Tuning of PID Controllers*. Instrument Society of America, Research Triangle Park, NC.

Åström, K.J. and Hägglund, T. (1995) *PID Controllers: Theory, Design and Tuning*. Instrument Society of America, Research Triangle Park, NC.

Chien, I.L. and Fruehauf, P.S. (1990) Consider IMC tuning to improve controller performance. *Chemical Engineering Progress*, October, 33–41.

Hang, C.C., Åström, K.J. and Wang, Q.G. (2002) Relay feedback auto-tuning of process controllers – a tutorial review. *J. Proc. Cont.*, **12**, 143–162.

Hang, C.C., Åström, K.J. and Ho, W.K. (1991) Refinement of the Ziegler–Nichols tuning formula. *IEE Proceedings-D*, **138**, 111–118.

Hang, C.C., Lee, T.H. and Ho, W.K. (1993) *Adaptive Control*. Instrument Society of America, Research Triangle Park, NC.

Harriott, P. (1964) *Process Control*. McGraw-Hill, New York.

Huang, C.T. and Chou, C.J. (1994) Estimation of the underdamped second-order parameters from the system transient. *Ind. Eng. Chem. Res.*, **33**, 174–176.

Huang, C.T. and Huang, M.F. (1993) Estimating of the second-order parameters from process transient by simple calculation. *Ind. Eng. Chem. Res.*, **32**, 228–230.

Huang, C.T. and Clements Jr, W.C. (1982) Parameter estimation for the second-order-plus-dead-time model. *Ind. Eng. Chem. Process Des. Dev.*, **21**, 601–603.

Huang, H.P. and Jeng, J.C. (2002) Monitoring and assessment of control performance for single loop systems. *Ind. Eng. Chem. Res.*, **41**, 1297–1309.

Huang, H.P. and Jeng, J.C. (2003) Identification for monitoring and autotuning of PID controllers. *J. Chem. Eng. Japan*, **36**, 284–296.

Huang, H.P., Lee, M.W. and Chen, C.L. (2001) A system of procedures for identification of simple models using transient step response. *Ind. Eng. Chem. Res.*, **40**, 1903–1915.

Huang, H.P., Lee, M.W. and Chien, I.L. (2000) Identification of transfer-function models from the relay feedback tests. *Chem. Eng. Commun.*, **180**, 231–253.

Li, W., Eskinat, E. and Luyben, W.L. (1991) An improved autotune identification method. *Ind. Eng. Chem. Res.*, **30**, 1530–1541.

Luyben, W.L. (2001) Getting more information from relay-feedback tests. *Ind. Eng. Chem. Res.*, **40**, 4391–4402.

McMillan, G.K. (1994) *Tuning and Control Loop Performance, A Practitioner Guide*, 3rd edn. Instrument Society of America, Research Triangle Park, NC.

Oldenbourg, R.C. and Sartorius, H. (1948) The dynamics of automatic control. *Trans. ASME*, **77**, 75–79.

Shinskey, F. (1990) How good are our controllers in absolute performance and robustness? *Measurement and Control*, **23**, 114–121.

Sundareesan, K.R., Chandra Prasad, C. and Krishnaswamy, C. (1978) Evaluating parameters for process transients. *Ind. Eng. Chem. Process Des. Dev.*, **17**, 237–241.

Sung, S.W., Jungmin, O., Lee, I.B., Lee, J. and Yi, S.H. (1996) Automatic tuning of PID controller using second-order plus time delay model. *J. Chem. Eng. Japan*, **6**, 990–999.

Wang, Q.G., Guo, X. and Zhang, Y. (2001) Direct identification of continuous time delay systems from step response. *J. Proc. Contr.*, **11**, 531–542.

Yu, C.C. (1999) *Autotuning of PID Controllers*. Springer-Verlag, London.

9 Fuzzy Logic and Genetic Algorithm Methods in PID Tuning

Learning Objectives
9.1 Introduction
9.2 Fuzzy PID Controller Design
9.3 Multi-Objective Optimised Genetic Algorithm Fuzzy PID Control
9.4 Applications of Fuzzy PID Controllers to Robotics
9.5 Conclusions and Discussion
Acknowledgments
References

Learning Objectives

Conventional PID controllers are extensively used in industry due to their simplicity in design and tuning and effectiveness for general linear systems, with convenient implementation and low cost. However, the tuning of the gains is always a challenge in the state of the art of PID controller design. In this chapter, approaches for the design of optimal PID controllers using genetic algorithms with multi-criteria and fuzzy logic are presented.

The learning objectives for this chapter are to:

- Understand the structure of a fuzzy PID controller and its components.
- Understand how the concepts of fuzzification and defuzzification are used in a fuzzy PID controller
- Understand how to create a PID controller using genetic algorithms concepts.
- Appreciate the potential performance benefits of these controllers by looking at the results from several industrial applications.

9.1 Introduction

PID control refers to a family of controllers with various configurations of the Proportional, Integral and Derivative terms. It is well known that in general, the P-controller is used to reduce the detected tracking error, independent of the phase shift between the output and input; the I-controller is applied to reduce (or even eliminate) the steady state tracking error; and the D-controller can reduce the maximum overshoot (but may retain a steady state tracking error). A full assembly of the PID controller is usually expected to combine these merits; namely to have fast rise time and settling time, with small or no overshoot and oscillation, and with small or no steady state errors.

Conventional PID controllers have been extensively used in industry due to their simplicity in design and tuning and effectiveness for general linear systems, with convenient implementation and low cost. For example, it has been reported that more than 90% of control loops in Japanese industry are of PID type (Yamamoto and Hashimoto, 1991), and this is also believed to be true elsewhere (Swallow, 1991). Because of their practical value, conventional PID controllers continually attract attention from both academic and industrial researchers, with many new studies being published (Grimble, 2001; Liu et al., 2001; O'Dwyer, 2003, Silva et al., 2002; Suchomski, 2001; Tokuda and Yamamoto, 2002; Zheng et al., 2002).

Conventional PID controllers have been well developed for nearly a century (Bennett, 1993). Spurred by the rapid development of advanced micro-electronics and digital processors, these controllers have recently gone through a technological evolution from their early versions as pneumatic controllers implemented by analogue electronics (Bennett, 2001) to the current versions as microprocessors implemented by digital circuits.

Although PID controllers can be analytically designed and pre-tuned for precisely given lower-order linear systems, they have to be manually operated for most practical systems that involve higher-order components, nonlinearity and uncertainties. To find easy ways of choosing suitable control gains in these controllers, Ziegler and Nichols (1942; 1943) and Cohen and Coon (1953) of the Taylor Instrument Company initiated the now well-known heuristic rules for experimental design and tuning methods.

However, tuning the gains is always a challenge in the state of the art of PID controller design. This problem becomes more important and critical, in particular, when issues including specifications, stability and performance are considered. In this chapter, the use of two distinct intelligent approaches to this problem, fuzzy logic and genetic algorithms, is reviewed.

Fuzzy logic as a model to mimic the experienced operator is adapted to the design of autotuning PID. Besides the direct combination of fuzzy logic and PID controllers, some non-conventional PID controllers employing fuzzy logic have also been developed (Chen, 1996; Chen and Pham, 2000; Ying et al., 1990).

Genetic algorithms (GA) work on the Darwinian principle of natural selection. They possess an intrinsic flexibility and freedom to search for a desirable solution according to the design specifications. Whether the specifications are nonlinear, constrained or multimodal, GA are entirely equal to the challenge. In addition, they have the distinct advantage of being able to solve the class of multi-objective problems to which controller design often belongs.

9.2 Fuzzy PID Controller Design

As has been known, and pointed out explicitly by Driankov et al. (1995), conventional PID controllers are generally insufficient to control processes with additional complexities such as large time delays, significant oscillatory behaviour, parameter variations, nonlinearities, and MIMO plants. To improve conventional PID controllers, fuzzy logic is adapted.

Fuzzy-logic-based PID controllers are a kind of intelligent autotuning PID controller (Åström, 1992; Åström et al., 1992). There is some active research on the topics of autotuning of PID control systems using fuzzy logic reported in the current literature. In addition to those briefly reviewed by Chen (1996), the following are some more recent approaches reported.

Xu et al. (1996) discussed the tuning of fuzzy PI controllers based on gain/phase margin specifications and the ITAE index. Furthermore, Xu et al. (2000) studied a parallel structure for fuzzy PID controllers with a new method for controller tuning. Better performance was obtained, and stability was analysed, in a way similar to the approach of Malki et al. (1994) and Chen and Ying (1997). A new method was proposed by Woo et al. (2000) for on-line tuning of fuzzy PID-type controllers via tuning of the scaling factors of the controllers, yielding better transient and steady state responses. Mann et al. (2001) discussed two-level fuzzy PID controller tuning – linear and nonlinear tuning at the lower and higher levels, respectively – thereby obtaining better performance results. Visioli (2001) presented a comparison among several different methods, all based on fuzzy logic, for tuning PID controllers, where the practical implementation issues of these controllers were also discussed. The simplest possible PI and PD controllers that have only two fuzzy sets in the input and three in the output paths have also been studied and reported upon (Patel and Mohan, 2002; Mohan and Patel, 2002). Design, performance evaluation and stability analyses of fuzzy PID controllers are all found in the work of Malki and Chen (1994) and Chen and Ying (1997). Kuo and Li (1999) combined the fuzzy PI and fuzzy PD controllers for an active suspension system, in which the fusion of genetic algorithms (GA) and fuzzy control provides much better performance; namely, a much more comfortable ride service to the passengers.

One alternative form of fuzzy logic-based PID controller is to combine fuzzy logic control with the PID controller to obtain a kind of behaviourally similar but better-performing PID controller.

There were some suggestions for designing a fuzzy PID controller by simply adding a few fuzzy IF–THEN rules to a conventional PID controller, but the improvement in control performance in such a design is not convincing. Further, such controllers cannot usually be used to directly replace a PID controller working in a plant or process, not to mention that these controllers do not have guaranteed stability, so this approach is generally not suitable for real-time applications.

Therefore a desirable goal has been to develop new PID controllers with the following properties and features:

1. Similar structures, and working in a similar way, to conventional PID controllers.

2. Precise control formulae for calculation of controller parameters (if needed) and guaranteed closed-loop stability properties by design.

3. Can be used to directly replace any commercial PID controller currently in use without modifying the original setup.

4. Use fuzzy logic (such as IF–THEN rules) only for the design step (so as to improve the control performance), not throughout the entire real-time control operation and activity.

5. Cost only a little more than PC-operated PID controllers but perform much better, so as to satisfy the industrial consumer specification and trade-off.

It turns out that this goal can be achieved to a certain degree of satisfaction, and the resulting fuzzy PID controllers have almost all the properties listed above. They include the blue print fuzzy PD controller (Malki et al., 1994), fuzzy PI controller, (Ying et al., 1990) and the fuzzy PID controller (Misir et al., 1996, Tang et al., 2001a), among other combinations (Carvajal et al., 2000; Sooraksa and Chen, 1998; Tang et al., 2001b). Computer simulations have already shown that these fuzzy PID controllers work equally as well as conventional PID controllers for low- (first- and second-) order linear plants, and yet have significant improvement over conventional controllers for high-order and time-delayed linear systems, and outperform conventional PID controllers on many nonlinear plants (Malki et al., 1994; Chen and Pham, 2000). Furthermore, it has been proven that the closed-loop stability of fuzzy PD controllers is guaranteed to within a sufficient condition (Chen and Ying, 1997; Chen and Pham, 2000).

9.2.1 Fuzzy PI Controller Design

A conventional structure for a PI+D control system is shown in Figure 9.1. This has the PI term acting on the closed-loop tracking error signal $e = e(t) = r(t) - y(t)$ and the D term acts only on the process output $y = y(t)$. As explained in Chapter 1, the relocation of the D term in the feedback path is to prevent derivative kick occurring.

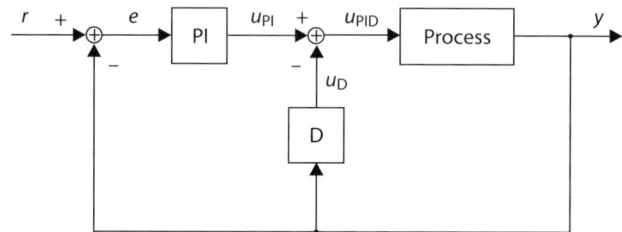

Figure 9.1 Conventional continuous-time PI+D control system.

Within the framework of Figure 9.1, the output signal of the conventional analogue PI controller in s-domain notation is given by

$$U_{PI}(s) = \left(K_p^c + \frac{K_i^c}{s} \right) E(s)$$

This equation for the control signal $U_{PI}(s)$ can be converted into the z-domain by applying the usual bilinear transformation given by

$$s = \left(\frac{2}{T}\right)\left(\frac{z-1}{z+1}\right) = \left(\frac{2}{T}\right)\left(\frac{1-z^{-1}}{1+z^{-1}}\right)$$

where $T > 0$ is the sampling period. The bilinear transformation has the advantage that it maps the left half-plane of the s-domain onto the unit circle in the z-domain. Hence stable poles map to stable locations in the z-domain and left half-plane zeros map to within the unit circle in the z-domain. Applying the bilinear transformation to the signal $U_{PI}(s)$ results in the following analysis:

$$U_{PI}(z) = \left(K_p^c + \frac{K_i^c T}{2}\left(\frac{1+z^{-1}}{1-z^{-1}}\right) \right) E(z) = \left(K_p^c + \frac{K_i^c T}{2}\left(\frac{-(1-z^{-1})+2}{1-z^{-1}}\right) \right) E(z)$$

and hence

$$U_{PI}(z) = \left(K_p^c - \frac{K_i^c T}{2} + \frac{K_i^c T}{1-z^{-1}} \right) E(z)$$

Define two new gains:

$$K_p = K_p^c - \frac{K_i^c T}{2},\ K_i = K_i^c$$

Then

$$(1-z^{-1})U_{PI}(z) = (K_p(1-z^{-1}) + K_i T)E(z)$$

Take the inverse z-transform to obtain the following difference equation:

$$u_{PI}(nT) - u_{PI}(nT-T) = K_p[e(nT) - e(nT-T)] + K_i T e(nT)$$

This difference equation can be rearranged as follows:

$$u_{PI}(nT) = u_{PI}(nT-T) + T\left[K_p\left(\frac{e(nT) - e(nT-T)}{T}\right) + K_i e(nT)\right]$$

to give

$$u_{PI}(nT) = u_{PI}(nT-T) + T\Delta u_{PI}(nT)$$

where

$$\Delta u_{PI}(nT) = K_p e_v(nT) + K_i e_p(nT), e_p(nT) = e(nT) \text{ and } e_v(nT) = \frac{e(nT) - e(nT-T)}{T}$$

Thus within this simple time-domain update equation for the PI control signal $u_{PI}(nT)$, the term $\Delta u_{PI}(nT)$ is the incremental control output of the PI controller, $e_p(nT)$ is the error signal and $e_v(nT)$ is the rate of change of the error signal.

To facilitate the entry to a fuzzy control structure, the term $T\Delta u_{PI}(nT)$ is replaced by a fuzzy control action $K_{uPI}\Delta u_{PI}(nT)$, where K_{uPI} is a constant gain to be determined, the PI output then becomes

$$u_{PI}(nT) = u_{PI}(nT-T) + K_{uPI}\Delta u_{PI}(nT)$$

where the fuzzy PI constant control gain is K_{uPI}.

Summary of Fuzzy PI Controller Equation Set

PI controller update: $u_{PI}(nT) = u_{PI}(nT-T) + K_{uPI}\Delta u_{PI}(nT)$

Incremental control output: $\Delta u_{PI}(nT) = K_p e_v(nT) + K_i e_p(nT)$

Error signal: $e_p(nT) = e(nT)$

Error signal rate of change: $e_v(nT) = \dfrac{e(nT) - e(nT-T)}{T}$

9.2.2 Fuzzy D Controller Design

The D control signal in the PI+D control system as shown in Figure 9.1 can be given the s-domain expression

$$U_D(s) = sK_d^c Y(s)$$

where K_d^c is the derivative gain and $Y(s)$ is the process output signal.

The z-domain form of $U_D(s)$ from the bilinear transform is

$$U_D(z) = K_d^c \frac{2}{T}\left(\frac{1-z^{-1}}{1+z^{-1}}\right)Y(z)$$

Introduce the new gain definition

$$K_d = \frac{2K_d^c}{T}$$

Then the expression for $U_D(z)$ rearranges as

$$(1+z^{-1})U_D(z) = K_d(1-z^{-1})Y(z)$$

so that the discrete time-domain difference equation is given by

$$u_D(nT) + u_D(nT-T) = K_d[y(nT) - y(nT-T)]$$

and thence to

$$u_D(nT) = -u_D(nT-T) + T\Delta u_D(nT)$$

where $\Delta u_D(nT) = K_d \Delta y(nT)$ is the incremental control output of the D controller and

$$\Delta y(nT) = \frac{y(nT) - y(nT-T)}{T}$$

is the rate of change of the output $y(t)$.

The entry to the fuzzy controller is achieved by replacing the term $T\Delta u_D(nT)$ in the above equation by a fuzzy control action term $K_{uD}\Delta u_D(nT)$ to give

$$u_D(nT) = -u_D(nT-T) + K_{uD}\Delta u_D(nT)$$

where K_{uD} is a constant gain to be determined.

To enable better performance of this D controller, namely, to avoid drastic changes of signal flows due to the differentiation on non-smooth internal signals, the equation giving the incremental controller output can be slightly modified by adding the signal $Ky_d(nT)$ to the right-hand side, so as to obtain

$$\Delta u_D(nT) = K_d \Delta y(nT) + Ky_d$$

where $y_d(nT) = y(nT) - r(nT) = -e(nT)$.

Summary of Fuzzy D Controller Equation Set

D controller update: $u_D(nT) = -u_D(nT-T) + K_{uD}\Delta u_D(nT)$

Incremental controller output: $\Delta u_D(nT) = K_d \Delta y(nT) + Ky_d(nT)$

Output rate of change: $\Delta y(nT) = \dfrac{y(nT) - y(nT-T)}{T}$

Addition to controller output: $y_d(nT) = y(nT) - r(nT) = -e(nT)$

9.2.3 Fuzzy PID Controller Design

The fuzzy PID control law can now be completed by algebraically summing the fuzzy PI control law update and fuzzy D control law update together; namely from Figure 9.1:

$$u_{PID}(nT) = u_{PI}(nT) - u_D(nT)$$

Hence the fuzzy PID control law equivalence will be

$$u_{PID}(nT) = u_{PI}(nT-T) + K_{uPI}\Delta u_{PI}(nT) + u_D(nT-T) - K_{uD}\Delta u_D(nT)$$

To this end, the configuration of the fuzzy PID controller is shown in Figure 9.2, where the above fuzzy PID control law has been implemented along with the control laws of the PI and D fuzzy controller equation sets above (Sections 9.2.1 and 9.2.2 respectively).

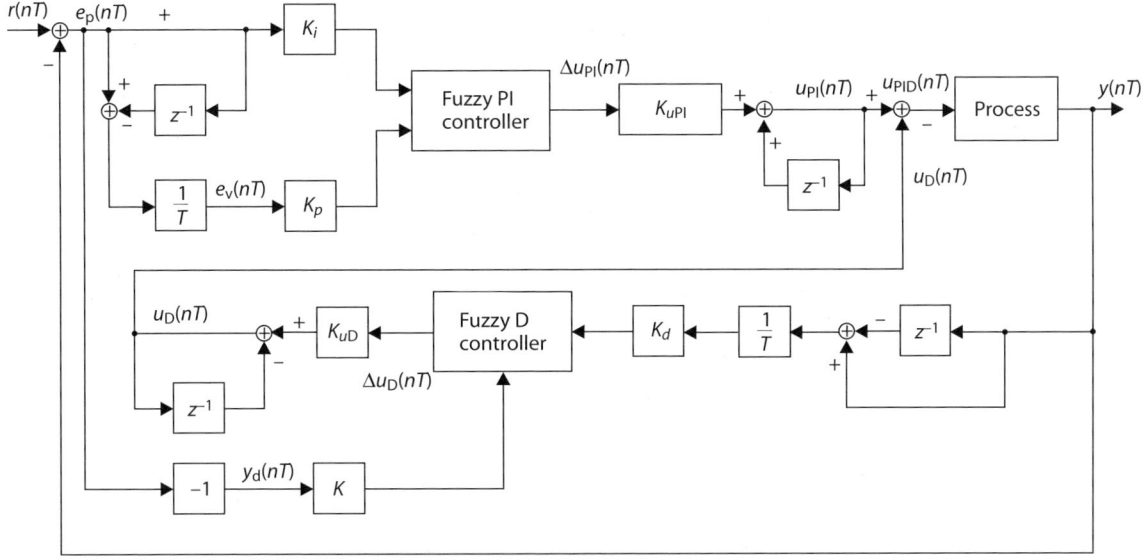

Figure 9.2 Configuration of the fuzzy PID controller.

9.2.4 Fuzzification

Having established the fuzzy PID control law the required fuzzification of the PI and D components can be made separately. The input and output membership functions of the PI component are shown in Figure 9.3, while those for the D component are in Figure 9.4.

It should be noted that, for simplicity, the number of membership functions is minimal, but the performance is still very good, as demonstrated later. To tune the membership function a single constant L is used since the inputs and outputs will be weighted by gains $K_p, K_i, K_d, K, K_{uPI}$ and K_{uD}. The value of L can be determined manually or optimally by using genetic algorithms as given in this chapter.

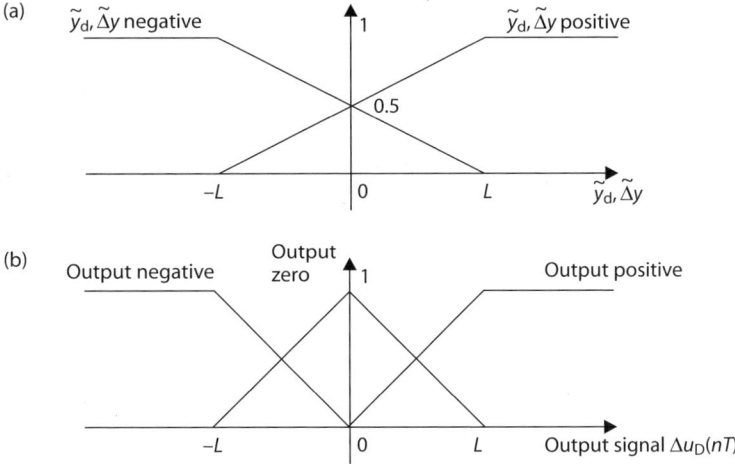

Figure 9.3 Membership functions of PI component. (a) Input membership functions; (b) Output membership functions.

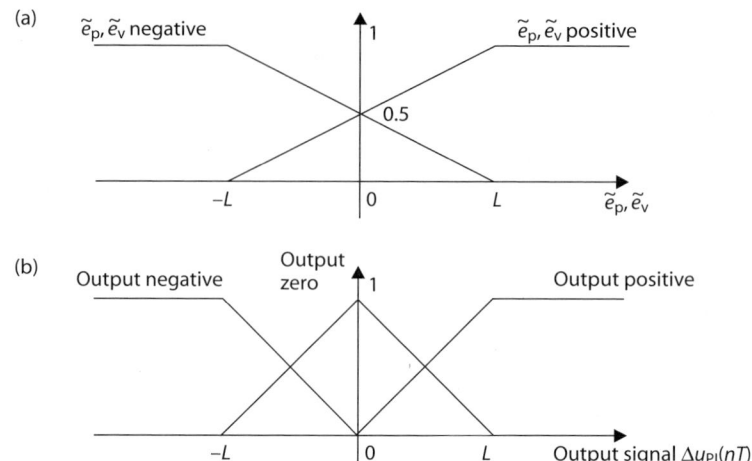

Figure 9.4 Membership functions for D component. (a) Input membership functions; (b) Output membership functions.

9.2.5 Fuzzy Control Rules

Using the aforementioned membership functions, the following four control rules are established for the fuzzy PI controller:

(R1)   IF $\tilde{e}_p = \tilde{e}_p \bullet n$ AND $\tilde{e}_v = \tilde{e}_v \bullet n$, THEN $\Delta u_{PI}(nT) = o \bullet n$

(R2)   IF $\tilde{e}_p = \tilde{e}_p \bullet n$ AND $\tilde{e}_v = \tilde{e}_v \bullet p$, THEN $\Delta u_{PI}(nT) = o \bullet z$

(R3)   IF $\tilde{e}_p = \tilde{e}_p \bullet p$ AND $\tilde{e}_v = \tilde{e}_v \bullet n$, THEN $\Delta u_{PI}(nT) = o \bullet z$

(R4)   IF $\tilde{e}_p = \tilde{e}_p \bullet p$ AND $\tilde{e}_v = \tilde{e}_v \bullet p$, THEN $\Delta u_{PI}(nT) = o \bullet p$

where $\Delta u_{PI}(nT)$ is the PI controller output, "$\tilde{e}_p = \tilde{e}_p \bullet p$" means "error positive", "$o \bullet p$" means output positive etc., and the logical "AND" takes the minimum.

Similarly, from the membership functions of the fuzzy D controller, the following control rules are used:

(R5)   IF $\tilde{y}_d = \tilde{y}_d \bullet p$ AND $\tilde{\Delta} y = \tilde{\Delta} y \bullet p$ THEN $\Delta u_D(nT) = o \bullet z$

(R6)   IF $\tilde{y}_d = \tilde{y}_d \bullet p$ AND $\tilde{\Delta} y = \tilde{\Delta} y \bullet n$ THEN $\Delta u_D(nT) = o \bullet p$

(R7)   IF $\tilde{y}_d = \tilde{y}_d \bullet n$ AND $\tilde{\Delta} y = \tilde{\Delta} y \bullet p$ THEN $\Delta u_D(nT) = o \bullet n$

(R8)   IF $\tilde{y}_d = \tilde{y}_d \bullet n$ AND $\tilde{\Delta} y = \tilde{\Delta} y \bullet n$ THEN $\Delta u_D(nT) = o \bullet z$

where $\Delta u_D(nT)$ is the output of fuzzy D controller and the other terms are defined similarly to the PI component.

9.2.6 Defuzzification

The input–output block diagram structures for the fuzzy PI and D controller components are shown in Figure 9.5. It is the task of defuzzification to use the input values to produce the incremental control output of the fuzzy PID control.

9.2 Fuzzy PID Controller Design

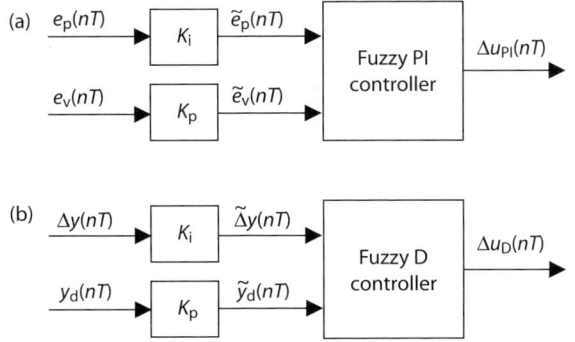

Figure 9.5 Block diagram structure for (a) the fuzzy PI and (b) the D controller components.

In the task of defuzzification for both fuzzy PI and D controllers (as shown in Figure 9.5) the incremental control of the fuzzy PID control is obtained using the centroid formula as follows:

$$\Delta u(nT) = \frac{\Sigma \text{ membership value of input} \times \text{output corresponding to the membership value of input}}{\Sigma \text{ membership values of input}}$$

A key diagram for the defuzzification task is given in Figure 9.6. This shows the input combinations (ICs) for the two fuzzy controller components, and the explanation for the use of these diagrams follows in the next two sections.

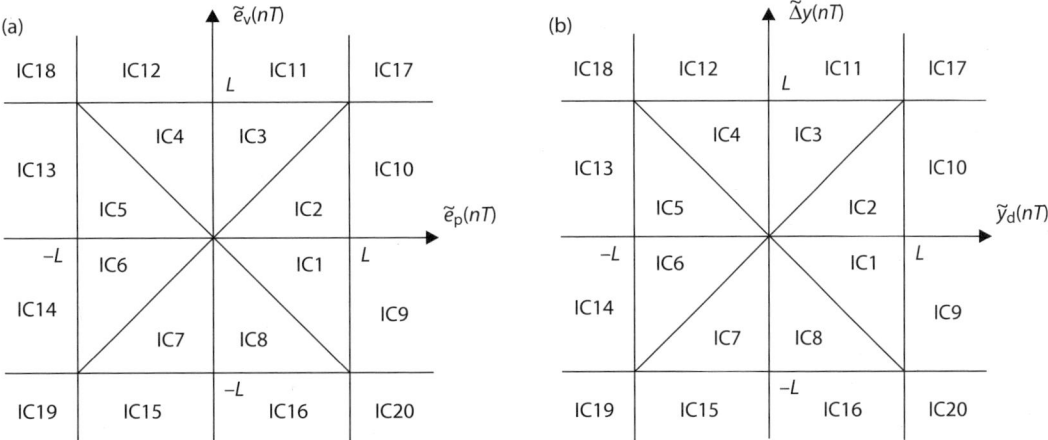

Figure 9.6 Regions of (a) fuzzy PI and (b) D controller input combination values.

Defuzzification of Fuzzy PI Controller

For the defuzzification PI controller, the value ranges of the two inputs, the error signal $\tilde{e}_p(nT) = K_i e_p(nT)$ and the rate of change of the error signal $\tilde{e}_v(nT) = K_p e_v(nT)$ are actually decomposed into 20 adjacent input combination (IC) regions, as shown in Figure 9.6(a). This figure is understood as follows:

1. Place the membership function of the error $\tilde{e}_p$ (Figure 9.3(a)) over the horizontal axis in Figure 9.6(a).

2. Place the membership function of the rate of change of error $\tilde{e}_v$ (given by the same curve in Figure 9.3(a)) over the vertical axis in Figure 9.6(a).

3. Combining 1 and 2 above will form a three-dimensional picture (albeit hidden in Figure 9.6(a)) over the two-dimensional regions shown in Figure 9.6(a).

The appropriate fuzzy PI control laws for each IC region can be executed by combining the fuzzy PI control rules (R1–R4), membership functions (Figure 9.3(a)) and IC regions (Figure 9.6(a)).

As an example, consider the situation of the location of error $\tilde{e}_p$ and the rate $\tilde{e}_v$ in the region IC1 and IC2. For the region of IC1, $\tilde{e}_p$ is within the region $[0, L]$ and $\tilde{e}_v$ in $[-L, 0]$. For the two signals, Figure 9.3(a) indicates the rule as $(\tilde{e}_v \bullet n > 0.5 > \tilde{e}_p \bullet n)$. According to R1, where the logical "AND" is used, this leads to:

"error $= \tilde{e}_p \bullet n$ AND rate $= \tilde{e}_v \bullet n$" $= \min\{\tilde{e}_p \bullet n, \tilde{e}_v \bullet n\} = \tilde{e}_p \bullet n$

Similarly, rules R1–R4 yield

$R1 \begin{cases} \text{the selected input membership value is } [\tilde{e}_p \bullet n]; \\ \text{the corresponding output membership value is } [o \bullet n] \end{cases}$

$R2 \begin{cases} \text{the selected input membership value is } [\tilde{e}_p \bullet n]; \\ \text{the corresponding output membership value is } [o \bullet z] \end{cases}$

$R3 \begin{cases} \text{the selected input membership value is } [\tilde{e}_v \bullet n]; \\ \text{the corresponding output membership value is } [o \bullet z] \end{cases}$

$R4 \begin{cases} \text{the selected input membership value is } [\tilde{e}_v \bullet n]; \\ \text{the corresponding output membership value is } [o \bullet p] \end{cases}$

It can be verified that the above rules are true for regions IC1 and IC2. Thus it follows from the centroid defuzzification formula that

$$\Delta u_{Pi}(nT) = \frac{\tilde{e}_p \bullet n \times o \bullet n + \tilde{e}_p \bullet n \times o \bullet z + \tilde{e}_v \bullet n \times o \bullet z + \tilde{e}_v \bullet n \times o \bullet p}{\tilde{e}_p \bullet n + \tilde{e}_p \bullet n + \tilde{e}_v \bullet n + \tilde{e}_{vp} \bullet p}$$

To this end, by applying $o \bullet p = L, o \bullet n = -L, o \bullet z = 0$ from Figure 9.3(b) and the following straight line formulae from the geometry of the membership functions associated with Figure 9.6(a):

$$\tilde{e}_p \bullet p = \frac{K_i e_p(nT) + L}{2L}; \quad \tilde{e}_p \bullet p = \frac{-K_i e_p(nT) + L}{2L}$$

$$\tilde{e}_v \bullet p = \frac{K_i e_v(nT) + L}{2L}; \quad \tilde{e}_v \bullet p = \frac{-K_i e_v(nT) + L}{2L}$$

Then

$$\Delta u_{PI}(nT) = \frac{L}{2(2L - K_i e_p(nT))}[K_i e_p(nT) + K_p e_v(nT)]$$

It should be noted that $e_p(nT) \geq 0$ in regions IC1 and IC2. In the same way, it can be verified that region IC5 and IC6 will yield

$$\Delta u_{PI}(nT) = \frac{L}{2(2L + K_i e_p(nT))}[K_i e_p(nT) + K_p e_v(nT)]$$

where $e_p(nT) \leq 0$ in the region IC5 and IC6.

Hence by combining the above two formulae the following result is obtained for the four regions IC1, IC2, IC5 and IC6:

$$\Delta u_{PI}(nT) = \frac{L[K_i e_p(nT) + K_p e_v(nT)]}{2[2L - K_i |e_p(nT)|]}$$

Working through all the regions in the same way, the overall formulae for the 20 regions are given in Table 9.1.

Table 9.1 Formulae for IC regions of fuzzy PI controllers.

IC Regions	$\Delta u_{PI}(nT)$
1,2,5,6	$\dfrac{L[K_i e_p(nT) + K_p e_v(nT)]}{2[2L - K_i |e_p(nT)|]}$
3,4,7,8	$\dfrac{L[K_i e_p(nT) + K_p e_v(nT)]}{2[2L - K_p |e_v(nT)|]}$
9,10	$\dfrac{1}{2}[K_p e_v(nT) + L]$
11,12	$\dfrac{1}{2}[K_i e_p(nT) + L]$
13,14	$\dfrac{1}{2}[K_p e_v(nT) - L]$
15,16	$\dfrac{1}{2}[K_i e_p(nT) - L]$
18,20	0
17	L
19	$-L$

Defuzzification for Fuzzy D Controller

Similarly, defuzzification of the fuzzy D controller follows the same procedure as described for the PI component, except that the input signals in this case are different (see Figure 9.5(b)). The IC combinations of these two inputs are decomposed into 20 similar regions as shown in Figure 9.6(b). The formulae for the fuzzy D controller for these IC regions are given in Table 9.2.

9.2.7 A Control Example

Consider a nonlinear process with the following simple model:

$$\dot{y}(t) = 0.0001|y(t)| + u(t)$$

It is known that the conventional PID controller is not able to track the setpoint, no matter how its parameters are changed. However, assuming that the sampling time $T = 0.1$ and tuning $K_p = 1.5$,

Table 9.2 Formulae for IC regions of fuzzy D controllers.

IC Regions	$\Delta u_D(nT)$
1,2,5,6	$\dfrac{L[Ky_d(nT) - K_d \Delta y(nT)]}{2[2L - |e_p(nT)|]}$
3,4,7,8	$\dfrac{L[Ky_d(nT) - K_d \Delta y(nT)]}{2[2L - K_d|\Delta y(nT)|]}$
9,10	$\dfrac{1}{2}[-K_d \Delta y(nT) + L]$
11,12	$\dfrac{1}{2}[Ky_d(nT) - L]$
13,14	$\dfrac{1}{2}[-K_d \Delta y(nT) - L]$
15,16	$\dfrac{1}{2}[Ky_d(nT) + L]$
17,19	0
18	$-L$
20	L

$K_d = 0.1, K_i = 2.0, K = 1.0, K_{uPI} = 0.11, K_{uD} = 1.0$ and $L = 45.0$, the output response, as depicted in Figure 9.7, can be accomplished by the proposed fuzzy PID controller.

9.3 Multi-Objective Optimised Genetic Algorithm Fuzzy PID Control

Despite the success of the fuzzy PID controller design, the main drawback is the selection of a set of parameters (K_p, K_i, K_d, K, K_{uD}, K_{uPI}, L) for the controller to reach an optimal performance. However, a detailed design and performance analysis of the optimal tuning of the controller gains for the fuzzy PID controller can be obtained using a multi-objective Genetic Algorithm (GA) method (Tang et al., 2001a) and this is the GA application described in this section.

GA is an optimisation technique that is inspired by the mechanism of natural selection. In the method, each individual potential solution is encoded into a string, which is also known as a chromosome. According to the problem specifications, a fitness value can be assigned to each chromosome in order to reflect its goodness for solving the problem. Starting from a random population of individuals, new offspring are generated by the mating of selected parental chromosomes. Usually, an individual with higher fitness value has a higher chance to be chosen as the parent. A mutation process then occurs to increase the genetic variation in the chromosomes. The offspring are finally reinserted into the original population for next generation. A similar generational cycle is repeated until the solution is found or some criteria are fulfilled. Figure 9.8 depicts a typical genetic cycle.

9.3 Multi-Objective Optimised Genetic Algorithm Fuzzy PID Control

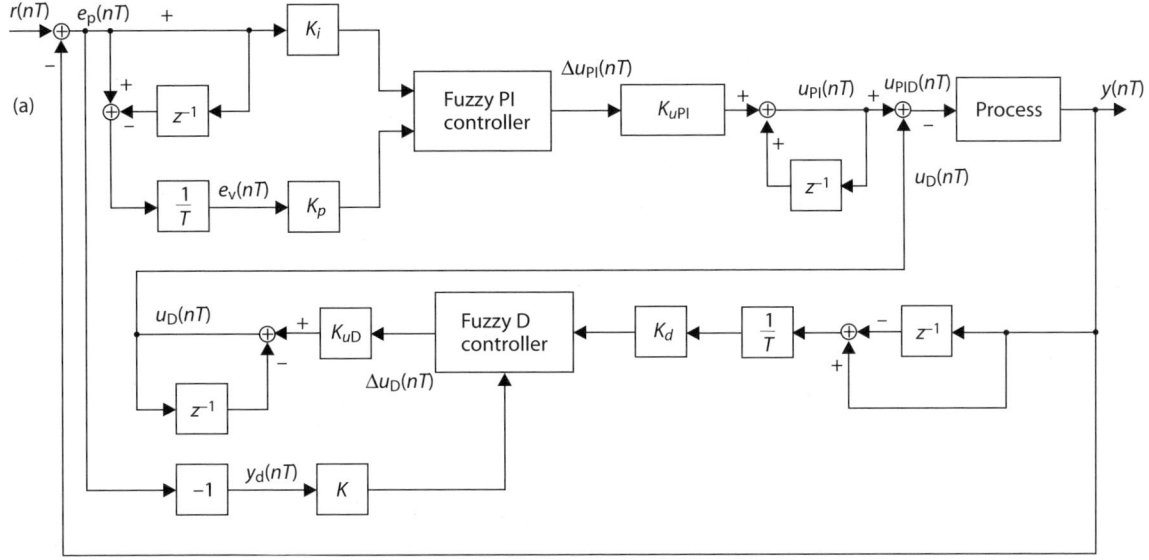

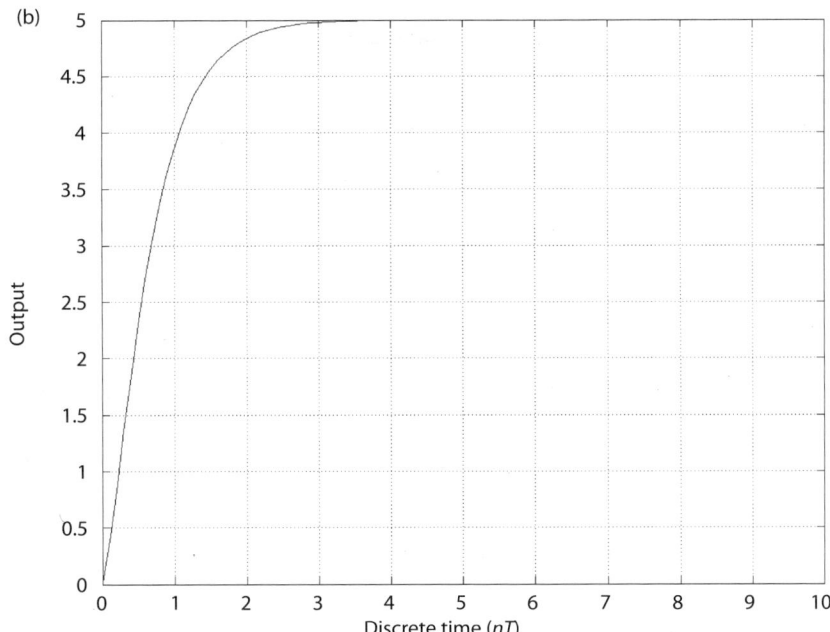

Figure 9.7 Control of a nonlinear process with the fuzzy PID controller (a) Simulink model; (b) the controlled output response.

9.3.1 Genetic Algorithm Methods Explained

The direct approach for the use of GA in finding the optimal gains of conventional PID controller is to construct the chromosome in a binary or real number representation (Jones *et al.*, 1996; Krohling *et al.*, 1997; Ota and Omatu, 1996). A similar genetic approach has been applied to tuning digital PID controllers (Porter and Jones, 1992).

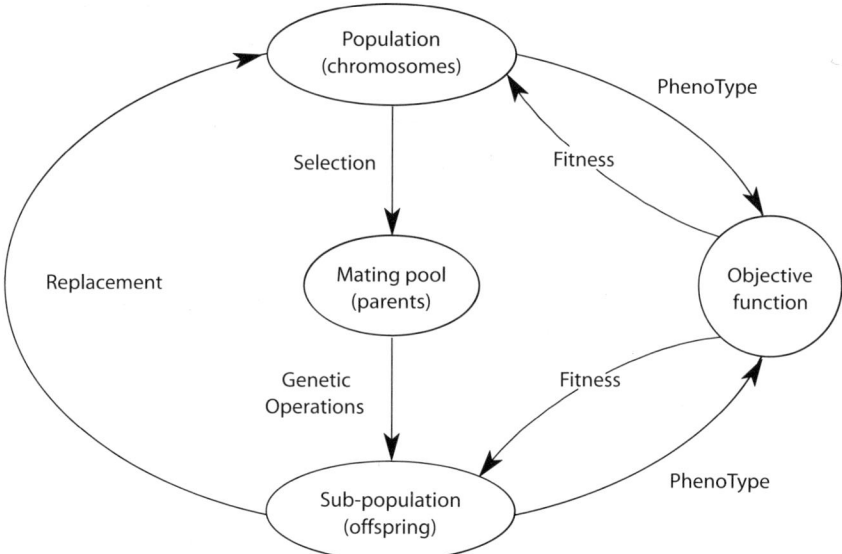

Figure 9.8 The genetic cycle.

To generate an appropriate design various fitness functions have been adopted: for example, a design with gain margin and phase margin specification (Jones et al., 1996), or alternatively a design that minimises the integral of squared error (ISE) or an improved version using the integral of time multiplied-square error (ITSE) in the frequency domain (Krohling et al., 1997). Classical specifications of rise time, overshoot and steady state error have also been used (Tang et al., 2001a).

Recently, the development of the multi-objective optimisation approach in GA makes it possible to have multiple criteria. For a general control problem, a GA can be used to optimise a number of system performance measures: for example, overshoot, settle time and rise time. Instead of aggregating them with a weighting function, the approach of Pareto-based fitness assignment (Fonseca and Fleming, 1998) is adopted.

The rank of the chromosome I is defined as:

$$rank(I) = I + p$$

where I is dominated by other p chromosomes. The criterion to have u dominated by v for a minimisation problem is that

$$f_i(u) \geq f_i(v) \quad \text{for all } i = 1, 2, \ldots, n$$

and

$$f_j(u) > f_j(v) \quad \text{for some } j = 1, 2, \ldots, n$$

A desirable function for industrial automation is the autotuning of PID control parameters, mainly comprising the control gains and perhaps also some scaling parameters used in the controller, according to the changes of the systems (plants, processes) and their working environments. However, due to the convergence of the population, a restart of the optimisation cycle is usually needed for a direct approach (van Rensburg et al., 1998).

From the presentation in Section 9.2, there are seven control parameters to be found for the fuzzy PID controller. Thus, the chromosome I can be defined using a real-number representation of the seven parameters as $I = \{K_p, K_i, K_d, K, K_{uD}, K_{uPI}, L\}$.

The fundamental genetic cycle is shown in Figure 9.8. The specialised genetic operations developed in the evolution program GENECOP (Michalewicz, 1996) for real-number chromosome representations are adopted. For crossover, the jth gene of the offspring I' can be determined by

$$I'_j = \alpha I_j^{(x)} + (1-\alpha) I_j^{(y)}$$

where $\alpha \in [0,1]$ is a uniformly distributed random number and $I^{(x)}, I^{(y)}$ are the selected parents. Mutation is performed within the confined region of the chromosome by applying Gaussian noise to the genes.

For the general control problem, it is desirable to optimise a number of different system performance measures. Consider a step input $r(t) = r$ and the output response $y(t)$; then the following objectives can be specified (Tang et al., 2001a):

1. Minimising the maximum overshoot of the output: $O_1 = \min\{OS\}$ where the overshoot is defined as $OS = |\max\{y(t)\}| - r$

2. Minimise the settle time of the output, $t_s(\pm 2\%)$, $O_2 = \min\{t_s\}$, where $0.98r \leq y(t) \leq 1.02r \ \forall t \geq t_s$

3. Minimise the rise time of the output, $t_r(10\%, 90\%)$, $O_3 = \min\{t_r\}$, where $t_r = t_1 - t_2$, $y(t_1) = 0.1r$ and $y(t_2) = 0.9r$

Instead of aggregating the objectives with a weighting function, a multi-objective approach based on Pareto ranking can be applied. Pareto-based ranking can correctly assign all non-dominated chromosomes with the same fitness. However, the Pareto set may not be uniformly sampled. Usually, the finite populations will converge to only one or some of these, due to stochastic errors in the selection process. This phenomenon is known as genetic drift. Therefore fitness sharing (Fonseca and Fleming, 1998; Goldberg and Richardson, 1987) can be adopted to prevent the drift and promote the sampling of the whole Pareto set by the population. An individual is penalised due to the presence of other individuals in its neighbourhood. The number of neighbours governed by their mutual distance in objective spaces is counted and the raw fitness value of the individual is then weighted by this niche count. Eventually, the total fitness in the population is redistributed, favouring those regions with fewer chromosomes located.

9.3.2 Case study A: Multi-Objective Genetic Algorithm Fuzzy PID Control of a Nonlinear Plant

Recall the example in Section 9.2.7, where a nonlinear process was represented by the following model:

$$\dot{y}(t) = 0.0001|y(t)| + u(t)$$

The multi-objective genetic algorithm (MOGA) method was used to select the optimal gains for the fuzzy PID controller. The output performance of the fuzzy PID controllers with the optimal gain located by MOGA and with the previous manually tuned gain is shown in Figure 9.9.

The figure clearly reveals that the fuzzy PI+D controller, after optimisation via the GA, has generally better steady state and transient responses due to the multi-objective optimal criteria formulated for the setpoint tracking control tasks. The objective values of the rank 1 solutions obtained by the MOGA are tabulated in Table 9.3.

The corresponding output responses of the rank 1 solutions are plotted in Figure 9.10. The table and the set of graphs clearly demonstrate the trade-off between the specified objectives and the ability of MOGA to find the solutions along the Pareto set.

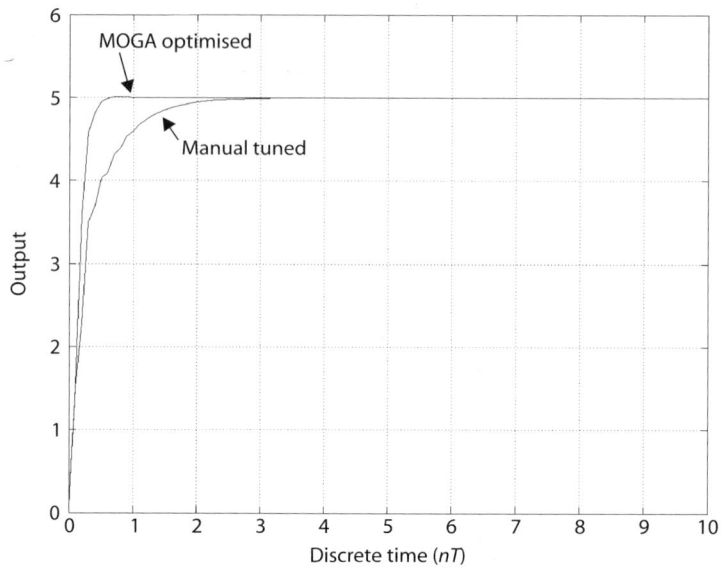

Figure 9.9 Output responses of the Fuzzy PID controller.

Table 9.3 Objective values of the rank 1 solutions for $K = 1$ in all cases.

Solutions	Overshoot (%)	Settle time (nT)	Rise time (nT)	K_p	K_i	K_d	K_{uD}	K_{uPI}	L
1	0	0.9	0.5	0.4076	1.4354	1.1935	1.8884	8.7155	31.8446
2	0.008	0.8	0.5	0.3319	1.1667	1.2036	1.7506	5.9769	21.1941
3	0.010	0.6	0.5	0.3319	1.1667	1.1938	1.7311	5.9769	21.1941
4	0.020	0.6	0.4	0.3514	1.2448	1.1938	1.5748	5.9769	20.2176
5	0.064	0.6	0.2	0.4076	1.7479	1.1935	1.4197	8.7155	31.8446
6	0.092	0.5	0.4	0.3319	1.2448	1.1938	1.5748	5.9769	20.2176
7	0.124	0.4	0.2	0.5070	2.3947	1.1975	1.7311	5.9769	146.0702
8	12.164	0.5	0.1	0.5621	3.9321	0.9614	2.3547	5.6675	250.9499

9.3.3 Case study B: Control of Solar Plant

To reinforce the practical use of the fuzzy PID controller design, the same controller structure is applied to a solar plant at Tabernas in Almeria, Spain (Camacho et al., 1988). This solar plant consists of 480 distributed solar ACUREX collectors arranged in rows forming 10 parallel loops. The collector uses parabolic mirrors to reflect solar radiation onto a pipe to heat up the circulating oil inside.

A sunlight tracking system is installed to drive the mirrors to revolve around the pipes to achieve the maximum benefit from the sunlight. The cold inlet oil is pumped from the bottom of the storage tank

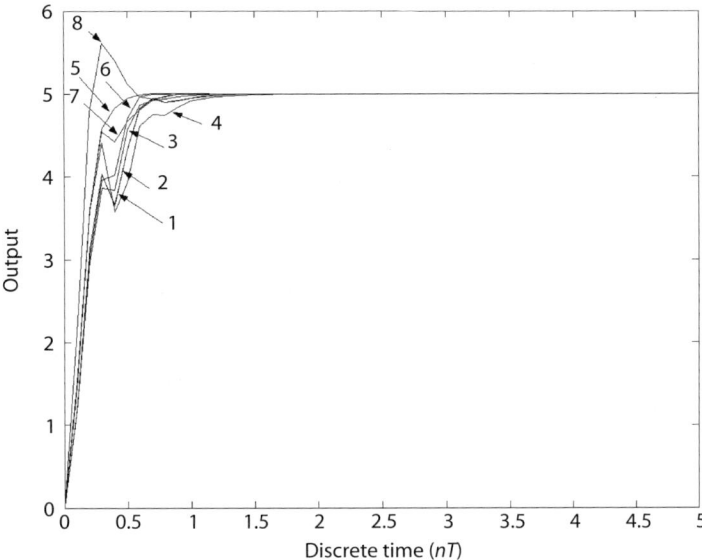

Figure 9.10 Output responses of the rank 1 solutions.

Table 9.4 Genetically tuned fuzzy PID parameters.

K_p	K_i	K_d	K	K_{uD}	K_{uPI}	L
3.3004	2.7463	3.1141	0.3077	1.2086	0.0348	185.1126

and passed through the field inlet. The heated oil is then transferred to a storage tank for use to generate the electrical power. The system has a three-way valve located in the field outlet to allow the oil to be recycled in the field until its outlet temperature is adequately heated for entering the top of the storage tank.

The most important design objective of this control system is to maintain the outlet oil temperature at a constant desirable level despite significant variation in process disturbances; an example is shown in Figure 9.11. The disturbances may be caused by changes in solar radiation level, mirror reflectivity and inlet oil temperature.

With the use of the genetically tuning fuzzy PID controller design (Tang *et al.*, 2000), a set of controller parameters $\{K_p, K_i, K_d, K, K_{uD}, K_{uPI}, L\}$ is obtained, as shown in Table 9.4.

As a result, the outlet temperature of the controlled solar plant for a given step setpoint demand of 180 °C is shown in Figure 9.12. It can be observed that the output temperature tracks the reference temperature well within the fluctuation in temperature of under 0.6 °C and with an overshoot limited to 3 °C.

9.4 Applications of Fuzzy PID Controllers to Robotics

There are many reports and suggestions about real-world applications of various fuzzy PID controllers. A comprehensive review of this subject is far beyond the scope of this chapter. Even a survey of

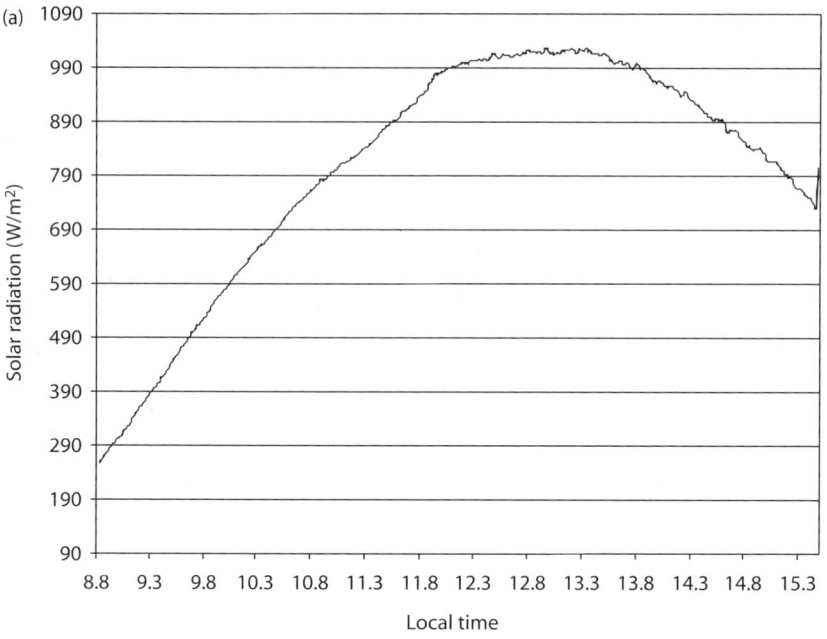

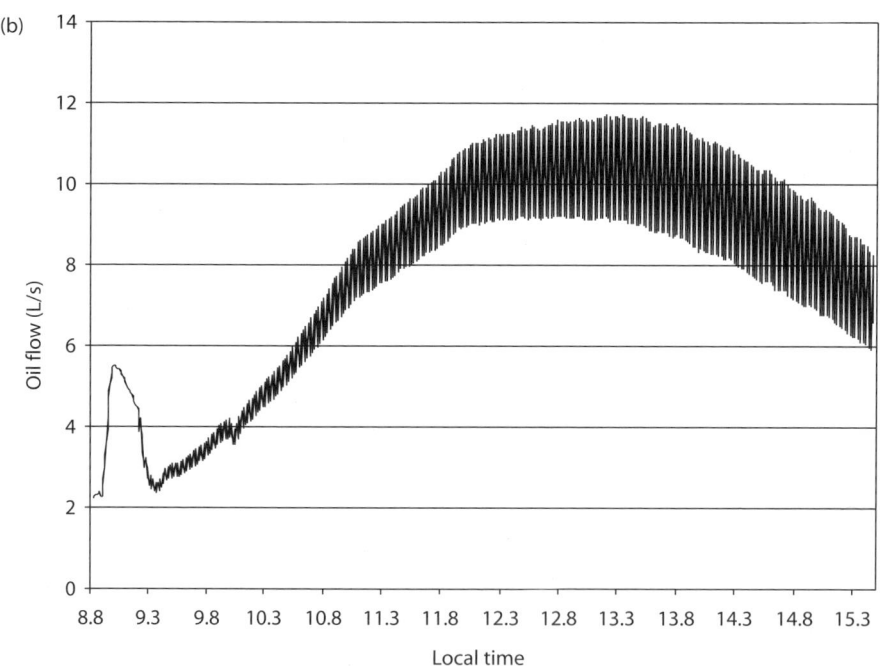

Figure 9.11 Disturbance variation of solar plant. (a) Variation in daily solar radiation; (b) Variation in daily oil flow.

comprehensive applications of fuzzy PID controllers seems to be impossible for a chapter of this modest size. Therefore, it is only mentioned here the recent studies of applications of those fuzzy PID

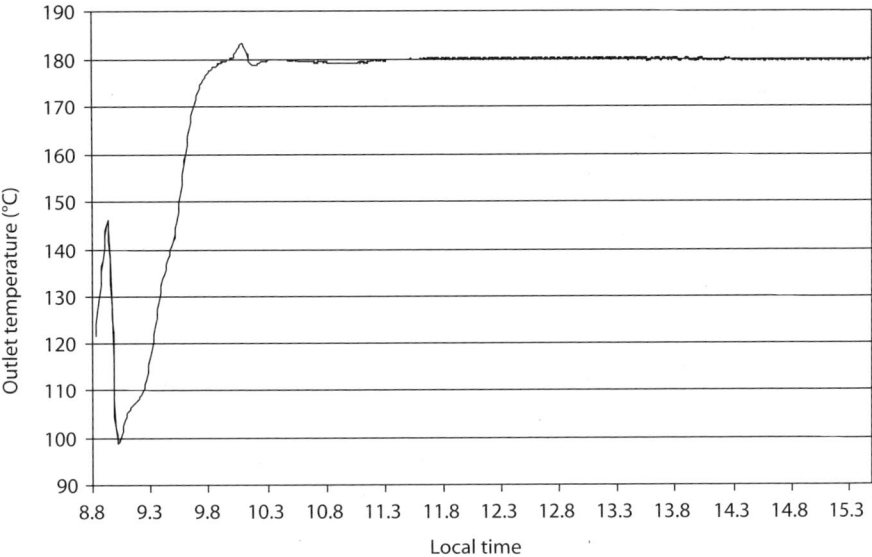

Figure 9.12 Output response of the outlet oil temperature.

controllers in the area of robotics, which include the investigations of Malki *et al.* (1997), Pattaradej *et al.* (2001a,b), Sooraksa and Chen (1998), Sooraksa *et al.* (2002) and Tang *et al.* (2001b).

These fuzzy PID controller types, such as the fuzzy PI, PD, PI+D and PD+I controllers and some of their variants, have been implemented via hardware on various robotic machines. Some typical applications include (Chen, 2000):

- The double inverted Pendubot with varying loads, implemented by Sanchez and his team at the INVESTAV, Guadalajara, Mexico.

- The RobotGuard, a six-legged insect robot, a bicycle robot and robot soccer players, implemented by Sooraksa and his team at the King Mongkut's Institute of Technology Ladkrabang, Bangkok, Thailand.

It is particularly worth mentioning that the proposed fuzzy PID controllers have been implemented by microchips by Lara-Rojo and Sanchez, with applications in robotics and didactic level systems control in Mexico (Lara-Rojo *et al.*, 2001; Sanchez *et al.*, 1998). Some other variants of fuzzy PID controller design, analysis and applications are also available in the literature (Hu *et al.*, 2001; Li and Tso, 2000; Li and Chen, 2001; Li *et al.*, 2001; Mann *et al.*, 1999; Xu *et al.*, 2001).

9.5 Conclusions and Discussion

It has been the experience of many researchers that fuzzy PID controllers generally outperform conventional PID controllers. Yet it must be emphasised that these two types of PID controller are not exactly comparable; in fact, it is generally impossible to have exactly the same conditions for a fair comparison of these controllers. The point to make here is that as long as fuzzy PID controllers work for some control problems on some systems and processes where the conventional PID controller does work or does not work so well, then fuzzy PID control has merit and should be further developed and applied.

Although various fuzzy PID types of controller seem quite promising for real-world control applications with complex systems, further development and improvement remains a challenge that calls for new efforts and endeavours from the research and engineering communities. Nevertheless, it is believed that fuzzy PID controllers have a bright future in the continually developing technology of PID control and in modern industrial applications.

On the other hand, the genetic algorithm method works as a powerful optimisation tool in order to choose the best parameters for controllers. It is typically useful in the sense that the functions for optimisation are usually nonlinear and multimodal. In addition, the multi-objective approach makes it a very good candidate for handling the multi-criteria that exist in most controller design problems.

Acknowledgments

The authors are very grateful for the support of Hong Kong Research Grants Council under Grants 9040604. It has been the great pleasure of authors, K.S. Tang, G.R. Chen, K.F. Man and S. Kwong to work with Hao Ying, Heidar A. Malki, Edgar N. Sanchez, Pitikhate Sooraksa and P.C.K. Luk on fuzzy PID controllers and genetic algorithms over recent years.

References

Andreiev, N. (1981) A new dimension: a self-tuning controller that continually optimises PID constants. *Control Eng.*, **28**, 84–85.

Åström, K.J. (1992) Intelligent tuning. In *Adaptive Systems in Control and Signal Processing* (eds. L. Dugard, M. M'Saad and I.D. Landau), Pergamon Press, Oxford, pp. 360–370.

Åström, K.J., Hang, C.C., Persson, P. and Ho, W.K. (1992) Towards intelligent PID control. *Automatica*, **28**, 1–9.

Bennett, S. (1993) Development of the PID controller. *IEEE Control Systems Magazine*, December, 58–65.

Bennett, S. (2001) The past of PID controllers. *Annual Reviews in Control*, **25**, 43–53.

Camacho, E.F., Rubio, F.R. and Gutierrez, J.A. (1988) Modelling and simulation of a solar power plant system with a distributed collector system. *Proc. Int. IFAC Symp. Power Systems Modelling and Control Applications*, pp. 11.3.1–11.3.5.

Carvajal, J., Chen, G. and Ogmen, H. (2000) Fuzzy PID controller: design, performance evaluation, and stability analysis, *Int. J. Inform. Sci.*, **123**, 249–270.

Chen, G. (1996) Conventional and fuzzy PID controllers: an overview. *Int. J. Intelligent Control Systems*, **1**, 235–246.

Chen, G. and Ying, H. (1997) BIBO stability of nonlinear fuzzy PI control systems. *J. Intelligence and Fuzzy Systems*, **5**, 245–256.

Chen, G. and Pham, T.T. (2000) *Introduction to Fuzzy Sets, Fuzzy Logic and Fuzzy Control Systems*. CRC Press, Boca Raton, FL.

Chen, G. (2002) http://www.ee.cityu.edu.hk/~gchen/fuzzy.html.

Cohen, G.H. and Coon, G.A. (1953) Theoretical condition of retarded control. *Trans. ASME*, **75**, 827–834.

Driankov, D., Hellendoorn, H. and Palm, R. (1995) Some research directions in fuzzy control. In *Theoretical Aspects of Fuzzy Control* (ed. H.T. Nguyen, M. Sugeno, R. Tong and R.R. Yager), Chapter 11. John Wiley, Chichester.

Fonseca, C.M. and Fleming, P.J. (1998) Multi-objective optimisation and multiple constraint handling with evolutionary algorithms Part I: A unified formulation. *IEEE Trans. on Systems, Man, and Cybernetics, Part A: Systems and Humans*, **28**(1), 26–38.

Goldberg, E. and Richardson, J. (1987) Genetic algorithms with sharing for multimodal function optimisation. *Proc. 2nd Int. Conf. on Genetic Algorithms* (ed. J. J. Grefenstette), pp. 41–49. Lawrence Erlbaum, New York.

Grimble, M.J. (2001) LQG optimisation of PID structured multi-model process control systems: one DOF tracking and feedforward control. *Dynamics and Control*, **11**, 103–132.

Hu, B.-G., Mann, G.K.I. and Gosine, R.G. (2001) A systematic study of fuzzy PID controllers – function-based evaluation approach. *IEEE Trans. Fuzzy Systems*, **9**, 699–712.

Jones, A.H., Ajlouni, N. and Uzam, M. (1996) Online frequency domain identification and genetic tuning of PID controllers. *IEEE Conf. Emerging Technologies and Factory Automation*, Vol. 1, pp. 261–266.

Krohling, R.A., Jaschek, H. and Rey, J.P. (1997) Designing PI/PID controllers for a motion control system based on genetic algorithms. *Proc. IEEE Int. Sym. Intelligent Control*, Turkey, pp. 125–130.

Kuo, Y.-P. and Li, T.-H.S. (1999) GA-based fuzzy PI/PD controller for automotive active suspension system. *IEEE Trans. Indus. Electr.*, **46**, 1051–1056.

Lara-Rojo, F., Sanchez, E.N. and Navarro, D.Z. (2001) Real-time fuzzy microcontroller for didactic level system. *Proc. ELECTRO'2001*, Instituto Tecnologico de Chihuahua, Mexico, pp. 153–158.

Li, H.-X. and Tso, S.K. (2000) Quantitative design and analysis of fuzzy proportional-integral-derivative control – a step towards autotuning. *Int. J. Sys. Sci.*, **31**, 545–553.

Li, H.-X. and Chen, G. (2001) Dual features of conventional fuzzy logic control. *Acta Automatica Sinica*, **27**, 447–459.

Li, W., Chang, X.G., Farrell, J. and Wahl, F.W. (2001) Design of an enhanced hybrid fuzzy P+ID controller for a mechanical manipulator. *IEEE Trans. Sys. Man Cybern.-B*, **31**, 938–945.

Liu, G.P., Dixon, R. and Daley, S. (2001) Design of stable proportional-integral-plus controllers. *Int. J. Control*, **74**, 1581–1587.

Malki, H.A., Li, H. and Chen, G. (1994) New design and stability analysis of fuzzy proportional-derivative control systems. *IEEE Trans. Fuzzy Systems*, **2**, 245–254.

Malki, H.A., Misir, D., Feigenspan, D. and Chen, G. (1997) Fuzzy PID control of a flexible-joint robot arm with uncertainties from time-varying loads. *IEEE Trans. Contr. Sys. Tech.*, **5**, 371–378.

Mann, G.K.I., Hu, B.-G. and Gosine, R.G. (1999) Analysis of direct action fuzzy PID controller structures. *IEEE Trans. Sys. Man Cybern.-B*, **29**, 371–388.

Mann, G.K.I., Hu, B.-G. and Gosine, R.G. (2001) Two-level tuning of fuzzy PID controllers. *IEEE Trans. Sys. Man Cybern.-B*, **31**, 263–269.

Michalewicz, Z. (1996) *Genetic Algorithms + Data Structures = Evolution Programmes*, 3rd edn. Springer-Verlag, London.

Misir, D., Malki, H.A. and Chen, G. (1996) Design and analysis of a fuzzy proportional-integral-derivative controller. *Fuzzy Sets and Systems*, **79**, 297–314.

Mohan, B.M. and Patel, A.V. (2002) Analytic structures and analysis of the simplest fuzzy PD controllers. *IEEE Trans. Sys. Man Cybern.-B*, **32**, 239–248.

O'Dwyer, A. (2003) *Handbook of PI and PID Controller Tuning Rules*. World Scientific, Singapore.

Ota, T. and Omatu, S. (1996) Tuning of the PID control gains by GA. *IEEE Conf. Emerging Technologies and Factory Automation*, Vol. 1, pp. 272–274.

Patel, A.V. and Mohan, B.M. (2002) Analytic structures and analysis of the simplest fuzzy PI controllers. *Automatica*, **38**, 981–993.

Pattaradej, T., Sooraksa, P. and Chen, G. (2001a) Implementation of a modified fuzzy proportional-integral computer-aided control for DC servo motors. *Proc. Intelligent Technologies*, Thailand, 27–29 November, pp. 85–90.

Pattaradej, T., Chandang, P. and Chen, G. (2001b) A speed evaluation for conventional PID and fuzzy controllers. *Proc. Intelligent Technologies*, Thailand, 27–29 November, pp. 91–96.

Porter, B. and Jones, A.H. (1992) Genetic tuning of digital PID controllers. *Electronic Letters*, **28**, 843–844.

Sanchez, E.N., Nuno, L.A., Hsu, Y.-C. and Chen, G. (1998) Real time fuzzy control for an under-actuated robot. *Proc. 4th Joint Conf. on Inform. Sci.*, Research Triangle Park, NC.

Silva, G.J., Datta, A. and Bhattacharyya, S.P. (2002) New results on the synthesis of PID controllers. *IEEE Trans. Auto. Control*, **47**, 241–252.

Sooraksa, P. and Chen, G. (1998) Mathematical modelling and fuzzy control of flexible robot arms. *Math. Comput. Modelling*, **27**, 73–93.

Sooraksa, P., Luk, B.L., Tso, S.K. and Chen, G. (2002) Multi-purpose autonomous robust carrier for hospitals (MARCH): Principles, design, and implementation. *Proc. M2VIP'2002*, Thailand (on CD-ROM).

Suchomski, P. (2001) Robust PI and PID controller design in delta domain. *IEE Proc. Control Theory and Applications*, **148**, 350–354.

Swallow, J.P. (1991) The best of the best in advanced process control. *Keynote speech* at the 4th Int. Chemical Process Control Conf., Padre Island, TX.

Tang, K.S., Man, K.F. and Chen, G. (2000) Solar plant control using genetic fuzzy PID controller. *Proc. IECON'2000*, Nagoya, Japan, Oct, pp. 1686–1691.

Tang, K.S., Man, K.F., Chen, G. and Kwong, S. (2001a) An optimal fuzzy PID controller. *IEEE Trans. Indust. Elect.*, **48**, 757–765.

Tang, W.M., Chen, G. and Lu, R.D. (2001b) A modified fuzzy PI controller for a flexible-joint robot arm with uncertainties. *Int. J. Fuzzy Sets and Systems*, **118**, 109–119.

Tokuda, M. and Yamamoto, T. (2002) A neural-net based controller supplementing a multi-loop PID control system. *IEICE Trans. Fundamentals*, **E85-A**, 256–261.

van Rensburg, P.J., Shaw, I.S. and van Wyk, J.D. (1998) Adaptive PID-control using a genetic algorithm. *Proc. Int. Conf. Knowledge-based Intelligent Electronic Systems*, pp. 21–23.

Visioli, A. (2001) Tuning of PID controllers with fuzzy logic. *IEE Proc. Control Theory and Applications*, **148**, 1–8.

Xu, J.-X., Liu, C. and Hang, C.-C. (1996) Tuning of fuzzy PI controllers based on gain/phase margin specifications and ITAE index. *ISA Trans.*, **35**, 79–91.

Xu, J.-X., Hang, C.-C. and Liu, C. (2000) Parallel structure and tuning of a fuzzy PID controller. *Automatica*, **36**, 673–684.

Xu, J.-X., Xu, J. and Cao, W.-J. (2001) A PD type fuzzy logic learning control approach for repeatable tracking control tasks. *Acta Automatica Sinica*, 7, 434–446.

Woo, Z.-W., Chung, H.-Y. and Lin, J.-J. (2000) A PID type fuzzy controller with self-tuning scaling factors. *Fuzzy Sets and Systems*, **115**, 321–326.

Yamamoto, S. and Hashimoto, I. (1991) Present status and future needs: the view from Japanese industry. *Proc. 4th Int. Chemical Process Control Conf.*, Padre Island, TX, pp. 1–28.

Ying, H., Siler, W. and Buckley, J.J. (1990) Fuzzy control theory: a nonlinear case. *Automatica*, **26**, pp. 513–552.

Zheng, F., Wang, Q.G. and Lee, T.H. (2002) On the design of multivariable PID controllers via LMI approach. *Automatica*, **38**, 517–526.

Ziegler, J.G. and Nichols, N.B. (1942) Optimum settings for automatic controllers. *Trans. ASME*, **64**, 759–768.

Ziegler, J.G. and Nichols, N.B. (1943) Process lags in automatic control circuits. *Trans. ASME*, **65**, 433–444.

10 Tuning PID Controllers Using Subspace Identification Methods

Learning Objectives

10.1 Introduction

10.2 A Subspace Identification Framework for Process Models

10.3 Restricted Structure Single-Input, Single-Output Controllers

10.4 Restricted-Structure Multivariable Controller Characterisation

10.5 Restricted-Structure Controller Parameter Computation

10.6 Simulation Case Studies

References

Learning Objectives

Model-free control methods make no use of a mathematical model to capture the system information. In Chapter 3, the method of iterative feedback tuning used the system directly to provide the responses necessary to compute the cost function gradients for controller tuning. The method of subspace identification is also heavily dependent on system data, but the data is organised into a linear model representation. The linear model representation can be used directly, as in the method of this chapter, or can be used to generate state space and transfer function models. For this reason, the method provides a bridge between model-free methods and parametric model techniques that use state space and transfer function models to represent the system.

A second feature of this chapter is the idea of restricted structure controllers. The controller framework adopted is generically multivariable and the controller tuning derives from an optimisation problem – in this case, an integral linear quadratic cost function. The optimisation problem is solved for a low-order pre-specified controller structure. The class of PID controllers is one type of restricted structure controller. Thus the tuning formulation has the potential for tuning a larger class of controllers than just the PID controller set. This will have value in the future when the full potential of digital control is exploited in the same way that PID control is standard today.

The learning objectives for this chapter are to:

- Briefly introduce the concepts of subspace identification.

- Set up the controller tuning problem in an optimisation framework.

- Demonstrate how restricted structures are formed in the cases of SISO and MIMO PID controllers.

- Use the subspace identification model in a solution of the optimisation problem.

- Demonstrate the algorithm devised on a simulated wastewater industry process.

10.1 Introduction

An optimisation formulation for the design and computation of a process controller invariably leads to a controller whose order is at least that of the process model used. When weighting function are used in the optimisation problem to inculcate closed loop performance specifications into the controller computed, the controller order may be much in excess of the process model order. Restricted structure controllers are those whose structure has been selected independently of the plant order. In some cases, this restriction may extend to the allowable range of numerical values that the controller parameters can take. In general, restricted structure controllers are of a lower order than the process model used to represent the plant under control. Typical examples of such controllers that are commonly employed in industry are from the family of classical controllers; phase lead, phase lag, phase lead–lag and industrial PID controllers.

Design methods for these controllers use a wide range of control theory and computational approaches. Therefore it is quite difficult to provide a classification of all the design techniques developed. This chapter will consider the broad classifications of model-based and model-free methods. Model-based methods can be further categorised into parametric model methods and nonparametric model methods. The parametric model methods can use transfer function models, state space models, or, as in this chapter, data-based identified subspace models. Nonparametric methods usually use one or two frequency response data points on which to base the controller design. On the other hand, model-free methods manage to use plant responses directly without the need for an intervening model description.

The general problem presented in this chapter is the tuning of conventional controllers, for example those of PID type, such that their performance is as close as possible to that obtainable from a full-order optimal LQG controller. In the most general form, the solution procedure involves optimised restricted structure multivariable controllers and models from the subspace identification paradigm. The restriction of the multivariable structure of a process controller introduces a number of options into the formulation. Decisions have to be made for (a) the input–output structure of the multivariable controller, (b) the type of controller elements to be used within the controller structure, (c) the number of controller parameters within the individual controller elements and (d) the numerical range of the controller parameters.

The optimisation problem formulation to compute the controller has origins in the longstanding integral quadratic cost function optimisation paradigm. The novel contribution is to join this to models from the subspace identification framework. The subspace identification method involves the use of closed-loop plant data and subspace identification techniques to develop a linear process model in matrix form. Generically these subspace models are multivariable and subsume the single input, single-output system case in a completely natural way.

The chapter opens with the subspace framework employed to develop the data-driven process plant models; this is briefly introduced in Section 10.2. This is followed by two sections (10.3 and 10.4) that relate how single-input, single-output and then multivariable restricted structure controllers are formulated. These use the PID controller as the restricted structure controller to be implemented and the multivariable case provides a more general perspective on setting up this type of controller. The solution of the optimisation problem and the associated regularity conditions are presented in Section 10.5. The novel aspect here is the use of the subspace model in the optimisation to compute the PID or restricted controller parameters (Sánchez et al., 2003, 2004). Several simulation case studies of applications of the PID control methods to a wastewater treatment plant are given in Section 10.6.

10.2 A Subspace Identification Framework for Process Models

The subspace framework used in this controller tuning method is an intermediate system representation between no model at all and the full state space matrix representation. It uses data taken from the closed-loop system and knowledge of the controller in the loop to give an input–output system model. The organisation and manipulation of system data uses the economically convenient notation of matrices and vectors. The section concludes with the presentation of two cases of incremental models for use in the control tuning method.

10.2.1 The Subspace Identification Framework

The closed-loop system is assumed to be under the control of a digital controller. Thus data collected from the system occurs at regular discrete time steps. The model representation for this in the discrete time domain has the discrete-time controller output passing through a zero-order hold providing a piecewise continuous input to the process. Thus, the zero-order hold plus continuous-time process can be represented as a sampled data process so that the discrete-time block diagram can be given as in Figure 10.1.

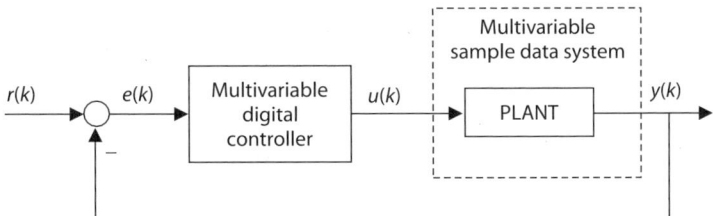

Figure 10.1 Sampled data closed loop system block diagram.

As in Figure 10.1, consider a plant represented by the set of state space model equations and in closed loop with a controller also represented by a set of state space equations.

Process Model State Space Equations

State equation: $\quad x(k+1) = Ax(k) + Bu(k) + Kv(k)$

Output equation: $\quad y(k) = Cx(k) + Du(k) + v(k)$

In these equations, $u(k) \in \Re^m$, $y(k) \in \Re^m$ and $x(k) \in \Re^n$ are the process input, output and state vectors respectively sampled at time instant $t = kT$, where T denotes the sample interval. Note that the process is assumed to have m inputs and m outputs. In this model, K is the Kalman filter gain and $v(k)$ is an unknown innovation sequence with covariance $E = [v(k)v^T(k)] = S$.

Controller State Space Equations

Controller state equation: $x_c(k+1) = A_c x_c(k) + B_c[r(k) - y(k)]$

Controller output equation: $u(k) = C_c x_c(k) + D_c[r(k) - y(k)]$

In these equations, $r(k) \in \Re^m$ and $y(k) \in \Re^m$ are the process reference input and output vectors respectively. Vector $x_c(k) \in \Re^l$ is the controller state vector and the controller output; equivalently, the process input is $u(k) \in \Re^m$. All the vectors are sampled at time instant $t = kT$, where T denotes the sample interval.

The problem can then be formulated as how to identify the plant parameters using closed-loop data and knowledge from the controller. Although there are several subspace identification methods of solving this problem (for example, the method presented by Verhaegen (1993)), the method due to van Overschee and De Moor (1996a) and also reported by Favoreel et al. (1998) has been adopted in this study, and is described below.

Consider a sufficiently large amount of input and output data $\{u(k)\}$ and $\{y(k)\}$ and knowledge of the controller parameters, so that the past and future block Hankel matrices for $u(k)$ and $y(k)$ can be constructed. For the case of input data $\{u(k)\}$, a past and future horizon of dimension N is assumed and the data is organised in matrix form as

Past input data: $U_p = \dfrac{1}{\sqrt{j}} \begin{bmatrix} u_0 & u_1 & \cdots & u_{j-1} \\ u_1 & u_2 & \cdots & u_j \\ \vdots & \vdots & \ddots & \vdots \\ u_{N-1} & u_N & \cdots & u_{N+j-2} \end{bmatrix}$

Future input data: $U_f = \dfrac{1}{\sqrt{j}} \begin{bmatrix} u_N & u_{N+1} & \cdots & u_{N+j-1} \\ u_{N+1} & u_{N+2} & \cdots & u_{N+j} \\ \vdots & \vdots & \ddots & \vdots \\ u_{2N-1} & u_{2N} & \cdots & u_{2N+j-2} \end{bmatrix}$

where j is the number of columns. Note the notation on the large matrix U: subscript p representing *past* data and subscript f representing *future or forward* data. The factor $1/\sqrt{j}$ has been added for statistical purposes. Clearly, the output data can be similarly organised to give matrices Y_p and Y_f.

In general, subspace identification assumes that there are long time series of data ($j \to \infty$), and that the data is ergodic. Due to this, the expectation operator E (average over a finite number of experiments) can be replaced with a different operator E_j applied to the sum of variables. The E_j operator is defined as

$$E_j = \lim_{j \to \infty} \frac{1}{j}[\bullet]$$

and the factor $1/\sqrt{j}$ has the function of pre-processing the data matrices.

Recursive substitution of process state space equations and the controller state space equations followed by organising the results into the data formats of Y_f, X_f, U_f, E_f and U_f, X_f^c, R_f, Y_f yield the matrix input–output equations for the plant and controller as

Matrix IO process equation: $Y_f = \Gamma_N X_f + H_N U_f + H_N^s E_f$

Matrix IO controller equation: $U_f = \Gamma_N^c X_f^c + H_N^c (R_f - Y_f)$

The matrices H_N, Γ_N, H_N^c and Γ_N^c are the lower block triangular Toeplitz matrices and the extended observability matrices for the process and the controller respectively, and are defined as follows:

10.2 A Subspace Identification Framework for Process Models

Process Toeplitz matrix:
$$H_N = \begin{bmatrix} D & 0 & \cdots & 0 \\ CB & D & \cdots & 0 \\ \vdots & \vdots & \ddots & \vdots \\ CA^{N-2} & CA^{N-3} & \cdots & D \end{bmatrix}$$

Process observability matrix:
$$\Gamma_N = [C \quad CA \quad CA^2 \quad \cdots \quad CA^{n-1}]$$

Controller Toeplitz matrix:
$$H_N^c = \begin{bmatrix} D_c & 0 & \cdots & 0 \\ C_c B_c & D_c & \cdots & 0 \\ \vdots & \vdots & \ddots & \vdots \\ C_c A_c^{N-2} B_c & C_c A_c^{N-3} B_c & \cdots & D_c \end{bmatrix}$$

Controller observability matrix:
$$\Gamma_N^c = [C_c \quad C_c A_c \quad C_c A_c^2 \quad \cdots \quad C_c A_c^{n-1}]$$

Eliminating the input matrix U_f between the matrix IO process equation and the matrix IO controller equation gives the input–output expression for the system operating in closed loop as

$$Y_f = T_N \Gamma_N X_f + T_N H_N N_f + T_N H_N^s E_f$$

where $T_N^{-1} = I + H_N H_N^c$ and $N_f = \Gamma_N^c X_f^c + H_N^c R_f$.

Using the equation pair

$$U_f = \Gamma_N^c X_f^c + H_N^c (R_f - Y_f) \quad N_f = \Gamma_N^c X_f^c + H_N^c R_f$$

yields straightforwardly

$$N_f = U_f + H_N^c Y_f$$

noting that N_f is uncorrelated with E_f since $X_f^c E_f^T = 0$ and $R_f E_f^T = 0$. This is due to the assumption that E_f is a Hankel matrix of white noise which is uncorrelated to the future reference input matrix R_f and the future controller states X_f^c (van Overschee and De Moor, 1996a). Therefore employing a linear predictor the output prediction $\hat{Y}_f$ can be estimated when $j \to \infty$ as the closed-loop input–output model,

$$\hat{Y}_f = L_w^c W_p + L_u^c N_f$$

where the composite past data matrix W_p is defined by

$$W_p = \begin{bmatrix} Y_p \\ U_p \end{bmatrix}$$

Within the equation for the output predictions $\hat{Y}_f$, the term $L_w^c W_p$ is a bank of Kalman filters, as proven by van Overschee and De Moor (1996b, pp. 69–72); therefore $\hat{Y}_f$ is considered to be the best possible prediction (estimate) of Y_f. The computation of the matrices L_w^c and L_u^c is achieved by minimising the Frobenius norm expression

$$\min_{L_w^c, L_u^c} \left\| Y_f - [L_w^c \quad L_u^c] \begin{bmatrix} W_p \\ N_f \end{bmatrix} \right\|_F^2$$

where the Frobenius norm of matrix $A \in \Re^{m \times n}$ is defined as

$$\|A\|_F = \sqrt{\sum_{i=1}^m \sum_{j=1}^n a_{ij}^2}$$

To find the closed-loop matrices L_w^c and L_u^c, the following numerical RQ decomposition is used:

$$\begin{bmatrix} W_p \\ U_f \\ Y_f \end{bmatrix} = \begin{bmatrix} R_{11} & 0 & 0 \\ R_{21} & R_{22} & 0 \\ R_{31} & R_{32} & R_{33} \end{bmatrix} \begin{bmatrix} Q_1^T \\ Q_2^T \\ Q_3^T \end{bmatrix}$$

With this decomposition it is possible to prove that matrices L_w^c and L_u^c can be calculated as (van Overschee and De Moor, 1996b; Ruscio, 2000)

$$[L_w^c \quad L_u^c] = [R_{31} \quad R_{32}] \begin{bmatrix} R_{11} & 0 \\ R_{21} & R_{22} \end{bmatrix}^\dagger$$

where † denotes the Moore–Penrose pseudo-inverse.

If the closed-loop model is compared with the open-loop model, the closed-loop matrices L_w^c and L_u^c are related to the open-loop matrices L_w and L_u by the equations

$$L_u^c = T_N L_u$$

and

$$L_w^c = T_N L_w$$

Recall that matrix T_N satisfies $T_N = [1 + H_N H_N^c]^{-1}$, and hence the open-loop matrices L_w and L_u can be found as

$$L_u = L_u^c (I - H_N^c L_u^c)^{-1} \quad \text{and} \quad L_w = (I + L_u H_N^c) L_w^c$$

The matrix L_w must be approximated to a rank n deficient matrix, where n is selected using the Singular Value Decomposition:

$$L_w = [U_1 \quad U_2] \begin{bmatrix} S_n & 0 \\ 0 & 0 \end{bmatrix} \begin{bmatrix} V_1^T \\ V_2^T \end{bmatrix}$$

Returning to vector forms, the plant prediction model is then given as a function of the future input vector $\hat{u}_f$ and the past input–output vector w_p as

$$\hat{y}_f = L_w w_p + L_u \hat{u}_f$$

where $\hat{u}_f = [u_1 \ \ldots \ u_N]^T$ and $w_p [y_{-N+1} \ \ldots \ y_0 \ u_{-N+1} \ \ldots \ u_0]^T$.

Note that the identification method is valid even if the signals are generated by purely deterministic systems. This, however, can lead to rank-deficient Hankel matrices, which can produce numerical problems depending on the decomposition algorithm employed. This phenomenon is produced when a straightforward implementation of the Schur algorithm is used to compute the R factor in a fast RQ decomposition using the Hankel structure. As stated by van Overschee and De Moor (1996b, p.163), this is not often the case in practice; however, systems with many outputs or with heavily coloured input signals can generate Hankel matrices that are nearly rank-deficient.

10.2.2 Incremental Subspace Representations

The key vector input–output data model, $\hat{y}_f = L_w w_p + L_u \hat{u}_f$ gives the best prediction of the output $\hat{y}_f$ for the future inputs $\hat{u}_f$ and past outputs and inputs in composite vector w_p. However, it is sometimes more useful to have a model defined in terms of the *change* in the control signal rather than the control signal *per se*. To achieve this modification, several different approaches have been followed in the literature, such as that due to Ruscio (1997) or to Kadali *et al.* (2003). Both approaches yield the same formulae; however, they consider different signal frameworks. Ruscio (1997) considers a deterministic approach while Kadali *et al.* (2003) follow a stochastic framework. Both approaches are presented in the next two subsections.

Deterministic Process Model Case

If the basic process is considered deterministic, then the process state space equations will have the noise term $v(k)$ set to zero. To achieve a control change equation, introduce a new state variable $z(k)$ such that $z(k) = x(k) - x(k-1)$. By straightforward substitution, it is easily shown that the system state space equations become

$$z(k+1) = Az(k) + B\Delta u(k)$$
$$\Delta y(k) = Cz(k) + D\Delta u(k)$$

where $\Delta u(k) = u(k) - u(k-1)$ and $\Delta y(k) = y(k) - y(k-1)$.

Clearly, the structure of the transformed process state space equation is identical to the original process state space equations, and consequently the process matrix input–output equation is

$$\Delta Y_f = \Gamma_N Z_f + H_N \Delta U_f$$

Note that the extended observability matrix and the block lower triangular Toeplitz matrix have not changed, since the system matrices are the same. Using the new process equations recursively the output prediction $\hat{y}$ at instant $k+N$ is given by

$$\hat{y}(k+N) = y(k) + (CA^{N-1} + \ldots + C)z(k+1)$$
$$+ (CA^{N-2}B + \ldots + CB + D)\Delta u(k+1) + \ldots$$
$$+ (CB + D)\Delta u(k+N-1) + D\Delta u(k+N)$$

Therefore in matrix–vector form the output prediction is

$$\hat{y}_f = y_t + \Gamma_N^\Delta z(k+1) + H_N^\Delta \Delta \hat{u}_f$$

where

$$\Gamma_N^\Delta = \begin{bmatrix} C \\ CA + C \\ \vdots \\ CA^{N-1} + \ldots + C \end{bmatrix} \text{ and } H_N^\Delta = \begin{bmatrix} D & 0 & \ldots & 0 \\ CB + D & D & \ldots & 0 \\ CAB + D & CB + D & \ldots & 0 \\ \vdots & \vdots & \ddots & \vdots \\ CA^{N-2}B + \ldots + D & CA^{N-3}B + \ldots + D & \ldots & D \end{bmatrix}$$

By comparing matrices Γ_N^Δ and H_N^Δ with matrices Γ_N and H_N, it is simple to verify the relations

$$L_u^\Delta = L_u \begin{bmatrix} I_{mN} & 0 & 0 & 0 \\ I_{mN} & I_{mN} & \ldots & 0 \\ \vdots & \vdots & \ddots & \vdots \\ I_{mN} & I_{mN} & \ldots & I_{mN} \end{bmatrix} \text{ and } L_w^\Delta = \begin{bmatrix} l_{w_1} \\ l_{w_1} + l_{w_2} \\ \ldots \\ \sum_{i=1}^N l_{w_i} \end{bmatrix}$$

where l_{w_i} is the ith row block of dimension m of L_w.

Consequently, the incremental form of the input–output data prediction system model is given by

$$\hat{y}_f = y_t + L_w^\Delta \Delta w_p + L_u^\Delta \Delta \hat{u}_f$$

Stochastic Process Model Case

Begin from the stochastic process state space equation:

State equation: $\quad x(k+1) = Ax(k) + Bu(k) + Kv(k)$

Output equation: $\quad y(k) = Cx(k) + Du(k) + v(k)$

The noise process captured by $v(k)$ is given the integrating white noise model:

$$v(k+1) = v(k) + a(k) \text{ with } v(k) = \frac{a(k)}{\Delta}$$

Introduce a new state variable new state variable $z(k)$ such that $z(k) = x(k) - x(k-1)$; then by straightforward substitution it is easily shown that the incremental stochastic process state space model is described by the equations

$$z(k+1) = Az(k) + B\Delta u(k) + Ka(k)$$
$$\Delta y(k) = Cz(k) + D\Delta u(k) + a(k)$$

Since the structure of these new model equations is exactly that of the original model equation, the input–output matrix equation is found as

$$\Delta Y_f = \Gamma_N Z_f + H_N \Delta U_f + H_N^s A_f$$

The output prediction at sampling instant $k + N$ can then be calculated as

$$\begin{aligned}\hat{y}(k+N) = {}& y(k) + (CA^{N-1} + \ldots + C)z(k+1) \\ & + (CA^{N-2}B + \ldots + CB + D)\Delta u(k+1) + \ldots \\ & + (CB+D)\Delta u(k+N-1) + D\Delta u(k+N) \\ & + (a(k) + \ldots + a(k+N))\end{aligned}$$

This leads to precisely the same equations as derived for the deterministic case in the preceding subsection. Thus, ultimately the best output prediction can be calculated using the equation

$$\hat{y}_f = y_t + L_w^\Delta \Delta w_p + L_u^\Delta \Delta \hat{u}_f$$

where

$$L_u^\Delta = L_u \begin{bmatrix} I_{mN} & 0 & 0 & 0 \\ I_{mN} & I_{mN} & \cdots & 0 \\ \vdots & \vdots & \ddots & \vdots \\ I_{mN} & I_{mN} & \cdots & I_{mN} \end{bmatrix}, \quad L_w^\Delta = \begin{bmatrix} l_{w_1} \\ l_{w_1} + l_{w_2} \\ \cdots \\ \sum_{i=1}^N l_{w_i} \end{bmatrix}$$

and l_{w_i} is the ith row block of dimension m of L_w.

10.3 Restricted Structure Single-Input, Single-Output Controllers

An introduction to the problem of the parameterisation and definition of the structure for a single-input, single-output (SISO) process is presented in this section. The parameterisation of a typical PID-type discrete controller is used to exemplify the problem of characterising the restricted structure controller. This enables the general controller structure for a SISO-type system to be more readily understood. Subsequently, a more complete and general description of these topics is given for multivariable controller characterisation in Section 10.4.

10.3.1 Controller Parameterisation

A SISO closed-loop discrete system has a block diagram closely related to Figure 10.1. The signals of the loop are scalar and for the subspace methods to be applied an extensive database of the loop signals r, y and u is required. The SISO closed-loop discrete system block diagram is shown in Figure 10.2.

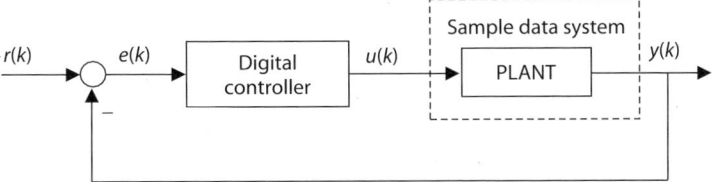

Figure 10.2 Closed-loop SISO discrete system model block diagram.

A discrete PID control action can be defined as

$$u(k) = k_P e(k) + k_I \sum_{n=1}^{k} e(n) + k_D[e(k) - e(k-1)]$$

and the incremental control action by

$$\Delta \hat{u}(k) = u(k) - u(k-1)$$

where k_P, k_I and k_D are the proportional, integral and derivative gains respectively.

Using the discrete PID controller formula with the incremental control action, the time expression for the PID control action increment is obtained as

$$\Delta \hat{u}(k) = \rho^{(1)} e(k) + \rho^{(2)} e(k-1) + \rho^{(3)} e(k-2)$$

where $\rho^{(1)} = k_P + k_I + k_D$, $\rho^{(2)} = -k_P - 2k_D$ and $\rho^{(3)} = k_D$.

To introduce generality into the incremental PID control expression, use

$$\Delta \hat{u}(k) = [e(k) \quad e(k-1) \quad e(k-2)] \begin{bmatrix} \rho^{(1)} \\ \rho^{(2)} \\ \rho^{(3)} \end{bmatrix}$$

or in compact notation

$$\Delta \hat{u}(k) = [e_k \quad e_{k-1} \quad e_{k-2}] \rho$$

where $\rho \in \Re^3$.

To comply with a digital PID structure the controller parameter vector $\rho \in \Re^3$ must comply with the following linear constraint:

$$\begin{bmatrix} -1 & 0 & 0 \\ 0 & 1 & 0 \\ 0 & 0 & -1 \end{bmatrix} \rho \leq \begin{bmatrix} 0 \\ 0 \\ 0 \end{bmatrix}$$

or equivalently

$$\phi \cdot \rho \leq 0 \in \Re^3$$

where $\phi \in \Re^{3 \times 3}$ and $\rho \in \Re^3$.

Using the compact notation for the incremental control action, the future incremental control action can be written as

$$\Delta \hat{u}_f = \begin{bmatrix} \Delta \hat{u}_1 \\ \Delta \hat{u}_2 \\ \vdots \\ \Delta \hat{u}_N \end{bmatrix} = \begin{bmatrix} e_1 & | & e_0 & | & e_{-1} \\ e_2 & | & e_1 & | & e_0 \\ \vdots & | & \vdots & | & \vdots \\ e_N & | & e_{N-1} & | & e_{N-2} \end{bmatrix} \rho$$

and in compact form as

$$\Delta \hat{u}_f = \varepsilon(\rho)\rho$$

where $\varepsilon(\rho) = [\xi_1 \ \xi_2 \ \xi_3] \in \Re^{N \times 3}, \xi_i \in \Re^N, i = 1, 2, 3$ and $\rho \in \Re^3$.

10.3.2 Controller Structure and Computations

To compute the vector of future control actions $\Delta \hat{u}_f \in \Re^N$ it is still necessary to calculate the matrix of closed-loop errors $\varepsilon(\rho) \in \Re^{N \times 3}$. This matrix can be directly derived from the error definition, $e_k = r_k - y_k$ where $k = -\infty, \ldots, -1, 0, 1, \ldots, \infty$. Since r_f is set to zero, the matrix ε can be evaluated as

$$\varepsilon = [\xi_1 \ \xi_2 \ \xi_3]$$

and

$$\varepsilon = -[T_{f_1}\hat{y}_f \ \ T_{f_2}\hat{y}_f \ \ T_{f_3}\hat{y}_f] - [T_{p_1}y_p \ \ T_{p_2}y_p \ \ T_{p_3}y_p]$$

where $T_{f_1} = I_N, T_{f_2} = \begin{bmatrix} 0 & \ldots & 0 & 0 & 0 \\ 1 & \ldots & 0 & 0 & 0 \\ \vdots & \ddots & \vdots & \vdots & \vdots \\ 0 & \ldots & \ddots & 0 & 0 \\ 0 & \ldots & \ldots & 1 & 0 \end{bmatrix}$ and $T_{f_3} = \begin{bmatrix} 0 & \ldots & 0 & 0 & 0 \\ 0 & \ldots & 0 & 0 & 0 \\ 1 & \ldots & 0 & 0 & 0 \\ \vdots & \ddots & \vdots & \vdots & \vdots \\ 0 & \ldots & 1 & 0 & 0 \end{bmatrix}$

with $T_{p_1} = 0_N, T_{p_2} = \begin{bmatrix} 0 & \ldots & 0 & 0 & 1 \\ 0 & \ldots & 0 & 0 & 0 \\ \vdots & \ddots & \vdots & \vdots & \vdots \\ 0 & \ldots & 0 & 0 & 0 \\ 0 & \ldots & 0 & 0 & 0 \end{bmatrix}$ and $T_{p_3} = \begin{bmatrix} 0 & \ldots & 0 & 1 & 0 \\ 0 & \ldots & 0 & 0 & 1 \\ 0 & \ldots & 0 & 0 & 0 \\ \vdots & \ddots & \vdots & \vdots & \vdots \\ 0 & \ldots & 0 & 0 & 0 \end{bmatrix}$

As will be discussed in the next section, the matrices T_f and T_p can be directly calculated from the controller structure. The discussion in this section is limited to a SISO controller and the controller structure is of a simple (default) type characterised by the scalar structure matrix $\Lambda_S = \{[1]\}$. Later in the chapter, when multivariable controllers are considered, a full discussion regarding the controller structure and its importance in the controller design is presented.

By bringing together these equations:

$$\hat{y}_f = y_t + L_w^\Delta \Delta w_p + L_u^\Delta \Delta \hat{u}_f$$
$$\Delta \hat{u}_f = \varepsilon(\rho)\rho$$
$$\varepsilon(\rho) = -[T_{f_1}\hat{y}_f \ \ T_{f_2}\hat{y}_f \ \ T_{f_3}\hat{y}_f] - [T_{p_1}y_p \ \ T_{p_2}y_p \ \ T_{p_3}y_p]$$

and defining $\Phi = r_f - y_t - L_w^\Delta \Delta w_p$, a composite equation is obtained as

$$\Omega(\rho)\begin{bmatrix}\xi_1\\\xi_2\\\xi_3\end{bmatrix}=\omega$$

where

$$\Omega(\rho)=\begin{bmatrix}I_N+\rho^{(1)}L_u^\Delta & \rho^{(2)}L_u^\Delta & \rho^{(3)}L_u^\Delta\\ \rho^{(1)}T_{f_2}L_u^\Delta & I_N+\rho^{(2)}T_{f_2}L_u^\Delta & \rho^{(3)}T_{f_2}L_u^\Delta\\ \rho^{(1)}T_{f_3}L_u^\Delta & \rho^{(2)}T_{f_3}L_u^\Delta & I_N+\rho^{(3)}T_{f_3}L_u^\Delta\end{bmatrix},\;\omega=\begin{bmatrix}\Phi\\ T_{f_2}\Phi-T_{p_2}y_p\\ T_{f_3}\Phi-T_{p_3}y_p\end{bmatrix}\text{ and}$$

$\xi_i\in\Re^N,\,i=1,2,3$

It is interesting to observe that the left-hand side of this equation $\Omega(\rho)$ involves signals to be predicted over the forward (future) horizon, while the right-hand side ω involves signals previously recorded in the past.

10.4 Restricted-Structure Multivariable Controller Characterisation

The following section is divided into two parts. The first part considers the characterisation of a restricted-structure controller in terms of a finite number of parameters and an incremental control action. The second part examines the structure of a multivariable controller structure and establishes definitions for this structure. The results of the section generalise those of the previous section, which only considered the case of a single-input, single-output controller.

10.4.1 Controller Parameterisation

For the multivariable discrete multivariable controller, the method assumes that the controller can be parameterised and expressed in an incremental control action form. A general form for this type of parameterisation is,

$$\Delta\hat{u}(k)=\Xi(k)\rho$$

where the control increment satisfies $\Delta\hat{u}(k)\in\Re^m$ for all time indices, and $\rho\in\Re^{n_c}$ is a vector of n_c controller parameters.

This situation is very common and many industrial controllers can be written in this way. Since the plant has m inputs and m outputs, let $\Delta\hat{u}_{i,j}(k+1)$ represent the ith incremental control action due to the error sequence of the jth output, where $i=1,\ldots,m$; $j=1,\ldots,m$. Then the effective incremental control action for the ith input, denoted $\Delta\hat{u}_i(k+1)$, can be calculated as a sum of partial contributions:

$$\Delta\hat{u}_i(k+1)=\sum_{j=1}^m\Delta\hat{u}_{i,j}(k+1)$$

This parameterisation can be used to demonstrate that the incremental control actions for all the plant inputs at sampling instant $k+1$ can be written in the general form

$$\Delta\hat{u}(k+1)=E(k+1)\rho$$

where

$$\rho=[\rho_{1,1}^{(1)}\;\rho_{1,1}^{(2)}\;\rho_{1,1}^{(3)}\;\cdots\;\rho_{1,m}^{(1)}\;\rho_{1,m}^{(2)}\;\rho_{1,m}^{(3)}\;|\;\cdots$$
$$\cdots\;|\;\rho_{m,1}^{(1)}\;\rho_{m,1}^{(2)}\;\rho_{m,1}^{(3)}\;\cdots\;\rho_{m,m}^{(1)}\;\rho_{m,m}^{(2)}\;\rho_{m,m}^{(3)}]^T$$

$$E(k+1) = \begin{bmatrix} \varepsilon(k+1) & \cdots & 0 \\ \vdots & \ddots & \vdots \\ 0 & \cdots & \varepsilon(k+1) \end{bmatrix}$$

and

$$\varepsilon(k+1) = [e_1(k+1) \quad e_1(k) \quad e_1(k-1) \quad \cdots \quad e_m(k+1) \quad e_m(k) \quad e_m(k-1)]$$

The indices in the notation $\rho_{i,j}^{(k)}$ indicate that the coefficient is the kth parameter of the controller acting over the ith input using the error sequence from the jth output. Note that if it is assumed that all the controllers have the same number of parameters p, then $\rho \in \Re^{pm^2}$, $E(k+1) \in \Re^{m \times pm^2}$ and $\varepsilon(k+1) \in \Re^{1 \times pm}$. In a more involved presentation, the number of controller parameters p for each individual controller could be set to different values. For the specific case of all the controller elements being PID controllers, the common value of $p = 3$ accrues.

10.4.2 Multivariable Controller Structure

The controller characterisation leads to the definition of the structure of the multivariable controller. The definition of the multivariable controller structure allows the specification of the output errors that are used to compute each input. The following definition in conjunction with the parameterisation completely defines the controller.

Definition 10.1: Multivariable controller structure matrix

Let a controller structure be defined by the matrix $\Lambda \in \Re^{m \times m}$ which links the output error of the plant to its inputs through a restricted-structure multivariable controller, where

$$\Lambda = \begin{bmatrix} a_{1,1} & \cdots & a_{1,m} \\ \vdots & \ddots & \vdots \\ a_{m,1} & \cdots & a_{m,m} \end{bmatrix} \text{ with } a_{i,j} \in \{0,1\} \text{ for all pairs from } i=1,\ldots,m;\ j=1,\ldots,m$$

For a connection between the ith input and the jth output set $a_{i,j} = 1$; for no connection between the ith input and the jth output set $a_{i,j} = 0$.

□

The matrix Λ can then be decomposed into the sum of f matrices, where f is the number of non-zero elements in Λ, as

$$\Lambda = \begin{bmatrix} a_{1,1} & \cdots & 0 \\ \vdots & \ddots & \vdots \\ 0 & \cdots & 0 \end{bmatrix} + \ldots + \begin{bmatrix} 0 & \cdots & 0 \\ \vdots & \ddots & \vdots \\ 0 & \cdots & a_{m,m} \end{bmatrix}$$

or compactly:

$$\Lambda = \Lambda_{1,1} + \ldots + \Lambda_{m,m}$$

Thus it is also possible to characterise any multivariable controller structure enumerated in set Λ_S, where $\Lambda_S = \{\Lambda_{1,1}, \Lambda_{1,2}, \ldots, \Lambda_{m,m}\}$.

10.5 Restricted-Structure Controller Parameter Computation

The approach developed computes a set of controller parameters that minimise a finite horizon LQG cost index. The algorithm uses the data-based input–output plant model described by

$\hat{y}_f = y_t + L_w^\Delta \Delta w_p + L_u^\Delta \Delta \hat{u}_f$, the incremental control law uses the relation $\Delta \hat{u}(k+1) = E(k+1)\rho$ and a structure characterised by the set $\Lambda_S = \{\Lambda_{1,1}, \Lambda_{1,2}, \ldots, \Lambda_{m,m}\}$. The method ensures that if the plant is linear and has no unstable hidden modes and the optimisation converges then the solution is a controller that stabilises the closed-loop process plant. The details of the computation to find the restricted structure controller and to comply with the conditions of closed-loop stability are given in this section.

10.5.1 Cost Index

The closed-loop multivariable system framework is that of Figure 10.1. It is assumed that the controller is in the loop, and that sufficient data has been accumulated of the system signals r, y and u. This data is used in the subspace routines to produce a model of the form $\hat{y}_f = y_t + L_w^\Delta \Delta w_p + L_u^\Delta \Delta \hat{u}_f$. The desired controller performance is specified through the use of a finite horizon LQG cost index:

$$J(\rho) = (r_f - \hat{y}_f)^T Q (r_f - \hat{y}_f) + \Delta \hat{u}_f^T R \Delta \hat{u}_f$$

where $y_f, \Delta \hat{u}_f \in \Re^{mN}$, $Q^T = Q \geq 0 \in \Re^{mN \times mN}$ is the error weighting matrix and $R^T = R > 0 \in \Re^{mN \times mN}$ is the control increment weighting matrix. The problem is to find a set of controller parameters such that the cost function $J(\rho) \in \Re_+$ is minimised.

To simplify the numerical problem, it is assumed that the system is regulated around an operating point and all the signals are normalised. By using the equation $\Delta \hat{u}(k+1) = E(k+1)\rho$, the future incremental control output can be written as

$$\Delta \hat{u}_f = \phi \rho$$

where

$$\Delta \hat{u}_f \in \Re^{mN}, \phi = \begin{bmatrix} E(k+1) \\ \vdots \\ E(k+N) \end{bmatrix} \in \Re^{mn \times pm^2} \text{ and } \rho \in \Re^{pm^2}$$

Thus by substitution of $\Delta \hat{u}_f = \phi \rho$ and $\hat{y}_f = y_t + L_w^\Delta \Delta w_p + L_u^\Delta \Delta \hat{u}_f$ and defining $\Phi = r_f - y_t - L_w^\Delta \Delta w_p$, the cost function $J(\rho) \in \Re_+$ expands to

$$J(\rho) = \rho^T (\phi^T L_u^{\Delta T} Q L_u^\Delta \phi + \phi^T R \phi) \rho - 2(\Phi^T Q L_u^\Delta \phi) \rho + \Phi^T Q \Phi$$

This expression for the cost function in terms of the controller parameter vector $\rho \in \Re^{pm^2}$ has the quadratic form $\rho^T A \rho + b^T \rho + c$, which can be minimised with respect to $\rho \in \Re^{pm^2}$ either by using efficient numerical methods or by directly computing the derivative of $J(\rho)$ with respect to ρ. However, the problem is not straightforward. There are still the problems of computing the matrix ϕ, which is ρ-dependent, and also the provision of conditions so that the computed multivariable controller yields a closed-loop stable system.

10.5.2 Formulation as a Least-Squares Problem

The matrices $\phi \in \Re^{mN \times pm^2}$ and L_u^Δ are usually large and ill-conditioned; therefore the choice of a robust numerical algorithm to calculate the minimum of cost index $J(\rho)$ is of vital importance. An efficient solution to this problem is to use a least-squares approach. Since the error weighting matrix Q is symmetric positive semi-definite and the control increment weighting matrix R is symmetric positive definite, it is possible to find matrices S_Q and S_R such that

$$Q = S_Q^T S_Q \text{ and } R = S_R^T S_R$$

Consequently, it is relatively simple to prove that the minimisation of $J(\rho)$ is equivalent to the minimum norm problem given by

$$\min_{\rho \in \Re^{pm^2}} \left\| \begin{bmatrix} S_Q(L_u^\Delta \phi \rho - \Phi) \\ S_R \phi \rho \end{bmatrix} \right\|^2$$

where $\Phi = r_f - y_t - L_w^\Delta \Delta w_p$ and $\rho \in \Re^{pm^2}$.

Thus the vector of controller parameters $\rho \in \Re^{pm^2}$ is the least-squares solution to

$$\begin{bmatrix} S_Q L_u^\Delta \phi \\ S_R \phi \end{bmatrix} \rho = \begin{bmatrix} S_Q \Phi \\ 0 \end{bmatrix}$$

10.5.3 Computing the Closed-Loop System Condition

In the least-squares solution procedure, the problem of how to calculate the Hankel matrix of errors $\phi \in \Re^{mN \times pm^2}$ such that the closed-loop system is stable remains to be considered. A solution to this problem is to calculate the future errors based on past data by satisfying the closed-loop equations. To do so the matrix $\phi \in \Re^{mN \times pm^2}$ can be decomposed into pm^2 column vectors $\xi_{i,j}^{(k)} \in \Re^{mN}$, where the indices i, j, k have the same meaning given previously to $\rho_{i,j}^{(k)}$, so that

$$\phi = [\xi_{1,1}^{(1)} \quad \xi_{1,1}^{(2)} \quad \xi_{1,1}^{(3)} \quad \ldots \quad \xi_{1,m}^{(p-2)} \quad \xi_{1,m}^{(p-1)} \quad \xi_{1,m}^{(p)} \quad | \quad \ldots$$
$$\ldots \quad | \quad \xi_{m,1}^{(1)} \quad \xi_{m,1}^{(2)} \quad \xi_{m,1}^{(3)} \quad \ldots \quad \xi_{1,1}^{(p-2)} \quad \xi_{1,1}^{(p-1)} \quad \xi_{1,1}^{(p)}]$$

Using the closed-loop system equation, each error vector $\xi_{i,j}^{(k)}$ can be calculated for $i = 1, \ldots, m$, $j = 1, \ldots, m$ and $k = 1, \ldots, p$ by solving the equation

$$\xi_{i,j}^{(k)} = -T_{f_{i,j}^{(k)}} \{L_u^\Delta \phi \rho - \Phi\} - T_{p_{i,j}^{(k)}} y_p$$

The resulting system of equations is

$$\Omega(\rho) \upsilon = \omega$$

where

$$\Omega(\rho) = \begin{bmatrix} I + \rho_{1,1}^{(1)} T_{f_{1,1}^{(1)}} L_u^\Delta & \rho_{1,1}^{(2)} T_{f_{1,1}^{(1)}} L_u^\Delta & \ldots & \rho_{m,m}^{(p)} T_{f_{1,1}^{(1)}} L_u^\Delta \\ \rho_{1,1}^{(1)} T_{f_{1,1}^{(2)}} L_u^\Delta & I + \rho_{1,1}^{(2)} T_{f_{1,1}^{(2)}} L_u^\Delta & \ldots & \rho_{m,m}^{(p)} T_{f_{1,1}^{(2)}} L_u^\Delta \\ \vdots & \vdots & \ddots & \vdots \\ \rho_{1,1}^{(1)} T_{f_{m,m}^{(p)}} L_u^\Delta & \rho_{1,1}^{(2)} T_{f_{m,m}^{(p)}} L_u^\Delta & \ldots & I + \rho_{m,m}^{(p)} T_{f_{m,m}^{(p)}} L_u^\Delta \end{bmatrix}$$

$$\upsilon = \begin{bmatrix} \xi_{1,1}^{(1)} \\ \xi_{1,1}^{(2)} \\ \vdots \\ \xi_{m,m}^{(p)} \end{bmatrix} \quad \text{and} \quad \omega = \begin{bmatrix} T_{f_{1,1}^{(1)}} \Phi - T_{P_{1,1}^{(1)}} y_p \\ T_{f_{1,1}^{(2)}} \Phi - T_{P_{1,1}^{(2)}} y_p \\ \vdots \\ T_{f_{m,m}^{(p)}} \Phi - T_{P_{m,m}^{(p)}} y_p \end{bmatrix}$$

The problem now relies on finding the matrices $T_{f_{i,j}}^{(k)}$ and $T_{P_{i,j}}^{(k)}$. The construction of these matrices is simple and comes as an immediate result from the controller structure definition. The controller structure is defined as a combination (or a subset) of the matrices of the set $\Lambda_S = \{\Lambda_{1,1}, \Lambda_{1,2}, \ldots, \Lambda_{m,m}\}$. The following lemma states how the matrices $T_{f_{i,j}}^{(k)}$ and $T_{P_{i,j}}^{(k)}$ are constructed based on the controller structure.

Lemma 10.1

Let $\Lambda_S = \{\Lambda_{1,1}, \Lambda_{1,2}, \ldots, \Lambda_{m,m}\}$ be the set that generates all the possible combinations of controller structures. Let B be a multivariable controller structure defined as a subset of Λ_S where $B \subseteq \Lambda_S$ and with $B_{i,j} \in B \subseteq \Lambda_S$.

The matrix $T_{f_{i,j}}^{(k)}$ can be calculated as

$$T_{f_{i,j}}^{(k)} = \left[\begin{array}{c|c|c} 0 & I & 0 \\ \hline - & - & - \\ \beta & I & 0 \end{array}\right] \text{ where } \beta = \begin{bmatrix} B_{i,j} & \cdots & 0 \\ \vdots & \ddots & \vdots \\ 0 & \cdots & B_{i,j} \end{bmatrix}_{m(N-k+1)}$$

The matrices $T_{p_{i,j}}^{(k)}$ can be calculated as

$$T_{p_{i,j}}^{(k)} = \left[\begin{array}{c|c|c} 0 & I & \gamma \\ \hline - & - & - \\ 0 & I & 0 \end{array}\right] \text{ where } \gamma = \begin{cases} \begin{bmatrix} B_{i,j} & \cdots & 0 \\ \vdots & \ddots & \vdots \\ 0 & \cdots & B_{i,j} \end{bmatrix} & \text{for } k=2,\ldots,p \\ 0_m & \text{for } k=1 \end{cases}$$

Proof

This lemma can be verified by constructing the error vectors $\xi_{i,j}^{(k)}$ from the definition $\xi_{i,j}^{(k)} = -T_{f_{i,j}}^{(k)} \hat{y}_f - T_{p_{i,j}}^{(k)} y_p$. □

10.5.4 Closed-Loop Stability Conditions

The problem of closed-loop system stability can be addressed using a result presented by Giovanini and Marchetti (1999). In this paper, it was demonstrated that to ensure exponential stability of restricted-structure digital controllers it is sufficient to comply with the following condition

$$|\Delta \hat{u}(k+N)_i| \leq \sigma \quad \text{for } i=1,\ldots,m$$

This condition transforms into the constraint

$$\Theta \rho \leq \Psi$$

where

$$\Theta = \begin{bmatrix} E(k+N) \\ -E(k+N) \end{bmatrix} \text{ and } \Psi = \begin{bmatrix} \sigma \\ \vdots \\ \sigma \end{bmatrix}$$

This closed-loop stability constraint has to be solved simultaneously with the closed-loop computation $\Omega(\rho)\upsilon = \omega$.

The use of this stability condition in the optimisation problem will produce a controller for which the system is closed-loop stable; however, it might be possible that this condition cannot be met, and in that case the optimisation will be infeasible. Some suggestions to avoid infeasible optimisations are to enlarge either the cost index time horizon N or the domain of attraction σ, or both.

10.5.5 The Controller Tuning Algorithm

The steps discussed and specified in the Sections 10.2–10.5 can be combined to give a restricted-structure controller tuning algorithm based on subspace identification for the process model input, and finite time quadratic cost function optimisation for the controller tuning.

Algorithm 10.1: Subspace multivariable restricted structure tuning algorithm

Step 1 With the process in closed loop operation, record sufficient input–output data of $y(k)$, $u(k)$ and normalise the data with respect to the operating point.

Step 2 Select the forward–backward horizon N based on the settling time required.

Step 3 Construct the data Hankel matrices

$$W_p = \begin{bmatrix} Y_p \\ U_p \end{bmatrix}$$

and use the controller model to construct the controller Toeplitz matrix H_N^c.

Step 4 Calculate $N_f = U_f + H_N^c Y_f$.

Step 5 Calculate L_u^c and L_w^c by performing the RQ-decomposition and then use the equation

$$[L_w^c \quad L_u^c] = [R_{31} \quad R_{32}] \begin{bmatrix} R_{11} & 0 \\ R_{21} & R_{22} \end{bmatrix}^\dagger$$

Step 6 Calculate L_u and L_w using the equations

$$L_u = L_u^c (I - H_N^c L_u^c)^{-1} \quad \text{and} \quad L_w = (I + L_u H_N^c) L_w^c$$

Step 7 Calculate L_u^Δ and L_w^Δ using the equations

$$L_u^\Delta = L_u \begin{bmatrix} I_{mN} & 0 & 0 & 0 \\ I_{mN} & I_{mN} & \cdots & 0 \\ \vdots & \vdots & \ddots & \vdots \\ I_{mN} & I_{mN} & \cdots & I_{mN} \end{bmatrix}, \quad L_w^\Delta = \begin{bmatrix} l_{w_1} \\ l_{w_1} + l_{w_2} \\ \vdots \\ \sum_{i=1}^N l_{w_i} \end{bmatrix}$$

where l_{w_i} is the ith row block of dimension m of L_w.

Step 8 Define the number of controller parameters per controller element p.

Step 9 Define the controller structure as a subset of $\Lambda_S = \{\Lambda_{1,1}, \Lambda_{1,2}, \ldots, \Lambda_{m,m}\}$.

Step 10 If required, define the controller parameter numerical ranges and express as an inequality constraint.

Step 11 Calculate the matrices T_f and T_p using Lemma 10.1.

Step 12 Select values for the cost index weighting matrices Q and R.

Step 13 Select the stability condition variable σ.

Step 14 Select values for the controller parameter vector $\rho \in \mathfrak{R}^{pm^2}$.

Step 15 Solve

$$\begin{bmatrix} S_Q L_u^\Delta \phi \\ S_R \phi \end{bmatrix} \rho = \begin{bmatrix} S_Q \Phi \\ 0 \end{bmatrix}$$

in the least-squares sense, subject to $\Omega(\rho)\upsilon = \omega$ (closed-loop system condition) and $\Theta \rho \leq \Psi$ (closed-loop stability condition), and any additional constraints arising from controller parameter numerical range restrictions.

Algorithm end

10.6 Simulation Case Studies

The multivariable restricted structure controller tuning algorithm is tested in an investigation for the regulation of dissolved oxygen in a benchmark simulation for an activated sludge wastewater treatment plant.

Four different controller structures containing three-term controller elements are used as examples to show the efficiency of the algorithm and to demonstrate the complexity of control structures that may be assessed.

A COST Action research group developed the wastewater treatment plant (WWTP) benchmark simulation used in the control studies of this chapter. COST was founded in 1971 as an intergovernmental framework for European Co-operation in the field of Scientific and Technical Research. COST Action 682 *Integrated Wastewater Management* (1992–1998) focused on biological wastewater treatment processes and the optimisation of the design and operation based of dynamic process models. COST Action 624 was dedicated to the optimisation of performance and cost-effectiveness of wastewater management systems. The COST research group for Actions 624 and 682 developed the wastewater treatment plant (WWTP) benchmark simulation used here (Copp, 2002).

The benchmark simulation is a fully defined simulation protocol for evaluating activated sludge wastewater treatment control strategies. The simulation is platform-independent and has been tested with several programming languages. The SIMBA implementation of the benchmark simulation was used for these control case studies.

10.6.1 Activated Sludge Wastewater Treatment Plant Layout

The benchmark process is composed of five cascade biological reactors and a 10-layer non-reactive secondary settling tank. The plant layout is presented in Figure 10.3. The plant is fully defined and has the following characteristic features:

1. Five biological cascade reactor tanks with a secondary settler

2. Total biological volume of 5999 m^3
    2.1 Reactor tanks 1 and 2, each of 1000 m^3
    2.2 Reactor tanks 3, 4 and 5 each of 1333 m^3
    2.3 Reactor tanks 1 and 2 unaerated, but fully mixed

3. Aeration of reactor tanks 3, 4 and 5 achieved using a maximum of $K_{La} = 360$ d^{-1}

4. DO saturation level of 8 mg/l

5. Non-reactive secondary settler with a volume of 6000 m^3
    5.1 Area of 1500 m^2
    5.2 Depth of 4 m^2
    5.3 Subdivided into 10 layers
    5.4 Feed point to the settler at 2.2 m from the bottom

6. Two internal recycles
    6.1 Nitrate recycle from Reactor tank 5 to Reactor tank 1, default flow rate of 55338 m^3/day
    6.2 RAS recycle from the underflow of the secondary settler to the front end of the plant at a default flow rate of 18446 m^3/day

7. WAS is continuously pumped from the secondary settler underflow at a default rate of 385 m^3/day

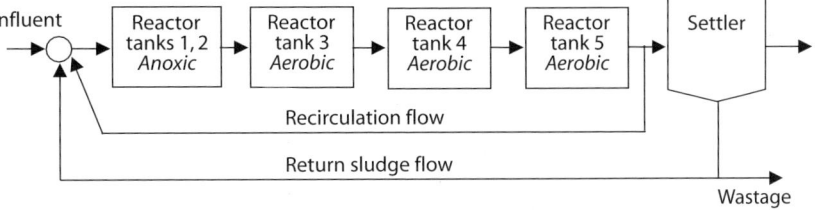

Figure 10.3 Activated sludge wastewater treatment plant layout.

Process Models and Controllers

The biological reactor tanks (aerated and unaerated) are modelled using ASM1 (Henze *et al.*, 1987). The settler is modelled using a double-exponential settling velocity function (Takcs *et al.*, 1991). The first two reactor tanks are anoxic, while the last three tanks are aerobic. The model also has a recirculation flow and a return sludge flow, as shown in Figure 10.3. The dissolved oxygen sensor used in the simulations has a 1 minute time delay and 1 minute sampling time. Actuators have been modelled as physical limitations in the air supply equivalent to a maximum oxygen transfer $K_{La} = 360$ d^{-1}. As a source of oxygen, air is introduced into the appropriate reactors through blowers (actuators) which are under PID control. The benchmark simulation also provides three files of dynamic influent data for dry, rain and storm conditions, and a file of constant influent data used to stabilise the plant.

The case studies examine the design of four possible controller configurations for the treatment plant.

10.6.2 Case study 1: Single-Input, Single-Output Control Structure

This case study shows the use of the tuning algorithm for a SISO plant type. The benchmark simulation is used by considering that only the last reactor tank has controlled aeration. The input to the model is the airflow rate scaled to a base of 10 and the output is the oxygen concentration in the reactor tank outflow. The airflow is introduced into the reactor tank through blowers (actuators) which are controlled by a PID controller. Figure 10.4 shows details of the control loop for the aerobic Reactor tank 5. The control loop also accounts for unmeasured disturbances as changes in the plant load. These disturbances are included as the signal $d(k)$. The identification considers a 1 minute sampling rate with an oxygen sensor with 1 minute time delay. The initial controller, with which the plant was identified in closed loop, is a PI with parameters shown in Table 10.1.

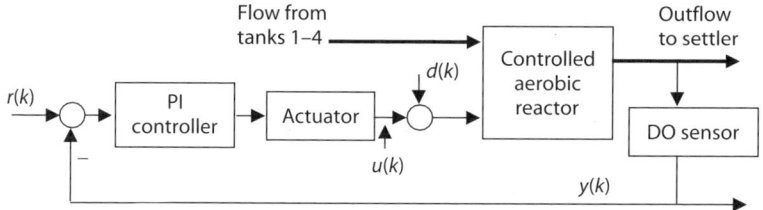

Figure 10.4 Case study 1: DO control structure.

Table 10.1 Initial and the optimal controller parameters.

	$\rho^{(1)}$	$\rho^{(2)}$	$\rho^{(3)}$
Initial parameters	1	−0.9308	0
Optimal parameters	8.2673	−12.2142	4.5825

For the identification, 1200 points of data were collected with the system excited by a pseudo-random binary signal (PRBS) of zero mean and size ±5 mg/l about a setpoint $r = 1$ mg/l. The forward and backward horizons were set to $N = 100$. The system is approximated by the first five singular values ($n = 5$) in the singular value decomposition of the identification algorithm. The algorithm was implemented in MATLAB. Table 10.1 shows the initial and final optimal controller parameters.

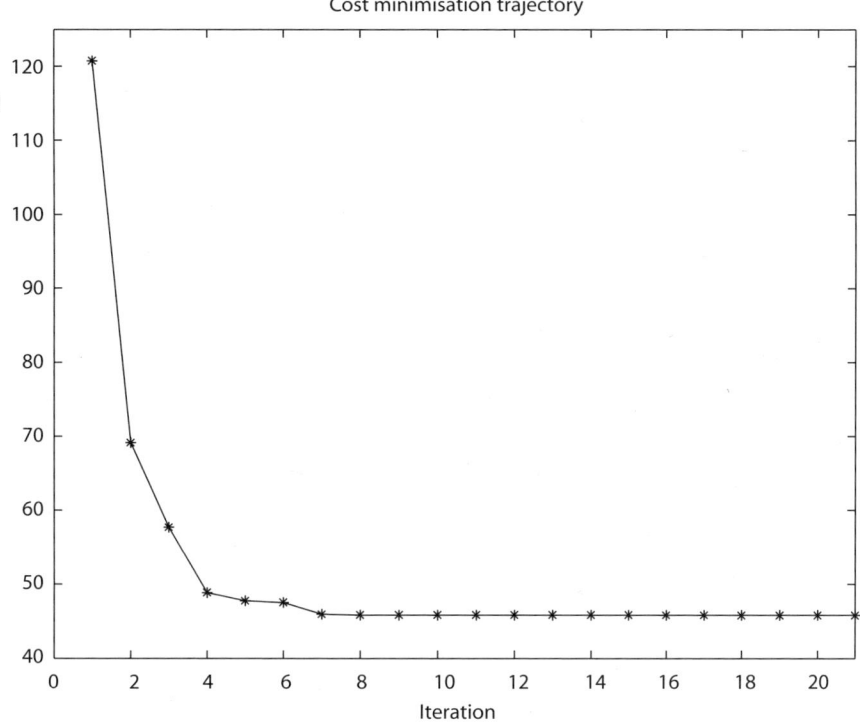

Figure 10.5 Cost index values per iteration.

The solution took from 10 to 20 iterations depending on the stringency of convergence conditions, as shown in Figure 10.5.

The cost index and the stability constraint specifications are given in Table 10.2. The optimal control parameters are given in Table 10.1 and Figure 10.6 shows the responses of the initial and optimised controllers for a unit step.

Table 10.2 Optimisation problem specifications.

N	M	R	Q	σ
100	100	8	5	0.01

10.6.3 Case Study 2: Control of Two Reactors with a Lower Triangular Controller Structure

This case study illustrates the design of a control system for Reactor tanks 4 and 5. Thus input–output data from both reactor tanks is required, as well as details of the Reactor tank 4 and 5 PID controllers. One day of input–output data was collected at a sampling interval of 1 minute (1440 samples). The system was excited with a pseudo-random binary signal perturbation of ±1 mg/l about the setpoint $r = 2$ mg/l. The initial PID controllers for the three control elements were set to be the same and these are given in Table 10.1.

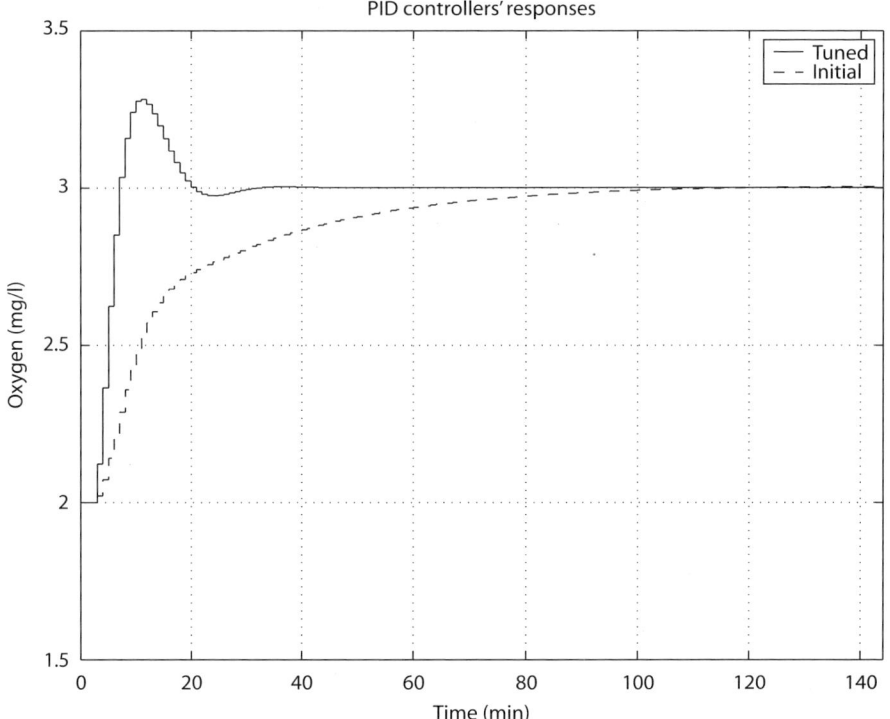

Figure 10.6 Comparison between the initial and optimally tuned PID controller responses.

The controller structure was selected to be lower triangular as specified through the structure matrix

$$B = \begin{bmatrix} 1 & 0 \\ 1 & 1 \end{bmatrix}$$

This structure defines an interaction between the error signal in Reactor tank 4 and the control signal in Reactor tank 5. Simulation results were performed for three different cases as presented in Table 10.3, where n is the order approximation in the identification SVD.

Figures 10.7 and 10.8 present the simulation results.

Notice that the response in Case 1 is much faster and has an acceptable overshoot compared with Cases 2 and 3. However, the effect of the coupling controller produces a much more aggressive response in the oxygen concentration in Reactor tank 5.

Table 10.3 Optimisation problem specifications.

	R	Q	N	n	σ
Case 1	diag(10 10)	diag(10 10)	60	5	0.01
Case 2	diag(10 10)	diag(10 20)	60	5	0.01
Case 3	diag(10 20)	diag(10 20)	60	5	0.01

10.6 Simulation Case Studies

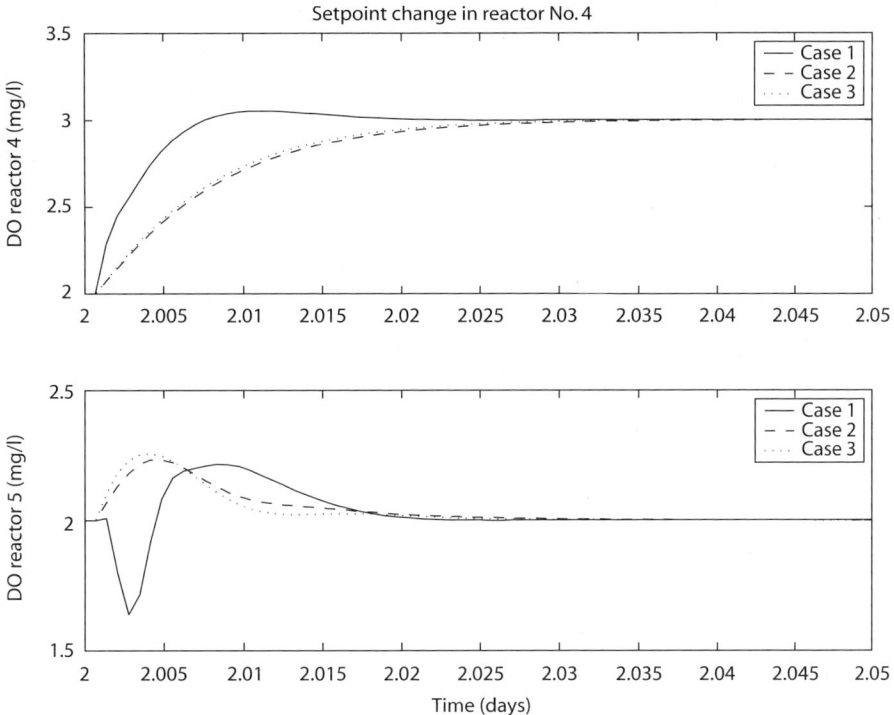

Figure 10.7 Effects of a setpoint change in Reactor tank 4.

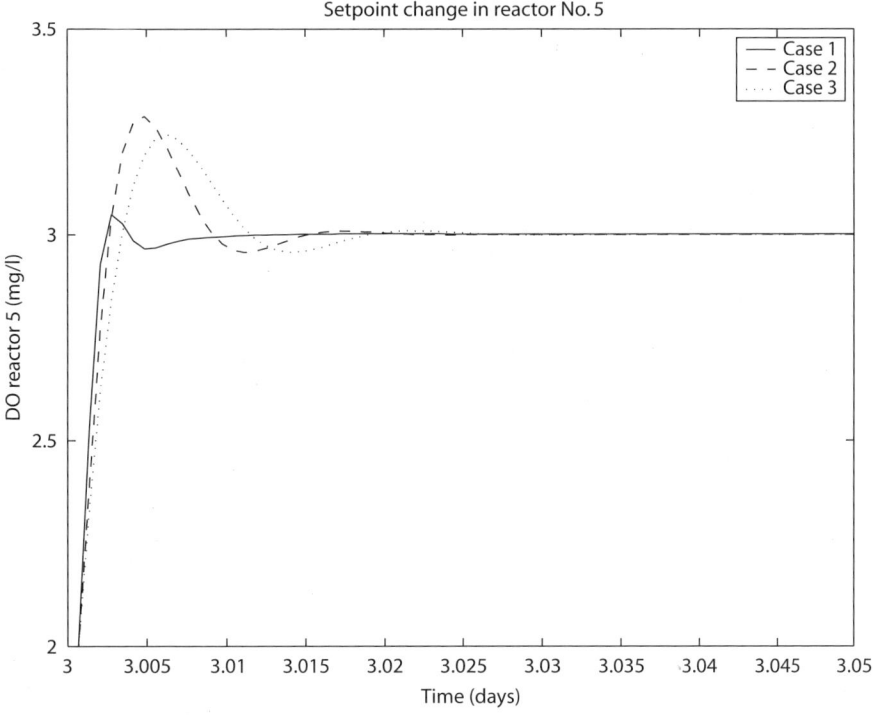

Figure 10.8 Effects of a setpoint change in Reactor tank 5.

10.6.4 Case Study 3: Control of Three Reactors with a Diagonal Controller Structure

This example considers the tuning of the three PID controllers with a diagonal controller structure. The same amount of data was collected as in Case Study 2 and using the same procedure. The controller structure was selected to be diagonal as specified through the structure matrix $B = \text{diag}\{1,1,1\}$. In this example, however, simulations were performed for prediction horizons of $N = 40$ and $N = 60$. For each horizon, four cases of different weights were considered. Table 10.4 summarises the different cases, where the weighting matrices Q and R have been chosen such that all their elements are the same. For $N = 40$, simulation results are presented in Figures 10.9–10.11 and for $N = 60$ in Figures 10.12–10.14.

Table 10.4 Optimisation problem specifications.

	R	Q	n	σ
Case 1	1	1	5	0.01
Case 2	1	3	5	0.01
Case 3	3	1	5	0.01
Case 4	10	10	5	0.01

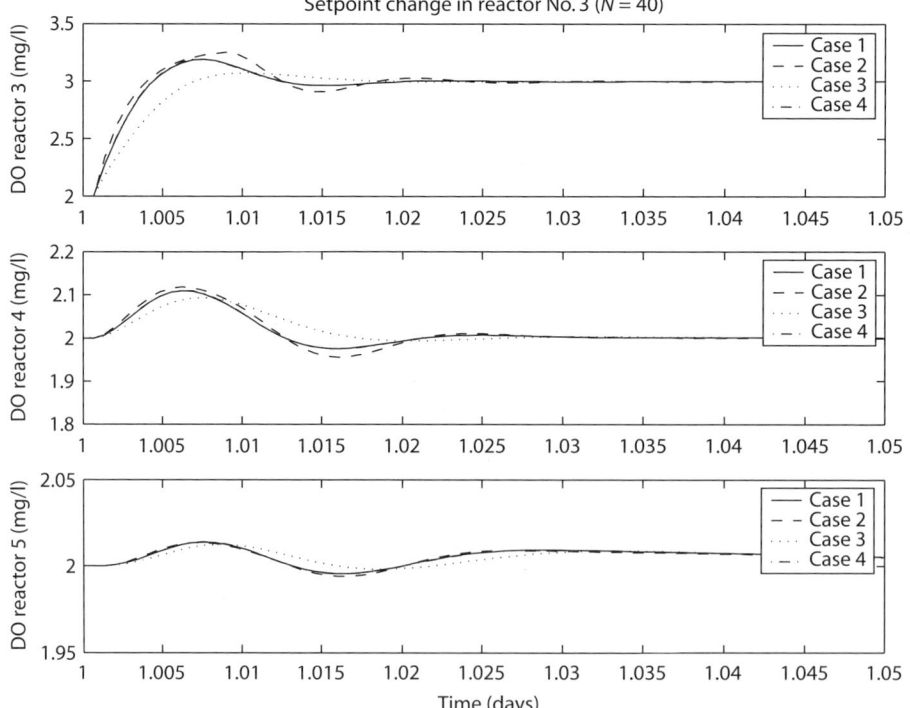

Figure 10.9 Effects of a setpoint change in Reactor tank 3 with $N = 40$.

10.6 Simulation Case Studies

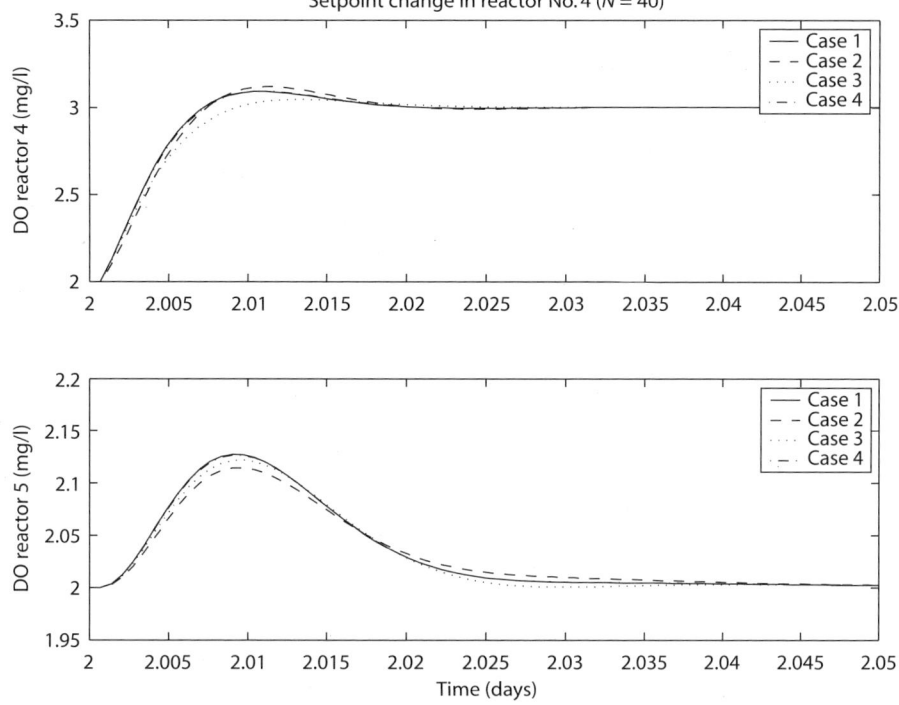

Figure 10.10 Effects of a setpoint change in Reactor tank 4 with $N = 40$.

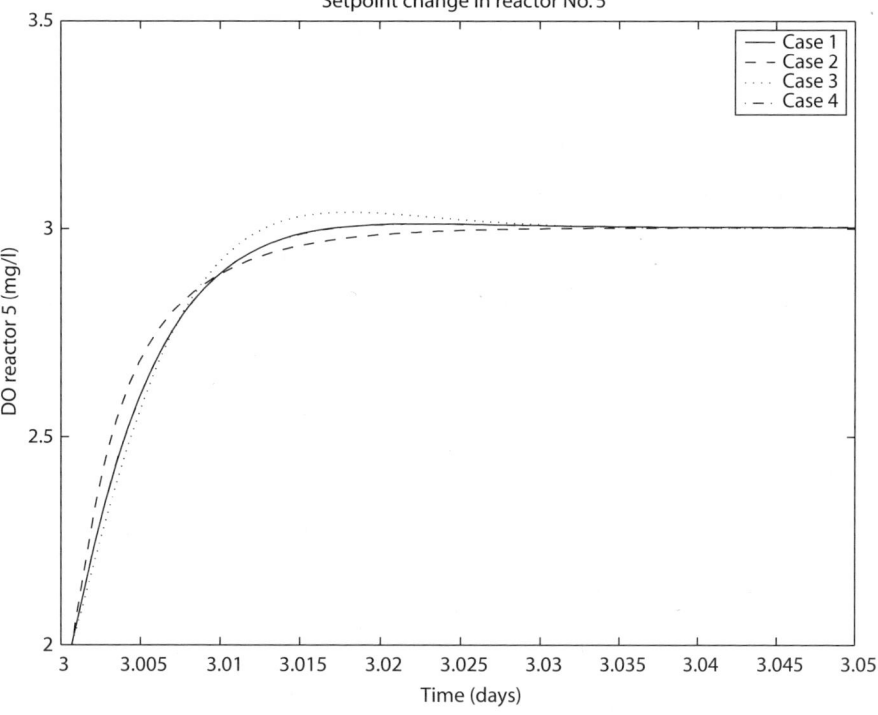

Figure 10.11 Effects of a setpoint change in Reactor tank 5 with $N = 40$.

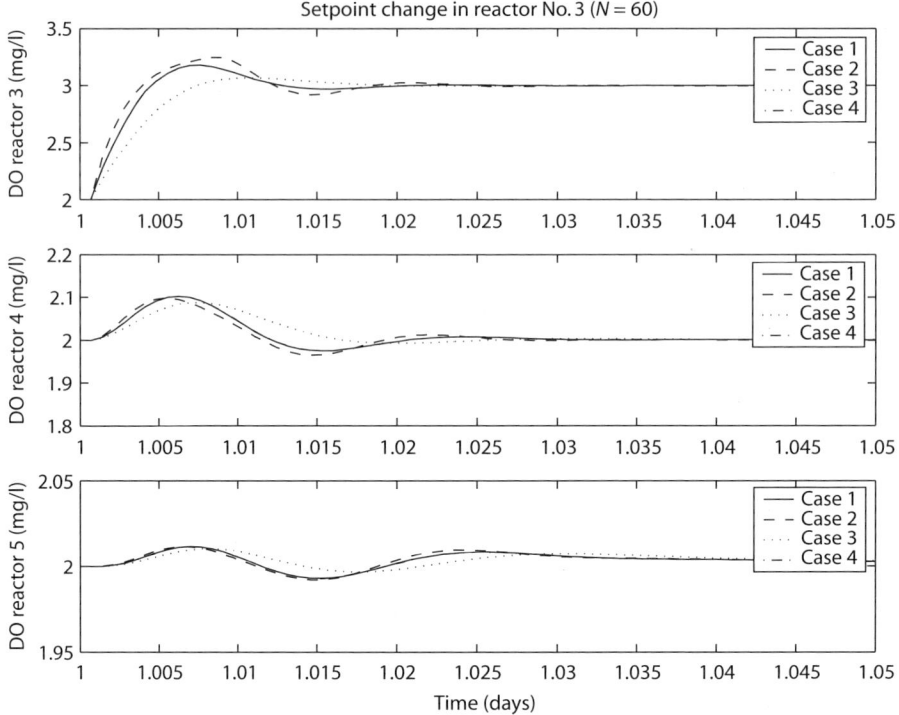

Figure 10.12 Effects of a setpoint change in Reactor tank 3 with $N = 60$.

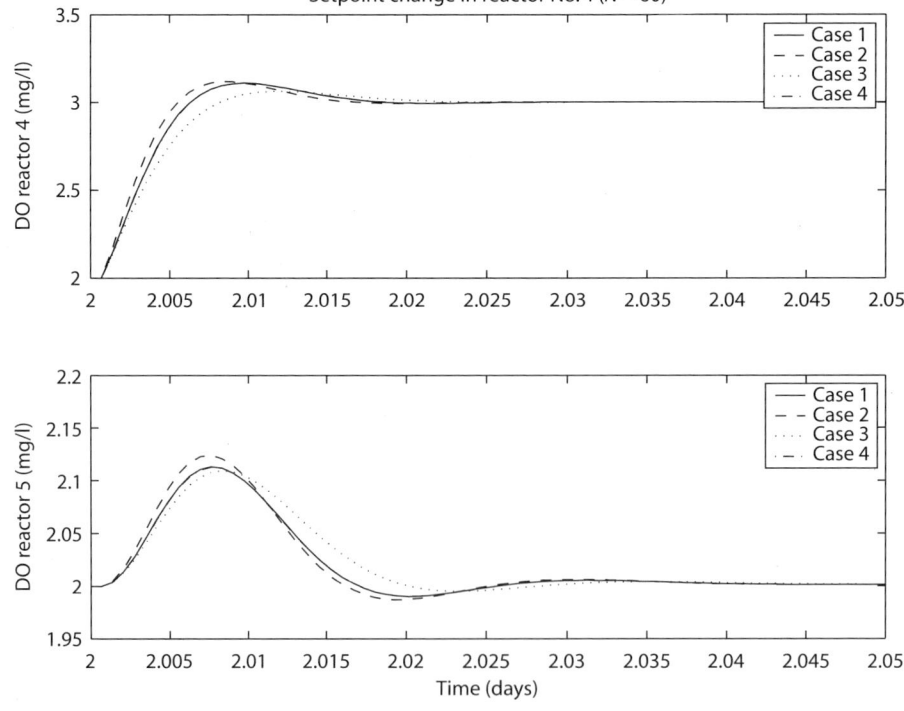

Figure 10.13 Effects of a setpoint change in Reactor tank 4 with $N = 60$.

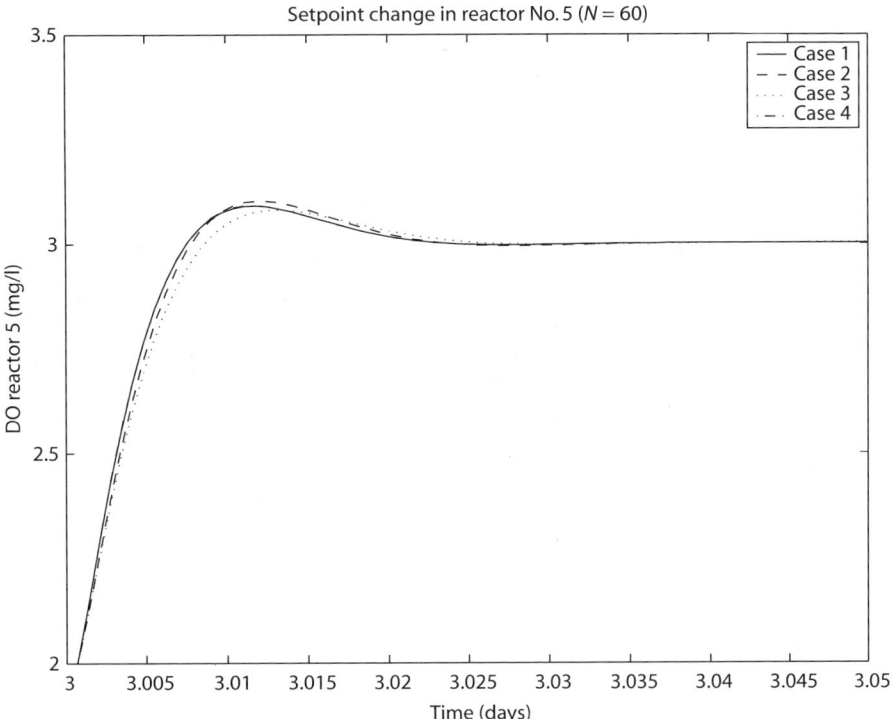

Figure 10.14 Effects of a setpoint change in Reactor tank 5 with $N = 60$.

The most significant observation in this example is that there was no major difference between the simulations with different prediction horizons. This suggests that longer horizons do not contribute significantly to the optimality of the solution; however, longer prediction horizons certainly increase the computation requirements.

10.6.5 Case Study 4: Control of Three Reactors with a Lower Triangular Controller Structure

This last case study examines the design of a lower triangular controller, similar to that of Case Study 2 but involving all three aerobic reactor tanks. The controller structure was selected to be lower triangular as specified through the structure matrix

$$B = \begin{bmatrix} 1 & 0 & 0 \\ 1 & 1 & 0 \\ 1 & 1 & 1 \end{bmatrix}$$

Two simulations for different sets of weights are presented. Table 10.5 gives the details of the optimisation problem specifications.

Table 10.5 Optimisation problem specifications.

	R	Q	n	σ
Case 1	diag(3 3 3)	diag(3 3 3)	5	0.01
Case 2	diag(10 20 30)	diag(10 20 30)	5	0.01

386 Tuning PID Controllers Using Subspace Identification Methods

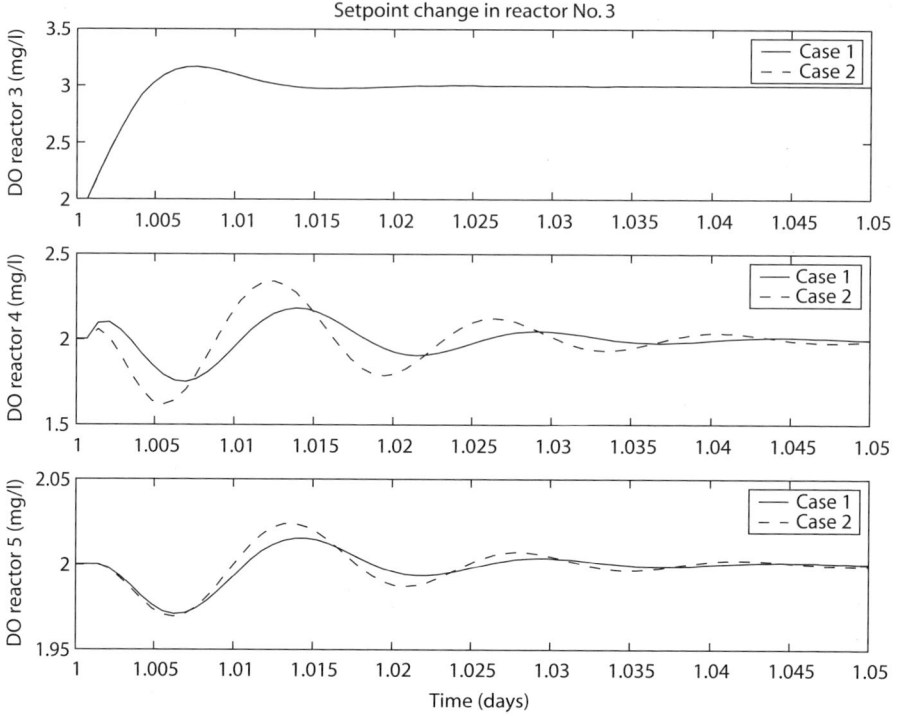

Figure 10.15 Effects of a setpoint change in Reactor tank 3.

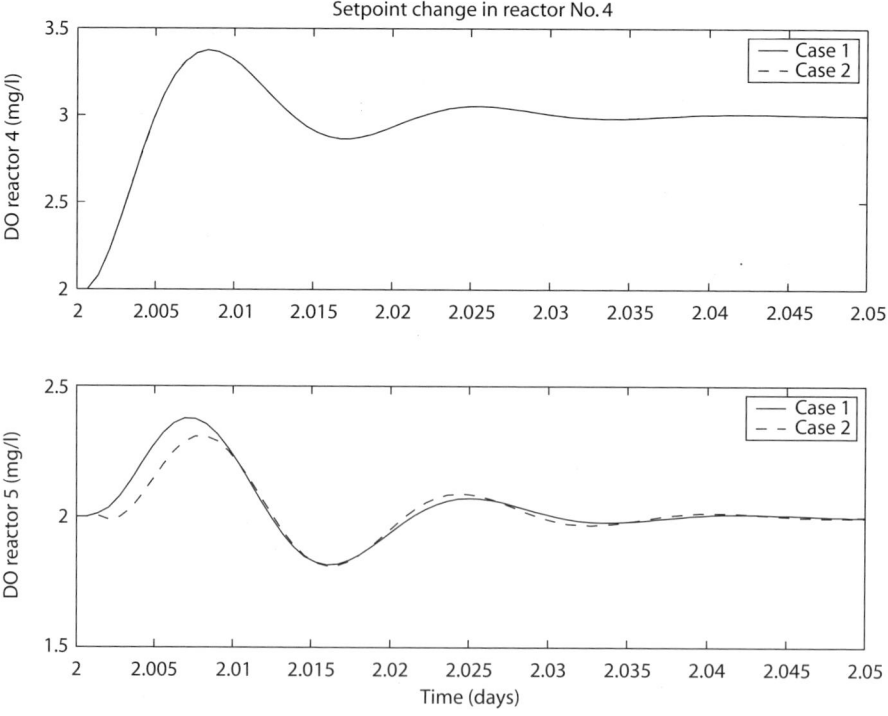

Figure 10.16 Effects of a setpoint change in Reactor tank 4.

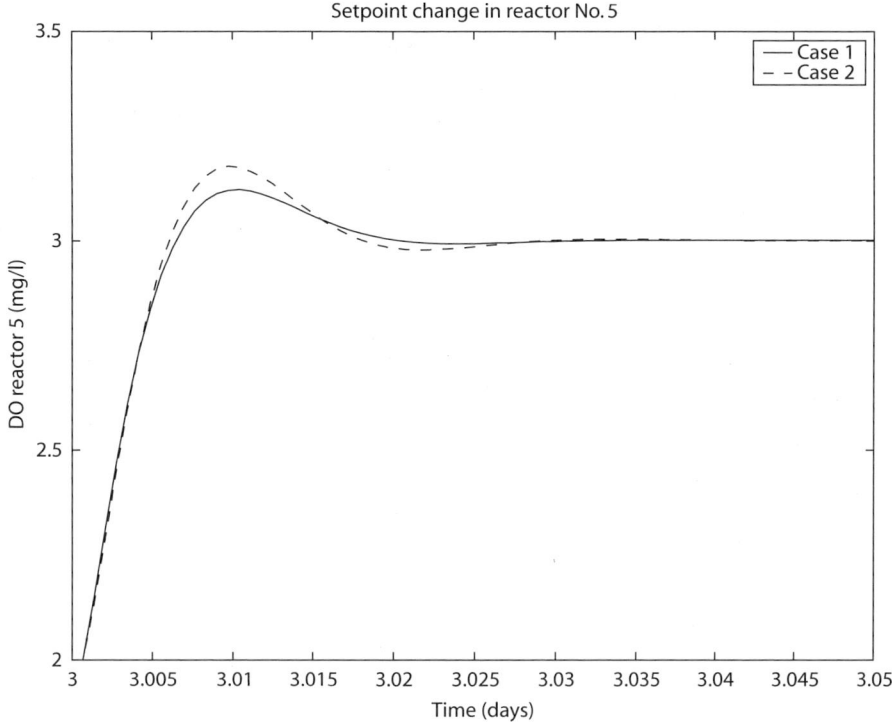

Figure 10.17 Effects of a setpoint change in Reactor tank 5.

Both simulations assume a prediction horizon of $N = 40$ and simulation responses are presented in Figures 10.15–10.17. Notice that when the weights are increased the system responds faster and becomes more oscillatory, especially when changing the setpoint in reactor 3 due to the higher weights in reactors 4 and 5 and the strong couplings with the first reactor. The effect in reactors 3 and 4 is, however, attenuated when changing the setpoints in reactors 4 and 5, since the couplings in the reverse direction are much smaller.

References

Copp, J. (ed.) (2002) COST *Action 624 – The COST Simulation Benchmark: Description and Simulation Manual*. European Commission – The COST Programme.

Favoreel, W., De Moor, B., Gevers, M. and Van Overschee, P. (1998) Closed loop model-free subspace-based LQG-design. *Technical Report ESAT-SISTA/TR 1998-108*. Departement Elecktrotechniek, Katholieke Universiteit Leuven, Belgium.

Giovanini, L. and Marchetti, J. (1999) Shaping time-domain responses with discrete controllers. *Ind. Eng. Chem. Res.*, **38**, 4777–4789.

Henze, M., Grady, C.P.L., Gujer, W., Marais, G.v.R. and Matsuo, T. (1987) Activated Sludge Model No. 1. *IAWQ Scientific And Technical Report No. 1*.

Kadali, R., Huang, B. and Rossiter, A. (2003) A data driven approach to predictive controller design. *Control Engineering Practice*, **11**(3), 261–278.

Ruscio, D.D. (1997) Model based predictive control: an extended state space approach. *Proc. 36th Conference on Decision and Control*, San Diego, CA, pp. 3210–3217.

Ruscio, D.D. (2000) *Subspace System Identification: Theory and Applications*. Lecture Notes, 6th edn, Telemark Institute of Technology, Porsgrunn, Norway.

Sánchez, A., Katebi, M.R. and Johnson, M.A. (2003) Subspace identification based PID control tuning. *Proc. IFAC Symposium on System Identification*, Rotterdam, The Netherlands.

Sánchez, A., Katebi, M.R. and Johnson, M.A. (2004) A tuning algorithm for multivariable restricted structure control systems using subspace identification. *Int. J. Adapt. Control Signal Process.*, **18**, 745–770.

Takcs, I., Patry, G.G. and Nolasco, D. (1991) A dynamic model of the clarification thickening process. *Water Research*, **25**(10), 1263–1271.

van Overschee, P. and De Moor, B. (1996a) Closed-loop subspace system identification. *Technical Report ESAT-SISTA/TR 1996-521*, Departement Elecktrotechniek, Katholieke Universiteit Leuven, Belgium.

van Overschee, P. and De Moor, B. (1996b) *Subspace Identification for Linear Systems*. Kluwer Academic Publishers, New York.

Verhaegen, M. (1993) Application of a subspace model identification technique to identify LTI systems operating in closed-loop. *Automatica*, **29**(4), 1027–1040.

11 Design of Multi-Loop and Multivariable PID Controllers

Learning Objectives

11.1 Introduction

11.2 Multi-Loop PID Control

11.3 Multivariable PID Control

11.4 Conclusions

References

Learning Objectives

In this chapter, a first major assumption is that the designer has a multi-input, multi-output transfer function model for the process to be controlled. This will be used to show the deep characteristics of multivariable systems and the difficulty in multivariable control arising from the interactions between the multiple process elements. The dictum of the methods to be discussed is that the less interacting the system, the more it behaves like independent SISO loops.

For multi-loop PID control, the BLT tuning method will be examined first. This method does not assume much knowledge of the multivariable system and may be used to give preliminary controller settings. Better settings may be obtained from the dominant pole placement method with the help of some computer graphs, and this new multi-loop dominant pole assignment method will be given a full explanation and derivation in this part of the chapter.

For multivariable PID control, decoupling design is recommended for the simplification of design and the improvement of performance. Trial and error methods have been found hard to use and a systematic method to achieve a reasonable engineering design is presented. One aspect of this procedure is the formal use of de-tuning rules to achieve near-optimal control performance.

Design of Multi-Loop and Multivariable PID Controllers

The learning objectives for this chapter are to:

- Help the reader to understand the nature of multivariable systems and the difficulty in multivariable control.
- Design multi-loop PID controllers.
- Design multivariable PID controllers with decoupling specification.

11.1 Introduction

11.1.1 Multivariable Systems

Processes with inherently more than one variable at the output to be controlled are frequently encountered in industry and are known as multi-input, multi-output (MIMO) or multivariable processes. Interactions usually exist between the control loops of multivariable processes, which accounts for their renowned difficulty in control when compared with the control of single-input, single output (SISO) processes.

For illustration, consider a system with two inputs, u_1 and u_2, and two outputs, y_1 and y_2, whose dynamics is given by

$$y_1 = \frac{1}{s+1}u_1 + \frac{\alpha}{s+1}u_2$$
$$y_2 = \frac{\beta}{s+1}u_1 + \frac{1}{s+1}u_2$$

This process can be seen in Figure 11.1, where the process interaction terms are highlighted and two potential control loops sketched in position. Note that the interaction terms are linked to the parameters α and β.

Let r_1 and r_2 denote the process setpoints (reference signals) for outputs y_1 and y_2, respectively. If input u_1 controls output y_1 using a PI-controller given by

$$u_1 = k_1\left(1 + \frac{1}{s}\right)(r_1 - y_1)$$

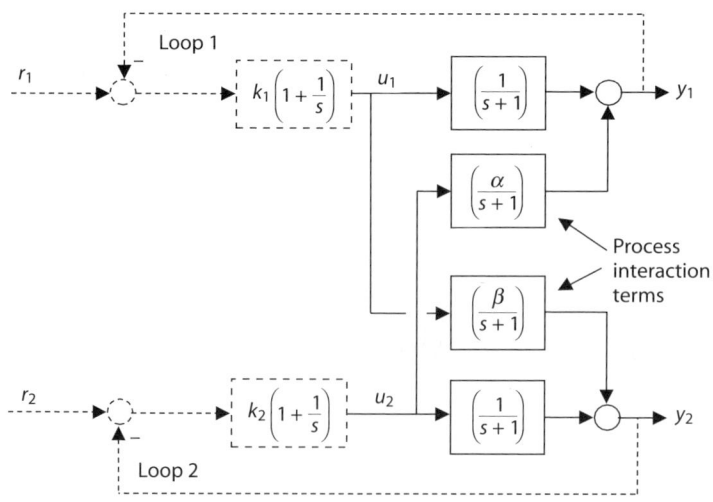

Figure 11.1 Interactive process and two possible control loops.

and input u_2 is set to zero, then the closed-loop transfer function is

$$\frac{y_1}{r_1} = \frac{k_1}{s+k_1}$$

and this is seen to be a stable transfer function for any $k_1 > 0$. Similarly, if the system is controlled only using Loop 2 such that input u_2 controls output y_2 using a PI-controller given by

$$u_2 = k_2\left(1+\frac{1}{s}\right)(r_2 - y_2)$$

and input u_1 is set to zero, then the closed-loop transfer function is

$$\frac{y_2}{r_2} = \frac{k_2}{s+k_2}$$

and this is stable whenever $k_2 > 0$. By way of contrast, consider what happens if both PI controllers operate together so that both loops are closed. Let $k_1 = 1$ and $k_2 = 1$ and operate the process with both Loop 1 and Loop 2 closed. When both loops are closed, the transfer function from r_1 to y_1 becomes

$$\frac{y_1}{r_1} = \frac{s+1-\alpha\beta}{s^2+2s+1-\alpha\beta}$$

which is unstable if $\alpha\beta > 1$. This shows that the interaction can destabilise a system of individually stabilised multi-loops.

The goal of controller design to achieve satisfactory loop performance for multivariable systems with interactions has hence posed a great challenge in the area of control design.

11.1.2 Multivariable Control

In the process industries, multi-loop controllers or decentralised controllers are often more favoured than fully cross-coupled multivariable controllers because they have a simpler structure with fewer tuning parameters to determine. In addition, in the case of actuator or sensor failure it is relatively easy to stabilise the process with a multi-loop controller manually (Skogestad and Morari, 1989). Like the SISO case, controllers of the Proportional plus Integral (PI) structure are employed in a multi-loop fashion for many multivariable processes in industry. The PI controller is unquestionably the most commonly used control algorithm in the process control industries (Åström et al., 1998). PID control is always preferred to more advanced controllers in practical applications unless evidence is given or accumulated which shows that PID control is not adequate and cannot meet the performance specifications. The reliability of the family of three-term controllers has been proven by decades of success and they enjoy a high level of acceptance among practising engineers.

There have been elegant theorems on general decentralised control of delay-free linear systems using state space and matrix fraction approaches (Ozguler, 1994). However, the decentralised PID control case is less developed; for example, the stability problem of decentralised PID control systems is difficult to address and still remains open (Åström and Hägglund, 1995; Tan et al., 1999). It can be shown that the problem of closed-loop stability of PID control is equivalent to that of output feedback stabilisation (Zheng et al., 2001) and this problem is NP hard. As a result, mainstream PID control research is directed at devising methods that meet certain performance specifications on the closed-loop time domain response or in the frequency domain, and leave closed-loop stability to be post-checked. Besides, largely due to rapid advances in computer technology, there is also an increasing use of parameter optimisation for the design of decentralised controllers with a specific structure such as a fixed order (usually low-order or PID type). If the problem is formulated in the frequency domain the algorithms can be greatly simplified, but then stability cannot be ensured; on the other hand, stability may be ensured using Youla parameterisations, but then it is no longer possible to guarantee a simple structure in the resulting controller (Boyd et al., 1988).

Multi-loop control cannot change the interactive nature of a given multivariable system and thus may limit achievable performance in many cases. A fully cross-coupled multivariable controller has the potential for a performance improvement over the decentralised controller and general multivariable control theory has developed over recent years (Luyben, 1990; Maciejowski, 1989; Skogestad and Postlethwaite, 1996). It may be noted that most multivariable design techniques, such as state space, optimisation, Inverse Nyquist Array and characteristic locus methods, are unlikely to lead to PID-type controllers. For multivariable PID control, Koivo and Tanttu (1991) gave a survey of recent tuning techniques. The survey showed that these techniques mainly aimed to decouple the plant at certain frequencies. A software package that includes several methods for tuning MIMO PI/PID controllers as well as an expert system to assist the user has also been developed (Lieslehto et al., 1993). Recently, Wang et al. (1997) have proposed a fully cross-coupled multivariable PID controller tuning method with decoupling design.

11.1.3 Scope of the Chapter and Some Preliminary Concepts

Firstly, consider multi-loop control. Since an independent SISO loop design may fail when applied to an interactive multivariable system, a simple way to cope with it is to de-tune all the loops together for stability and performance. If the system is stable then the multi-loop feedback system can always be made stable provided the loop gains are de-tuned to be sufficiently small. In Section 11.2, a de-tuning method called the Biggest Log-Modulus Tuning (BLT) is introduced. To reduce the conservativeness in tuning the common gain and to get better performance, it is proposed and demonstrated that the loop gains should be tuned separately to fit the different characteristics of individual loops.

It is noted that control performance is largely influenced by closed-loop pole positions. For time delay systems there are an infinite number of closed-loop poles, and full pole placement is not possible with conventional output feedback control. Thus the dominant pole method becomes popular for both analysis and design of delay control systems. This method has the twin advantages of addressing closed-loop stability as well as performance. The SISO dominant pole placement design method (Åström and Hägglund, 1995) is known to work well for most SISO processes. Given the popularity of this SISO design method and the use of decentralised control in practice, the dominant pole placement method is extended to the decentralised case and presented in this chapter. For a two-input, two-output (TITO) process controlled by a multi-loop PI controller, two pairs of desired poles can be specified for the closed-loop system, and the design problem is to find the parameters of the multi-loop PI controller such that the roots of the closed-loop characteristic equation are placed at the desired positions. Unlike its SISO counterpart, where the controller parameters can be obtained analytically, the multi-loop version involves solving some coupled nonlinear equations with complex coefficients. This is not surprising, as the coupling between loops makes the closed-loop characteristic equation depend on both unknown controllers in a nonlinear way and hence the problem is difficult to solve analytically. In Section 11.2, a novel approach is presented to solve the closed-loop characteristic equation using a "root trajectory" method. The design procedure is given and simulation examples are provided to show the effectiveness of the method.

Consider next multivariable control. Modern industrial processes are becoming more complex. This complexity can be caused by the dynamics of the individual channels, but more often it is caused by the strong interaction existing between the control loops, especially when the number of controlled variables becomes large and the individual loops contain significant time delays. Such interactions can give rise to extremely complicated dynamics when the loops are closed, and multi-loop control is inadequate, leading only to poor performance. In Section 11.3, a systematic method is presented for the design of multivariable PID controllers with the objectives of achieving fast loop responses with acceptable overshoot and minimum loop interactions. The design is based on the fundamental relations that describe the decoupling of a multivariable process and the characterisation of the unavoidable time delays and non-minimum phase zeros that are inherent in the decoupled feedback loops.

To conclude this section, the basic control system and some assumptions for use in subsequent sections are introduced. The plant and controller framework used in the chapter is exclusively the conventional unity feedback control system shown in Figure 11.2, where $G(s)$ represents the transfer matrix of the plant and $C(s)$ the controller.

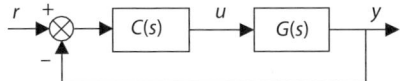

Figure 11.2 Unity feedback control system.

It is assumed that the process or plant has a known model representation in the form of a square and non-singular transfer matrix:

$$G(s) = \begin{bmatrix} g_{11}(s) & \cdots & g_{1m}(s) \\ \vdots & \ddots & \vdots \\ g_{m1}(s) & \cdots & g_{mm}(s) \end{bmatrix}$$

where the elements $g_{ij}(s)$ have the form $g_{ij}(s) = g_{ij0}(s)e^{-L_{ij}s}$, and each term $g_{ij0}(s)$ is a strictly proper, stable scalar real rational function and the parameter L_{ij} is a non-negative constant representing the time delay in the process model element.

The controller $C(s)$ is an $m \times m$ transfer matrix:

$$C(s) = \begin{bmatrix} c_{11}(s) & \cdots & c_{1m}(s) \\ \vdots & \ddots & \vdots \\ c_{m1}(s) & \cdots & c_{mm}(s) \end{bmatrix}$$

where all the controller elements $c_{ij}(s)$ are of PID type. As a result, the controller always has integral control. If it is assumed that the plant has no zero at the origin, then the integral control action present will lead to zero steady state errors in the time response to step inputs provided that the closed loop is stable.

Section 11.2 focuses on multi-loop or decentralised controller structure where

$$c_{ij}(s) = 0, \text{ if } i \neq j$$

Thus the controller transfer function has a diagonal structure:

$$C(s) = \begin{bmatrix} c_{11}(s) & 0 & \cdots & 0 \\ 0 & c_{22}(s) & \cdots & 0 \\ \vdots & \vdots & \ddots & \vdots \\ 0 & 0 & \cdots & c_{mm}(s) \end{bmatrix}$$

In Section 11.3, a general controller $C(s)$ will be considered where all the controller elements are generally non-zero.

It is also assumed throughout the chapter that a proper input–output pairing has been made for the multivariable process $G(s)$ with a tool such as the Relative Gain Array (RGA). This assumed structural rearrangement will ensure that none of the m principal minors of $G(s)$ is zero (the ith principal minor of $G(s)$ is the determinant of an $(m-1) \times (m-1)$ matrix obtained by deleting the ith row and ith column of $G(s)$).

11.2 Multi-Loop PID Control

Multi-loop control is widely used in industry. Its advantage lies in its simplicity for implementation and manual tuning. The disadvantages are a lack of flexibility for interaction adjustment and few powerful tools for its design compared with general multivariable control. This section will introduce a popular tuning method (the Biggest Log-Modulus Tuning) and a newly developed tuning method (dominant pole placement tuning) and close with some comparative examples of the two methods.

11.2.1 Biggest Log-Modulus Tuning Method

Control engineers know how to tune a SISO PID controller. To capitalise on this knowledge, one simple way to tune a multi-loop PID controller is first to tune each individual loop one after another, totally ignoring loop interactions; that is, tune the ith loop PID controller $c_{ii}(s)$ for the plant transfer diagonal element $g_{ii}(s)$ only. Then de-tune all the loops together with a common factor such that the overall system is stable and gives acceptable load disturbance rejection responses.

The biggest log-modulus tuning (BLT) procedure (Luyben, 1990) provides such a standard tuning methodology. Its objective is to arrive at reasonable controller settings with only a small amount of engineering and computational effort. It may not necessarily produce the best results. But the method is easy to use and understand by control engineers, and leads to settings that compare very favourably with the empirical settings found by exhaustive and expensive trial-and-error tuning methods used in many studies. The method should be viewed in the same light as the classical SISO Ziegler–Nichols method. It gives reasonable settings that provide a starting point for further tuning and a benchmark for comparative studies. The BLT tuning method consists of four steps, as given in the Algorithm 11.1.

Algorithm 11.1: BLT tuning method

Step 1 Initial Ziegler–Nichols PID settings

Calculate the Ziegler–Nichols settings for each individual loop. The ultimate gain and ultimate frequency of each transfer function matrix diagonal element $g_{ii}(s)$ are calculated in the classical SISO way. There are several ways to find the ultimate data settings, including the method of sustained oscillations, the relay experiment or even straightforward Nyquist plotting using software like MATLAB. Denote the ultimate frequency by ω_{ui} from which the ultimate period is $P_{ui} = 2\pi/\omega_{ui}$, and denote the ultimate gain by K_{ui}.

The PI controllers have the assumed representation:

$$c_{ZNi}(s) = K_{ZNi}\left(1 + \frac{1}{\tau_{ZNi} s}\right)$$

The Ziegler–Nichols settings are

$K_{ZNi} = K_{ui}/2.2, \quad i = 1, \ldots, m$
$\tau_{ZNi} = P_{ui}/1.2, \quad i = 1, \ldots, m$

Step 2 The de-tuning step

A common detuning factor F is assumed. The de-tuning factor F should always be greater than unity. Typical values lie in the range $1.5 < F < 4$.

The de-tuned PI controllers have the assumed representation

$$c_{ii}(s) = K_{ci}\left(1 + \frac{1}{\tau_{Ii} s}\right)$$

The parameters of all the feedback controllers $c_{ii}(s)$ are de-tuned as follows:

The gains: $K_{ci} = K_{ZNi}/F$, $i = 1,\ldots,m$

The reset times: $\tau_{Ii} = \tau_{ZNi} \times F$

The F factor can be considered as a detuning factor that is applied to all loops. The larger the value of F, the more stable the system will be, but the more sluggish the setpoint and load disturbance rejection responses will be. The method yields settings that give a reasonable compromise between stability and performance in multivariable systems.

Step 3 The log modulus plot

Using the guessed value of F and the resulting controller settings, a multivariable Nyquist plot of the scalar function $W(s) = -1 + \det[I + G(s)C(s)]$ is made. The closer this contour is to the $(-1,0)$ point, the closer the system is to instability. Therefore the quantity $W/(1+W)$ will be similar to the closed-loop servo transfer function $GC/(1+GC)$ for a SISO system. Thus, based on intuition and empirical grounds, a multivariable closed-loop log modulus function L_{cm} is defined as

$$L_{cm} = 20 \log_{10} \left| \frac{W}{1+W} \right|$$

The peak in the plot of L_{cm} over the entire frequency range is the biggest log modulus: L_{cm}^{max}.

Step 4 Re-selecting the F factor

The F factor is varied until L_{cm}^{max} is equal to $2m$, where m is the number of outputs of the system. For $m = 1$, the SISO case, the familiar +2 dB maximum closed-loop log modulus criterion emerges. For a 2×2 system, a +4 dB value of L_{cm}^{max} is used; for a 3×3 system a +6 dB value of L_{cm}^{max} is used; and so on.

Algorithm end

This empirically determined criterion has been tested on a large number of cases and gives reasonable performance, which is a little on the conservative side. The method weights each loop equally, so that each loop is equally detuned. If it is important to keep tighter control of some output variables than others, the method can be easily modified by using different weighting factors for different output variables. The less important loops being detuned more than the more important loops.

11.2.2 Dominant Pole Placement Tuning Method

The presentation for the dominant pole placement method is specialised to the decentralised control of a general two-input, two-output process. It begins from a 2×2 transfer function model given by

$$\begin{bmatrix} y_1(s) \\ y_2(s) \end{bmatrix} = \begin{bmatrix} g_{11}(s) & g_{12}(s) \\ g_{21}(s) & g_{22}(s) \end{bmatrix} \begin{bmatrix} u_1(s) \\ u_2(s) \end{bmatrix}$$

The process is to be controlled in a conventional negative feedback configuration (Figures 11.1 and 11.2) by the multi-loop or decentralised control law:

$$\begin{bmatrix} u_1(s) \\ u_2(s) \end{bmatrix} = \begin{bmatrix} c_1(s) & 0 \\ 0 & c_2(s) \end{bmatrix} \begin{bmatrix} e_1(s) \\ e_2(s) \end{bmatrix}$$

where the error signal is given as

$$\begin{bmatrix} e_1(s) \\ e_2(s) \end{bmatrix} = \begin{bmatrix} r_1(s) \\ r_2(s) \end{bmatrix} - \begin{bmatrix} y_1(s) \\ y_2(s) \end{bmatrix} = \begin{bmatrix} r_1(s) - y_1(s) \\ r_2(s) - y_2(s) \end{bmatrix}$$

The characteristic equation for this 2×2 closed-loop process is given by

$$\rho_{CL}(s) = \det[I + G(s)C(s)] = 0$$

where I is the 2×2 identity matrix.

It is well known that the roots of the characteristic equation are the poles of the resultant closed-loop system, and for this 2×2 closed-loop process it follows that

$$\rho_{CL}(s) = \det[I + G(s)C(s)] = 1 + g_{11}(s)c_1(s) + g_{22}(s)c_2(s) + \Delta(s)c_1(s)c_2(s) = 0$$

where $\Delta(s) = g_{11}(s)g_{22}(s) - g_{12}(s)g_{21}(s)$.

Let the multi-loop PI controller be parameterised as

$$c_l(s) = k_{Pl} + \frac{k_{Il}}{s}, \quad l = 1,2$$

where $k_{Pl}, l = 1, 2$ are the proportional gains and $k_{Il}, l = 1, 2$ are the integral gains. This multi-loop PI controller has four adjustable parameters $k_{Pl}, k_{Il}, l = 1, 2$; hence two pairs of desired closed-loop dominant pole locations can be specified. Thus define the pre-specified pole placement locations as

$$p_i = \sigma_i + j\omega_i = -\omega_{0i}\zeta_i + j\omega_{0i}\sqrt{1-\zeta^2}, \quad i=1,2$$

and

$$\bar{p}_i = \sigma_i - j\omega_i = -\omega_{0i}\zeta_i - j\omega_{0i}\sqrt{1-\zeta^2}, \quad i=1,2$$

The Pole Placement Design Equations

The controller parameters will be determined so that two pre-specified pairs of complex numbers are the roots of the characteristic equation. It is assumed that the desired closed-loop pole locations $p_i, \bar{p}_i, i=1,2$, are not already poles of $G(s)$, and as a result the numbers $g_{ij}(p_i)$, $g_{ij}(\bar{p}_i)$, $\Delta(p_i)$, $\Delta(\bar{p}_i)$, $i=1,2$ are all finite. Thus the pole placement requirement at p_i and $\bar{p}_i, i=1,2$ is equivalent to the following set of equations:

$$\rho_{CL}(p_i) = 1 + g_{11}(p_i)c_1(p_i) + g_{22}(p_i)c_2(p_i) + \Delta(p_i)c_1(p_i)c_2(p_i) = 0, \quad i=1,2$$

$$\rho_{CL}(\bar{p}_i) = 1 + g_{11}(\bar{p}_i)c_1(\bar{p}_i) + g_{22}(\bar{p}_i)c_2(\bar{p}_i) + \Delta(\bar{p}_i)c_1(\bar{p}_i)c_2(\bar{p}_i) = 0, \quad i=1,2$$

$$c_l(p_i) = k_{Pl} + \frac{k_{Il}}{\sigma_i + j\omega_i} = k_{Pl} + \frac{k_{Il}(\sigma_i - j\omega_i)}{\sigma_i^2 + \omega_i^2}, \quad l=1,2; \; i=1,2$$

$$c_l(\bar{p}_i) = k_{Pl} + \frac{k_{Il}}{\sigma_i - j\omega_i} = k_{Pl} + \frac{k_{Il}(\sigma_i + j\omega_i)}{\sigma_i^2 + \omega_i^2}, \quad l=1,2; \; i=1,2$$

But for a real rational transfer function matrix $G(s)$ and controller $C(s)$, the properties $\overline{g_{ij}(s)} = g_{ij}(\bar{s})$ and $\overline{c_{ij}(s)} = c_{ij}(\bar{s})$ hold $(i=1,2; j=1,2)$ so that if $\rho_{CL}(p_i) = 0$ and p_i is a closed-loop pole then $\overline{\rho_{CL}(p_i)} = \rho_{CL}(\bar{p}_i) = 0$ and $\bar{p}_i$ will also be a closed-loop pole and vice versa. This enables the above set of equations to be reduced to

$$\rho_{CL}(p_i) = 1 + g_{11}(p_i)c_1(p_i) + g_{22}(p_i)c_2(p_i) + \Delta(p_i)c_1(p_i)c_2(p_i) = 0, \quad i=1,2$$

$$c_l(p_i) = k_{Pl} + \frac{k_{Il}}{\sigma_i + j\omega_i} = k_{Pl} + \frac{k_{Il}(\sigma_i - j\omega_i)}{\sigma_i^2 + \omega_i^2}, \quad l=1,2; \; i=1,2$$

with $\Delta(p_i) = g_{11}(p_i)g_{22}(p_i) - g_{12}(p_i)g_{21}(p_i)$.

In this reduced set of equations, the four unknowns are the controller parameters $k_{Pl}, k_{Il}, l = 1, 2$. One route to a solution would be to eliminate the quantities $c_l(p_i) = \tilde{c}_{li}(k_{Pl}, k_{Il}), l=1,2; i=1,2$ from the closed-loop characteristic equations to give two complex equations:

$$\rho_{CL}(p_i) = \tilde{\rho}_{CLi}(k_{P1}, k_{I1}, k_{P2}, k_{I2}) = 0, \quad i=1,2$$

11.2 Multi-Loop PID Control

Since these two closed-loop characteristic equations are complex equations, this would give four real nonlinear equations from which to find solutions for $(k_{P1}, k_{I1}, k_{P2}, k_{I2})$. However, a different solution strategy that eventually leads to an innovative graphical procedure is followed instead. Essentially this alternative solution procedure uses the characteristic equations to solve for relationships between potential values of $c_l(p_i), l = 1,2; i = 1,2$ and uses the controller equations to solve for $(k_{P1}, k_{I1}, k_{P2}, k_{I2})$.

To avoid the pole placement problem from becoming trivial it is assumed for the rest of this section that $g_{12}(p_i)g_{21}(p_i) \neq 0$ or $\Delta(p_i) \neq g_{11}(p_i)g_{22}(p_i)$. Otherwise, if $g_{12}(p_i)g_{21}(p_i) = 0$ or $\Delta(p_i) = g_{11}(p_i)g_{22}(p_i)$ so that the transfer function matrix $G(s)$ evaluated at p_i is diagonal or triangular, then the closed-loop characteristic condition reduces to two trivial SISO pole placement problems specified through

$$\rho_{CL}(p_i) = [1 + g_{11}(p_i)c_1(p_i)][1 + g_{22}(p_i)c_2(p_i)] = 0$$

The required pole placement characteristic equation can be used to solve for relationships between potential values of $c_l(p_i)$. This is achieved by factoring $\rho_{CL}(p_i)$ for the controller $c_2(s)$ as follows:

$$\rho_{CL}(p_i) = [\Delta(p_i)c_1(p_i) + g_{22}(p_i)]c_2(p_i) + [1 + g_{11}(p_i)c_1(p_i)] = 0; \quad i = 1,2$$

where for $i = 1,2, [\Delta(p_i)c_1(p_i) + g_{22}(p_i)] \neq 0$ since $g_{12}(p_i)g_{21}(p_i)c_1(p_i) \neq 0$. Hence solving for $c_2(p_i)$ yields

$$c_2(p_i) = -\frac{[1 + g_{11}(p_i)c_1(p_i)]}{[\Delta(p_i)c_1(p_i) + g_{22}(p_i)]}, \quad i = 1,2$$

The second part of the solution strategy is to obtain relationships from the controller equations. Evaluating the controllers $c_l(s)$ for $s = p_1 = \sigma_1 + j\omega_1$ yields:

$$c_l(p_1) = k_{Pl} + \frac{k_{Il}}{\sigma_1 + j\omega_1} = k_{Pl} + \frac{k_{Il}(\sigma_1 - j\omega_1)}{\sigma_1^2 + \omega_1^2}, \quad l = 1,2$$

$$\text{Re}\{c_l(p_1)\} = k_{Pl} + \frac{k_{Il}\sigma_1}{\sigma_1^2 + \omega_1^2}, \quad l = 1,2$$

$$\text{Im}\{c_l(p_1)\} = \frac{-k_{Il}\omega_1}{\sigma_1^2 + \omega_1^2}, \quad l = 1,2$$

Then solving for $k_{Pl}, k_{Il}, l = 1, 2$ gives

$$k_{Pl} = \text{Re}\{c_l(p_1)\} + \frac{\sigma_1}{\omega_1}\text{Im}\{c_l(p_1)\}, \quad l = 1,2$$

$$k_{Il} = -\left(\frac{\sigma_1^2 + \omega_1^2}{\omega_1}\right)\text{Im}\{c_l(p_1)\}, \quad l = 1,2$$

Continuing with relationships from the controller equations, a relation between $c_l(p_1)$ and $c_l(p_2)$ can be established as follows:

$$c_l(p_2) = k_{Pl} + \frac{k_{Il}}{\sigma_2 + j\omega_2} = k_{Pl} + \frac{k_{Il}(\sigma_2 - j\omega_2)}{\sigma_2^2 + \omega_2^2}, \quad l = 1,2$$

and substituting for $k_{Pl}, k_{Il}, l = 1,2$ yields

$$c_l(p_2) = \left(\text{Re}\{c_l(p_1)\} + \frac{\sigma_1}{\omega_1}\text{Im}\{c_l(p_1)\}\right) + \left(-\left(\frac{\sigma_1^2 + \omega_1^2}{\omega_1}\right)\text{Im}\{c_l(p_1)\}\right)\frac{(\sigma_2 - j\omega_2)}{\sigma_2^2 + \omega_2^2}$$

$$c_l(p_2) = \text{Re}\{c_l(p_1)\} + \left[\frac{\sigma_1}{\omega_1} - \left(\frac{\sigma_1^2 + \omega_1^2}{\omega_1}\right)\frac{(\sigma_2 - j\omega_2)}{(\sigma_2^2 + \omega_2^2)}\right]\text{Im}\{c_l(p_1)\}$$

where $l = 1, 2$.

It is useful at this point to summarise the equation set whose solution will lead to PI multi-loop control as:

$$c_2(p_i) = -\frac{[1 + g_{11}(p_i)c_1(p_i)]}{[\Delta(p_i)c_1(p_i) + g_{22}(p_i)]} \quad i = 1, 2$$

$$\Delta(p_i) = g_{11}(p_i)g_{22}(p_i) - g_{12}(p_i)g_{21}(p_i) \quad i = 1, 2$$

$$c_l(p_2) = \text{Re}\{c_l(p_i)\} + \left[\frac{\sigma_1}{\omega_1} - \left(\frac{\sigma_1^2 + \omega_1^2}{\omega_1}\right)\frac{(\sigma_2 - j\omega_2)}{(\sigma_2^2 + \omega_2^2)}\right]\text{Im}\{c_l(p_i)\} \quad l = 1, 2$$

$$k_{Pl} = \text{Re}\{c_l(p_i)\} + \frac{\sigma_1}{\omega_1}\text{Im}\{c_l(p_i)\} \quad l = 1, 2$$

$$k_{Il} = -\left(\frac{\sigma_1^2 + \omega_1^2}{\omega_1}\right)\text{Im}\{c_l(p_i)\} \quad l = 1, 2$$

Further Equation Simplification

To simplify and reduce the equation set, some properties of complex numbers are used. The complex number z satisfies the following identities:

$$\text{Re}\{z\} = \frac{z + \bar{z}}{2} \quad \text{and} \quad \text{Im}\{z\} = \frac{z - \bar{z}}{2}$$

where $\bar{z}$ is the complex conjugate of z. If this property is applied to the controller equation, it can be shown that

$$c_l(p_2) = \alpha_1 c_l(p_1) + \alpha_2 \overline{c_l(p_1)}$$

where

$$\alpha_1 = \frac{1}{2} + \left[\sigma_1 - \frac{(\sigma_1^2 + \omega_1^2)(\sigma_2 - j\omega_2)}{(\sigma_2^2 + \omega_2^2)}\right]\left(\frac{1}{2j\omega_1}\right)$$

$$\alpha_2 = \frac{1}{2} - \left[\sigma_1 - \frac{(\sigma_1^2 + \omega_1^2)(\sigma_2 - j\omega_2)}{(\sigma_2^2 + \omega_2^2)}\right]\left(\frac{1}{2j\omega_1}\right)$$

Based on the above relationship, a complex function f is introduced where

$$f(z) = \alpha_1 z + \alpha_2 \bar{z}$$

and α_1, α_2 are known complex numbers. Then the controller equation becomes

$$c_l(p_2) = f(c_l(p_1)), \quad l = 1, 2$$

Thus the new equation subset is:

$$c_2(p_1) = -\frac{[1 + g_{11}(p_1)c_1(p_1)]}{[\Delta(p_1)c_1(p_1) + g_{22}(p_1)]}$$

$$c_2(p_2) = -\frac{[1 + g_{11}(p_2)c_1(p_2)]}{[\Delta(p_2)c_1(p_2) + g_{22}(p_2)]}$$

$$c_1(p_2) = f(c_1(p_1))$$

$$c_2(p_2) = f(c_2(p_1))$$

Reduction to a single equation in terms of $c_1(p_1)$ occurs as follows:

$$f(c_2(p_1)) = f\left(-\frac{[1+g_{11}(p_1)c_1(p_1)]}{[\Delta(p_1)c_1(p_1)+g_{22}(p_1)]}\right)$$

$$= c_2(p_2)$$

$$= -\frac{[1+g_{11}(p_2)c_1(p_2)]}{[\Delta(p_2)c_1(p_2)+g_{22}(p_2)]}$$

$$= -\frac{[1+g_{11}(p_2)f(c_1(p_1))]}{[\Delta(p_2)f(c_1(p_1))+g_{22}(p_2)]}$$

and the final equation set is:

$$f\left(\frac{[1+g_{11}(p_1)c_1(p_1)]}{[\Delta(p_1)c_1(p_1)+g_{22}(p_1)]}\right) = \frac{[1+g_{11}(p_2)f(c_1(p_1))]}{[\Delta(p_2)f(c_1(p_1))+g_{22}(p_2)]}$$

$$c_2(p_1) = -\frac{[1+g_{11}(p_1)c_1(p_1)]}{[\Delta(p_1)c_1(p_1)+g_{22}(p_1)]}$$

$$\Delta(p_1) = g_{11}(p_1)g_{22}(p_1) - g_{12}(p_1)g_{21}(p_1)$$

$$k_{Pl} = \text{Re}\{c_l(p_1)\} + \frac{\sigma_1}{\omega_1}\text{Im}\{c_l(p_1)\} \qquad l=1,2$$

$$k_{Il} = -\left(\frac{\sigma_1^2+\omega_1^2}{\omega_1}\right)\text{Im}\{c_l(p_1)\} \qquad l=1,2$$

It is clear that the nonlinear equation of the solution set will admit a solution for the unknown $c_1(p_1)$ if and only if the original problem of placing two pre-specified poles at p_i, $\bar{p}_i$, $i=1,2$ for the process $G(s)$ is solvable by a PI multi-loop controller. It should also be noted that this is a nonlinear equation with complex coefficients. It may have multiple solutions. It is thus desirable to have an effective way to determine whether the equation admits solutions, and how to find them if they exist. In case of multiple solutions, a best one should be chosen for $c_1(p_1)$. Having found candidate solutions for $c_1(p_1)$, the equations in the set are then evaluated in sequence; $c_2(p_1)$ can then be obtained and subsequently the parameters of the PI controller can be computed. Thus the main issue is to solve the fundamental nonlinear equation, and in what follows a novel approach will be presented.

Solving the Nonlinear Equation

The nonlinear equation to be solved is

$$f\left(\frac{[1+g_{11}(p_1)c_1(p_1)]}{[\Delta(p_1)c_1(p_1)+g_{22}(p_1)]}\right) = \frac{[1+g_{11}(p_2)f(c_1(p_1))]}{[\Delta(p_2)f(c_1(p_1))+g_{22}(p_2)]}$$

where $f(z) = \alpha_1 z + \alpha_2 \bar{z}$ and α_1, α_2 are the known complex numbers.

A complex number can always be expressed equivalently in its polar form. Thus let $c_1(p_1) = re^{j\theta}$, where $r > 0$ and θ are real numbers; then the nonlinear equation becomes

$$\frac{[1+g_{11}(p_1)c_1(p_1)]}{[\Delta(p_1)c_1(p_1)+g_{22}(p_1)]} + \alpha_2 \frac{[1+\overline{g_{11}(p_1)}\overline{c_1(p_1)}]}{[\overline{\Delta(p_1)}\overline{c_1(p_1)}+\overline{g_{22}(p_1)}]} = \frac{[1+g_{11}(p_2)(\alpha_1 c_1(p_1)+\alpha_2 \overline{c_1(p_1)})]}{[\Delta(p_2)(\alpha_1 c_1(p_1)+\alpha_2 \overline{c_1(p_1)})+g_{22}(p_2)]}$$

and hence

$$\frac{[1+g_{11}(p_1)re^{j\theta}]}{[\Delta(p_1)re^{j\theta}+g_{22}(p_1)]} + \alpha_2 \frac{[1+\overline{g_{11}(p_1)}re^{-j\theta}]}{[\overline{\Delta(p_1)}re^{-j\theta}+\overline{g_{22}(p_1)}]} = \frac{[1+g_{11}(p_2)(\alpha_1 e^{j\theta}+\alpha_2 e^{-j\theta})r]}{[\Delta(p_2)(\alpha_1 e^{j\theta}+\alpha_2 e^{-j\theta})r+g_{22}(p_2)]}$$

The above equation can be rearranged into a third-order polynomial in variable r, namely

$$(\beta_7 e^{-j\theta} + \beta_6 e^{j\theta})r^3 + (\beta_5 e^{-j\theta} + \beta_4 + \beta_3 e^{j\theta})r^2 + (\beta_2 e^{-j\theta} + \beta_1 e^{j\theta})r + \beta_0 = 0$$

where $\beta_i, i = 0, 1, 2, \ldots, 7$ are known complex numbers given by

$$\beta_7 = \alpha_2[\alpha_1 g_{11}(p_1)\overline{\Delta(p_1)}\Delta(p_2) + \alpha_2 \overline{g_{11}(p_1)}\Delta(p_1)\Delta(p_2) - g_{11}(p_2)|\Delta(p_1)|^2]$$

$$\beta_6 = \beta_7 \alpha_1 / \alpha_2$$

$$\beta_5 = \alpha_2[\alpha_2 \overline{g_{11}(p_1)} g_{22}(p_1)\Delta(p_2) + \alpha_1 \overline{\Delta(p_1)}\Delta(p_2) - g_{11}(p_2)g_{22}(p_1)\overline{\Delta(p_1)}]$$

$$\beta_4 = \alpha_1^2 \overline{\Delta(p_1)}\Delta(p_2) + \alpha_2^2 \Delta(p_1)\Delta(p_2) - |\Delta(p_1)|^2 + \alpha_1 g_{11}(p_1)g_{22}(p_2)\overline{\Delta(p_1)}$$
$$+ \alpha_2 \overline{g_{11}(p_1)} g_{22}(p_2)\Delta(p_1) - \alpha_1 g_{11}(p_2)g_{22}(p_1)\overline{\Delta(p_1)}$$
$$- \alpha_2 g_{11}(p_2)g_{22}(p_1)\Delta(p_2) + 2\alpha_1 \alpha_2 \operatorname{Re}[\overline{g_{11}(p_1)} g_{22}(p_1)]\Delta(p_2)$$

$$\beta_3 = \alpha_1[\alpha_1 g_{11}(p_1)\overline{g_{22}(p_1)}\Delta(p_2) + \alpha_2 \Delta(p_1)\Delta(p_2) - g_{11}(p_2)\overline{g_{22}(p_1)}\Delta(p_1)]$$

$$\beta_2 = -\overline{\Delta(p_1)} g_{22}(p_1) + \alpha_2 \overline{g_{11}(p_1)} g_{22}(p_1) g_{22}(p_2) - \alpha_2 g_{11}(p_2)|g_{22}(p_1)|^2$$
$$+ \alpha_2^2 g_{22}(p_1)\Delta(p_2) + \alpha_1 \alpha_2 \overline{g_{22}(p_1)}\Delta(p_2) + \alpha_1 g_{22}(p_2)\overline{\Delta(p_1)}$$

$$\beta_1 = -\Delta(p_1)\overline{g_{22}(p_1)} + \alpha_1 g_{11}(p_1)\overline{g_{22}(p_1)} g_{22}(p_2) - \alpha_1 g_{11}(p_2)|g_{22}(p_1)|^2$$
$$+ \alpha_1^2 \overline{g_{22}(p_1)}\Delta(p_2) + \alpha_1 \alpha_2 g_{22}(p_1)\Delta(p_2) + \alpha_2 g_{22}(p_2)\Delta(p_1)$$

$$\beta_0 = \alpha_1 \overline{g_{22}(p_1)} g_{22}(p_2) - |g_{22}(p_1)|^2 + \alpha_2 g_{22}(p_1)g_{22}(p_2)$$

The third-order polynomial in unknown variable r is also a function of the unknown variable θ; a careful analysis of the roles in the polynomial of θ and then r is presented to derive a solution procedure for this equation.

Feasible Regions for Variable θ

Firstly, it is noted from the controller equation that

$$\theta = \arg\{c_1(p_1)\} = \arg\left\{k_{P1} + \frac{k_{I1}}{\sigma_1 + j\omega_1}\right\}$$

For non-negative parameters, the PI controller $c_1(s)$ evaluated at $s = p_1$, can only contribute a phase angle θ in the interval

$$\left(-\pi + \arctan\left(\frac{\sqrt{1-\zeta_1^2}}{\zeta_1}\right), 0\right)$$

for positive parameters or

$$\left(\arctan\left(\frac{\sqrt{1-\zeta_1^2}}{\zeta_1}\right), \pi\right)$$

for negative parameters. The θ value on the two endpoints of the intervals corresponds to either pure integral control or pure proportional control. The sign of the controller parameters of $c_1(s)$ should be determined such that a negative feedback configuration is formed. Maciejowski (1989) has shown that the equivalent process faced by $c_1(s)$ can be derived as

$$g_1(s) = g_{11}(s) - \frac{g_{12}(s)g_{21}(s)c_2(s)}{1 + g_{22}(s)c_2(s)}$$

Since integral control action is assumed in each loop this gives

$$g_1(0) = g_{11}(0) - \frac{g_{12}(0)g_{21}(0)}{g_{22}(0)}$$

If $g_1(0)$ is positive, then the controller parameters of $c_1(s)$ should be chosen to be positive to form a negative feedback system and vice versa. Thus θ is confined in the interval Θ given by

$$\Theta = \begin{cases} \left[\left(-\pi + \arctan\left(\frac{\sqrt{1-\zeta_1^2}}{\zeta_1}\right)\right), 0 \right] & \text{if } g_1(0) > 0 \\ \left(\arctan\left(\frac{\sqrt{1-\zeta_1^2}}{\zeta_1}\right), \pi\right) & \text{if } g_1(0) < 0 \end{cases}$$

Root Trajectory Solutions in Variable r

For each given $\theta \in \Theta$, the polynomial equation in pair (r, θ) simply becomes a third-degree polynomial equation in r, which has three roots. As θ varies in Θ, the solutions for r form a trajectory (containing three branches) in the complex plane, and this *root trajectory* can be formally defined as

$$L \equiv r \in \mathbb{C} \mid (\beta_7 e^{-j\theta} + \beta_6 e^{j\theta})r^3 + (\beta_5 e^{-2j\theta} + \beta_4 + \beta_3 e^{2j\theta})r^2 + (\beta_2 e^{-j\theta} + \beta_1 e^{j\theta})r + \beta_0 = 0, \ \theta \in \Theta$$

Each root trajectory of L can be plotted on the complex plane. An intersection of the trajectories with the positive real axis means a set of values for $r > 0$ and $\theta \in \Theta$ which solves the polynomial equation. Thus the trajectories L will intersect with the positive real axis at as many points as the number of solutions to the polynomial equation with $r > 0$ and $\theta \in \Theta$, and if no such solution exists then there are no intersections. Let the set of all such intersection points be denoted by Z.

Identifying the Feasible Solution Set Z

The original complex nonlinear equation given by

$$f\left(\frac{[1 + g_{11}(p_1)c_1(p_1)]}{[\Delta(p_1)c_1(p_1) + g_{22}(p_1)]}\right) = \frac{[1 + g_{11}(p_2)f(c_1(p_1))]}{[\Delta(p_2)f(c_1(p_1)) + g_{22}(p_2)]}$$

has been transformed to a quasi-polynomial equation:

$$(\beta_7 e^{-j\theta} + \beta_6 e^{j\theta})r^3 + (\beta_5 e^{-j\theta} + \beta_4 + \beta_3 e^{j\theta})r^2 + (\beta_2 e^{-j\theta} + \beta_1 e^{j\theta})r + \beta_0 = 0$$

and then solved by a root trajectory method for a set of solutions Z. However, it should be noted that although each element in Z is a solution to the quasi-polynomial equation, the corresponding $z = re^{j\theta}$ may not necessarily be a solution for the unknown $c_1(p_1)$ in the nonlinear equation. This is because an extra solution can be introduced when transforming from the nonlinear equation form to the quasi-polynomial form where a multiplication of the common denominator is performed, and the solution set may be enlarged. It can be shown that if

$$\arg\left\{\frac{-g_{22}(p_1)}{\Delta(p_1)}\right\} \in \Theta$$

then

$$z = \frac{-g_{22}(p_1)}{\Delta(p_1)} \in Z$$

and that this particular $z \in Z$ is *not* a solution to the original nonlinear equation and hence should be discarded from Z. The remaining elements in Z are all the solutions for $c_1(p_1)$ in the nonlinear equation with $\arg\{c_1(p_1)\} \in \Theta$ and each value of $c_1(p_1)$ can be used to compute the value of $c_1(p_1)$ and hence provides a set of multi-loop PI controller settings $(k_{P1}, k_{I1}, k_{P2}, k_{I2})$ which achieves the placement of the poles at $(p_i, \bar{p}_i, i = 1, 2)$.

Selecting the Controller Parameters ($k_{P1}, k_{I1}, k_{P2}, k_{I2}$)

As the performance of a control system may not be solely determined by the dominant poles but also be affected by other factors, it is desirable to choose from the possibly multiple feasible solutions the one that gives the best performance apart from achieving the requirement of pole-placement. It is proposed that the controller parameters selected should give the most stable and robust closed-loop system.

The stability and robustness of the closed-loop system can be examined by plotting the Nyquist curve of the closed-loop characteristic function (Luyben, 1990),

$$F(s) = \det[I + G(s)C(s)] = 1 + g_{11}(s)c_1(s) + g_{22}(s)c_2(s) + \Delta(s)c_1(s)c_2(s)$$

where the argument s traverses the usual Nyquist contour D. To have a stable closed-loop, the curve must avoid encircling the origin, and further, the closer the plot to the origin, then the less stable and robust the system. Therefore, among all possible multiple feasible solutions the selected one should give the largest value of the criterion

$$\min_{s \in D} |F(s)|$$

Simulation studies have shown that this way of discarding inferior solutions is appropriate.

A Multi-Loop Controller Design Algorithm

Thus far, a method for multi-loop PI controller design that places two pairs of pre-specified complex numbers as the closed-loop poles has been developed. It is thought that the proposed procedure may be too complicated to be used in on-line autotuning but this presentation has shown that the multi-loop problem essentially needs iteration. The solution presented is technical and avoids the need for iteration with a graphical method. A question from user's viewpoint is how to specify the dominant poles, that is the selection of ($p_i, \bar{p}_i, i=1,2$), or equivalently the selection of $\omega_{0_i}, \zeta_i, i=1,2$. The parameters basically determine the damping of the control system in a way much like the SISO case and are recommended to be in the range $0.5 \leq \zeta_i \leq 0.9$, $i=1,2$, with the default of $\zeta_i = 0.707$, $i=1,2$. The parameters ω_{0_i}, $i=1,2$ can be viewed as tuning parameters for the response speed. Selecting small values of parameters ω_{0_i} gives a slow closed-loop system response and vice versa. Generally, $g_{ij}(s)$, $i=1,2$; $j=1,2$ may be of high order and/or have time delays consequently the closed-loop system will have more than four poles. The condition that the poles ($p_i, \bar{p}_i, i=1,2$) are dominant will thus give an admissible range of the parameters ω_{0_i}, $i=1,2$. Further discussion on this issue has been given by Åström and Hägglund (1995).

It should be noted that the multi-loop PI controller for a 2×2 process has four adjustable parameters and thus is generically, but may not be always, capable of positioning two pairs of complex conjugate closed-loop poles. It is interesting to know what theoretical conditions the process and the specified poles should satisfy to have a solution for such a pole placement problem. Obviously, the feasibility of solving the problem amounts to the solvability of the nonlinear complex equation, or equivalently, the existence of an intersection point between the root trajectories defined by the quasi-polynomial equation and the positive real axis. An analytical feasibility condition would be desirable but remains an open question at present. Fortunately, from the numerical algorithm, it is always possible to post-check the solution feasibility given the process data and the specified poles, thus avoiding the difficulty which other iterative design methods may suffer (for example, searching for a solution which may not exist at all). In the case that the proposed method indicates no feasible solution, it is suggested to seek a feasible solution with closed-loop dominant poles adjusted using $p_i = \gamma(\sigma_i + j\omega_i)$ and $\bar{p}_i = \gamma(\sigma_i - j\omega_i)$, $i=1,2$ where $0 < \gamma < 1$. A search may also be performed on γ to optimise an objective function. Normally, it is useful to maximise γ subject to obtaining a feasible solution; the user can also add other constraints. For example, it is possible to specify that $\min_{s \in D} |F(s)|$ be greater than a certain threshold to ensure a desired level of stability robustness. An alternative is to specify that $c_1(s)$ stabilises $g_{11}(s)$ (and/or $c_2(s)$ stabilises

$g_{22}(s)$) so that the control system remains stable in face of Loop 1 and/or Loop 2 failure so that a kind of control loop failure sensitivity is handled by the solution.

It is useful to conclude this section by presenting the full design algorithm.

Algorithm 11.2: Dominant pole placement multi-loop PI design

Step 1 Dominant pole placement specification

Select $p_i, \bar{p}_i, i=1,2$

Case (a) $p_i = \sigma_i + j\omega_i$, $\bar{p}_i = \sigma_i - j\omega_i$, $i=1,2$;

Case (b) $\begin{cases} p_i = -\omega_{0i}\zeta_i + j\omega_{0i}\sqrt{1-\zeta^2}, i=1,2 \\ \bar{p}_i = -\omega_{0i}\zeta_i - j\omega_{0i}\sqrt{1-\zeta^2}, i=1,2 \end{cases}$

Select $0.5 \leq \zeta_i \leq 0.9$, $i=1,2$, with ω_{0_i}, $i=1,2$ tuneable.

Step 2 Compute alpha constants

Compute α_1, α_2

$$\alpha_1 = \frac{1}{2} + \left[\sigma_1 - \frac{(\sigma_1^2 + \omega_1^2)(\sigma_2 - j\omega_2)}{(\sigma_2^2 + \omega_2^2)}\right]\left(\frac{1}{2j\omega_1}\right)$$

$$\alpha_2 = \frac{1}{2} - \left[\sigma_1 - \frac{(\sigma_1^2 + \omega_1^2)(\sigma_2 - j\omega_2)}{(\sigma_2^2 + \omega_2^2)}\right]\left(\frac{1}{2j\omega_1}\right)$$

Step 3 Compute beta constants

Compute β_i, $i=7,6,\ldots,0$

$\beta_7 = \alpha_2[\alpha_1 g_{11}(p_1)\overline{\Delta(p_1)}\Delta(p_2) + \alpha_2 \overline{g_{11}(p_1)}\Delta(p_1)\Delta(p_2) - g_{11}(p_2)|\Delta(p_1)|^2]$

$\beta_6 = \beta_7 \alpha_1 / \alpha_2$

$\beta_5 = \alpha_2[\alpha_2 \overline{g_{11}(p_1)}g_{22}(p_1)\Delta(p_2) + \alpha_1 \overline{\Delta(p_1)}\Delta(p_2) - g_{11}(p_2)g_{22}(p_1)\overline{\Delta(p_1)}]$

$\beta_4 = \alpha_1^2 \overline{\Delta(p_1)}\Delta(p_2) + \alpha_2^2 \Delta(p_1)\Delta(p_2) - |\Delta(p_1)|^2 + \alpha_1 g_{11}(p_1)g_{22}(p_2)\overline{\Delta(p_1)}$
$\quad + \alpha_2 \overline{g_{11}(p_1)}g_{22}(p_2)\Delta(p_1) - \alpha_1 g_{11}(p_2)g_{22}(p_1)\overline{\Delta(p_1)}$
$\quad - \alpha_2 g_{11}(p_2)\overline{g_{22}(p_1)}\Delta(p_2) + 2\alpha_1 \alpha_2 \text{Re}[\overline{g_{11}(p_1)}g_{22}(p_1)]\Delta(p_2)$

$\beta_3 = \alpha_1[\alpha_1 g_{11}(p_1)g_{22}(p_1)\Delta(p_2) + \alpha_2 \Delta(p_1)\Delta(p_2) - g_{11}(p_2)\overline{g_{22}(p_1)}\Delta(p_1)]$

$\beta_2 = -\overline{\Delta(p_1)}g_{22}(p_1) + \alpha_2 g_{11}(p_1)g_{22}(p_1)g_{22}(p_2) - \alpha_2 g_{11}(p_2)|g_{22}(p_1)|^2$
$\quad + \alpha_2^2 g_{22}(p_1)\Delta(p_2) + \alpha_1 \alpha_2 \overline{g_{22}(p_1)}\Delta(p_2) + \alpha_1 g_{22}(p_2)\overline{\Delta(p_1)}$

$\beta_1 = -\Delta(p_1)\overline{g_{22}(p_1)} + \alpha_1 g_{11}(p_1)\overline{g_{22}(p_1)}g_{22}(p_2) - \alpha_1 g_{11}(p_2)|g_{22}(p_1)|^2$
$\quad + \alpha_1^2 \overline{g_{22}(p_1)}\Delta(p_2) + \alpha_1 \alpha_2 g_{22}(p_1)\Delta(p_2) + \alpha_2 g_{22}(p_2)\Delta(p_1)$

$\beta_0 = \alpha_1 \overline{g_{22}(p_1)}g_{22}(p_2) - |g_{22}(p_1)|^2 + \alpha_2 g_{22}(p_1)g_{22}(p_2)$

Step 4 Calculate Θ, range for θ

Compute $g_1(0) = g_{11}(0) - \dfrac{g_{12}(0)g_{21}(0)}{g_{22}(0)}$

$$\text{Determine } \Theta = \begin{cases} \left(-\pi + \arctan\left(\frac{\sqrt{1-\zeta_1^2}}{\zeta_1}\right), 0\right) & \text{if } g_1(0) > 0 \\ \left(\arctan\left(\frac{\sqrt{1-\zeta_1^2}}{\zeta_1}\right), \pi\right) & \text{if } g_1(0) < 0 \end{cases}$$

Step 5 *Plot root trajectory and determine intersections*
Fix $\theta \in \Theta$
Plot roots of

$$(\beta_7 e^{-j\theta} + \beta_6 e^{j\theta})r^3 + (\beta_5 e^{-j\theta} + \beta_4 + \beta_3 e^{j\theta})r^2 + (\beta_2 e^{-j\theta} + \beta_1 e^{j\theta})r + \beta_0 = 0$$

Find set of intersection points with real axis Z.

Step 6 *Find feasible solutions in Z*

$$\text{Compute } Test = \arg\left\{\frac{-g_{22}(p_1)}{\Delta(p_1)}\right\}$$

If $Test \in \Theta$, then delete $z = \frac{-g_{22}(p_1)}{\Delta(p_1)}$ from Z

Step 7 *Compute multi-loop PI controllers*
For each $z \in Z$, set $c_1(p_1) = z$

$$c_2(p_1) = -\frac{[1 + g_{11}(p_1)c_1(p_1)]}{[\Delta(p_1)c_1(p_1) + g_{22}(p_1)]}$$
$$\Delta(p_1) = g_{11}(p_1)g_{22}(p_1) - g_{12}(p_1)g_{21}(p_1)$$

Compute, for $l = 1, 2$, the following:

$$k_{Pl} = \text{Re}\{c_l(p_1)\} + \frac{\sigma_1}{\omega_1}\text{Im}\{c_l(p_1)\}$$

$$k_{Il} = -\left(\frac{\sigma_1^2 + \omega_1^2}{\omega_1}\right)\text{Im}\{c_l(p_1)\}$$

Step 8 *Multi-loop PI controller assessment*
For each set of multi-loop PI controller settings $(k_{P1}, k_{I1}, k_{P2}, k_{I2})$
Step 8.1 *Stability and robustness assessment*
Compute $F(s) = \det[I + G(s)C(s)] = 1 + g_{11}(s)c_1(s) + g_{22}(s)c_2(s) + \Delta(s)c_1(s)c_2(s)$
Choose $(k_{P1}, k_{I1}, k_{P2}, k_{I2})$ which maximises $\min_{s \in D}|F(s)|$
Step 8.2 *Time-domain assessment*
Run simulations to review time domain reference tracking and disturbance rejection performance.
Algorithm end

11.2.3 Examples

In the following examples, Algorithm 11.2 for dominant pole placement design is demonstrated. Time domain results from the BLT method are used as a basis for a comparison of performance.

Example 11.1

Consider the well-known Wood/Berry binary distillation column plant model (Wood and Berry, 1973):

$$\begin{bmatrix} y_1(s) \\ y_2(s) \end{bmatrix} = \begin{bmatrix} \dfrac{12.8e^{-s}}{16.7s+1} & \dfrac{-18.9e^{-3s}}{21.0s+1} \\ \dfrac{6.60e^{-7s}}{10.9s+1} & \dfrac{-19.4e^{-3s}}{14.4s+1} \end{bmatrix} \begin{bmatrix} u_1(s) \\ u_2(s) \end{bmatrix} + \begin{bmatrix} \dfrac{3.8e^{-8s}}{14.9s+1} \\ \dfrac{4.9e^{-3s}}{13.2s+1} \end{bmatrix} d(s)$$

The application of the dominant pole placement algorithm follows.

Step 1: Dominant pole placement specification

It is noted that the input data for the proposed multi-loop PI design with closed-loop dominant pole placement are $(p_i, \bar{p}_i, i=1, 2)$ or $\zeta_i, \omega_{0_i}, i=1, 2$ and $\{G(p_i) \in \mathcal{R}^{2\times 2}, i=1, 2\}$.

Let $\omega_{0_1} = 0.28, \omega_{0_2} = 0.1$ and $\zeta_i = 0.707, i=1, 2$; hence $p_1 = -0.198 + j0.198$ and $p_2 = -0.0707 + j0.0707$. These pole positions yield $\sigma_1 = -0.198, \omega_1 = 0.198, \sigma_2 = -0.0707$ and $\omega_2 = 0.0707$. After $g_{11}(p_i), g_{22}(p_i)$ and $\Delta(p_i), i=1, 2$ are calculated, the design steps of the proposed multi-loop PI controller design procedure can then be activated as follows.

Step 2: Compute alpha constants

Constants α_1, α_2 are calculated as $\alpha_1 = 1.9 - j0.8997$ and $\alpha_2 = -0.9 + j0.8997$

Step 3: Compute beta constants

The coefficients $\beta_i, i = 7, 6, \ldots, 0$ are computed as

$$\beta_7 = (66.17 + j5.973) \times 10^3 \qquad \beta_6 = (-9.962 - j4.605) \times 10^4$$
$$\beta_5 = (-16.05 + j5.545) \times 10^3 \qquad \beta_4 = (3.305 - j3.039) \times 10^4$$
$$\beta_3 = (-0.8348 + j25.45) \times 10^3 \qquad \beta_2 = (3.989 + j34.94) \times 10^2$$
$$\beta_1 = (-1.542 - j1.836) \times 10^3 \qquad \beta_0 = (1.643 - j0.08934) \times 10^2$$

Step 4: Calculate Θ, range for θ

Compute $g_1(0)$ and test; hence $g_1(0) = 6.37 > 0$ gives

$$\Theta = (-\pi + \arctan((\sqrt{1-\zeta_1^2})/\zeta_1), 0) = (-3\pi/4, 0)$$

Step 5: Plot root trajectory and determine intersections

For each $\theta \in \Theta$ the quasi-polynomial in variable r is solved and the resultant root trajectories are plotted as L in Figure 11.3.

From the graph, two intersection points between L and the positive real axis are found, namely

$$z_1 = 0.5521 - j0.2237$$
$$z_2 = -0.0842 - j0.1317$$

These form the entries in the set Z.

Step 6: Find feasible solutions in Z

Since

$$\text{Test} = \arg\left\{\frac{-g_{22}(p_1)}{\Delta(p_1)}\right\} = -2.14 \in \Theta$$

there must be one false solution in the set $Z = \{z_1, z_2\}$. Actually, the intersection corresponding to z_2 is almost equal to

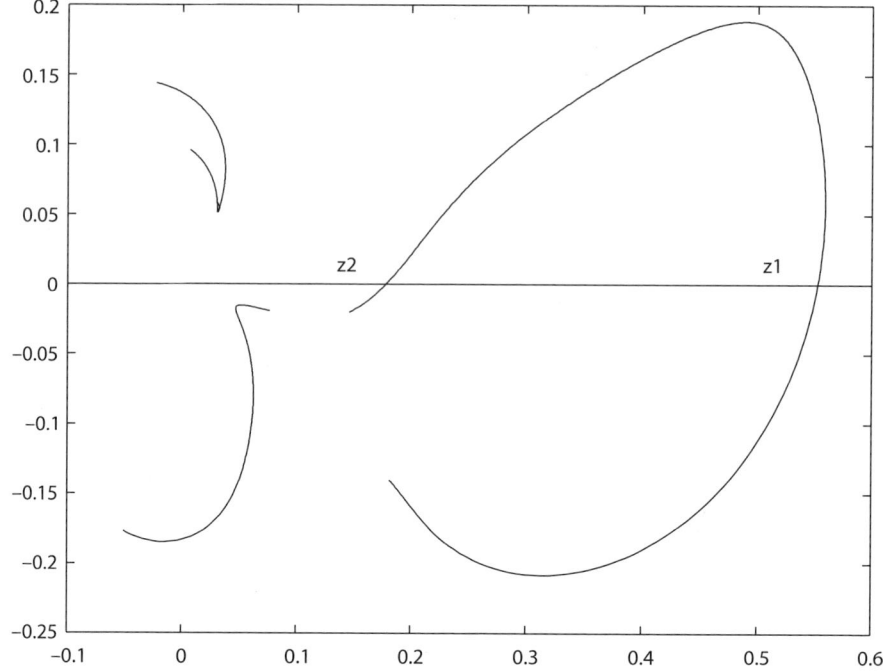

Figure 11.3 Root trajectory L for Example 11.1.

$$\frac{-g_{22}(p_1)}{\Delta(p_1)} = -0.0843 - j0.1321$$

and is a false solution. The small error is caused by the numerical computation during the search for the graph intersection and hence the set $Z = \{z_1, z_2\}$ is reduced to $Z = \{z_1\}$.

Step 7: Compute multi-loop PI controllers
Set $c_1(p_1) = z_1 = 0.5521 - j0.2237$

Compute $c_2(p_1) = -\dfrac{[1 + g_{11}(p_1)c_1(p_1)]}{[\Delta(p_1)c_1(p_1) + g_{22}(p_1)]} = 0.006881 + j0.03572$

To compute the controller parameters for $l = 1, 2$ use the formulae:

$$k_{Pl} = \text{Re}\{c_l(p_1)\} + \frac{\sigma_1}{\omega_1}\text{Im}\{c_l(p_1)\}$$

$$k_{Il} = -\left(\frac{\sigma_1^2 + \omega_1^2}{\omega_1}\right)\text{Im}\{c_l(p_1)\}$$

Compute

$k_{P1} = 0.7757 \quad k_{I1} = 0.0886$
$k_{P2} = -0.0288 \quad k_{I2} = -0.0141$

Step 8: Multi-loop PI controller assessment
The step responses of the resultant feedback system to unit setpoint changes followed by load disturbance change of -0.2 are shown in Figure 11.4 with solid lines. The corresponding step responses using

the BLT method (Luyben, 1986a) with $k_{P1} = 0.375$, $k_{I1} = 0.375/8.29$ and $k_{P2} = -0.075$, $k_{I2} = -0.075/23.6$ are given with dashed lines in the figure. It is observed that the proposed dominant pole placement method gives better loop performance and shorter settling time for the couplings, especially for the slow (second) loop.

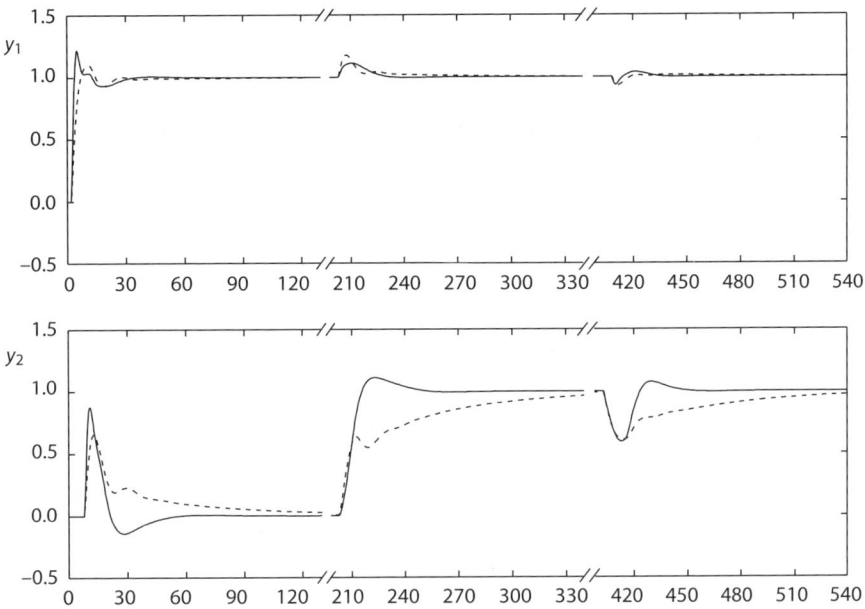

Figure 11.4 Step responses for Example 11.1 (solid line: proposed; dashed line: BLT).

Example 11.2

The process studied by Luyben (1986a) has the following transfer function matrix:

$$\begin{bmatrix} y_1(s) \\ y_2(s) \end{bmatrix} = \begin{bmatrix} \dfrac{0.126e^{-6s}}{60s+1} & \dfrac{-0.101e^{-12s}}{(45s+1)(48s+1)} \\ \dfrac{0.094e^{-8s}}{38s+1} & \dfrac{-0.12e^{-8s}}{35s+1} \end{bmatrix} \begin{bmatrix} u_1(s) \\ u_2(s) \end{bmatrix}$$

Let $\omega_{01} = 0.1$, $\omega_{02} = 0.08$ and $\zeta_1 = \zeta_2 = 0.707$, so that $p_1 = -0.0707 + j0.0707$ and $p_2 = -0.0566 + j0.0566$. The resultant loci are plotted as L in Figure 11.5.

From the graph of Figure 11.5, three intersection points, namely $z_1 = 29.34 - j16.99$, $z_2 = 17.96 - j12.83$ and $z_3 = 9.286 - j6.018$ are found so that $Z = \{z_1, z_2, z_3\}$. As $\arg\{-g_{22}(p_1)/\Delta(p_1)\}$, there must be one false solution in the set $Z = \{z_1, z_2, z_3\}$. Actually, the intersection corresponding to z_2 is almost equal to $-g_{22}(p_1)/\Delta(p_1) = 18.0106 - j12.7857$ and is a false solution.

After elimination of the false solution there are two feasible solutions, namely z_1 and z_3, for $c_1(p_1)$ and two cases arise:

1. Case (a): For $c_1(p_1) = z_1$, the controller parameters are obtained as $k_{P1} = 46.34$, $k_{I1} = 2.403$ and $k_{P2} = -7.895$, $k_{I2} = -0.6214$.

2. Case (b): For $c_1(p_1) = z_3$, the controller parameters are obtained as $k_{P1} = 15.30$, $k_{I1} = 0.8509$ and $k_{P2} = -21.30$, $k_{I2} = -0.8453$.

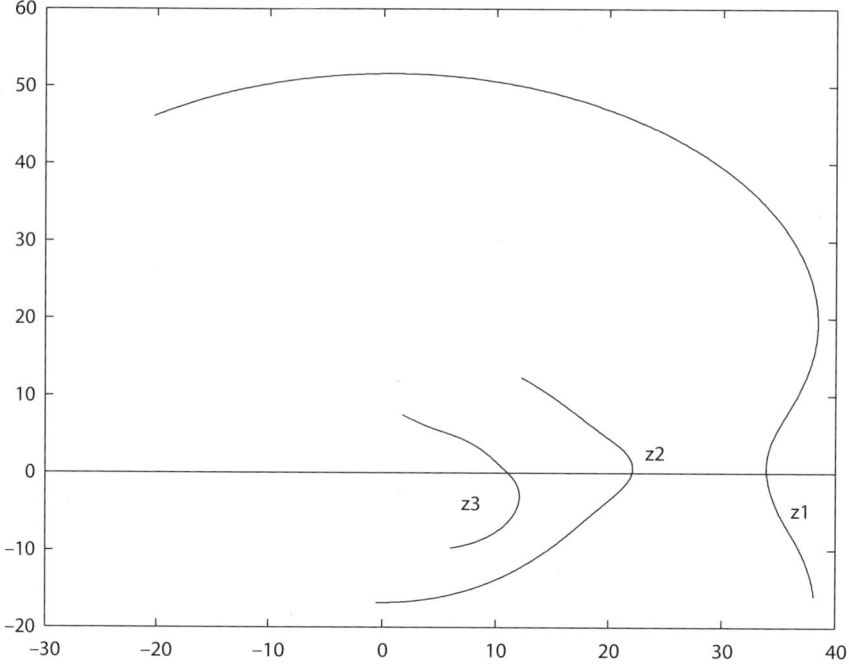

Figure 11.5 Root trajectory L for Example 11.2.

It is found that the controller setting of Case (b) which corresponds to z_3 yields a larger value for the $\min_{s \in D} |F(s)|$ criterion and thus is taken as the multi-loop PI setting. The step responses of the resultant feedback system to unit setpoint changes followed by load disturbance changes are shown in Figure 11.6 with solid lines. As there is no load disturbance model for this process, load disturbances are applied directly on the two process inputs. Again, the proposed controller exhibits better loop and decoupling performance than the BLT method.

11.3 Multivariable PID Control

In this section on multivariable PID control design, decoupling of the closed-loop system will be set as a design objective. The reasons for this requirement are:

1. In practical situations, at least in the process control industry, decoupling is usually required to ease process operations (Kong, 1995).

2. According to our experience, poor decoupling in closed loop could, in many cases, lead to poor diagonal loop performance; in other words, good decoupling is helpful for good loop performance.

3. It should be noted that even if the decoupling requirement is relaxed to allow a limited amount of coupling, it will still lead to near decoupling if the number of inputs/outputs is large. Roughly, if the number is 10, total couplings from all other loops are limited to 30%. Then each off-diagonal element will have a relative gain less than 3%, so the system is almost decoupled.

4. The decoupling property also simplifies the design procedure. In fact, it enables element-wise controller design; as will be seen subsequently. The alternative is the extremely difficult problem of designing the controller matrix as a whole if the decoupling is removed or relaxed. However, it

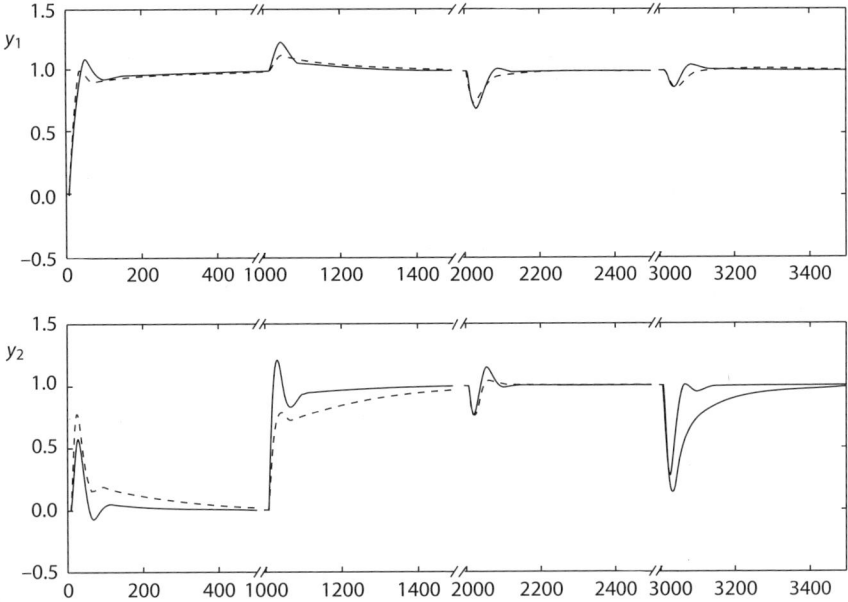

Figure 11.6 Step responses for Example 11.2 (solid line: proposed; dashed line: BLT).

should be emphasised that there are cases where decoupling should not be used; instead, couplings are deliberately employed to boost performance. The design of these control systems is not considered in this chapter.

11.3.1 Decoupling Control and Design Overview

Consider the conventional feedback system as shown in Figure 11.2. The design task in this configuration is to find a controller matrix $C(s)$ such that the closed-loop transfer matrix between the output vector $y(s)$ and the reference vector $r(s)$ given by

$$H(s) = G(s)C(s)[I_m + G(s)C(s)]^{-1}$$

is decoupled; that is $H(s)$ is diagonal and non-singular. If $H(s)$ is inverted then it follows that

$$\begin{aligned}
[H(s)]^{-1} &= [G(s)C(s)[I_m + G(s)C(s)]^{-1}]^{-1} \\
&= [I_m + G(s)C(s)][G(s)C(s)]^{-1} \\
&= [G(s)C(s)]^{-1} + I_m
\end{aligned}$$

Thus the closed-loop $H(s)$ is decoupled if and only if the open loop $G(s)C(s)$ is decoupled, namely

$$G(s)C(s) = Q(s) = diag\{q_{11}(s), q_{22}(s), \ldots, q_{mm}(s)\}$$

Denote the ith column of controller $C(s)$ as $c_{\bullet i}(s)$; then each column of $G(s)C(s)$ may be written as

$$G(s)c_{\bullet i}(s) = q_{ii}(s)e_i$$

or

$$c_{\bullet i}(s) = G(s)^{-1} q_{ii}(s) e_i$$

where the standard basis vectors are written as $e_i \in \Re^m$.

Let G^{ij} be the cofactor corresponding to g_{ij} in $G(s)$. It follows from linear algebra (Noble, 1969) that the inverse of $G(s)$ can be given as

$$G^{-1} = \frac{\text{adj}(G)}{\det[G]}$$

where $\text{adj}(G) = [G^{ij}]^T = [G^{ji}]$. Thus, for $i = 1, \ldots, m$, the controller columns $c_{\bullet i}(s)$ are given by

$$c_{\bullet i}(s) = G(s)^{-1} q_{ii}(s) e_i = \frac{[G^{ij}]^T}{\det[G]} q_{ii}(s) e_i$$

$$= [G^{i1}, G^{i2}, \ldots, G^{im}]^T \frac{q_{ii}(s)}{\det[G]}$$

and the controller elements are

$$c_{ji}(s) = G^{ij} \frac{q_{ii}(s)}{\det[G]} \quad j \neq i$$

$$c_{ii}(s) = G^{ii} \frac{q_{ii}(s)}{\det[G]} \quad j = i$$

Eliminating between these two relations results in

$$c_{ji}(s) = \frac{G^{ij}}{G^{ii}} c_{ii}(s) = \psi_{ji} c_{ii}(s), \quad j \neq i, \ i = 1, \ldots, m; \ j = 1, \ldots, m$$

and the decoupled open loop transfer function matrix is given by

$$G(s)C(s) = \text{diag}\{\tilde{g}_{ii}(s) c_{ii}(s)\} = \text{diag}\left\{\frac{\det[G(s)]}{G^{ii}(s)} c_{ii}(s)\right\}, \ i = 1, \ldots, m$$

where $\tilde{g}_{ii}(s)$ can be viewed as an equivalent SISO process for the ith loop.

The above analysis shows that in order for $G(s)C(s)$ to be decoupled, the off-diagonal elements of the controller $C(s)$ are uniquely determined by the diagonal elements of the controller matrix and that the resulting diagonal element of $G(s)C(s)$ contains only the controller's diagonal element $c_{ii}(s)$.

Example 11.3 Part 1: Cofactors and related computations

For an illustration of the computations involved, consider the process given by

$$G(s) = \begin{bmatrix} \dfrac{e^{-2s}}{s+2} & \dfrac{-e^{-6s}}{s+2} \\ \dfrac{(s-0.5)e^{-3s}}{(s+2)^2} & \dfrac{(s-0.5)^2 e^{-8s}}{2(s+2)^3} \end{bmatrix}$$

Simple calculations give

$$\det[G(s)] = \frac{2(s-0.5)(s+2)e^{-9s} + (s-0.5)^2 e^{-10s}}{2(s+2)^4}$$

$$G^{11} = \frac{(s-0.5)^2 e^{-8s}}{2(s+2)^3} \quad G^{21} = \frac{e^{-6s}}{s+2}$$

$$G^{22} = \frac{e^{-2s}}{s+2} \quad G^{12} = -\frac{(s-0.5)e^{-3s}}{(s+2)^2}$$

It follows that the decoupled loops have their respective equivalent processes as

$$\tilde{g}_{11}(s) = \frac{\det[G(s)]}{G^{11}(s)} = \frac{2(s+2)e^{-s} + (s-0.5)^2 e^{-2s}}{(s-0.5)(s+2)}$$

$$\tilde{g}_{22}(s) = \frac{\det[G(s)]}{G^{22}(s)} = \frac{2(s-0.5)(s+2)e^{-7s} + (s-0.5)^2 e^{-8s}}{2(s+2)^3}$$

The *psi*-transfer functions linking the off-diagonal controller elements to the diagonal controller elements are given by

$$\psi_{21}(s) = \frac{G^{12}(s)}{G^{11}(s)} = \frac{2(s+2)e^{5s}}{(s-0.5)}$$

$$\psi_{12}(s) = \frac{G^{21}(s)}{G^{22}(s)} = e^{-4s}$$

At first glance, a potential design procedure is to design each diagonal controller $c_{ii}(s)$ using the apparent process $\tilde{g}_{ii}(s)$, and then the off-diagonal controllers can be determined from using the diagonal controller elements and the *psi*-transfer functions $\psi_{ij}(s)$. It is well known that the performance limitations are imposed by the non-minimum phase part of the system. However, taking into account non-minimum phase part of $\tilde{g}_{ii}(s)$ is not enough to design the ith loop. This is because a realisable $c_{ii}(s)$ based on $\tilde{g}_{ii}(s)$ only may result in unrealisable or unstable off-diagonal controllers $c_{ji}(s)$, $j \neq i$. In other words, realisable $c_{ji}(s)$, $j \neq i$ may necessarily impose additional time delay and non-minimum phase zeros on $c_{ii}(s)$. Thus it is necessary to derive a characterisation of the unavoidable time delays and non-minimum phase zeros that are inherent in any feedback loop of the decoupled plant. Furthermore, it is noted from the above derivation that the exact decoupling controller is likely to be very complex and thus not practical for industrial implementation. For industrial implementation, a controller of PID type that attains near-decoupling performance would be much preferred. These considerations lead to the following three-stage decoupling design algorithm.

Algorithm 11.3: Multivariable PID controller design

Step 1 Based on the process characteristics, derive $h_{ii}(s)$, the objective closed-loop transfer functions from the ith setpoint $r_i(s)$ to the ith output $y_i(s)$ for each decoupled closed loop. The ith objective open-loop transfer function $q_{ii}(s)$ is then given by

$$q_{ii}(s) = \frac{h_{ii}(s)}{1 - h_{ii}(s)}$$

Step 2 Match the actual loop $\tilde{g}_{ii}(s)c_{ii}(s)$ to the objective or target loop $q_{ii}(s)$ to find the ideal diagonal elements of the controller as

$$c_{ii}^{\text{IDEAL}}(s) = \tilde{g}_{ii}^{-1}(s)q_{ii}(s) = \frac{G^{ii}(s)}{\det[G(s)]}q_{ii}(s), \ i = 1,\ldots,m$$

and the corresponding ideal decoupling off-diagonal elements $c_{ji}(s)$ of the controller given by

$$c_{ji}^{\text{IDEAL}}(s) = \frac{G^{ij}(s)}{G^{ii}(s)}c_{ii}^{\text{IDEAL}}(s) = \frac{G^{ij}(s)}{\det[G(s)]}q_{ii}(s)$$

Step 3 For each element, approximate the ideal controller transfer functions $c_{ij}^{\text{IDEAL}}(s)$, $i = 1,\ldots,m$; $j = 1,\ldots,m$ with PID controllers. Detune $h_{ii}(s)$ if necessary, so that the PID controller approximation can reach the given accuracy.

Algorithm end

11.3.2 Determination of the Objective Loop Performance

The achievable closed-loop performance is closely related to the process open-loop characteristics, and thus cannot be specified arbitrarily. To reflect various key factors in the process dynamics that limit the achievable performance, let the ith decoupled loop control specifications be expressed in terms of the following closed-loop transfer function:

$$H(s) = diag\{h_{ii}(s)\}$$

where

$$h_{ii}(s) = e^{-L_i s} f_i(s) \prod_{z \in Z^+_{det[G]}} \left(\frac{z-s}{z+s}\right)^{n_i(z)}, \; i = 1, 2, \ldots, m$$

Time delays and non-minimum phase zeros are inherent characteristics of a process that cannot be altered by any feedback control. They are reflected in the above closed loop expression through the terms $e^{-L_i s}$ and

$$\prod_{z \in Z^+_{det[G]}} \left(\frac{z-s}{z+s}\right)^{n_i(s)}$$

respectively. The ith loop filter $f_i(s)$ is chosen throughout this section simply as

$$f_i(s) = \frac{1}{\tau_i s + 1}, \; i = 1, 2, \ldots, m$$

where τ_i is the only tuning parameter used for each loop to achieve the appropriate compromise between performance and robustness and to keep the action of the manipulated variable within bounds.

In what follows, a characterisation of the time delays and non-minimum phase zeros for the decoupling controller and the resulting decoupled loops is developed. For the non-singular process $G(s)$ whose elements have the delay form $g_{ij}(s) = g_{ij0}(s)e^{-L_{ij}s}$, $i = 1, \ldots, m$; $j = 1, \ldots, m$, a general expression for the components $\tilde{g}_{ii}(s)$ in the decoupled forward path transfer $G(s)C(s)$ will be

$$\phi(s) = \frac{\sum_{k=0}^{M} n_k(s) e^{-\alpha_k s}}{d_0(s) + \sum_{l=1}^{N} d_l(s) e^{-\beta_l s}}$$

where $n_k(s)$ and $d_l(s)$ are non-zero scalar polynomials of s with $\alpha_0 < \alpha_1 < \ldots < \alpha_M$ and $0 < \beta_1 < \beta_2 < \ldots < \beta_N$.

Time Delay Limitation Conditions

Define the time delay for the nonzero transfer function $\phi(s)$ using the operator γ, where

$$\gamma(\phi(s)) = \alpha_0$$

For any nonzero transfer functions $\phi_1(s)$, $\phi_2(s)$ and $\phi(s)$, it is easy to verify the properties:

(i) $\gamma(\phi_1(s)\phi_2(s)) = \gamma(\phi_1(s)) + \gamma(\phi_2(s))$

and

(ii) $\gamma(\phi^{-1}(s)) = -\gamma(\phi(s))$

From the definition of the operator γ, if $\gamma(\phi(s)) \geq 0$ then this measures the minimum time required for $\phi(s)$ to have a nonzero output in response to a step input. If $\gamma(\phi(s)) < 0$ then the system output will depend on the future values of the input. It is obvious that for any realisable and nonzero $\phi(s)$, $\gamma(\phi(s))$ cannot be negative. Therefore, by extension, a realisable controller $C(s)$ requires

$$\gamma(c_{ji}(s)) \geq 0, \ i=1,2,\ldots,m; \ j \in J_i$$

and $J_i = \{j=1,2,\ldots,m \,|\, G^{ij} \neq 0\}$. Recall the desired relationship for the off-diagonal controller elements in the decoupling controller as

$$c_{ji}(s) = \frac{G^{ij}}{G^{ii}} c_{ii}(s)$$

Applying the delay relation yields

$$\gamma(c_{ji}(s)) = \gamma\left(\frac{G^{ij}}{G^{ii}} c_{ii}(s)\right) = \gamma(G^{ij}) - \gamma(G^{ii}) + \gamma(c_{ii}(s)), \ i=1,2,\ldots,m; \ j \in J_i$$

and this gives

$$\gamma(c_{ii}(s)) \geq \gamma(G^{ii}) - \gamma(G^{ij}), \ i=1,2,\ldots,m; \ j \in J_i$$

or

$$\gamma(c_{ii}(s)) \geq \gamma(G^{ii}) - \widetilde{\gamma}_i, \ i=1,2,\ldots,m$$

where

$$\widetilde{\gamma}_i = \min_{j \in J_i} \gamma(G^{ij}), \ i=1,2,\ldots,m.$$

This relation is a characterisation of the controller diagonal elements in terms of their time delays, which indicates the minimum time delay that the ith diagonal controller elements must contain.

Implications for the Diagonal Closed-Loop System

The resulting closed-loop ith diagonal element of the closed-loop system is given by

$$h_{ii}(s) = \frac{\widetilde{g}_{ii}(s) c_{ii}(s)}{1 + \widetilde{g}_{ii}(s) c_{ii}(s)}$$

and this will then satisfy the following conditions:

$$\gamma(h_{ii}(s)) \geq \gamma(\widetilde{g}_{ii}(s)) + \gamma(c_{ii}(s)) \geq \gamma(\widetilde{g}_{ii}(s)) + \gamma(G^{ii}) - \widetilde{\gamma}_i, \ i=1,2,\ldots,m$$

But since

$$\widetilde{g}_{ii}(s) = \frac{\det[G(s)]}{G^{ii}(s)}$$

we have

$$\gamma(\widetilde{g}_{ii}(s)) = \gamma\left(\frac{\det[G(s)]}{G^{ii}(s)}\right) = \gamma(\det[G(s)]) - \gamma(G^{ii}(s))$$

Hence by substitution

$$\gamma(h_{ii}(s)) \geq \gamma(\det[G(s)]) - \widetilde{\gamma}_i, \ i=1,2,\ldots,m$$

The above bound relation characterises the decoupled ith closed-loop transfer function in terms of time delays. The minimum time delay that the ith decoupled closed-loop transfer function must contain is indicated by the bound. Unavoidable time delay in excess of this bound is harmful to control performance; thus the time delays in desired closed-loop diagonal form are selected as

$$L_i = \gamma(\det[G(s)]) - \tilde{\gamma}_i, \ i = 1, 2, \ldots, m$$

Example 11.3 Part 2: Time-delay characterisations

The results presented here continue Example 11.3. The process transfer function matrix is recalled as

$$G(s) = \begin{bmatrix} \dfrac{e^{-2s}}{s+2} & \dfrac{-e^{-6s}}{s+2} \\ \dfrac{(s-0.5)e^{-3s}}{(s+2)^2} & \dfrac{(s-0.5)^2 e^{-8s}}{2(s+2)^3} \end{bmatrix}$$

Simple computations lead to

$$\det[G(s)] = \frac{2(s-0.5)(s+2)e^{-9s} + (s-0.5)^2 e^{-10s}}{2(s+2)^4}$$

$$G^{11} = \frac{(s-0.5)^2 e^{-8s}}{2(s+2)^3} \quad G^{21} = \frac{e^{-6s}}{s+2}$$

$$G^{22} = \frac{e^{-2s}}{s+2} \quad G^{12} = -\frac{(s-0.5)e^{-3s}}{(s+2)^2}$$

It follows from the definition of the operator γ that the time delays for the respective functions are

$$\gamma(G^{11}) = 8 \quad \gamma(G^{12}) = 3$$
$$\gamma(G^{21}) = 6 \quad \gamma(G^{22}) = 2$$

and $\gamma(\det[G(s)]) = 9$.

The formula for $\tilde{\gamma}_i$ is given by

$$\tilde{\gamma}_i = \min_{j \in J_i} \gamma(G^{ij}), \ i = 1, 2, \ldots, m$$

Hence we obtain

$$\tilde{\gamma}_1 = \min\{\gamma(G^{11}), \gamma(G^{12})\} = \min\{8, 3\} = 3$$
$$\tilde{\gamma}_2 = \min\{\gamma(G^{11}), \gamma(G^{12})\} = \min\{6, 2\} = 2$$

Using the controller time delay bound $\gamma(c_{ii}(s)) \geq \gamma(G^{ii}) - \tilde{\gamma}_i, \ i = 1, 2, \ldots, m$ it follows that $c_{22}(s)$ and $c_{11}(s)$ are characterised for their time delays by

$$\gamma(c_{11}(s)) \geq \gamma(G^{11}) - \tilde{\gamma}_1 = 8 - 3 = 5$$
$$\gamma(c_{22}(s)) \geq \gamma(G^{22}) - \tilde{\gamma}_2 = 2 - 2 = 0$$

Also, $h_{11}(s)$ and $h_{22}(s)$ must meet the conditions:

$$\gamma(h_{ii}(s)) \geq \gamma(\det[G(s)]) - \tilde{\gamma}_i, \ i = 1, 2, \ldots, m$$

Hence one concludes that

$$\gamma(h_{11}(s)) \geq \gamma(\det[G(s)]) - \tilde{\gamma}_1 = 9 - 3 = 6$$
$$\gamma(h_{22}(s)) \geq \gamma(\det[G(s)]) - \tilde{\gamma}_2 = 9 - 2 = 7$$

Non-Minimum Phase Zero Limitation Conditions

Consider now the performance limitation due to the non-minimum phase zeros. Denote by C^+ the closed right half of the complex plane (RHP). For the general nonzero transfer function $\phi(s)$, let Z_ϕ^+ be the set of all the RHP zeros of $\phi(s)$, that is

$$Z_\phi^+ = \{z \in C^+ \mid \phi(z) = 0\}$$

Let $\eta_z(\phi)$ be an integer v such that

$$\lim_{s \to z} \left(\frac{\phi(s)}{(s-z)^v} \right)$$

exists and is nonzero. Thus $\phi(s)$ has $\eta_z(\phi)$ zeros at $s = z$ if $\eta_z(\phi) > 0$, or $-\eta_z(\phi)$ poles if $\eta_z(\phi) < 0$, or neither poles nor zeros if $\eta_z(\phi) = 0$. For any nonzero transfer functions $\phi_1(s), \phi_2(s)$ and $\phi(s)$ it is easy to verify the following two properties:

(i) $\eta(\phi_1(s)\phi_2(s)) = \eta(\phi_1(s)) + \eta(\phi_2(s))$

(ii) $\eta(\phi^{-1}(s)) = -\eta(\phi(s))$

It can be shown that a nonzero transfer function $\phi(s)$ is stable if and only if $\eta_z(\phi) \geq 0$ for each $z \in C^+$. Thus a stable controller matrix $C(s)$ requires

$$\eta_z(c_{ji}(s)) \geq 0, \ i = 1,2,\ldots,m; \ j \in J_i$$

where $J_i = j = 1,2,\ldots,m \mid G^{ij} \neq 0$.

Using the expression for $c_{ji}(s)$ given in terms of the controller diagonal elements gives

$$\eta_z(c_{ji}(s)) = \eta_z\left(\frac{G^{ij}}{G^{ii}} c_{ii}(s)\right) = \eta_z(G^{ij}) - \eta_z(G^{ii}) + \eta_z(c_{ii}(s)), \ i = 1,2,\ldots,m; \ j \in J_i$$

Using $\eta_z(c_{ji}(s)) \geq 0$ leads to the bound relations

$$\eta_z(c_{ii}(s)) \geq \eta_z(G^{ii}) - \eta_z(G^{ij})$$

and

$$\eta_z(c_{ii}(s)) = \eta_z(G^{ii}) - \tilde{\eta}_i(z)$$

where

$$\tilde{\eta}_i(z) = \min_{j \in J_i} \eta_z(G^{ij}), \ i = 1,2,\ldots,m$$

Since the process transfer function matrix $G(s)$ is assumed stable, so are the co-factors $G^{ij}(s)$. Consequently

$$\tilde{\eta}_i(z) = \min_{j \in J_i} \eta_z(G^{ij}) \geq 0 \text{ for } i = 1,2,\ldots,m \text{ and all } z \in C^+$$

Thus for $i = 1,2,\ldots,m$ and all $z \in C^+$, it follows that

$$\eta_z(G^{ii}) \geq \eta_z(G^{ii}) - \tilde{\eta}_i(z) = \eta_z(G^{ii}) - \min_{j \in J_i} \eta_z(G^{ij}) \geq \eta_z(G^{ii}) - \eta_z(G^{ij})|_{j=i} = 0$$

This yields the important bound relation:

$$\eta_z(G^{ii}) \geq \eta_z(G^{ii}) - \tilde{\eta}_i(z) \geq 0 \text{ for } i = 1,2,\ldots,m \text{ and all } z \in C^+$$

This bound implies that $c_{ii}(s)$ need not have any non-minimum phase zeros except at $z \in Z^+_{G^{ii}} = \{z \in C^+ | G^{ii}(z) = 0\}$. Therefore $c_{ii}(s)$ is characterised for its non-minimum phase zeros by

$$\eta_z(c_{ii}(s)) \geq \eta_z(G^{ii}) - \eta_i(z), \quad i = 1, 2, \ldots, m; \quad z \in Z^+_{G^{ii}}$$

Implications for the Diagonal Closed-Loop System

The resulting closed-loop ith diagonal element of the closed-loop system is given by

$$h_{ii}(s) = \frac{\tilde{g}_{ii}(s) c_{ii}(s)}{1 + \tilde{g}_{ii}(s) c_{ii}(s)}$$

Thus for $i = 1, 2, \ldots, m$ and all $z \in C^+$ this will then satisfy the following conditions:

$$\eta_z(h_{ii}(s)) \geq \eta_z(\tilde{g}_{ii}(s)) + \eta_z(c_{ii}(s)) \geq \eta_z(\tilde{g}_{ii}(s)) + \eta_z(G^{ii}) - \tilde{\eta}_i(z)$$

But since

$$\tilde{g}_{ii}(s) = \frac{\det[G(s)]}{G^{ii}(s)}$$

it follows that

$$\eta_z(\tilde{g}_{ii}(s)) = \eta_z\left(\frac{\det[G(s)]}{G^{ii}(s)}\right) = \eta_z(\det[G(s)]) - \eta_z(G^{ii}(s))$$

Hence for $i = 1, 2, \ldots, m$ and all $z \in C^+$, direct substitution gives

$$\eta_z(h_{ii}(s)) \geq \eta_z(\det[G(s)]) - \tilde{\eta}_i(z)$$

It is readily seen that for $i = 1, 2, \ldots, m$ and all $z \in C^+$

$$\eta_z(\det[G(s)]) \geq \eta_z(\det[G(s)]) - \tilde{\eta}_i(z)$$

and

$$\eta_z(\det[G(s)]) - \tilde{\eta}_i(z) = \eta_z\left(\frac{\det[G(s)]}{(s-z)^{\tilde{\eta}_i(z)}}\right)$$

$$= \eta_z\left(\frac{1}{(s-z)^{\tilde{\eta}_i(z)}} \sum_{j=1}^m g_{ij}(s) G^{ij}(s)\right)$$

$$= \eta_z\left(\sum_{j=1}^m g_{ij}(s) \frac{G^{ij}(s)}{(s-z)^{\tilde{\eta}_i(z)}}\right) \geq 0$$

leading to the bound relation:

$$0 \leq \eta_z(\det[G(s)]) - \tilde{\eta}_i(z) \leq \eta_z(\det[G(s)]) \text{ for } i = 1, 2, \ldots, m \text{ and all } z \in C^+$$

The bound relationship also indicates that the ith closed-loop transfer function $h_{ii}(s)$ need not have any non-minimum phase zeros except at $z \in Z^+_{\det[G(s)]}$. Therefore the characterisations of the non-minimum-phase zeros for the $h_{ii}(s)$ elements are given by

$$\eta_z(h_{ii}(s)) \geq \eta_z(\det[G(s)]) - \tilde{\eta}_i(z) \geq 0 \text{ for } i = 1, 2, \ldots, m \text{ and all } z \in Z^+_{\det[G(s)]}$$

Thus in the generic expression for $h_{ii}(s)$ given by

$$h_{ii}(s) = e^{-L_i s} f_i(s) \prod_{z \in Z^+_{\det[G]}} \left(\frac{z-s}{z+s}\right)^{n_i(z)}, \quad i = 1, 2, \ldots, m$$

the integers $n_i(z)$ are chosen as

$$n_i(z) = \eta_z(\det[G(s)]) - \tilde{\eta}_i(z) \text{ for } i = 1, 2, \ldots, m \text{ and all } z \in Z^+_{\det[G(s)]}$$

Example 11.3 Part 3: Non-minimum phase characterisations

This example is a continuation of Example 11.3. Using the process transfer function matrix simple computation led to

$$\det[G(s)] = \frac{2(s-0.5)(s+2)e^{-9s} + (s-0.5)^2 e^{-10s}}{2(s+2)^4}$$

$$G^{11} = \frac{(s-0.5)^2 e^{-8s}}{2(s+2)^3} \quad G^{21} = \frac{e^{-6s}}{s+2}$$

$$G^{12} = -\frac{(s-0.5)e^{-3s}}{(s+2)^2} \quad G^{22} = \frac{e^{-2s}}{s+2}$$

Direct inspection of the above elements gives:

$$Z^+_{G^{11}} = \{0.5\}, \quad Z^+_{G^{22}} = \emptyset, \quad Z^+_{\det[G(s)]} = \{0.5\}$$

with

$$\eta_z(G^{11})|_{z=0.5} = 2 \quad \eta_z(G^{12})|_{z=0.5} = 1$$
$$\eta_z(G^{21})|_{z=0.5} = \emptyset \quad \eta_z(G^{22})|_{z=0.5} = \emptyset$$

$$\eta_z(\det[G(s)])|_{z=0.5} = 1$$

Computing the indices $\tilde{\eta}_i(z) = \min_{j \in J_i} \eta_z(G^{ij}), i = 1, 2, \ldots, m$ yields:

$$\tilde{\eta}_1(z)|_{z=0.5} = \min\{\eta_z(G^{11})|_{z=0.5}, \eta_z(G^{12})|_{z=0.5}\} = \min\{2,1\} = 1$$
$$\tilde{\eta}_2(z)|_{z=0.5} = \min\{\eta_z(G^{21})|_{z=0.5}, \eta_z(G^{22})|_{z=0.5}\} = \min\{0,0\} = 0$$

Controller elements $c_{11}(s)$ and $c_{22}(s)$ should satisfy

$$\eta_z(c_{ii}(s)) = \eta_z(G^{ii}) - \tilde{\eta}_i(z) \geq 0, \; i = 1, 2; \; z \in Z^+_{G^{ii}}$$

giving $i = 1; z \in Z^+_{G^{11}}$, whence $z = 0.5$ so that

$$\eta_z c_{11}(s) = \eta_z(G^{11})|_{z=0.5} - \tilde{\eta}_1(z)|_{z=0.5} = 2 - 1 = 1$$

For $i = 2; z \in Z^+_{G^{22}}, Z^+_{G^{22}} = \emptyset$ and there will be no constraint on the non-minimum phase zeros of $c_{22}(s)$.
The characterisation of the non-minimum-phase zeros for the $h_{ii}(s)$ elements is given by

$$\eta_z(h_{ii}(s)) \geq \eta_z(\det[G(s)]) - \tilde{\eta}_i(z) \geq 0 \text{ for } i = 1, 2 \text{ and } z \in Z^+_{\det[G(s)]}$$

So that in this case

$$\eta_z(h_{11}(s))|_{z=0.5} \geq \eta_z(\det[G(s)])|_{z=0.5} - \tilde{\eta}_1(z)|_{z=0.5} = 1 - 1 = 0$$
$$\eta_z(h_{22}(s))|_{z=0.5} \geq \eta_z(\det[G(s)])|_{z=0.5} - \tilde{\eta}_2(z)|_{z=0.5} = 1 - 0 = 1$$

Thus the first loop need not contain any non-minimum phase zero and the second loop must contain a non-minimum phase zero at $z = 0.5$ of multiplicity 1.

Example 11.3 Part 4: A look at possible controllers

The computations performed in Parts 1, 2 and 3 of Example 11.3 lead to the table of characterisations of Table 11.1.

Table 11.1 Controller and closed-loop characterisations.

Controller	$c_{11}(s)$	$c_{22}(s)$
Time delay	$\gamma(c_{11}(s)) \geq 5$	$\gamma(c_{22}(s)) \geq 0$
Non-minimum phase	$z = 0.5$ and $\eta_z(c_{11}(s)) = 1$	$Z^+_{G^{22}} = \emptyset$; no constraint
Diagonal closed loop	$h_{11}(s)$	$h_{22}(s)$
Time delay	$\gamma(h_{11}(s)) \geq 6$	$\gamma(h_{22}(s)) \geq 7$
Non-minimum phase	$\eta_z(h_{11}(s))|_{z=0.5} \geq 0$	$\eta_z(h_{22}(s))|_{z=0.5} \geq 1$

Using these characterisations, choose the controller diagonal elements as

$$c_{11}(s) = \frac{(s-0.5)e^{-5s}}{s+\rho} \quad \text{and} \quad c_{22}(s) = \frac{1}{s+\rho}$$

where $\rho > 0$ and the lower bounds shown in the table are exactly met. The off-diagonal controllers are then obtained using the formula:

$$c_{ji}(s) = \frac{G^{ij}}{G^{ii}} c_{ii}(s), \quad j \neq i, \quad i = 1, 2; \quad j = 1, 2$$

and the complete controller is

$$C(s) = \begin{bmatrix} \dfrac{(s-0.5)e^{-5s}}{s+\rho} & \dfrac{e^{-4s}}{s+\rho} \\ \dfrac{-2(s+2)}{s+\rho} & \dfrac{1}{s+\rho} \end{bmatrix}$$

This controller is both realisable and stable, and the resulting forward path transfer function matrix $G(s)C(s)$ is calculated as

$$G(s)C(s) = \begin{bmatrix} \left(\dfrac{2s+4+(s-0.5)e^{-s}}{(s+2)(s+\rho)}\right)e^{-6s} & 0 \\ 0 & (s-0.5)\left(\dfrac{2s+4+(s-0.5)e^{-s}}{2(s+2)^3(s+\rho)}\right)e^{-7s} \end{bmatrix}$$

It can be seen that the diagonal elements are compatible with the previous characterisation for time delays and non-minimum phase zeros, namely (Table 11.1) the time delays for the decoupled loops are no less than 6 and 7, respectively, and loop two must contain a non-minimum phase zero at $s = 0.5$ of multiplicity 1.

In order to make a comparison, if the controller diagonal elements are chosen as

$$c_{11}(s) = c_{22}(s) = \frac{1}{s+\rho}$$

which violates the characteristics, this gives rise to an off-diagonal controller element:

$$c_{21}(s) = \frac{-2(s+2)e^{5s}}{(s-0.5)(s+\rho)}$$

which is neither realisable nor stable. The resultant diagonal closed-loop element $h_{11}(s)$ is given by

$$h_{11}(s) = \tilde{g}_{11}(s)c_{11}(s) = \frac{2s+4+(s+0.5)e^{-s}}{(s-0.5)(s+2)(s+\rho)}$$

which is unstable too. Clearly such incompatible controllers should not be used.

Unstable Pole–Zero Cancellations in Decoupling Control

A quite special phenomenon in such a decoupling control is that there might be unstable pole–zero cancellations in forming $h_{ii}(s) = \tilde{g}_{ii}(s)c_{ii}(s)$. In contrast to normal intuition, this will not cause instability. For example, notice in the above example that $\tilde{g}_{11}(s)$ has a pole at $s = 0.5$ and $c_{11}(s)$ has at least one zero at the same location, since $\eta_z(c_{11}(s))|_{z=0.5} \geq 1$. Hence an unstable pole–zero cancellation at $s = 0.5$ occurs in forming $h_{11}(s) = \tilde{g}_{11}(s)c_{11}(s)$. However, the resulting closed-loop system is still stable because all the controller diagonal and off-diagonal elements are stable. In fact, as far as $G(s)$ and $C(s)$ are concerned, there is no unstable pole–zero cancellation, since both $G(s)$ and $C(s)$ are stable.

Selecting the Tuning Parameters τ_i, $i = 1, ..., m$

Recall the target diagonal closed loop as

$$H(s) = diag\{h_{ii}(s)\}$$

where

$$h_{ii}(s) = e^{-L_i s} f_i(s) \prod_{z \in Z^+_{det[G]}} \left(\frac{z-s}{z+s}\right)^{n_i(z)}, \quad i = 1, 2, ..., m$$

Within this the ith loop filter $f_i(s)$ is chosen as

$$f_i(s) = \frac{1}{\tau_i s + 1}, \quad i = 1, 2, ..., m$$

where τ_i is the only tuning parameter used for each loop to achieve the appropriate compromise between performance and robustness and to meet any input constraints that might exist in the process. The selection of the tuning parameters τ_i, $i = 1, ..., m$ will be made to meet a desired phase margin which is given in degrees as ϕ_{PM}. To achieve phase margin ϕ_{PM}, the gain crossover frequency, ω_{gci} of the ith ideal open-loop transfer function

$$(GC)_{ii}(s) = \frac{h_{ii}(s)}{1 - h_{ii}(s)}$$

where

$$|(GC)_{ii}(j\omega_{gci}, \tau_i)| = \left|\frac{h_{ii}(j\omega_{gci}, \tau_i)}{1 - h_{ii}(j\omega_{gci}, \tau_i)}\right| = 1$$

should meet

$$\arg\{(GC)_{ii}(j\omega_{gci},\tau_i)\} = \arg\left\{\frac{h_{ii}(j\omega_{gci},\tau_i)}{1-h_{ii}(j\omega_{gci},\tau_i)}\right\} \geq -180° + \phi_{PM}$$

Empirical studies by the author suggest that $\phi_{PM} = 65°$ will usually be a good choice for most 2×2 industrial processes, whereas a larger phase margin, such as $\phi_{PM} = 70°$, will be necessary for 3×3 or even 4×4 processes.

To consider the performance limitation imposed by input constraints, a frequency-by-frequency analysis is used. Assume that at each frequency $|u_i(j\omega)| \leq \bar{u}_i$ and that the model has been appropriately scaled such that the reference signal in each channel satisfies $|r_i(j\omega)| \leq 1$. Now consider the presence of a reference signal in a channel at a time the ideal controlled variable $y(s) \in \Re^m(s)$ would be $y(s) = [0 \;\; \ldots \;\; h_{ii}(s)r_i(s) \;\; \ldots \;\; 0]^T$, and thus the corresponding manipulated variable becomes

$$u(s) = G(s)^{-1} y(s) = G(s)^{-1} [0 \;\; \ldots \;\; h_{ii}(s)r_i(s) \;\; \ldots \;\; 0]^T$$

Hence

$$u_l(s) = \frac{G^{il}(s)}{\det[G(s)]} h_{ii}(s) r_i(s)$$

Considering the worst-case references $|r_i(j\omega)| = 1$ then it is required that

$$|u_l(j\omega)| = \left|\frac{G^{il}(j\omega)}{\det[G(j\omega)]} h_{ii}(j\omega)\right| \leq \bar{u}_i$$

or

$$|h_{ii}(j\omega)| \leq \min_{l\in J_i}\left\{\left|\frac{\det[G(j\omega)]}{G^{il}(j\omega)}\right|\bar{u}_i\right\}$$

To derive an inequality on τ_i imposed by input constraints, note that at the bandwidth frequency $\omega_{BW} = 1/\tau_i$

$$|h_{ii}(j\omega_{BW})| = \left|e^{-L_{ij}j\omega_{BW}}\right|\left|\left(\frac{1}{\tau_i(j\omega_{BW})+1}\right)\right| \prod_{z\in Z^+_{\det[G]}} \left|\left(\frac{z-j\omega_{BW}}{z+j\omega_{BW}}\right)\right|^{n_i(z)} = \frac{1}{\sqrt{2}}, \; i=1,2,\ldots,m$$

Hence

$$\frac{1}{\sqrt{2}} \leq \min_{l\in J_i}\left\{\left|\frac{\det[G(j\omega_{BW})]}{G^{il}(j\omega_{BW})}\right|\bar{u}_i\right\} \text{ where } \omega_{BW} = 1/\tau_i$$

Thus, the right-hand side of the above equation can be plotted to determine when the inequality holds to find the suitable range of τ_i.

In order to have the fastest response while having a given stability margin and meeting input constraints, the tuning parameter τ_i in the filter should be chosen to be the smallest that simultaneously satisfies the equation set:

$$|(GC)_{ii}(j\omega_{gci},\tau_i)| = \left|\frac{h_{ii}(j\omega_{gci},\tau_i)}{1-h_{ii}(j\omega_{gci},\tau_i)}\right| = 1$$

$$\arg\{(GC)_{ii}(j\omega_{\text{gci}},\tau_i)\} = \arg\left\{\frac{h_{ii}(j\omega_{\text{gci}},\tau_i)}{1-h_{ii}(j\omega_{\text{gci}},\tau_i)}\right\} \geq -180°+\phi_{\text{PM}}$$

$$\frac{1}{\sqrt{2}} \leq \min_{l\in J_i}\left\{\left\|\frac{\det[G(j\omega_{\text{BW}})]}{G^{il}(j\omega_{\text{BW}})}\bar{u}_i\right\|\right\}, \quad \omega_{\text{BW}} = 1/\tau_i$$

11.3.3 Computation of PID Controller

The analysis of the foregoing sections has led to a target multivariable controller matrix $C(s)$ which decouples the multivariable system and meets certain performance and robustness specifications. However, the implementation of this ideal controller will use a multivariable PID controller in the form:

$$\hat{C}(s) = \{\hat{c}_{ij}(s)\} \quad \text{where } \hat{c}_{ij}(s) = k_{Pij} + \frac{k_{Iij}}{s} + k_{Dij}s$$

Thus the task is to find PID parameters so as to match controller elements $\hat{c}_{ij}(s)$ to the obtained ideal controller elements $c_{ij}(s)$ obtained as closely as possible. This objective can be realised by solving a frequency domain least square fitting problem. This requires minimising one loss function for each element in the controller matrix over a specified frequency range. For the ijth controller element minimise the loss function:

$$\min_{\hat{c}_{ij}(s)} J_{ij} = \min_{\hat{c}_{ij}(s)} \sum_{l=1}^{L} |\hat{c}_{ij}(j\omega_l) - c_{ij}(j\omega_l)|^2$$

subject to the same sign for parameters $\hat{k}_{Pij}$, $\hat{k}_{Iij}$, $\hat{k}_{Dij}$. This problem can be solved by standard non-negative least squares (Lawson and Hanson, 1974) to give the optimal PID parameters. Here, the fitting frequencies $(\omega_1, \omega_2, ..., \omega_L)$ are chosen as $\omega_l = l \times \omega_{\text{BW max}}/100$, $l = 1, ..., 100$, where $\omega_{\text{BW max}}$ is the maximum closed-loop bandwidth among all the loops, that is

$$\omega_{\text{BW max}} = \max_i\{\omega_{\text{BW}i}\} = \max_i\left\{\frac{1}{\tau_i}\right\}$$

Detuning to Moderate Excessive Interactions

In MIMO systems, it might happen that the loop performance is satisfactory while the interactions between loops are too large to be accepted. A detuning technique is introduced to fine-tune the design towards acceptable levels of interaction. Let the open-loop transfer function be $Q = GC$ for the ideal controller C and $\hat{Q} = G\hat{C}$ for the reduced controller $\hat{C}$.

Define a frequency interval $\Omega = [0, \omega_L]$; then to ensure both decoupling and loop performance it is required that

$$ERR_{oi} = \max_{\omega\in\Omega}\left|\frac{\hat{q}_{ii}(j\omega)-q_{ii}(j\omega)}{q_{ii}(j\omega)}\right| \leq \varepsilon_{oi}, \quad i=1,2,...,m$$

$$ERR_{di} = \max_{\omega\in\Omega}\frac{\Sigma_{j\neq i}|\hat{q}_{ji}(j\omega)|}{|\hat{q}_{ii}(j\omega)|} \leq \varepsilon_{di}, \quad i=1,2,...,m$$

where ε_{oi} and ε_{di} are the specified performance requirements on the loop performance and loop interactions, respectively.

Usually, both ε_{oi} and ε_{di} may be set as 10% and if both of the above ERR bounds hold then the design is complete. Otherwise, it is always possible to detune PID by relaxing the specification on the performance filter $f_i(s)$, namely increasing the filter time constant τ_i. An appropriate detuning rule is

$$\tau_i^{k+1} = \tau_i^k + \eta_i^k \max\{\gamma(\det[G(s)]) - \gamma_i, \min_{z \in Z_{\det[G]}^+}\{\text{Re}(z)\}\}$$

where k is an iteration counter and η_i is an adjustable factor of the ith loop reflecting both the approximation accuracy and the decoupling performance of the kth iteration and is set as

$$\eta_i = \max(\eta_{oi}, \eta_{di})$$

where

$$\eta_{oi} = \begin{cases} 0 & \text{if } ERR_{oi} \leq 10\% \\ 0.25 & \text{if } 10\% \leq ERR_{oi} \leq 50\% \\ 0.5 & \text{if } 50\% \leq ERR_{oi} \leq 100\% \\ 1 & \text{if } ERR_{oi} > 100\% \end{cases} \quad \eta_{di} = \begin{cases} 0 & \text{if } ERR_{di} \leq 10\% \\ 0.5 & \text{if } 10\% < ERR_{di} \end{cases}$$

The iteration continues until both of the above ERR bounds are satisfied.

11.3.4 Examples

Some simulation examples are given here to demonstrate the devised PID tuning algorithm and compare it with the multivariable PID controller design used by Dong and Brosilow (1997). They used a PID controller in the following form:

$$\tilde{C}(s) = \tilde{K}_P + \tilde{K}_I\left(\frac{1}{s}\right) + \tilde{K}_D\left(\frac{s}{\alpha s + 1}\right)$$

where matrix coefficients satisfy $\tilde{K}_P, \tilde{K}_I, \tilde{K}_D \in \Re^{(m \times m)}$ with

$$\alpha = \max_{i,j}((\tilde{K}_P^{-1}\tilde{K}_D)(i,j)/20)$$

provided that $\tilde{K}_P^{-1}$ exists. The ideal PID controller used in the approximation analysis above:

$$\hat{C}(s) = \{\hat{c}_{ij}(s)\} \text{ where } \hat{c}_{ij}(s) = k_{Pij} + \frac{k_{Iij}}{s} + k_{Dij}s$$

is not physically realisable and thus is replaced by

$$\hat{c}_{ij}(s) = k_{Pij} + \frac{k_{Iij}}{s} + \frac{k_{Dij}s}{|k_{Dij}/20|s + 1}$$

Example 11.4: Luyben's process model

Consider the process model given by Luyben (1986b):

$$G(s) = \begin{bmatrix} \dfrac{-2.2e^{-s}}{7s+1} & \dfrac{1.3e^{-0.3s}}{7s+1} \\ \dfrac{-2.8e^{-1.8s}}{9.5s+1} & \dfrac{4.3e^{-0.35s}}{9.2s+1} \end{bmatrix}$$

Following the procedure of this section, the target diagonal closed-loop process is

$$H(s) = \begin{bmatrix} \dfrac{e^{-1.05s}}{\tau_1 s + 1} & 0 \\ 0 & \dfrac{e^{-s}}{\tau_2 s + 1} \end{bmatrix}$$

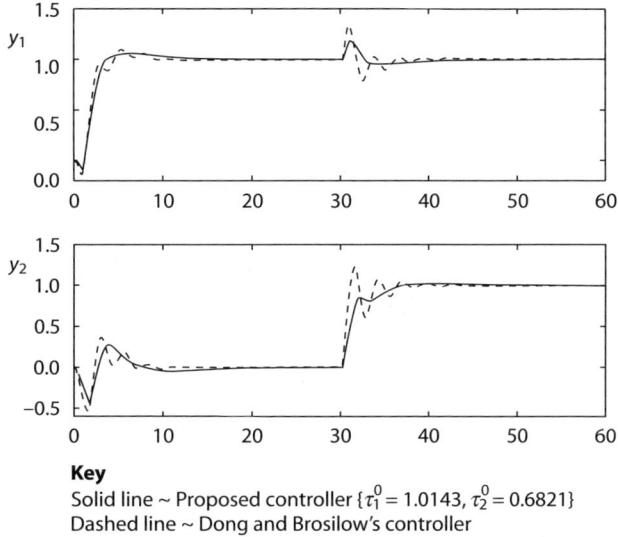

Figure 11.7 Step responses for Example 11.4: the first controllers obtained.

The tuning parameters $\{\tau_1, \tau_2\}$ are to be chosen to meet a desired phase margin specification and an input saturation constraint. The resulting initial values are obtained as

$$\{\tau_1^0 = 1.0143, \tau_2^0 = 0.6821\}$$

The proposed method in this section is used to obtain an ideal controller followed by a frequency domain fitting of the multivariable PID controller. The resulting overall controller is

$$\hat{C}(s) = \begin{bmatrix} -1.833 - \dfrac{0.434}{s} - 1.042s & 0.663 + \dfrac{0.169}{s} + 0.610s \\ -0.646 - \dfrac{0.306}{s} & 1.398 + \dfrac{0.265}{s} \end{bmatrix}$$

with

$ERR_{o1} = 31.76\%$ $ERR_{d1} = 42.14\%$
$ERR_{o2} = 93.23\%$ $ERR_{d2} = 43.31\%$

Dong and Brosilow's method generates the PID controller parameters as

$$\tilde{K}_P = \begin{bmatrix} -2.418 & 1.873 \\ -1.312 & 3.151 \end{bmatrix}, \tilde{K}_I = \begin{bmatrix} -0.367 & 0.216 \\ -0.239 & 0.366 \end{bmatrix}, \tilde{K}_D = \begin{bmatrix} 0.188 & -0.558 \\ 1.040 & -1.102 \end{bmatrix}$$

Since the derivative terms obtained using Dong and Brosilow's method have different signs from other terms in a single PID, PI control is used instead of a PID controller for this case, as suggested by Dong and Brosilow (1997). The step responses in Figure 11.7 show that both the proposed PID controller and Dong's PI controller cannot yield good loop performance.

Fine-Tuning the Responses

Based on the proposed iterative procedure, the tuning parameters $\{\tau_1, \tau_2\}$ are increased until the approximation error and the decoupling index are small enough to satisfy the loop performance and

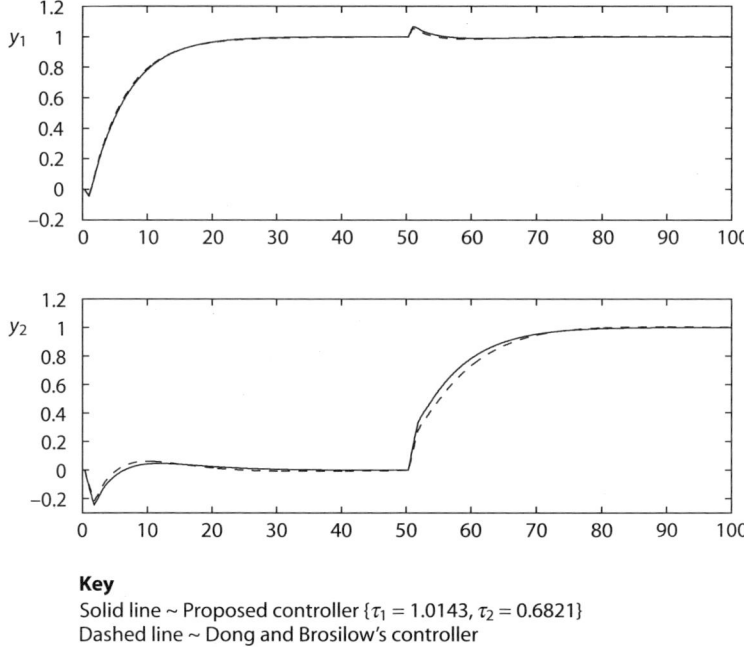

Key
Solid line ~ Proposed controller $\{\tau_1 = 1.0143, \tau_2 = 0.6821\}$
Dashed line ~ Dong and Brosilow's controller

Figure 11.8 Step responses for Example 11.4: the fine-tuned controllers.

loop interactions error bounds respectively. This results in new values for $\{\tau_1, \tau_2\}$, $\hat{C}(s)$ and the *ERR* bounds as respectively:

$$\{\tau_1 = 6.0143, \tau_2 = 5.9321\}$$

$$\hat{C}(s) = \begin{bmatrix} -0.669 - \dfrac{0.106}{s} - 0.507s & 0.270 + \dfrac{0.033}{s} + 0.428s \\ -0.317 - \dfrac{0.069}{s} & 0.426 + \dfrac{0.055}{s} \end{bmatrix}$$

$ERR_{o1} = 9.05\%$ $ERR_{d1} = 7.70\%$
$ERR_{o2} = 6.38\%$ $ERR_{d2} = 5.58\%$

Although Dong and Brosilow (1997) did not suggest any detuning rule, their method with the same new values of $\{\tau_1, \tau_2\}$ yields,

$$\tilde{K}_P = \begin{bmatrix} -0.676 & 0.306 \\ -0.365 & 0.515 \end{bmatrix}, \tilde{K}_I = \begin{bmatrix} -0.105 & 0.036 \\ -0.069 & 0.060 \end{bmatrix}, \tilde{K}_D = \begin{bmatrix} 0.172 & -0.107 \\ 0.362 & -0.206 \end{bmatrix}$$

The closed-loop step responses are shown in Figure 11.8 and are quite smooth.

Example 11.5: Vasnani's process model

Consider the 3×3 plant presented by Vasnani (1994):

11.3 Multivariable PID Control

$$G(s) = \begin{bmatrix} \dfrac{119e^{-5s}}{21.7s+1} & \dfrac{40e^{-5s}}{337s+1} & \dfrac{-2.1e^{-5s}}{10s+1} \\ \dfrac{77e^{-5s}}{50s+1} & \dfrac{76.7e^{-3s}}{28s+1} & \dfrac{-5e^{-5s}}{10s+1} \\ \dfrac{93e^{-5s}}{50s+1} & \dfrac{-36.7e^{-5s}}{166s+1} & \dfrac{-103.3e^{-4s}}{23s+1} \end{bmatrix}$$

Following the procedure of this section, the target diagonal closed-loop process is

$$H(s) = \begin{bmatrix} \dfrac{e^{-7.3s}}{\tau_1 s+1} & 0 & 0 \\ 0 & \dfrac{e^{-3.8s}}{\tau_2 s+1} & 0 \\ 0 & 0 & \dfrac{e^{-7.9s}}{\tau_3 s+1} \end{bmatrix}$$

The guideline for tuning the filters gives $\tau_1 = 6.85$, $\tau_2 = 4.90$, $\tau_3 = 6.60$ and this results in the proposed multivariable PID controller:

$$\hat{C}(s) = 10^{-3} \times \begin{bmatrix} 21.8 + \dfrac{0.9}{s} + 26.8s & -1.4 - \dfrac{0.1}{s} - 0.9s & 0.0002 + \dfrac{0.04}{s} \\ -9.8 - \dfrac{0.9}{s} & 59.3 + \dfrac{2.1}{s} + 40.4s & -4.1 - \dfrac{0.1}{s} \\ 9.3 + \dfrac{0.7}{s} + 1.5s & -3.2 - \dfrac{0.3}{s} & -27.8 - \dfrac{1.1}{s} - 24.7s \end{bmatrix}$$

with

$ERR_{o1} = 4.33\%$ $ERR_{d1} = 4.58\%$
$ERR_{o2} = 2.64\%$ $ERR_{d2} = 3.41\%$
$ERR_{o3} = 1.95\%$ $ERR_{d3} = 5.95\%$

Dong and Brosilow's method yields

$$\tilde{K}_P = \begin{bmatrix} -0.144 & 0.365 & -0.010 \\ 0.153 & -0.301 & 0.007 \\ -0.260 & 0.578 & -0.036 \end{bmatrix}, \quad \tilde{K}_I = 10^{-3} \times \begin{bmatrix} 1.10 & -0.80 & 0.01 \\ -1.00 & 2.40 & -0.10 \\ 1.30 & -1.60 & -0.90 \end{bmatrix},$$

$$\tilde{K}_D = \begin{bmatrix} 79.312 & -180.405 & 4.583 \\ -76.487 & 173.189 & -4.409 \\ 119.610 & -269.632 & 6.804 \end{bmatrix}$$

Unfortunately, the Dong and Brosilow controller leads to an unstable closed loop. When one of the three parameters that has a different sign from the other two elements in $\tilde{c}_{ij}(s)$ is eliminated, a stable step response can be obtained, as shown in Figure 11.9. In the same figure, it can be seen that the proposed method shows much better performance than Dong and Brosilow's design.

Generally, the controller needed to produce an equivalent decoupled process of a significantly interactive multivariable process becomes highly complicated. Although it is possible to detune the PID controller to a sufficient extent to generate a stable closed loop, a sluggish response must result. This is because a PID controller is too simple to reshape complicated dynamics satisfactorily over a large frequency range for high performance. In such a case, a more complex controller than PID is necessary if higher performance is required.

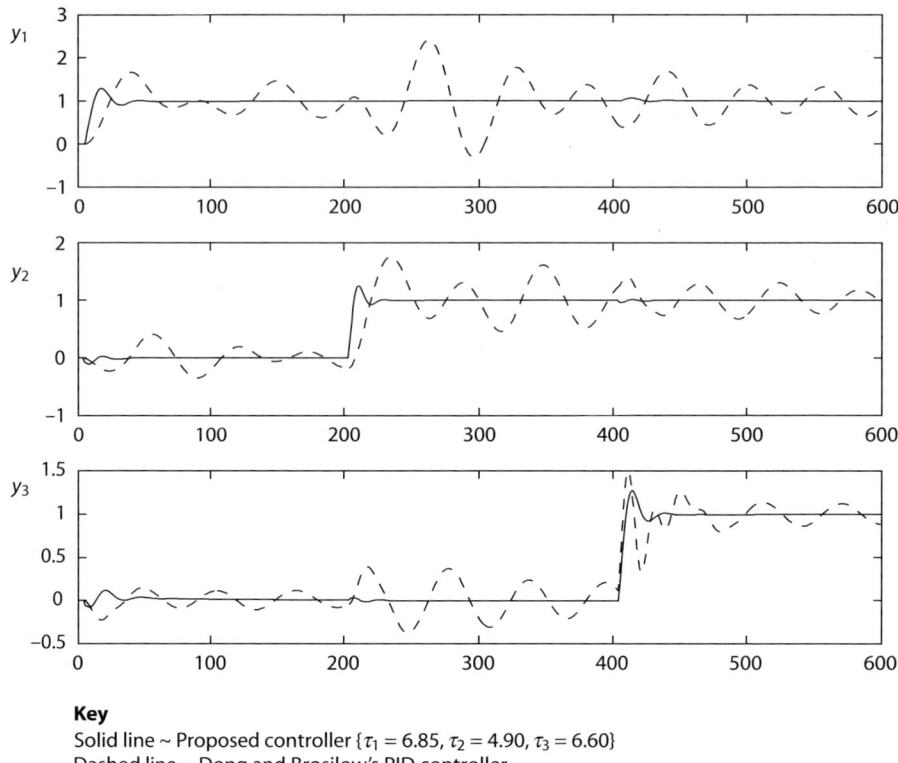

Key
Solid line ~ Proposed controller $\{\tau_1 = 6.85, \tau_2 = 4.90, \tau_3 = 6.60\}$
Dashed line ~ Dong and Brosilow's PID controller

Figure 11.9 Step responses for Example 11.5.

11.4 Conclusions

The BLT tuning method should be viewed as giving preliminary controller settings that can be used as a benchmark for comparative studies. Note that the procedure guarantees that the system is stable with all controllers on automatic and also that each individual loop is stable if all others are on manual (the F factor is limited to values greater than one, so the settings are always more conservative than the Ziegler–Nichols values). Thus a portion of the integrity question is automatically answered. However, further checks of stability would have to be made for other combinations of manual/automatic operation.

The dominant pole placement method places two pairs of pre-specified complex numbers as the closed-loop poles. Unlike its SISO counterpart, where the controller parameters can be obtained analytically, the multi-loop version amounts to solving a nonlinear equation with complex coefficients. A novel approach was presented to solve the equation using a "root trajectory" method. The design procedure is simple and simulations show that the method is effective.

Possible future research work can deal with the design of multi-loop PID controllers to place two more real poles. It should be pointed out that the method shares the same problems as in the SISO counterpart: that is, the placed poles may not necessarily be as dominant as expected and the closed-loop stability cannot be ensured; they must be post-checked. Research in these directions will be welcome and valuable.

The multivariable PID design is based on the fundamental relations for the decoupling of a multivariable process and the characterisation of the resultant closed-loop systems in terms of their unavoidable time delays and non-minimum phase zeros. The proposed design tackles stability, decoupling and loop performance as a whole, and can achieve, as close as PID control will allow, decoupling and best achievable loop performance with the given detuning formula. This truly reflects what many industrial control systems need. Examples have been presented to illustrate that these novel approaches give very satisfactory loop and decoupling performance. Although the method is somewhat computationally involved, some simplification may be introduced without much performance loss. There are few de-tuning rules for multivariable PID control and this could be a interesting topic for future research. After all, exact decoupling is not usually necessary in practice and designs that have allowable specified interactions would be a logical next move.

References

Åström, K.J. and Hägglund, T. (1995) *PID Controllers: Theory, Design, and Tuning*, 2nd edn. Instrument Society of America. Research Triangle Park, NC.

Åström, K.J., Panagopoulos, H. and Hägglund, T. (1998) Design of PI controllers based on non-convex optimisation. *Automatica*, **34**(5), 585–601.

Boyd, S.P., Balakrishnan, V. and Baratt, C.H. (1988) A new CAD method and associated architectures for linear controllers. *IEEE Trans. Automatic Control*, **23**, 268–283.

Dong, J. and Brosilow, C.B. (1997) Design of robust multivariable PID controllers via IMC. *Proc. American Control Conference*, Vol. 5. Albuquerque, New Mexico, pp. 3380–3384.

Koivo, H.N. and Tanttu, J.T. (1991) Tuning of PID controllers: survey of SISO and MIMO techniques. *Preprints IFAC International Symposium on Intelligent Tuning and Adaptive Control*.

Kong, K.Y. (1995) *Feasibility Report on a Frequency-Domain Adaptive Controller*. Department of Electrical Engineering. National University of Singapore, Singapore.

Lawson, C.L. and Hanson, R.J. (1974) *Solving Least Squares Problems*. Prentice Hall, Englewood Cliffs, NJ.

Lieslehto, J., Tanttu, J.T. and Koivo, H.N. (1993) An expert system for multivariable controller design. *Automatica*, **29**, 953–968.

Luyben, W.L. (1986) Simple method for tuning SISO controllers in multivariable systems. *Industrial and Engineering Chemistry: Process Design and Development*, **25**, 654–660.

Luyben, W.L. (1990). *Process Modelling, Simulation, and Control for Chemical Engineers*. McGraw-Hill, New York.

Maciejowski, J.M. (1989) *Multivariable Feedback Design*. Addison-Wesley, Reading, MA.

Noble, B. (1969) *Applied Linear Algebra*. Prentice Hall, Englewood Cliffs, NJ.

Ozguler, A.B. (1994) *Linear Multichannel Control: a System Matrix Approach*. Prentice Hall, New York.

Skogestad, S. and Postlethwaite, I. (1996) *Multivariable Feedback Control: Analysis and Design*. John Wiley & Sons, New York.

Skogestad, S. and Morari, M. (1989) Robust performance of decentralised control systems by independent designs. *Automatica*, **25**(1), 119–125.

Tan, K.K., Wang, Q.G. and Hang, C.C. (1999) Advances in PID Control. Springer-Verlag, London.

Vasnani, V.U. (1994) Towards relay feedback auto-tuning of multi-loop systems. *PhD Thesis*, National University of Singapore.

Wang, Q.G., Zou, B., Lee, T.H. and Bi, Q. (1997) Auto-tuning of multivariable PID controllers from decentralised relay feedback. *Automatica*, **33**(3), 319–330.

Wood, R.K. and Berry, M.W. (1973) Terminal composition control of a binary distillation column. *Chemical Engineering Science*, **28**, 1707–1717.

Zheng, F., Wang, Q.-G. and Lee, T.-H. (2002) On the design of multivariable PID controllers via LMI approach. *Automatica*, **38**(3), 517–526.

12 Restricted Structure Optimal Control

Learning Objectives

12.1 Introduction to Optimal LQG Control for Scalar Systems

12.2 Numerical Algorithms for SISO System Restricted Structure Control

12.3 Design of PID Controllers Using the Restricted Structure Method

12.4 Multivariable Optimal LQG Control: an Introduction

12.5 Multivariable Restricted Structure Controller Procedure

12.6 An Application of Multivariable Restricted Structure Assessment – Control of the Hotstrip Finishing Mill Looper System

12.7 Conclusions

Acknowledgements

References

Learning Objectives

Industrial control applications usually involve simple well-justified control structures for multivariable problems and use various fixed-order controller structures, for example PID controllers, to minimise the engineering expertise overhead needed to install and tune process control hardware. Scalar and multivariable optimal restricted controller procedures can help with both of these real industrial problems. In particular, this chapter shows how to use the well-developed optimal Linear Quadratic Gaussian (LQG) control framework to design and tune PID compensators and provide advice on the controller structural issues arising in the control of complex systems.

The learning objectives for this chapter are to:

- Help the reader to understand how to set up and solve a scalar LQG optimisation problem to provide an optimal controller whose properties are to be replicated in a restricted structure controller, for example, a PID controller.

- Present the numerical algorithms to compute the optimal restricted controller.

- Describe the application of the restricted structure concept to the design of an industrial PID controller.

- Help the reader to extend the restricted controller method to multivariable systems and appreciate the additional issues that arise in design multivariable controllers.

- Present the necessary multivariable optimal LQG control ideas needed for as a basis for the benchmark control solution.

- Help the reader to understand the derivation of the numerical algorithms to required to compute the optimal multivariable restricted controllers and the associated cost function values.

- Describe an extended example from the steel industry which illustrates how the optimal multivariable restricted controllers and the associated cost function values can be used in resolving the different structural issues in industrial multivariable control.

12.1 Introduction to Optimal LQG Control for Scalar Systems

The theory of optimal control is well developed in both the state space (Kwakernaak and Sivan, 1972; Anderson and Moore, 1972) and polynomial settings (Kucera, 1979; Grimble and Johnson, 1988a,b). In both cases, the controller that results from the optimisation is of an order at least equal to the plant order. However, industrial control applications exploit various controller structures that are of fixed order regardless of the plant order. These include PID controllers or Lead/Lag, which are the focus of this book. The question considered here is how to use the optimal control framework to design and tune PID compensators.

Model reduction via Balanced Truncation or Hankel Norm Approximation (Zhou *et al.*, 1996) is one way to circumvent the difficulty of generating a low-order optimal controller, by reducing the order of the plant such that the characteristics of the original plant are retained in some sense. An optimal controller will then be of the same order as this reduced plant. The problem considered in this chapter is how to minimise a Linear Quadratic Gaussian (LQG) cost function for a high-order plant where the structure for the controller is fixed (Grimble, 2000a). The PID controller is a special case of this more general problem. The aim is to simplify the design of such controllers whilst retaining the well-established and trusted structures.

The strategy followed here is to minimise an LQG criterion in such a way that the controller is of the desired form and is causal. A simple analytic solution cannot be obtained, as in the case where the controller structure is unconstrained. However, a relatively straightforward direct optimisation problem can be established which provides the desired numerical solution. An obvious and necessary assumption is that, for the given controller structure, a stabilising controller law exists. Given that a stabilising solution exists then the proposed method should enable the optimal gain values to be found.

The main application area for this first technique is in tuning PID controllers for industrial applications. It will be necessary to provide a representative linear model obtained by step tests, PRBS or other methods. However, once the model is obtained the LQG design is very simple, involving the choice of one or more scalars that determine the bandwidth/speed of response (Grimble, 1994).

There is rather more freedom in computing the reduced-order/PID controller which minimises the chosen LQG criterion than might be expected. This freedom arises because the frequency domain version of the cost function involves an optimisation over all frequencies. However, the most important frequency range for stability and robustness is a decade above and below the unity-gain crossover frequency. Thus an option is to minimise the frequency domain criterion over a limited selected frequency range. This facility is particularly valuable for very high-order systems where the low-order controller is required, since it provides an additional tuning facility.

12.1.1 System Description

The scalar process is assumed to be linear, continuous-time and single-input, single-output, as shown in Figure 12.1. The external white noise sources drive colouring filters that represent the reference $W_r(s)$ and disturbance $W_d(s)$ subsystems. The system equations are:

Output	$y(s) = W(s)u(s) + d(s)$
Input disturbance	$d(s) = W_d(s)\xi_d(s)$
Reference	$r(s) = W_r(s)\xi_r(s)$
Tracking error	$e(s) = r(s) - y(s)$
Control signal	$u(s) = C_0(s)e(s) = C_0(s)(r(s) - y(s))$

The system transfer functions are all assumed to be functions of the Laplace transform variable s, and for notational simplicity the arguments are omitted in $W(s)$ and the other disturbance and reference models.

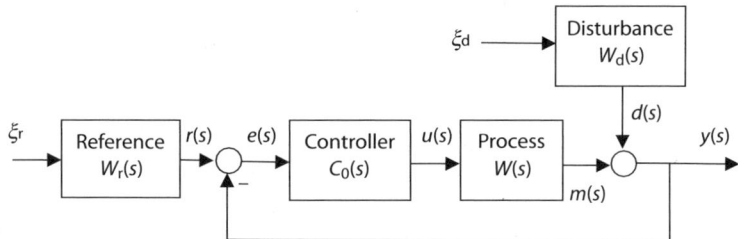

Figure 12.1 Single degree of freedom unity feedback control system.

Process Assumptions

1. The white noise sources ξ_r and ξ_d are zero-mean and mutually statistically independent. The intensities for these signals are without loss of generality taken to be unity.

2. The plant model W is assumed free of unstable hidden modes and the reference W_r and disturbance W_d subsystem models are asymptotically stable.

Signals and Sensitivity Transfer Functions

The following expressions are easily derived for the output, error and control signals:

$$y(s) = WMr(s) + Sd(s)$$
$$e(s) = (1 - WM)r(s) - Sd(s)$$

and

$$u(s) = SC_0(r(s) - d(s))$$

These equations involve the following sensitivity transfer functions:

Sensitivity $\quad S(s) = (I + WC_0)^{-1}$

Control sensitivity $\quad M(s) = C_0(s)S(s) = C_0(I + WC_0)^{-1}$

The Process in Polynomial System Form

The system shown in Figure 12.1 may be represented in polynomial form (Kailath, 1980), where the system transfer functions are written as:

Plant model $\quad W = A^{-1}B$

Reference generator $\quad W_r = A^{-1}E$

Input disturbance $\quad W_d = A^{-1}C_d$

The various polynomials $A(s), B(s), C_d(s), E(s)$ are not necessarily coprime but the plant transfer function is assumed to be free of unstable hidden modes.

12.1.2 Cost Function and Optimisation Problem

The LQG time domain cost function to be minimised is defined as

$$J = \lim_{T \to \infty} E\left\{ \frac{1}{2T} \int_{-T}^{T} \{[(H_q * e)(t)]^2 + [(H_r * u)(t)]^2\} dt \right\}$$

where $f * g$ denotes convolution of the functions f and g.

The transfer function form of the cost function may be stated as (Youla et al., 1976)

$$J = \frac{1}{2\pi j} \oint_D [Q_c(s)\Phi_{ee}(s) + R_c(s)\Phi_{uu}(s)] ds$$

where Q_c and R_c represent dynamic weighting elements acting on the spectrum of the error $\Phi_{ee}(s)$ and on the spectrum of the feedback control signal $\Phi_{uu}(s)$, respectively. The R_c weighting term is assumed to be positive definite and Q_c is assumed to be positive semi-definite on the D contour of the s-plane.

Let the Laplace transform of the adjoint of W be denoted W^*, where $W^*(s) = W(-s)$. Then the spectra $\Phi_{ee}(s)$ and $\Phi_{uu}(s)$ may be expanded as

$$\Phi_{ee}(s) = (1 - WM)\Phi_{rr}(1 - WM)^* + (1 - WM)\Phi_{dd}(1 - WM)^*$$

giving

$$\Phi_{ee}(s) = (1 - WM)\Phi_{ff}(1 - WM)^*$$

Similarly,

$$\Phi_{uu}(s) = M\Phi_{ff}(s)M^*$$

The spectrum of the composite signal $f(s) = (r(s) - d(s))$ is denoted by $\Phi_{ff}(s)$, and a generalised spectral factor Y_f may be defined from this spectrum using:

$$Y_f Y_f^* = \Phi_{ff} = \Phi_{rr} + \Phi_{dd}$$

In polynomial form, the generalised spectral factor can shown to be $Y_f = A^{-1}D_f$. The disturbance model is assumed to be such that D_f is strictly Hurwitz and satisfies

$$D_f D_f^* = EE^* + C_d C_d^*$$

The cost function weightings have transfer function forms which can be written in polynomial form as:

Error signal weight $\quad Q_c = Q_n/(A_q^* A_q)$

Control signal weight $\quad R_c = R_n/(A_r^* A_r)$

In these weighting expressions A_q is a Hurwitz polynomial and A_r is a strictly Hurwitz polynomial.

Restricted Structure Problem and Solution Strategy

The restricted structure controller problem assumes that the control engineer has captured the desirable feature of the control design through the specification of the LQG cost function definition. The problem then is to minimise the LQG criterion given by

$$J = \frac{1}{2\pi j} \oint_D \{Q_c(s)\Phi_{ee}(s) + R_c(s)\Phi_{uu}(s)\} ds$$

whilst the controller has a specified restricted structure such as:

Reduced order controller $\quad C_0(s) = \dfrac{c_{n0} + c_{n1}s + \ldots + c_{np}s^p}{c_{d0} + c_{d1}s + \ldots + c_{nv}s^v}$

where $v \geq p$ is less than the order of the system (plus weightings).

Lead–lag controller $\quad C_0(s) = \dfrac{(c_{n0} + c_{n1}s)(c_{n2} + c_{n3}s)}{(c_{d0} + c_{d1}s)(c_{d2} + c_{d3}s)}$

PID controller $\quad C_0(s) = k_0 + \dfrac{k_1}{s} + k_2 s$

The terminology *restricted structure* controller is used because the controller is sub-optimal and does not have the same number of coefficients that appears in the full LQG optimal controller. In solving this optimisation, an assumption must be made that a stabilising control law exists for the assumed controller structure. The optimal control solution is required to be causal. It will be assumed that the controller structure is consistent with the choice of error weighting in the sense that if A_q includes a j-axis zero then the controller denominator $C_{0d}(s)$ also includes such a zero. In practice, this simply recognises that when the term $1/A_q$ in the error weighting function denotes an integrator, the optimal controller will also include an integrator.

The strategy for a solution is to solve the scalar LQG control problem using polynomial system methods. This leads to the optimisation of a reduced integral cost function expression. One route from this integral leads to the full optimal solution, but a second route is to introduce the restricted structure controller into the formulation and devise a numerical method to find the optimal coefficients of the desired restricted structure controller.

Steps to an Optimal LQG Controller Solution
Recall the LQG error and control weighted criterion as

$$J = \frac{1}{2\pi j} \oint_D \{Q_c(s)\Phi_{ee}(s) + R_c(s)\Phi_{uu}(s)\} ds$$

with expressions for the weightings as $Q_c = Q_n/(A_q^* A_q)$ and $R_c = R_n/(A_r^* A_r)$, and for the spectra $\Phi_{ee}(s)$ and $\Phi_{uu}(s)$ as

$$\Phi_{ee}(s) = (1 - WM)\Phi_{ff}(1 - WM)^* \text{ and } \Phi_{uu}(s) = M\Phi_{ff}(s)M^*$$

Then substitution of the error and control input power spectra into the cost criterion gives

$$J = \frac{1}{2\pi j}\oint_D \{(W^*Q_cW + R_c)M\Phi_{ff}M^* - Q_cWM\Phi_{ff} - Q_cW^*M^*\Phi_{ff} + Q_c\Phi_{ff}\}ds$$

To make further manipulation of the cost function possible it is necessary to introduce the control and filter spectral factors into the analysis. Define the control spectral factor $Y_c(s)$ using

$$Y_c^*Y_c = W^*Q_cW + R_c = [D_cA^{-1}]^*[D_cA^{-1}]$$

so that $Y_c = [D_cA^{-1}]$ and then recall the filter spectral factor Y_f as satisfying

$$Y_f Y_f^* = \Phi_{ff} = \Phi_{rr} + \Phi_{dd}$$

where $Y_f = A^{-1}D_f$.

If further substitutions are made using these generalised spectral factors, the resulting cost is

$$J = \frac{1}{2\pi j}\oint_D \left\{(Y_cMY_f)(Y_cMY_f)^* - (Y_cMY_f)\frac{Q_cW\Phi_{ff}}{Y_cY_f} - (Y_c^*M^*Y_f^*)\frac{Q_cW^*\Phi_{ff}}{Y_c^*Y_f^*} + Q_c\Phi_{ff}\right\}ds$$

Using a completing-the-squares argument in the cost function leads to

$$J = \frac{1}{2\pi j}\oint_D \left\{\left(Y_cMY_f - \frac{Q_cW^*\Phi_{ff}}{Y_c^*Y_f^*}\right)\left(Y_cMY_f - \frac{Q_cW^*\Phi_{ff}}{Y_c^*Y_f^*}\right)^* - \frac{W^*WQ_c^*Q_c\Phi_{ff}^*\Phi_{ff}}{Y_c^*Y_cY_fY_f^*} + Q_c\Phi_{ff}\right\}ds$$

The cost function is now in a form such that the final two terms of the integrand are independent of the feedback controller $C_0(s)$. The remaining difficulty is that the product within the integrand must be decomposed into guaranteed stable and unstable components, because

$$\frac{Q_cW^*\Phi_{ff}}{Y_c^*Y_f^*}$$

is certainly unstable but Y_cMY_f stability depends upon the plant.

This decomposition can be achieved using two Diophantine equations, but first the product terms must be expanded into the desired form by substituting the various system polynomial definitions into the integrand; this gives the result

$$\left(Y_cMY_f - \frac{Q_cW^*\Phi_{ff}}{Y_c^*Y_f^*}\right) = \frac{(B^*A_r^*Q_nD_f)BA_rC_{0n} + (A^*A_q^*R_nD_f)AA_qC_{0n}}{D_c^*A_c(AC_{0d} + BC_{0n})} - \frac{B^*A_r^*Q_nD_f}{D_c^*AA_q}$$

Using the Diophantine equation pair below:

$$D_c^*G_0 + F_0AA_q = B^*A_r^*Q_nD_f$$

$$D_c^*H_0 - F_0BA_r = A^*A_q^*R_nD_f$$

it is then trivial to show, under the assumptions on system models, that

$$Y_cMY_f - \frac{Q_cW^*\Phi_{ff}}{Y_c^*Y_f^*} = \frac{C_{0n}H_0A_q - C_{0d}G_0A_r}{A_w(AC_{0d} + BC_{0n})} - \frac{F_0}{D_c^*} = T_1^+ + T_1^-$$

where

12.1 Introduction to Optimal LQG Control for Scalar Systems

$$T_1^+ = \frac{C_{0n}H_0A_q - C_{0d}G_0A_r}{A_w(AC_{od} + BC_{on})} \quad \text{and} \quad T_1^- = -\frac{F_0}{D_c^*}$$

The first term $T_1^+(s)$ is strictly stable and the second term $T_1^-(s)$ is strictly unstable. This decomposition enables the cost function integral to be written as

$$J = \frac{1}{2\pi j}\oint_D \{(T_1^+ + T_1^-)(T_1^+ + T_1^-)^* + \Phi_0\}\,ds$$

The T_1^- and Φ_0 terms are independent of the controller C_0; hence the optimal control problem reduces to finding C_0 such that

$$J = \frac{1}{2\pi j}\oint_D \{T_1^+ T_1^{+*}\}\,ds$$

is minimised. In the LQG optimal controller case, optimality is accomplished by

$$T_1^+ = \frac{C_{0n}H_0A_q - C_{0d}G_0A_r}{A_w(AC_{od} + BC_{on})} = 0$$

and the optimal controller is obtained as

$$C_0(s) = \frac{C_{0n}}{C_{0d}} = \frac{G_0 A_r}{H_0 A_q}$$

The results above can be collected together and the following theorem stated.

Theorem 12.1: LQG optimal control – single degree of freedom process
For the given LQG error and control weighted criterion

$$J = \frac{1}{2\pi j}\oint_D \{Q_c(s)\Phi_{ee}(s) + R_c(s)\Phi_{uu}(s)\}\,ds$$

with cost function weights defined as

$$Q_c = Q_n/(A_q^* A_q)$$
$$R_c = R_n/(A_r^* A_r)$$

compute the Hurwitz filter and control spectral factors $D_f(s)$ and $D_c(s)$ from

$$D_f D_f^* = EE^* + C_d C_d^*$$
$$D_c^* D_c = B^* A_r^* Q_n A_r B + A^* A_q^* R_n A_q A$$

Solve the simultaneous Diophantine equations for (G_0, H_0, F_0) with F_0 of minimum degree:

$$D_c^* G_0 + F_0 A A_q = B^* A_r^* Q_n D_f$$
$$D_c^* H_0 - F_0 B A_r = A^* A_q^* R_n D_f$$

Then the optimal LQG controller $C_0(s) = C_{0n}/C_{0d}$ minimises the causal component in the cost function term:

$$J_1 = \frac{1}{2\pi}\int_{-\infty}^{\infty} T_1^+(j\omega) T_1^+(-j\omega)\,d\omega$$

where

$$T_1^+ = \frac{C_{0n}H_0A_q - C_{0d}G_0A_r}{A_w(AC_{od} + BC_{on})}$$

In which case the optimal LQG controller is obtained as

$$C_0(s) = \frac{C_{0n}}{C_{0d}} = \frac{G_0A_r}{H_0A_q} \quad \text{and} \quad J_1^{\min} = 0$$

12.2 Numerical Algorithms for SISO System Restricted Structure Control

The results of Theorem 12.1 can be exploited to give a solution to the restricted structure optimal controller problem. In these results optimality originated from minimising the cost function component

$$J_1 = \frac{1}{2\pi} \int_{-\infty}^{\infty} T_1^+(j\omega)T_1^+(-j\omega)d\omega$$

where

$$T_1^+ = \frac{C_{0n}H_0A_q - C_{0d}G_0A_r}{A_w(AC_{od} + BC_{on})}$$

It is a small but critical step to replace the full order controller by a restricted structure controller $C_0^{RS}(s) = C_{0n}^{RS}/C_{0d}^{RS}$ and minimise the frequency domain integral with respect to the parameters of the restricted structure controller. This is the essence of the numerical procedures described next. It should be noted that if the controller has restricted structure then the minimum of the cost term

$$J_1 = \frac{1}{2\pi} \int_{-\infty}^{\infty} T_1^+(j\omega)T_1^+(-j\omega)d\omega$$

will satisfy $J_1^{\min} \geq 0$, whilst for an unconstrained optimal LQG solution the minimum is achieved when $T_1^+ = 0$ and then $J_1^{\min} = 0$.

12.2.1 Formulating a Restricted Structure Numerical Algorithm

It has been explained in Theorem 12.1 that the computation of the optimal feedback controller C_0 reduces to minimisation of the term J_1 given by

$$J_1 = \frac{1}{2\pi j} \oint_D \{T_1^+ T_1^{+*}\} ds = \frac{1}{2\pi} \int_{-\infty}^{\infty} T_1^+(j\omega)T_1^+(-j\omega)d\omega$$

Recall the expression for T_1^+ as

$$T_1^+ = \frac{C_{0n}H_0A_q - C_{0d}G_0A_r}{A_w(AC_{od} + BC_{on})}$$

If the polynomials $H_0, A_q, G_0, A_r, A_w, A$ and B are all known then T_1^+ can be written as

$$T_1^+ = [C_{0n}L_1 - C_{0d}L_2]/[C_{0d}L_3 + C_{0n}L_4]$$

12.2 Numerical Algorithms for SISO System Restricted Structure Control

where $C_0(s) = C_{0n}(s)/C_{0d}(s)$ has a specified restricted structure.

Assume, for example, that $C_0(s)$ has a modified PID structure of the form

$$C_0(s) = k_P + \frac{k_I}{s} + \frac{k_D s}{\tau s + 1} = \frac{C_{0n}(s)}{C_{0d}(s)}$$

so that the controller numerator and denominator are respectively

$$C_{0n}(s) = k_P(1+\tau s)s + k_I(1+\tau s) + k_D s^2$$

and

$$C_{0d}(s) = s(1+\tau s)$$

Here T_1^+ is obviously nonlinear in k_P, k_I and k_D, making it difficult to minimise the J_1 cost function directly. However, an iterative solution is possible if the values of k_P, k_I and k_D in the denominator of T_1^+ are assumed known, since a minimisation can then be performed using unknown k_P, k_I and k_D values in the numerator (linear terms) part of T_1^+. Thus to set up this problem let

$$T_1^+ = C_{0n} L_{n1} - C_{0d} L_{n2}$$

where

$$L_{n1} = L_1/(C_{0n}L_3 + C_{0d}L_4)$$
$$L_{n2} = L_2/(C_{0n}L_3 + C_{0d}L_4)$$

Setting $s = j\omega$, the controller numerator and denominator can be written in terms of real and imaginary parts as

$$C_{0n}(j\omega) = C_{0n}^r + jC_{0n}^i$$
$$C_{0d}(j\omega) = C_{0d}^r + jC_{0d}^i$$

where superscripts r and i denote the real and imaginary components.

The controller numerator term may be split into frequency-dependent components through the following steps. Recall the controller numerators as

$$C_{0n}(s) = k_P(1+\tau s)s + k_I(1+\tau s) + k_D s^2$$

Hence setting $s = j\omega$ yields

$$C_{0n}(j\omega) = -k_P \omega^2 \tau + k_I - k_D \omega^2 + j(k_P \omega + k_I \omega \tau)$$

and $C_{0n}(j\omega) = C_{0n}^r + jC_{0n}^i$ leads to the identifications

$$C_{0n}^r = -k_P \omega^2 \tau + k_I - k_D \omega^2$$

and

$$C_{0n}^i = k_P \omega + k_I \omega \tau$$

Similarly, for the denominator term

$$C_{0d}(j\omega) = -\omega^2 \tau + j\omega$$

and using $C_{0d}(j\omega) = C_{0d}^r + jC_{0d}^i$ yields

$$C_{0d}^r = -\omega^2 \tau \text{ and } C_{0d}^i = \omega$$

Reduction of $T_1^+(s)$

Recall the equation for $T_1^+(s)$ as

$$T_1^+(s) = C_{0n}(s)L_{n1}(s) - C_{0d}(s)L_{n2}(s)$$

Setting $s = j\omega$ yields

$$T_1^+(j\omega) = C_{0n}(j\omega)L_{n1}(j\omega) - C_{0d}(j\omega)L_{n2}(j\omega)$$

and substituting for real and imaginary parts of $C_{0n}(j\omega)$, $L_{n1}(j\omega)$, $C_{0d}(j\omega)$, $L_{n2}(j\omega)$ we obtain

$$T_1^+ = (C_{0n}^r + jC_{0n}^i)(L_{n1}^r + jL_{n1}^i) - (C_{0d}^r + jC_{0d}^i)(L_{n2}^r + jL_{n2}^i)$$
$$= C_{0n}^r L_{n1}^r - C_{0n}^i L_{n1}^i - C_{0d}^r L_{n2}^r + C_{0d}^i L_{n2}^i + j(C_{0n}^i L_{n1}^r + C_{0n}^r L_{n1}^i - C_{0d}^i L_{n2}^r - C_{0d}^r L_{n2}^i)$$

and substitution for $C_{0n}^r, C_{0n}^i, C_{0d}^r, C_{0n}^i$ gives

$$T_1^+ = k_0(-\omega^2 \tau L_{n1}^r - \omega L_{n1}^i + j(\omega L_{n1}^r - \omega^2 \tau L_{n1}^i))$$
$$+ k_1(L_{n1}^r - \omega\tau L_{n1}^i + j(\omega\tau L_{n1}^r + L_{n1}^i)) + k_2(-\omega^2 L_{n1}^r - j\omega^2 L_{n1}^i)$$
$$+ \omega^2 \tau L_{n2}^r + \omega L_{n2}^i + j(\omega^2 \tau L_{n2}^i - \omega L_{n2}^r)$$

where the controller coefficient identification

$$\begin{bmatrix} k_0 \\ k_1 \\ k_2 \end{bmatrix} = \begin{bmatrix} k_P \\ k_I \\ k_D \end{bmatrix}$$

is used.

The real and imaginary parts of $T_1^+(j\omega)$ may therefore be written as

$$T_1^+(j\omega) = T_1^{+r} + jT_1^{+i}$$

and it follows that

$$|T_1^+|^2 = (T_1^{+r})^2 + (T_1^{+i})^2$$

The above equations can be written in matrix–vector form as

$$\begin{bmatrix} T_1^{+r} \\ T_1^{+i} \end{bmatrix} = [F]\begin{bmatrix} k_0 \\ k_1 \\ k_2 \end{bmatrix} - L = [F]\theta - L$$

where

$$F(\omega) = \begin{bmatrix} (-\omega^2 \tau L_{n1}^r - \omega L_{n1}^i) & (L_{n1}^r - \omega\tau L_{n1}^i) & (-\omega^2 L_{n1}^r) \\ (\omega L_{n1}^r - \omega^2 \tau L_{n1}^i) & (\omega\tau L_{n1}^r + L_{n1}^i) & (-\omega^2 L_{n1}^i) \end{bmatrix}$$

$F: \Re^+ \to \Re^{2\times 3}$

$$L(\omega) = \begin{bmatrix} -\omega^2 \tau L_{n2}^r - \omega L_{n2}^i \\ -\omega^2 \tau L_{n2}^i + \omega L_{n2}^r \end{bmatrix}$$

$L: \Re^+ \to \Re^{2\times 1}$

and $\theta = [k_0 \quad k_1 \quad k_2]^T \in \Re^3$

A simple iterative solution can be obtained if the integral is approximated (Yukitomo et al., 1998) by a summation with a sufficient number of frequency points $\{\omega_1, \omega_2, ..., \omega_N\}$. The optimisation can then be performed by minimising the sum of squares at each of the frequency points. The minimisation of the cost term J_0 is therefore required, where

$$J_0 = \sum_{i=1}^{N} (F(\omega_i)\theta - L(\omega_i))^T (F(\omega_i)\theta - L(\omega_i)) = (b - A\theta)^T (b - A\theta)$$

and

$$A = \begin{bmatrix} F(\omega_1) \\ \vdots \\ F(\omega_N) \end{bmatrix} \in \Re^{2N \times 3}, \quad b = \begin{bmatrix} L(\omega_1) \\ \vdots \\ L(\omega_1) \end{bmatrix} \in \Re^{2N \times 1}, \quad \theta = \begin{bmatrix} k_p \\ k_i \\ k_d \end{bmatrix} \in \Re^3$$

Assuming that the matrix $[A^T A] \in \Re^{3 \times 3}$ is non-singular, then the least squares optimal solution (Noble, 1969) follows as

$$\theta = [A^T A]^{-1} A^T b$$

The assumption is made that θ is already known when defining the denominator of T_1^+, and this is a case when the method of successive approximation (Luenberger, 1969) is used. This involves a transformation T such that $\theta_{k+1} = T(\theta_k)$. Under appropriate conditions, the sequence $\{\theta_k\}$ converges to a solution of the original equation. Since this optimisation problem is nonlinear there may not be a unique minimum. However, the algorithm presented in the next subsection does always appear to converge to an optimal solution for a large number of industrial examples.

12.2.2 Iterative Solution for the SISO Restricted Structure LQG Controller

The following successive approximation algorithm (Luenberger, 1969) can be used to compute the restricted structure LQG controller.

Algorithm 12.1: SISO restricted structure LQG controller

Step 1 *Spectral factor and Diophantine equation solutions*
Set up optimal control problem by choosing appropriate weightings.
Solve for the spectral factors D_c, D_f.
Solve the Diophantine equations for G_0, H_0, F_0.
Compute full order controller from $C_0(s) = (G_0 A_r)/(H_0 A_q)$

Step 2 *Restricted structure controller, PID*
Set $\alpha_0 = (1 + s/\theta_D)s$, $\alpha_1 = (1 + s/\theta_D)$, $\alpha_2 = s^2$
Set $L_1 = H_0 A_q$, $L_2 = G_0 A_r$, $L_3 = A_w B$, $L_4 = A_w A$

Step 3 *Initial controller*

$$\text{Set values for } \begin{bmatrix} k_0 \\ k_1 \\ k_2 \end{bmatrix} = \begin{bmatrix} k_P \\ k_I \\ k_D \end{bmatrix}$$

Step 4 *Controller functions*
Compute $C_{0n}(s) = \alpha_0(s)k_0 + \alpha_1(s)k_1 + \alpha_2(s)k_2$ and $C_{0d}(s) = (1 + s/\theta_D)s$
Compute $L_{n1}(s) = L_1/(C_{0n}L_3 + C_{0d}L_4)$ and $L_{n2}(s) = L_2/(C_{0n}L_3 + C_{0d}L_4)$

Step 5 *Compute F matrix elements*
For all chosen frequencies
Compute $L_{n1}^r(\omega), L_{n1}^i(\omega), L_{n2}^r(\omega), L_{n2}^i(\omega)$

Compute $\alpha_0^r(\omega), \alpha_0^i(\omega), \alpha_1^r(\omega), \alpha_1^i(\omega), \alpha_2^r(\omega), \alpha_2^i(\omega)$
Compute

$$f_{11}^r(\omega) = \alpha_0^r(\omega)L_{n1}^r(\omega) - \alpha_0^i(\omega)L_{n1}^i(\omega) \quad f_{12}^r(\omega) = \alpha_1^r(\omega)L_{n1}^r(\omega) - \alpha_1^i(\omega)L_{n1}^i(\omega)$$
$$f_{13}^r(\omega) = \alpha_2^r(\omega)L_{n1}^r(\omega) - \alpha_2^i(\omega)L_{n1}^i(\omega) \quad f_{11}^i(\omega) = \alpha_0^i(\omega)L_{n1}^r(\omega) + \alpha_0^r(\omega)L_{n1}^i(\omega)$$
$$f_{12}^i(\omega) = \alpha_1^i(\omega)L_{n1}^r(\omega) + \alpha_1^r(\omega)L_{n1}^i(\omega) \quad f_{13}^i(\omega) = \alpha_2^i(\omega)L_{n1}^r(\omega) + \alpha_2^r(\omega)L_{n1}^i(\omega)$$

Step 6 *Compute L matrix elements*
For all chosen frequencies
Compute $C_{0n}^r(\omega), C_{0n}^i(\omega), C_{0d}^r(\omega), C_{0d}^i(\omega)$
Compute

$$L_{11}^r(\omega) = C_{0d}^r(\omega)L_{n2}^r(\omega) - C_{0d}^i(\omega)L_{n2}^i(\omega)$$
$$L_{11}^i(\omega) = C_{0d}^r(\omega)L_{n2}^i(\omega) + C_{0d}^i(\omega)L_{n2}^r(\omega)$$

Step 7 *Assemble the frequency-dependent matrices*

$$A = \begin{bmatrix} f_{11}^r(\omega_1) & f_{12}^r(\omega_1) & f_{13}^r(\omega_1) \\ f_{11}^i(\omega_1) & f_{12}^i(\omega_1) & f_{13}^i(\omega_1) \\ \vdots & \vdots & \vdots \\ f_{11}^r(\omega_N) & f_{12}^r(\omega_N) & f_{13}^r(\omega_N) \\ f_{11}^i(\omega_N) & f_{12}^i(\omega_N) & f_{13}^i(\omega_N) \end{bmatrix} \quad b = \begin{bmatrix} L_{11}^r(\omega_1) \\ L_{11}^i(\omega_1) \\ \vdots \\ L_{11}^r(\omega_N) \\ L_{11}^i(\omega_N) \end{bmatrix}$$

Step 8 *Optimal solution*

$$\text{Compute } \theta = \begin{bmatrix} k_0 \\ k_1 \\ k_2 \end{bmatrix} = [A^T A]^{-1} A^T b$$

Step 9 *Convergence test*
If the cost is lower than the previous cost, repeat Steps 4–8 using the new $C_{0n}(s)$.
If the cost decrease is less than the convergence tolerance, compute the controller using

$$C_{0n}(s) = \alpha_0(s)k_0 + \alpha_1(s)k_1 + \alpha_2(s)k_2 \text{ and } C_0(s) = C_{0n}(s)/C_{0d}(s) \text{ and STOP}$$

Algorithm end

Remarks 12.1
1. The stopping criterion is not important, and in fact experience reveals that a small fixed number of steps, say 10, can be used. More sophisticated numerical methods such as gradient algorithms would give even faster convergence.

2. The functions in Steps 6 and 7 can be evaluated for each frequency and stored in vectors. The Hadamard product of two vectors $X \circ Y = [x_i y_i]$ can then be used to form the element-by-element products, and a compact matrix solution can be generated.

12.2.3 Properties of the Restricted Structure LQG Controller

The following lemma summarises the properties of the LQG controller with a restricted structure.

Lemma 12.1: Restricted structure LQG controller solution properties
The characteristic polynomial, which determines stability, and the implied equation are respectively

$$\rho_c = AC_{0d} + BC_{0n} \text{ and } AA_qH_0 + BA_rG_0 = D_cD_f$$

The minimum of the cost function with a controller of restricted structure is:

$$J_{min} = \frac{1}{2\pi j}\oint_D \{T_1^+T_1^{+*} + T_1^-T_1^{-*} + \Phi_0\}\, ds$$

$$= \frac{1}{2\pi j}\oint_D \left\{T_1^+T_1^{+*} + \frac{F_0F_0^* + Q_nD_fD_f^*R_n}{D_c^*D_c}\right\} ds$$

Proof

The proof of this theorem, which is mainly algebraic, is due to Grimble (2000a). ☐

Remarks 12.2

1. For some problems, the required solutions for the pair H_0, G_0 may be computed from the so-called implied equation $AA_qH_0 + BA_rG_0 = D_cD_f$ rather than solving the simultaneous Diophantine equation pair.

2. The cost relationship given in the Lemma is interesting since for the full-order LQG cost $T_1^+(s) = 0$ and hence the LQG optimal cost may be defined as

$$J_{LQG}^{min} = \frac{1}{2\pi j}\oint_D \{T_1^-T_1^{-*} + \Phi_0\}\, ds$$

Thus the restricted structure cost can be written

$$J_{min} = \frac{1}{2\pi j}\oint_D \{T_1^+T_1^{+*}\}\, ds + J_{LQG}^{min}$$

This relation reveals the increase in cost which occurs by restricting the controller structure, namely

$$\Delta J_{min} = J_{min} - J_{LQG}^{min} = \frac{1}{2\pi j}\oint_D \{T_1^+T_1^{+*}\}\, ds$$

Enhancing Restricted Controller Robustness

There are two general approaches to robustness improvement:

1. The robustness of the full-order design can be improved with the expectation that the reduced order solution will also be better.

2. The restricted structure controller can be retuned to directly improve robustness properties.

The former approach involves adding sensitivity terms in the criterion to try to improve both full and reduced order solutions. The second approach could involve retuning using the Quantitative Feedback Theory (QFT) approach. That is, the LQG restricted structure approach can be used to provide a starting point for QFT designs (Horowitz, 1979), which is particularly helpful for open-loop unstable and non-minimum phase systems. The main advantage of the QFT approach is the ability to improve robustness. However, there are some problems in obtaining optimum stochastic properties and a merger of the approaches may provide benefits over both of these methods.

12.3 Design of PID Controllers Using the Restricted Structure Method

The approach of the presentation of the restricted structure method took a route that emphasised the generality of the concepts involved. This next section will specialise in the restricted structure controller concept for the design of PID controllers and will demonstrate the performance of the numerical algorithm on a marine controller example.

12.3.1 General Principles for Optimal Restricted Controller Design

An important assumption for the restricted structure controller concept is that the requirements of the desired controller are captured in the specification of the LQG cost function. This means that the issue of weight selection for the LQG cost function must be resolved. To inform the discussion to follow, it is useful to recall the cost function and weighting definitions as:

LQG cost function $\quad J = \dfrac{1}{2\pi j} \oint_D \{Q_c(s)\Phi_{ee}(s) + R_c(s)\Phi_{uu}(s)\}\, ds$

Error weighting $\quad Q_c(s) = \dfrac{Q_n(s)}{A_q^*(s)A_q(s)}$

Control weighting $\quad R_c(s) = \dfrac{R_n(s)}{A_r^*(s)A_r(s)}$

where $A_q(s)$ is a Hurwitz polynomial and $A_r(s)$ is a strictly Hurwitz polynomial.

The LQG controller should be designed in such a way that it is consistent with the restricted controller structure of interest. For example, $A_q(s)$ should approximate a differentiator if near integral action is required. The assumption made in deriving Theorem 12.1 was that the controller structure is compatible with the choice of error weighting, and if $1/A_q(s)$ includes a j-axis pole then this will be included in the chosen controller. In fact, the usual situation will be that the designer decides that the controller should include integral action and the weighting $1/A_q(s)$ will be chosen as an integrator. The control weighting $1/A_r(s)$ is not so critical, but if, for example, a PID structure is to be used, then the point at which the differential (lead term) comes in can help to determine the $A_r(s)$ weighting.

Clearly, there is no point in designing an LQG controller that has an ideal response, in some sense, but which cannot be approximated by the chosen controller structure. Thus the cost weightings should be selected so that the closed-loop properties are satisfactory but taking into consideration the limitations of the controller structure required.

The choice of frequency range and spacing between frequency points used in the cost equation approximation J_0, namely the selection of frequency points $\{\omega_1, \omega_2, \ldots, \omega_N\}$, is another design issue of note. This is because the restricted structure controller must incorporate the salient features of the full-order frequency response. Inspection of this response allows the engineer to select the important regions so that any integral or derivative action is included in the restricted controller along with other relevant frequency characteristics.

The structure of the low-order controller itself is a design consideration, as certain structures may be more suited to some problems than others. This will not be covered here, however, as only the PID controller form is of interest.

With regard to PID control, one further design factor is the choice of derivative filter. In practical situations, the D term will have some high-frequency filtering in order to prevent exogenous noise from causing excessive actuator variations. Depending upon the amount of high-frequency gain in the full order controller, the derivative filter may be tuned to provide a better match between restricted and high-order designs.

The above design ideas are demonstrated in the following simple example.

12.3.2 Example of PID Control Design

The following scalar design example is based upon the dynamic position (DP) control of a ship. Results are obtained for a full-order LQG controller and for a modified PID restricted structure LQG controller.

The system model is decomposed into the polynomial data needed for restricted controller algorithm computations.

Plant Polynomials

$$A = s(s+1)(s+0.000509)$$
$$B = 6.5 \times 10^{-5}(s+1)$$
$$C_d = 16s(s+0.000509)$$
$$E = 0.1(s+1)(s+0.000509)$$

Cost Function Definition

$$Q_c = \frac{Q_n}{A_q^* A_q} = \frac{0.01}{-s^2}$$

$$R_c = \frac{R_n}{A_r^* A_r} = \frac{-10^{-3} s^2}{(s+1)(-s+1)}$$

An explanation of the design decisions made to determine the weightings can be found in Grimble (1994). In the sequel, the comparison between the full-order LQG and the solution with the restricted structure will be highlighted. However, note that the error weight $Q_c(s)$ contains an integrator so that integral action appears in the controller and the control weight $R_c(s)$ contains a filtered differential so that the amount of derivative action incorporated is not excessive.

The full-order unrestricted LQG controller for this system can be calculated as

$$C_{0f} = \frac{3.56s^4 + 7.138s^3 + 3.602s^2 + 0.02179s + 9.994 \times 10^{-6}}{s^4 + 0.319s^3 + 0.0508s^2 + 0.0043s}$$

$$= \frac{3.56(s+1.003)(s+1)(s+0.0056)(s+0.0005)}{s(s+0.171)(s+0.0742+j0.141)(s+0.0742-j0.141)}$$

The frequency response of the full-order unrestricted LQG controller is shown in Figure 12.2.

Modified PID Controller Structure

The assumed controller structure is modified PID; that is, a PID controller with a high-frequency filter on the differential term. The assumed structure and the computed controller using the iterative algorithm are as follows:

$$C_0 = k_0 + \frac{k_1}{s} + \frac{k_2 s}{1+s/\theta_D} = (k_0 \alpha_0 + k_1 \alpha_1 + k_2 \alpha_2)/C_{0d}$$

where $\alpha_0 = (1+s/\theta_D)s$, $\alpha_1 = (1+s/\theta_D)$, $\alpha_2 = s^2$, $C_{0d} = (1+s/\theta_D)s$.

On inspection of Figure 12.2, the frequency interval of interest is selected as $[10^{-5} \text{ rad s}^{-1}, 10 \text{ rad s}^{-1}]$. This is divided into 20 frequency points. With some experimentation, it is found that $\theta_D = 0.0667$ for the D-term filter gives a good approximation to the high-frequency characteristics and correlation with the time response of the full-order controller. Computation of the restricted structure controller results in gains $k_P = 4.89$, $k_I = 0.0023$ and $k_D = 866$, and the ensuing controller is

$$C_{0r} = \frac{62.6s^2 + 0.328s + 0.000154}{s^2 + 0.0667s} = \frac{62.6(s+0.047)(s+0.005)}{s(s+0.0667)}$$

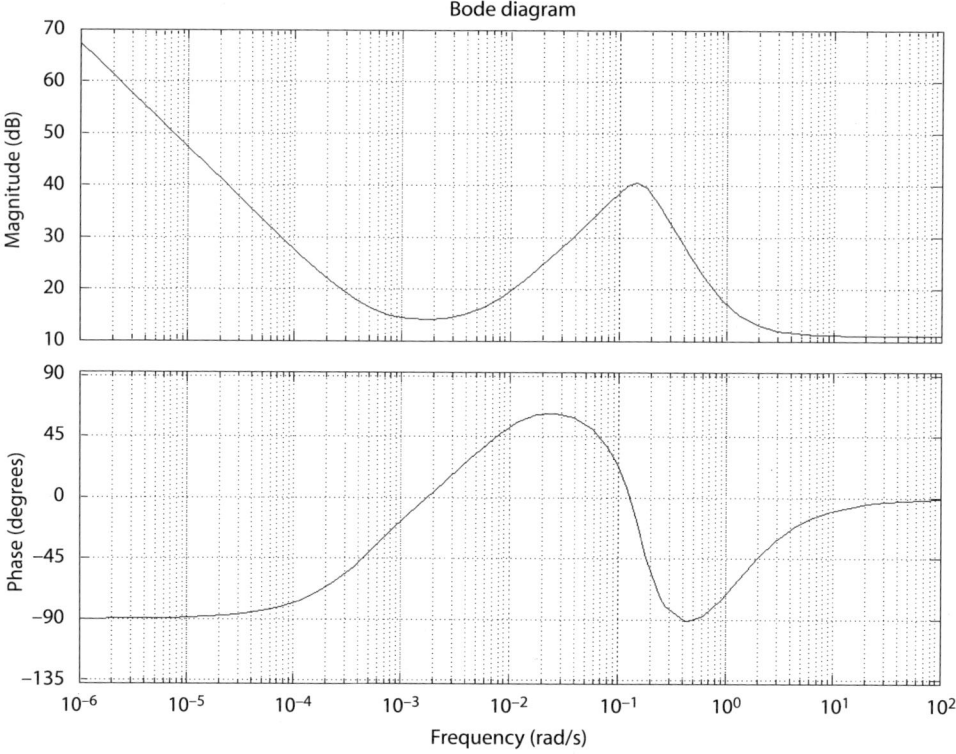

Figure 12.2 Full-order controller Bode plot.

Clearly, the restricted structure controller is simpler than the true LQG controller. However, relative to the full-order design, the step and frequency response results will not be as good for this limited structure controller. The question to be explored is how much worse the results become.

Time Response Results

The unit step position response for the ship is shown in Figure 12.3. The restricted structure LQG controller results are surprisingly good relative to the alternative higher-order design. In fact, the overshoot and rise time are only slight greater, and the settling time is about the same as for the full-order designs.

Frequency Response Results

The controller magnitude response for each of the designs is shown in Figure 12.4. The resulting open loop frequency response is shown in Figure 12.5. In both of these figures, the classical and restricted LQG controller gains are very similar at low frequencies. The closed-loop frequency responses are shown in Figure 12.6. Clearly, there is excellent low-frequency following, and near the roll-off point the differences are not too great. These results clearly indicate that there is little penalty to be paid for use of the restricted structure controller.

12.4 Multivariable Optimal LQG Control: An Introduction

The introduction of multivariable system restricted structure to optimal LQG control involves the interplay of a number of interesting ideas, some of which are not present in the scalar system problem.

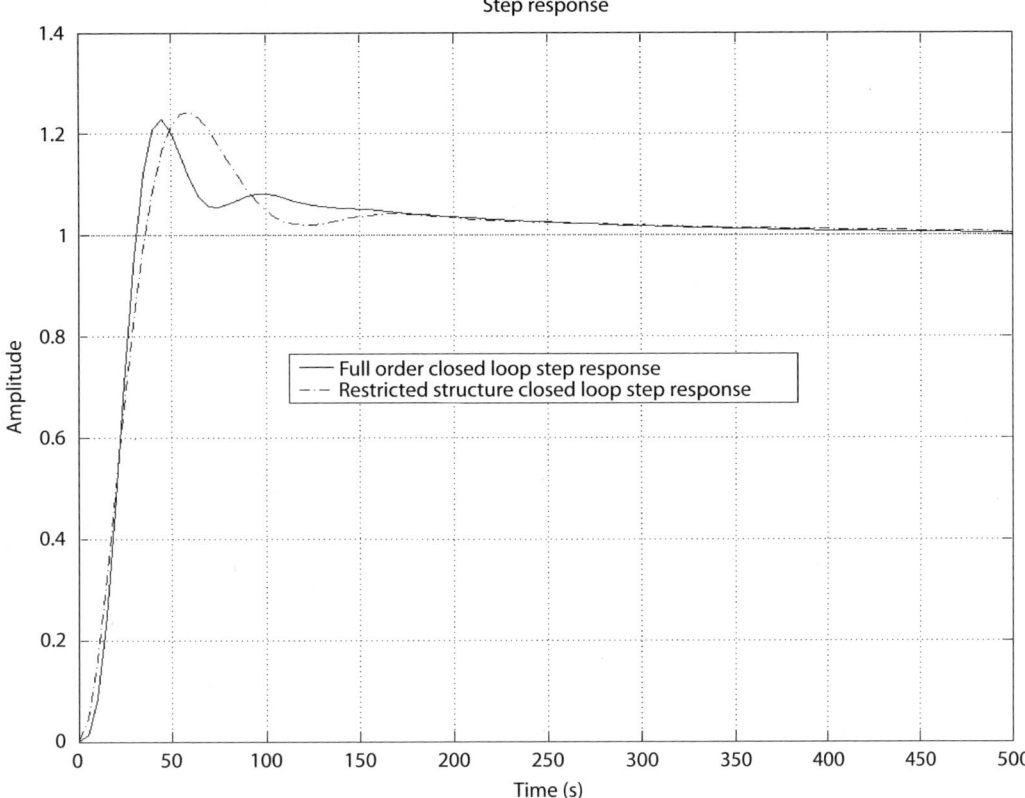

Figure 12.3 Step responses for full optimal and PID controllers

Benchmark Design and Optimal Solution

A key assumption for the use of the restricted structure idea is that the optimal control solution is a desirable controller solution. The properties of the optimal controller then become a benchmark against which to compare different restricted structure controllers. The computation of the restricted structure controller will be based on the full optimal controller solution, and by definition the restricted structure controller will be sub-optimal. Another aspect of the design issue is that optimal benchmark LQG designs should be easily set up. For optimal control, this has been a longstanding development process, but it is now considered that extensive know-how is available (Grimble and Johnson, 1988a,b; Grimble, 1994, 2001a).

Multivariable Control Structures

In creating a multivariable control solution for an industrial system, an important consideration will be the question of which system inputs should be used to control which process outputs. Tools like the Relative Gain Array (Skogestad and Postlethwaite, 1996; Albertos and Sala, 2004) coupled with some physical insight about the process usually settle this question.

With an input–output structure determined for the controller, a typical industrial multivariable controller might be given by

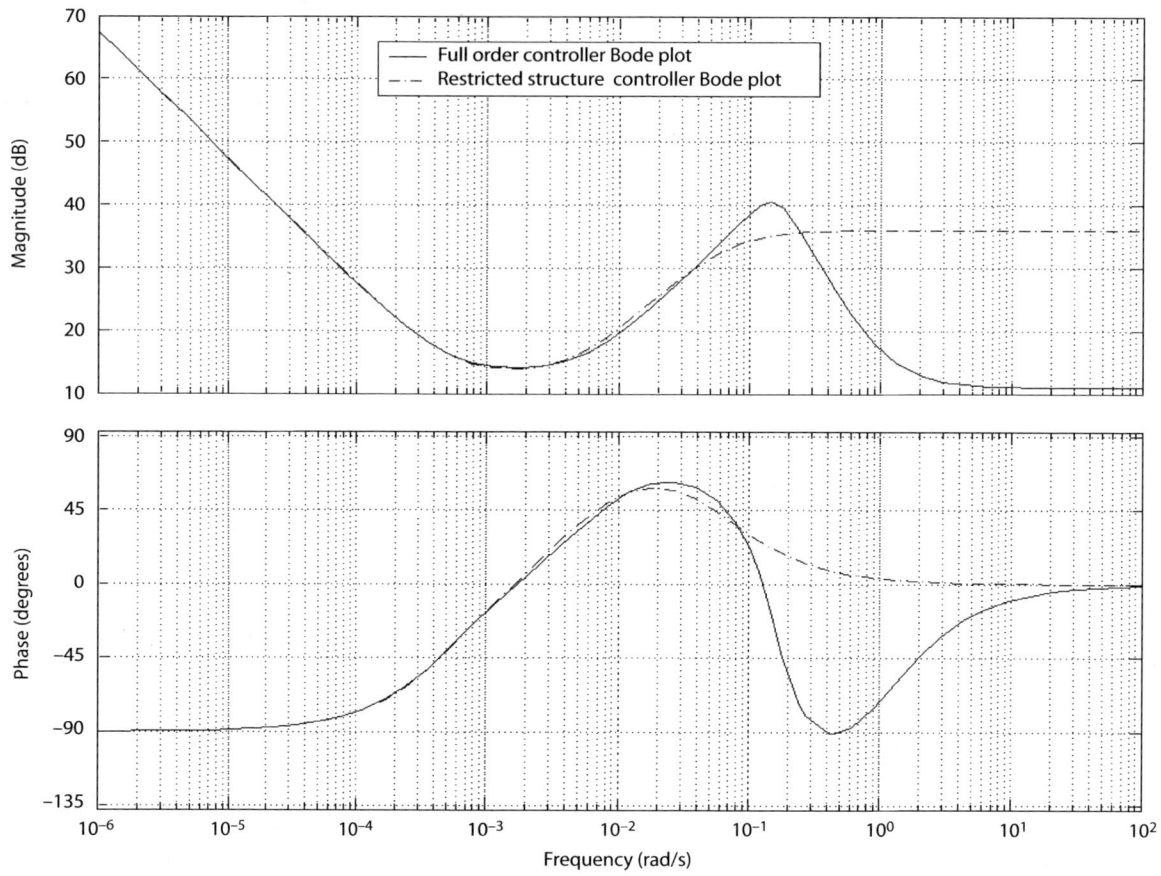

Figure 12.4 Comparison of the controller Bode plots.

$$u(s) = \begin{bmatrix} K_{11}(s) & K_{12}(s) & K_{13}(s) \\ K_{21}(s) & K_{22}(s) & K_{23}(s) \\ K_{31}(s) & K_{32}(s) & K_{33}(s) \end{bmatrix} e(s) = \begin{bmatrix} k_{11}^P + \dfrac{k_{11}^I}{s} & 0 & k_{13}^P + \dfrac{k_{13}^I}{s} \\ k_{21}^P + k_{21}^D s & k_{22}^P & 0 \\ 0 & 0 & k_{33}^P + \dfrac{k_{33}^I}{s} + k_{33}^D s \end{bmatrix} e(s)$$

It is easily seen that there are two more structural issues present in such a controller: a macro- and a micro-structure for the multivariable controller.

Macro-Structure of the Controller

It can be seen that some controller elements $K_{ij}(s)$ are completely zero. If this pattern of zero controllers still yields good control performance, this is a real bonus since the number of controllers which have to be implemented is significantly reduced from the m^2 elements of the full controller matrix. Sometimes there are sound physical system reasons why a controller element can be set to zero. Industrial engineers are often mightily encouraged when a plausible underlying physical system justification is given for a macro-controller structure. The major problem in complex systems is, of course, finding and justifying these latent structures. Restricted structure controller assessment can assist with this problem.

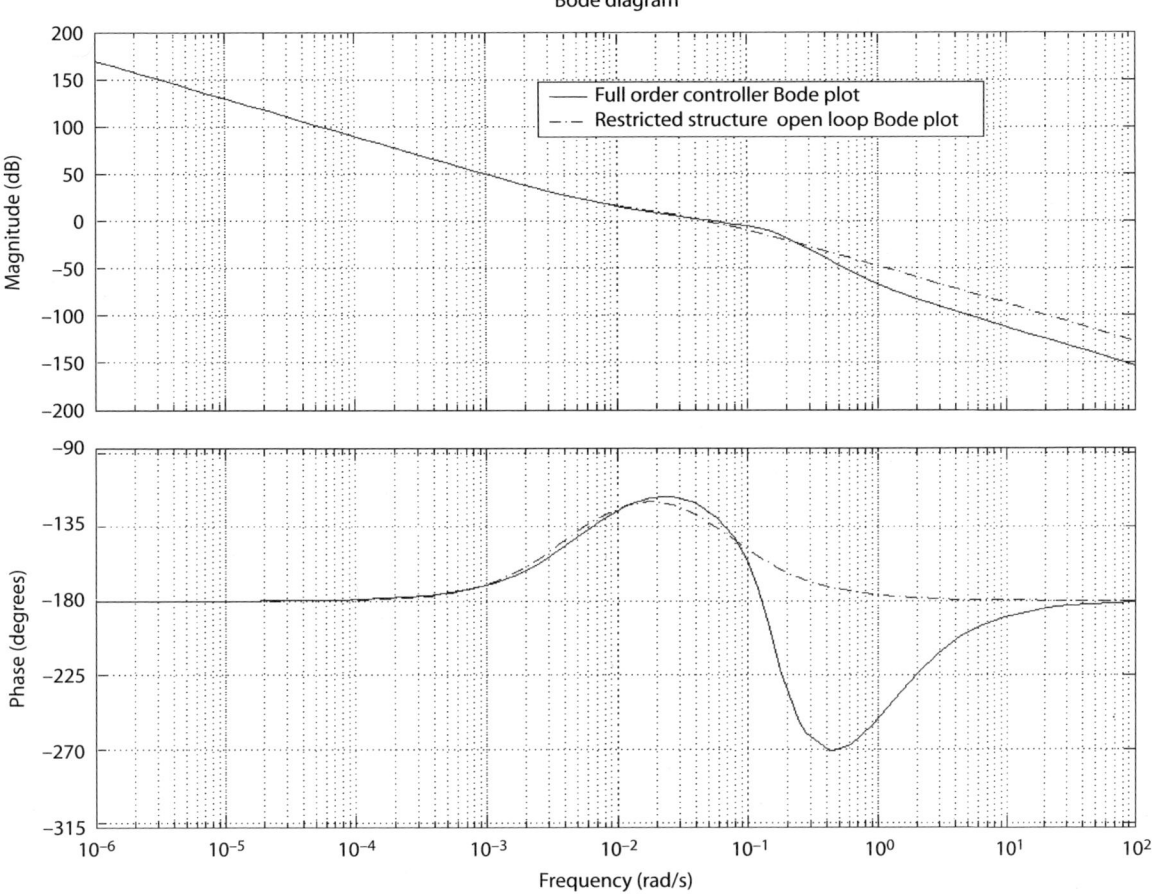

Figure 12.5 Comparison of open-loop forward path Bode plots.

Micro-Structures in the Controller Structure

Looking again at the multivariable controller, some controller elements are P, some are PI, some are PD, and some are PID. To emphasise the micro-controller issue the controller elements are written as:

$$K_{11}(s) = \text{PID}(k_{11}^P, k_{11}^I, 0) \quad K_{12}(s) = \text{PID}(0, 0, 0) \quad K_{13}(s) = \text{PID}(k_{13}^P, k_{13}^I, 0)$$
$$K_{21}(s) = \text{PID}(k_{21}^P, 0, k_{21}^D) \quad K_{22}(s) = \text{PID}(k_{22}^P, 0, 0) \quad K_{23}(s) = \text{PID}(0, 0, 0)$$
$$K_{31}(s) = \text{PID}(0, 0, 0) \quad K_{32}(s) = \text{PID}(0, 0, 0) \quad K_{33}(s) = \text{PID}(k_{33}^P, k_{33}^I, k_{33}^D)$$

The problem of the microstructure within a multivariable PID controller is to determine what terms in the PID controller elements should be retained whilst still yielding satisfactory control performance. For example, questions like should element $K_{ij}(s)$ be a P, PI or PID controller element have to be answered. If the controller parameters are all augmented into a single vector $\theta \in \Re^{n_c}$, then the controller may be denoted as $K(s) = K_{RS}(s, \theta)$. The index n_c is the total number of possible controller parameters given the totality of possible control structures within a given fixed framework. For example, in the above example nine PID controllers are possible, so with potentially three parameters per controller $n_c = 27$ in all. A restricted structure controller assessment procedure based on optimisation can provide a basis for both macro- and micro-structure assessment for multivariable controllers.

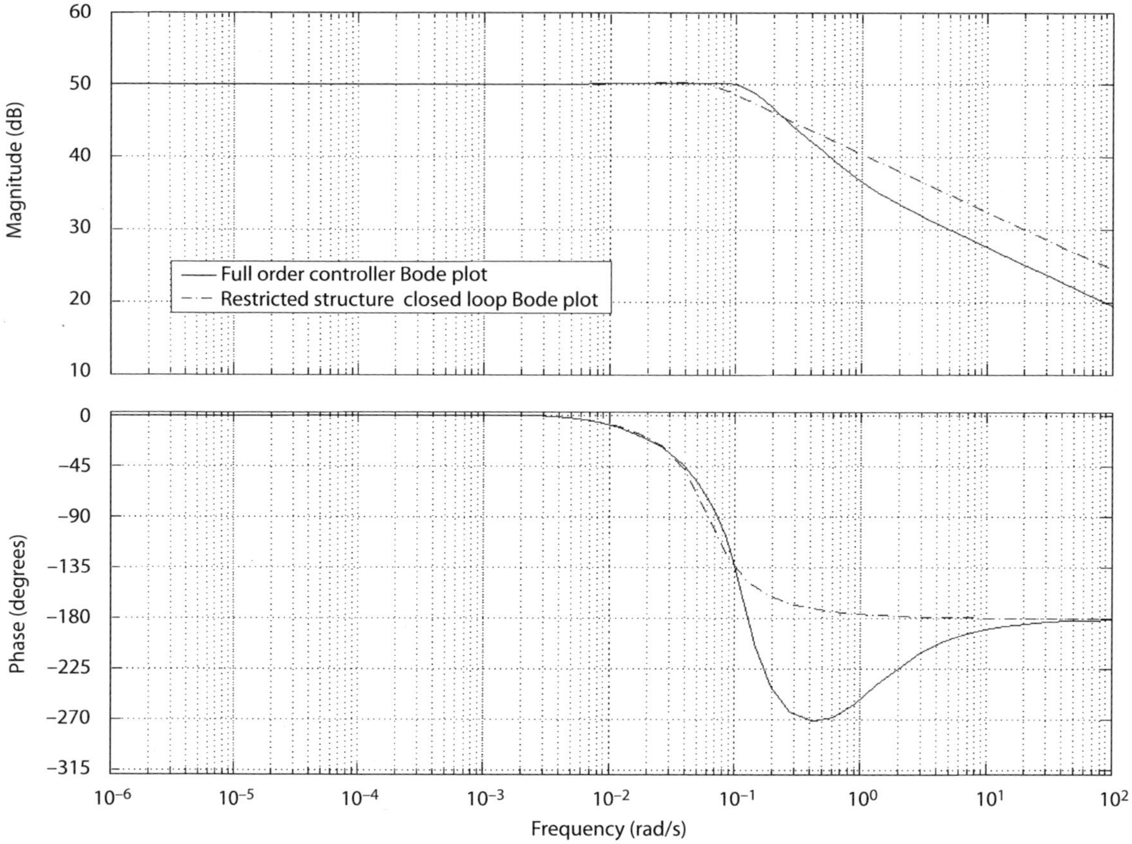

Figure 12.6 Comparison of closed-loop Bode plots.

The first task is to establish the multivariable LQG optimal control results and the associated design scheme on which the restricted structure control formulation will be based.

12.4.1 Multivariable Optimal LQG Control and Cost Function Values

The results to be presented are based on standard multivariable LQG optimal control theory adapted to the control structure of the plant as shown in Figure 12.7.

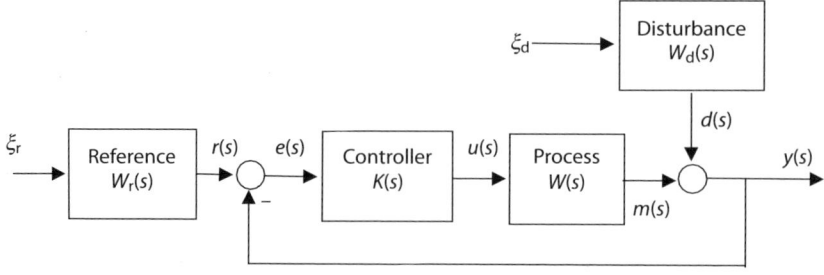

Figure 12.7 Multivariable system model configuration.

12.4 Multivariable Optimal LQG Control: An Introduction

Using Figure 12.7, the multivariable system equations can be defined as:

Process model $\quad y(s) = W(s)u(s) + d(s)$

Disturbance model $\quad d(s) = W_d(s)\xi_d(s)$

Reference model $\quad r(s) = W_r(s)\xi_r(s)$

Controller $\quad u(s) = K(s)e(s)$

Error equation $\quad e(s) = r(s) - y(s)$

where $y(s), e(s), r(s), u(s), d(s) \in \Re^m(s)$, $W(s), W_r(s), K(s) \in \Re^{m \times m}(s)$ and $W_d(s) \in \Re^{m \times d}(s)$. The white noise signals ξ_r and ξ_d are zero-mean with unit covariance matrix, and are mutually statistically independent. The system transfer functions have a common denominator, are coprime and may be written in polynomial matrix form as

$$[W \quad W_d \quad W_r] = A^{-1}[B \quad D \quad E]$$

The LQG cost function J_{LQG} is used in a form appropriate for polynomial system methods and analysis (Grimble and Johnson, 1988b). Thus the steady state stochastic LQG quadratic cost function is given by

$$J_{LQG} = \frac{1}{2\pi j} \oint_D \{trace\{Q_c(s)\Phi_{ee}(s)\} + trace\{R_c(s)\Phi_{uu}(s)\}\} ds$$

where $\Phi_{ee}(s) \in \Re^{m \times m}(s)$ and $\Phi_{uu}(s) \in \Re^{m \times m}(s)$ are real rational matrix spectral density transfer functions for the error $e(s) \in \Re^m(s)$ and the control $u(s) \in \Re^m(s)$. The integration is around the usual D contour. The multivariable dynamic weights are denoted $Q_c(s) \in \Re^{m \times m}$ and $R_c(s) \in \Re^{m \times m}$; these are used by the control engineer to imbue the controller and the closed-loop response with desired features.

The first result to be presented is a decomposition of the LQG cost function which establishes the framework for an LQG optimal control solution and subsequently a restricted structure procedure. To enable this result to be stated, various polynomial system definitions are given.

Cost Function Weighting Matrices
The polynomial matrix representations of the weights are

Error weighting $\quad Q_c(s) = A_q^{-*} B_q^* B_q A_q^{-1} = Q_c^*(s)$

Control weighting $\quad R_c(s) = A_r^{-*} B_r^* B_r A_r^{-1} = R_c^*(s)$

where the adjoint operation is defined through $[X(s)]^* = [X^T(-s)]$ and the polynomial matrices A_q, B_q, A_r, B_r are chosen to make $Q_c \geq 0$ and $R_c \geq 0$ on the D-contour.

The Weighted System Transfer Function
This is useful in algebraic manipulations and is given for the weighted system. It is defined as

$$W_{WT} = A_q^{-1} A^{-1} B A_r = B_1 A_1^{-1}$$

Spectral Factors
The control and filter spectral factors are defined. The control spectral factor $Y_c(s)$ is given as

$$Y_c^* Y_c = W^* Q_c W + R_c = [D_c(A_r A_1)^{-1}]^* [D_c(A_r A_1)^{-1}]$$

with the polynomial matrix control spectral factor $D_c(s)$ being computed as

$$D_c^* D_c = B_1^* B_q^* B_q B_1 + A_1^* B_r^* B_r A_1$$

The total spectrum of the signal $e(s) \in \Re^m(s)$ is denoted by $\Phi_{TT}(s) \in \Re^{m \times m}$ and a generalised spectral factor $Y_f(s) \in \Re^{m \times m}(s)$ is defined from this spectrum, namely

$$\Phi_{TT} = \Phi_{rr} + \Phi_{dd} = Y_f Y_f^* = [A^{-1} D_f][A^{-1} D_f]^*$$

where the polynomial matrix filter spectral factor $D_f(s)$ is

$$D_f D_f^* = EE^* + DD^*$$

Diophantine Equations

The following Diophantine equation solutions are needed in the optimal solution. Solve for triple (H_0, G_0, F_0) with F_0 of smallest row degree

$$D_c^* G_0 + F_0 A_2 = B_1^* B_q^* B_q D_2$$
$$D_c^* H_0 - F_0 B_2 = A_1^* B_r^* B_r D_3$$

with right coprime decompositions

$$A_2 D_2^{-1} = D_f^{-1} A A_q$$
$$B_2 D_3^{-1} = D_f^{-1} B A_r$$

Theorem 12.2: Multivariable LQG cost function decomposition

The LQG cost function J_{LQG} may be decomposed into two terms as

$$J_{LQG} = J_A + J_{BC}$$

where

$$J_A = \frac{1}{2\pi j} \oint_D trace\{X_A^* X_A\} ds$$

$$J_{BC} = \frac{1}{2\pi j} \oint_D trace\{X_B^* X_B + X_C\} ds$$

and

$$X_A(s) = [H_0 D_3^{-1} A_r^{-1} K_n - G_0 D_2^{-1} A_q^{-1} K_d](A K_d + B K_n)^{-1} D_f, \; X_B(s) = -D_C^{-*}(s) F_0(s) \text{ and}$$
$$X_C(s) = [Q_c - Q_c W^*(Y_c^* Y_c)^{-1} W Q_c] \Phi_{TT}$$

Proof

The proof follows the standard polynomial optimal LQG techniques as given by Grimble and Johnson (1988a, b). Greenwood (2003) gives full details of the steps for this particular version of the optimality result.

□

Corollary 12.1: Optimal multivariable LQG control and cost values

The optimal value of the multivariable LQG cost function is given by

$$J_{LQG}^{min} = J_{BC}$$

and the associated optimal controller is

$$K_{opt}(s) = A_r(s) D_3(s) H_0^{-1}(s) G_0(s) D_2^{-1}(s) A_q^{-1}(s)$$

Proof

The cost function $J_{LQG} = J_A + J_{BC}$ has only the term J_A available for optimisation with respect to the controller parameters. Since $J_A \geq 0$, the minimal value obtains when $J_A = 0$ and

$$X_A(s) = [H_0 D_3^{-1} A_r^{-1} K_n K_d^{-1} - G_0 D_2^{-1} A_q^{-1}](A + BK_n K_d^{-1})^{-1} D_f = 0$$

whence the optimal controller

$$K_{opt}(s) = A_r(s) D_3(s) H_0^{-1}(s) G_0(s) D_2^{-1}(s) A_q^{-1}(s)$$

results. In this case $J_{LQG} = 0 + J_{BC} = J_{LQG}^{min}$ follows. ❏

From the results of this theorem and corollary it is possible to capture the computation of the optimal controller as Algorithm 12.2.

Algorithm 12.2: Calculation of the optimal multivariable LQG controller

Step 1 *Define the plant, disturbance and reference models*
   Define and compute $[W \quad W_d \quad W_r] = A^{-1}[B \quad D \quad E]$

Step 2 *Define dynamic weightings for the design*
   Define $A_q(s), B_q(s)$ for error weighting $Q_c(s) = A_q^{-*} B_q^* B_q A_q^{-1}$
   Define $A_r(s), B_r(s)$ for control weighting $R_c(s) = A_r^{-*} B_r^* B_r A_r^{-1}$

Step 3 *Compute matrices and spectral factors*
   Calculate matrices $A_1(s), B_1(s)$
   Calculate $D_c(s)$ from the spectral factorisation of $D_c^* D_c = B_1^* B_q^* B_q B_1 + A_1 B_r^* B_r A_1$
   Calculate $D_f(s)$ from the spectral factorisation of $D_f D_f^* = EE^* + DD^*$
   Calculate matrices D_2, A_2, D_3 and B_2

Step 4 *Compute Diophantine equation solution*
   Simultaneously solve the pair

$$D_c^* G_0 + F_0 A_2 = B_1^* B_q^* B_q D_2$$

$$D_c^* H_0 - F_0 B_2 = A_1^* B_r^* B_r D_3$$

Step 5 *Optimal controller*
   Compute $K_{opt}(s) = A_r(s) D_3(s) H_0^{-1}(s) G_0(s) D_2^{-1}(s) A_q^{-1}(s)$

End Algorithm

With reference to the discussion to come, the important equations from the above results are:

1. For restricted structure procedure the LQG cost function is $J_{LQG} = J_A + J_{LQG}^{min}$, where

$$J_A = \frac{1}{2\pi j} \oint_D trace\{X_A^* X_A\} ds \text{ and } X_A(s) = [H_0 D_3^{-1} A_r^{-1} K_n K_d^{-1} - G_0 D_2^{-1} A_q^{-1}](A + BK_n K_d^{-1})^{-1} D_f$$

2. The optimal LQG multivariable controller can be computed from

$$K_{opt}(s) = A_r(s) D_3(s) H_0^{-1}(s) G_0(s) D_2^{-1}(s) A_q^{-1}(s)$$

12.4.2 Design Procedures for an Optimal LQG Controller

The use of an optimal multivariable LQG controller as a benchmark against which to assess and compute restricted structure controllers requires some straightforward guidelines for the setup of a cost function to achieve desirable controller features. It is the structure of the equation for the optimal controller that indicates how these rules should be established.

Recall the optimal controller equation as

$$K_{opt}(s) = A_r(s)D_3(s)H_0^{-1}(s)G_0(s)D_2^{-1}(s)A_q^{-1}(s)$$

and also the cost function weight matrices chosen in terms of polynomial matrices A_q, B_q, A_r, B_r as equations:

Error weighting $\qquad Q_c(s) = A_q^{-*}B_q^*B_q A_q^{-1}$

Control weighting $\qquad R_c(s) = A_r^{-*}B_r^*B_r A_r^{-1}$

It can be seen that the optimal controller contains A_q as a "denominator" polynomial matrix and A_r as a "numerator" polynomial matrix. This is easily seen in a 2×2 controller, where the polynomial matrices A_q and A_r are taken to be diagonal, for example

$$K_{opt}(s) = \begin{bmatrix} a_{r1}(s) & 0 \\ 0 & a_{r2}(s) \end{bmatrix} D_3 H_0^{-1} G_0 D_2^{-1} \begin{bmatrix} \dfrac{1}{a_{q1}(s)} & 0 \\ 0 & \dfrac{1}{a_{q2}(s)} \end{bmatrix}$$

Thus A_q and A_r can be used to introduce integral and differential action into the optimal controller respectively.

An alternative interpretation can be given by looking at weighted error and control variables as they appear in the cost function, namely

$$\tilde{e}(s) = B_q(s)A_q^{-1}(s)e(s)$$
$$\tilde{u}(s) = B_r(s)A_r^{-1}(s)u(s)$$

Thus by using simple (diagonal) matrix structure and simple low-order polynomials in the elements of A_q, B_q, A_r, B_r the frequency dependence of the weightings can be used to weight different frequency ranges in the error and control signal respectively.

The usual weight function correlation is that speed of response is related to the overall numerical balance between the error and control weights. Hence it is usual to include a scalar weight $\rho^2 > 0$ in the control weighting to implement the speed of response tuning. As is well known, as $\rho^2 \to 0$ the global speed of response increases, corresponding to cheap optimal control (Grimble and Johnson, 1988a).

Frequency-Dependent Error Weight $Q_c(s) = A_q^{-*}B_q^*B_q A_q^{-1}$

Simple frequency-dependent weighting in the error weight can be obtained using diagonal polynomial matrices where the ith relation is given by

$$\tilde{e}_i(s) = \tilde{q}_{ci}(s)e_i(s)$$

with

$$\tilde{q}_{ci}(s) = \frac{b_{qi}(s)}{a_{qi}(s)} = \frac{s + \omega_{qi}}{s} = 1 + \frac{\omega_{qi}}{s}$$

and hence

$$\tilde{e}_i(s) = \tilde{q}_{ci}(s)e_i(s) = \left(1 + \frac{\omega_{qi}}{s}\right)e_i(s)$$

so that ω_{qi} determines the onset of integral action in the weighted error signal $\tilde{e}_i(s)$.

As an example, let $m = 2$; then

$$B_q A_q^{-1} = \begin{bmatrix} \dfrac{s+\omega_{q1}}{s} & 0 \\ 0 & \dfrac{s+\omega_{q2}}{s} \end{bmatrix} = \begin{bmatrix} (s+\omega_{q1}) & 0 \\ 0 & (s+\omega_{q2}) \end{bmatrix} [sI_2]^{-1}$$

Frequency-Dependent Control Weighting $R_c(s) = A_r^{-*} B_r^* B_r A_r^{-1}$

Simple frequency-dependent weighting in the control weight can be obtained using diagonal polynomial matrices where the ith relation is given by

$$\tilde{u}_i(s) = \tilde{r}_{ci}(s) u_i(s)$$

with

$$\tilde{r}_{ci}(s) = \dfrac{b_{ri}(s)}{a_{ri}(s)} = \rho_i \left(1 + \dfrac{s}{\omega_{ri}}\right)$$

and then

$$\tilde{u}_i(s) = \tilde{r}_{ci}(s) u_i(s) = \rho_i \left(1 + \dfrac{s}{\omega_{ri}}\right) u_i(s)$$

Thus the onset of high-frequency weighting in the control signal is determined by ω_{ri}.

As an example, let $m = 2$:

$$B_r A_r^{-1} = \begin{bmatrix} \rho_1\left(1+\dfrac{s}{\omega_{r1}}\right) & 0 \\ 0 & \rho_2\left(1+\dfrac{s}{\omega_{r2}}\right) \end{bmatrix} = \begin{bmatrix} (s+\omega_{r1}) & 0 \\ 0 & (s+\omega_{r2}) \end{bmatrix} \begin{bmatrix} \dfrac{\omega_{r1}}{\rho_1} & 0 \\ 0 & \dfrac{\omega_{r2}}{\rho_2} \end{bmatrix}^{-1}$$

This second form ensures that the polynomials in the weightings remain monic.

Response Speed Weighting ρ_i^2, $i = 1, \ldots, m$

Looking at the ith element of the LQG cost function yields

$$\tilde{e}_i^* \tilde{e}_i + \tilde{u}_i^* \tilde{u}_i = \tilde{q}_{ci}^* \tilde{q}_{ci} e_i^* e_i + \rho_i^2 \tilde{r}_{ci}^* \tilde{r}_{ci} u_i^* u_i$$

Thus the set of parameters ρ_i^2, $i = 1, \ldots, m$ maintain the weighting balance between the weighted error and control terms and the individual parameters ρ_i^2 can be used to affect the speed of response in the ith output variable.

These simple frequency-dependent weighting matrix structures provide the basis for a simple weight selection procedure. For example, Grimble (2000a) suggests the following starting points for the weighing functions based on empirical correlations for SISO systems: $\omega_{qi} = \omega_{gco}/10$, $\omega_{ri} = 10\omega_{gco}$, $i = 1, \ldots, m$, where the unity gain crossover frequency is denoted ω_{gco}. More detailed procedures have been devised by Grimble (1994, 2001b).

12.5 Multivariable Restricted Structure Controller Procedure

In most circumstances, the multivariable LQG optimal control method will produce a controller of a high order. On industrial plants, the multivariable controller structure and the orders of the individual controller elements are fixed *a priori*, either due to practical considerations such as the hardware utilised, or to keep the controller simple. The industrial controller design is therefore of a reduced order in relation to the optimal multivariable solution. Here a new multivariable LQG restricted structure controller procedure is introduced, where the controller is based on the minimisation of the LQG cost function with respect to the fixed reduced control structure.

12.5.1 Analysis for a Multivariable Restricted Structures Algorithm

The analysis opens by assuming that an LQG optimal control problem has been formulated which captures the desired optimal controller design properties. It is then assumed that the Algorithm 12.2 has been used to compute the matrices and the full optimal LQG control solution. The procedure is based on minimising the J_A term of the full cost function with respect to a parameterised restricted structure controller, where the full cost function value is given by $J_{LQG} = J_A + J_{LQG}^{min}$, with

$$J_A = \frac{1}{2\pi j} \oint_D trace\{X_A^* X_A\} ds$$

and

$$X_A(s) = [H_0 D_3^{-1} A_r^{-1} K_n K_d^{-1} - G_0 D_2^{-1} A_q^{-1}](A + BK_n K_d^{-1})^{-1} D_f$$

The J_A term of the full cost function is zero in the case of the optimal solution $K^{opt}(s) = K_n K_d^{-1}$. The analysis for the development of a numerical procedure follows next.

To emphasise the restricted structure nature of the controller, rename the J_A term of the full cost function as J_{RS}, where

$$J_{RS} = \frac{1}{2\pi j} \oint_D trace\{X_A^* X_A\} ds$$

This cost function is evaluated round the D contour, given a frequency-domain form with $s = j\omega$ and defined in terms of the real and imaginary parts of complex matrix $X_A(j\omega)$, namely

$$J_{RS} = \frac{1}{2\pi j} \oint_D trace\{X_A^* X_A\} ds = \frac{1}{2\pi} \int_{-\infty}^{\infty} trace\{X_A^T(-j\omega) X_A(j\omega)\} d\omega$$

and

$$J_{RS} = \frac{1}{2\pi} \int_{-\infty}^{\infty} trace\{X_R^T X_R + X_I^T X_I\} d\omega$$

where $X_A(j\omega) = X_R(\omega) + jX_I(\omega)$.

The assumption that the full optimal LQG solution is known implies that all the following matrices are known and available: $H_0(s), D_3(s), A_r(s), G_0(s), D_2(s), A_q(s), D_f(s)$. Define therefore the following components of the matrix $X_A(s)$:

$$\alpha(s) = H_0 D_3^{-1} A_r^{-1} \text{ and } \beta(s) = G_0 D_2^{-1} A_q^{-1}$$

Then $X_A(s)$ may be written

$$X_A(s) = [\alpha(s) K_n(s) K_d^{-1}(s) - \beta(s)](A + BK_n(s) K_d^{-1}(s))^{-1} D_f$$

Let the controller be identified as

$$K(s) = K_n(s) K_d^{-1}(s)$$

12.5 Multivariable Restricted Structure Controller Procedure

and then

$$X_A(s) = [\alpha(s)K(s) - \beta(s)](A + BK(s))^{-1} D_f$$

The objective of the analysis is an iterative scheme that minimises the cost function J_{RS} with respect to the parameters of the restricted structure controller. Denote the unknown restricted structure controller as $K(s) = K_{RS}(s,\theta)$, where the associated vector of unknown controller parameters is $\theta \in \Re^{n_c}$. Within the iterative scheme to be developed, at the kth iteration denote the vector of controller parameters as $\theta^k \in \Re^{n_c}$ and the associated restricted structure controller as $K(s) = K_{RS}(s,\theta^k)$. Thus, at the kth iteration the term $X_A(s)$ may be written

$$X_A(s) = [\alpha(s)K_{RS}(s,\theta) - \beta(s)](A + BK_{RS}(s,\theta^k))^{-1} D_f$$

where $K_{RS}(s,\theta)$ is considered unknown and $K_{RS}(s,\theta^k)$ is known.

Introduce a new quantity that can be computed from the kth value of the restricted structure controller $K_{RS}(s,\theta^k)$ as

$$\gamma(s) = (A + BK_{RS}(s,\theta^k))^{-1} D_f$$

Thus $X_A(s)$ becomes

$$X_A(s) = [\alpha(s)K_{RS}(s,\theta) - \beta(s)]\gamma(s)$$

It is now possible to consider finding the real and imaginary parts of $X_A(j\omega)$, namely $X_R(\omega)$, $X_I(\omega)$. Set $s = j\omega$ and omitting to denote the frequency dependency for brevity, define real and imaginary parts of α, β, γ, K as:

$$\alpha(j\omega) = \alpha_R + j\alpha_I$$
$$\beta(j\omega) = \beta_R + j\beta_I$$
$$\gamma(j\omega) = \gamma_R + j\gamma_I$$
$$K(j\omega,\theta) = K_R + jK_I$$

These are substituted into the complex form of term $X_A(j\omega)$:

$$X_A(j\omega) = [\alpha(j\omega)K_{RS}(j\omega,\theta) - \beta(j\omega)]\gamma(j\omega)$$

so that

$$X_A(j\omega) = [(\alpha_R + j\alpha_I)(K_R + jK_I) - (\beta_R + j\beta_I)](\gamma_R + j\gamma_I)$$

and then identify with

$$X_A(j\omega) = X_R(\omega) + jX_I(\omega)$$

Thus, multiplying out and collecting the real and imaginary terms gives:

$$X_R = \alpha_R K_R \gamma_R - \alpha_I K_R \gamma_I - \alpha_I K_I \gamma_R - \alpha_R K_I \gamma_I - \beta_R \gamma_R + \beta_I \gamma_I$$
$$X_I = \alpha_I K_R \gamma_R + \alpha_R K_R \gamma_I + \alpha_R K_I \gamma_R - \alpha_I K_I \gamma_I - \beta_I \gamma_R - \beta_R \gamma_I$$

Parameterisation of the Restricted Structure Controller

The step to simplify X_R, X_I is obtained by expanding the terms K_R, K_I within the controller expression $K(j\omega,\theta) = K_R + jK_I$. This step is demonstrated by a simple example. Consider the case of a 2× 2 PID controller given by

$$K(s,\theta) = \begin{bmatrix} k_{11}^P + \dfrac{k_{11}^I}{s} + k_{11}^D s & k_{12}^P + \dfrac{k_{12}^I}{s} + k_{12}^D s \\ k_{21}^P + \dfrac{k_{21}^I}{s} + k_{21}^D s & k_{22}^P + \dfrac{k_{22}^I}{s} + k_{22}^D s \end{bmatrix}$$

where $\theta = [k_{11}^P \ k_{11}^I \ k_{11}^D \ \ldots \ k_{22}^P \ k_{22}^I \ k_{22}^D]^T \in \Re^{12}$.

Evaluate the controller for $s = j\omega$:

$$K(j\omega,\theta) = \begin{bmatrix} k_{11}^P - j\dfrac{k_{11}^I}{\omega} + jk_{11}^D \omega & k_{12}^P - j\dfrac{k_{12}^I}{\omega} + jk_{12}^D \omega \\ k_{21}^P - j\dfrac{k_{21}^I}{\omega} + jk_{21}^D \omega & k_{22}^P - j\dfrac{k_{22}^I}{\omega} + jk_{22}^D \omega \end{bmatrix}$$

and collect the real and imaginary parts:

$$K(j\omega,\theta) = \begin{bmatrix} k_{11}^P & k_{12}^P \\ k_{21}^P & k_{22}^P \end{bmatrix} + j\begin{bmatrix} -\dfrac{k_{11}^I}{\omega} + jk_{11}^D \omega & -\dfrac{k_{12}^I}{\omega} + jk_{12}^D \omega \\ -\dfrac{k_{21}^I}{\omega} + jk_{21}^D \omega & -\dfrac{k_{22}^I}{\omega} + jk_{22}^D \omega \end{bmatrix}$$

Then identify with

$$K(j\omega,\theta) = K_R + jK_I$$

Thus extract:

$$K_R = E_1 k_{11}^P + E_2 k_{12}^P + E_3 k_{21}^P + E_4 k_{22}^P$$
$$K_I = E_5 k_{11}^I + E_6 k_{11}^D + \ldots + E_{11} k_{22}^I + E_{12} k_{22}^D$$

where $E_1, E_2, \ldots, E_{12} \in \Re^{2\times 2}$ are simple matrices delineating the participation of the individual controller parameters in the real and imaginary parts K_R, K_I of $K(j\omega,\theta)$. Thus the controller expression is decomposed into a linear sum of matrices each associated with only one scalar parameter in the restricted structure controller. The actual pattern and number of component matrices will depend on the complexity of the restricted structure controller. This decomposition is written generically as

$$K_R = \sum_{i=1}^{n_c} E_{Ri}\theta_i \ \text{ and } \ K_I = \sum_{i=1}^{n_c} E_{Ii}\theta_i$$

where $E_{Ri}, E_{Ii} \in \Re^{m\times m}$, $i = 1, \ldots, n_c$, $\theta \in \Re^{n_c}$ and n_c is the number of restricted controller parameters.

Linear Forms for the Terms X_R, X_I

It is now possible to put together the two sets of equations for X_R, X_I and K_R, K_I, namely

$$X_R = \alpha_R K_R \gamma_R - \alpha_I K_R \gamma_I - \alpha_I K_I \gamma_R - \alpha_R K_I \gamma_I - \beta_R \gamma_R + \beta_I \gamma_I$$
$$X_I = \alpha_I K_R \gamma_R + \alpha_R K_R \gamma_I + \alpha_R K_I \gamma_R - \alpha_I K_I \gamma_I - \beta_I \gamma_R - \beta_R \gamma_I$$

with

$$K_R = \sum_{i=1}^{n_c} E_{Ri}\theta_i \ \text{ and } \ K_I = \sum_{i=1}^{n_c} E_{Ii}\theta_i$$

to obtain by direct substitution

12.5 Multivariable Restricted Structure Controller Procedure

$$X_R = \sum_{i=1}^{n_c} \delta_i \theta_i + X_{R0} \in \Re^{m \times m} \text{ and } X_I = \sum_{i=1}^{n_c} \xi_i \theta_i + X_{I0} \in \Re^{m \times m}$$

where

$$\delta_i = \alpha_R E_{Ri} \gamma_R - \alpha_I E_{Ri} \gamma_I - \alpha_I E_{Ii} \gamma_R - \alpha_R E_{Ii} \gamma_I \text{ and } X_{R0} = -\beta_R \gamma_R + \beta_I \gamma_I$$

$$\xi_i = \alpha_I E_{Ri} \gamma_R + \alpha_R E_{Ri} \gamma_I + \alpha_R E_{Ii} \gamma_R - \alpha_I E_{Ii} \gamma_I \text{ and } X_{I0} = -(\beta_I \gamma_R + \beta_R \gamma_I), \; i=1,\ldots,n_c$$

Evaluating the Restricted Structure Cost Function J_{RS}

Recall the restricted structure cost function as

$$J_{RS} = \frac{1}{\pi} \int_0^\infty trace\{X_R^T X_R + X_I^T X_I\} d\omega$$

for which

$$X_R = \sum_{i=1}^{n_c} \delta_i \theta_i + X_{R0} \in \Re^{m \times m} \text{ and } X_I = \sum_{i=1}^{n_c} \xi_i \theta_i + X_{I0} \in \Re^{m \times m}$$

Firstly, evaluating the kernel within the trace expression yields

$$X_R^T X_R + X_I^T X_I = \sum_{i=1}^{n_c} \sum_{j=1}^{n_c} (\delta_i^T \delta_j + \xi_i^T \xi_j) \theta_i \theta_j$$
$$+ \sum_{i=1}^{n_c} (\delta_i^T X_{R0} + X_{R0}^T \delta_i + \xi_i^T X_{R0} + X_{R0}^T \xi_i) \theta_i + X_{R0}^T X_{R0} + X_{I0}^T X_{I0}$$

Applying the trace operation and using standard trace results gives

$$trace\{X_R^T X_R + X_I^T X_I\} = \theta^T \widetilde{P} \theta + 2\widetilde{q}^T \theta + \widetilde{c}$$

where:

$$\widetilde{p}_{ij} = trace\{\delta_i^T \delta_j + \xi_i^T \xi_j\}$$
$$2\widetilde{q}_i = trace\{\delta_i^T X_{R0} + X_{R0}^T \delta_i + \xi_i^T X_{I0} + X_{I0}^T \xi_i\}$$
$$\widetilde{c} = trace\{X_{R0}^T X_{R0} + X_{I0}^T X_{I0}\}$$

and $i=1,\ldots,n_c; \; j=1,\ldots,n_c$.

Note that $\widetilde{p}_{ij} = trace\{\delta_i^T \delta_j + \xi_i^T \xi_j\} = trace\{\delta_j^T \delta_i + \xi_j^T \xi_i\} = \widetilde{p}_{ji}$. Thus in finding the elements of $\widetilde{P}$ it is sufficient to find $\widetilde{p}_{ij}$ for $i=1,\ldots,n_c$ and $j=i, i+1, \ldots, n_c$.

The above theory is exact and frequency-dependent, and has been defined using the kth estimate of $\theta \in \Re^{n_c}$ so that

$$J_{RS} = \frac{1}{2\pi} \int_{-\infty}^{\infty} trace\{X_R^T(\omega) X_R(\omega) + X_I^T(\omega) X_I(\omega)\} d\omega$$
$$= \frac{1}{2\pi} \int_{-\infty}^{\infty} (\theta^T \widetilde{P}(\omega) \theta + 2\widetilde{q}^T(\omega) \theta + \widetilde{c}(\omega)) d\omega$$

and

$$J_{RS} = \theta^T P \theta + 2q^T \theta + c = J_{RS}(\theta | \theta^k)$$

where

$$P = \frac{1}{2\pi} \int_{-\infty}^{\infty} \tilde{P}(\omega) d\omega, \quad q = \frac{1}{2\pi} \int_{-\infty}^{\infty} \tilde{q}(\omega) d\omega \text{ and } c = \frac{1}{2\pi} \int_{-\infty}^{\infty} \tilde{c}(\omega) d\omega$$

As can be seen, the frequency dependence is integrated out and the resulting restricted structure cost function is a quadratic form in the controller parameter vector $\theta \in \Re^{n_c}$ which can be minimised using the standard result so that the unconstrained minimum is $\theta = -P^{-1}q$. This new value for $\theta \in \Re^{n_c}$ has been calculated by embedding $\theta^k \in \Re^{n_c}$ into the calculation, so in terms of an iterative routine this can be expressed as $\theta^{k+1} = -P(\theta^k)^{-1} q(\theta^k)$ and the whole procedure can be repeated until satisfactory convergence is achieved.

Remarks 12.3
1. In effecting the optimisation, different approaches can be taken to evaluating the integrals within the finite dimensional cost function, namely calculating

$$P = \frac{1}{2\pi} \int_{-\infty}^{\infty} \tilde{P}(\omega) d\omega, \quad q = \frac{1}{2\pi} \int_{-\infty}^{\infty} \tilde{q}(\omega) d\omega \text{ and } c = \frac{1}{2\pi} \int_{-\infty}^{\infty} \tilde{c}(\omega) d\omega$$

These integrals can be computed over a specific finite frequency range or over the full frequency range.

2. Majecki (2002) has noted that the above derivation can be given using the Kronecker or tensor product with the advantage of enhanced brevity of presentation.

12.5.2 Multivariable Restricted Structure Algorithm and Nested Restricted Structure Controllers

An algorithm for the design and computation of an LQG restricted structure controller based on the above analysis follows next.

Algorithm 12.3: Multivariable restricted structure LQG control

Step 1 Set up and solve the LQG problem
Record $H_0(s), D_3(s), A_r(s), G_0(s), D_2(s), A_q(s), D_f(s)$
Compute $\alpha(s) = H_0 D_3^{-1} A_r^{-1}, \beta(s) = G_0 D_2^{-1} A_q^{-1}$

Step 2 Define restricted controller structure
Define the $\theta \in \Re^{n_c}$ vector of restricted structure parameters
Define matrices $E_{Ri}(\omega), E_{Ii}(\omega)$ for $i = 1, \ldots, n_c$

Step 3 Initialisation
Set counter $k = 0$; define $K(s, \theta^0)$
Determine frequency interval and step-size $0 \leq \omega_0 < \ldots < \omega_N$
Frequency index is $i_1 = 0(1)N$

Step 4 Loop step
For $i_1 = 0(1)N$
Step 4.1 Using Step 1 data, compute:
$\alpha_R(\omega_{i_1})$ and $\alpha_I(\omega_{i_1})$ from $\alpha(j\omega_{i_1})$
$\beta_R(\omega_{i_1})$ and $\beta_I(\omega_{i_1})$ from $\beta(j\omega_{i_1})$
$\gamma(j\omega_{i_1}) = [A(j\omega_{i_1}) + B(j\omega_{i_1})K(j\omega_{i_1}, \theta^k)]^{-1} D_f(j\omega_{i_1})$
$\gamma_R(\omega_{i_1})$ and $\gamma_I(\omega_{i_1})$ from $\gamma(j\omega_{i_1})$

12.5 Multivariable Restricted Structure Controller Procedure

$E_{R_i}(\omega_{i_1})$ and $E_{I_i}(\omega_{i_1})$, $i = 1, 2, \ldots, n_c$

Step 4.2 Compute for $i = 1, 2, \ldots, n_c$

$$\delta_i = \alpha_R E_{Ri}\gamma_R - \alpha_I E_{Ri}\gamma_I - \alpha_I E_{Ii}\gamma_R - \alpha_R E_{Ii}\gamma_I \quad X_{R0} = -\beta_R \gamma_R + \beta_I \gamma_I$$
$$\xi_i = \alpha_I E_{Ri}\gamma_R + \alpha_R E_{Ri}\gamma_I + \alpha_R E_{Ii}\gamma_R - \alpha_I E_{Ii}\gamma_I \quad X_{I0} = -(\beta_I \gamma_R + \beta_R \gamma_I)$$

Step 4.3 Compute for $i = 1, 2, \ldots, n_c$, $j = i, i+1, \ldots, n_c$

$$\tilde{p}_{ij} = trace\{\delta_i^T \delta_j + \xi_i^T \xi_j\}$$
$$\tilde{q}_i = trace\{X_{R0}^T \delta_i + X_{I0}^T \xi_i\}$$

Step 5 Integration and optimisation

Use matrix $\tilde{P} = [\tilde{p}_{ij}(\omega_{i_1})]$, $i_1 = 1, \ldots, N$ and vector $\tilde{q} = [\tilde{q}_i(\omega_{i_1})]$, $i_1 = 1, \ldots, N$

$$\text{Compute } P = \frac{1}{2\pi} \int_{-\infty}^{\infty} \tilde{P}(\omega) d\omega, \quad q = \frac{1}{2\pi} \int_{-\infty}^{\infty} \tilde{q}(\omega) d\omega$$

Compute $\theta^{k+1} = -P^{-1} q$

Check convergence – if $\|\theta^{k+1} - \theta^k\| \leq tol$ STOP with $\theta^{opt} = \theta^{k+1}$
Otherwise $k := k + 1$ goto Step 4

Algorithm end

The Restricted Structure Controller Parameterised – a Decoupled Controller Example

A key step of the above procedure occurs when the restricted structure controller is decomposed and written in terms of the simple matrices $E_{Ri}, E_{Ii} \in \Re^{m \times m}$, $i = 1, \ldots, n_c$ and parameters $\theta \in \Re^{n_c}$, where n_c is the number of restricted controller parameters. It is useful to examine how this step is performed for a decoupled 2×2 PI controller. A typical example of such a controller is given by

$$K(s, \theta) = \begin{bmatrix} k_{P1} + \dfrac{k_{I1}}{s} & 0 \\ 0 & k_{P2} + \dfrac{k_{I2}}{s} \end{bmatrix}$$

where $n_c = 4$ and $\theta = [\theta_1 \ \theta_2 \ \theta_3 \ \theta_4]^T = [k_{P1} \ k_{I1} \ k_{P2} \ k_{I2}]^T \in \Re^4$.

Evaluate the controller for $s = j\omega$ and collect into the real and imaginary parts. Then

$$K(j\omega, \theta) = \begin{bmatrix} k_{P1} - j\dfrac{k_{I1}}{\omega} & 0 \\ 0 & k_{P2} - j\dfrac{k_{I2}}{\omega} \end{bmatrix}$$

gives

$$K(j\omega, \theta) = \begin{bmatrix} k_{P1} & 0 \\ 0 & k_{P2} \end{bmatrix} + j \begin{bmatrix} -\dfrac{k_{I1}}{\omega} & 0 \\ 0 & -\dfrac{k_{I2}}{\omega} \end{bmatrix}$$

which is then compared with the identity

$$K(j\omega, \theta) = K_R + jK_I$$

Forming the linear sum of matrices for the real part of the decoupled PI controller

$$K_R = \begin{bmatrix} 1 & 0 \\ 0 & 0 \end{bmatrix} k_{P1} + \begin{bmatrix} 0 & 0 \\ 0 & 0 \end{bmatrix} k_{I1} + \begin{bmatrix} 0 & 0 \\ 0 & 1 \end{bmatrix} k_{P2} + \begin{bmatrix} 0 & 0 \\ 0 & 0 \end{bmatrix} k_{I2}$$

or generically

$$K_R = E_{R1}\theta_1 + E_{R2}\theta_2 + E_{R3}\theta_3 + E_{R4}\theta_4$$

For the imaginary part:

$$K_I = \begin{bmatrix} 0 & 0 \\ 0 & 0 \end{bmatrix} k_{P1} + \begin{bmatrix} -1/\omega & 0 \\ 0 & 0 \end{bmatrix} k_{I1} + \begin{bmatrix} 0 & 0 \\ 0 & 0 \end{bmatrix} k_{P2} + \begin{bmatrix} 0 & 0 \\ 0 & -1/\omega \end{bmatrix} k_{I2}$$

or generically

$$K_I = E_{I1}\theta_1 + E_{I2}\theta_2 + E_{I3}\theta_3 + E_{I4}\theta_4$$

Thus the simple matrices $E_{Ri}, E_{Ii} \in \Re^{2\times 2}, i=1,\ldots,4$ are found.

Nested Restricted Structure Controllers

The restricted structure controller derives from the following optimisation problem:

$$\min_{\theta \in \Re^{n_c}} \{J_{LQG}(\theta)\}$$

where $J_{LQG}(\theta) = J_{RS}(\theta) + J_{LQG}^{min}$ and assuming a solution which stabilises the closed loop exists. The restricted structure of the controller is characterised by the vector $\theta \in \Re^{n_c}$, where n_c is taken to be the totality of the number of controller parameters. Within this vector, reduced structures can be introduced into the controller by adjoining a zero controller parameter constraint. Consider the example where four restricted structure controllers might be specified though the following notation:

$$S_f = \{\theta \in \Re^{n_c}\}, \; S_1 = \{\theta \in \Re^{n_c}; \theta_1 = 0\}$$

$$S_2 = \{\theta \in \Re^{n_c}; \theta_2 = 0\} \; \text{and} \; S_{12} = \{\theta \in \Re^{n_c}; \theta_1 = 0; \theta_2 = 0\}$$

Note that the structures satisfy

$$S_{12} \subseteq S_1 \subseteq S_f \; \text{and} \; S_{12} \subseteq S_2 \subseteq S_f$$

Then assuming minima exist for which the optimising $\theta \in \Re^{n_c}$ are closed-loop stable, the condition of optimality will give

$$J_{RS}^{min}(\theta(S_f)) \leq J_{RS}^{min}(\theta(S_1)) \leq J_{RS}^{min}(\theta(S_{12}))$$

and

$$J_{RS}^{min}(\theta(S_f)) \leq J_{RS}^{min}(\theta(S_2)) \leq J_{RS}^{min}(\theta(S_{12}))$$

This nesting property of the optimal cost can be used to assess different levels of restricted structure controllers. Retain the structural notation and define the relationship between LQG and RS optimality as

$$0 \leq J_{LQG}^{min} \leq J_{LQG}^{min}(\theta(S))$$

where S denotes a particular structure and

$$J_{LQG}^{min}(\theta(S)) = J_{RS}^{min}(\theta(S)) + J_{LQG}^{min}$$

Hence introduce a performance assessment index:

$$PAI(S) = \frac{J_{LQG}^{\min}}{J_{RS}^{\min}(\theta(S)) + J_{LQG}^{\min}} \quad \text{where} \quad 0 \leq PAI(S) \leq 1$$

In the case of the set of structures

$$0 \leq \frac{J_{LQG}^{\min}}{J_{RS}^{\min}(\theta(S_{12})) + J_{LQG}^{\min}} \leq \frac{J_{LQG}^{\min}}{J_{RS}^{\min}(\theta(S_1)) + J_{LQG}^{\min}} \leq \frac{J_{LQG}^{\min}}{J_{RS}^{\min}(\theta(S_f)) + J_{LQG}^{\min}} \leq 1$$

yields

$$0 \leq PAI(S_{12}) \leq PAI(S_1) \leq PAI(S_f) \leq 1$$

In the above relations, the optimal cost quantities are normalised so that the performance assessment index whose value is closer to unity indicates that the restricted controller has near optimal performance. Such measures have the percentage form

$$PAI(S)\% = \frac{J_{LQG}^{\min}}{J_{RS}^{\min}(\theta(S)) + J_{LQG}^{\min}} \times 100\% \quad \text{where} \quad 0 \leq PAI(S)\% \leq 100\%$$

These performance indices can be computed using Algorithm 12.3 for each structure in the combinatorial set. Thus the algorithm can also be used to assess families of different restricted structure controllers, as will be shown in the next section on a steel industry application.

12.6 An Application of Multivariable Restricted Structure Assessment – Control of the Hotstrip Finishing Mill Looper System

The hotstrip mill is the first stage in the manufacture of steel strip. The steel strip produced is typically used in consumer white goods and car bodies. The whole hotstrip mill process takes place with the steel strip at temperatures of 1000 °C to 350 °C. The hotstrip mill comprises a sequence of strip processing units. These begin with the casting shop and the reheating furnaces and then move on to the preliminary rolling in the roughing mills. The major dimensional transformation to steel strip takes place in the hotstrip finishing mill. After the finishing mill, strip is cooled to about 350 °C on run-out tables and then coiled ready for transportation to the cold mill for further processing. The multivariable control problem for the application of a controller performance assessment is the control of the interstand looper in the finishing mill.

12.6.1 The Hotstrip Finishing Mill Looper System

The hotstrip finishing mill is a tandem mill of six or seven stands. Each rolling stand is usually a four-high rolling stand. The outer two rolls are large diameter backup rolls that provide stiffness for the two inner work rolls. The smaller diameter work rolls are in contact with the hot steel strip. The purpose of the roll stands is to effect a gauge (thickness) reduction in the strip as it travels through the mill. Between each pair of stands is a looper arm as shown in Figure 12.8.

During the rolling operation, the looper arm is raised or lowered to control the tension in the steel strip. The steady operational looper arm angle is about 15°, but this angle varies in reaction to tension disturbances and weight changes. Control of the looper angle and strip tension is a multivariable problem with considerable process interaction.

Looper System Model

Johnson *et al.* (1999) gave a full description of the modelling of the finishing mill inter-stand looper system. In that work, a small signal linear looper system model was given in transfer function form:

462 Restricted Structure Optimal Control

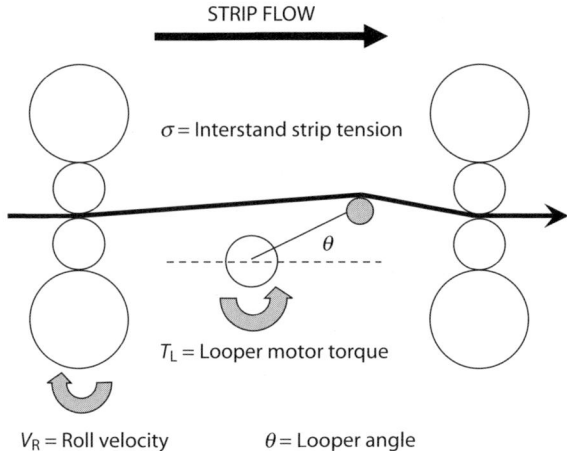

Figure 12.8 Hotstrip finishing mill looper system.

$$\begin{bmatrix} \Delta\sigma \\ \Delta\theta \end{bmatrix} = \begin{bmatrix} G_{11}(s) & G_{12}(s) \\ G_{21}(s) & G_{22}(s) \end{bmatrix} \begin{bmatrix} \Delta V_R \\ \Delta T_L \end{bmatrix}$$

where the process variables are defined as: interstand strip tension σ, looper angle θ, roll velocity V_R and loop motor torque T_L.

A decentralised looper control structure and system block diagram is shown in Figure 12.9.

A decentralised control structure (Figure 12.9) which used two controllers $K_1(s)$ and $K_2(s)$ was investigated by Johnson *et al.* (1999). The designs in that study finally settled on PI and PID forms for the controllers $K_1(s)$ and $K_2(s)$ respectively and the tuning used standard Ziegler–Nichols methods. However, the controller assessment theory developed in this chapter now provides the tools for

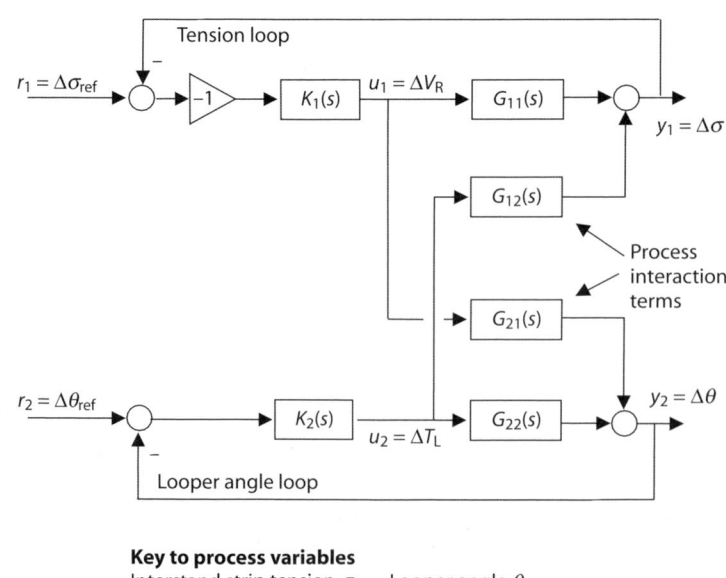

Figure 12.9 Decentralised looper control structure and system block diagram.

12.6 An Application of Multivariable Restricted Structure Assessment

comparing the performance of different classical controllers of increasing complexity. This method uses an optimal control solution as a benchmark against which to assess the performance of different control structures based on different selections of classical PID controllers.

The Looper Model – Numerical Issues
The physical variables under control in the looper system are strip tension and looper arm angle. The tension is measured in Nm^{-2} with numerical values of the order 10^5, while angle is measured in degrees with a step *change* being typically $\pm 1.5°$. The zero frequency gains of the full-order transfer function models reported by Johnson et al. (1999) lie in a range from 3.1×10^{-3} to 10^8. This range of scaling was found to cause difficulties for the numerical polynomial procedures for spectral factorisation and simultaneous Diophantine equation solution. Spurious effects and incorrect results occurred so that it was necessary to construct a demonstration system based on the observed physical behaviour of the full system but without the extreme numerical range of the full transfer function model. The results presented here use this demonstration looper system represented by the model $G_{LS}(s)$ and given as

$$G_{LS}(s) = \begin{bmatrix} G_{11}(s) & G_{12}(s) \\ G_{21}(s) & G_{22}(s) \end{bmatrix} = \begin{bmatrix} \dfrac{-120}{s+1.2} & \dfrac{100s+9}{s^2+9s+9} \\ \dfrac{24}{s+1.2} & \dfrac{0.9s+9}{s^2+9s+9} \end{bmatrix}$$

12.6.2 An Optimal Multivariable LQG Controller for the Looper System

Completing the LQG System Description
To complete the LQG system description, the basic system model $G_{LS}(s)$ was augmented with additional models.

(a) Reference model
To reflect the process requirements for performance with step reference changes, the reference model was selected as

$$W_R(s) = diag\left\{\dfrac{0.05}{s}, \dfrac{0.05}{s}\right\}$$

(b) Disturbance model
Both the tension and angle variables are subject to low-frequency disturbances, and the composite disturbance model was selected as

$$W_D(s) = diag\left\{\dfrac{1}{10s+1}, \dfrac{0.1}{10s+1}\right\}$$

LQG Cost Function Weight Selection
The main requirement in the controller was for integral action and a balance between the cost weights to give a good speed of response. Consequently, the following weight function polynomial were selected:

$B_q(s) = diag\{s+9, s+11\}$
$A_q(s) = diag\{s, s\}$
$B_r(s) = diag\{s+40, s+10\}$
$A_r(s) = diag\{40, 10\}$

The above weighting function polynomials are based on the simple design rules given in Section 12.4.2 and implement the following parameter values:

$$\omega_{r1} = 40 \quad \omega_{q1} = 9 \quad \rho_1 = 0.95$$
$$\omega_{r2} = 30 \quad \omega_{q2} = 11 \quad \rho_2 = 0.05$$

Algorithm 12.2 was used to compute the full LQG optimal controller for the modified looper system, which was calculated as

$$K_{opt}(s) = \begin{bmatrix} K_{opt}^{(1,1)} & K_{opt}^{(1,2)} \\ K_{opt}^{(2,1)} & K_{opt}^{(2,2)} \end{bmatrix}$$

where

$$K_{opt}^{(1,1)} = \frac{-0.02649s^5 - 1.598s^4 - 35.20s^3 - 244.40s^2 - 302.80s - 69.07}{0.1s^6 + 10.31s^5 + 368.1s^4 + 5030s^3 + 3015s^2 + 246.5s}$$

$$K_{opt}^{(1,2)} = \frac{1.351s^5 + 80.29s^4 + 1727s^3 + 6891s^2 + 8102s + 2620}{0.1s^6 + 10.31s^5 + 368.1s^4 + 5030s^3 + 3015s^2 + 246.5s}$$

$$K_{opt}^{(2,1)} = \frac{5.175s^4 + 293.3s^3 + 2283s^2 + 2365s + 143.8}{0.1s^6 + 10.31s^5 + 368.1s^4 + 5030s^3 + 3015s^2 + 246.5s}$$

$$K_{opt}^{(2,2)} = \frac{0.002594s^5 + 18.53s^4 + 1207s^3 + 9997s^2 + 1.355s + 4187}{0.1s^6 + 10.31s^5 + 368.1s^4 + 5030s^3 + 3015s^2 + 246.5s}$$

How Optimal LQG Controls the Looper

In order to make a comparison, the controllers of the decentralised PI system were tuned as individual loops using the refined Ziegler–Nichols rules (Wilkie et al., 2002). Figures 12.10 and 12.11 show the responses to a looper angle step change $\Delta\theta_{ref} = 2.5°$ for the Ziegler–Nichols tuned decentralised control system and the optimal LQG multivariable controller solution respectively.

The interesting feature of the responses is the different treatment of interaction between the two loops. In the response from the Ziegler–Nichols tuned decentralised control system for a looper angle reference change of 2.5°, observe that as the looper angle step change occurs the strip tension changes in an impulsive manner (Figure 12.10). This happens because as the looper angle increases step-wise, the

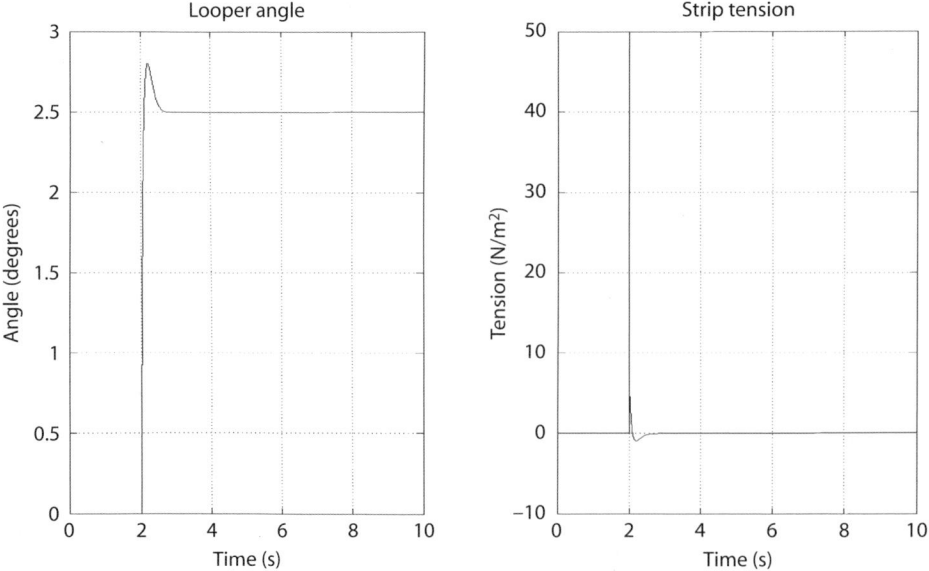

Figure 12.10 Ziegler–Nichols tuned decentralised control system responses for $\Delta\theta_{ref} = 2.5°$.

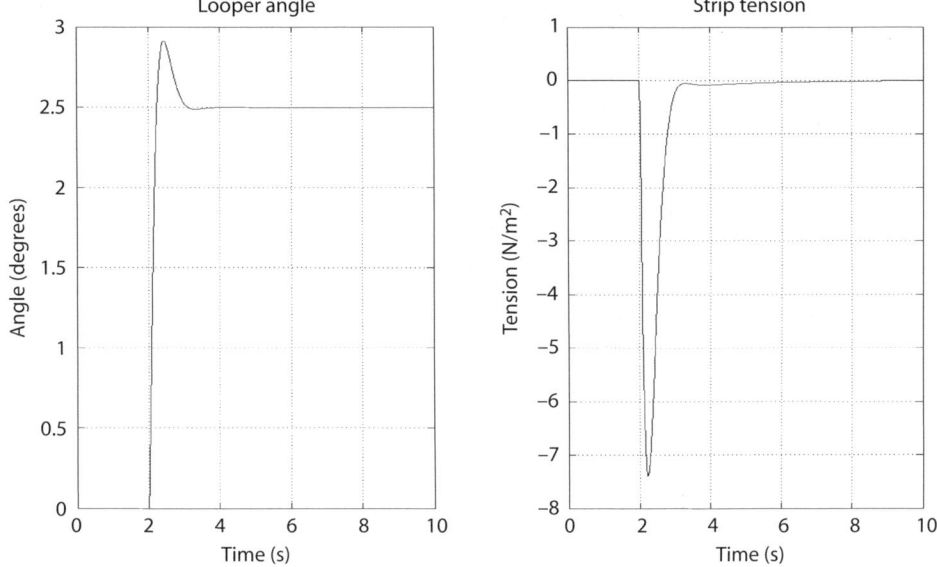

Figure 12.11 Optimal LQG multivariable controlled system responses for $\Delta\theta_{\text{ref}} = 2.5°$.

strip speed (ΔV_R) remains unchanged and the strip is stretched over the increased distance that it now has to travel, and the tension increases sharply as a result. The decoupled controllers regain control of the steady state but only after the strip has experienced the tension impulse, which might have broken the strip and caused a serious mill wreck.

A look at the optimal LQG multivariable controller results for the looper angle reference change of 2.5° shows a different approach to controlling the change (Figure 12.11). In the optimal solution, as the looper angle goes positive the tension decreases. This allows slack in the strip that is taken up safely by the change in looper angle. The tension reduction is relatively small compared with the impulsive tension change occurring in the decentralised solution. The optimal control solution produces these coordinated responses because both loops in the system are cross-coupled and the controller exploits the coupling beneficially.

12.6.3 A Controller Assessment Exercise for the Hotstrip Looper System

Restricted Structure Algorithm Convergence Results

The restricted structure algorithm was implemented using the optimal multivariable LQG controller solution as the benchmark of desired performance. Table 12.1 lists some typical structures and orders of the controllers tested with Algorithm 12.3. The table shows the number of control parameters n_c for each structure and the last two columns cover the convergence details. In particular, the table shows the number of iterations for the restricted structure algorithm to converge to a value below a pre-specified convergence tolerance threshold.

The convergence tolerance was set at a value of 0.1 and, as can be seen in the table, the greater the number of parameters in the controller for a *particular* macro-structure, the same or fewer iterations were needed to meet this convergence tolerance. This seems to reflect the notion that as the controller within a given macro-structure has more parameters there is more freedom to minimise the cost and convergence occurs with less iteration. Thus convergence occurs in families of patterns depending on the macro-structure of the controllers. The analysis of the controller *PSIs* will show if a similar behavioural pattern governs the indices.

Table 12.1 Typical restricted structure algorithm convergence results.

Case	Control structure	No. controller parameters, n_c	No. of RS algorithm iterations	Convergence tolerance $\|\theta^{k+1}-\theta^k\| \leq tol$
Limited interactive restricted structure				
8	$\begin{bmatrix} K_{PI} & 0 \\ K_{PI} & K_{PI} \end{bmatrix}$	6	8	0.0305
9	$\begin{bmatrix} K_{PI} & 0 \\ K_{PI} & K_{PID} \end{bmatrix}$	7	8	0.0165
10	$\begin{bmatrix} K_{PID} & 0 \\ K_{PID} & K_{PID} \end{bmatrix}$	9	8	0.0139
11	$\begin{bmatrix} K_{PI} & K_{PI} \\ 0 & K_{PI} \end{bmatrix}$	6	22	0.0864
12	$\begin{bmatrix} K_{PI} & K_{PI} \\ 0 & K_{PID} \end{bmatrix}$	7	18	0.0871
13	$\begin{bmatrix} K_{PID} & K_{PID} \\ 0 & K_{PID} \end{bmatrix}$	9	17	0.0914
14	$\begin{bmatrix} 0 & K_{PI} \\ K_{PI} & 0 \end{bmatrix}$	4	4	0.0497
Full interactive restricted structure				
15	$\begin{bmatrix} K_{PI} & K_{PI} \\ K_{PI} & K_{PI} \end{bmatrix}$	8	4	0.0221
16	$\begin{bmatrix} K_{PI} & K_{PI} \\ K_{PI} & K_{PID} \end{bmatrix}$	9	4	0.0218
17	$\begin{bmatrix} K_{PID} & K_{PID} \\ K_{PID} & K_{PID} \end{bmatrix}$	12	4	0.0227

Restricted Structure Controller Performance Assessment

The controller performance assessment is based on the results from the multiple application of Algorithm 12.3 to the restricted structure combinations chosen. The performance assessment index (PAI) for the restricted structure controllers is computed using

$$PAI = \frac{J_{LQG}^{\min}}{J_{RS}^{\min}(\theta) + J_{LQG}^{\min}}$$

which satisfies the bound

12.6 An Application of Multivariable Restricted Structure Assessment

$$0 \leq PAI = \frac{J_{LQG}^{\min}}{J_{RS}^{\min}(\theta) + J_{LQG}^{\min}} \leq 1$$

The performance assessment index for those control design obtained using other tuning methods (for example, the decentralised Ziegler–Nichols designs) is computed as

$$PAI = \frac{J_{LQG}^{\min}}{J_{LQG}^{actual}}$$

which also satisfies the bound

$$0 \leq PAI = \frac{J_{LQG}^{\min}}{J_{LQG}^{actual}} \leq 1$$

The value of J_{LQG}^{actual} is simply the LQG cost function evaluated for the simulated responses obtained using the non-optimal control design. Percentage forms of all the above performance assessment indices are obtained by multiplying the fractional *PAI* by 100%.

Table 12.2 lists the cost values and the computed performance assessment indices (*PAIs*) for all the restricted structure and Ziegler–Nichols control designs undertaken. The results are in order of worst performance to best performance; better performance occurs as *PAI* tends closer to 100%. The one exception is the first case, which is the optimal LQG solution with the default value PAI = 100%.

The table of assessment can be used in the following different ways.

Looking at Non-Optimal and Optimal Designs

The key assumption is that the LQG formulation has captured the specification of what is considered to be desirable control for the hotstrip looper system. From this it is very clearly seen that there is an order of magnitude difference in the performance indices for the non-optimal designs (*PAI* ~ 2% to 3%) when compared to the lowest values of the performance index for the optimal designs (*PAI* > 47%). This can be interpreted as meaning that non-optimal designs of decentralised Ziegler–Nichols type can at best attain only 3% of possible optimal performance.

Looking at Macro-Structure in Controllers

The table is grouped into the macro-structure controller sets: decentralised restricted structure, limited interactive restricted structure and full interactive restricted structure. The results show some interesting findings that are best surveyed by group, as shown in Table 12.3.

This reveals the very strong performance obtainable by using a skew-diagonal optimal decentralised restricted structure, which is of the order of 80% of full optimal performance. This skew-diagonal performance spills over into the upper triangular optimal limited interactive restricted structure performance, which is at around 73% levels. However, the skew-diagonal performance does not transfer well into the lower triangular optimal limited interactive restricted structure performance, which is only at around 49% levels. Such results would prompt the control engineer to look more closely at this good controller structure and understand the physical reasons for this performance.

Looking at Micro-Structure in Controllers

The question being examined here is what order or type should be chosen for the controller elements within a given controller macro-structure. As an example, Table 12.4 extracts the results for three cases within the optimal limited interactive restricted structure lower triangular set of controllers.

Clearly, the question then relates to the difference in performance obtainable between using PI controllers and mixtures of PI and PID controllers. Note that as the number of controller parameters

Table 12.2 Performance assessment index values.

Case	Control structure	No. of controller parameters, n_c	Cost function value	Performance index, PI (%)
Full optimal structure				
1	$\begin{bmatrix} K_{LQG} & K_{LQG} \\ K_{LQG} & K_{LQG} \end{bmatrix}$	47	937.346	100
Decentralised structure Z–N tuning				
2	$\begin{bmatrix} K_{PI} & 0 \\ 0 & K_{PI} \end{bmatrix}$	4	50018.35	1.87
3	$\begin{bmatrix} K_{PID} & 0 \\ 0 & K_{PID} \end{bmatrix}$	6	37364.35	2.51
Decentralised restricted structure				
4	$\begin{bmatrix} K_{PI} & 0 \\ 0 & K_{PI} \end{bmatrix}$	4	1987.746	47.16
5	$\begin{bmatrix} K_{PID} & 0 \\ 0 & K_{PI} \end{bmatrix}$	5	1988.446	47.14
6	$\begin{bmatrix} K_{PI} & 0 \\ 0 & K_{PID} \end{bmatrix}$	5	1970.246	47.58
7	$\begin{bmatrix} K_{PID} & 0 \\ 0 & K_{PID} \end{bmatrix}$	6	1969.846	47.58
Limited interactive restricted structure				
8	$\begin{bmatrix} K_{PI} & 0 \\ K_{PI} & K_{PI} \end{bmatrix}$	6	1927.816	48.62
9	$\begin{bmatrix} K_{PI} & 0 \\ K_{PI} & K_{PID} \end{bmatrix}$	7	1905.544	49.19
10	$\begin{bmatrix} K_{PID} & 0 \\ K_{PID} & K_{PID} \end{bmatrix}$	9	1905.155	49.20
11	$\begin{bmatrix} K_{PI} & K_{PI} \\ 0 & K_{PI} \end{bmatrix}$	6	1291.538	72.58
12	$\begin{bmatrix} K_{PI} & K_{PI} \\ 0 & K_{PID} \end{bmatrix}$	7	1282.085	73.11

12.6 An Application of Multivariable Restricted Structure Assessment

Table 12.2 (continued)

Case	Control structure	No. of controller parameters, n_c	Cost function value	Performance index, PI (%)
13	$\begin{bmatrix} K_{PID} & K_{PID} \\ 0 & K_{PID} \end{bmatrix}$	9	1281.751	73.13
14	$\begin{bmatrix} 0 & K_{PI} \\ K_{PI} & 0 \end{bmatrix}$	4	1175.139	79.76
Full interactive restricted structure				
15	$\begin{bmatrix} K_{PI} & K_{PI} \\ K_{PI} & K_{PI} \end{bmatrix}$	8	1110.798	84.38
16	$\begin{bmatrix} K_{PI} & K_{PI} \\ K_{PI} & K_{PID} \end{bmatrix}$	9	1110.797	84.38
17	$\begin{bmatrix} K_{PID} & K_{PID} \\ K_{PID} & K_{PID} \end{bmatrix}$	12	1110.787	84.39

increases from $n_c = 6$ to $n_c = 9$ the percentage performance increases from 48.62% to 49.20%, which is in line with the idea of nested decreasing optimal cost function values and nested increasing *PAI* values as the number of controller parameter increases. However, the increase in performance from Case 8, which only uses PI controllers, by adding in extra D-terms to the controller elements is only one or two per cent at best. Hence there seems little justification for using PID controllers in any of these structures and the conclusion is to use only the PI controller elements as in the structure of Case 8.

Table 12.3 Performance assessment of macro-structure in RS controllers.

Controller macro-structure grouping	Range of *PAI*'s percentages
Optimal decentralised restricted structure *Leading diagonal*	~ 40%
Optimal limited interactive restricted structure *Lower triangular*	~ 49%
Optimal limited interactive restricted structure *Upper triangular*	~ 73%
Optimal decentralised restricted structure *Skew diagonal*	~ 80%
Optimal full interactive restricted structure	~ 84%
Optimal LQG controller	Exactly 100%

Table 12.4 Micro-structure in a controller family.

Case	Structure	n_c	Cost	PAI
8	$\begin{bmatrix} K_{PI} & 0 \\ K_{PI} & K_{PI} \end{bmatrix}$	6	1927.816	48.62
9	$\begin{bmatrix} K_{PI} & 0 \\ K_{PI} & K_{PID} \end{bmatrix}$	7	1905.544	49.19
10	$\begin{bmatrix} K_{PID} & 0 \\ K_{PID} & K_{PID} \end{bmatrix}$	9	1905.155	49.20

The conclusions coming from this type of controller assessment can be used to guide a more in-depth assessment of particular controller structures involving system simulation, case studies and simply the control engineer interrogating the system setup to gain insight into the possible physical mechanisms at work in identified good or bad multivariable control structures. In the case of multivariable hotstrip looper control, such an interrogative analysis has been given by Greenwood (2003).

Note: In Table 12.2, Case 5 is just marginally out of line with expected trends; look at Cases 4, 5 and 6 together. This is believed to be a small numerical error. Unfortunately, due to the relentless march of time neither author (DG or MAJ) has been able to revisit this computation.

12.7 Conclusions

This chapter has considered the design of optimal controllers that minimise an LQG criterion, but where the structure of the controller is chosen to be of a restricted form, PID in particular. The technique provides a simple method of generating low-order controllers, which should be valuable in practice, and the results suggest that a low-order controller will often give a satisfactory response. Of course, a low-order controller is also easier to implement and is less subject to numerical inaccuracies.

Some interesting design issues occur, since it should be possible to choose the LQG weighting parameters to achieve a design that satisfies performance requirements and is compatible with the structure chosen for implementation. Thus, for example, if integral action is required the error weighting should include an integrator, so that both the full-order and restricted order LQG controllers will include an integrator. The desired controller structure can be chosen and then the LQG design should use weighting selections that give the required performance and frequency response compatible with the structural choice. If these steps are not consistent, it is an indication that either the performance requirements are unrealistic or the required structure should be changed.

From a theoretical viewpoint there are many obvious scalar system extensions, including discrete-time domain procedures (Grimble, 2001a), multi-model (Grimble, 2000b) and predictive versions (Grimble, 2004) of these results. However, the main value of the work is in providing a relatively simple method of generating PID-type controllers for industrial applications.

The introduction of multivariable system restricted structure to optimal LQG control involves the interplay of a number of interesting ideas, some of which are not present in the scalar system problem. However, the main storyline, which leads to the key LQG optimal multivariable control results and then on to the algorithm for computing the optimal restricted structure controllers, follows that of the scalar formulations described in the earlier sections of the chapter. Some extra issues arising from the multivariable nature of the system reported in this chapter were as follows.

Benchmark Design and Optimal Solution

A key assumption for the use of the optimal restricted structure idea is that the optimal control solution is a desirable controller solution. The full optimal controller then becomes a benchmark from which to compute and against which to compare different restricted structure controllers. A further aspect of the design issue is that optimal benchmark LQG designs should be easily setup and a brief outline of the selection procedure for multivariable weighting functions was given.

Multivariable Control Structures

Often the first question to be settled in creating a multivariable control solution for an industrial system is that of which system inputs will be used to control which process outputs. Tools like the Relative Gain Array coupled with some physical insight about the process can be used to resolve this question. However, with an input–output structure determined for the controller, new structural questions appear. A typical industrial multivariable controller might be given by

$$u(s) = \begin{bmatrix} K_{11}(s) & K_{12}(s) & K_{13}(s) \\ K_{21}(s) & K_{22}(s) & K_{23}(s) \\ K_{31}(s) & K_{32}(s) & K_{33}(s) \end{bmatrix} e(s) = \begin{bmatrix} k_{11}^P + \dfrac{k_{11}^I}{s} & 0 & k_{13}^P + \dfrac{k_{13}^I}{s} \\ k_{21}^P + k_{21}^D s & k_{22}^P & 0 \\ 0 & 0 & k_{33}^P + \dfrac{k_{33}^I}{s} + k_{33}^D s \end{bmatrix} e(s)$$

and it is easily seen that there are two more structural issues present in such a controller: a macro- and a micro-structure for the multivariable controller.

Macro-Structure of the Controller

It can be seen that some of the controller elements are completely zero and sometimes there are sound inherent physical system reasons why a controller element can be set to zero. Industrial engineers are often mightily encouraged when a plausible underlying physical system justification can be given for a macro-controller structure. The major problem in complex systems is, of course, finding and justifying these hidden structures. The restricted structure assessment for multivariable controllers was proposed as a tool to find such interesting structures. A steel industry control example was used to demonstrate this aspect of performance assessment.

Micro-Structures in the Controller Structure

A second look at the multivariable controller shows that some control elements are P, some are PI, some are PD and some are PID. Thus the problem of the micro-structure within a multivariable controller is to determine what terms in the PID controller elements should be retained whilst still yielding satisfactory control performance. Again, it was shown how this problem also fits in with the restricted structure assessment philosophy with an illustration from the same steel industry example.

Acknowledgements

The authors, M.J. Grimble, M.A. Johnson, P. Martin and D. Greenwood, acknowledge the kind financial support of the United Kingdom's Engineering and Physical Science Research Council under grants GR/L98237 (Steel project) and the Platform Grant (GR/R04683-01).

References

Albertos, P. and Sala, A. (2004) *Multivariable Control Systems*. Springer-Verlag, London.

Anderson, B.D.O and Moore, J.B. (1972) *Linear Optimal Control*. Prentice Hall, Englewood Cliffs, NJ.

Greenwood, D.R. (2003) *PhD Thesis*. Industrial Control Centre, University of Strathclyde, Glasgow.

Greenwood, D., Johnson, M.A. and Grimble, M.J. (2002) Multivariable LQG optimal control – restricted structure control for benchmarking and tuning. *European Control Conference*, Cambridge, UK.

Grimble, M.J. (1994) *Robust Industrial Control*. Prentice Hall, Hemel Hempstead.

Grimble, M.J. (2000a) Restricted-structure LQG optimal control for continuous-time systems. *IEE Proc. D, Control Theory Appl.*, **147**(2), 185–195.

Grimble, M.J. (2000b) Restricted structure linear estimators for multiple-model systems. *IEE Proceedings on Vision, Image and Signal Processing*, **147**, 193–204.

Grimble, M.J. (2001a) Restricted structure optimal linear pseudo-state filtering for discrete-time systems. *IEEE Transactions on Signal Processing*, **49**, 2522–2535.

Grimble, M.J. (2001b) Restricted structure controller performance assessment and benchmarking. *Proc. American Control Conference*, pp. 2718–2723.

Grimble, M.J. (2004) Restricted structure predictive optimal control. *Optimal Control Applications and Methods* (in press).

Grimble, M.J. and Johnson, M.A. (1988a) *Optimal Control and Stochastic Estimation*, Vol. 1. John Wiley & Sons, Chichester.

Grimble, M.J. and Johnson, M.A. (1988b) *Optimal Control and Stochastic Estimation*, Vol. 2. John Wiley & Sons, Chichester.

Horowitz, I.M. (1979) Quantitative synthesis of uncertain multiple-input, multiple-output feedback systems. *Int. J. Control*, **30**(1), 81–106.

Johnson, M.A., Hearns, G. and Lee, T. (1999) *The Hot Strip Rolling Mill Looper: a Control Case Study, Mechatronic Systems, Techniques And Applications* (ed. C. T. Leondes). Gordon & Breach Science Publishers, New Jersey. Volume 1: Industrial Manufacturing, Chapter 4, pp. 169–253.

Kailath, T. (1980) *Linear Systems*. Prentice Hall, Englewood Cliffs, NJ.

Kucera, V. (1979) *Discrete Linear Control*. John Wiley & Sons, New York.

Kwakernaak, H. and Sivan, R. (1972) *Linear Optimal Control Systems*. John Wiley & Sons, New York.

Luenberger, D. (1969) *Optimisation by Vector Space Methods*. John Wiley & Sons, New York.

Majecki, P. (2002) Personal communication.

Noble, B. (1969) *Applied Linear Algebra*. Prentice Hall, Englewood Cliffs, NJ.

Skogestad, S. and Postlethwaite, I. (1996) *Multivariable Feedback Control*. John Wiley & Sons, Chichester.

Wilkie, J., Johnson, M.A. and Katebi, M.R. (2002) *Control Engineering – an Introductory Course*. Palgrave, Basingstoke.

Youla, D.C., Jabr, H.A. and Bongiorno, J.J. (1976) Modern Wiener–Hopf design of optimal controllers, part II: the multivariable case. *IEEE Trans. Auto. Control*, **AC-21**(3), 319–338.

Yukitomo, M., Iino, Y., Hino, S., Takahashi, K. and Nagata, K. (1998) A new PID controller tuning system and its application to a flue gas temperature control in a gas turbine power plant. *IEEE Conference on Control Applications*, Trieste, Italy

Zhou, K., Doyle, J.C. and Glover, K. (1996) Robust and Optimal Control. Prentice Hall, Upper Saddle River, NJ.

Ziegler, J.G. and Nichols, N.B. (1942) Optimum settings for automatic controllers. *Trans. ASME*, **64**, 759–768.

13 Predictive PID Control

Learning Objectives

13.1 Introduction

13.2 Classical Process Control Model Methods

13.3 Simple Process Models and GPC-Based Methods

13.4 Control Signal Matching and GPC Methods

Acknowledgements

Appendix 13.A

References

Learning Objectives

Processes with a long time delay are difficult to stabilise with PID controllers. Can the Smith Predictor structure offer a solution to this PID control design problem? If future knowledge of the reference is known, how can this knowledge be used with PID control to improve the control quality? Generalised Predictive Control is very popular, but can it be used to tune PID controllers? There have been several attempts to use the control signal from an advanced control method as a benchmark signal for PID control design and performance. Is there a systematic method for achieving optimal control signal matching using PID control? This last chapter looks at all these questions and provides some answers and demonstrations of how the different predictive control approaches work.

The learning objectives for this chapter are to:

- Appreciate the context and categories of predictive PID control approaches.
- Study the theory of the Hägglund Smith Predictor-based technique.
- Investigate a worked example to demonstrate the effectiveness of the additional delay-estimate tuning parameter incorporated into the PI controller structure.

- Understand the use of simple process models with GPC methods to derive a predictive PID controller.

- Appreciate the control quality benefits that accrue from predictive action.

- Study the design of SISO and MIMO predictive PID controllers using the control signal matching approach.

- Achieve an appreciation of the wide application potential of the M-step ahead predictive PID controller method.

13.1 Introduction

A number of researchers have used advanced control techniques such as optimal control and GPC to design PID controllers. Many such methods restrict the structure of these controllers to retrieve the PID controller. Rivera *et al.* (1986) introduced an IMC based PID controller design for a first-order process model and Chien (1988) extended IMC-PID controller design to cover the second-order process model. The limitation of these works is that the tuning rules are derived for the delay-free system. A first-order Padé approximation or a first-order Taylor series expansion was used to remove the delay from the system. Morari and Zafiriou (1989) and Morari (1994) have shown that IMC leads to PID controllers for virtually all models common in industrial practice. In Wang *et al.* (2000) a least squares algorithm was used to compute the closest equivalent PID controller to an IMC design and a frequency response approach is adopted. However, the design is still ineffective when applied to time-delay and unstable systems.

Prediction in PID control design comprises two distinct contributions, both of which take a model based approach.

Classical Process Control Model Methods
The control design for processes which have long time delays usually begins from the Smith predictor and the Internal Model Control paradigm. The approach of Hägglund (1992, 1996) which is presented in this chapter does not involve optimisation but is straightforward judicious engineering based on the Smith predictor principle and the use of simple classical process control models.

Optimisation-Based Predictive Control
One of the successful developments of modern control is the group of optimisation methods of model-based predictive control, typically generalised predictive control using the receding horizon principle. They have found important application in the process industries, particularly at the supervisory level of complex plant where there is often some knowledge available about the future reference signals of processes and the ability to handle operational constraints is a very useful facility. In these advanced routines, the optimisation of a quadratic cost function is the common route to a control computation. When process control constraints are added, the computation step becomes a constrained optimisation, often with a concomitant increase in the numerical processing required for a solution. A process model is important in these approaches since it is needed to predict the future output response to possible future control trajectories. Controls are usually computed for a future finite time horizon and then an assumption about how the current and future controls will be used is invoked. In a receding horizon control philosophy only the current control is implemented; the time horizon moves forward by one step and the whole process repeated.

Moving from such a successful predictive control design method to the design of PID controllers requires some ingenuity. Two optimisation approaches are discussed and presented in this chapter.

- *Restricted models methods*

  This route uses the general model-based predictive control problem formulation but introduces restrictions on the process model so that the predictive control law which emerges is PID in form. In this chapter, the presentation is based on a solution devised by Tan *et al.* (2000, 2002b).

- *Control signal matching methods*

  Control signal matching methods also use the full model-based predictive control problem to yield a solution. The difference is the introduction of a modified PID control law which has only one set of controller coefficients but is able to generate future control moves. The PID controller gains are then chosen to give a close match between the predictive control signal and the predictive PID control signal. In this chapter the approach of Katebi and Moradi (2001) is described.

The classifications of the various approaches to predictive PID control are shown in Figure 13.1.

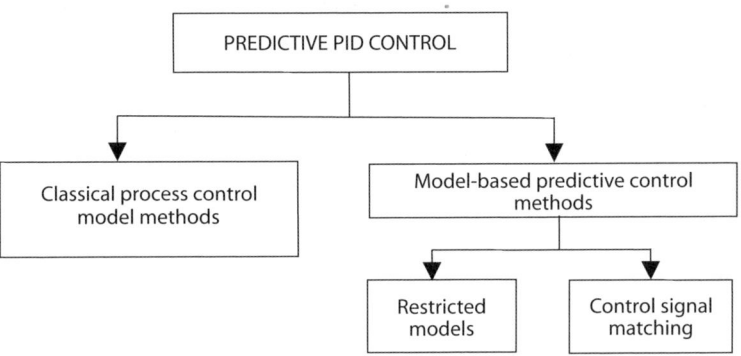

Figure 13.1 Approaches to predictive PID control.

13.2 Classical Process Control Model Methods

Simplicity and effectiveness are often key attributes of methods adopted for use in industrial applications. In many cases a technique is used because there is a good balance between the effort needed to persuade industrial engineers to introduce a new control method and the adequate but improved performance achieved by the method. Sometimes, this consideration is more important than trying to use a more complex control solution that attains the ultimate performance possible. PID control is successful because it is simple and effective for an important class of industrial control problems.

The method in this section of the chapter uses the simplicity of the PID control structure to achieve enhanced performance for processes with long dead-times. It will be seen that the method due to Hägglund (1992, 1996) introduces an additional tuning input to the PI control structure to give an incremental technological solution that very nicely meets the criteria of simplicity and effectiveness.

13.2.1 Smith Predictor Principle

Like the Ziegler–Nichols PID tuning rules, the Smith Predictor is a survivor from the early years of the development of electronic control systems. In this case the concept dates from the late 1950s (Smith, 1957, 1958) and is sufficiently fundamental to still be generating applications and research interest. The principle is captured by Figure 13.2 which shows a process model $G_p(s) = \widetilde{G}(s)e^{-Ls}$ containing a process dead time $L > 0$. The controller is given by $G_c(s)$ and there is an internal loop involving a process model representation $G_m(s) = \widetilde{G}_m(s)e^{-L_m s}$.

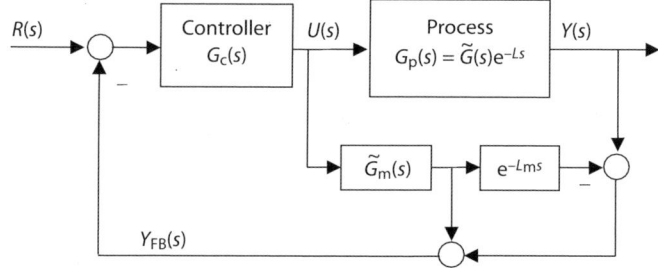

Figure 13.2 Smith predictor block diagram.

Analysis

Using Figure 13.2, the feedback signal $Y_{FB}(s)$ is given by

$$Y_{FB}(s) = \tilde{G}_m(s)U(s) + (\tilde{G}(s)e^{-Ls} - \tilde{G}_m(s)e^{-L_m s})U(s)$$

Thus if there is a perfect match between model and process so that $\tilde{G}_m(s) = \tilde{G}(s)$ and $L_m = L$, then the second term in the expression for $Y_{FB}(s)$ is zero and $Y_{FB}(s) = \tilde{G}_m(s)U(s)$. The condition of perfect model match yields more insight into the nature of the feedback signal, since

$$Y_{FB}(s) = \tilde{G}_m(s)U(s)$$

but

$$U(s) = \frac{1}{\tilde{G}(s)e^{-Ls}}Y(s) \text{ and } \tilde{G}_m(s) = \tilde{G}(s)$$

and hence

$$Y_{FB}(s) = \tilde{G}_m(s)\frac{1}{\tilde{G}(s)e^{-Ls}}Y(s) = e^{Ls}Y(s)$$

Clearly, in the time domain $y_{FB}(t) = y(t + L)$ and the feedback signal is the *prediction* of the output signal.

To complete the closed-loop analysis, solve

$$Y(s) = \tilde{G}(s)e^{-Ls}U(s)$$
$$U(s) = G_c(s)(R(s) - Y_{FB}(s))$$
$$Y_{FB}(s) = e^{Ls}Y(s)$$

to obtain

$$Y(s) = \left[\frac{\tilde{G}(s)e^{-Ls}G_c(s)}{1 + \tilde{G}(s)G_c(s)}\right]R(s)$$

As can be seen in this complementary sensitivity expression, the closed-loop *stability* design for controller $G_c(s)$ can proceed as though the plant is delay-free, but the *tracking performance* is still dependent on the presence of the delay.

Closed-Loop System Performance

To gain a further insight into the closed-loop system performance, the system should be augmented with the load disturbance signal $D_L(s)$ and the measurement noise input $N(s)$ as in Figure 13.3.

13.2 Classical Process Control Model Methods

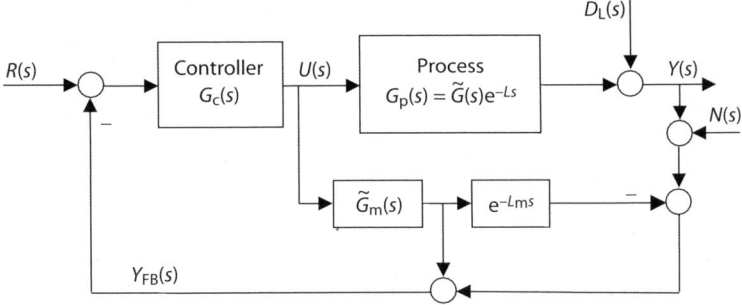

Figure 13.3 Smith predictor with disturbance inputs.

The disturbance and measurement noise rejection properties of the Smith predictor can be investigated through the usual transfer function relationship derived as

$$Y(s) = \left[\frac{\tilde{G}(s)e^{-Ls}G_c(s)}{1+\tilde{G}(s)G_c(s)}\right](R(s)-N(s)) + \left[\frac{1+\tilde{G}(s)G_c(s)(1-e^{-Ls})}{1+\tilde{G}(s)G_c(s)}\right]D_L(s)$$

and

$$Y(s) = [T(s)](R(s)-N(s)) + [S(s)]D_L(s)$$

where $T(s)$ is the complementary sensitivity function and $S(s)$ is the sensitivity function.

As can be seen, even in this ideal situation of perfect model matching, whilst the stability design relation will be delay-free, the reference tracking performance and the load disturbance rejection will still depend on the delay properties of the system. These properties are subject to the standard sensitivity relationship $S(s) + T(s) = 1$. Some interesting material on the properties of Smith predictors has been given by Morari and Zafiriou (1989).

13.2.2 Predictive PI With a Simple Model

The approach established by Hägglund (1992, 1996) and subsequently extended by Matauŝek and Micić (1996) was based on the Smith predictor principle and concerned the problem of tuning a PI controller for a process plant with a long time delay. The process was modelled by the standard FOPDT model:

$$G_p(s) = \left[\frac{K_m e^{-L_m s}}{T_m s + 1}\right]$$

where the process static gain is $K_m > 0$, the process time constant is $T_m > 0$, the process time delay is $L_m > 0$ and $L_m \gg T_m$. In the process industries the FOPDT model is fairly commonly used for non-oscillatory processes and is often identified using a step response test to find the three parameters K_m, T_m and L_m. The controller is assumed to be of PI form:

$$G_c(s) = k_P\left(1 + \frac{1}{\tau_i s}\right)$$

where the controller proportional gain is k_P and the integral time constant is τ_i.

Hägglund's contention was that a Smith predictor framework incorporating a PI controller could be used to provide effective control of a process with a long time delay. However, with this form of solution there would be some five parameters to tune: three parameters (K_m, T_m, L_m) from the FOPDT model and two parameters (k_P, τ_i) from the PI controller. Hägglund's analysis sought to find simple rules to reduce the tuning task and retain a philosophy of manual tuning for the predictor.

Analysis

The Smith predictor network of Figure 13.2 is filled with the details of the FOPDT model and the PI controller to yield the new Figure 13.4.

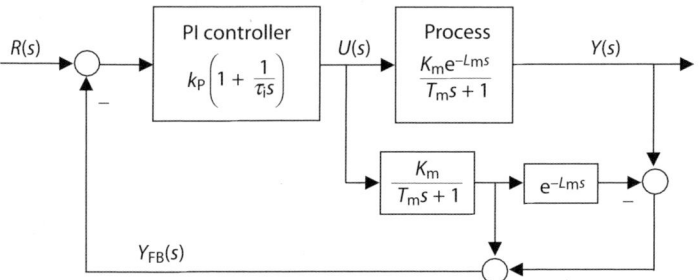

Figure 13.4 PI control with a Smith predictor.

The control signal $U(s)$ may be found from Figure 13.4 as

$$U(s) = k_P\left(1 + \frac{1}{\tau_i s}\right)[R(s) - Y_{FB}(s)]$$

and, after substitution,

$$U(s) = k_P\left(1 + \frac{1}{\tau_i s}\right)\left[R(s) - \left(\frac{K_m}{T_m s + 1}\right)(U(s) - e^{-L_m s}U(s)) - Y(s)\right]$$

Using $E(s) = R(s) - Y(s)$ this rearranges to

$$U(s) = k_P\left(1 + \frac{1}{\tau_i s}\right)E(s) - k_P\left(\frac{\tau_i s + 1}{\tau_i s}\right)\left(\frac{K_m}{T_m s + 1}\right)(U(s) - e^{-L_m s}U(s))$$

If the parameterisation $K_m = \alpha/k_P$, $T_m = \beta\tau_i$ is introduced, then the control signal $U(s)$ can be written in terms of the controller parameters k_P, τ_i and the two selectable parameters α, β as

$$U(s) = k_P\left(1 + \frac{1}{\tau_i s}\right)E(s) - \alpha\left(\frac{1 + \tau_i s}{\tau_i s(1 + \beta\tau_i s)}\right)(U(s) - e^{-L_m s}U(s))$$

Remarks 13.1
1. The form of the control signal now comprises a PI controller acting on the reference error $E(s) = R(s) - Y(s)$ and the low-pass filtering of the component of the control concerned with prediction.

2. It is useful to note that once a selection rule for parameters α, β has been determined, the PI controller parameters can be found as $k_P = \alpha/K_m$ and $\tau_i = T_m/\beta$. These rules could be used if a prior FOPDT model is determined.

Selecting the parameters α, β

It has already been shown in the previous section that the closed-loop expression for the Smith predictor is given by

$$Y(s) = \left[\frac{\widetilde{G}(s)e^{-Ls}G_c(s)}{1+\widetilde{G}(s)G_c(s)} \right] R(s)$$

Closed-loop stability will be determined by the polynomial equation

$$\rho_{CL}(s) = \widetilde{d}(s)d_c(s) + \widetilde{n}(s)n_c(s) = 0$$

where the transfer functions have the representations

$$\widetilde{G}(s) = \left[\frac{\widetilde{n}(s)}{\widetilde{d}(s)} \right] \text{ and } G_c(s) = \left[\frac{n_c(s)}{d_c(s)} \right]$$

Using the particular forms

$$\widetilde{G}(s) = \left[\frac{K_m}{T_m s + 1} \right] \text{ and } G_c(s) = \left[\frac{k_P(\tau_i s + 1)}{\tau_i s} \right]$$

gives

$$\rho_{CL}(s) = T_m \tau_i s^2 + (1 + K_m k_P)\tau_i s + K_m k_P = 0$$

The open-loop system model has a pole at $s = -1/T_m$ and Hägglund selects a critically damped closed-loop stability specification so that the closed-loop system has a repeated pole at the open-loop pole location. This gives a target closed-loop polynomial equation

$$\rho_{CL}^{\text{Target}}(s) = T_m^2 s^2 + 2T_m s + 1 = 0$$

Thus equating coefficients yields the controller design equations as

$$k_P = 1/K_m$$
$$\tau_i = T_m$$

Equivalently, the selectable parameters α, β have been chosen as $\alpha = \beta = 1$.

With such a choice, the control signal $U(s)$ reduces to

$$U(s) = k_P\left(1 + \frac{1}{\tau_i s}\right)E(s) - \left(\frac{1}{\tau_i s}\right)(U(s) - e^{-L_m s}U(s))$$

The critically damped specification for the second-order closed-loop characteristic polynomial $\rho_{CL}(s)$ is motivated by a desire to have a fast-acting reference tracking performance with no overshoot. The target closed-loop pole locations equal to the open loop pole position are in contrast with the tuning targets of the classical rule-based methods like Ziegler–Nichols, which attempt to have controller computations dependent on the size of the time delay L_m. An interesting point is that other second-order closed-loop specifications could also be set and investigated.

In the process industries reference tracking performance may not be of primary concern; often it is the load disturbance rejection which is considered to be more important. Hägglund (1992) has presented some analysis which shows that the integral absolute error (*IAE*) criterion satisfies the following relationship:

$$IAE_{\text{PredPI}} = \left(\frac{T_m + L_m}{2L_m}\right) IAE_{\text{PI}}$$

where $IAE = \int |e(\tau)|\,d\tau$, IAE_{PredPI} is evaluated for the Smith predictor-based PI controller, IAE_{PI} has a PI controller designed for critical damping, and the process model time constant and delay time are respectively T_m, L_m. Thus in the case of long dead times, where $L_m \gg T_m$, the value of IAE_{PredPI} approaches 50% of IAE_{PI}. Hägglund cited simulation evidence that this reduction can fall to 30% for other tuning rules, but even this is nonetheless still an improvement.

The extension to this methodology found in the contribution of Matausek and Mićić (1996) was basically three-fold:

1. To give an extension of the methodology to integrating processes.
2. To modify the Smith predictor framework to improve the disturbance rejection capabilities.
3. To retain the simplicity of controller tuning by finding clear physical interpretations for the tuning parameters; parameters which were different from those of the Hägglund method.

13.2.3 Method Application and an Example

The emphasis in this section is to use the predictive PI controller as it might be used in an industrial application. Firstly, a discussion of implementation is given along with a short procedure for use, and then some simulation results are presented.

Implementation Aspects

Assuming that the selectable parameters α, β have been chosen as recommended, namely $\alpha = \beta = 1$, this gives a reduced control signal $U(s)$ as

$$U(s) = k_P \left(1 + \frac{1}{\tau_i s}\right) E(s) - \left(\frac{1}{\tau_i s}\right)(U(s) - e^{-L_m s} U(s))$$

where $E(s) = R(s) - Y(s)$. To avoid proportional kick, Hägglund suggests that the control law be modified to

$$U(s) = k_P \left(-Y(s) + \left(\frac{1}{\tau_i s}\right) E(s)\right) + \left(\frac{1}{\tau_i s}\right)(e^{-L_m s} U(s) - U(s))$$

A block diagram showing this implementation is given as Figure 13.5.

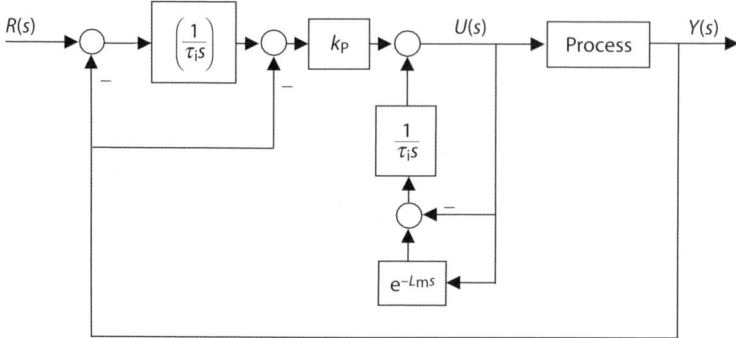

Figure 13.5 Predictor PI control implementation.

Clearly relatively small changes have to be made to the standard PID controller to implement the predictive PI controller. The values for controller parameters k_P, τ_i and L_m are required to run the controller; however, a degree of manual tuning flexibility has been retained. A tuning procedure is given next as Algorithm 13.1.

Algorithm 13.1: Predictor PI control
Step 1 Open loop step test
        Run an open loop step response test
        Estimate values
             Static gain $\hat{K}_m$
             Dominant time constant $\hat{T}_m$
             Delay time $\hat{L}_m$
Step 2 Controller parameters
        Compute $k_P = 1/\hat{K}_m$
        Set $\tau_i = \hat{T}_m$
        Set $L_m = \hat{L}_m$
Step 3 Test and fine tune
        Run step response
        Observe disturbance rejection performance
        Fine tune if necessary
Algorithm end

Example 13.1

The process is represented by an FOPDT model:

$$G_p(s) = \frac{3.0}{2.0s + 1} e^{-8s}$$

so that model static gain is $K_m = 3.0$, process model time constant is $T_m = 2.0$ and process delay time is $L_m = 8.0$. The case studies have been chosen to demonstrate some performance characteristics and the manual tuning facility of the new controller. The simulations were performed using Simulink.

Case Study 13.1.1: Perfect Model Match – Reference Step Response
This is a straightforward step response test with perfect model matching so that the controller is designed with perfect parameter knowledge of the model (see Figure 13.6).
    The unit step reference input was applied at time $t = 0$ and the response traces show the control signal acting immediately whilst the output responds after the process delay time has passed. The critically damped output response shows a smooth rise to the steady state level of unity.

Case Study 13.1.2: Perfect Model Match – Reference Step Plus Load disturbance response
In this case study, the controller is designed with perfect model matching. The unit step reference input was applied at time $t = 0$ and at $t = 22$ a +10% load disturbance occurs in the process output. The response traces are given in Figure 13.7.
    The output response to the reference step rises as before to the steady state level and the process output trace shows the 10% load disturbance at occurring at $t = 22$. The control signal reacts immediately to the output change, but the process output remains unaffected for the duration of the process delay and then the control brings the output back to the desired unity reference level.

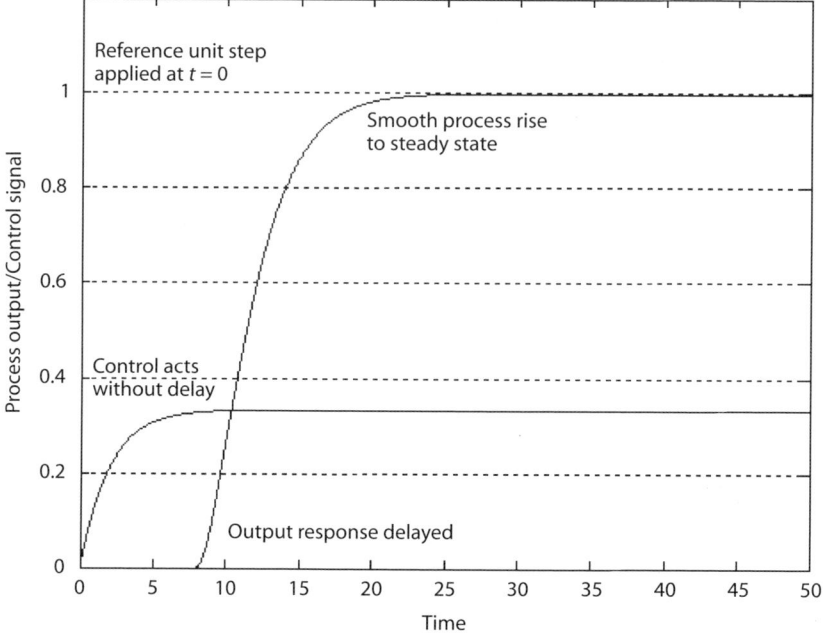

Figure 13.6 Perfect model match: reference step response signals.

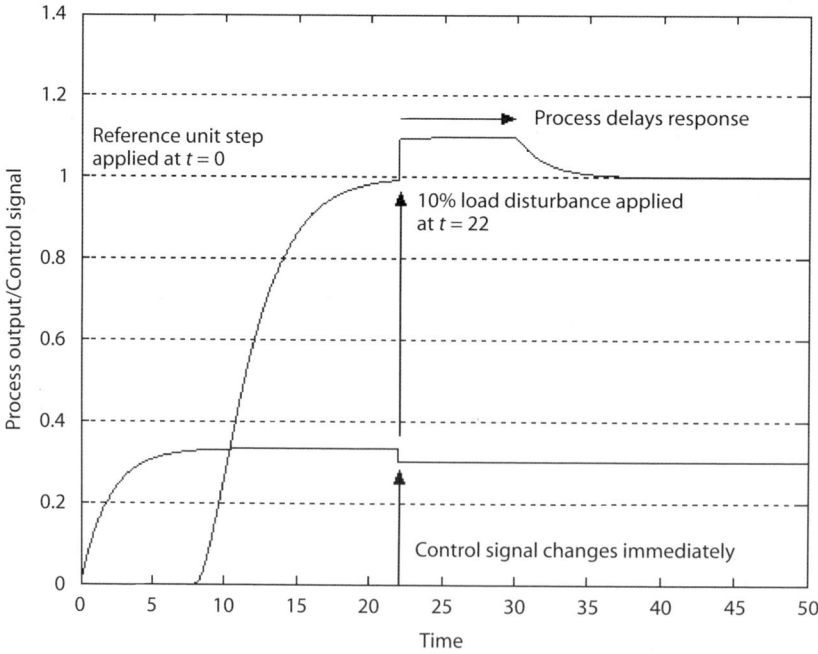

Figure 13.7 Perfect model match: reference step and load disturbance. responses (unit step reference applied at $t = 0$ and +10% load disturbance applied at $t = 22$).

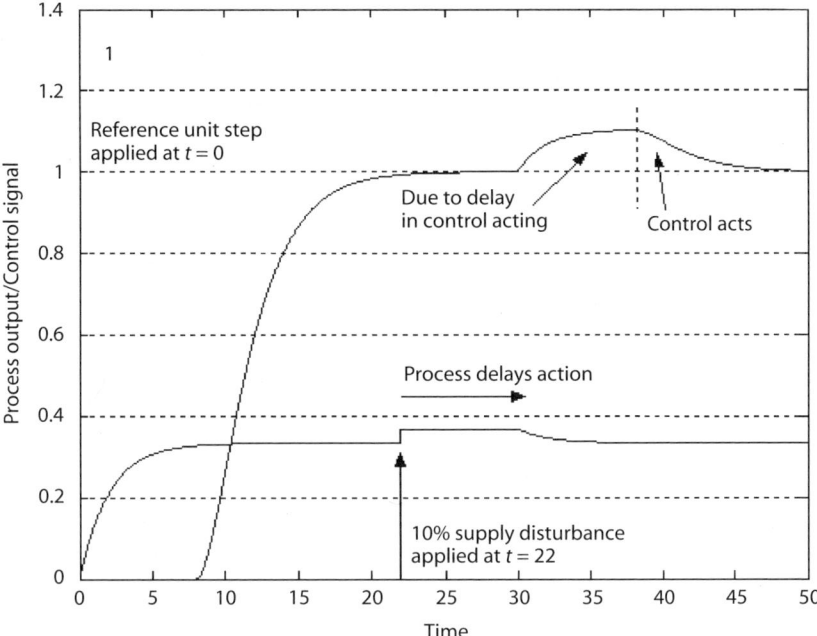

Figure 13.8 Perfect model match: reference step and 10% supply disturbance response (unit step reference applied at $t = 0$ and +10% supply disturbance applied at $t = 22$).

Case Study 13.1.3: Perfect Model Match – Reference Step Plus Supply disturbance response

Once more the controller is designed with perfect model matching. The unit step reference input was applied at time $t = 0$ and at $t = 22$ a +10% supply disturbance occurred at the process input. The resulting response traces are given in Figure 13.8.

The output response to the reference step rises as before to the steady state level and the control signal shows the 10% supply disturbance at occurring at $t = 22$. The 10% supply disturbance is at the process input and nothing happens until the disturbance passes through the process delay. The eventual result is an additional response feature in the process output. Once the process output change has managed to go round the feedback loop, the controller takes action to bring the process output back to the desired unity reference level.

Case Study 13.1.4: Model Mismatch – Reference Step Response

In this case study, the controller was designed with model parameters at 110% of nominal value. This meant that the estimated static gain was $\hat{K}_m = 3.3$, estimated time constant was $\hat{T}_m = 2.2$ and estimated delay time was $\hat{L}_m = 8.8$. This case study was designed to replicate the likely outcome of using Algorithm 13.1 where a prior estimation step with an approximate FOPDT model is standing duty for high-order process behaviour. A reference unit step was applied at time $t = 5$ and the response traces are shown in Figure 13.9.

The responses of Figure 13.9 should be compared with those of the perfect model matching case in Figure 13.6. As can be seen, the output responses for the mismatched model-based controller are much slower and no longer have the smooth rise to the steady state level that previously existed in Figure 13.6. Eventually, steady state offset elimination occurs by virtue of the presence of the integral term in the controller.

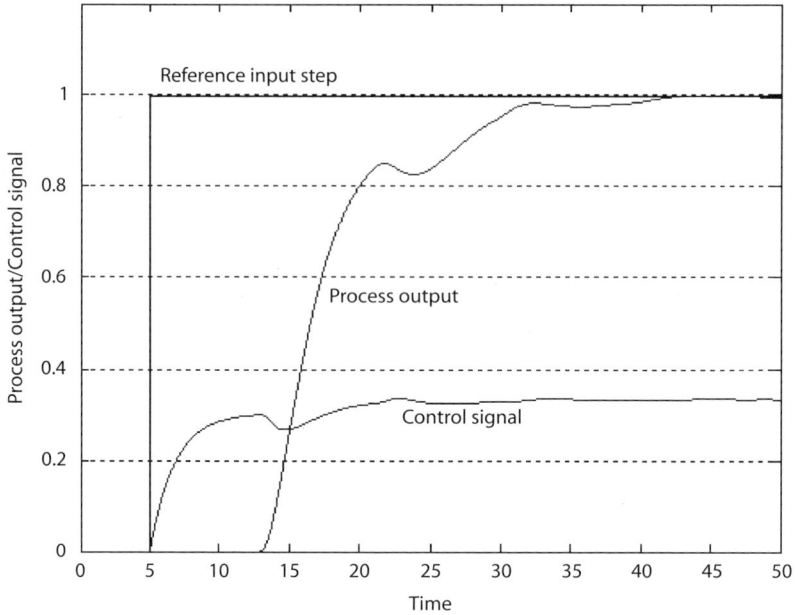

Figure 13.9 Model mismatch: reference step response (model parameters +10% incorrect and unit reference step applied at $t = 5$).

However, the inherent strength of this new PI controller structure is the tuning input available to change the estimated process delay $\hat{L}_m$. Figure 13.10 shows the effect of only altering the estimated process delay whilst the rest of the controller parameters remain unchanged having been computed using the mismatch model parameters.

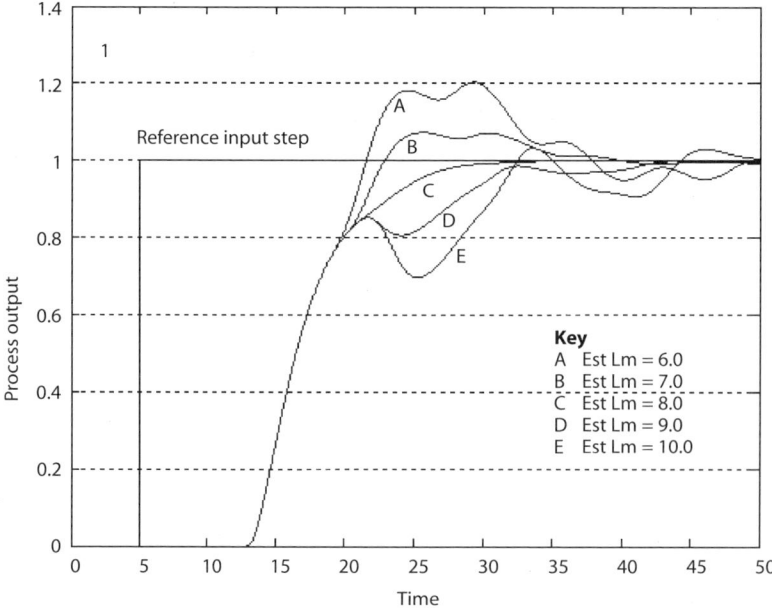

Figure 13.10 Process output responses obtained by tuning the estimated process delay $\hat{L}_m$.

It is useful to re-examine Figure 13.5 to see that re-tuning the process delay could be considered as a simple extra tuning input to the controller that does not interfere with the computation of the terms of the PI controller. As the responses in Figure 13.10 show, this can be a highly effective tuning device for improving speed and shape of the process response.

13.3 Simple Process Models and GPC-Based Methods

Of the modern control methods, Generalised Predictive Control is popular because it is easy to understand, easy to implement as a computer-based control technology, and effective and flexible, being able to accommodate process operational constraints fairly easily. The method described in this section uses the general model-based predictive control problem formulation, but introduces restrictions on the process model so that the predictive control law which emerges can be given error-actuated PID form. The presentation to follow is based on a solution devised by Tan et al. (2000, 2002b). In this presentation, the analysis retains the future reference setpoint information so that a pre-act predictive trim signal is derived and this is explored further in the numerical example given.

13.3.1 Motivation for the Process Model Restriction

The key to the method to be presented lies in a careful consideration of the requirements for the PID control law. The PID law may be given in continuous time as

$$u(t) = k_\mathrm{P} e(t) + k_\mathrm{I} \int^t e(\tau) \mathrm{d}\tau + k_\mathrm{D} \frac{\mathrm{d}e}{\mathrm{d}t}$$

If the sampling interval is T, then a sampled version of this law can be obtained. Introduce the time point $t = kT$ and define sampled signals:

Error signal $\qquad e_k = e(kT)$

Control signals $\qquad u_k = u(kT)$

Define

$$\Phi_k = \frac{1}{T} \int^{kT} e(\tau) \mathrm{d}\tau$$

then the recurrent relation for the integral term is

$$\Phi_{k+1} = \Phi_k + e_k$$

Approximate the derivative term as

$$\frac{\mathrm{d}e}{\mathrm{d}t} = \frac{e_k - e_{k-1}}{T}$$

Then the sampled PID control law is

$$u_k = \tilde{k}_\mathrm{P} e_k + \tilde{k}_\mathrm{I} \Phi_k + \tilde{k}_\mathrm{D}(e_k - e_{k-1})$$

where $\tilde{k}_\mathrm{P} = k_\mathrm{P}$, $\tilde{k}_\mathrm{I} = k_\mathrm{I} \times T$ and $\tilde{k}_\mathrm{D} = k_\mathrm{D}/T$.

A simple re-ordering of the sampled PID law reveals a simple linear combination of two error states e_{k-1}, e_k and one integrator state Φ_k, namely

$$u_k = K(1) e_{k-1} + K(2) e_k + K(3) \Phi_k$$

where $K(1) = -\tilde{k}_D$, $K(2) = (\tilde{k}_P + \tilde{k}_D)$ and $K(3) = \tilde{k}_I$.

As can be seen from the above equations, the integrator state Φ_k also has a recurrent relationship dependent on the error state e_k. With just two error states coming from the process this motivates a process model restriction to a second-order process. Consider then a general second-order process model with additional input delay of $h \geq 0$ given in z-transform form as

$$\frac{Y(z)}{U(z)} = G(z) = \frac{(b_1 z + b_2) z^{-h}}{z^2 + a_1 z + a_2}$$

Such a process model could originate from an s-domain transform model given by

$$G_p(s) = \frac{(as + b)}{s^2 + cs + d} e^{-Ls}$$

and this would, as Tan et al. (2002b) indicate, cover a wide spectrum of process model behaviour, including process time delay, underdamped and overdamped responses, minimum and non-minimum phase effects, integrating and unstable dynamics. The tacit assumption is that the process time delay L has translated into a whole integer number of sample intervals, so that $hT = L$ exactly.

13.3.2 Analysis for a GPC PID Controller

Motivated by the PID law, which can be written as a linear control law:

$$u_k = (-\tilde{k}_D) e_{k-1} + (\tilde{k}_P + \tilde{k}_D) e_k + (\tilde{k}_I) \Phi_k$$

or

$$u_k = K(1) e_{k-1} + K(2) e_k + K(3) \Phi_k$$

a dynamic system description is sought for the three state variables e_k, e_{k-1}, Φ_k. Recall the z-domain definition of the system error and the system description as:

$$E(z) = R(z) - Y(z)$$

$$Y(z) = \frac{(b_1 z + b_2) z^{-h}}{z^2 + a_1 z + a_2} U(z)$$

Together these give the sampled time domain equation:

$$e_{k+2} + a_1 e_{k+1} + a_2 e_k = r_{k+2} + a_1 r_{k+1} + a_2 r_k - (b_1 u_{k-h+1} + b_2 u_{k-h})$$

Hence rearranging and shifting the time counter by -1 yields

$$e_{k+1} = -a_2 e_{k-1} - a_1 e_k - b_1 \left(u_{k-h} + \frac{b_2}{b_1} u_{k-h-1} \right) + r_{k+1} + a_1 r_k + a_2 r_{k-1}$$

Introduce variables $\tilde{u}_{k-h}$ and $\tilde{r}_k$ defined by

$$\tilde{u}_{k-h} = \left(u_{k-h} + \frac{b_2}{b_1} u_{k-h-1} \right)$$

and

$$\tilde{r}_k = r_{k+1} + a_1 r_k + a_2 r_{k-1}$$

then

$$e_{k+1} = -a_2 e_{k-1} - a_1 e_k + (-b_1) \tilde{u}_{k-h} + \tilde{r}_k$$

To obtain a state space representation of the error process, introduce the state vector

$$X_k = \begin{bmatrix} e_{k-1} \\ e_k \\ \Phi_k \end{bmatrix} \in \Re^3$$

Then the state equation follows as

$$X_{k+1} = \begin{bmatrix} e_k \\ e_{k+1} \\ \Phi_{k+1} \end{bmatrix} = \begin{bmatrix} e_k \\ -a_2 e_{k-1} - a_1 e_k + (-b_1)\tilde{u}_{k-h} + \tilde{r}_k \\ \Phi_k + e_k \end{bmatrix}$$

and

$$X_{k+1} = \begin{bmatrix} 0 & 1 & 0 \\ -a_2 & -a_1 & 0 \\ 0 & 1 & 1 \end{bmatrix} \begin{bmatrix} e_{k-1} \\ e_k \\ \Phi_k \end{bmatrix} + \begin{bmatrix} 0 \\ -b_1 \\ 0 \end{bmatrix} \tilde{u}_{k-h} + \begin{bmatrix} 0 \\ 1 \\ 0 \end{bmatrix} \tilde{r}_k$$

The required state space system is

$$X_{k+1} = F X_k + B \tilde{u}_{k-h} + E \tilde{r}_k$$

where $X_{k+1}, X_k \in \Re^3$; $\tilde{u}_{k-h}, \tilde{r}_k \in \Re$ and

$$F = \begin{bmatrix} 0 & 1 & 0 \\ -a_2 & -a_1 & 0 \\ 0 & 1 & 1 \end{bmatrix} \in \Re^{3\times 3} \quad B = \begin{bmatrix} 0 \\ -b_1 \\ 0 \end{bmatrix} \in \Re^3 \quad E = \begin{bmatrix} 0 \\ 1 \\ 0 \end{bmatrix} \in \Re^3$$

Setting Up the GPC Cost Function

An assumption for the formulation of the GPC problem is that future reference signals are known and that the controller is to be designed with this knowledge. In the model derivation above the modified reference signal was given as $\tilde{r}_k = r_{k+1} + a_1 r_k + a_2 r_{k-1}$, so clearly future reference information would enable the computation of the sequence $\{\tilde{r}_k\}$. Introduce the future cost horizon for the system state as N_0 and the control cost horizon as N_c. It is usual for the cost horizons to satisfy $0 \leq h \leq N_c \leq N_0$, where h is the system delay.

The GPC cost function is defined as

$$J = \sum_{l=1}^{N_0} X_{k+l}^T Q_l X_{k+l} + \sum_{j=1}^{N_c} \tilde{u}_{k-h+j-1}^T R_j \tilde{u}_{k-h+j-1}$$

where weight functions satisfy $Q_l \in \Re^{3\times 3}$ and $R_j \in \Re$. The components in the cost function are briefly described as follows.

1. *State signal costs*: the forward values of the state vector $X_k \in \Re^3$ are weighted for time indices from $t = k+1$ to $t = k + N_0$, where N_0 is the state cost horizon. The state cost horizon is selected to satisfy $N_0 \geq N_c$. The weighting at each discrete time step is $Q_l \in \Re^{3\times 3}$, where for $l = 1, \ldots, N_0$, $Q_l^T = Q_l \geq 0 \in \Re^{3\times 3}$.

2. *Control signal costs*: the first state vector to be weighted is at time $t = k+1$ and after taking into account that system delay of h time steps, the first control which affects this state is at time $t = k - h$; consequently the forward values of the scalar control signal vector $\tilde{u}_{k-h}$ are weighted for N_c time

steps, from $t=k-h$ to $t=k-h+N_c-1$. Control cost horizon N_c is selected to be larger than the system delay, h. The weighting at each discrete time step is $R_j \in \Re$, where for $j=1,\ldots,N_c, R_j > 0 \in \Re$.

Figure 13.11 gives a diagrammatic interpretation of the relationship between time indices and the cost intervals.

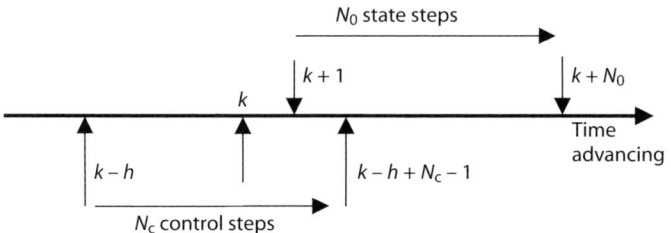

Figure 13.11 GPC cost time indices and the cost intervals.

The GPC cost function can be written in stacked form as

$$J = [X_{k+1}^T \ \cdots \ X_{k+N_0}^T] \begin{bmatrix} Q_1 & 0 & 0 \\ 0 & \ddots & 0 \\ 0 & 0 & Q_{N_0} \end{bmatrix} \begin{bmatrix} X_{k+1} \\ \vdots \\ X_{k+N_0} \end{bmatrix} + [\tilde{u}_{k-h} \ \cdots \ \tilde{u}_{k-h+N_c-1}] \begin{bmatrix} R_1 & 0 & 0 \\ 0 & \ddots & 0 \\ 0 & 0 & R_{N_c} \end{bmatrix} \begin{bmatrix} \tilde{u}_{k-h} \\ \vdots \\ \tilde{u}_{k-h+N_c-1} \end{bmatrix}$$

so that

$$J = \tilde{X}^T Q \tilde{X} + \tilde{U}^T R \tilde{U}$$

where

$$\tilde{X} = \begin{bmatrix} X_{k+1} \\ \vdots \\ X_{k+N_0} \end{bmatrix} \in \Re^{3N_0}, \quad Q = \begin{bmatrix} Q_1 & 0 & 0 \\ 0 & \ddots & 0 \\ 0 & 0 & Q_{N_0} \end{bmatrix} \in \Re^{3N_0 \times 3N_0}$$

$$\tilde{U} = \begin{bmatrix} \tilde{u}_{k-h} \\ \vdots \\ \tilde{u}_{k-h+N_c-1} \end{bmatrix} \in \Re^{N_c} \text{ and } R = \begin{bmatrix} R_1 & 0 & 0 \\ 0 & \ddots & 0 \\ 0 & 0 & R_{N_c} \end{bmatrix} \in \Re^{N_c \times N_c}$$

System Responses

The next step is to generate the stacked vector $\tilde{X} = \in \Re^{3N_0}$. Recall the state space system equation as $X_{k+1} = FX_k + B\tilde{u}_{k-h} + E\tilde{r}_k$; then this is recurrently advanced from time index $t=k$ to $t=k+N_0-1$. However, it should be noted that $N_0 \geq N_c$; consequently it is necessary to make an assumption about the control values $\tilde{u}_i$ whenever $i > k-h+N_c-1$. In this derivation, it is assumed that $\tilde{u}_i = \tilde{u}_{k-h+N_c-1}$ whenever $i > k-h+N_c-1$. The following state responses are found:

Time	State response
$k+1$	$X_{k+1} = FX_k + B\tilde{u}_{k-h} + E\tilde{r}_k$
$k+2$	$X_{k+2} = F^2 X_k + FB\tilde{u}_{k-h} + B\tilde{u}_{k-h+1} + FE\tilde{r}_k + E\tilde{r}_{k+1}$
$k+3$	$X_{k+3} = F^3 X_k + F^2 B\tilde{u}_{k-h} + FB\tilde{u}_{k-h+1} + B\tilde{u}_{k-h+2} + F^2 E\tilde{r}_k + FE\tilde{r}_{k+1} + E\tilde{r}_{k+2}$

$$k+p \quad X_{k+p} = F^p X_k + F^{p-1}B\tilde{u}_{k-h} + F^{p-2}B\tilde{u}_{k-h+1} + \ldots + B\tilde{u}_{k-h+p-1}$$
$$+ F^{p-1}E\tilde{r}_k + F^{p-2}E\tilde{r}_{k+1} + \ldots + E\tilde{r}_{k+p-1}$$

$$\vdots$$

$$k+N_c \quad X_{k+N_c} = F^{N_c}X_k + F^{N_c-1}B\tilde{u}_{k-h} + \ldots + B\tilde{u}_{k-h+N_c-1} + F^{N_c-1}E\tilde{r}_k + \ldots + E\tilde{r}_{k+N_c-1}$$

$$k+N_c+1 \quad X_{k+N_c+1} = F^{N_c+1}X_k + F^{N_c}B\tilde{u}_{k-h} + \ldots + (FB+B)\tilde{u}_{k-h+N_c-1} + F^{N_c}E\tilde{r}_k + \ldots + E\tilde{r}_{k+N_c}$$

$$\vdots$$

$$k+N_0 \quad X_{k+N_0} = F^{N_0}X_k + F^{N_0-1}B\tilde{u}_{k-h} + \ldots + (F^{N_0-N_c}B + \ldots + B)\tilde{u}_{k-h+N_c-1} + F^{N_0-1}E\tilde{r}_k + \ldots$$
$$+ E\tilde{r}_{k+N_0-1}$$

Augmenting these system responses into stacked vectors leads to the equation

$$\tilde{X} = GFX_k + \tilde{A}\tilde{U} + \tilde{E}\tilde{R}$$

with the details

Vectors:
$$\tilde{X} = \begin{bmatrix} X_{k+1} \\ \vdots \\ X_{k+N_0} \end{bmatrix} \in \Re^{3N_0}, \quad \tilde{U} = \begin{bmatrix} \tilde{u}_{k-h} \\ \vdots \\ \tilde{u}_{k-h+N_c-1} \end{bmatrix} \in \Re^{N_c}, \quad \tilde{R} = \begin{bmatrix} \tilde{r}_k \\ \vdots \\ \tilde{r}_{k+N_0-1} \end{bmatrix} \in \Re^{N_0}$$

Matrices:
$$F \in \Re^{3\times 3}, \quad G = \begin{bmatrix} 1 \\ F \\ \vdots \\ F^{N_0-1} \end{bmatrix} \in \Re^{3N_0 \times 3}, \quad \tilde{E} = \begin{bmatrix} E & 0 & \ldots & 0 \\ FE & E & 0 & \vdots \\ \vdots & \vdots & \vdots & 0 \\ F^{N_0-1}E & \ldots & \ldots & E \end{bmatrix} \in \Re^{3N_0 \times N_0}$$

$$\tilde{A} = \begin{bmatrix} B & 0 & \ldots & 0 \\ FB & B & 0 & \vdots \\ \vdots & \vdots & \vdots & 0 \\ F^{N_c-1}B & \ldots & \ldots & B \\ F^{N_c}B & \ldots & \ldots & S_1 \\ \vdots & \vdots & \vdots & \vdots \\ F^{N_0-1}B & \ldots & \ldots & S_{N_0-N_c} \end{bmatrix} \in \Re^{3N_0 \times N_c}$$

where $S_j = F^j B + F^{j-1}B + \ldots + B$, $j = 1, \ldots, N_0 - N_c$.

GPC Optimal Control

The GPC control optimisation problem can be stated as

$$\min_{\text{wrt}\tilde{U}} J = \tilde{X}^T Q \tilde{X} + \tilde{U}^T R \tilde{U}$$

where $\tilde{X} = GFX_k + \tilde{A}\tilde{U} + \tilde{E}\tilde{R}$.

Direct substitution gives

$$J = (GFX_k + \tilde{A}\tilde{U} + \tilde{E}\tilde{R})^T Q (GFX_k + \tilde{A}\tilde{U} + \tilde{E}\tilde{R}) + \tilde{U}^T R \tilde{U}$$

The unconstrained optimality condition is

$$\frac{\partial J}{\partial \widetilde{U}} = 2\widetilde{A}^T Q(GFX_k + \widetilde{A}\widetilde{U}_{opt} + \widetilde{E}\widetilde{R}) + 2R\widetilde{U}_{opt} = 0 \in \Re^{N_c}$$

which is solved as

$$\widetilde{U}_{opt} = -[\widetilde{A}^T Q \widetilde{A} + R]^{-1} \widetilde{A}^T Q(GFX_k + \widetilde{E}\widetilde{R})$$

Using the receding horizon philosophy, only the first control step is applied. Thus set

$$D = [1 \quad 0 \quad \ldots \quad 0] \in \Re^{1 \times N_c}$$

Then

$$D\widetilde{U}_{opt} = \widetilde{u}_{k-h} = [-D[\widetilde{A}^T Q \widetilde{A} + R]^{-1} \widetilde{A}^T Q GF]X_k + [-D[\widetilde{A}^T Q \widetilde{A} + R]^{-1} \widetilde{A}^T Q \widetilde{E}]\widetilde{R}(k:k + N_0 - 1)$$

and the receding horizon GPC control law is given by

$$\widetilde{u}_{k-h} = K_{GPC} X_k + K_{ref} \widetilde{R}(k:k + N_0 - 1)$$

where

$$K_{GPC} = [-D[\widetilde{A}^T Q \widetilde{A} + R]^{-1} \widetilde{A}^T Q GF] \in \Re^{1 \times 3}$$

and

$$K_{ref} = [-D[\widetilde{A}^T Q \widetilde{A} + R]^{-1} \widetilde{A}^T Q \widetilde{E}] \in \Re^{1 \times N_0}$$

As can be seen from the control law, the control input at time $t = k - h$ is given in terms of the system state at time which is in the future. Two different cases can be developed from here.

13.3.3 Predictive PID Control: Delay-Free System $h = 0$

In the delay-free case, the control law is given as

$$\widetilde{u}_k = K_{GPC} X_k + K_{ref} \widetilde{R}(k:k + N_0 - 1)$$

where

$$K_{GPC} = [-D[\widetilde{A}^T Q \widetilde{A} + R]^{-1} \widetilde{A}^T Q GF] \in \Re^{1 \times 3}$$

and

$$K_{ref} = [-D[\widetilde{A}^T Q \widetilde{A} + R]^{-1} \widetilde{A}^T Q \widetilde{E}] \in \Re^{1 \times N_0}$$

It is observed that both $K_{GPC} \in \Re^{1 \times 3}$ and $K_{ref} \in \Re^{1 \times N_0}$ are constant gain matrices; further the control law comprises two components, a state feedback law and a control trim based on future setpoint information.

The PID Control Component

From the full control law identify the PID component as

$$\widetilde{u}_k^{PID} = K_{GPC} X_k$$

Resolving the state feedback into element form

$$\tilde{u}_k^{\text{PID}} = K_{\text{GPC}}(1)X_k(1) + K_{\text{GPC}}(2)X_k(2) + K_{\text{GPC}}(3)X_k(3)$$
$$= K_{\text{GPC}}(1)e_{k-1} + K_{\text{GPC}}(2)e_k + K_{\text{GPC}}(3)\Phi_k$$
$$= (K_{\text{GPC}}(1) + K_{\text{GPC}}(2))e_k + K_{\text{GPC}}(3)\Phi_k + (-K_{\text{GPC}}(1))(e_k - e_{k-1})$$
$$= \tilde{k}_P e_k + \tilde{k}_I \Phi_k + \tilde{k}_D(e_k - e_{k-1})$$

yields the following identifications for the PID control law coefficients:

Proportional	Integral	Derivative
$\tilde{k}_P = K_{\text{GPC}}(1) + K_{\text{GPC}}(2)$	$\tilde{k}_I = K_{\text{GPC}}(3)$	$\tilde{k}_D = -K_{\text{GPC}}(1)$
$k_P = \tilde{k}_P$	$k_I = \tilde{k}_I / T$	$k_D = \tilde{k}_D \times T$

PID control law $\quad u(t) = k_P e(t) + k_I \int^t e(\tau)\mathrm{d}\tau + k_D \dfrac{\mathrm{d}e}{\mathrm{d}t}$

Sampled control law $\quad u_k = \tilde{k}_P e_k + \tilde{k}_I \Phi_k + \tilde{k}_D(e_k - e_{k-1})$

where the sampling interval is T.

The structure of the control law arising from the GPC-type of formulation is shown in Figure 13.12.

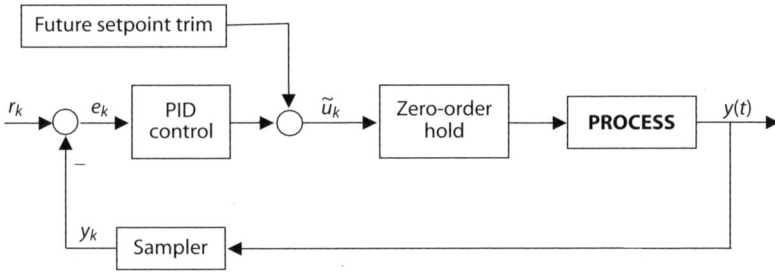

Figure 13.12 Structure of the control law for delay-free systems.

13.3.4 Predictive PID Control: Systems with Delay $h > 0$

The control law for systems with delay is given as

$$\tilde{u}_{k-h} = K_{\text{GPC}} X_k + K_{\text{ref}} \tilde{R}(k:k + N_0 - 1)$$

where

$$K_{\text{GPC}} = [-D[\tilde{A}^T Q \tilde{A} + R]^{-1} \tilde{A}^T Q G F] \in \Re^{1 \times 3}$$

and

$$K_{\text{ref}} = [-D[\tilde{A}^T Q \tilde{A} + R]^{-1} \tilde{A}^T Q \tilde{E}] \in \Re^{1 \times N_0}$$

It is observed at this point that both $K_{\text{GPC}} \in \Re^{1 \times 3}$ and $K_{\text{ref}} \in \Re^{1 \times N_0}$ are constant gain matrices; further the control law comprises two components, a state feedback law and a control trim based on future setpoint information. Following Tan et al. (2002b) the analysis assumes that the setpoint trim has been set to zero and concentrates on deriving the PID law for the dynamic component of the control law, namely:

$$\tilde{u}_{k-h} = K_{\text{GPC}} X_k \quad \text{where} \quad K_{\text{GPC}} = [-D[\tilde{A}^T Q \tilde{A} + R]^{-1} \tilde{A}^T Q G F] \in \Re^{1 \times 3}$$

The required PID control law is to have the form

$$\tilde{u}_k = K_{GPC}X_k = \tilde{k}_p e_k + \tilde{k}_i \Phi_k + \tilde{k}_d(e_k - e_{k-1})$$

Updating the GPC law yields

$$\tilde{u}_k = K_{GPC}X_{k+h}$$

and it can be seen that the control input depends on future state information.

To describe the state X_{k+h} in terms of the state X_k, noting that the control law has constant gains in matrix K_{GPC}, Tan et al. (2002b) give the following results.

Case (a): the Time Index Satisfies $0 \leq k < h$

From time $t = k$ to time $t = h$ the system free response relation gives state $X_h = F^{h-k}X_k$. At $t = h$ until $t = k + h$ the system is in closed loop for k time steps and $X_{k+h} = (F + BK_{GPC})^k X_h$. Together, these give $X_{k+h} = (F + BK_{GPC})^k F^{h-k} X_h$ whenever $0 \leq k < h$.

Figure 13.13(a) shows the intuitive nature of the result.

Case (b): the Time Index Satisfies $h \leq k$

The system is under closed-loop control for h steps from state X_k; hence the system response is given by $X_{k+h} = (F + BK_{GPC})^h X_k$ whenever $k \geq h$. The intuitive nature of the result is shown in Figure 13.13(b).

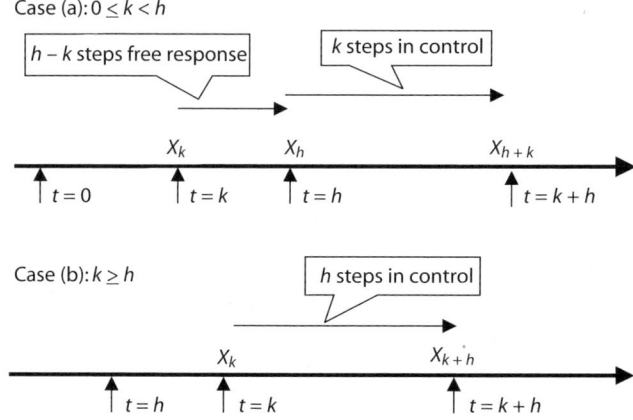

Figure 13.13 System state relationships.

The PID Controller Parameters

The PID controller parameters take on values according to the two cases discussed above, and this classification is used once more.

Case (a): the Time Index Satisfies $0 \leq k < h$

The control law used to identify the PID component is

$$\tilde{u}_k = K_{GPC}X_{k+h}$$

Substituting for $X_{k+h} = (F + BK_{GPC})^k F^{h-k} X_h$ yields

$$\tilde{u}_k^{PID} = K_{GPC}(F + BK_{GPC})^k F^{h-k} X_h$$
$$= K_k X_k$$

where $K_k = K_{GPC}(F + BK_{GPC})^k F^{h-k}$.

Resolving the state feedback into element form

$$\tilde{u}_k^{PID} = K_k(1)X_k(1) + K_k(2)X_k(2) + K_k(3)X_k(3)$$
$$= K_k(1)e_{k-1} + K_k(2)e_k + K_k(3)\Phi_k$$
$$= (K_k(1) + K_k(2))e_k + K_k(3)\Phi_k + (-K_k(1))(e_k - e_{k-1})$$

and

$$\tilde{u}_k^{PID} = \tilde{k}_P(k)e_k + \tilde{k}_I(k)\Phi_k + \tilde{k}_D(k)(e_k - e_{k-1})$$

yields the following identifications for the PID control law coefficients:

Proportional Integral Derivative
$\tilde{k}_P(k) = K_k(1) + K_k(2)$ $\tilde{k}_I(k) = K_k(3)$ $\tilde{k}_D(k) = -K_k(1)$

Time-varying sampled PID control law:

$$u_k = \tilde{k}_P(k)e_k + \tilde{k}_I(k)\Phi_k + \tilde{k}_D(k)(e_k - e_{k-1})$$

where the sampling interval is T and

$$K_k = K_{GPC}(F + BK_{GPC})^k F^{h-k}$$

Case (b): the Time Index Satisfies $h \leq k$

The control law used to identify the PID component is

$$\tilde{u}_k = K_{GPC}X_{k+h}$$

Substituting for $X_{k+h} = (F + BK_{GPC})^h X_k$ yields

$$\tilde{u}_k^{PID} = K_{GPC}(F + BK_{GPC})^h X_k$$
$$= KX_k$$

where $K = K_{GPC}(F + BK_{GPC})^h$.

Resolving the state feedback into element form, as in Case (a), yields the following identifications for the PID control law coefficients:

Proportional Integral Derivative
$\tilde{k}_P = K(1) + K(2)$ $\tilde{k}_I = K(3)$ $\tilde{k}_D = -K(1)$

Time-invariant sampled PID control law:

$$u_k = \tilde{k}_P e_k + \tilde{k}_I \Phi_k + \tilde{k}_P(e_k - e_{k-1})$$

where the sampling interval is T and

$$K = K_{GPC}(F + BK_{GPC})^h, h \leq k$$

13.3.5 Predictive PID Control: An Illustrative Example

To demonstrate some of the properties of this predictive PID method, a very simple but illuminating example is used; this is the predictive PI control of a delay-free first-order system. The process is assumed to have a model:

$$Y(s) = G_p(s)U(s)$$

where

$$G_p(s) = \frac{K}{\tau s + 1}$$

A sample interval size T is selected and the discrete model (including zero-order hold inputs) is then

$$y_k = \left[\frac{b}{q-a}\right] u_k$$

where $a = e^{-T/\tau}$ and q is the usual one-step time advance operator.

The continuous time PI control law is given by

$$u(t) = k_P e(t) + k_I \int^t e(\tilde{\tau}) d\tilde{\tau}$$

Following the sampling procedure of Section 13.4.3, sampling at $t = kT$ leads to

$$u_k = k(1) e_k + k(2) \theta_k$$

with

$$k(1) = k_P, \quad k(2) = k_I \times T, \quad \theta_k = \frac{1}{T} \int^{kT} e(\tilde{\tau}) d\tilde{\tau}$$

where $\theta_{k+1} = \theta_k + e_k$. This motivates a state representation for the GPC analysis where

$$X_k = \begin{bmatrix} e_k \\ \theta_k \end{bmatrix} \in \Re^2$$

If the procedure of the above sections is followed, the state space system describing the closed-loop system is derived as

$$X_{k+1} = AX_k + Bu_k + E\tilde{r}_k$$

where

$$X_k = \begin{bmatrix} e_k \\ \theta_k \end{bmatrix} \in \Re^2, \quad u_k, \tilde{r}_k \in \Re, \quad \tilde{r}_k = r_{k+1} - ar_k, \quad A = \begin{bmatrix} a & 0 \\ 1 & 1 \end{bmatrix}, \quad B = \begin{bmatrix} -b \\ 0 \end{bmatrix} \text{ and } E = \begin{bmatrix} 1 \\ 0 \end{bmatrix}$$

GPC Cost Function and Solution

Recall the full description of the GPC cost function as

$$J = \sum_{l=1}^{N_0} X_{k+l}^T Q_l X_{k+l} + \sum_{j=1}^{N_c} \tilde{u}_{k-h+j-1}^T R_j \tilde{u}_{k-h+j-1}$$

where future cost horizon for the system state is N_0 steps and the control cost horizon is N_c steps satisfying $0 \leq h \leq N_c \leq N_0$, where $0 \leq h$ is the system delay and the weight functions are $Q_l \geq 0 \in \Re^{2 \times 2}$ and $R_j > 0 \in \Re$.

For the example of this section, the full panoply of options available in the general GPC cost function are reduced as follows. There is no system delay; hence $h = 0$ and the cost horizons are set to be equal with $N_0 = N_c = N$. For simplicity, the state weightings are all set to $Q_l = I_2$, $l = 1, \ldots, N$ and the control weightings are all set to $R_j = \rho^2$, $j = 1, \ldots, N$. Thus the cost function becomes

$$J = \sum_{l=1}^{N} X_{k+l}^T X_{k+l} + \rho^2 \sum_{j=1}^{N} u_{k+j-1}^T u_{k+j-1}$$

and this can be given stacked form as

$$J = X^T X + \rho^2 U^T U$$

where $X \in \Re^{2N}$ and $U \in \Re^N$.

The closed-loop state space model $X_{k+1} = AX_k + Bu_k + E\tilde{r}_k$ is used straightforwardly to yield

$$X = \tilde{A} X_k + \tilde{B} U + \tilde{C} \tilde{R}$$

with the details:

Vectors: $X = \begin{bmatrix} X_{k+1} \\ \vdots \\ X_{k+N} \end{bmatrix} \in \Re^{2N},\ U = \begin{bmatrix} u_k \\ \vdots \\ u_{k+N-1} \end{bmatrix} \in \Re^N,\ \tilde{R} = \begin{bmatrix} \tilde{r}_k \\ \vdots \\ \tilde{r}_{k+N-1} \end{bmatrix} \in \Re^N$

Matrices: $\tilde{A} = \begin{bmatrix} A \\ A^2 \\ \vdots \\ A^N \end{bmatrix} \in \Re^{2N \times 2},\ \tilde{B} = \begin{bmatrix} B & 0 & \cdots & 0 \\ AB & B & 0 & \vdots \\ \vdots & \vdots & \vdots & 0 \\ A^{N-1}B & \cdots & \cdots & B \end{bmatrix} \in \Re^{2N \times N}$

and $\tilde{C} = \begin{bmatrix} E & 0 & \cdots & 0 \\ AE & E & 0 & \vdots \\ \vdots & \vdots & \vdots & 0 \\ A^{N-1}E & \cdots & \cdots & E \end{bmatrix} \in \Re^{2N \times N}$

The GPC Solution

Substitution of the stacked equation $X = \tilde{A}X_k + \tilde{B}U + \tilde{C}\tilde{R}$ into the reduced GPC cost function $J = X^T X + \rho^2 U^T U$ and optimisation with respect to the control $U \in \Re^N$ yields

$$U = K_{\text{GPC}} X_k + K_{\text{GPC}}^R \tilde{R}$$

where $K_{\text{GPC}} = -[\tilde{B}^T \tilde{B} + \rho^2 I_N]^{-1} \tilde{B}^T \tilde{A}$ and $K_{\text{GPC}}^R = -[\tilde{B}^T \tilde{B} + \rho^2 I_N]^{-1} \tilde{B}^T \tilde{C}$.

From this optimal vector expression, the application of the first control step yields the control law as

$$u_k = K_{\text{GPC}}(1,1) e_k + K_{\text{GPC}}(1,2) \theta_k + \sum_{i=1}^{N} K_{\text{GPC}}^R(1,i) \tilde{r}_{k+i}$$

At this point it is possible to return to the PI control law and write the PI controller gains as $k_P = K_{\text{GPC}}(1,1)$ and $k_I = K_{\text{GPC}}(1,2)/T$. Notice also that the Predictive PI control law has a predictive trim signal

$$\text{TRIM}_k = \sum_{i=1}^{N} K_{\text{GPC}}^R(1,i) \tilde{r}_{k+i}$$

dependent on future reference signals; this will be investigated in the numerical simulations to follow.

Selection of Control Weighting ρ^2

There are two methods for choosing control weighting parameter ρ^2. One direct method simply involves a simulation of the step response and selecting ρ^2 when the response is considered acceptable.

The second involves computing the closed-loop pole positions and selecting ρ^2 on the basis of these locations. The closed-loop analysis is based on the set of system equations:

$$\text{System } y_k = \left[\frac{b}{q-a}\right] u_k$$

Integral action $\theta_{k+1} = \theta_k + e_k$

Control law $u_k = K_{\text{GPC}}(1,1) e_k + K_{\text{GPC}}(1,2) \theta_k$

A z-domain analysis gives a closed-loop characteristic equation

$$\rho_{\text{CL}} = z^2 + (bK_{\text{GPC}}(1,1) - a - 1)z + a - b(K_{\text{GPC}}(1,1) - K_{\text{GPC}}(1,2)) = 0$$

Thus for a range of values of ρ^2 gains $K_{\text{GPC}}(1,1)$, $K_{\text{GPC}}(1,2)$ can be calculated and then closed-loop pole positions computed and viewed for acceptability.

The different stages of this example can be collected together as a procedure as follows.

Algorithm 13.2: Predictive PI control – first-order process

Step 1   Process model computations

For process model $G_p(s) = \dfrac{K}{\tau s + 1}$ select parameters K and τ

Select sample interval parameter T

For model $y_k = \left[\dfrac{b}{q-a}\right] u_k$, compute $a = e^{-T/\tau}, b = K(1-a)$

For model $X_{k+1} = AX_k + Bu_k + E\tilde{r}_k$ compute

$$A = \begin{bmatrix} a & 0 \\ 1 & 1 \end{bmatrix}, \; B = \begin{bmatrix} -b \\ 0 \end{bmatrix} \text{ and } E = \begin{bmatrix} 1 \\ 0 \end{bmatrix}$$

Step 2   GPC cost function

Select cost horizon parameter N

For the model $X = \tilde{A}X_k + \tilde{B}U + \tilde{C}R$ compute

$$\tilde{A} = \begin{bmatrix} A \\ \vdots \\ A^N \end{bmatrix} \in \Re^{2N \times 2}, \; \tilde{B} = \begin{bmatrix} B & \cdots & 0 \\ \vdots & \vdots & 0 \\ A^{N-1}B & \vdots & B \end{bmatrix} \in \Re^{2N \times N} \text{ and } \tilde{C} = \begin{bmatrix} E & \cdots & 0 \\ \vdots & \vdots & 0 \\ A^{N-1}E & \vdots & E \end{bmatrix} \in \Re^{2N \times N}$$

Step 3   GPC solution

Select the control weighting parameter ρ^2

Compute

$$K_{\text{GPC}} = -[\tilde{B}^T \tilde{B} + \rho^2 I_N]^{-1} \tilde{B}^T \tilde{A}$$
$$K_{\text{GPC}}^R = -[\tilde{B}^T \tilde{B} + \rho^2 I_N]^{-1} \tilde{B}^T \tilde{C}$$

Step 4   Solution assessment

Simulate closed-loop step response and if acceptable STOP else goto Step 3

Compute the closed-loop pole positions using

$$\rho_{\text{CL}} = z^2 + (bK_{\text{GPC}}(1,1) - a - 1)z + a - b(K_{\text{GPC}}(1,1) - K_{\text{GPC}}(1,2)) = 0$$

If pole positions acceptable STOP else goto Step 3
Algorithm end

Example 13.2: Numerical case study

For process model

$$G_p(s) = \frac{K}{\tau s + 1}$$

the parameters K and τ were chosen as $K = 3.7$ and $\tau = 0.5$. The sample interval was set as $T = 0.1$ and the costing horizon for the cost functions was $N = 10$. Algorithm 13.2 was followed using MATLAB programmes.

Case Study 13.2.1: Selecting Control Weighting ρ^2

As in Step 4, Algorithm 13.2, s-domain closed-loop pole locations were computed and tabulated as in Table 13.1.

Table 13.1 Computed values for a range of ρ^2 parameters.

Case No.	ρ^2 value	$K_{GPC}(1,1)$	$K_{GPC}(1,2)$	s-domain closed-loop poles
(a)	0.1	1.9622	0.8245	$s = -18.032, -10.859$
(b)	1.0	1.4058	0.5367	$s = -7.233 \pm j4.4469$
(c)	10.0	0.7583	0.2390	$s = -3.770 \pm j3.0089$
(d)	100.0	0.3315	0.0868	$s = -2.119 \pm j1.642$
(e)	1000.0	0.0683	0.0170	$s = -1.6503, -0.7791$

Case (c) seems to have satisfactory s-domain pole locations and the unit step response showed about 12% overshoot, and this case was used in the subsequent case studies.

Case Study 13.2.2: System Dynamic Response Without Prediction

The simulation was run to produce the closed-loop dynamic response to only the PI control part of the GPC control law. Thus the GPC formulation was being used to compute suitable PI controller parameters. Using Figure 13.14 the output response looks satisfactory, but there is significant proportional kick on the control signal. It would be interesting to explore whether standard PID structures could be introduced into this GPC methodology and compare the differences in the responses obtained.

Case Study 13.2.3: System Dynamic Response With PI Control and Prediction Trim

In this case study, the full Predictive PI control law derived from the GPC formulation is implemented. The responses are shown in Figure 13.15.

The graphs of Figure 13.15 clearly show the predictive trim signal acting before the step reference change occurs. This is also reflected in the control signal, which also acts prior to the reference step being applied. There is much less aggression in the control signal; compare the control peak of nearly 0.8 (Figure 13.14) with a peak of around 0.4 in the case when prediction is used. The output in the Predictive PI control response case looks more like an approximation to the step reference signal and smoothly follows the path of the reference step change.

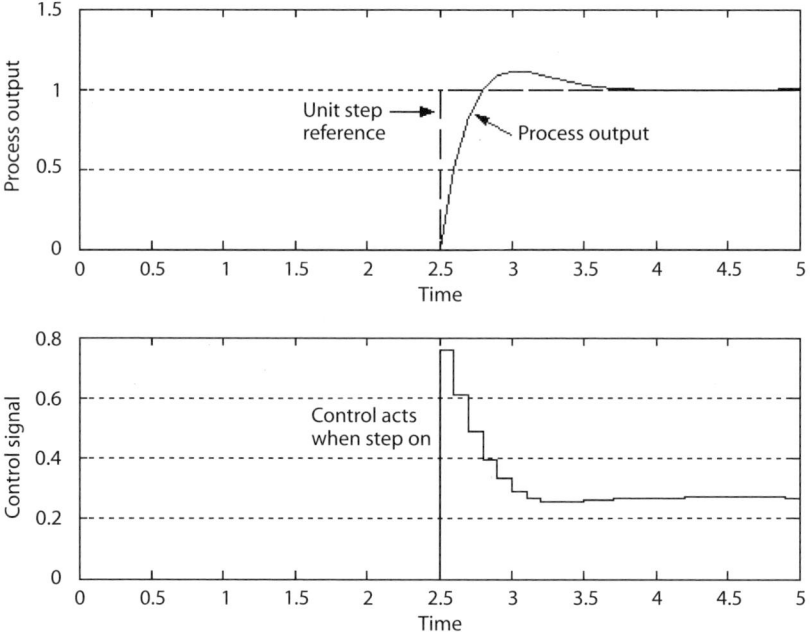

Figure 13.14 Output responses and control signal for the GPC designed PI controller.

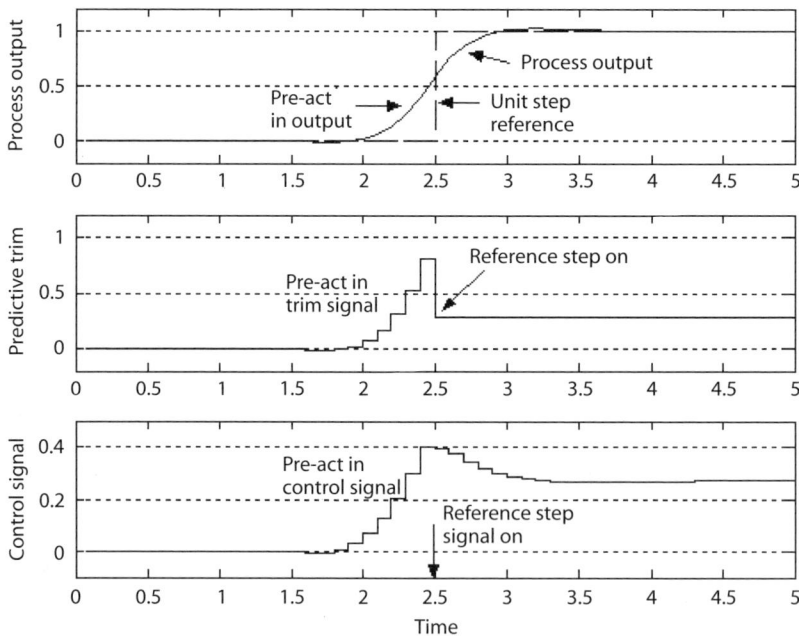

Figure 13.15 System dynamic response with PI control and prediction trim.

Case Study 13.2.4: The Benefits of Predictive PI Control

To view the benefits of Predictive PI control graphs of the reference errors for the two cases were produced as in Figure 13.16. These are simply plots of $e(t) = r(t) - y(t)$ for the case of PI control without (Plot (a)) and with (Plot (b)) prediction active.

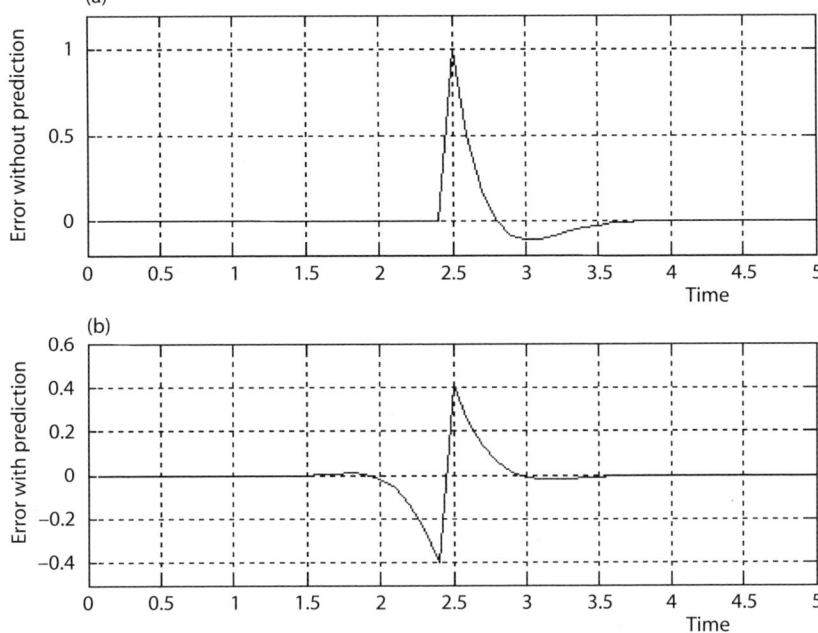

Figure 13.16 Graphs of the error signal for PI control without (a) and with (b) predictive action.

It is useful to be reminded of the benefits of predictive control in redistributing the error in the process output signal. In this example, the absolute magnitude of the error is reduced by better than 50% when predictive PI control is used. Restructuring the PI controller could possibly reduce this figure for the non-predictive PI controller, but the predictive PI control responses have the benefit of before-and-after distribution of error in the signal; this could be a significant advantage in manufacturing processes where quality is statistically monitored.

Summary Comments

A true Predictive PID control action is seen when the predictive trim signal is implemented. Tan and his colleagues have explored a number of avenues using this type of simple model coupled to a GPC analysis. Tan et al. (2002b) exploited the freedom to select cost function weightings to achieve a close specification of the desired closed-loop system parameters. Smith predictor design was also given a new look using this GPC approach (Tan et al., 2002a). It would be interesting to investigate three further issues:

1. *Cost weighting selection*: Tan et al. (2002b) have initiated this investigation and there seems to be scope for more guidelines in this area. Automated setup procedures would reduce the expertise needed to specify the weights to achieve different closed-loop system specification objectives.

2. *Robustness*: a key issue is that for high-order and uncertain processes an identified low-order process model will be used by this predictive PID control method. There is a need for studies to determine the extent of model mismatch permissible before the control designs deteriorate.

3. *Implementation issues*: simple simulation studies show the considerable benefits that accrue if prediction is used. It would be interesting to investigate how to incorporate this for routine use in an industrial PID controller unit.

13.4 Control Signal Matching and GPC Methods

As a method of computing PID controller coefficients, combining the theory of optimal control and model-based predictive control with PID control laws has been of considerable interest. The optimal control route was important in the Iterative Feedback Tuning method (Chapter 3), in the subspace model method (Chapter 10) and in the method of restricted structures (Chapter 12). Other contributions can be found in the literature, for example Rusnak (1999, 2000) used linear quadratic regulator theory to formulate optimal tracking problems and showed how some cases have a solution as PID control. This approach avoided heuristics and gave a systematic explanation of the good performance of the PID controllers. Rusnak (2000) also introduced a generalised PID structure that was applied to up to fifth-order systems. Combining model-based predictive control with PID control was previously presented in Section 13.3, where the method due to Tan *et al.* (2002b) used reduced-order process models to give a PID control solution; an additive trim signal produced the necessary predictive action in this controller. However, in this section of the chapter a different method for designing and computing predictive PID controllers is presented. The method exploits a simple idea where the PID controller is defined by using a bank of M parallel conventional PID controllers with M as the prediction horizon. Computation of the appropriate PID control law gains uses a control signal matching concept. The presentation is for both single-input, single-output and multi-input, multi-output systems.

13.4.1 Design of SISO Predictive PID Controllers

The method is first presented for the case of a single-input, single-output system. A prediction horizon of M steps is used to define the PID controller as a bank of M parallel conventional PID controllers. The predictive PID controller signal is then matched to the optimal GPC control signal and a method for the computation of the PID controller gains follows.

A Predictive Form for the PID Controller

The form of predictive PID controller used in this method is defined as

$$u(k) = \sum_{i=0}^{M} \left\{ k_P e(k+i) + k_I \sum_{j=1}^{k} e(j+i) + k_D [e(k+i) - e(k+i-1)] \right\}$$

where M is the prediction horizon. The controller consists of M parallel PID controllers as shown in Figure 13.17(a). For $M = 0$, the controller is identical to the conventional discrete PID control law. For $M > 0$, the proposed controller has a predictive capability similar to Model Predictive Control. The prediction horizon M will be selected to find the best approximation to a GPC control signal solution. The total predictive PID control signal $u(k)$ can be decomposed into M control signal components:

$$u(k) = \widetilde{u}(k) + \widetilde{u}(k+1) + \ldots + \widetilde{u}(k+M)$$

where

$$\widetilde{u}(k+i) = k_P e(k+i) + k_I \sum_{j=1}^{k} e(j+i) + k_D (e(k+i) - e(k+i-1)), \ i = 0, \ldots, M$$

For the reference error signal, it is assumed that $r = 0$ so that $e = r - y = -y$. If the incremental form of control signal is considered, then define $\Delta u(k) = u(k) - u(k-1)$ and the incremental control signal $\Delta u(k)$ can be written as

$$\Delta u(k) = \Delta \widetilde{u}(k) + \Delta \widetilde{u}(k+1) + \ldots + \Delta \widetilde{u}(k+M)$$

where

13.4 Control Signal Matching and GPC Methods

(a)
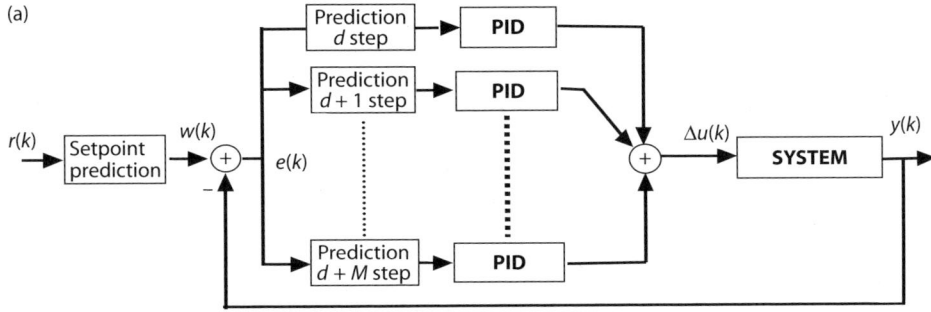

(b)
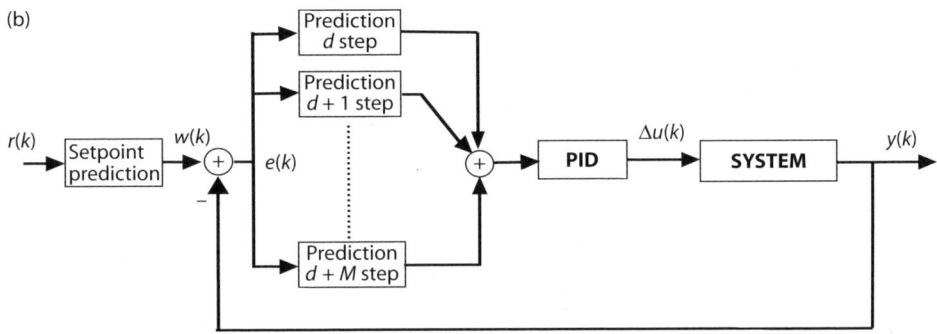

(c)
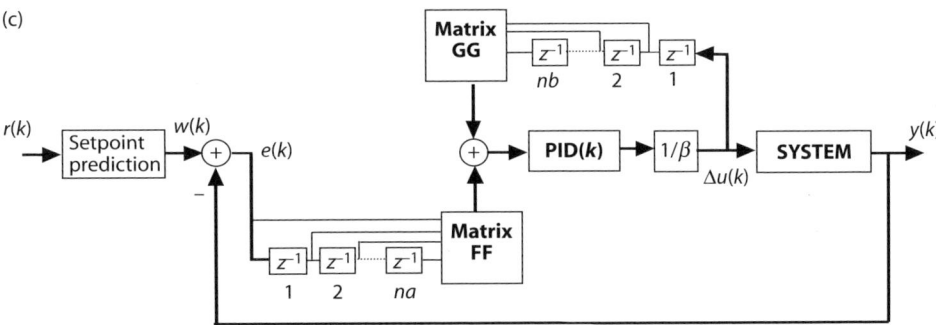

Figure 13.17 Block diagram interpretations of M-horizon predictive PID control. (a) M level of PID controller; (b) M horizon prediction; (c) structure of controller.

$$\Delta \tilde{u}(k+i) = -\{k_P[y(k+i) - y(k+i+1)] + k_I y(k+i) + k_D[y(k+i) - 2y(k+i-1) + y(k+i-2)]\}$$

In compact form, the components $\Delta \tilde{u}(k+i)$ can be rewritten as

$$\Delta \tilde{u}(k+i) = -KY(k+i)$$

where

$$K = [k_D \quad -k_P - 2k_D \quad k_P + k_I + k_D] \in \Re^{1 \times 3}$$

and

$$Y(k+i) = [y(k+i-2) \quad y(k+i-1) \quad y(k+i)]^T \in \Re^{3 \times 1}$$

Using the above equations, substitution gives

$$\Delta u(k) = \Delta \tilde{u}(k) + \Delta \tilde{u}(k+1) + \ldots + \Delta \tilde{u}(k+M)$$
$$= -K[Y(k) + Y(k+1) + \ldots + Y(k+M)]$$

This implies that the current control signal value is a linear combination of the future predicted outputs and a relation to determine these future outputs is required.

Output Predictor for the Predictive PID Controller

A CARIMA model for a SISO process can be expressed as

$$A(z^{-1})y(k) = B(z^{-1})\Delta u(k-1)$$

where z^{-1} is interpreted as the one-step time delay

$$A(z^{-1}) = 1 + a_1 z^{-1} + \ldots + a_{n_a} z^{-n_a} \text{ and } B(z^{-1}) = b_0 + b_1 z^{-1} + \ldots + b_{n_b} z^{-n_b}$$

The ith step-ahead prediction of the output can be obtained from the following equation (Camacho and Bordons, 1999):

$$y(k+i) = [g_{i-1} \quad g_{i-2} \quad \cdots \quad g_0] \begin{bmatrix} \Delta u(t) \\ \Delta u(t+1) \\ \vdots \\ \Delta u(t+i-1) \end{bmatrix} + [f_{i1} \quad f_{i2} \quad \cdots \quad f_{i(n_a+1)}] \begin{bmatrix} y(t) \\ y(t-1) \\ \vdots \\ y(t-n_a) \end{bmatrix}$$

$$+ [g'_{i1} \quad g'_{i2} \quad \cdots \quad g'_{in_b}] \begin{bmatrix} \Delta u(t-1) \\ \Delta u(t-2) \\ \vdots \\ \Delta u(t-n_b) \end{bmatrix}$$

Therefore, combining expressions for the control components $\Delta \tilde{u}(k+i) = -KY(k+i)$ with the above predictor relations gives the output prediction for the ith PID controller as

$$Y(k+i) = G_i \Delta \tilde{u}(k) + F_i y_0(k) + G'_i \Delta u_0(k)$$

where

$$G_i = \begin{bmatrix} g_{i-3} & \cdots & g_0 & 0 & 0 \\ g_{i-2} & g_{i-3} & \cdots & g_0 & 0 \\ g_{i-1} & g_{i-2} & g_{i-3} & \cdots & g_0 \end{bmatrix} \quad F_i = \begin{bmatrix} f_{(i-2)1} & f_{(i-2)2} & \cdots & f_{(i-2)(n_a+1)} \\ f_{(i-1)1} & f_{(i-1)2} & \cdots & f_{(i-1)(n_a+1)} \\ f_{i1} & f_{i2} & \cdots & f_{i(n_a+1)} \end{bmatrix}$$

$$G'_i = \begin{bmatrix} g'_{(i-2)1} & g'_{(i-2)2} & \cdots & g'_{(i-2)n_b} \\ g'_{(i-1)1} & g'_{(i-1)2} & \cdots & g'_{(i-1)n_b} \\ g'_{i1} & g'_{i2} & \cdots & g'_{in_b} \end{bmatrix}$$

$$\Delta \tilde{u}(k) = [\Delta u(k) \quad \Delta u(k+1) \quad \ldots \quad \Delta u(k+N_u-1)]^T$$
$$\Delta u_0(k) = [\Delta u(k-1) \quad \Delta u(k-2) \quad \ldots \quad \Delta u(k-n_b)]^T$$
$$y_0(k) = [y(k) \quad y(k-1) \quad \ldots \quad y(k-n_a)]^T$$

In the above equation, future control increments $\Delta u(k+i)$, $i = 1, \ldots, N_u - 1$ are needed to calculate the current control signal. Using the output prediction equation in the expression for the total kth control increment yields

$$\Delta u(k) = -K \sum_{i=0}^{M} Y(k+i) = -K \sum_{i=0}^{M} G_i \Delta \tilde{u}(k) + K \sum_{i=0}^{M} F_i y_0(k) + \sum_{i=0}^{M} G_i' \Delta u_0(k)$$

In compact form, the control signal for the PID predictive controller is obtained as

$$\Delta u(k) = -K\{\alpha \Delta \tilde{u}(k) + F_f y_0(k) + G_g \Delta u_0(k)\}$$

where

$$\alpha = \sum_{i=0}^{M} G_i \quad F_f = \sum_{i=0}^{M} F_i \quad G_g = \sum_{i=0}^{M} G_i'$$

In the real world, where systems are causal, the control signal cannot depend on an unknown future control signal, so it is assumed that $\Delta u(k+i) = 0$ for time indices $i = 1, \ldots, N_u - 1$ and the control increment is then evaluated as

$$\Delta u(k) = -(1 + K\alpha)^{-1} K[F_f y_0(k) + G_g \Delta u(k)]$$

Thus the output of predictive PID controller $\Delta u(k)$ depends on the details of the process specification captured as α, F_f, G_g and the PID controller gains K.

Modifications for Systems with a Time Delay

For a system with time delay d, the output of the process will not be affected by $\Delta u(k)$ until the time instant $t = (k + d + 1)T$ and the previous outputs will be a part of the system free response; consequently, there is no point in considering them as part of the cost function. In this case the first PID predicts d steps ahead and the last PID predicts $d + M$ steps ahead. This is given diagrammatic interpretation in Figure 13.17(b).

The control signal can be written as $\Delta u(k - d) = -(1 + K\alpha)^{-1} K[F_f y_0(k) + G_g \Delta u_0(k)]$.

Shifting the control signal for d steps ahead gives

$$\Delta u(k) = -(1 + K\alpha_d)^{-1} K[F_{fd} y_0(k) + G_{gd} \Delta u_0(k)]$$

where the coefficient matrices are

$$\alpha_d = \sum_{i=0}^{d+M} G_i \quad F_{fd} = \sum_{i=0}^{d+M} F_i \quad G_{gd} = \sum_{i=0}^{d+M} G_i$$

13.4.2 Optimal Values of Predictive PID Controller Gains

To obtain the optimal values of the predictive PID controller gains, the Generalised Predictive Control (GPC) algorithm is used. This theory obtains a control sequence by optimising a cost function of the form

$$J_{GPC} = \sum_{i=N_1}^{N_2} [y(k+d+i) - w(k+d+i)]^2 + \sum_{i=1}^{N_u} \lambda(i)[\Delta u(k+i-1)]^2$$

For process control applications, default settings of output cost horizon, namely $N_1 = 1$ and $N_2 = N$, and the control cost horizon $N_u = 1$ can be used in the GPC method to give reasonable performance (Clarke et al., 1987). Thus GPC consists of applying a control sequence that minimises the following cost function:

$$J_{GPC} = \sum_{i=1}^{N} [y(k+d+i) - w(k+d+i)]^2 + \lambda[\Delta u(k)]^2$$

The minimum of J_{GPC} (assuming there are no constraints on the control signals) is found using the usual gradient analysis, which leads to (Camacho and Bordons, 1999):

$$\Delta u(k) = (G^TG + \lambda I)^{-1} G^T[w - Fy_0(k) - G\Delta u_0(k)]$$

This form can expanded (assuming the future set point $w(k+i) = 0$) as

$$\Delta u(k) = -K_{GPC}[F \quad G'] \begin{bmatrix} y_0(k) \\ \Delta u_0(k) \end{bmatrix} - K_0 \begin{bmatrix} y_0(k) \\ \Delta u_0(k) \end{bmatrix}$$

where

$$K_0 = K_{GPC}[F \quad G'] \qquad \Delta u_0(k) = [\Delta u(k-1) \quad \Delta u(k-2) \quad \ldots \quad \Delta u(k-n_b)]^T$$
$$y_0(k) = [y(k) \quad y(k-1) \quad \ldots \quad y(k-n_a)]^T \quad K_{GPC} = (G^TG + \lambda I)^{-1} G^T$$

To compute the optimal values of predictive control PID gains the concept of control signal matching is used. In this case the PID controller gains are to be selected to make the PID control signal as close as possible to the GPC control signal. Noting that the GPC gain is subsumed into the gain K_0 and the predictive PID gain is defined through the gain matrix K, the following minimum norm optimisation problem is to be solved.

Control Signal Matching Optimisation Problem

$$\min_{\substack{wrt\ K \in K_{PID}^S \\ \text{and horizon } M}} J(K, K_0)$$

where the norm expression for the control signal matching is given by

$$J(K, K_0) = \|-(1 + K\alpha)^{-1} K[F_f \quad G_g] Z(K) + K_0 Z(K_0)\|_2$$

the set of stability gains for the PID controller is K_{PID}^S, the predictive controller prediction horizon is M and

$$Z(\cdot) = \begin{bmatrix} y_0(k) \\ \Delta u_0(k) \end{bmatrix}$$

depends on the particular control gains used, K or K_0.

Computing the Optimal Predictive PID Gains

In order to obtain explicit formulae for predictive PID control, the solution procedure assumes that the PID control $Z(K)$ can be written in terms of a perturbation about the GPC control value for $Z_0(K)$, namely write $Z(K) = Z(K_0) + \Delta Z$, then the norm expression $J(K, K_0)$ is given bounds as follows:

$$J(K, K_0) = \|-(1 + K\alpha)^{-1} K[F_f \quad G_g] Z(K) + K_0 Z(K_0)\|_2$$
$$= \|-(1 + K\alpha)^{-1} K[F_f \quad G_g](Z(K_0) + \Delta Z) + K_0 Z(K_0)\|_2$$
$$\leq \|[-(1 + K\alpha)^{-1} K[F_f \quad G_g] + K_0] Z(K_0)\|_2 + \|-(1 + K\alpha)^{-1} K[F_f \quad G_g] \Delta Z\|_2$$

and

$$J(K, K_0) \leq \|-(1 + K\alpha)^{-1} K[F_f \quad G_g] + K_0\|_2 \|Z(K_0)\|_2 + \|-1(1 + K\alpha)^{-1} K[F_f \quad G_g]\|_2 \|\Delta Z\|_2$$

Thus it is sought to minimise this bound as follows:

1. The value of the PID controller gain matrix K is selected to make $\|-(1+K\alpha)^{-1} K[F_f \quad G_g] + K_0\|_2 = 0$, so that $(1 + K\alpha)^{-1} K[F_f \quad G_g] = K_0$.

2. It is assumed that it is possible to find suitable predictive horizon and PID controller gains K so that $\|\Delta Z\|_2 = \|Z(K) - Z(K_0)\|_2$ is suitably small.

An explicit solution formula for K can be found in terms of K_0 as follows.
Define $S_0 = [F_f \quad G_g]$; then

$$K_0 = (1 + K\alpha)^{-1} K[F_f \quad G_g] = (1 + K\alpha)^{-1} KS_0$$
$$(1 + K\alpha)K_0 = KS_0$$

and

$$K(S_0 - \alpha K_0) = K_0$$

A unique solution to this equation always exists and takes the form

$$K = K_0(S_0 - \alpha K_0)^T[(S_0 - \alpha K_0)(S_0 - \alpha K_0)^T]^{-1}$$

After the calculation of K, the PID gains are calculated as:

$$k_P = -[K(2) + 2K(1)]$$
$$k_I = K(1) + K(2) + K(3)$$
$$k_D = K(1)$$

The Setpoint Rebuilt

In the GPC algorithm, information about the future horizon of N setpoint values are used in the calculation of the control sequence to be applied. In this control signal matching method the new setpoint is generated to save the information about the future setpoint. The new setpoint $r(k)$ is calculated from the previous setpoint $w(k)$ using

$$r(1) = \sum_{i=1}^{N} w(i)/N$$
$$r(i) = r(i-1) + w(i + N - 1) - w(i - 1) \quad i = 2, \ldots, n$$

The rebuilt setpoint is the average of setpoint values in the next N steps. The predictive PID controller uses these generated setpoints to achieve GPC performance.

An alternative for rebuilding a new set point is to use

$$r(k) = K_{GPC} W(k), \text{ for } k = 1, \ldots, n$$

where the future setpoint of the system is $W(k) = [w(k) \quad w(k+1) \quad \ldots \quad w(k+N)]$ and the GPC gain matrix is calculated from $K_{GPC} = (G^T G + \lambda I)^{-1} G^T$.

Predictive PID Control Algorithm

The calculations detailed above for the predictive PID controller can be implemented using the following algorithm.

Algorithm 13.3: Predictive PID controller for SISO processes

Step 1 *Initialisation*
Find a discrete system model and calculate the polynomials $A(z^{-1})$ and $B(z^{-1})$.
Choose the value of PID controller prediction horizon M and define the future setpoint values $\{r(k)\}_{k=1}^{\infty}$.

Step 2 *Off-line calculation*
Calculate the matrices α, F_f, G_g and $S_0 = [F_f \quad G_g]$
Calculate the GPC gains K_{GPC} and K_0.

Calculate the optimal value of predictive PID gains

$$K = K_0(S_0 - \alpha K_0)^T[(S_0 - \alpha K_0)(S_0 - \alpha K_0)^T]^{-1}$$

Iterate over the value of M to minimise the cost function.

Step 3 *On-line calculation*

Calculate the signals:
(a) $[F_f \quad G_g]Z(K)$
(b) $R(k) = [r(k-2) \quad r(k-1) \quad r(k)]$

Calculate the control increment:
$$u(k) = u(k-1) + (I + K\alpha)^{-1}K[R(k) - [F_g \quad G_g]Z(K)]$$

Step 4 *Fine tuning step*

Apply the control signal and validate the closed-loop performance.
Fine tune the PID gains if necessary.

Algorithm end

Stability Study for the Predictive PID Control Method

To study the stability of a predictive PID controlled system, the closed-loop transfer function of the system as shown in Figure 13.17(a) is computed. Writing the predictive PID control law in matrix form, retaining the setpoint, gives

$$\Delta u(k) = -K\alpha \Delta u(k) + F_f \theta(k) + G_g \Delta U(k)$$

where:

$$G_i = \begin{bmatrix} g_{i-3} \\ g_{i-2} \\ g_{i-1} \end{bmatrix}, \quad G_i' = \begin{bmatrix} g'_{(i-2)1} & g'_{(i-2)2} & \cdots & g'_{(i-2)n_b} \\ g'_{(i-1)1} & g'_{(i-1)2} & \cdots & g'_{(i-1)n_b} \\ g'_{i1} & g'_{i2} & \cdots & g'_{in_b} \end{bmatrix}, \quad F_i = \begin{bmatrix} f_{(i-2)1} & f_{(i-2)2} & \cdots & f_{(i-2)(n_a+1)} \\ f_{(i-1)1} & f_{(i-1)2} & \cdots & f_{(i-1)(n_a+1)} \\ f_{i1} & f_{i2} & \cdots & f_{i(n_a+1)} \end{bmatrix}$$

$$\alpha = \sum_{i=0}^{M} G_i \quad F_f = \sum_{i=0}^{M} F_i \quad G_g = \sum_{i=0}^{M} G_i'$$

$$\theta(k) = [-e(k) \quad -e(k-1) \quad \cdots \quad -e(k-n_a)]^T$$

$$\Delta U(k) = [\Delta u(k-1) \quad \Delta u(k-2) \quad \cdots \quad \Delta u(k-n_b)]^T$$

$$e(k-i) = w(k-i) - y(k-i) \quad i = 0, \ldots, n_a$$

Rewrite the control law in compact form; then some straightforward algebra gives

$$A_c(z^{-1})y = w^* - B_c(z^{-1})\Delta u$$

where

$$A_c(z^{-1}) = KF_f Z_y, \quad B_c(z^{-1}) = G^* Z_u, \quad w^* = k_{ref} w, \quad k_{ref} = KF_f,$$

$$Z_y = [1 \quad z^{-1} \quad \cdots \quad z^{-na}]^T, \quad G^* = [1 + K\alpha \quad KG_g] \text{ and } Z_u = [1 \quad z^{-1} \quad \cdots \quad z^{-na}]^T$$

A CARIMA model for system is given by

$$\widetilde{A}(z^{-1})y(k) = z^{-d}B(z^{-1})\Delta u(k-1) = z^{-d-1}B(z^{-1})\Delta u(k)$$

where $\widetilde{A}(z^{-1}) = D(z^{-1})A(z^{-1}) = (1 - z^{-1})A(z^{-1})$ and $d \geq 0$ is the time delay of the system.

Inserting the control equation into the process system model gives the closed-loop transfer function of the system as

$$y = \frac{B(z^{-1})z^{-d-1}}{A_c(z^{-1})B(z^{-1})z^{-d-1} + B_c(z^{-1})\tilde{A}(z^{-1})} w^*$$

and the closed-loop poles are the roots of characteristic equation

$$A_c(z^{-1})B(z^{-1})z^{-d-1} + B_c(z^{-1})\tilde{A}(z^{-1}) = 0$$

The above equations show that the poles of closed-loop system, which are the roots of characteristic equation, depend on PID gains. On other hand, the PID gains are affected by selection of the prediction horizon of the proposed method.

Predictive PID Control Applied to Second-Order System Models

A discrete model for a second-order system with control increments is given by

$$\tilde{A}(z^{-1})y(k) = z^{-d}B(z^{-1})\Delta u(k-1)$$

where

$$D(z^{-1}) = 1 - z^{-1}, \quad A(z^{-1}) = 1 + a_1 z^{-1} + a_2 z^{-2}, \quad B(z^{-1}) = b_1 + b_2 z^{-1},$$
$$\tilde{A}(z^{-1}) = D(z^{-1})A(z^{-1}) = 1 + (a_1 - 1)z^{-1} + (a_2 - a_1)z^{-2} - a_2 z^{-3}$$

and $d \geq 0$ is the time delay of the system. The equivalent time domain difference equation is

$$y(k+1) = -a_1' y(k) - a_2' y(k-1) - a_3' y(k-2) + b_1 \Delta u(k-1-d) + b_2 \Delta u(k-2-d)$$

where $a_1' = a_1 - 1, a_2' = a_2 - a_1$ and $a_3' = -a_2$. Using the redefinition of the control signal

$$v(k-1) = u(k-1) + \frac{b_2}{b_1} u(k-2)$$

the above system equation can be rewritten as

$$y(k+1) = -a_1' y(k) - a_2' y(k-1) - a_3' y(k-2) + b_1 \Delta v(k-1-d)$$

If the predictive PID controller design procedure is applied to the above system, the optimisation is respect to the new manipulated signal Δv. Once $\Delta v(k)$ has been calculated then the control increment $\Delta u(k)$ can be computed as

$$\Delta u(k) = \Delta v(k) - \Delta u(k-1)$$

or

$$\Delta u(k) = \sum_{i=0}^{k-1} \left(\frac{b_2}{b_1}\right)^i \Delta v(k-i) \quad \text{for} \quad k = 1, 2, \ldots$$

Using the GPC cost function with an output prediction horizon of N steps and a control costing horizon of $N_u = 1$, give the GPC control signal as

$$\Delta v(k) = -K_0 \begin{bmatrix} y(k) \\ y(k-1) \\ y(k-2) \end{bmatrix}$$

For the predictive PID control method with $M = 0$ the control signal is

$$\Delta v(k-d) = KY(k) \quad \text{where} \quad K = [k_P + k_I + k_D \quad -k_P - 2k_D \quad k_D]$$

To achieve GPC performance the control signal of predictive PID should be equal to the GPC control signal. Thus if $d = 0$, after some algebra we obtain

$$K = K_{GPC} F = K_0$$

so that $k_P = -K_0(2) - 2K_0(3)$ and $k_I = K_0(1) + K_0(2) + K_0(3)$ and $k_D = K_0(3)$. Therefore, GPC for a second-order delay-free system is equal to a PID controller with the above coefficients. For a second-order system with delay, the control signal will be

$$\Delta v(k) = KY(k+d)$$

This equation can be written as

$$\Delta v(k) = -(1 + K\alpha)^{-1} K[\widetilde{F}][Y(k)]$$

where

$$\widetilde{F} = \begin{bmatrix} f_{d1} & f_{d2} & f_{d3} \\ f_{(d-1)1} & f_{(d-1)2} & f_{(d-1)3} \\ f_{(d-2)1} & f_{(d-2)2} & f_{(d-2)3} \end{bmatrix} \text{ and } \alpha = [g_{d-1} \quad g_{d-2} \quad g_{d-3}]$$

To achieve the same result as GPC the PID control signal should be equal to the GPC control signal, so that $K_0(1 + K\alpha) = K\widetilde{F}$ and $K(\widetilde{F} - \alpha K_0) = K_0$. In this equation a unique solution for K exists and the required PID gains will be

$$k_P = -K(2) - 2K(3)$$
$$k_I = K(1) + k(2) + k(3)$$
$$k_D = K(3)$$

where

$$K = K_0(\widetilde{F}_k - \alpha_k K_0)^{-1} \quad k < d$$
$$K = K_0(\widetilde{F} - \alpha K_0)^{-1} \quad k > d$$

$$\widetilde{F}_K = \begin{bmatrix} f_{d(k-d+1)} & f_{d(k-d+2)} & f_{d(k-d+3)} \\ f_{(d-1)(k-d+1)} & f_{(d-1)(k-d+2)} & f_{(d-1)(k-d+3)} \\ f_{(d-2)(k-d+1)} & f_{(d-2)(k-d+2)} & f_{(d-2)(k-d+3)} \end{bmatrix}, \alpha = \begin{bmatrix} g_{k-1} \\ g_{k-2} \\ g_{k-3} \end{bmatrix} \text{ and } f_{ij} = 0; \, i < 0 \text{ or } j > 0$$

Remark 13.2 For second-order system models, one level of PID ($M = 0$) is enough to achieve the equivalent GPC performance. For higher order systems, $M > 0$ must be selected to find the best approximation to the GPC solution.

Constrained Predictive PID Controller Design

All real-world control systems must be able to accommodate system constraints. These constraints can originate from amplitude limits in the control signal, slew rate limits of the actuator dynamics, and output signal limits. Typical system constraints can be expressed in condensed form as (Camacho and Bordons, 1999):

$$Ru \leq c$$

To absorb different types of system constraints into the design of the predictive PID controller two methods are investigated:

- Using an anti-windup integrator in the controller
- Calculation of optimal values for constrained Predictive PID controller gains

Using an Anti-Windup Integrator in the Predictive PID Controller

Windup problems were originally encountered when using PI/PID controllers for controlling linear systems with saturating control inputs (see Chapter 1). To implement anti-windup procedures in the predictive controller, the following changes in calculation of the control signal should be applied.

Using an anti-windup integrator (as shown in Figure 13.18), the constrained control signal will be

$$\Delta u(k) = \Delta \tilde{u}(k) + \Delta \tilde{u}(k+1) + \ldots + \Delta \tilde{u}(k+M) + \frac{u^*(k) - u(k)}{T_t}$$

$$= \Delta u(k)_{un} + \frac{u^*(k) - u(k)}{T_t}$$

where the unconstrained control signal is $\Delta u(k)_{un}$ and for the nonlinear function $f: \Re \rightarrow \Re$, $u^*(k) = f(u(k))$. After some straightforward algebra:

$$u(k) = \frac{T_t}{1 + T_t}\left[u(k-1) + \Delta u(k)_{un} + \frac{u^*(k)}{T_t}\right]$$

$$= \frac{T_t}{1 + T_t}[u(k-1) + \Delta u(k)_{un}] + \frac{u^*(k)}{1 + T_t}$$

Thus it is clear that for unconstrained systems where $u'(k) = u(k)$, the above expressions will revert to unconstrained control signals.

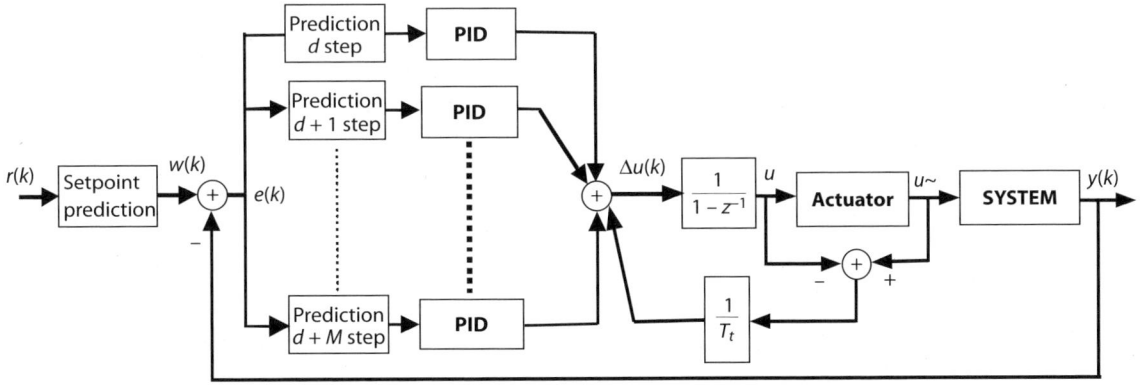

Figure 13.18 Predictive PID with an integral anti-windup facility.

Computing the Optimal Values for Constrained Predictive PID Controller Gains

To obtain the optimal values of the PID controller gains, the GPC cost function with constraints is used. Constrained GPC control consists of applying a control sequence that minimises the following cost function:

$$\min J(u) = \frac{1}{2} u^T H u + b^T u + f_0$$

subject to $Ru \leq c$.

It is clear that the implementation of GPC for processes with bounded signals requires the solution of a quadratic programming (QP) problem. There are many QP programs to choose from to compute the optimal values of GPC control signal $\Delta u(k)_{opt}$ for the above constrained cost function. After computing $\Delta u(k)_{opt}$, the control signal matching principle is applied to compute the optimal values of predictive PID gains. Thus, the predictive PID control signals should be made the same as the optimal GPC control signal. This means that the $\Delta u(k)_{opt}$ should be equal to

$$\Delta u(k)_{opt} = -(1 + K\alpha_d)^{-1} K[F_f \quad G_g] \begin{bmatrix} y_0 \\ \Delta u_0 \end{bmatrix}$$

Define

$$M_m = -[F_f \quad G_g] \begin{bmatrix} y_0 \\ \Delta u_0 \end{bmatrix}$$

Then

$$\Delta u(k)_{opt} = (1 + K\alpha_d)^{-1} K M_m$$

$$(1 + K\alpha) \Delta u(k)_{opt} = K M_m$$

and hence

$$K[M_m - \alpha \Delta u(k)_{opt}] = \Delta u(k)_{opt}$$

The value of the constrained predictive PID control gains can then be calculated as

$$K = (M_m - \alpha \Delta u(k)_{opt})^T [(M_m - \alpha \Delta u(k)_{opt})(M_m - \alpha \Delta u(k)_{opt})^T]^{-1} \Delta u(k)_{opt}$$

The values of the PID gains found in the above formula ensure that constraints are not violated. However, the gains are time-varying since the control signal calculated from the GPC algorithm changes with time. Hence an adaptive PID controller, as described later, can be formulated to handle system input constraints. The approach requires the solution of the constrained GPC control problem at each instant of time in order to find the constrained PID controller gains, and as such the question may arise as to why the GPC solution is not used directly. This is of course true, but this technique can be used in a supervisory role or can even be employed to replace the integral windup scheme in order to obtain a better solution. Furthermore, the algorithm can be used with existing industrial PID controllers without the need for major refurbishment of the control system. The adaptive PID algorithm is now stated.

Algorithm 13.4: Adaptive predictive PID controller for systems with input constraints

Step 1 Initialisation
   Set loop counter $t = 0$
   Follow Steps 1 and 2 of Algorithm 13.3
Step 2 On-line calculation
   If $Ru > c$ the constraints are violated so calculate the constrained PID gains.
   Follow Step 3 of the unconstrained Algorithm 13.3.
Step 3 Loop
   Set $t = t + 1$ and go to Step 2
Algorithm end

Case Studies

Some case studies of the stability and performance of different systems controlled by a predictive PID controller designed using the controls signal matching method will be presented.

Stability Studies

The effects of the prediction horizon M on the PID parameters and the variation of PID parameters on the stability region for three types of systems have been considered. The test systems are:

System 1: Non-minimum phase second-order system $G_1(s) = \dfrac{-10s+1}{40s^2+10s+1}$

System 2: Stable second-order system with time delay $G_2(s) = \dfrac{1}{40s^2+10s+1}e^{-3s}$

System 3: Unstable second-order system $G_3(s) = \dfrac{1}{40s^2+10s-1}$

The results showed that for a non-minimum phase system the stability region is closed and a larger value for M causes a smaller stability region. For stable systems the regions are not closed, but like non-minimum phase systems increasing the M-level of the PID controller decreases the size of the stability regions. For System 3, represented by $G_3(s)$, the stability regions are not closed and larger M values cause the stability regions to be bigger. These results are shown in Figure 13.19.

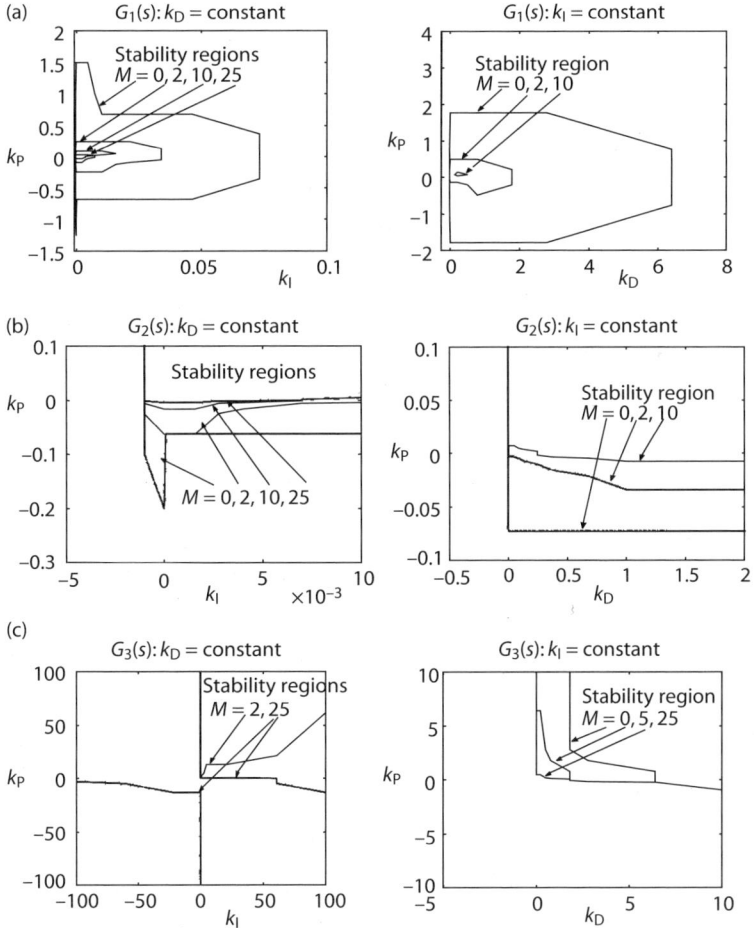

Figure 13.19 Closed-loop stability regions for three system types. (a) Second-order non-minimum phase system; (b) stable second-order system; (c) unstable second-order system.

Remark 13.3 Larger values of M decrease the PID coefficients, and for stable systems increase the possibility of stability. For non-minimum phase systems, larger M values decrease the possibility of stability, whilst for unstable systems increasing the value of M increases the possibility of stability.

Performance Study

For the study of performance six different systems were considered; the systems are listed as:

System 1 Second-order system
$$G_1(z) = \frac{0.3457}{z^2 - 0.8952z + 0.3679}$$

System 2 Second-order system with dead time
$$G_2(z) = \frac{0.3457}{z^2 - 0.8952z + 0.3679} z^{-4}$$

System 3 Second-order non-minimum phase
$$G_3(z) = \frac{0.09628z - 0.688}{z^2 - 0.8952z + 0.3679}$$

System 4 Third-order system
$$G_4(z) = \frac{0.1121z^2 - 0.3007z + 0.05282}{z^3 - 1.501z^2 + 0.9104z - 0.2231}$$

System 5 Boiler system (loop one)
$$G_5(z) = \frac{-0.007z + 0.007}{z^2 - 1.98z + 0.98}$$

System 6 Boiler system (loop two)
$$G_6(z) = \frac{z^2 - 1.96z + 0.96}{z^2 - 1.93z + 0.93}$$

GPC and predictive PID control methods were used to design the controller for each system when there is zero-mean white noise in the system. For GPC, the costing horizon of the output was $N = 25$ and the control costing horizon was $N_u = 1$. The control term cost weighting was $\lambda = 80$. The GPC gains and predictive PID control gains are shown in Table 13.2.

For stable second-order system the PID gains is calculated using the form from p. 507, namely $K = K_{GPC}F = K_0$, where the solution of predictive PID is exact. Thus the step responses of predictive PID control and the GPC methods are the same (see Figure 13.20). For the second-order system with dead time, the PID gains are calculated using the formula $K = K_0(\tilde{F} - \alpha K_0)^{-1}$ from p. 508. The closed-loop responses of the system to a step input for the predictive PID control and GPC methods are the same (see Figure 13.20). It should be noted that for the condition $k < d$ the predictive PID gains have been considered constant.

For System 3, represented by $G_3(z)$ the predictive PID gains are calculated using the full equation $K = K_0(S_0 - \alpha K_0)^T[(S_0 - \alpha K_0)(S_0 - \alpha K_0)^T]^{-1}$, which gives a predictive PID control gain approximation to GPC gains. From Figure 13.21, the step response of a closed-loop system obtained using the predictive PID controller is very close to the step response of the full GPC system response.

The closed-loop step responses for System 4, which is of third order, for the predictive PID control and GPC methods are shown in Figure 13.22. For this third-order unstable system it is necessary to increase the value of the prediction horizon M so that $M \geq 1$ to ensure closed stability. For $M = 1$ and $M = 2$, Figure 13.22 shows that the predictive PID control response are a reasonable approximation of those obtained from the full GPC gain matrix.

For the last example, the predictive PID controller design method has been applied to the loops of a boiler model (De Mello, 1991). The step responses of the closed-loop system for Loops 1 and 2 of the boiler model are shown in Figure 13.23. The results show that the method is applicable to this industrial boiler model with one and two levels of predictive PID control for Loops 1 and 2, respectively.

A comparison of the disturbance rejection properties of the predictive PID controller with GPC and conventional PID control for the second-order System 2 ($G_2(z)$) is shown in Figure 13.24(a). For the

Table 13.2 GPC and predictive PID control gains for six different systems.

System No.	Description	Level of PID (M)	GPC gain	Predictive PID control gain
1	Stable second-order system	0	[0.4128 −0.3696 0.1502]	[0.0692 0.1934 0.1502]
2	Second-order system with dead time	1	[0.408 0.365 0.15]	[0.081 0.225 0.175]
3	Non-minimum phase second-order system	1	[−0.536 0.483 −0.198 0.37]	[−0.087 −0.251 −0.198]
4	Stable third-order system	2 3	[1.383 −2.07 1.251 −0.305 0.482 0.072]	[−0.049 0.234 0.544] [0.041 0.156 0.393]
5	Boiler model; transfer function between throttle pressure and fuel/air	0	[0.1658 0.078 0.0287 0.1487]	[−0.029 −0.1004 0.0037]
6	Boiler model; transfer function between flow rate and control valve	1	[0.374 −0.129 0.0005 −0.11 0.069]	[0.128 0.25 −0.0005]

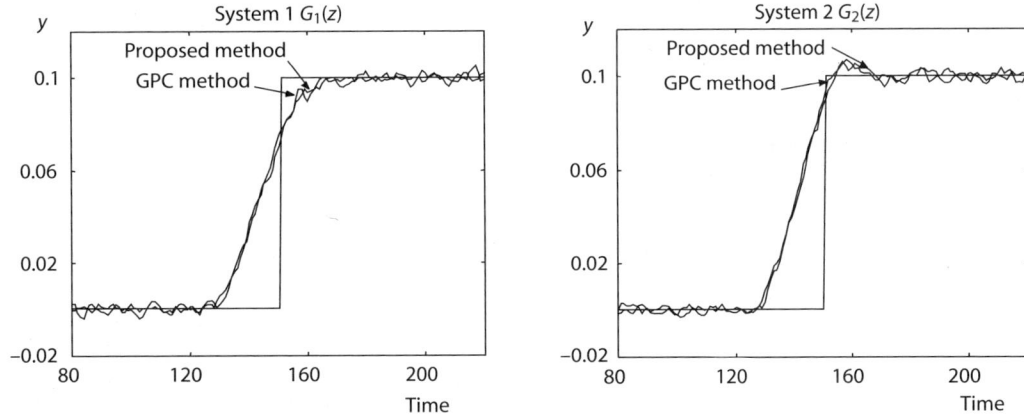

Figure 13.20 Comparison of GPC and predictive PID control for second-order systems.

value $M = 1$, the predictive PID controller rejects the disturbance much faster than the conventional PID at the expense of a larger overshoot. Figure 13.24(b) shows the simulation result for the case where the future setpoint is not known. The predictive PID response for $M = 1$ is almost critically damped compared to the conventional PID controller, where the response is overdamped. Figure 13.24(c) demonstrates the simulation for the case where the known setpoint changes are applied in advance for d steps, the period of the dead time. Again, better performance can be achieved using the predictive PID controller.

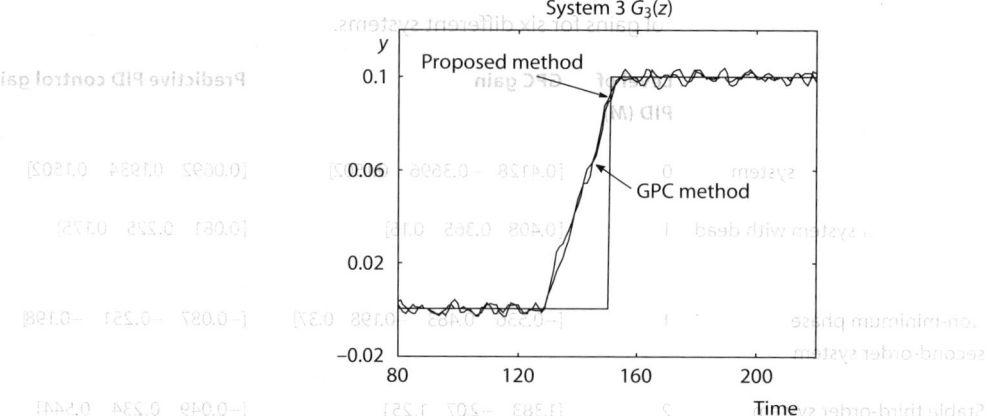

Figure 13.21 Output of GPC method and predictive PID control for System 3 with $M = 1$.

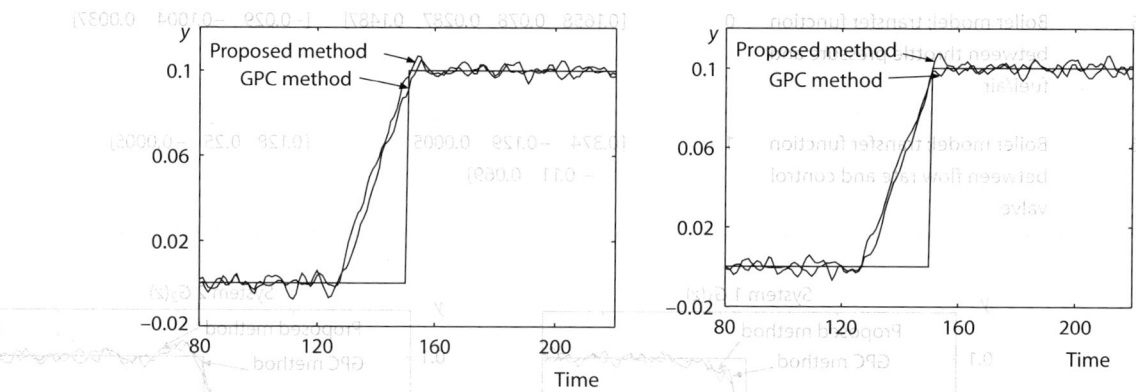

Figure 13.22 Output of GPC method and predictive PID control for System 4 with $M = 1, 2$.

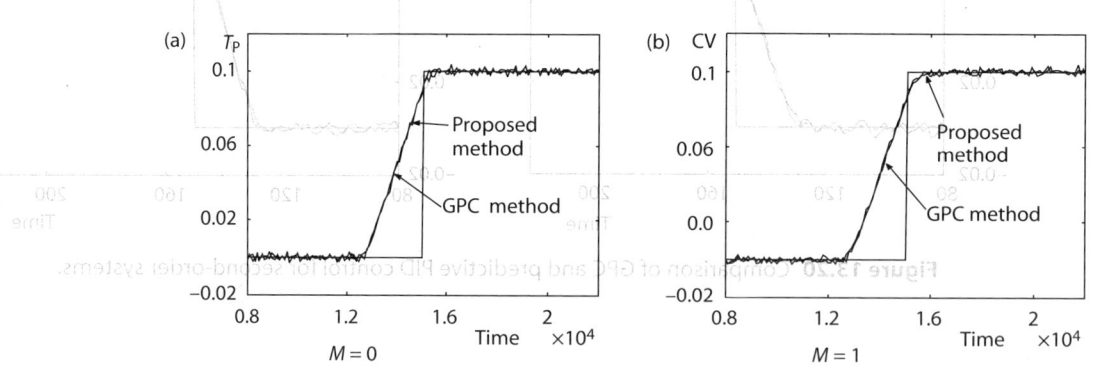

Figure 13.23 Comparison of GPC method and predictive PID control for an industrial boiler model: (a) Loop 1; (b) Loop 2.

Concluding Remarks

For any order of system, by changing the level parameter (M) of the predictive PID controller it is usually possible to achieve performance close to that obtainable from a GPC-controlled system. In the case of second-order systems, one level of PID ($M = 0$) is sufficient to achieve GPC performance. For

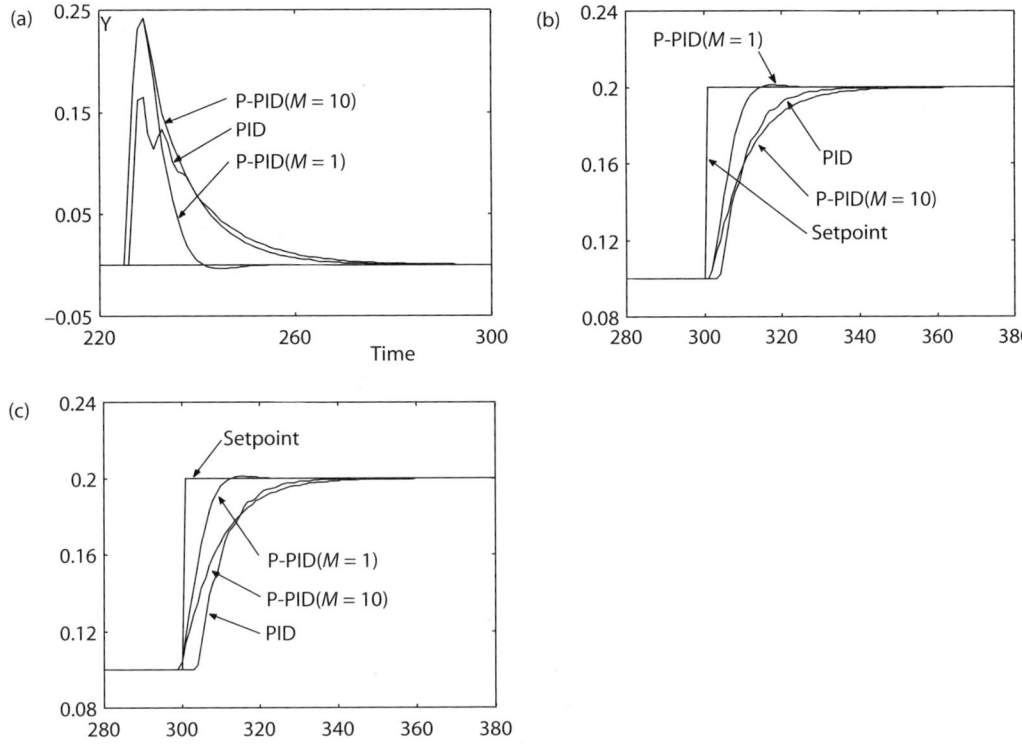

Figure 13.24 (a) Comparison of disturbance rejection of the conventional PID with predictive PID control. (b) Comparison of conventional PID with predictive PID control without future setpoint information. (c) Comparison of conventional PID with predictive PID control when the setpoint change is applied d seconds in advance.

higher-order systems, the prediction horizon parameter M has to be selected to find the best approximation to the GPC solution.

The new predictive PID controller structure based on the GPC approach was shown to be applicable to a wide variety of processes. In the proposed configuration, the system transient behaviour is managed by the predicted trajectories and closed-loop stability was assured by selecting an appropriate level for the predictive PID controller structure. The predictive PID controller reduces to the same structure as a conventional PI or PID controller for first- and second-order systems, respectively. It was shown how the optimal values of predictive PID controller gains can be found using a matching of the GPC control signal values. One of the advantages of the proposed controller is that it can be used with systems of any order and the PID tuning can be used to adjust the controller performance.

13.4.3 Design of MIMO Predictive PID controllers

The extension of scalar tuning methods to multivariable systems is not a straightforward issue. Optimisation methods represent an attractive route to the tuning of multivariable controllers and many different facets of the optimisation procedure have been exploited in the search for workable tuning methods. For example, Yusof et al. (1994) presented a multivariable self-tuning PID controller procedure, the objective of which was to minimise the variance of output. Wang and Wu (1995) have proposed a multi-objective optimisation method for calculating the parameters of PID controllers. Meanwhile,

Pegel and Engell (2001) presented a convex optimisation method to compute multivariable PID controllers based on the idea of the approximation of the optimal attainable closed-loop performance.

One family of approaches compute the PID parameters by minimising the error between the output of a PID controller and a more advanced controller such as one derived from optimal LQG theory, Model-based Predictive Control or even optimal H_∞ theory. This is the control signal matching approach that is used in this section where the GPC control method is used to provide the desired control signal and the predictive PID controller is designed to attain this level of optimal performance. Formally, the approach involves solving the optimisation problem

$$\min J = \|u_{\text{PID}}(k) - u_{\text{GPC}}(k)\|_p$$

where the optimal control signal to be attained is denoted $u_{\text{GPC}}(k)$, the predictive PID control signal is denoted $u_{\text{PID}}(k)$ and the norm index is p ($p = 1, 2, \ldots, \infty$).

An interesting potential of this approach is the ability to adjoin actuator constraints and other metrics representing closed-loop performance and stability to the optimisation. Depending on the order of the system and the time delay it is possible in some cases to obtain analytical solutions to the optimisation.

A predictive PID controller for SISO systems based on the control signal matching method was introduced in the last section. The technique was extended for MIMO systems with polynomial representation by Moradi et al. (2002). However, the presentation in this section uses the state space representation of multivariable predictive PID controllers.

State Space Representation of MIMO Predictive PID Controllers

The state space representation has the advantage that it is generically multivariable and can be used for multivariable processes in a straightforward manner. The control law is simply the feedback of a linear combination of state vector variables although the solution requires an observer if the states are not accessible. Both Morari (1994) and Maciejowski (2001) have presented useful state space formulations for model-based predictive control problems.

Process System Representation
Consider the deterministic state space representation of the plant as follows:

$$x_p(k+1) = A_p x_p(k) + B_p \Delta u(k)$$
$$y(k) = C_p x_p(k)$$

where the notation has the interpretations:

Process state vector	$x_p \in \Re^{N \times 1}$	State coefficient matrix	$A_p \in \Re^{N \times N}$
Incremental control vector	$\Delta u \in \Re^{L \times 1}$	Control coefficient matrix	$B_p \in \Re^{N \times L}$
Output vector	$y \in \Re^{L \times 1}$	Output coefficient matrix	$C_p \in \Re^{L \times N}$
Number of input and output variables	L	Number of state variables	N

The system triple (A_p, B_p, C_p) is assumed to be completely controllable and observable.

Output Prediction
The output of the model for step $(k + i)$, assuming that the state at step k and future control increments are known, can be computed by recursively applying the state space process equations to yield (Camacho and Bordons, 1999)

$$y(k+i) = C_p A_p^i x_p(k) + [C_p A_p^{i-1} B_p \quad C_p A_p^{i-2} B_p \quad \ldots \quad C_p B_p] \begin{bmatrix} \Delta u(k) \\ \Delta u(k+1) \\ \vdots \\ \Delta u(k+i-1) \end{bmatrix}$$

MIMO PID Controller Description

The conventional MIMO PID controller in discrete form for an L-square multivariable system can be represented by

$$u(k) = K_p e(k) + K_I \sum_{j=1}^{k} e(j) + K_D[e(k) - e(k-1)]$$

It is assumed that only L integral controllers are used, so matrix K_I is diagonal. The incremental MIMO PID controller can now be formulated as

$$\Delta u(k) = u(k) - u(k-1) = (K_p + K_I + K_D)e(k) + (-K_p - 2K_D)e(k-1) + K_D e(k-2)$$

In compact matrix form, the incremental MIMO PID controller can be written as

$$\Delta u(k) = KE(k) = K[R(k) - Y(k)]$$

where

$$\Delta u(k) = [\Delta u_1(k) \quad \Delta u_2(k) \quad \ldots \quad \Delta u_L(k)]^T \quad K = [K_D \quad -2K_D - K_p \quad K_D + K_p + K_I]$$
$$E(k) = [e(k-2)^T \quad e(k-1)^T \quad e(k)^T]^T \quad e(k) = [e_1(k) \quad e_2(k) \quad \ldots \quad e_L(k)]^T$$
$$Y(k) = [y(k-2)^T \quad y(k-1)^T \quad y(k)^T]^T \quad y(k) = [y_1(k) \quad y_2(k) \quad \ldots \quad y_L(k)]^T$$
$$R(k) = [r(k-2)^T \quad r(k-1)^T \quad r(k)^T]^T \quad r(k) = [r_1(k) \quad r_2(k) \quad \ldots \quad r_L(k)]^T$$
$$R, Y, E \in \Re^{3L \times 1} \quad K \in \Re^{L \times 3L} \quad r, y, \Delta u \in \Re^{L \times 1}$$

and $K_p, K_I, K_D \in \Re^{L \times L}$ are the proportional, integral and derivative gain matrices, respectively.

MIMO Predictive PID Controller

Using the MIMO PID controller equation, a predictive PID (PPID) controller is defined as follows:

$$\Delta u(k)_{PPID} = K \sum_{i=0}^{M} E(k+i) = K \sum_{i=0}^{M} R(k+i) - K \sum_{i=0}^{M} Y(k+i)$$

where M is the PID controller predictive horizon. The controller consists of M parallel PID controllers. For $M = 0$, the controller is identical to the conventional PID controller. For $M > 0$ the proposed controller has predictive capability similar to GPC. The predictive PID controller signal may be decomposed as

$$\Delta u(k)_{PPID} = K[E(k) + E(k+1) + \ldots + E(k+M)] = \Delta u(k) + \Delta u(k+1) + \ldots + E(k+M)$$

The input of the ith PID component at time depends on the error signal at time $(k+i)$. This implies that the current control signal value is a linear combination of the future predicted outputs. Thus the future outputs for M steps ahead are needed to calculate the control signal.

Lemma 13.1

For the system represented by the state space model:

$$x_p(k+1) = A_p x_p(k) + B_p \Delta u(k)$$
$$y(k) = C_p x_p(k)$$

and the predictive PID controller:

$$\Delta u(k)_{PPID} = K\sum_{i=0}^{M} E(k+i) = K\sum_{i=0}^{M} R(k+i) - K\sum_{i=0}^{M} Y(k+i)$$

the control signal of predictive PID will be given by

$$\Delta u_{PPID}(k) = (I + KH_t)^{-1}[KR_t(k) - KF_t x_p(k)]$$

where

$$H_t = \sum_{i=0}^{M} H_i', \quad R_t(k) = \sum_{i=0}^{M} R(k+i), \quad F_t = \sum_{i=0}^{M} F_i$$

Proof
See Appendix 13.A.1. ◻

The lemma shows that the control signal of a MIMO predictive PID controller is defined as a function of the process model (H_i', F_i), the future setpoint information $R_i(k)$ and the PID gains K.

Optimal Values of the Predictive PID Gains

To obtain the optimal values of the gains for the PID controller, the Generalised Predictive Control (GPC) algorithm is used as the ideal solution. For typical process control problems the default settings of the output cost horizon $[N_1:N_2] = [1:N]$ and the control cost horizon $N_u = 1$ can be used to obtain reasonable performance (Clarke *et al.*, 1987). This leads to a GPC control increment of (Camacho and Bordons, 1999)

$$\Delta u_{GPC}(k) = K_{GPC} W(k) - K_0 x_p(k)$$

where

$$K_{GPC} = [H^T H + \lambda I]^{-1} H^T, \quad K_0 = K_{GPC} F$$

$$W(k) = [w(k)^T \quad w(k+1)^T \quad \ldots \quad w(k+N)^T]^T, \quad w(k) = [w_1(k) \quad w_2(k) \quad \ldots \quad w_L(k)]^T$$

$$H = \begin{bmatrix} B_p & 0 & \ldots & 0 \\ A_p B_p & B_p & \ldots & 0 \\ \vdots & \vdots & \ddots & \vdots \\ A_p^{N-1} B_p & A_p^{N-2} B_p & \ldots & B_p \end{bmatrix}, \quad F = \begin{bmatrix} A_p \\ A_p^2 \\ \vdots \\ A_p^N \end{bmatrix}$$

Following the control signal matching method, the predictive PID control gains are chosen to minimise the norm difference between the predictive PID control signal and the GPC controller signal; thus the following optimisation problem has to be solved:

$$\min_{K \in K_{PID}^S, M} J(K, K_0) = \|\Delta u_{PPID}(K) - \Delta u_{GPC}(K_0)\|_2$$

where the set of stabilising PID gains is denoted K_{PID}^S and M is the predictive PID controller horizon. The predictive PID control signal and the GPC control signal both depend on the closed-loop response matrix $Z = x_p(k)$. But the matrix Z produced by each control law will be a function of the sequence of past control signals. To make the analysis tractable, it is assumed that Z arising from the application of predictive PID control law evolves closely to the outputs generated by the GPC algorithm, so that a first-order approximation of Z may be written as

$$Z(K) = Z(K_0) + \Delta Z$$

With this assumption the following theorem can be established.

Theorem 13.1
For the given predictive PID control law and the GPC control law, and assuming

$$Z_{\text{PPID}} = Z(K_0) + \Delta Z$$

then

$$J(K, K_0) = \|\Delta u_{\text{PID}}(K) - \Delta u_{\text{GPC}}(K_0)\|_2 \leq \| -(1 + KH_t)^{-1} KF_t\|_2 \|\Delta Z\|_2$$

with

$$(I + KH_i)^{-1} KF_i = K_0$$
$$(I + KH_i)^{-1} KR_i(k) = K_{\text{GPC}} W(k)$$

and where $\|\Delta Z\|_2$ is suitably small.

Proof
See Appendix 13.A.2 □

Using Theorem 13.1, the solution for the predictive PID control gain matrix K can now be found in terms of K_0 from the equation $(I + KH_i)^{-1} KF_i = K_0$, which implies

$$K(F_i - H_i K_0) = K_0$$

A unique solution to this equation always exists and takes the form (Levine, 1996)

$$K = K_0 (F_t - H_t K_0)^{\text{T}} [(F_t - H_t K_0)(F_t - H_t K_0)^{\text{T}}]^{-1}$$

Assuming the first, second and third L columns of K are denoted K_{1L}, K_{2L} and K_{3L} respectively, the predictive PID gain matrices can be calculated as

$$K_D = K_{1L}, \quad K_P = -K_{2L} - 2K_D \quad \text{and} \quad K_I = K_{3L} - K_D - K_P$$

Also, using the equation $(I + KH_i)^{-1} KR_i(k) = K_{\text{GPC}} W(k)$ from Theorem 13.1, the rebuilt future setpoint is calculated as

$$R_t(k) = [K^{\text{T}}(KK^{\text{T}})^{-1}](I + KH_t) K_{\text{GPC}} W(k)$$

Closed-Loop Stability Issues for MIMO Predictive PID Controllers
In this section, the closed-loop state space system representation for the proposed method will be found using the state space representation of a conventional PID controller. This closed-loop representation will be used for stability studies of the method. The discrete state space form for a conventional MIMO PID controller is

$$x_c(k+1) = A_c x_c(k) + B_c e(k)$$
$$\Delta u(k) = C_c x_c(k) + D_c e(k)$$

where A_c, B_c, C_c, D_c are controller state space coefficient matrices.

Lemma 13.2
Using the state space representation for a conventional PID controller and the predictive PID controller equation, the state space representation of the MIMO predictive PID controller is given by

$$x_c(k+1) = A_{cc}x_c(k) + B_{cc}\sum_{i=0}^{M} r(k+i) - A_{cp}x_p(k)$$

$$\Delta u(k) = C_{cc}x_c(k) + D'_c\sum_{i=0}^{M} r(k+i) - D_{cp}x_p(k)$$

Proof
See Appendix 13.A.3. □

Using the defined model of the system and the above state space predictive PID controller representation, the closed-loop state space system can be written as

$$\begin{bmatrix} x_p(k+1) \\ x(k+1) \end{bmatrix} = \begin{bmatrix} A - B_p D_{CP} & B_p C_{CC} \\ -A_{CP} & A_{CC} \end{bmatrix}\begin{bmatrix} x_p(k) \\ x(k) \end{bmatrix}$$

To verify the stability of the system, all the eigenvalues of the closed-loop system matrix should have less than unit modulus, where

$$A_{CL} = \begin{bmatrix} A_P - B_P D_{CP} & B_p C_{CC} \\ -A_{CP} & A_{CC} \end{bmatrix}$$

The MIMO predictive PID controller can be designed and implemented using the following algorithmic procedure.

Algorithm 13.5: Multivariable predictive PID controller computation

Step 1 Initialisation
  For the system model, find the discrete state space system matrices A_p, B_p, C_p.
  Select the prediction horizon M and compute the future setpoint vector W.

Step 2 PID gains calculation
  Compute the matrices H_t, F_t (Lemma 13.1).
  Calculate the GPC gain K_{GPC}.
  Calculate the optimal value of predictive PID gains.
  Iterate over the value of M to minimise the cost function and maximise closed-loop stability.

Step 3 Future setpoints
  Calculate the signals $F_t Z(K)$ and $R_t(K)$.
  Calculate the control signal $u(k) = u(k-1) + (I + KH_t)^{-1} K[R_t(k) - F_t Z(K)]$.

Step 4 Input constraints
  Repeat Step 2 but calculate the GPC gain K_{GPC} subject to the input constraints.

Algorithm end

The calculation for the constrained input case should be repeated at each sample time and results in an adaptive PID controller (Katebi and Moradi, 2001).

Case Studies

In this section, the stability and performance of multivariable predictive PID controllers are studied and compared to the GPC controller for two industrial systems.

Process 1: Stirred Tank Reactor
This linear model of a stirred tank reactor is (Camacho and Bordons, 1999):

$$G_1(s) = \begin{bmatrix} \dfrac{1}{1+0.7s} & \dfrac{5}{1+0.3s} \\ \dfrac{1}{1+0.5s} & \dfrac{5}{1+0.4s} \end{bmatrix}$$

where the manipulated variables $u_1(s)$ and $u_2(s)$ are the feed flow rate and the flow of coolant in the jacket respectively. The controlled output variables $Y_1(s)$ and $Y_2(s)$ are the effluent concentration and the reactor temperature respectively.

Process 2: A Boiler Model

The boiler model considered is due to Katebi et al. (2000) and is given by

$$G_2(s) = \begin{bmatrix} \dfrac{-0.007s + 0.00078}{s^2 + 0.0178s + 0.00055} & \dfrac{-0.005s + 7 \times 10^{-5}}{s^2 + 0.0178s + 0.00055} \\ \dfrac{-0.0067s + 0.00062}{s^2 + 0.017s + 0.00053} & \dfrac{s^2 - 0.043s + 0.00065}{s^2 + 0.067s + 0.0006} \end{bmatrix}$$

where the manipulated variables $u_1(s)$ and $u_2(s)$ are the feed/air demand and the control valve position respectively. The controlled output variables $Y_1(s)$ and $Y_2(s)$ are the throttle pressure and the steam flow respectively.

Stability Study

The closed-loop system matrix is used to study the stability issues for the two processes. The effect of the choice of prediction horizon parameter on the predictive PID controller parameters and the size of the stability region for systems are investigated.

For Process 1, Figure 13.25(a) shows that as parameter M increases, the stability region also increases in size. This is to be expected, as the system is minimum-phase and open-loop stable. For the non-minimum phase system of Process 2, the stability regions are bounded and as M increases the stability region decreases in size; see Figure 13.25(b). These results are difficult to generalise and the stability study should be repeated for each individual design case.

Performance Study

The GPC and predictive PID control methods were used to design multivariable controllers for the two processes $G_1(s)$ and $G_2(s)$. In the case of the GPC method, the output costing horizon of $N = 20$ and a control costing horizon of $N_u = 1$ were chosen. The controller gains for the two methods are shown in Table 13.3.

This table shows predictive PID controller gains for Process 1, which is a first-order 2I2O system having similar responses to the GPC controller. This occurred for $M = 0$. The step response of the closed-loop system of Process 1 for the two methods is shown in Figure 13.26.

For the boiler model of Process 2, setting $M = 1$ enables the predictive PID controller to approximate the GPC performance closely. The step response of the closed-loop system for two methods along with conventional MIMO PID control is shown in Figure 13.27. The results show that the predictive PID control performance is close to that of the GPC controller and is superior to that of conventional PID control.

The simulation results for Process 2 in the case where future setpoint information is not known are shown in Figure 13.28. The predictive PID control response is almost critically damped compared with conventional PID control, where the response is overdamped.

Concluding Remarks

Experience of simulations performed for various multivariable systems shows that for a first-order 2I2O system, conventional PID (namely, predictive PID with $M = 0$) is sufficient to attain GPC performance. For other processes, the variation in the step responses for different values of M shows that this parameter can be used to meet some additional time-domain control design specifications.

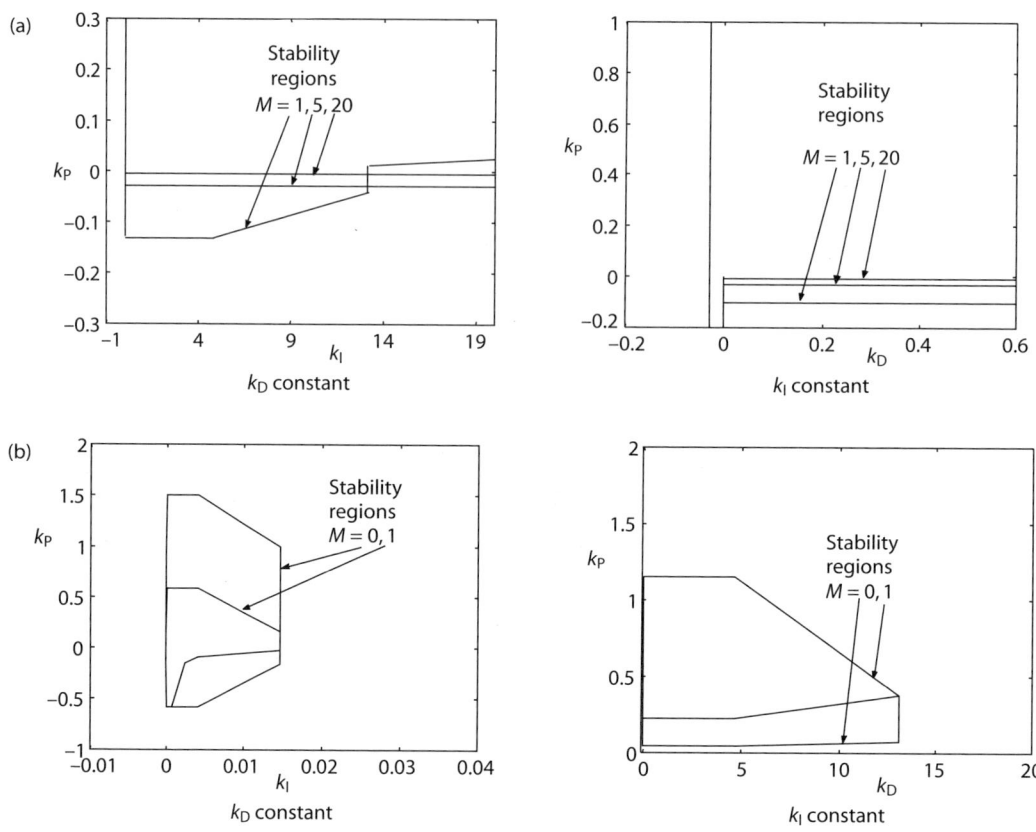

Figure 13.25 (a) Stability regions of predictive PID control for Process 1. (b) Stability regions of predictive PID control for Process 2, the boiler model.

Table 13.3 GPC and Predictive PID control design for processes $G_1(s)$ and $G_2(s)$.

	Level of PID (M)	GPC gain ($\times 10^{-3}$)	Predictive PID gain ($\times 10^{-3}$) $\begin{bmatrix} K_{pid_{11}} & K_{pid_{12}} \\ K_{pid_{21}} & K_{pid_{22}} \end{bmatrix}$
$G_1(s)$	0	$\begin{bmatrix} -72 & 285 & 19 & -62 & -1 & 3 & -20 & 10 \\ -266 & 18 & -64 & -3.8 & 2 & 0.2 & -8 & -273 \end{bmatrix}$	$\begin{bmatrix} -20 & -54 & -0.6 & 55 & 23 & 3 \\ -6 & 174 & 2 & 3.4 & 14 & 0.2 \end{bmatrix}$
	10		$\begin{bmatrix} -19 & -29 & -5 & 6.7 & 38 & 17 \\ -37 & -40 & 12 & -13 & -31 & 2.6 \end{bmatrix}$
$G_2(s)$	1	$\begin{bmatrix} -87 & 0 & 60 & 0.1 & 41 \\ -3 & 26 & -2 & 3 & -1.3 \\ 0.1 & 11 & 0 & 3 & 0 & 162 \\ 1 & -0.4 & -2 & -0.1 & 0 & -3 \\ 33 & 55 & 14 & 15 & 3 \\ 270 & -87 & 56 & -0.5 & -23 \end{bmatrix}$	$\begin{bmatrix} -50 & 140 & 40 & -0.3 & -0.3 & -0.2 \\ 10 & 30 & 10 & -3.4 & 13.5 & -5.8 \end{bmatrix}$
	5		$\begin{bmatrix} -4 & 50 & -13 & 0.5 & 1.5 & 0.2 \\ 0.5 & 1.5 & 0.2 & 17 & 47 & -61 \end{bmatrix}$

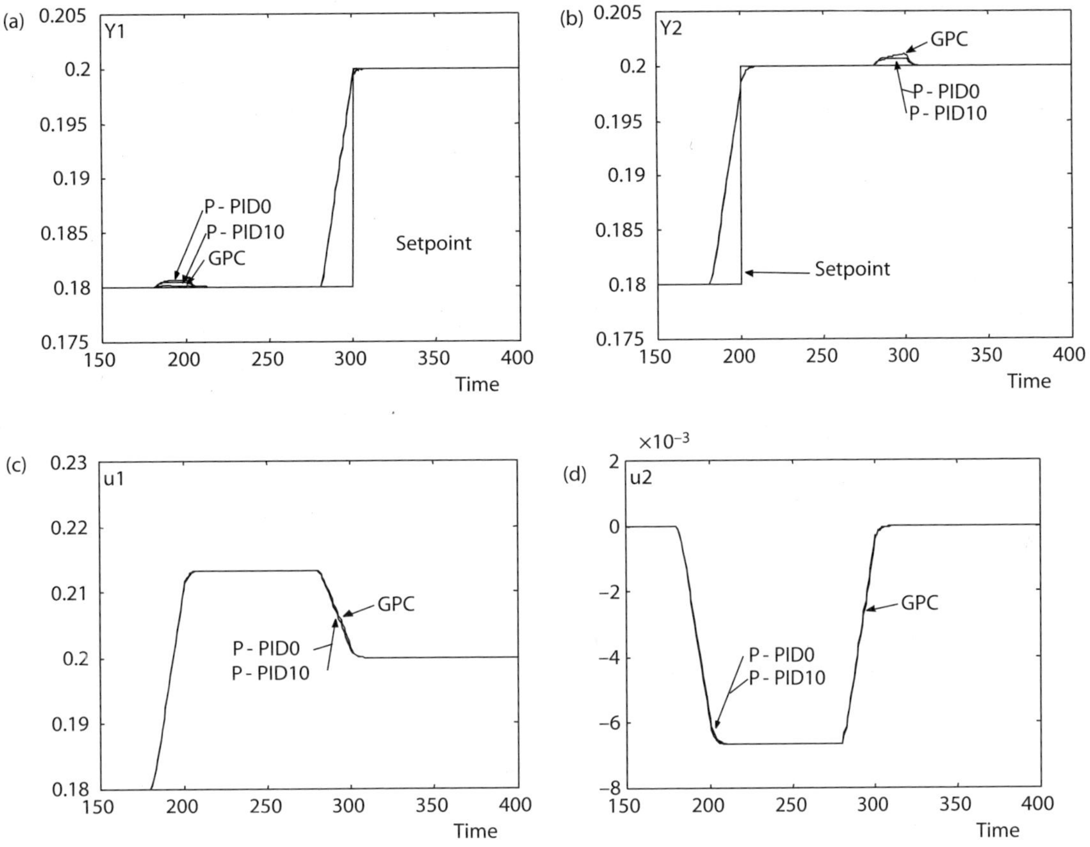

13.4 Control Signal Matching and GPC Methods

Figure 13.26 Comparison of GPC method with Predictive PID control for Process 1. (a) Output 1; (b) output 2; (c) control signal 1; (d) control signal 2.

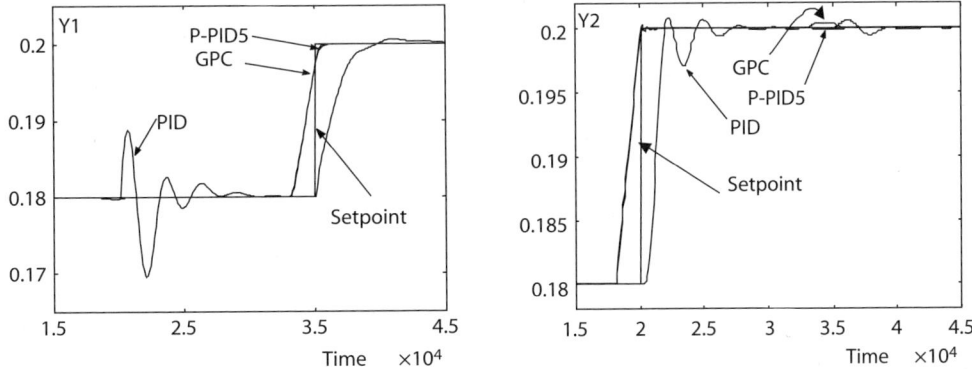

Figure 13.27 Comparison of Predictive PID control with GPC and conventional PID control for the boiler model of Process 2

The predictive PID method was proposed to calculate the PID gains for multivariable systems. The method is based on control signal matching between the PID and GPC controllers. The PPID controller has similar features to GPC controllers and can produce performance close to GPC using a PID control structure. The proposed controller can deal with future setpoints and results in an adaptive PID

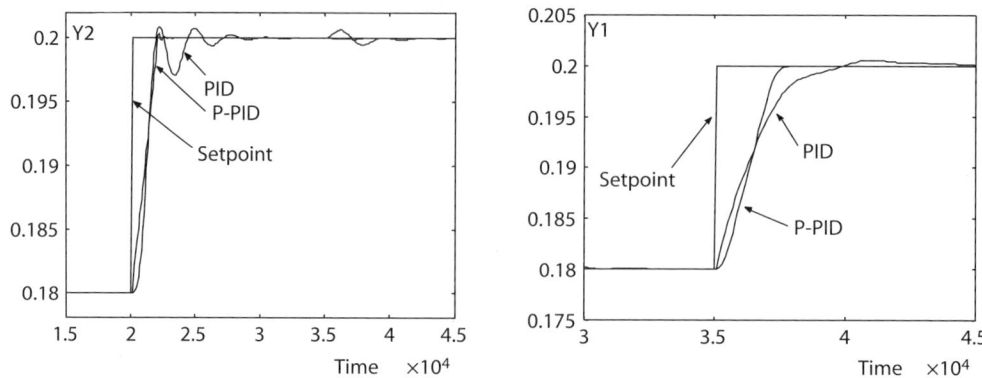

Figure 13.28 Comparison of predictive PID control with conventional PID control without future setpoint information.

controller when the input constraints are included. Simulation studies were presented to demonstrate the superior performance of the controller to the conventional PID controllers. The predictive PID controller is useful for processes where minimum changes to SCADA control systems are allowed. The method has the potential to accommodate advanced control design features such as input constraint handling and predictive tracking capability into the PID design methods. Furthermore, the implementation of the proposed predictive PID controllers requires only small changes to the process hardware and hence may be more attractive for some industrial applications. The method calculates optimal PID gains and hence may also be used as a tuning method to reduce tuning effort and commissioning time.

Acknowledgements

Dr Tore Hägglund (Lund University, Sweden) very kindly reviewed the section on his method for predictive controllers for processes with long time delays. Dr Kok Kiong Tan (National University of Singapore) kindly reviewed the section on his method of deriving predictive PID controllers using simple models and Generalised Predictive Control. Author Michael Johnson, who wrote these sections, would like to thank both Dr Hägglund and Dr Tan for their help.

Appendix 13.A

13.A.1 Proof of Lemma 13.1

The MIMO predictive PID controller was defined as

$$\Delta u(k)_{\text{PPID}} = K \sum_{i=0}^{M} E(k+i) = K \sum_{i=0}^{M} R(k+i) - K \sum_{i=0}^{M} Y(k+i)$$

Using the future output prediction, $Y(k+i)$ can be written as

$$Y(k+i) = \begin{bmatrix} y(k+i-2) \\ y(k+i-1) \\ y(k+i) \end{bmatrix} = \begin{bmatrix} C_p A_p^{i-2} \\ C_p A_p^{i-1} \\ C_p A_p^i \end{bmatrix} x_p(k)$$

$$+ \begin{bmatrix} C_p A_p^{i-3} B_p & \cdots & C_p B_p & 0 & 0 \\ C_p A_p^{i-2} B_p & \cdots & C_p A_p B_p & C_p B_p & 0 \\ C_p A_p^{i-1} B_p & \cdots & C_p A_p B_p & C_p A_p B_p & C_p B_p \end{bmatrix} \begin{bmatrix} \Delta u(k) \\ \Delta u(k+1) \\ \vdots \\ \Delta u(k+i-1) \end{bmatrix}$$

which can be given compact form as

$$Y(k+i) = F_i x_p(k) + H_i U_i(k)$$

with the following definitions:

$$Y(k+i) = \begin{bmatrix} y(k+i-2) \\ y(k+i-1) \\ y(k+i) \end{bmatrix} \quad F_i = \begin{bmatrix} C_p A_p^{i-2} \\ C_p A_p^{i-1} \\ C_p A_p^i \end{bmatrix} \quad A_p^i = \begin{cases} A_p^i & i \geq 0 \\ 0 & i < 0 \end{cases}$$

$$H_i = \begin{bmatrix} C_p A_p^{i-3} B_p & \cdots & C_p B_p & 0 & 0 \\ C_p A_p^{i-2} B_p & \cdots & C_p A_p B_p & C_p B_p & 0 \\ C_p A_p^{i-1} B_p & \cdots & C_p A_p B_p & C_p A_p B_p & C_p B_p \end{bmatrix} \quad U_i = \begin{bmatrix} \Delta u(k) \\ \Delta u(k+1) \\ \vdots \\ \Delta u(k+i-1) \end{bmatrix}$$

Using the above equations

$$\Delta u_{\text{PPID}}(k) = K \sum_{i=0}^{M} R(k+i) - K \left(\sum_{i=0}^{M} F_i \right) x_p(k) - K \left(\sum_{i=0}^{M} H_i' \right) U_M(k)$$

where

$$H_i' = \begin{cases} [H_i \quad Zero(3, M-I)] & i < M \\ H_i & i \geq M \end{cases}$$

If the control cost horizon N_u applies, the above equations can be written as

$$Y(k+i) = \begin{bmatrix} y(k+i-2) \\ y(k+i-1) \\ y(k+i) \end{bmatrix} \quad F_i = \begin{bmatrix} C_p A_p^{i-2} \\ C_p A_p^{i-1} \\ C_p A_p^i \end{bmatrix} \quad U_i = \begin{bmatrix} \Delta u(k) \\ \Delta u(k+1) \\ \vdots \\ \Delta u(k+i-1) \end{bmatrix}$$

$$H_i = \begin{bmatrix} C_p A_p^{i-3} B_p & \cdots & C_p B_p & 0 & 0 \\ C_p A_p^{i-2} B_p & \cdots & C_p A_p B_p & C_p B_p & 0 \\ C_p A_p^{i-1} B_p & \cdots & C_p A_p B_p & C_p A_p B_p & C_p B_p \end{bmatrix}$$

$$H_i' = \begin{cases} [H_i \quad Zero(3, M-i)] & i < N_u \\ H_i & i \geq N_u \end{cases}$$

If $N_u = 1$, then the matrices are

$$Y(k+i) = \begin{bmatrix} y(k+i-2) \\ y(k+i-1) \\ y(k+i) \end{bmatrix} \quad F_i = \begin{bmatrix} C_p A_p^{i-2} \\ C_p A_p^{i-1} \\ C_p A_p^i \end{bmatrix} \quad H_i = \begin{bmatrix} C_p A_p^{i-3} B_p \\ C_p A_p^{i-2} B_p \\ C_p A_p^{i-1} B_p \end{bmatrix}$$

and for the case $N_u = 1$ the equation reduces to

$$(I + KH_t)\Delta u_{\text{PPID}}(k) = KR_t - KF_t x_p(k)$$

where

$$H_t = \sum_{i=0}^{M} H_i' \quad R_t(k) = \sum_{i=0}^{M} R(k+i) \quad F_t = \sum_{i=0}^{M} F_i$$

From this the controller follows as

$$\Delta u_{\text{PPID}}(k) = (I + KH_t)^{-1}[KR_t(k) - KF_t x_p(k)]$$

13.A.2 Proof of Theorem 13.1

Recall the cost function as

$$J(K, K_0) = \|(I + K\alpha)^{-1} K[W - [F_f \quad G_g]Z(K)] - K_{\text{GPC}}w + K_0[Z(K_0)]\|_2$$

The quantities $Z(K), Z(K_0)$ depend on the particular control gain used and it is assumed that the relationship between them can be written as $Z(K) = Z(K_0) + \Delta Z$. Then the optimisation cost function can be developed as

$$J(K, K_0) = \|(1 - K\alpha)^{-1} K[R_t(k) - [F_f \quad G_g]Z(K)] - K_{\text{GPC}}w + K_0[Z(K_0)]\|_2$$
$$\leq \|(-(I - K\alpha)^{-1} K[F_f \quad G_g] + K_0)Z(K_0)\|_2$$
$$+ \|(-(I - K\alpha)^{-1} K[F_f \quad G_g]\Delta Z\|_2 + \|(1 + K\alpha)^{-1} KR_t(k) - K_{\text{GPC}}W(k)\|_2$$
$$\leq \|(-(I - K\alpha)^{-1} K[F_f \quad G_g] + K_0\|_2 \|Z(K_0)\|_2$$
$$+ \|(-(I - K\alpha)^{-1} K[F_f \quad G_g]\|_2 \|\Delta Z\|_2 + \|(I + K\alpha)^{-1} KR_t(k) - K_{\text{GPC}}W(k)\|_2$$

Thus the bound is minimised by selecting

$$(I + K\alpha)^{-1} K[F_f \quad G_g] = K_0$$

and

$$\|(I + K\alpha)^{-1} KR_t(k) - K_{\text{GPC}}W(k)\|_2 = 0$$

Hence it is assumed that it is possible to find suitable gain K so that $\|\Delta Z\|_2$ is suitably small, and that the second expression can be used to rebuild the future setpoint $R_t(k)$. These results lead to

$$J(K, K_0) = \|-(I - K\alpha)^{-1} K[F_f \quad G_g]\Delta Z\|_2 \leq \|-(I + K\alpha)^{-1} K[F_f \quad G_g]\|_2 \|\Delta Z\|_2$$

when

$$(I + K\alpha)^{-1} KR_t(k) = K_{\text{GPC}}W(k)$$

$$(I + K\alpha)^{-1} K[F_f \quad G_g] = K_0$$

13.A.3 Proof of Lemma 13.2

The process model is given by

$$x_p(k+1) = A_p x_p(k) + B_p \Delta u(k)$$
$$y(k) = C_p x_p(k)$$

The discrete representation of a MIMO PID controller can be written as

$$x_c(k+1) = A_c x_c(k) + B_c e(k)$$
$$\Delta u(k) = C_c x_c(k) + D_c e(k)$$

where the error signal is $e(k) = r(k) - y(k)$ and the setpoint is $r(k)$.

The closed-loop state space system can be then be written as

$$\begin{bmatrix} x_p(k+1) \\ x(k+1) \end{bmatrix} = \begin{bmatrix} A - B_p D_c C_p & B_p C_c \\ -B_c C_p & A_C \end{bmatrix} \begin{bmatrix} x_p(k) \\ x(k) \end{bmatrix}$$

$$A_{CL} = \begin{bmatrix} A - B_p D_c C_p & B_p C_c \\ -B_c C_p & A_C \end{bmatrix}$$

where A_{CL} is the closed-loop system matrix.

The M step ahead prediction of output in compact form is given by

$$Y_M(k) = F_M x_p(k) + H_M U_{M-1}(k)$$

where the matrices are defined as

$$Y_M(k) = \begin{bmatrix} y(k) \\ y(k+1) \\ y(k+2) \\ \vdots \\ y(k+M) \end{bmatrix} \quad F_M = \begin{bmatrix} C_p \\ C_p A_p \\ C_p A_p^2 \\ \vdots \\ C_p A_p^M \end{bmatrix} \quad U_{M-1}(k) = \begin{bmatrix} \Delta u(k) \\ \Delta u(k+1) \\ \vdots \\ \Delta u(k+M-1) \end{bmatrix}$$

$$H_i = \begin{bmatrix} 0 & 0 & 0 & 0 \\ C_p B_p & 0 & \cdots & 0 \\ C_p A_p B_p & C_p B_p & \cdots & 0 \\ \vdots & \vdots & \ddots & 0 \\ C_p A_p^{M-1} B_p & C_p A_p^{M-2} B_p & \cdots & C_p B_p \end{bmatrix}$$

The conventional state space representation for PID control is considered and the input to the controller is the sum of error signal from step k to step $(k+M)$, $\Sigma_{i=0}^M e(k+i)$ which depends on the future output of the system. Thus the controller is using $\Sigma_{i=0}^M e(k+i)$ at step k. Also, the states and outputs of the controller will be calculated using the future output of the system only. Hence

$$x_c(k+1) = A_c x_c(k) + B_c \sum_{i=0}^M r(k+i) - B_c \sum_{i=0}^M y(k+i)$$

$$\Delta u(k) = C_c x_c(k) + D_c \sum_{i=0}^M r(k+i) - D_c \sum_{i=0}^M y(k+i)$$

To calculate $\Delta u(k)$ in the above equation, it is necessary to predict $y(k+i)$ and to do this it is necessary to specify $\Delta u(k+i)$. In other words, the control signal at step k depends on future control signals. Since it is assumed that the system is causal, it is assumed that

$$\Delta u(k+i) = 0 \quad i > 0$$

With this assumption, it can be written that

$$y_t(k) = F_{Mt} x_p(k) + H_{Mt} \Delta u(k)$$

where

$$y_t(k) = \sum_{i=0}^{M} y(k+i) \quad F_{Mt} = C_p \sum_{i=0}^{M} A_p^i \quad H_{Mt} = C_p \left(\sum_{i=0}^{M-1} A_p^i \right) B_p$$

Inserting the above equations in the state space representation of the controller yields

$$x_c(k+1) = A_c x_c(k) + B_c \sum_{i=0}^{M} r(k+i) - B_c F_{Mt} x_p(k) - B_c H_{Mt} \Delta u(k)$$

$$\Delta u(k) = C_c x_c(k) + D_c \sum_{i=0}^{M} r(k+i) - D_c F_{Mt} x_p(k) - D_c H_{Mt} \Delta u(k)$$

Calculating $\Delta u(k)$ from the above equation gives

$$(I + D_c H_{Mt}) \Delta u(k) = C_c x_c(k) + D_c \sum_{i=0}^{M} r(k+i) - D_c F_{Mt} x_p(k)$$

and

$$\Delta u(k) = C'_c x_c(k) + D'_c \sum_{i=0}^{M} r(k+i) - D'_c F_{Mt} x_p(k)$$

where

$$C'_c = (I + D_c H_{Mt})^{-1} C_c \quad \text{and} \quad D'_c = (I + D_c H_{Mt})^{-1} D_c$$

Inserting $\Delta u(k)$ into the related equation gives

$$x_c(k+1) = (A_c - B_c H_{Mt} C'_c) x_c(k) + (B_c - B_c H_{Mt} D'_c) \sum_{i=0}^{M} r(k+i)$$
$$- (B_c F_{Mt} - B_c H_{Mt} D'_c F_{Mt}) x_p(k)$$

Rewriting this in compact form proves the lemma as

$$x_c(k+1) = A_{cc} x_c(k) + B_{cc} \sum_{i=0}^{M} r(k+i) - A_{cp} x_p(k)$$

$$\Delta u(k) = C_{cc} x_c(k) + D'_c \sum_{i=0}^{M} r(k+i) - D_{cp} x_p(k)$$

References

Camacho, E.F. and Bordons, C. (1999) *Model Predictive Control*. Springer-Verlag, London.

Chien, I.L. (1988) IMC-PID controller design – an extension. *IFAC Proceeding Series*, **6**, 147–152.

Clarke, D.W., Mohtadi, C. and Tuffs, P.S. (1987) Generalised predictive control, parts I & II. *Automatica*, **23**(2), 137–160.

De Mello, F.P. (1991) Boiler models for system dynamic performance. *IEEE Trans. Power Systems*, **61**, 66–74.

Hägglund, T. (1992) A predictive PI controller for processes with long dead times. *IEEE Control Systems Magazine*, February, 57–60.

Hägglund, T. (1996) An industrial dead-time compensating PI controller. *Control Engineering Practice*, **4**, 749–756.

Katebi, M.R. and Moradi, M.H. (2001) Predictive PID controllers. *IEE Proc. Control Theory Application*, **148**(6), 478–487.

Katebi, M.R., Moradi, M.H. and Johnson, M.A. (2000) Controller tuning methods for industrial boilers. *Proceedings IECON Conference*, Nagoya, Japan.

Levine, W. (1996) *The Control Handbook*. CRC Press, Boca Raton, FL.

Maciejowski, J.M. (2001) *Predictive Control with Constraints*. Prentice Hall, Chichester.

Matauŝek, M.R. and Mićic, A.D. (1996) A modified Smith predictor for controlling a process with an integrator and long deadtime. *IEEE Trans. Automatic Control*, **41**(8), 1199–1203.

Moradi, M.H. and Katebi, M.R. (2001) Predictive PID control for autopilot design. *IFAC Conference on Control Application in Marine Systems*, 18–20 July 2001, Glasgow.

Moradi, M.H., Katebi, M.R. and Johnson, M.A. (2001) Predictive PID control: a new algorithm. *Proceedings IECON 01 Conference*, December, Denver, CO.

Moradi, M.H. (2002) *PhD Thesis*. Industrial Control Centre, University of Strathclyde, Glasgow, UK.

Moradi, M.H., Katebi, M.R. and Johnson, M.A. (2002) The MIMO predictive PID controller design. *Asian Journal of Control*, **4**(4), 452–463.

Morari, M. (1994) *Advances in Model Predictive Control*. Oxford University Press, Oxford.

Morari, M. and Zafiriou, E. (1989) *Robust Process Control*. Prentice Hall International Editions, Chichester.

Pegel, S. and Engell, S. (2001) Multivariable PID controller design via approximation of the attainable performance. *Proceedings IECON 01 Conference*, December, Denver, CO, pp. 724–729.

Rivera, D.E., Skogestad, S. and Morari, M. (1986) Internal model control 4: PID controller design. *Ind. Eng. Chem. Proc. Des. & Dev.*, **25**, 252–265.

Rusnak, I. (1999) Generalised PID controllers. *Proceedings of the 7th IEEE Mediterranean Conference on Control & Automation, MED 99*, Haifa, Israel.

Rusnak, I. (2000) The generalised PID controller and its application to control of ultrasonic and electric motors. *IFAC Workshop PID'00*, Spain, pp. 125–130.

Smith, O.J.M. (1957) Close control of loops with dead time. *Chemical Engineering Progress*, **53**, 217–219.

Smith, O.J.M. (1958) *Feedback Control Systems*. McGraw-Hill, New York.

Tan, K.K., Wang, Q.-G. and Hang, C.C. (1999) *Advances in PID Control*. Springer-Verlag, London.

Tan, K.K., Huang, S.N. and Lee, T.H. (2000) Development of a GPC-based PID controller for unstable systems with dead time. *ISA Trans.*, **39**(1), 57–70.

Tan, K.K., Lee, T.H. and Leu, F.M. (2001) Predictive PI versus Smith control for dead-time compensation. *ISA Trans.*, **40**(1), 17–29.

Tan, K.K., Lee, T.H. and Leu, F.M. (2002a) Optimal Smith predictor design based on a GPC approach. *Industrial and Engineering Chemistry Research*, **41**, 1242–1248.

Tan, K.K., Lee, T.H., Huang, S.N. and Leu, F.M. (2002b) PID control design based on a GPC approach. *Industrial and Engineering Chemistry Research*, **41**, 2013–2022.

Wang, F. and Wu, T.Y. (1995) Multi-loop PID controllers tuning by goal attainment trade-off method. *Trans. Inst. Measurement and Control*, **17**(1), 27–34.

Wang, Q.-G., Hang, C.C. and Yang, X.P. (2000) Single loop controller design via IMC principles. *Proc. Asian Control Conference*, Shanghai, China.

Yusof, R., Omatu, S. and Khalid, M. (1994) Self-tuning PID control: a multivariable derivation and application. *Automatica*, **30**(12), 1975–1981.

About the Contributors

Contributors appear in alphabetical order of family name.

Guanrong (Ron) Chen

Guanrong (Ron) Chen received the MSc degree in Computer Science from Sun Yat-sen (Zhongshan) University, China, and the PhD degree in Applied Mathematics from Texas A&M University, USA. After working at Rice University and the University of Houston, Texas, USA for 15 years, he is currently a Chair Professor in the Department of Electronic Engineering, City University of Hong Kong. He became an IEEE Fellow in 1996, and has been conferred with more than 10 Honorary Guest-Chair Professorships from China and also an Honorary Professorship from the Central Queensland University of Australia. He is Chief, Advisory, Feature, and Associate Editor for eight IEEE Transactions and International Journals.

Professor Chen has (co)authored 15 research monographs and advanced textbooks, more than 300 journal papers, and about 180 conference papers published since 1981 in the field of nonlinear systems (both dynamics and controls) with applications to nonlinear dynamics, as well as intelligent and fuzzy control systems. His technical research monograph in this field is entitled *Introduction to Fuzzy Sets, Fuzzy Logic, and Fuzzy Control Systems* (CRC Press, 2000). He received the outstanding prize for the best journal paper award from the American Society of Engineering Education in 1998, and the best transactions annual paper award from the IEEE Aerospace and Electronic Systems Society in 2001.

James Crowe

James Crowe received the BSc and MPhil degrees from the University of Strathclyde, Glasgow, in 1981 and 1998, respectively. He completed his PhD degree in the field of process control systems at the University of Strathclyde in 2003. His supervisors were Professors Michael A. Johnson and Michael J. Grimble and the external examiner for his PhD thesis was Professor Derek Atherton, University of Sussex. His industrial experience has been gained in the steel, pharmaceutical and electricity generation industries. He has worked on applications as varied as the control of steel converters, continuous casting machines and continuous and batch chemical processes and the design and implementation of control and instrumentation systems for AGR reactors. He is a member of the Institution of Electrical Engineers (UK) and is a Chartered Engineer.

Raihana Ferdous

Raihana Ferdous is a postgraduate student doing her PhD with the National University of Singapore. Her research interests are in intelligent and robust PID control and process monitoring.

David Greenwood

David Greenwood graduated from the University of Sheffield in Electronic, Control and Systems Engineering (BEng) in 1997. Since then he has worked as a Research Assistant at the University of Strathclyde on an industrial and EPSRC-sponsored research project investigating applications of advanced control techniques to hotstrip rolling mills. The partners in this research included Corus plc and Alcan International. This was followed by an industrial internship with Alcan International based at Banbury, UK.

Promotion to a Research Fellowship at the Industrial Control Centre, University of Strathclyde, enabled the steel industry control research to be reinterpreted in a performance assessment and control benchmarking framework. He graduated with a PhD in July 2003, and his thesis was a culmination of the research carried out at the University of Strathclyde in collaboration with industrial partners. Currently, David Greenwood is employed by Network Rail, UK.

Mike J. Grimble

Michael Grimble was born in Grimsby, England. He was awarded a BSc (Coventry) in Electrical Engineering in 1971, and MSc, PhD and DSc degrees from the University of Birmingham in 1971, 1978 and 1982, respectively. Whilst in industry he started an Open University Degree studying mathematics and was awarded a BA in 1978. The University of Strathclyde, Glasgow, appointed him to the Professorship of Industrial Systems in 1981 and he developed the Industrial Control Centre, combining fundamental theoretical studies and real applications research.

As the Director of the Industrial Control Centre, he is concerned with industrial control problems arising in the automotive, aerospace, process control, manufacturing, wind energy and marine sectors. His interests in control theory are focused on robust control theory, multivariable design techniques, optimal control, adaptive control, estimation theory, performance assessment and condition monitoring. He has published five books and well over 100 papers.

He is the managing editor of two journals he founded, namely the *International Journal of Robust and Nonlinear Control* and the *International Journal of Adaptive Control and Signal Processing* (both published by John Wiley & Sons Ltd). He is the joint Series Editor of the Springer Verlag Monograph series on Advances in Industrial Control and is an Editor of the Springer Lecture Notes in Control and Information Sciences series. He is also the joint Series Editor of the Springer-Verlag textbook series on Control and Signal Processing. He was the editor of the Prentice Hall International series of books on Systems and Control Engineering and the Prentice Hall series on Acoustics Speech and Signal Processing.

Professor Grimble is a Fellow of the IEEE (1992), the IEE, the InstMC, the IMA and the Royal Society of Edinburgh (1999).

Hsiao-Ping Huang

Professor Huang's academic career started in 1970 when he was appointed as a lecturer in the department of chemical engineering at National Taiwan University. He has spent over 30 years in both teaching and research since then. He served as the department Chairman at NTU, editor-in-chief of the journal of the Chinese Institute of Chemical Engineering, programme director of the Chemical Engineering Division of the National Science Council in Taiwan, visiting associate professor at University of Wisconsin Madison, visiting scientist at MIT, and visiting professor at Kyoto University. He is currently a full-time professor at National Taiwan University, and an Associate Editor of *Journal of Process Control*. His work has concentrated on control engineering, theory and applications. He is the author or co-author of one book and more than 70 papers on system identifications predictive control, and controller design and assessment. He is chairing a standing committee of the PSE Asia, a small regional international conference, until 2005.

Jyh-Cheng Jeng

Jyh-Cheng Jeng received his PhD in Chemical Engineering from National Taiwan University in 2002, and is currently a post-doctoral researcher in the Department of Chemical Engineering, National Taiwan University. His current research interests include performance assessment and monitoring of control systems, controller autotuning, and identification of nonlinear systems.

Michael A. Johnson

Professor Johnson's academic career has concentrated on control engineering, theory and applications. He has significant industrial control applications and research experience. He is the author and co-author of books and papers on power generation, wastewater control, power transmission and control benchmarking. Previous European Union project experience includes the ESPRIT Projects MARIE and IMPROVE and a joint Italian, Spanish and UK THERMIE Project. He has had a number of student projects in the area of assistive technology and has successfully trained over 40 engineers to PhD level in the last 20 years. He is joint founding Series Editor of the Springer-Verlag monograph series Advances in Industrial Control and also of the graduate textbook series Advanced Textbooks in Control and Signal Processing series. He was an *Automatica* Associate Editor (1985–2002). His most recent textbook is *Control Engineering – An Introductory Course* (Palgrave, 2002), co-authored with colleagues Dr Jacqueline Wilkie and Dr Reza Katebi.

Another of Professor Johnson's interests is that of Assistive Technology. Collaboration with Dr Marion Hersh at the University of Glasgow has led to a set of targeted activities in this area. These included a new undergraduate degree course module, a new conference series and some research projects in the area. He is co-author and co-editor of the book *Assistive Technology for the Hearing-Impaired, Deaf and Deafblind* (Springer-Verlag, 2003). A companion volume, *Assistive Technology for Vision-Impaired, and Blind*, is currently in preparation.

Professor Johnson retired from the University of Strathclyde in December 2002 and was appointed Professor Emeritus of the University in April 2003. Professor Johnson is a Member of the IEEE, a Fellow of the IMA, a Chartered Mathematician, a member and Director of the charity Deafblind Scotland and a member of the British Retinitis Pigmentosa Society.

Reza Katebi

Reza Katebi completed a PhD in Control Engineering at the University of Manchester Institute of Science and Technology in 1984. From 1984 to 1989, he worked as a Research Fellow and a Senior Engineer on marine and steel rolling control systems at the University of Strathclyde and at Industrial Systems and Control Ltd. Dr Katebi joined the Department of Electronic and Electrical Engineering, University of Strathclyde, as Lecturer in 1989 and was promoted to a Senior Lectureship in 1994.

Dr Katebi's research interests have focused on the modelling and control of complex and nonlinear systems with applications to marine, steel and wastewater processes. He leads an EPSRC Marine Control Network and a research team of five engineers in wastewater systems control through the award of two Framework V European projects. Dr Katebi has been principal investigator or co-investigator in EPSRC and European research projects with a grant portfolio of more than £2 million, has published four books, 35 journal articles and book chapters, and 80 refereed conference papers, and has supervised 10 PhD students.

Dr Katebi has served on the International Programme Committee of several international conferences. He was deputy chairperson of the IFAC Conference on Control of Marine Systems and Applications in Glasgow, 2001. He is a Member of IEE and IEEE and chairperson of the IEEE Rolling Process group. He was the co-editor of a special issue of *IEEE Control System Technology* on rolling processes and is a member of the IFAC Technical Committees on Marine Control Systems Technology and on Control Education. Since 2002, he has been an Associate Editor of *Automatica*, the journal of the International Federation of Automatic Control.

Sam Kwong

Dr Sam Kwong received his BSc and MASc degrees in Electrical Engineering from the State University of New York at Buffalo, USA, and the University of Waterloo, Canada, in 1983 and 1985 respectively. He later obtained his PhD from the University of Hagen, Germany. From 1985 to 1987 he was a Diagnostic Engineer with Control Data Canada where he designed the diagnostic software to detect manufacturing faults in the VLSI chips of the Cyber 430 machine. He later joined Bell Northern Research Canada as a member of the scientific staff, where he worked on both the DMS-100 voice network and the DPN-100 data network projects. In 1990, he joined the City University of Hong Kong as a Lecturer in the Department of Electronic Engineering. He is currently an Associate Professor in the Department of Computer Science.

His research interests are in genetic algorithms, speech processing and recognition, digital watermarking, data compression and networking.

Tong Heng Lee

T. H. Lee received the BA degree with First Class Honours in the Engineering Tripos from Cambridge University, England, in 1980 and the PhD degree from Yale University in 1987. He is a Professor in the Department of Electrical and Computer Engineering at the National University of Singapore. He is also currently Head of the Drives, Power and Control Systems Group in this department, and Vice-President and Director of the Office of Research at the University.

Dr Lee's research interests are in the areas of adaptive systems, knowledge-based control, intelligent mechatronics and computational intelligence. He currently holds Associate Editor appointments in *Automatica*; the *IEEE Transactions in Systems, Man and Cybernetics*; *Control Engineering Practice* (an

IFAC journal); the *International Journal of Systems Science* (Taylor & Francis, London); and *Mechatronics* (Pergamon Press, Oxford).

Dr Lee was a recipient of the Cambridge University Charles Baker Prize in Engineering. He has also co-authored three research monographs and holds four patents (two of which are in the technology area of adaptive systems, and two in the area of intelligent mechatronics).

Kim-Fung Man

Kim-Fung Man received his PhD from Cranfield Institute of Technology, Bedfordshire, UK, in 1983. He is now a Professor with the Electronic Engineering Department at City University of Hong Kong. Dr Man is currently the Editor-in-chief of *Real-Time Systems Journal* and editorial member of *IEEE Transactions on Industrial Electronics*, *IEEE Transactions on Mobile Computing* and of the Advances in Industrial Control book series of Springer-Verlag. His research interests are in evolutionary computing and control engineering.

Peter Martin

Peter Martin earned a Bachelor's degree in Electrical and Electronic Engineering from University of Leicester, UK, in 1998, and a Master of Science degree in Advanced Control from the University of Manchester Institute of Science and Technology, UK, in 1999. Since that time, he has been studying for a PhD in Control Engineering with the University of Strathclyde, UK. In July 2003, he was seconded to BAE Systems Avionics in Edinburgh, UK, as a Research Fellow to work on a nonlinear servo systems problem. His research interests include optimal control using polynomial methods, and nonlinear systems. Peter Martin is a Member of the IEE (UK).

Mohammad H. Moradi

M. H. Moradi obtained the BSc and MSc degrees in 1991 and 1993 from the Sharif University of Technology and Tarbiat Modarres University, respectively. Dr Moradi joined the Bu-Ali Sina University in 1993 as a lecturer in the Department of Electronic and Electrical Engineering.

In 1999, he came to the University of Strathclyde to initiate a control engineering research programme at the Industrial Control Centre of the Department of Electronic and Electrical Engineering. One outcome has been the award of the PhD in control engineering in the Faculty of Engineering in 2002.

His current theoretical research interests include predictive control, system identification, robust control design, control through computer networks and information technology. His industrial interests are in the areas of large-scale systems especially power systems and the control of power plants and industrial automation. Dr Moradi has published a number of journal and conference papers in these areas.

Alberto Sanchez

Alberto Sanchez obtained his Engineering degree in 1999 from Escuela Politécnica Nacional in Quito-Ecuador. He later obtained his MSc degree from the University of Bradford, UK, in the field of power electronics and real-time control systems in 2000, and his PhD from the University of Strathclyde, UK, in 2004 in control systems.

Alberto Sanchez's background is in applied control systems to industrial processes and motor control. His industrial experience encompasses the implementation of supervision systems for petrochemical plants and instrumentation. In his academic research he has worked extensively in the implementation of field-vector control algorithms for induction motors and on different aspects of control theory. He is the author and co-author of several papers on mobile battery-power conversion control and the control of wastewater processes. Previous project experiences include the Interamerican Development Bank funded project (BID-085) for the development of commercial motor drives and the European Union project SMAC for the development and implementation of control strategies for wastewater treatment systems.

He is currently an Assistant Professor in the Department of Industrial Automation and Control at the Escuela Politécnica Nacional in Quito, Ecuador, and a development consultant for I&DE Ltd. (Ingeniería y Diseño Electrónico). His current research interests include subspace identification methods and embedded applications for mobile monitoring.

Kok Kiong Tan

K. K. Tan received his BEng in Electrical Engineering with honours in 1992 and PhD in 1995, all from the National University of Singapore. He was a Research Fellow with SIMTech, a national R&D institute spearheading the promotion of Research and Development in local manufacturing industries, from 1995–1996. During his SIMTech sojourn, he was involved in several industrial projects with various companies, including MC Packaging Pte Ltd, Panalpina World Transport, and SATS cargo.

He has been with the National University of Singapore since 1996, where he is now an Associate Professor. His current research interests are in the applications of advanced control techniques to industrial control systems and high precision mechatronic systems.

Kit S. Tang

Dr Tang obtained his BSc from the University of Hong Kong in 1988, and both MSc and PhD from the City University of Hong Kong in 1992 and 1996, respectively. He is currently an Associate Professor in the Department of Electronic Engineering, City University of Hong Kong. He has published over 40 journal papers and three book chapters, and co-authored two books, focusing on genetic algorithms and chaos theory. He is a member of IFAC Technical Committee on Optimal Control, a member of Intelligent Systems Committee in IEEE Industrial Electronics Society and the Associate Editor for *IEEE Transactions on Circuits and Systems II*.

Qing-Guo Wang

Qing-Guo Wang received a BEng in Chemical Engineering in 1982, and an MEng (1984) and a PhD (1987) in Industrial Automation, all from Zhejiang University, China. Since 1992, he has been with the Department of Electrical and Computer Engineering, National University of Singapore, where he is currently a Full Professor. He has published four research monographs and more than 100 international refereed journal papers, and holds four patents. He is an Associate Editor for the *Journal of Process Control* and *ISA Transactions*. His present research interests are mainly in robust, adaptive and multivariable control and optimisation with emphasis on their applications in process, chemical and environmental industries. He was the General Chair of the Fourth Asian Control Conference, Singapore, 2002, and the Fourth International Conference on Control and Automation, Montreal, Canada, 2003.

Yong Zhang

Yong Zhang received a BEng in both Automation and Applied Mathematics in 1994, and an MEng in Automation in 1997, all from Shanghai Jiaotong University, China, and a PhD in Electrical Engineering in 2001 from National University of Singapore. Since 2000, he has been with Global Research, General Electric Company, where he is currently Manager for the Real-Time/Power Controls Laboratory. He has published more than 20 international refereed journal and conference papers. His present research interests are mainly in modelling, system identification, controller design and real-time simulation and system implementation with emphasis on their applications in sensor system, process and power controls.

Yu Zhang

Yu Zhang received a BEng (1993) and an MEng (1996) in Automation from Shanghai Jiaotong University, China, and a PhD in Controls, Electrical Engineering, in 2000 from National University of Singapore. He worked for Hewlett-Packard Singapore from 1999 to 2000, and since 2000 has been with Global Research, General Electric Company, where he is currently a Senior Engineer and Programme Manager. He has published more than 20 international refereed journal and conference papers. His present research interests include multivariable system modelling and controls, real-time system, hardware-in-loop simulation, and renewable energy technology including wind generation.

Index

Bold page numbers indicate extended discussion or defining sections

A
ABCD state space model   51
actuator   3
anti-windup circuits   **26**, 28
autotune   34, 45, 220, **325**

B
benchmark metric   445, 465
biased relay   219
Biggest Log-Modulus Tuning (BLT)   392, **394**
block diagram   xxvi
   control loop (detailed)   4
boiler control model   521

C
cascade control system   **103**
   phase-locked loop method   295
   PID system   38, **105**
chromosome   352
closed loop identification *see* phase-locked loop method
closed loop stability   62, 68
commissioning, control loops   3
communications   3
control design *see* design
control design cycle   61
control signal matching methods   475
control spectral factors   434, 449
controller parameter cycling tuning   124, 131, 135
controller performance assessment   329
controller specification   5, **16**
controller structures
   1 degree of freedom   57, 88
   2 degree of freedom   57, 90
   3 degree of freedom   59
convolution integral model   **52**
coupled tank process   174
critical points of a system   185

D
DCS system   **35**, 43
decentralised control   101
   example   136
   phase-locked loop method   295
   PID design   **394**, **408**
   relay feedback method   184, **187**, 190
decoupling PID control design   **408**
   examples   422
defuzzification   346
DeltaV system   **40**
derivative control   **8**
   bandwidth-limited   18
   time constant   32
describing functions   215, 217
design
   Biggest Log-Modulus Tuning (BLT)   392, **394**
   control design cycle   61
   control signal matching methods   475
   control specification   5, **16**
   controller parameter cycling tuning   124, **131**, 135
   decoupling PID control design   **408**
   dominant pole-placement tuning   **395**   403
   Fourier method   **150**
   internal model control   43
   inverse model controller method   **320**
   iterative feedback tuning (IFT)   **110**, 122
   lambda-based tuning   43
   load disturbance rejection   **65**, 68
   multi-loop PID control   **394**, **408**
      design algorithm   403
      example   404
   PI control with Smith predictor   478
   reference tracking performance   **64**, 68
   restricted model methods   475
   Smith predictor   **475**
   supply disturbance rejection   **66**, 68
   sustained oscillation   152
   systematic methods   48
   trial and error   48
   Ziegler–Nichols rules   42

see also fuzzy PID control; GPC control signal matching, predictive PID control; GPC predictive PID control design; LQG optimal control; phase-locked loop method; predictive PI control using a simple model; process reaction curve method; relay feedback experiment; sine wave experiment; transient response method
diagonal band controller structure   102
diagonal controller structure   102, 393
differentiator   2
Diophantine equation   434, 463
dissolved-oxygen loop model   118
disturbances, input   61
   load   61
   noise   61, 68
   output   61
   supply   61
dominant pole-placement tuning   **395**, 403

F

filter spectral factors   434, 450
FOPDT model   56, 73, 84, 298, 477
Fourier method   **150**
Frobenius norm   365
function block   40
fuzzification   **345**
fuzzy control
   D term design   344
   defuzzification   346
   fuzzification   **345**
   genetic algorithm application   350
   multi-objectives design   353
   PI term design   342
   PID design   **340**
      example   349
   robotic applications   355
   solar plant application   354

G

GA see genetic algorithms; fuzzy control
gain margin   xxvi, **74**, 322
gain schedule   34
generalised predictive control (GPC)   485
genetic algorithms (GA)   350
golden section search   263
GPC control signal matching
   predictive PID control   **500**
      MIMO algorithm   **520**
      MIMO controller   517
      MIMO examples   520, 521
      MIMO optimal gains   518
      MIMO process response   516
      MIMO theory   **515**
      SISO algorithm   **505**
      SISO constrained case   508
      SISO example   512
      SISO second-order system case   507
      SISO theory   **500**
GPC cost function   487

GPC predictive PID control design   **486**
   algorithm   496
   benefits   498
   cost weight selection   499
   example   493, 497
   implementation   499
   robustness   499

H

Hessian matrix   113, 123
Hilbert transform   256
hotstrip finishing mill
   decentralised looper control   462
   looper control   **461**
   looper process model   461, 463
   optimal LQG control   463
   PID control, assessment   **465**

I

IAE cost function   329, 480
independent relay feedback method   184
industrial examples
   boiler model   521
   dissolved-oxygen loop model   118
   hotstrip mill, looper control   **461**
   robotic applications   355
   solar plant application   354
   stirred tank reactor   520
   wastewater treatment process   117, **376**
industrial PID control
   Emerson process management   **40**
   Siemens   **35**
input disturbances   61
input variables   50
integral control   **8**
integral time constant   32
integrator windup   **26**
integrator   1
Internal Model Control   43, 299, 474
inverse model controller method   **320**
inverse system response   29
iterative feedback tuning (IFT)   **110**, 122
   example   119
   multivariable systems   123

K

Kalman filter   244, 246
kick effects   18, **22, 24**

L

LabVIEW   176
LabVIEW module   236, 271, 272
lambda-based tuning   43
Laplace transform models   **53**
Levenberg–Marquardt procedure   134
load disturbances   6, 61
   induced errors in relay experiment   218
   rejection   **65**, 68, 178

Index   541

looper control model   463
lower triangular controller structure   101
LQG optimal control   **430**
   benchmark metric   445, 465
   control spectral factors   434, 449
   cost function   112, 432
   cost optimisation, subspace formulation   **372**
   Diophantine equation   434
   filter spectral factors   434, 450
   hotstrip mill, looper control   **461**
   multivariable optimal control   444, **448**
      controller algorithm (MIMO)   451
      controller design (MIMO)   **451**
      weight selection   451
   optimal controller   435, 450
   optimisation theory, scalar   432
   polynomial system models   449
   restricted controller structures
      algorithm (MIMO)   **458**
      algorithm (SISO)   436, 439
      controller theory (MIMO)   454
      control properties (SISO)   440
      nested costs   460, 465
      PID example (MIMO)   **461**
      PID example (SISO)   443
Luyben's process model   407, 422

M

marine control model   443
maximum sensitivity   xxvi, 75, **167**, 179
measurement noise   6, 18
model-based control design   5
model-free methods   **109**
model predictive control   474
models
   boiler model   521
   convolution integral model   **52**
   dynamic   49
   first-order plus dead time (FOPDT)   56, 73, 84, 298
   first-order transform   56
   framework for   49
   hotstrip mill, looper control   461, 463
   Laplace transform models   **53**
   Luyben's process   407, 422
   marine control   443
   polynomial system models   449
   second-order plus dead time (SOPDT)   56, 73, 85, 299
   second-order transform   56
   state space   **49**
      controller   364
      process   363
   static   48, 88
   stirred tank reactor   520
   three-input, three-output system   210
   Vasnani's process   424
   Wood and Berry process model   207, 405
   Zhuang and Atherton   136

multi-loop PID control design   394, **408**
   algorithm   403
   example   404
multivariable controllers   99, 393
   basic structures   372, **445**
   decentralised control   **101**
   diagonal band controller structure   102
   diagonal controller structure   102
   lower triangular controller structure   101
   upper triangular controller structure   101
multivariable PID control   391

N

noise disturbances   61, 68, 219
nonparametric identification   214

O

observability matrix   365
optimal control
   cost function gradient   112, 114, 124, 125
   Hessian matrix   **127**
   steepest descent optimisation   113
   see also LQG optimal control
optimal LQG controller   435, 450
oscillatory system modes   68
output disturbances   61
output variables   50

P

parameter stability margin   79, 80
peak resonance   262
phase frequency estimator   256
phase-locked loop methods   **151, 213**
   algorithm performance   236
   automated PI control design   270
      cascade control design   295
      decentralised control design   295
      gain and phase margin design   277
      phase margin and maximum sensitivity design   **286**
   basic theory   **229**
   closed loop identification
      unknown system, known controller   **268**
      unknown system, unknown controller   **265**
   comparison with relay method   236
   convergence theory   **232**
   digital module   230
   disturbance management methods   **248**
   implementation   **236**
   Kalman filter   244, 246
   LabVIEW module   236, 271, 272
   noise management methods   **242**, 246
   as phase margin estimator   260
   as second order system identifier   **261**
   voltage controlled oscillator (VCO)   229
phase margin estimator   260
phase margin   xxvi, 75, 322
PI control   15, 16, 33, 69, 91, 105

PID control
   controller nomenclature   32
   industrial controller structure   32
   parallel form   **9**
   series form   **12**
   with Smith predictor   478
   structure selection   15
   term tuning effects   14
   time constant form   **10**
post-commissioning, control loop   5
pre-act time   31
predictive PI control using a simple model   **477**
   algorithm   481
   example   **481**
process control unit   **33**
process models *see* models
process reaction curve method   **297**
process variable   32
proportional band   30
proportional control   **7**
proportional gain   32
proportional kick   18, **22**
proportional term   2

R

receding horizon principle   474
reference input   32
reference tracking performance   **64**, 68
Relative Gain Array (RGA)   393, 445
relay feedback experiment   **150**, 152
   adaptive delay method   161
   adaptive identification   311
   basic theory   **215**
   biased relay   219
   case studies   174
   coupled tank process   174
   decentralised relay feedback method   184, **187**, 190
   describing functions   215, 217
   FOPDT model determination   172
   general Nyquist point estimation   **161**, 177
   improved estimation accuracy   **155**, 176
   independent relay feedback method   184
   load disturbance induced errors   218
   load disturbance rejection   178
   maximum sensitivity specification   **167**, 179
   modified relay method   156
   multiple Nyquist point estimation   **164**, 177
   multivariable system methods
      bandwidth frequency   **202**
      bandwidth frequency algorithm   205
      case study   207, 210
      critical points   **186, 204**
      relay feedback algorithm   **201**
      three-input, three-output system   210
   noise disturbances   219
   noise rejection   154
   online tuning   **164**

   parametric model determination   **171**
   performance case studies   238, 239, 242
   problems arising from   **216**
   relay with hysteresis   220
   saturation relay   217
   sequential relay feedback method   184
   SOPDT model determination   174
   stability margin specifications   **170**, 179
   tuning for robustness specification   **166**, 179
   two-input, two-output (TITO) system
      case study   207
      critical points   185, 192
      PID control   191
      relay feedback algorithm   **201**
      steady state gains   192
      Wood and Berry process model   207
relay with hysteresis   154, 220
reset rate   31
reset time   31
restricted model methods   475
restricted structure controllers   **112**, 433
   MIMO controllers   371
   SISO controllers   368
reverse acting controllers   29
robotic applications   355
robustness measures   **73**, 166, 179
root trajectories   401

S

saturation relay   217
SCADA system   35, 43
self-tuning control   34
sequential relay feedback method   184
setpoint   32
sine wave experiment   **221**
   bisection method   222
   comparative study   227
   prediction method   **224**
sinusoidal autotune variation (SATV)   255
Smith predictor   178, **475**
solar plant application   354
stability margin specifications   **170**, 179
state space PID control
   2 dof control   **90**
   classical PI feedback   **95**
   integral error feedback   **91**
   reference error feedback   **88**
steepest descent optimisation   113
subspace identification method   **363**
   closed loop condition   374
   incremental model representation   366
   LQG cost optimisation   372
   wastewater process example   376
supervisory control   35
supply disturbance   6, 61
sustained oscillation   152, 213
   ultimate frequency   152, 154
   ultimate gain   152, 154

T

three-input, three-output system   210
Toeplitz matrix   365
transient response methods   **149**
  see also process reaction curve method
trend display   34, 44
tuning
  systematic methods   48
  trial and error   48
  see also design
two-input, two-output (TITO) system
  case study   207
  critical points   185, 192
  Wood and Berry process model   207

U

ultimate frequency   152, 154

ultimate gain   152, 154
upper triangular controller structure   101

V

Vasnani's process model   424
voltage-controlled oscillator (VCO)   229

W

wastewater process   117
windup *see* anti-windup circuits; integrator windup
Wood and Berry process model   207, 405

Z

Zhuang and Atherton process model   136
Ziegler–Nichols tuning   42, 84, 85, 87, 298, 394